Student Solutions Manual

Laurel Technical Services

Finite Mathematics

for Business, Economics, Life Sciences, and Social Sciences

Eighth Edition

Raymond A. Barnett

Michael R. Ziegler

Karl E. Byleen

PRENTICE HALL, Upper Saddle River, NJ 07458

Executive Editor: Sally Simpson
Managing & Supplements Editor: Gina M. Huck
Editorial Assistant: Sara Beth Newell
Special Projects Manager: Barbara A. Murray
Production Editor: Barbara A. Till
Supplement Cover Manager: Paul Gourhan
Supplement Cover Designer: Liz Nemeth
Manufacturing Buyer: Alan Fischer

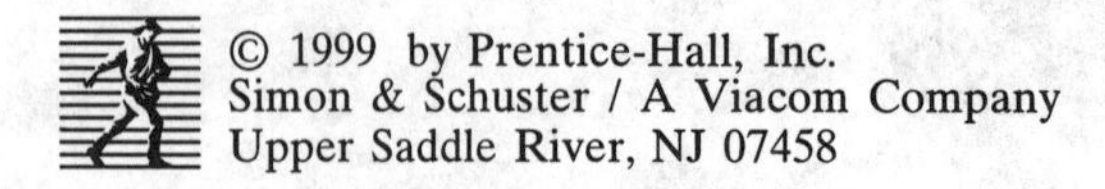

Printed in the United States of America

10 9 8 7 6 5 4 3 2 1

ISBN 0-13-932568-9

Prentice-Hall International (UK) Limited, *London*
Prentice-Hall of Australia Pty. Limited, *Sydney*
Prentice-Hall Canada, Inc., *London*
Prentice-Hall Hispanoamericana, S.A., *Mexico*
Prentice-Hall of India Private Limited, *New Delhi*
Prentice-Hall of Japan, Inc., *Tokyo*
Simon & Schuster Asia Pte. Ltd., *Singapore*
Editora Prentice-Hall do Brazil, Ltda., *Rio de Janeiro*

CONTENTS

1 A BEGINNING LIBRARY OF ELEMENTARY FUNCTIONS

EXERCISE 1-1

Things to remember:

1. A FUNCTION is a rule (process or method) that produces a correspondence between one set of elements, called a DOMAIN, and a second set of elements, called the RANGE, such that to each element in the domain there corresponds one and only one element in the range.

2. EQUATIONS AND FUNCTIONS:

   Given an equation in two variables. If there corresponds exactly one value of the dependent variable (output) to each value of the independent variable (input), then the equation defines a function. If there is more than one output for at least one input, then the equation does not define a function.

3. VERTICAL LINE TEST FOR A FUNCTION

   An equation defines a function if each vertical line in the coordinate system passes through at most one point on the graph of the equation. If any vertical line passes through two or more points on the graph of an equation, then the equation does not define a function.

4. AGREEMENT ON DOMAINS AND RANGES

   If a function is specified by an equation and the domain is not given explicitly, then assume that the domain is the set of all real number replacements of the independent variable (inputs) that produce real values for the dependent variable (outputs). The range is the set of all outputs corresponding to input values.

   In many applied problems, the domain is determined by practical considerations within the problem.

5. FUNCTION NOTATION—THE SYMBOL $f(x)$

   For any element x in the domain of the function f, the symbol $f(x)$ represents the element in the range of f corresponding to x in the domain of f. If x is an input value, then $f(x)$ is the corresponding output value. If x is an element which is not in the domain of f, then f is NOT DEFINED at x and $f(x)$ DOES NOT EXIST.

1. The table specifies a function, since for each domain value there corresponds one and only one range value.

3. The table does not specify a function, since more than one range value corresponds to a given domain value. (Range values 5, 6 correspond to domain value 3; range values 6, 7 correspond to domain value 4.)

5. This is a function.

7. The graph specifies a function; each vertical line in the plane intersects the graph in at most one point.

9. The graph does not specify a function. There are vertical lines which intersect the graph in more than one point. For example, the y-axis intersects the graph in three points.

11. The graph specifies a function.

13. $f(x) = 3x - 2$
$f(2) = 3(2) - 2 = 4$

15. $f(-1) = 3(-1) - 2$
$= -5$

17. $g(x) = x - x^2$
$g(3) = 3 - 3^2 = -6$

19. $f(0) = 3(0) - 2$
$= -2$

21. $g(-3) = -3 - (-3)^2$
$= -12$

23. $f(1) + g(2)$
$= [3(1) - 2] + (2 - 2^2) = -1$

25. $g(2) - f(2)$
$= (2 - 2^2) - [3(2) - 2]$
$= -2 - 4 = -6$

27. $g(3) \cdot f(0) = (3 - 3^2)[3(0) - 2]$
$= (-6)(-2)$
$= 12$

29. $\dfrac{g(-2)}{f(-2)} = \dfrac{-2 - (-2)^2}{3(-2) - 2} = \dfrac{-6}{-8} = \dfrac{3}{4}$

31. $y = f(-5) = 0$

33. $y = f(5) = 4$

35. $f(x) = 0$ at $x = -5, 0, 4$

37. $f(x) = -4$ at $x = 6$

39. domain: all real numbers or $(-\infty, \infty)$

41. domain: all real numbers except -4

43. $x^2 + 3x - 4 = (x + 4)(x - 1)$; domain: all real numbers except -4 and 1.

45. $7 - x \geq 0$ for $x \leq 7$; domain: $x \leq 7$ or $(-\infty, 7]$

47. $7 - x > 0$ for $x < 7$; domain: $x < 7$ or $(-\infty, 7)$

49. f is not defined at the values of x where $x^2 - 9 = 0$, that is, at 3 and -3; f is defined at $x = 2$, $f(2) = \dfrac{0}{-5} = 0$.

51. $g(x) = 2x^3 - 5$

53. $G(x) = 2\sqrt{x} - x^2$

55. Function f multiplies the domain element by 2 and subtracts 3 from the result.

57. Function F multiplies the cube of the domain element by 3 and subtracts twice the square root of the domain element from the result.

59. Given $4x - 5y = 20$. Solving for y, we have:

$$-5y = -4x + 20$$

$$y = \frac{4}{5}x - 4$$

Since each input value x determines a unique output value y, the equation specifies a function. The domain is R, the set of real numbers.

61. Given $x^2 - y = 1$. Solving for y, we have:

$$-y = -x^2 + 1 \quad \text{or} \quad y = x^2 - 1$$

This equation specifies a function. The domain is R, the set of real numbers.

63. Given $x + y^2 = 10$. Solving for y, we have:

$$y^2 = 10 - x$$

$$y = \pm\sqrt{10 - x}$$

This equation does not specify a function since each value of x, $x \le 10$, determines two values of y. For example, corresponding to $x = 1$, we have $y = 3$ and $y = -3$; corresponding to $x = 6$, we have $y = 2$ and $y = -2$.

65. Given $xy - 4y = 1$. Solving for y, we have:

$$(x - 4)y = 1 \quad \text{or} \quad y = \frac{1}{x - 4}$$

This equation specifies a function. The domain is all real numbers except $x = 4$.

67. Given $x^2 + y^2 = 25$. Solving for y, we have:

$$y^2 = 25 - x^2 \quad \text{or} \quad y = \pm\sqrt{25 - x^2}$$

Thus, the equation does not specify a function since, for $x = 0$, we have $y = \pm 5$, when $x = 4$, $y = \pm 3$, and so on.

69. Given $F(t) = 4t + 7$. Then:

$$\frac{F(3 + h) - F(3)}{h} = \frac{4(3 + h) + 7 - (4\cdot 3 + 7)}{h}$$

$$= \frac{12 + 4h + 7 - 19}{h} = \frac{4h}{h} = 4$$

71. Given $Q(x) = x^2 - 5x + 1$. Then:

$$\frac{Q(2 + h) - Q(2)}{h} = \frac{(2 + h)^2 - 5(2 + h) + 1 - (2^2 - 5\cdot 2 + 1)}{h}$$

$$= \frac{4 + 4h + h^2 - 10 - 5h + 1 - (-5)}{h} = \frac{h^2 - h - 5 + 5}{h}$$

$$= \frac{h(h - 1)}{h} = h - 1$$

73. Given $f(x) = 4x - 3$. Then:

$$\frac{f(a + h) - f(a)}{h} = \frac{4(a + h) - 3 - (4a - 3)}{h}$$

$$= \frac{4a + 4h - 3 - 4a + 3}{h} = \frac{4h}{h} = 4$$

75. Given $f(x) = 4x^2 - 7x + 6$. Then:

$$\frac{f(a + h) - f(a)}{h} = \frac{4(a + h)^2 - 7(a + h) + 6 - (4a^2 - 7a - 6)}{h}$$
$$= \frac{4(a^2 + 2ah + h^2) - 7a - 7h + 6 - 4a^2 + 7a - 6}{h}$$
$$= \frac{4a^2 + 8ah + 4h^2 - 7h - 4a^2}{h} = \frac{8ah + 4h^2 - 7h}{h}$$
$$= \frac{h(8a + 4h - 7)}{h} = 8a + 4h - 7$$

77. Given $f(x) = x^3$. Then:

$$\frac{f(a + h) - f(a)}{h} = \frac{(a + h)^3 - a^3}{h} = \frac{a^3 + 3a^2h + 3ah^2 + h^3 - a^3}{h}$$
$$= \frac{h(3a^2 + 3ah + h^2)}{h} = 3a^2 + 3ah + h^2$$

79. Given $f(x) = \sqrt{x}$. Then:

$$\frac{f(a + h) - f(a)}{h} = \frac{\sqrt{a + h} - \sqrt{a}}{h}$$
$$= \frac{\sqrt{a + h} - \sqrt{a}}{h} \cdot \frac{\sqrt{a + h} + \sqrt{a}}{\sqrt{a + h} + \sqrt{a}} \quad \text{(rationalizing the numerator)}$$
$$= \frac{a + h - a}{h(\sqrt{a + h} + \sqrt{a})} = \frac{h}{h(\sqrt{a + h} + \sqrt{a})} = \frac{1}{\sqrt{a + h} + \sqrt{a}}$$

81. Given $A = \ell w = 25$.

Thus, $\ell = \frac{25}{w}$. Now $P = 2\ell + 2w$

$$= 2\left(\frac{25}{w}\right) + 2w = \frac{50}{w} + 2w.$$

The domain is $w > 0$.

83. Given $P = 2\ell w + 2w = 100$ or $\ell + w = 50$ and $w = 50 - \ell$.
Now $A = \ell w = \ell(50 - \ell)$ and $A = 50\ell - \ell^2$.
The domain is $0 \le \ell \le 50$. [Note: $\ell \le 50$ since $\ell > 50$ implies $w < 0$.]

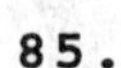
85.

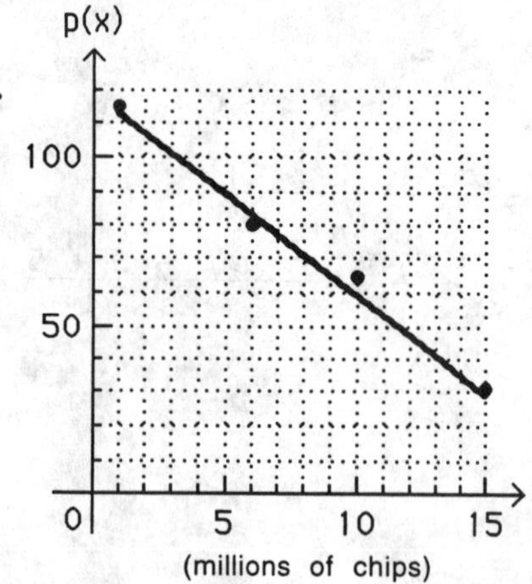

$p(8) = 71$ dollars per chip
$p(11) = 53$ dollars per chip

87. (A) $R(x) = xp(x) = x(119 - 6x)$
Domain: $1 \le x \le 15$

(B) Table 10 Revenue

x(millions)	$R(x)$ (millions)
1	\$113
3	303
6	498
9	585
12	564
15	435

(C)

89. (A) $P(x) = R(x) - C(x)$
$= x(119 - 6x) - (234 + 23x)$
$= -6x^2 + 96x - 234$ million dollars
Domain: $1 \le x \le 15$

(B) Table 12 Profit

x(millions)	$P(x)$ (millions)
1	-\$144
3	0
6	126
9	144
12	54
15	-144

(C)

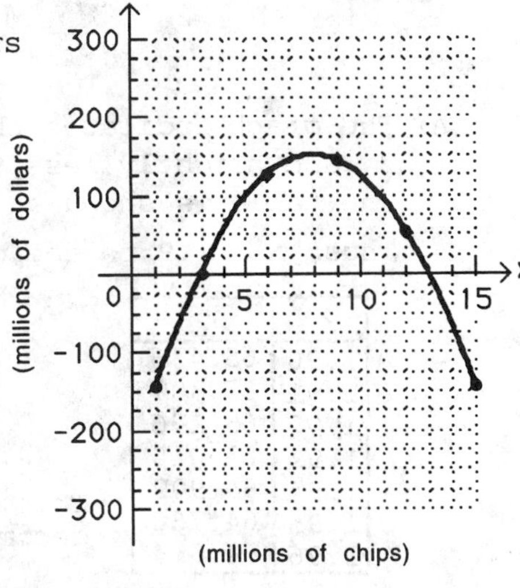

91.

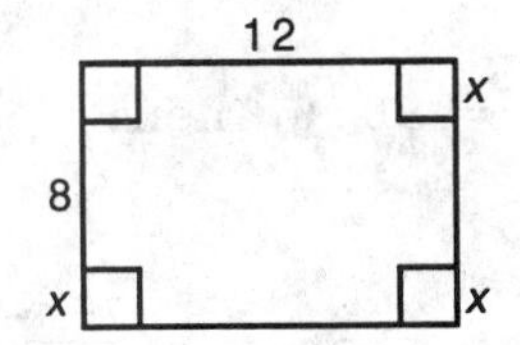

(A) $V =$ (length) (width) (height)
$V(x) = (12 - 2x)(8 - 2x)x$
$= x(8 - 2x)(12 - 2x)$

(B) Domain: $0 \le x \le 4$

(C) $V(1) = (12 - 2)(8 - 2)(1)$
$= (10)(6)(1) = 60$
$V(2) = (12 - 4)(8 - 4)(2)$
$= (8)(4)(2) = 64$
$V(3) = (12 - 6)(8 - 6)(3)$
$= (6)(2)(3) = 36$

Thus,

Volume

x	$V(x)$
1	60
2	64
3	36

(D)

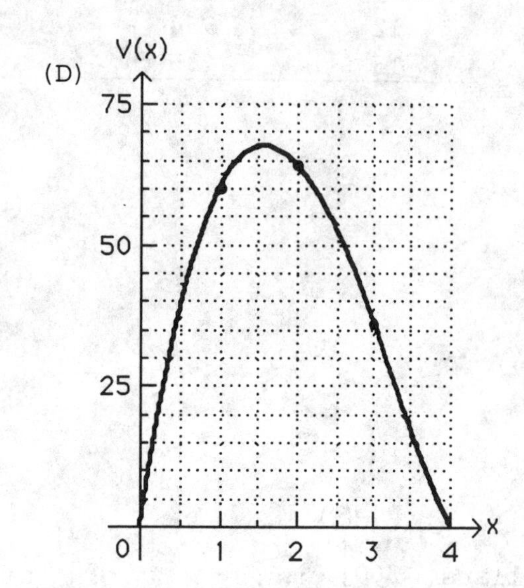

93. (A) The graph indicates that there is a value of x near 2, and slightly less than 2, such that $V(x) = 65$. The table is shown at the right.

Thus, $x = 1.9$ to one decimal place.

x	y_1
1.7	67.252
1.8	66.528
1.9	65.436
2.0	64.000

(B)

x	y_1
1.90	65.436
1.91	65.307
1.92	65.176
1.93	65.040
1.94	64.902
1.95	64.760
1.96	64.614

Thus, $x = 1.93$ to two decimal places.

95. Given $(w + a)(v + b) = c$. Let $a = 15$, $b = 1$, and $c = 90$. Then:
$(w + 15)(v + 1) = 90$
Solving for v, we have
$v + 1 = \frac{90}{w + 15}$ and $v = \frac{90}{w + 15} - 1 = \frac{90 - (w + 15)}{w + 15}$, so that $v = \frac{75 - w}{w + 15}$.
If $w = 16$, then $v = \frac{75 - 16}{16 + 15} = \frac{59}{31} \approx 1.9032$ cm/sec.

Things to remember:

1. LIBRARY OF ELEMENTARY FUNCTIONS

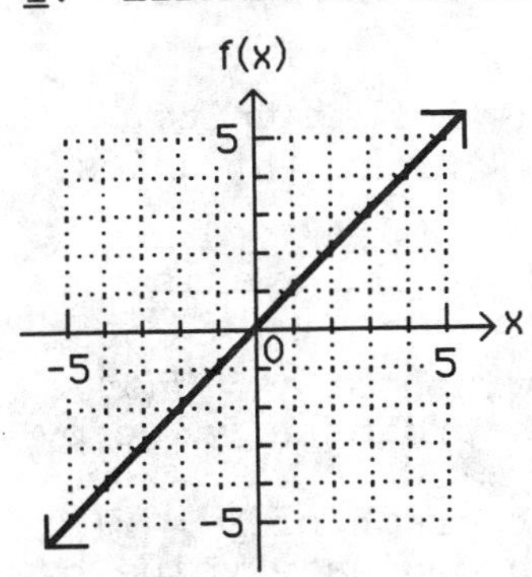

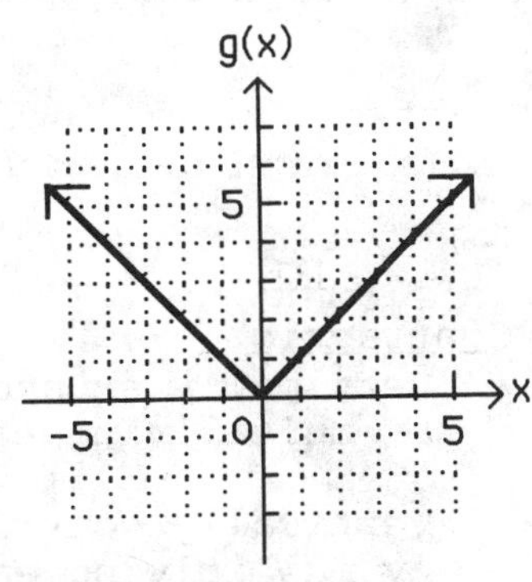

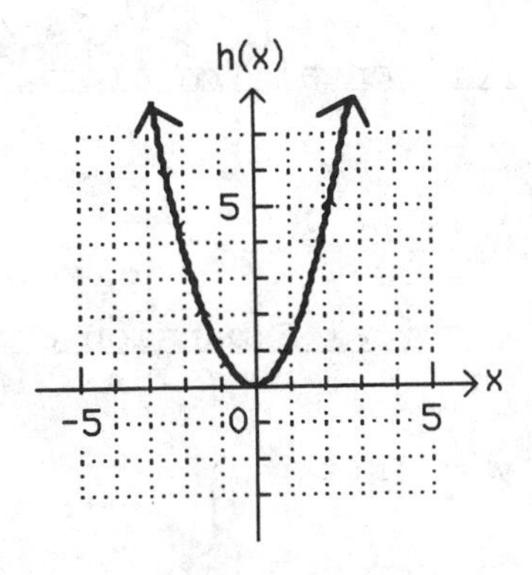

Identity Function
$f(x) = x$
Domain: All real numbers
Range: All real numbers
(a)

Absolute Value Function
$g(x) = |x|$
Domain: All real numbers
Range: $[0, \infty)$
(b)

Square Function
$h(x) = x^2$
Domain: All real numbers
Range: $[0, \infty)$
(c)

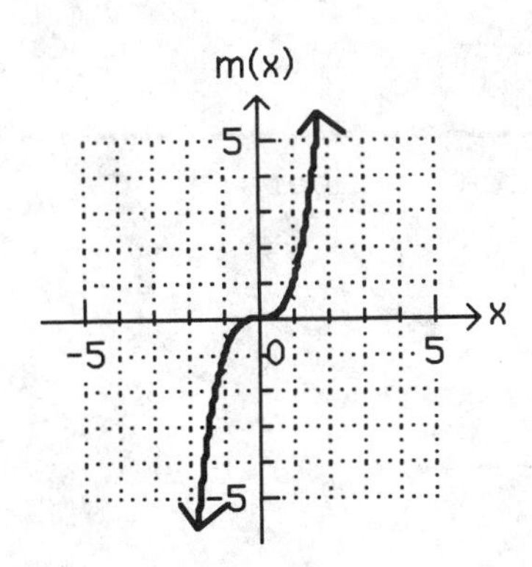

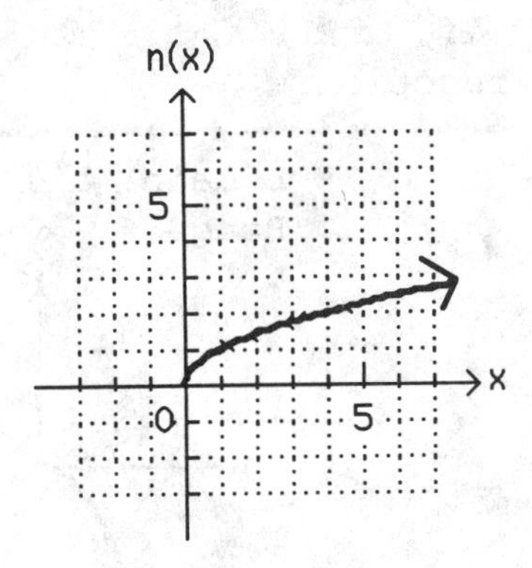

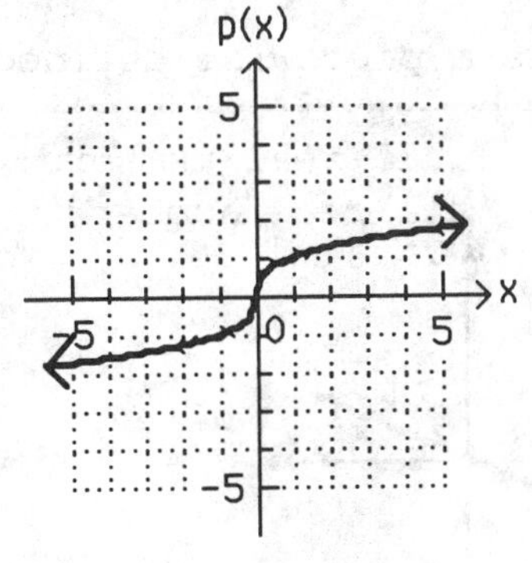

Cube Function
$m(x) = x^3$
Domain: All real numbers
Range: All real numbers
(d)

Square-Root Function
$n(x) = \sqrt{x}$
Domain: $[0, \infty)$
Range: $[0, \infty)$
(e)

Cube-Root Function
$p(x) = \sqrt[3]{x}$
Domain: All real numbers
Range: All real numbers
(f)

NOTE: Letters used to designate the above functions may vary from context to context.

2. GRAPH TRANSFORMATIONS SUMMARY

Vertical Translation:

$y = f(x) + k$ $\begin{cases} k > 0 & \text{Shift graph of } y = f(x) \text{ up } k \text{ units} \\ k < 0 & \text{Shift graph of } y = f(x) \text{ down } |k| \text{ units} \end{cases}$

Horizontal Translation:

$y = f(x + h)$ $\begin{cases} h > 0 & \text{Shift graph of } y = f(x) \text{ left } h \text{ units} \\ h < 0 & \text{Shift graph of } y = f(x) \text{ right } |h| \text{ units} \end{cases}$

Reflection:

$y = -f(x)$ Reflect the graph of $y = f(x)$ in the x axis

Vertical Expansion and Contraction:

$y = Af(x)$ $\begin{cases} A > 1 & \text{Vertically expand graph of } y = f(x) \text{ by multiplying each ordinate value by } A \\ 0 < A < 1 & \text{Vertically contract graph of } y = f(x) \text{ by multiplying each ordinate value by } A \end{cases}$

3. PIECEWISE-DEFINED FUNCTIONS

Functions whose definitions involve more than one rule are called PIECEWISE-DEFINED FUNCTIONS.

For example,

$$f(x) = |x| = \begin{cases} -x & \text{if } x < 0 \\ x & \text{if } x \geq 0 \end{cases}$$

is a piecewise-defined function.

---

1. Domain: all real numbers; Range: all real numbers

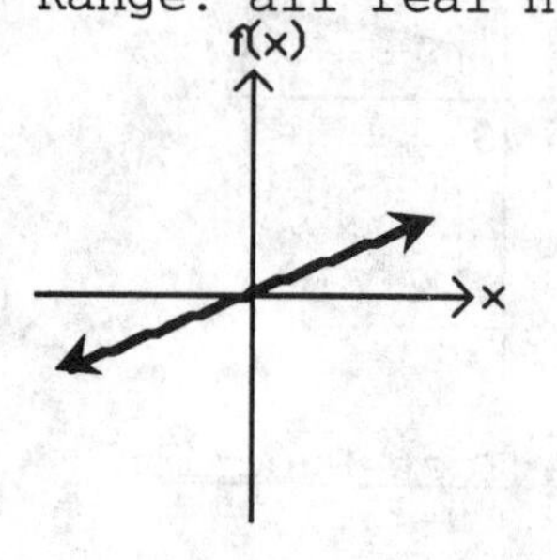

3. Domain: all real numbers; Range: $(-\infty, 0]$

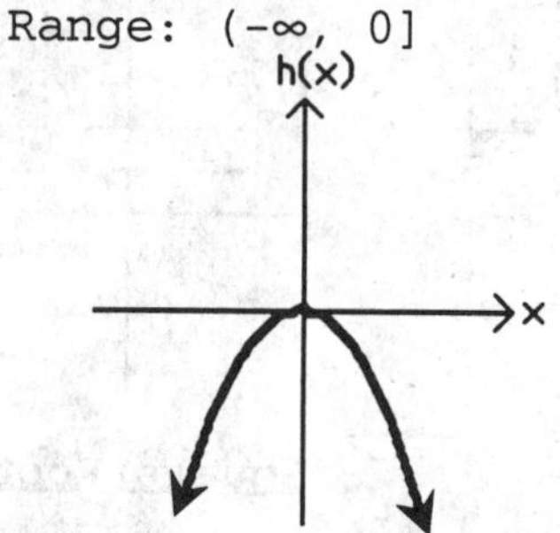

5. Domain: $[0, \infty)$; Range: $(-\infty, 0]$

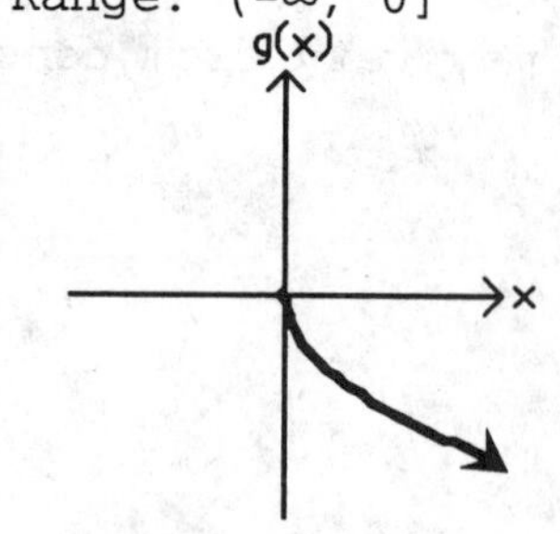

7. Domain: all real numbers; Range: all real numbers

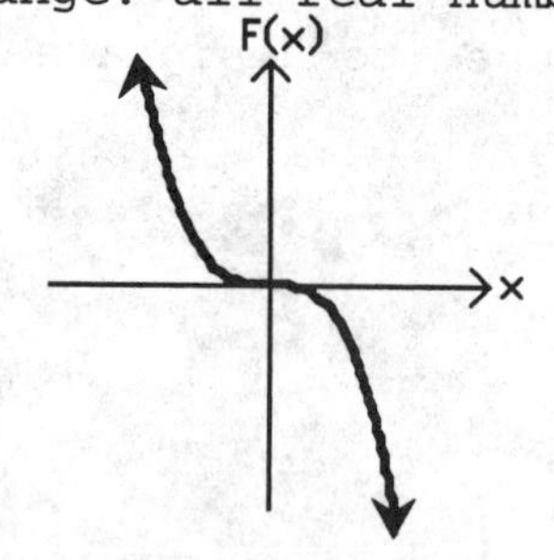

9.

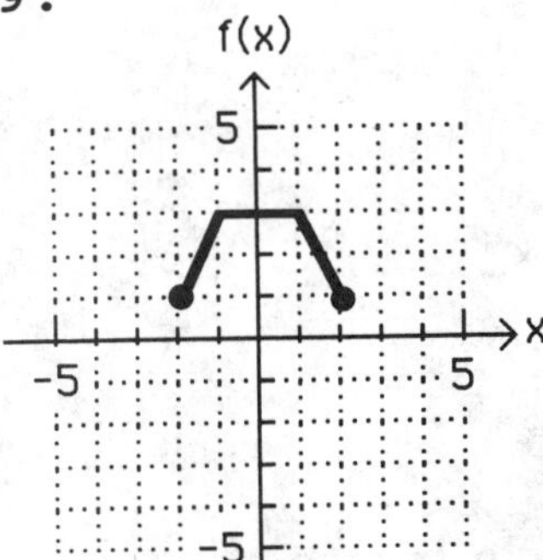

11.

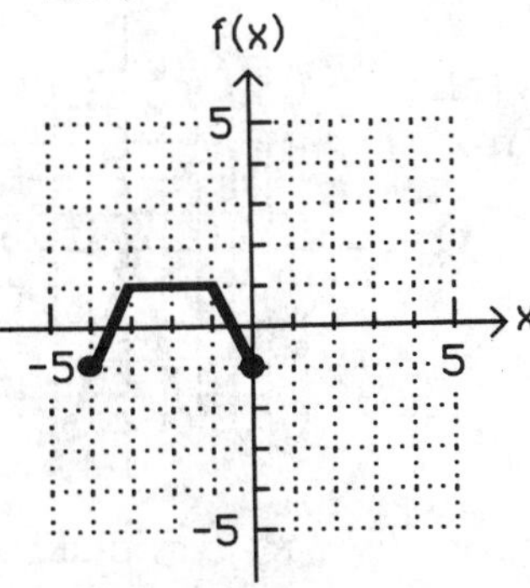

13.

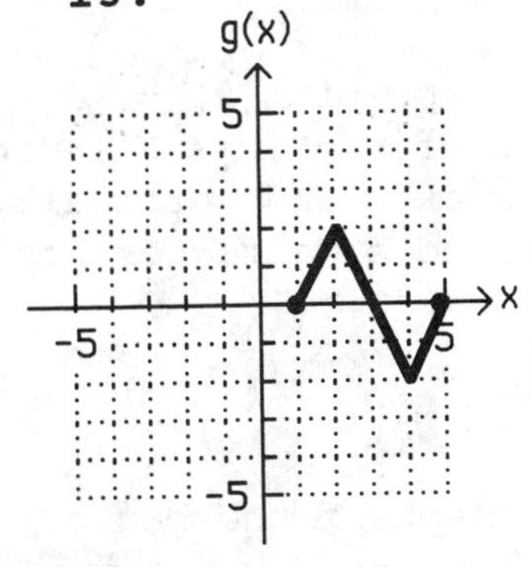

15.

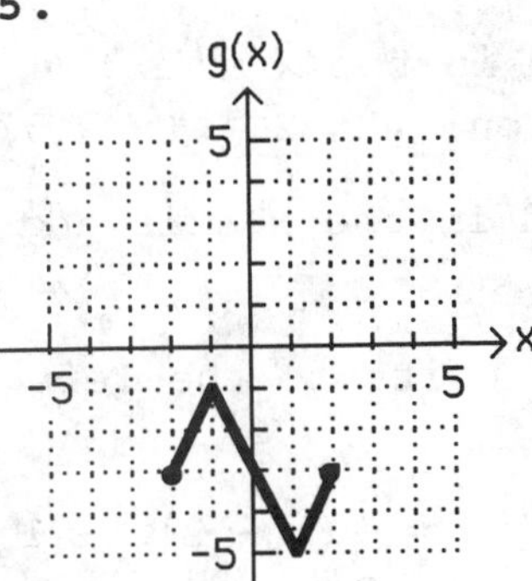

17.

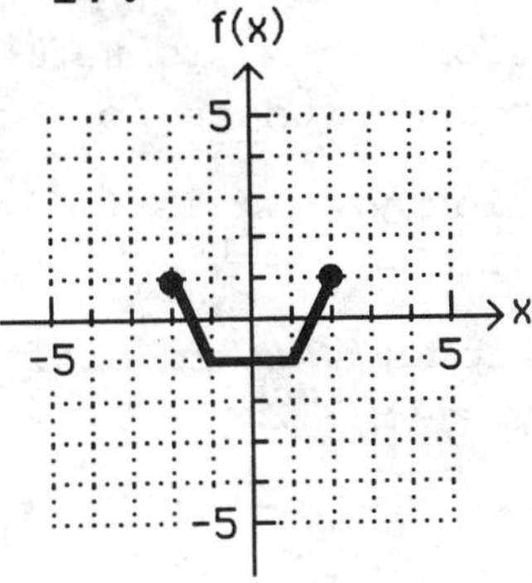

19.

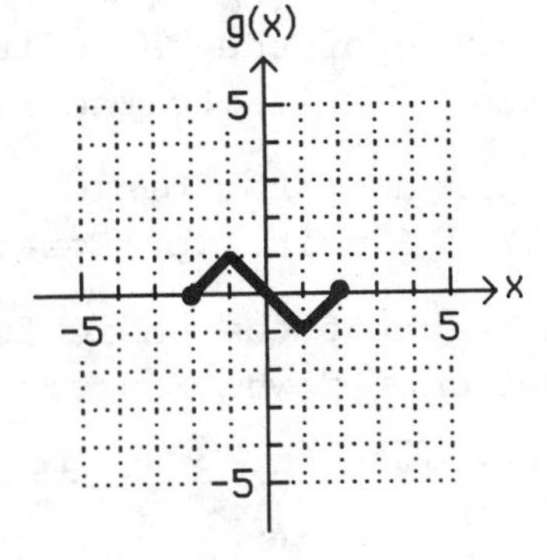

21. The graph of $g(x) = -|x + 3|$ is the graph of $y = |x|$ reflected in the x axis and shifted 3 units to the left.

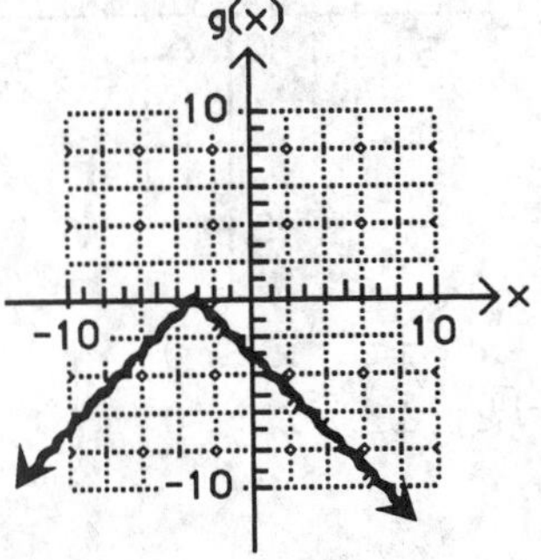

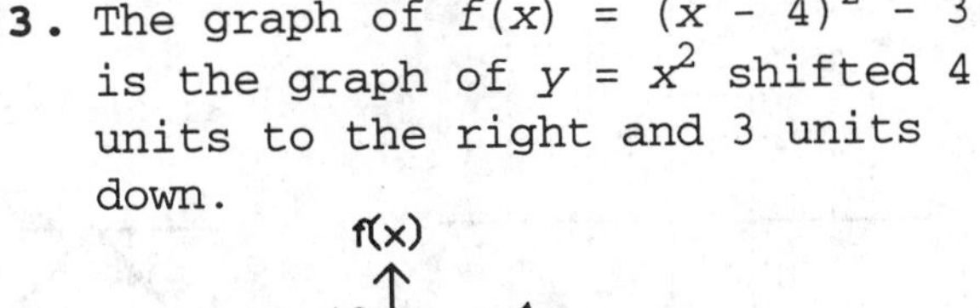

23. The graph of $f(x) = (x - 4)^2 - 3$ is the graph of $y = x^2$ shifted 4 units to the right and 3 units down.

25. The graph of $f(x) = 7 - \sqrt{x}$ is the graph of $y = \sqrt{x}$ reflected in the x axis and shifted 7 units up.

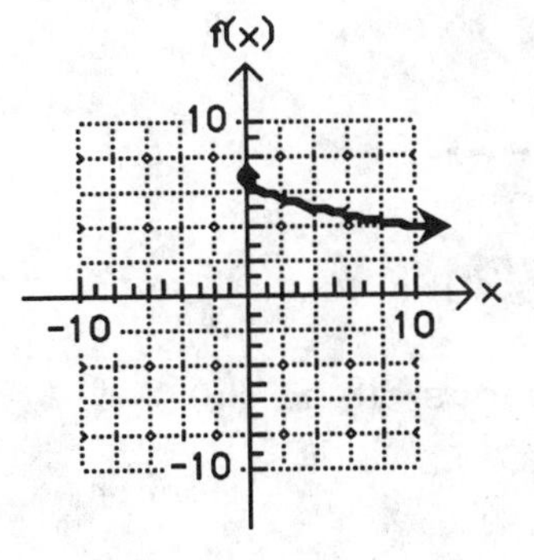

27. The graph of $h(x) = -3|x|$ is the graph of $y = |x|$ reflected in the x axis and vertically expanded by a factor of 3.

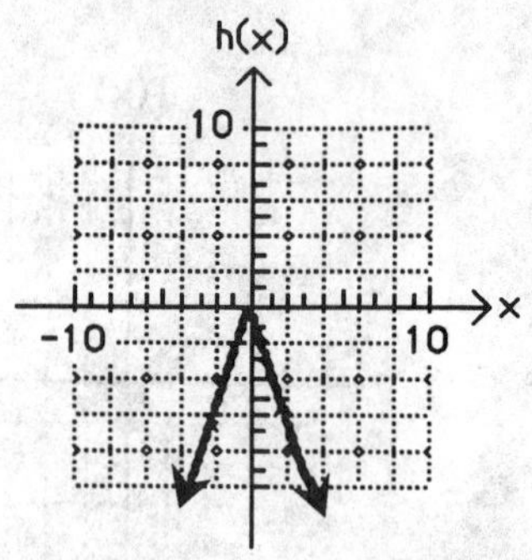

29. The graph of the basic function $y = x^2$ is shifted 2 units to the left and 3 units down. Equation: $y = (x + 2)^2 - 3$.

31. The graph of the basic function $y = x^2$ is reflected in the x axis, shifted 3 units to the right and 2 units up. Equation: $y = 2 - (x - 3)^2$.

33. The graph of the basic function $y = \sqrt{x}$ is reflected in the x axis and shifted 4 units up. Equation: $y = 4 - \sqrt{x}$.

35. The graph of the basic function $y = x^3$ is shifted 2 units to the left and 1 unit down. Equation: $y = (x + 2)^3 - 1$.

37. $g(x) = \sqrt{x - 2} - 3$

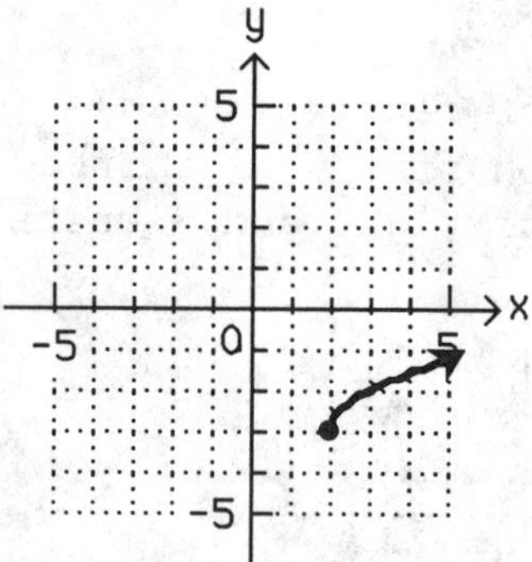

39. $g(x) = -|x + 3|$

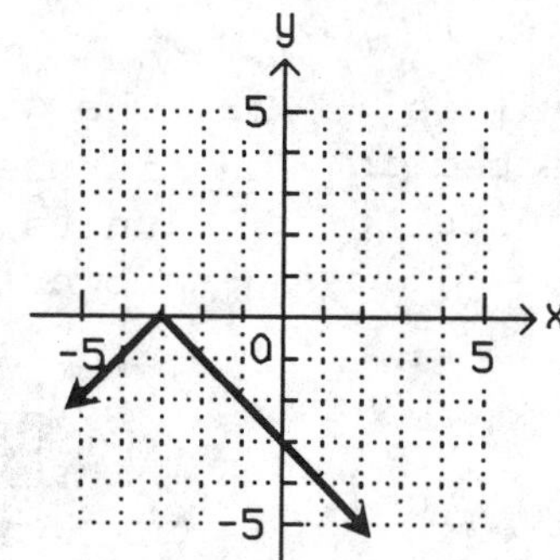

41. $g(x) = -(x - 2)^3 - 1$

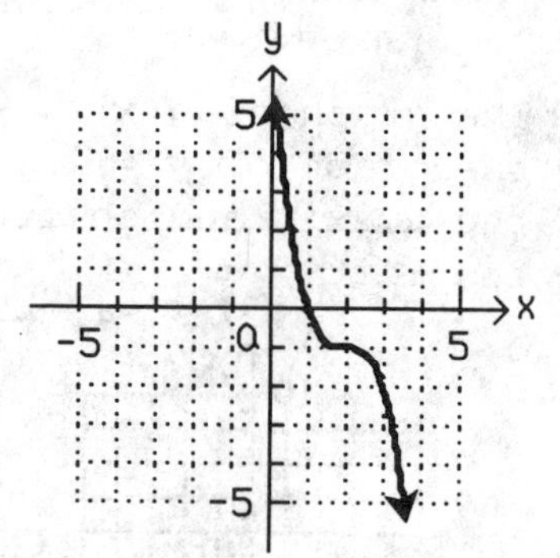

43. $f(x) = \begin{cases} x + 1 & \text{if } x < 0 \\ x - 1 & \text{if } x \geq 0 \end{cases}$

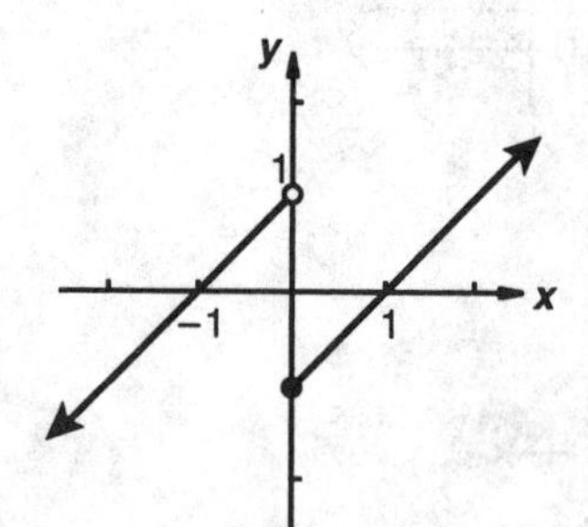

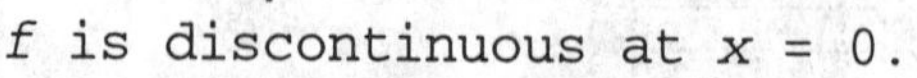
f is discontinuous at $x = 0$.

45. 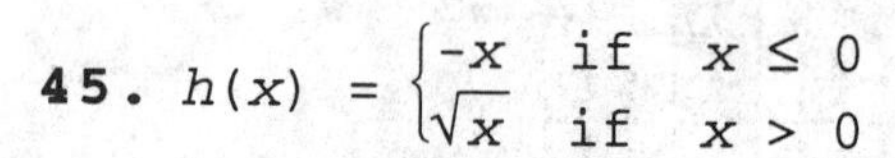
$h(x) = \begin{cases} -x & \text{if } x \leq 0 \\ \sqrt{x} & \text{if } x > 0 \end{cases}$

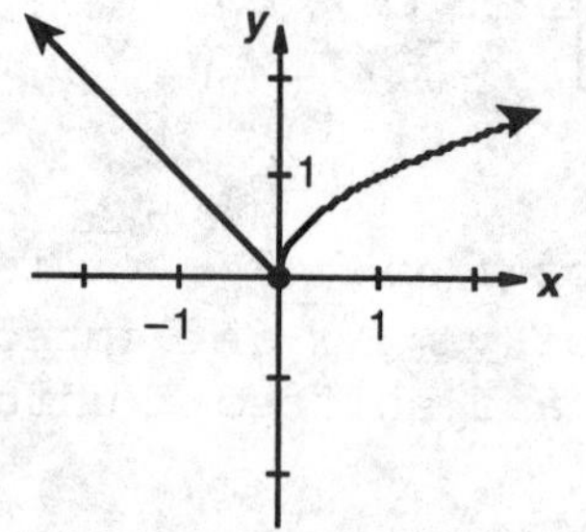

There are no points of discontinuity.

47. $p(x) = \begin{cases} -2x & \text{if } x \le -1 \\ x^2 & \text{if } -1 < x < 1 \\ \sqrt{x - 1} & \text{if } 1 \le x \end{cases}$

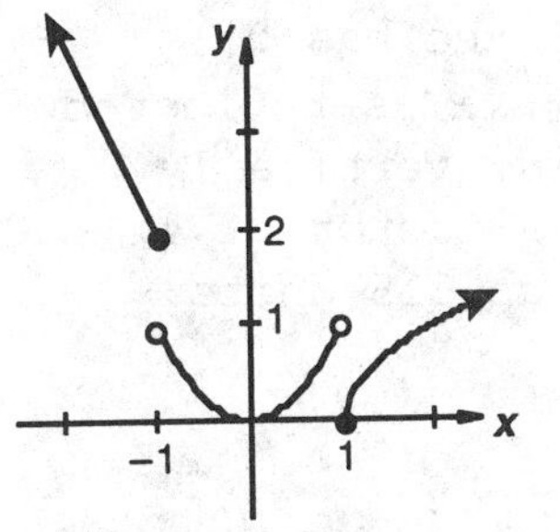

p is discontinuous at $x = -1$ and $x = 1$.

49. The graph of the basic function: $y = |x|$ is reflected in the x axis and has a vertical contraction by the factor 0.5. Equation: $y = -0.5|x|$.

51. The graph of the basic function $y = x^2$ is reflected in the x axis and is vertically expanded by the factor 2. Equation: $y = -2x^2$.

53. The graph of the basic function $y = \sqrt[3]{x}$ is reflected in the x axis and is vertically expanded by the factor 3. Equation: $y = -3\sqrt[3]{x}$.

55. Vertical shift, horizontal shift.
Reversing the order does not change the result. Consider a point (a, b) in the plane. A vertical shift of k units followed by a horizontal shift of h units moves (a, b) to $(a, b + k)$ and then to $(a + h, b + k)$.

In the reverse order, a horizontal shift of h units followed by a vertical shift of k units moves (a, b) to $(a + h, b)$ and then to $(a + h, b + k)$. The results are the same.

57. Vertical shift, reflection in the x axis.
Reversing the order can change the result. For example, let (a, b) be a point in the plane with $b > 0$. A vertical shift of k units, $k \neq 0$, followed by a reflection in the x axis moves (a, b) to $(a, b + k)$ and then to $(a, -[b + k]) = (a, -b - k)$.

In the reverse order, a reflection in the x axis followed by the vertical shift of k units moves (a, b) to $(a, -b)$ and then to $(a, -b + k)$; $(a, -b - k) \neq (a, -b + k)$ when $k \neq 0$.

59. Horizontal shift, reflection in y axis.
Reversing the order can change the result. For example, let (a, b) be a point in the plane with $a > 0$. A horizontal shift of h units followed by a reflection in the y axis moves (a, b) to the point $(a + h, b)$ and then to $(-[a + h], b) = (-a - h, b)$.

In the reverse order, a reflection in the y axis followed by the horizontal sift of h units moves (a, b) to $(-a, b)$ and then the $(-a + h, b)$, $(-a - h, b) \neq (-a + h, b)$ when $h \neq 0$.

61. (A) The graph of the basic function $y = \sqrt{x}$ is reflected in the x axis, vertically expanded by a factor of 4, and shifted up 115 units.

(B)

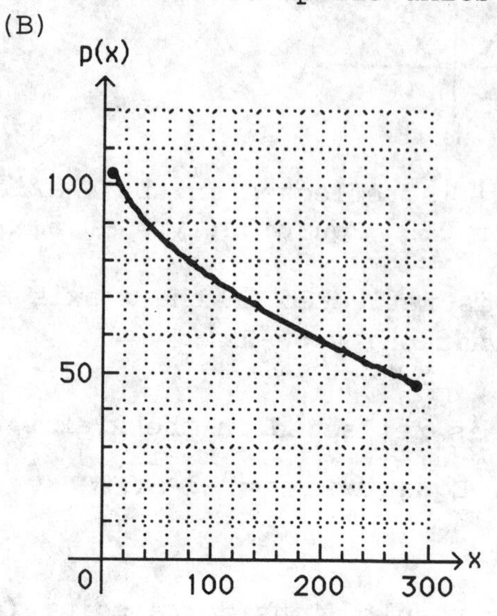

63. (A) The graph of the basic function $y = x^3$ is vertically contracted by a factor of 0.00048 and shifted right 500 units and up 60,000 units.

(B)

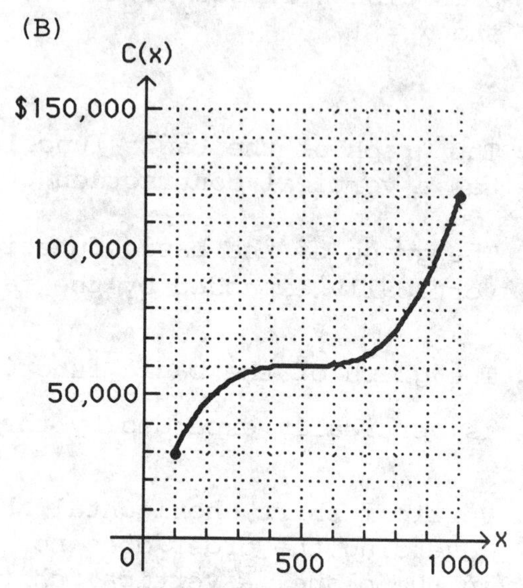

65. (A) According to the table, the price per box for x boxes is: \$10.06 for $1 \le x < 4$, \$8.52 for $4 \le x < 10$, and \$7.31 for $x \ge 10$. Thus,

$$C(x) = \begin{cases} 10.06x & \text{if} \quad 1 \le x < 4 \\ 8.52x & \text{if} \quad 4 \le x < 10 \\ 7.31x & \text{if} \quad 10 \le x \end{cases}$$

(B) The graph of C is:

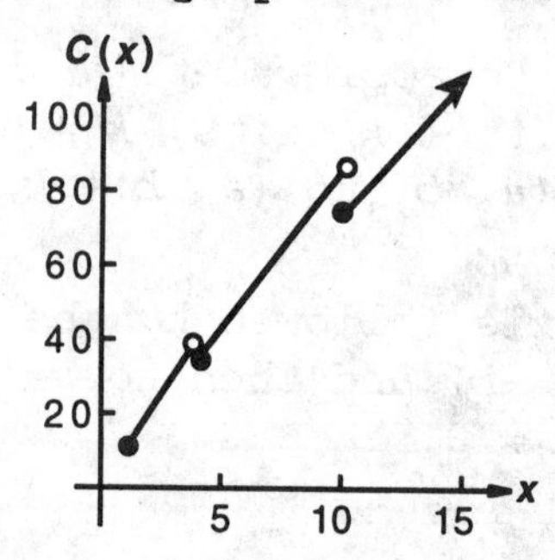

(C) $C(3) = 10.06(3) = \$30.18$ and $C(4) = 8.52(4) = \$34.08$. The customer would have to decide if a fourth box for an additional \$3.90 was desirable. However, $C(9) = 8.52(9) = \$76.68$ while $C(10) = 7.31(10) = \$73.10$ so the cost of 10 boxes is *cheaper* than cost of 9 boxes; it is to the customer's advantage to buy 10 boxes instead of 9.

67. (A) Let x = number of miles driven.
For $0 \le x \le 50$, the charge is $0.50x$.
At $x = 50$, the charge is \$25.

For $50 \le x \le 100$, the charge is $25 + 0.35(x - 50) = 7.5 + 0.35x$.
At $x = 100$, the charge is $7.5 + 0.35(100) = \$42.5$

For $x \ge 100$, the charge is: $42.5 + 0.20(x - 100) = 22.5 + 0.20x$.

Thus,

$$C(x) = \begin{cases} 0.50x & \text{if} \quad 0 \le x \le 50 \\ 7.5 + 0.35x & \text{if} \quad 50 < x \le 100 \\ 22.5 + 0.20x & \text{if} \quad 100 < x \end{cases}$$

(B)

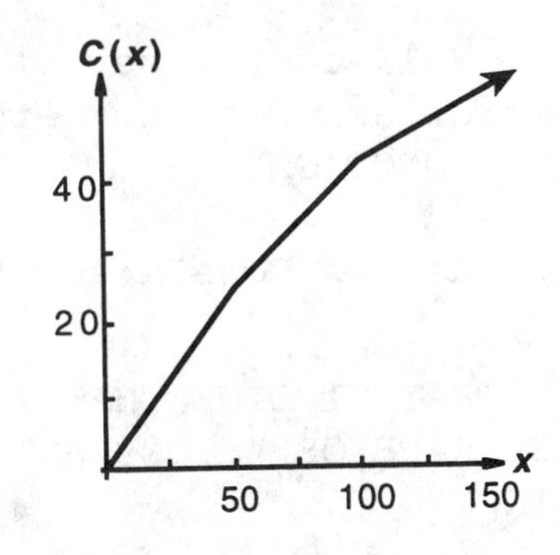

(C) The discount mileage rates do not apply to previous mileage. There are no breaks (gaps) in the graph at the points where the mileage rates change.

69. (A) The graph of the basic function $y = x$ is vertically expanded by a factor of 5.5 and shifted down 220 units.

(B)

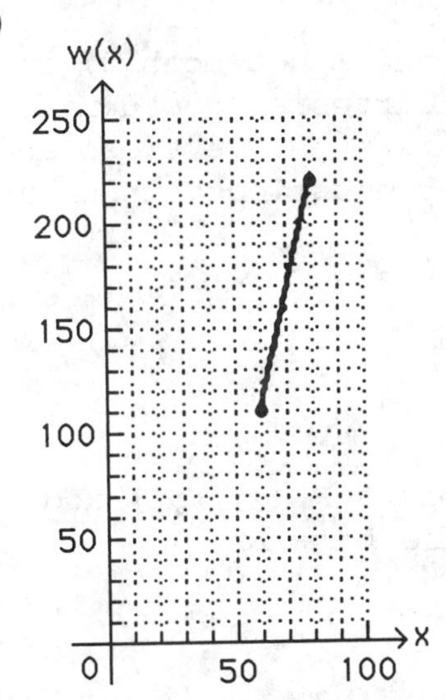

71. (A) The graph of the basic function $y = \sqrt{x}$ is vertically expanded by a factor of 7.08.

(B)

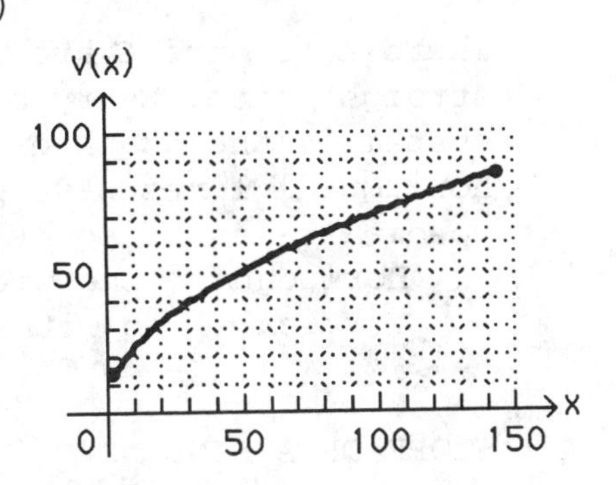

EXERCISE 1-3

Things to remember:

1. INTERCEPTS

   If the graph of a function f crosses the x axis at a point with x coordinate a, then a is called an **x intercept** of f. If the graph of f crosses the y axis at a point with y coordinate b, then b is called the **y intercept**. The x intercepts are the real solutions or roots of $f(x) = 0$; if f is defined at 0, then $f(0)$ is the y intercept.

2. LINEAR AND CONSTANT FUNCTIONS

A function f is a LINEAR FUNCTION if

$$f(x) = mx + b \qquad m \neq 0$$

where m and b are real numbers. The DOMAIN is the set of all real numbers and the RANGE is the set of all real numbers. If $m = 0$, then f is called a CONSTANT FUNCTION

$$f(x) = b$$

which has the set of all real numbers as its DOMAIN and the constant b as its RANGE.

THE GRAPH OF A LINEAR FUNCTION IS A STRAIGHT LINE THAT IS NEITHER HORIZONTAL NOR VERTICAL. THE GRAPH OF A CONSTANT FUNCTION IS A HORIZONTAL STRAIGHT LINE.

3. GRAPH OF A LINEAR EQUATION IN TWO VARIABLES

The graph of any equation of the form

$$AX + By = C \qquad \text{Standard Form} \qquad (5)$$

where A, B, and C are real constants (A and B not both 0) is a straight line. Every straight line in a Cartesian coordinate system is the graph of an equation of this type. Vertical and horizontal lines have particularly simple equations, which are special cases of equation (5):

Horizontal line with y intercept b: $y = b$

Vertical line with x intercept a: $x = a$

4. SLOPE OF A LINE

If a line passes through two distinct points $P_1(x_1, y_1)$ and $P_2(x_2, y_2)$, then its slope is given by the formula

$$m = \frac{y_2 - y_1}{x_2 - x_1} \qquad x_1 \neq x_2$$

$$= \frac{\text{Vertical change (rise)}}{\text{Horizontal change (run)}}$$

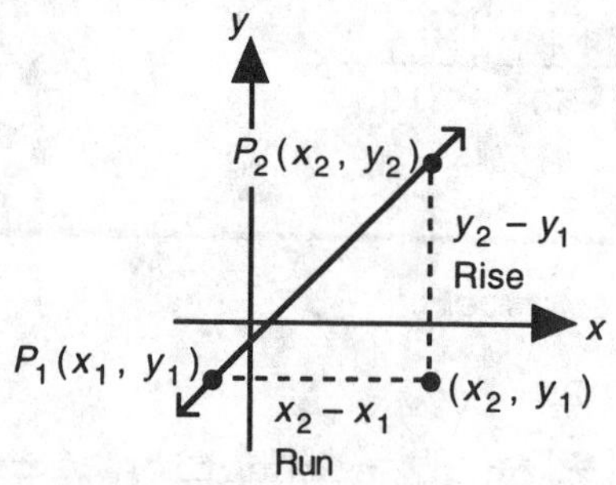

GEOMETRIC INTERPRETATION OF SLOPE

Line	Slope	Example
Rising as x moves from left to right	Positive	
Falling as x moves from left to right	Negative	
Horizontal	0	
Vertical	Not defined	

5. EQUATIONS OF A LINE

Standard form	$Ax + By = C$	A and B not both 0
Slope-intercept form	$y = mx + b$	Slope: m; y intercept: b
Point-slope form	$y - y_1 = m(x - x_1)$	Slope: m; Point: (x_1, y_1)
Horizontal line	$y = b$	Slope: 0
Vertical line	$x = a$	Slope: Undefined

1. (d)

3. (c); The slope is 0.

5. $y = 2x - 3$

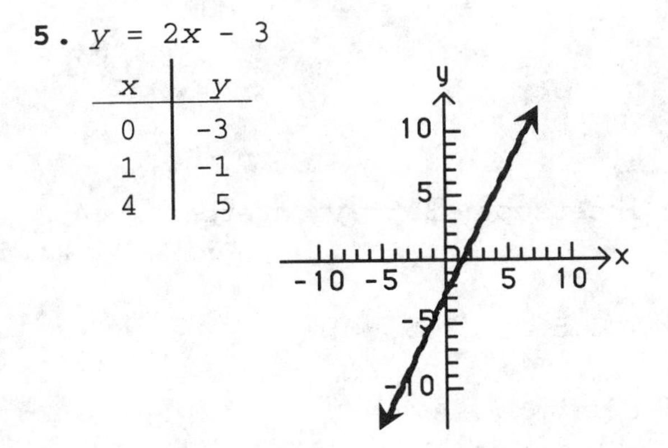

x	y
0	-3
1	-1
4	5

7. $2x + 3y = 12$

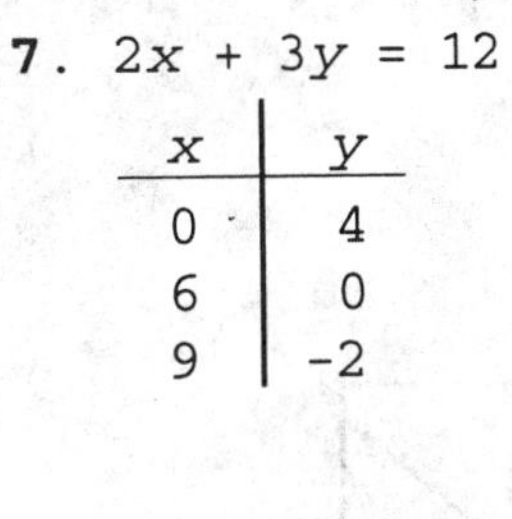

x	y
0	4
6	0
9	-2

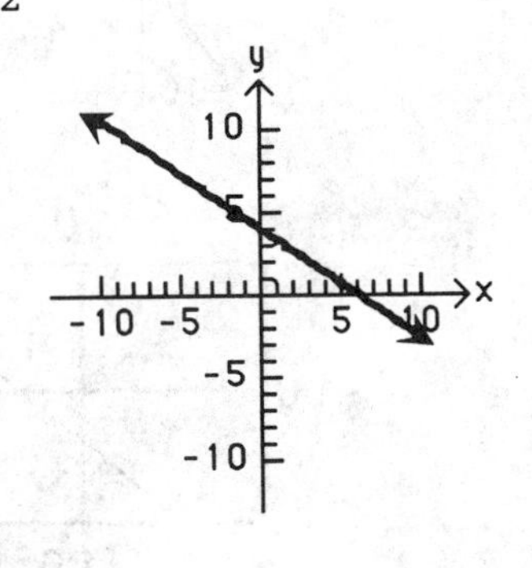

9. Slope $m = 2$

y intercept $b = -3$

11. Slope $m = -\frac{2}{3}$

y intercept $b = 2$

13. $m = -2$

$b = 4$

Using $\underline{5}$, $y = -2x + 4$.

15. $m = -\frac{3}{5}$

$b = 3$

Using $\underline{5}$, $y = -\frac{3}{5}x + 3$.

17. $y = -\frac{2}{3}x - 2$

$m = -\frac{2}{3}$, $b = -2$

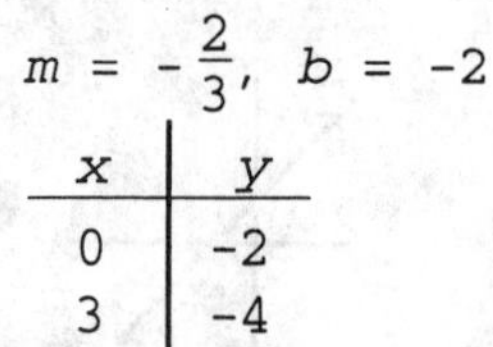

x	y
0	-2
3	-4
-3	0

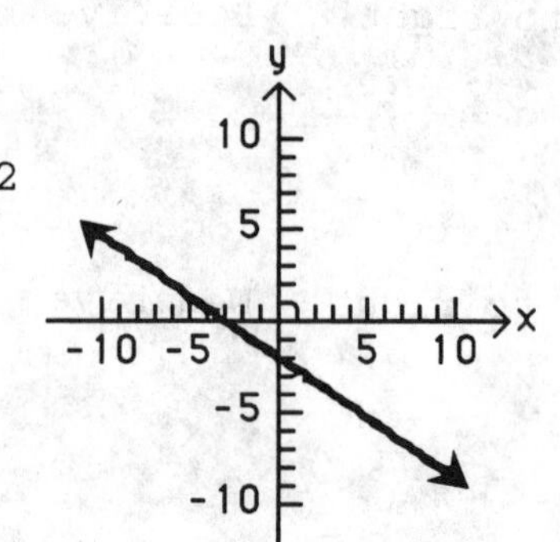

19. $3x - 2y = 10$

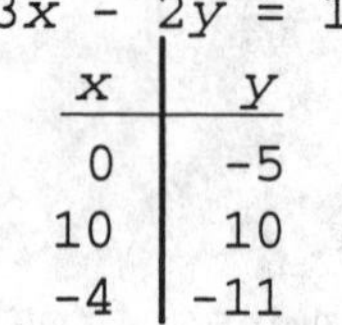

x	y
0	-5
10	10
-4	-11

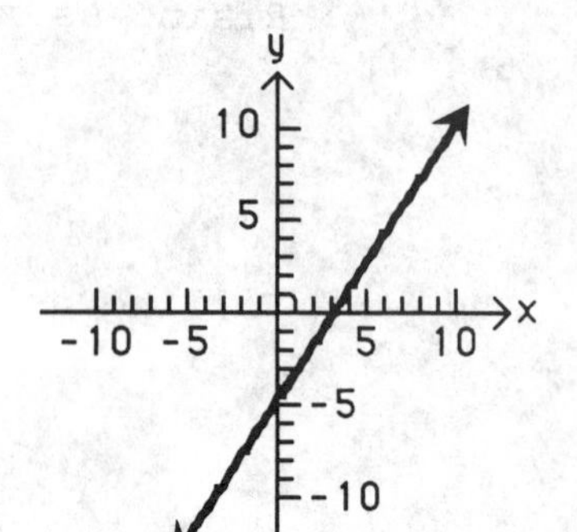

21.

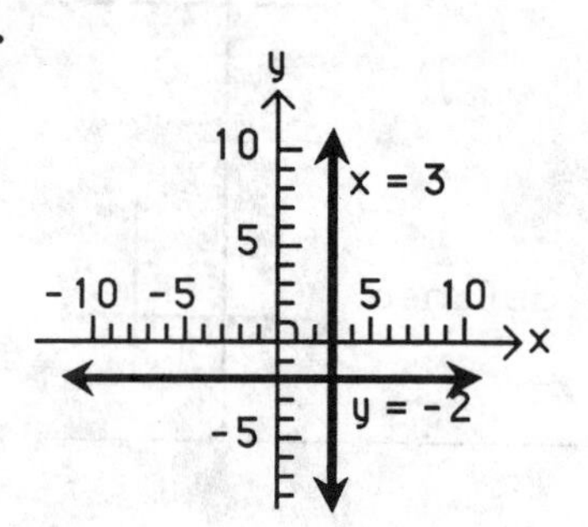

23. $3x + y = 5$

$y = -3x + 5$

$m = -3$ (using $\underline{5}$)

25. $2x + 3y = 12$

$3y = -2x + 12$

Divide both sides by 3:

$y = -\frac{2}{3}x + \frac{12}{3} = -\frac{2}{3}x + 4$

$m = -\frac{2}{3}$ (using $\underline{5}$)

27. (A)

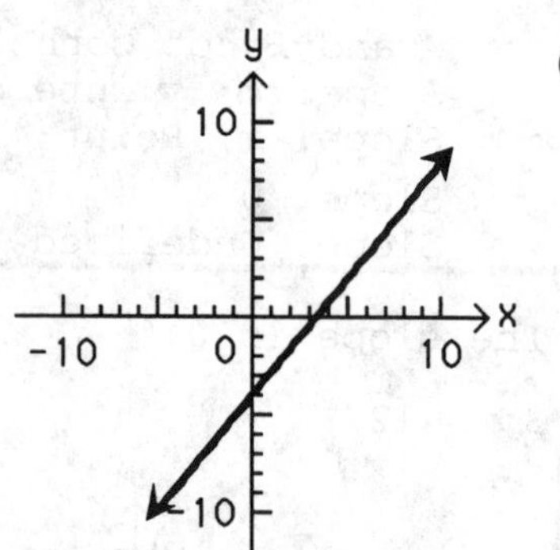

(B) x intercept--set $f(x) = 0$:

$1.2x - 4.2 = 0$

$x = 3.5$

y intercept--set $x = 0$:

$y = -4.2$

(C)

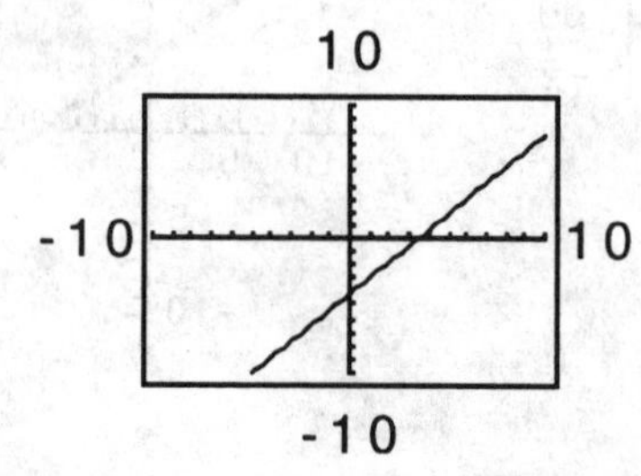

(D) x intercept: 3.5; y intercept: -4.2

(E) $x > 3.5$ or $(3.5, \infty)$

29. Using $\underline{3}$ with $a = 3$ for the vertical line and $b = -5$ for the horizontal line, we find that the equation of the vertical line is $x = 3$ and the equation of the horizontal line is $y = -5$.

31. Using 3 with $a = -1$ for the vertical line and $b = -3$ for the horizontal line, we find that the equation of the vertical line is $x = -1$ and the equation of the horizontal line is $y = -3$.

33. $m = -3$
For the point $(4, -1)$, $x_1 = 4$ and $y_1 = -1$. Using 6, we get:

$$y - (-1) = -3(x - 4)$$
$$y + 1 = -3x + 12$$
$$y = -3x + 11$$

35. $m = \frac{2}{3}$
For the point $(-6, -5)$, $x_1 = -6$ and $y_1 = -5$. Using 6, we get:

$$y - (-5) = \frac{2}{3}[x - (-6)]$$
$$y + 5 = \frac{2}{3}(x + 6)$$
$$y + 5 = \frac{2}{3}x + 4$$
$$y = \frac{2}{3}x - 1$$

37. $y - (-5) = 0(x - 3)$
$y + 5 = 0$ or $y = -5$
$y = 0$

39. The points are $(1, 3)$ and $(7, 5)$. Let $x_1 = 1$, $y_1 = 3$, $x_2 = 7$, and $y_2 = 5$. Using 4, we get:

$$m = \frac{5 - 3}{7 - 1} = \frac{2}{6} = \frac{1}{3}$$

41. Let $x_1 = -5$, $y_1 = -2$, $x_2 = 5$, and $y_2 = -4$. Using 4, we get:

$$m = \frac{-4 - (-2)}{5 - (-5)} = \frac{-4 + 2}{5 + 5} = \frac{-2}{10} = -\frac{1}{5}$$

43. $m = \frac{-3 - 7}{2 - 2} = \frac{-10}{0}$, the slope is not defined; the line through $(2, 7)$ and $(2, -3)$ is vertical.

45. $m = \frac{3 - 3}{-5 - 2} = \frac{0}{-7} = 0$

47. First, find the slope using 4:

$$m = \frac{y_2 - y_1}{x_2 - x_1} = \frac{5 - 3}{7 - 1} = \frac{2}{6} = \frac{1}{3}$$

Then, by using 6, $y - y_1 = m(x - x_1)$, where $m = \frac{1}{3}$ and $(x_1, y_1) = (1, 3)$ or $(7, 5)$, we get:

$y - 3 = \frac{1}{3}(x - 1)$ or $y - 5 = \frac{1}{3}(x - 7)$

These two equations are equivalent. After simplifying either one of these, we obtain:
$-x + 3y = 8$ or $x - 3y = -8$

49. First, find the slope using 4:

$$m = \frac{-4 - (-2)}{5 - (-5)} = \frac{-4 + 2}{5 + 5} = \frac{-2}{10} = -\frac{1}{5}$$

By using 6, and either one of these points, we obtain:

$$y - (-2) = -\frac{1}{5}[x - (-5)] \quad \text{[using } (-5, -2)\text{]}$$
$$y + 2 = -\frac{1}{5}(x + 5)$$
$$5(y + 2) = -x - 5$$
$$5y + 10 = -x - 5$$
$$x + 5y = -15$$

51. (2, 7) and (2, -3)
Since each point has the same
x coordinate, the graph of the line formed by these two points will be a *vertical line*. Then, using 3, with $a = 2$, we have $x = 2$ as the equation of the line.

53. (2, 3) and (-5, 3)
Since each point has the same
y coordinate, the graph of the line formed by these two points will be a *horizontal line*. Then, using 3, with $b = 3$, we have $y = 3$ as the equation of the line.

55. A linear function **57.** Not a function **59.** A constant function

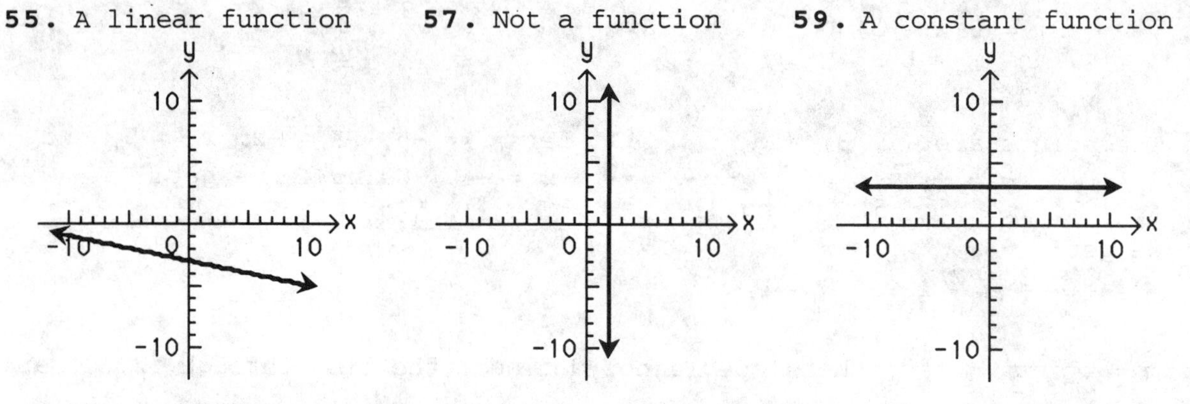

61. The graphs of $y = mx + 2$, m any real number, all have the same y intercept (0, 2); for each real number m, $y = mx + 2$ is a non-vertical line that passes through the point (0, 2).

65. The graph of $g(x) = |mx + b|$ coincides with the graph of $f(x) = mx + b$ for all x satisfying $mx + b \geq 0$. The graph of g is the reflection of the graph of f in the x axis for all x satisfying $mx + b < 0$. The function g is **never** a linear function.

67. We are given $A = 100(0.06)t + 100 = 6t + 100$

(A) At $t = 5$, we have $A = 6(5) + 100 = \$130$
At $t = 20$, we have $A = 6(20) + 100 = \$220$

(B)

A
$200
$100
0
10
20
t

(C) The equation $A = 6t + 100$ is in slope-intercept form. Thus, the slope is 6. Interpretation: The amount in the account is growing at the rate of $6 per year.

69. (A) We find an equation $C(x) = mx + b$ for the line passing through (0, 200) and (20, 3800).

$$m = \frac{3800 - 200}{20 - 0} = \frac{3600}{20} = 180$$

Also, since $C(x) = 200$ when $x = 0$, it follows that $b = 200$. Thus, $C(x) = 180x + 200$.

(B) The total costs at 12 boards per day are:

$C(x) = 180(12) + 200 = 2,360$ or \$2,360

(C)

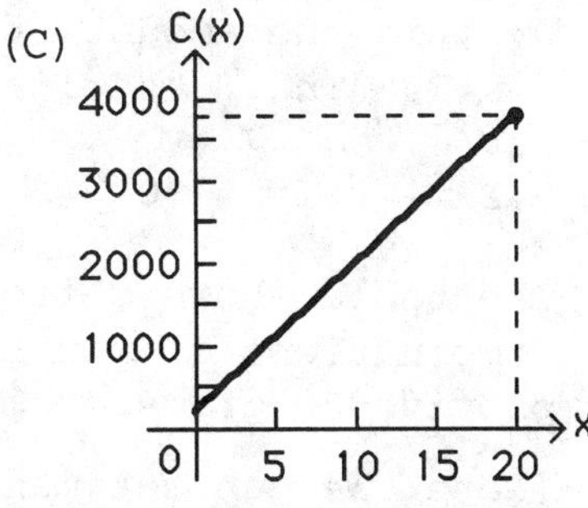

71. (A)

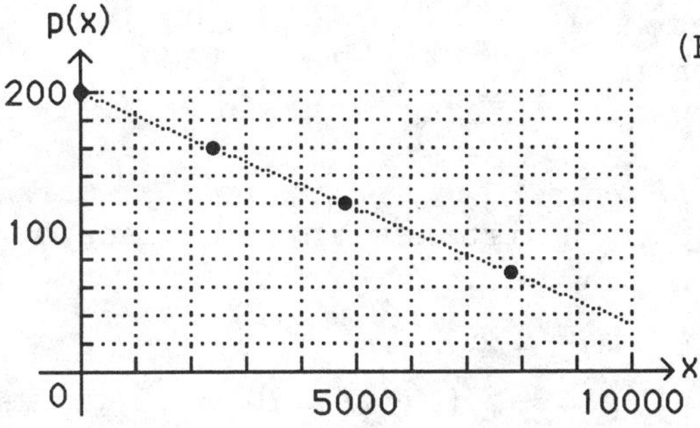

(B) slope: $m = \frac{160 - 200}{2,400 - 0} = \frac{-40}{2400} = -\frac{1}{60}$

y intercept: 200

equation: $p(x) = -\frac{1}{60}x + 200$

(C) $p(3000) = -\frac{1}{60}(3000) + 200 = -50 + 200 = \$150$

(D) The equation $p(x) = -\frac{1}{60}x + 200$ is in slope-intercept form. Thus the slope is $-\frac{1}{60} = -0.01666 \approx -0.02$. The price decreases \$0.02 (2 cents) for each unit increase in demand.

73. $f(x) = 5.74 + 0.97x$

(A)

x	0	1	2	3	4
Sales	5.9	6.5	7.7	8.6	9.7
$f(x)$	5.7	6.7	7.7	8.6	9.6

(B)

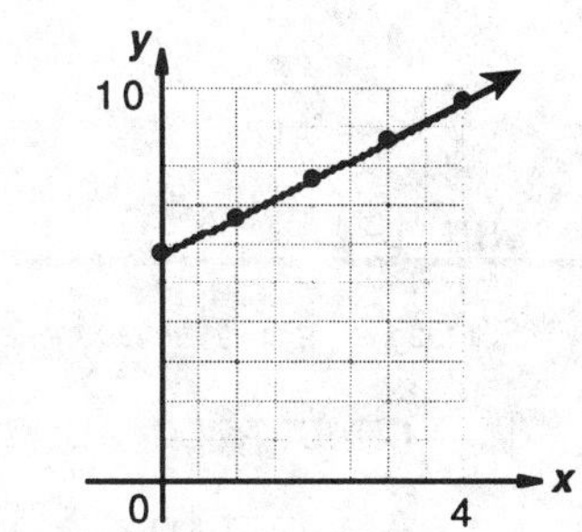

(C) The year 1993 corresponds to $x = 5$;

$f(5) = 5.74 + 0.97(5) = 5.74 + 4.85$
$= 10.59$

Estimated sales in 1993: \$10.6 billion.

The year 2000 corresponds to $x = 12$;

$f(12) = 5.74 + 0.97(12) = 5.74 + 11.64 = 17.38$

Estimated sales in 2000: \$17.4 billion.

(D) The sales are \$5.9 billion in 1988 and increase at approximately \$0.97 billion per year for the next four years.

75. Mix A contains 20% protein. Mix B contains 10% protein. Let x be the amount of A used, and let y be the amount of B used. Then $0.2x$ is the amount of protein from mix A and $0.1y$ is the amount of protein from mix B. Thus, the linear equation is:
$0.2x + 0.1y = 20$

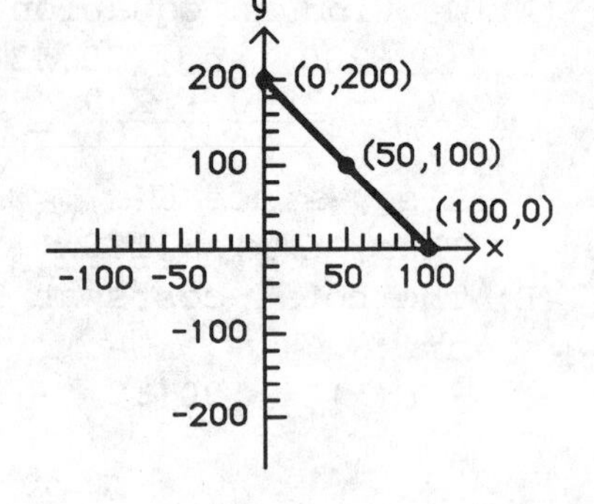

The table shows different combinations of mix A and mix B to provide 20 grams of protein.

Mix A x	Mix B y
100	0
0	200
50	100
10	180

[Note: We can get many more combinations. In fact, each point on the graph indicates a combination of mix A and mix B.]

77. $p = -\frac{1}{5}d + 70$, $30 \le d \le 175$, where d = distance in centimeters and p = pull in grams

(A) $d = 30$
$p = -\frac{1}{5}(30) + 70 = 64$ grams

$d = 175$
$p = -\frac{1}{5}(175) + 70 = 35$ grams

(B)

d	p
30	64
50	60
175	35

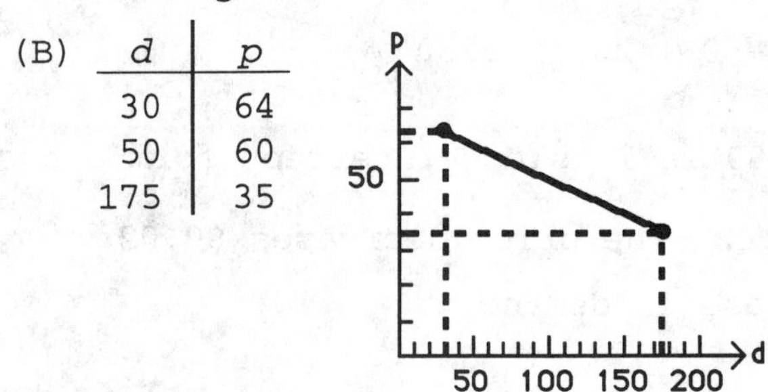

(C) Select two points (30, 64) and (50, 60) as (x_1, y_1) and (x_2, y_2), respectively, from part (B). Using 2:

$$\text{Slope } m = \frac{y_2 - y_1}{x_2 - x_1} = \frac{60 - 64}{50 - 30}$$
$$= -\frac{4}{20} = -\frac{1}{5}$$

EXERCISE 1-4

Things to remember:

1. QUADRATIC FUNCTION

A function f is a QUADRATIC FUNCTION if
$$f(x) = ax^2 + bx + c \qquad a \neq 0$$
where a, b, and c are real numbers. The domain of a quadratic function is the set of all real numbers.

2. PROPERTIES OF A QUADRATIC FUNCTION AND ITS GRAPH

Given a quadratic function

$f(x) = ax^2 + bx + c \quad a \neq 0$

and the form obtained by completing the square

$f(x) = a(x - h)^2 + k$

we summarize general properties as follows:

a. The graph of f is a parabola:

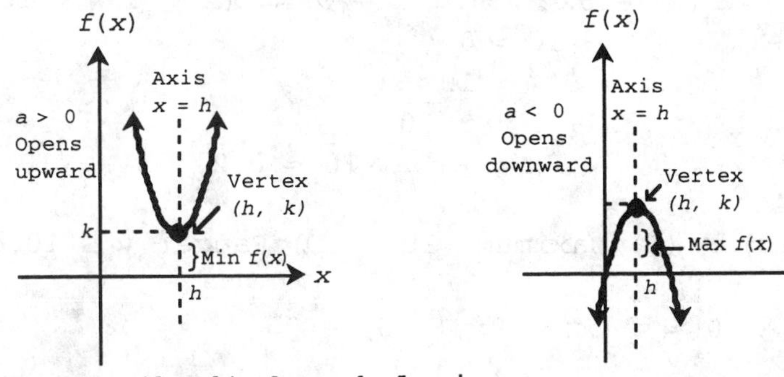

b. Vertex: (h, k) [parabola increases on one side of the vertex and decreases on the other]

c. Axis (of symmetry): $x = h$ (parallel to y axis)

d. $f(h) = k$ is the minimum if $a > 0$ and the maximum if $a < 0$

e. Domain: All real numbers
Range: $(-\infty, k]$ if $a < 0$ or $[k, \infty)$ if $a > 0$

f. The graph of f is the graph of $g(x) = ax^2$ translated horizontally h units and vertically k units.

---

1. (a), (c), (e), (f)

3. (A) m (B) g (C) f (D) n

5. (A) x intercepts: 1, 3; y intercept: -3 (B) Vertex: (2, 1)
(C) Maximum: 1 (D) Range: $y \leq 1$ or $(-\infty, 1]$
(E) Increasing interval: $x \leq 2$ or $(-\infty, 2]$
(F) Decreasing interval: $x \geq 2$ or $[2, \infty)$

7. (A) x intercepts: -3, -1; y intercept: 3 (B) Vertex: (-2, -1)
(C) Minimum: -1 (D) Range: $y \geq -1$ or $[-1, \infty)$
(E) Increasing interval: $x \geq -2$ or $[-2, \infty)$
(F) Decreasing interval: $x \leq -2$ or $(-\infty, -2]$

9. $f(x) = -(x - 2)^2 + 1 = -x^2 + 4x - 4 + 1 = -x^2 + 4x - 3 = -(x - 3)(x - 1)$
(A) x intercepts: 1, 3; y intercepts: -3 (B) Vertex: (2, 1)
(C) Maximum: 1 (D) Range: $y \leq 1$ or $(-\infty, 1]$

11. $M(x) = (x + 2)^2 - 1 = x^2 + 4x + 4 - 1 = x^2 + 4x + 3 = (x + 3)(x + 1)$
(A) x intercepts: -3, -1; y intercept 3 (B) Vertex: (-2, -1)
(C) Minimum: -1 (D) Range: $[-1, \infty)$

13. $y = -[x - (-2)]^2 + 5 = -(x + 2)^2 + 5$

15. $y = (x - 1)^2 - 3$

17. $f(x) = x^2 - 8x + 13 = x^2 - 8x + 16 - 3 = (x - 4)^2 - 3$

(A) x intercepts: $(x - 4)^2 - 3 = 0$

$$(x - 4)^2 = 3$$
$$x - 4 = \pm\sqrt{3}$$
$$x = 4 + \sqrt{3} \approx 5.7,\ 4 - \sqrt{3} \approx 2.3$$

y intercept: 13

(B) Vertex: $(4, -3)$ (C) Minimum: -3 (D) Range: $y \geq -3$ or $[-3, \infty)$

19. $M(x) = 1 - 6x - x^2 = -(x^2 + 6x + 9) + 1 + 9 = -(x + 3)^2 + 10$

(A) x intercepts: $-(x + 3)^2 + 10 = 0$

$$(x + 3)^2 = 10$$
$$x + 3 = \pm\sqrt{10}$$
$$x = -3 + \sqrt{10} \approx 0.2,\ -3 - \sqrt{10} = -6.2$$

y intercept: 1

(B) Vertex: $(-3, 10)$ (C) Maximum: 10 (D) Range: $y \leq 10$ or $(-\infty, 10]$

21. $G(x) = 0.5x^2 - 4x + 10 = \frac{1}{2}(x^2 - 8x + 16) + 2$

$$= \frac{1}{2}(x - 4)^2 + 2$$

(A) x intercepts: none, since $G(x) = \frac{1}{2}(x - 4)^2 + 2 \geq 2$ for all x;
y intercept: 10

(B) Vertex: $(4, 2)$ (C) Minimum: 2 (D) Range: $y \geq 2$ or $[2, \infty)$

23. $f(x) = 0.3x^2 - x - 8$

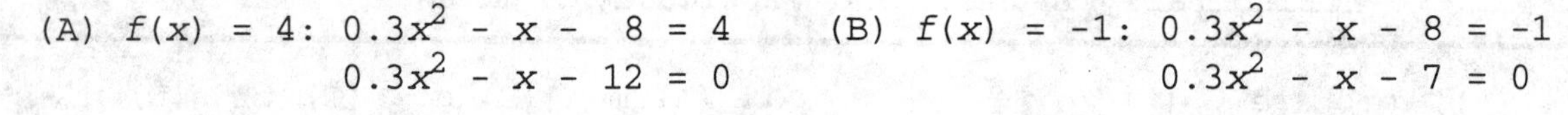

(A) $f(x) = 4$: $0.3x^2 - x - 8 = 4$
$0.3x^2 - x - 12 = 0$

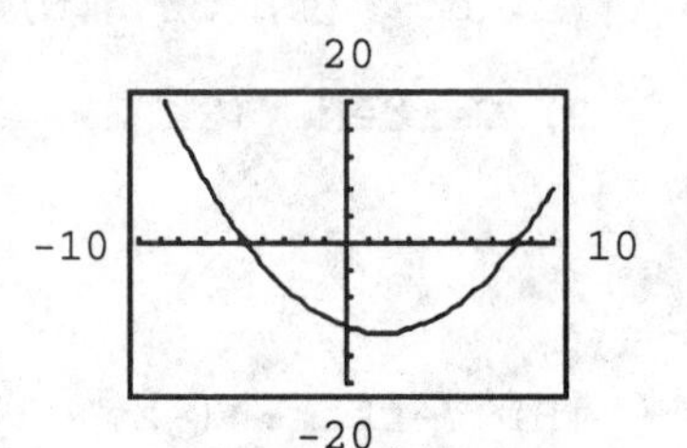

$x = -4.87,\ 8.21$

(B) $f(x) = -1$: $0.3x^2 - x - 8 = -1$
$0.3x^2 - x - 7 = 0$

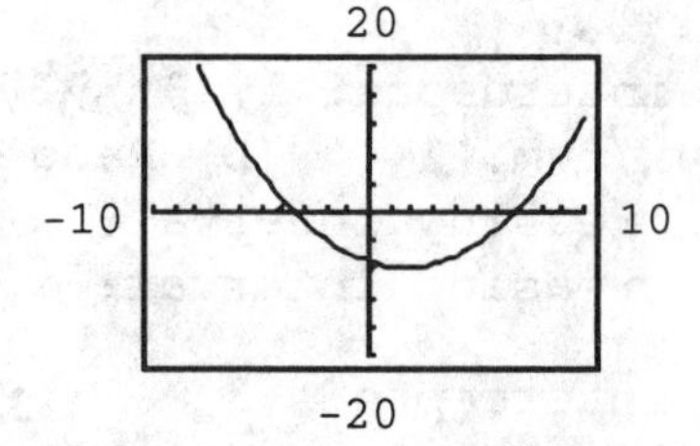

$x = -3.44,\ 6.78$

(C) $f(x) = -9$: $0.3x^2 - x - 8 = -9$
$0.3x^2 - x + 1 = 0$

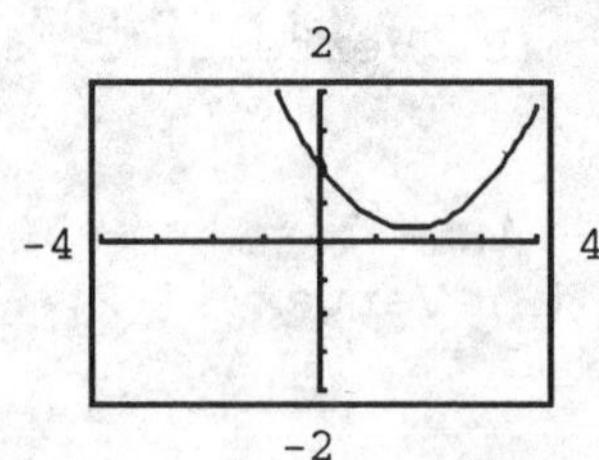

No solutions.

25. The vertex of the parabola is on the x axis.

27. $g(x) = 0.25x^2 - 1.5x - 7 = 0.25(x^2 - 6x + 9) - 2.25 - 7$
$= 0.25(x - 3)^2 - 9.25$

(A) x intercepts: $0.25(x - 3)^2 - 9.25 = 0$
$(x - 3)^2 = 37$
$x - 3 = \pm\sqrt{37}$
$x = 3 + \sqrt{37} \approx 9.1,\ 3 - \sqrt{37} \approx -3.1$

y intercept: -7

(B) Vertex: $(3, -9.25)$ (C) Minimum: -9.25

(D) Range: $y \geq -9.25$ or $[-9.25, \infty)$

29. $f(x) = -0.12x^2 + 0.96x + 1.2$
$= -0.12(x^2 - 8x + 16) + 1.92 + 1.2$
$= -0.12(x - 4)^2 + 3.12$

(A) x intercepts: $-0.12(x - 4)^2 + 3.12 = 0$
$(x - 4)^2 = 26$
$x - 4 = \pm\sqrt{26}$
$x = 4 + \sqrt{26} \approx 9.1,\ 4 - \sqrt{26} \approx -1.1$

y intercept: 1.2

(B) Vertex: $(4, 3.12)$ (C) Maximum: 3.12

(D) Range: $y \leq 3.12$ or $(-\infty, 3.12]$

31.

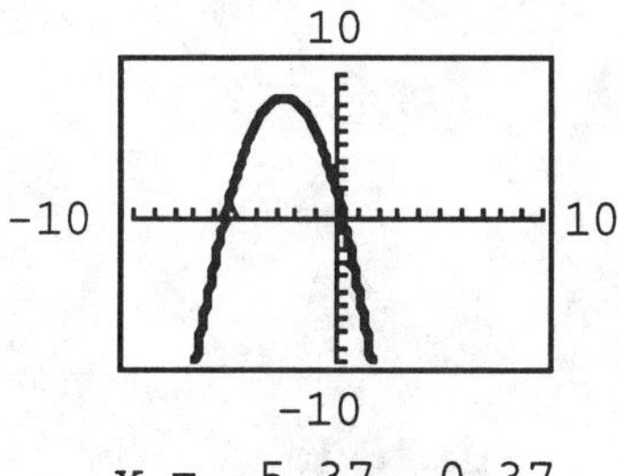

$x = -5.37, 0.37$

33.

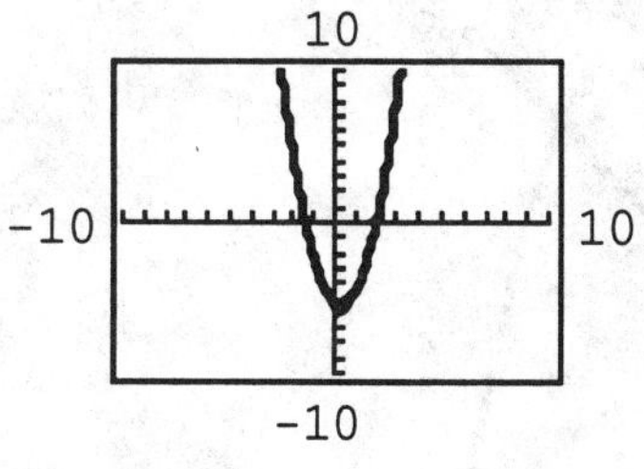

$-1.37 < x < 2.16$

35.

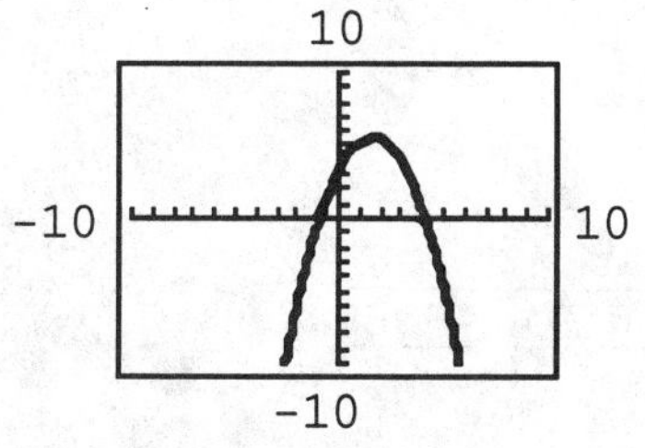

10

-10 10

-10

$x \leq -0.74$ or $x \geq 4.19$

37. f is a quadratic function and min $f(x) = f(2) = 4$
Axis: $x = 2$
Vertex: $(2, 4)$
Range: $y \geq 4$ or $[4, \infty)$
x intercepts: None

39. (A)

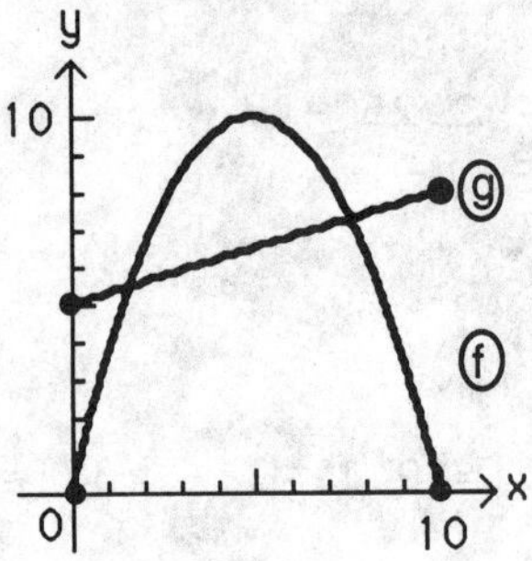

(B) $f(x) = g(x)$

$$-0.4x(x - 10) = 0.3x + 5$$
$$-0.4x^2 + 4x = 0.3x + 5$$
$$-0.4x^2 + 3.7x = 5$$
$$-0.4x^2 + 3.7x - 5 = 0$$
$$x = \frac{-3.7 \pm \sqrt{3.7^2 - 4(-0.4)(-5)}}{2(-0.4)}$$
$$x = \frac{-3.7 \pm \sqrt{5.69}}{-0.8} \approx 1.64,\ 7.61$$

(C) $f(x) > g(x)$ for $1.64 < x < 7.61$

(D) $f(x) < g(x)$ for $0 \le x < 1.64$ or $7.61 < x \le 10$

41. (A)

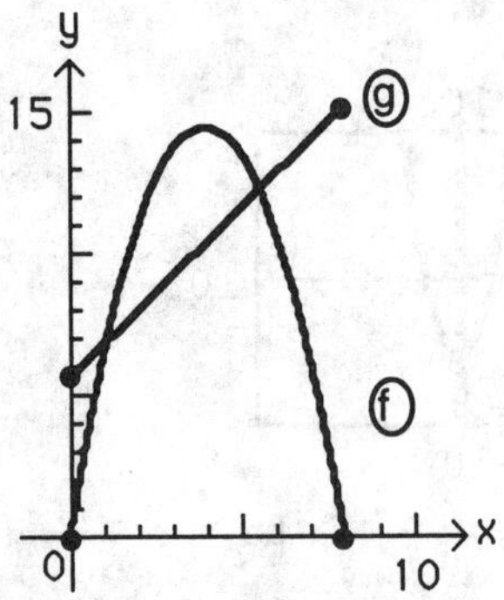

(B) $f(x) = g(x)$

$$-0.9x^2 + 7.2x = 1.2x + 5.5$$
$$-0.9x^2 + 6x = 5.5$$
$$-0.9x^2 + 6x - 5.5 = 0$$
$$x = \frac{-6 \pm \sqrt{36 - 4(-0.9)(-5.5)}}{2(-0.9)}$$
$$x = \frac{-6 \pm \sqrt{16.2}}{-1.8} \approx 1.1,\ 5.57$$

(C) $f(x) > g(x)$ for $1.10 < x < 5.57$

(D) $f(x) < g(x)$ for $0 \le x < 1.10$ or $5.57 < x \le 8$

43. $f(x) = x^2 + 1$ and $g(x) = -(x - 4)^2 - 1$ are two examples. Their graphs are:

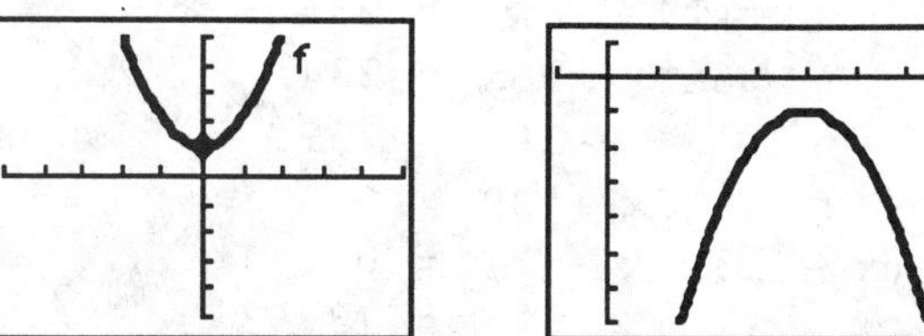

Their graphs do not intersect the x axis.

45. Mathematical model: $f(x) = -0.518x^2 + 33.3x - 481$

(A)

x	28	30	32	34	36
Mileage	45	52	55	51	47
$f(x)$	45.3	51.8	54.2	52.4	46.5

(B)

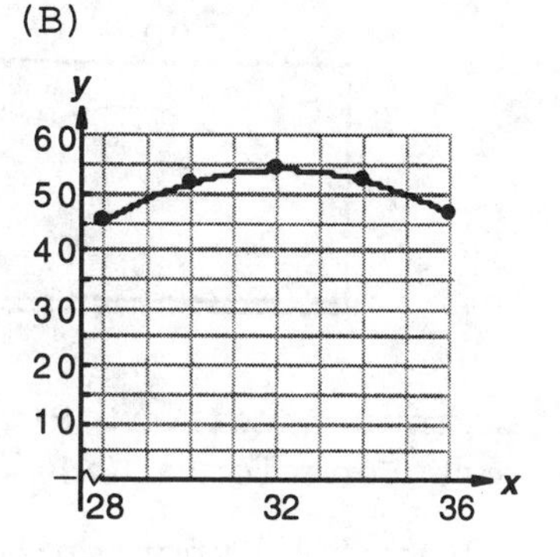

(C) $x = 31$: $f(31) = -0.518(31)^2 + 33.3(31) - 481 = 53.502$
$f(31) \approx 53.50$ thousand miles
$x = 35$: $f(35) = -0.518(35)^2 + 33(35) - 481 \approx 49.95$ thousand miles

(D) The maximum mileage is achieved at 32 lb/in^2 pressure. Increasing the pressure or decreasing the pressure reduces the mileage.

47. (A)

R(x)
700
0
15
x

(B) $R(x) = x(119 - 6x)$

$= -6x^2 + 119x = -6\left(x^2 - \frac{119}{6}x\right)$

$= -6(x^2 - 19.833x + 98.340) + 590.042$

$= -6(x - 9.917)^2 + 590.042$

Output: 9.917 million chips, i.e. 9,917,000 chips

Maximum revenue: 590.042 million dollars, i.e. \$590,042,000

(C)

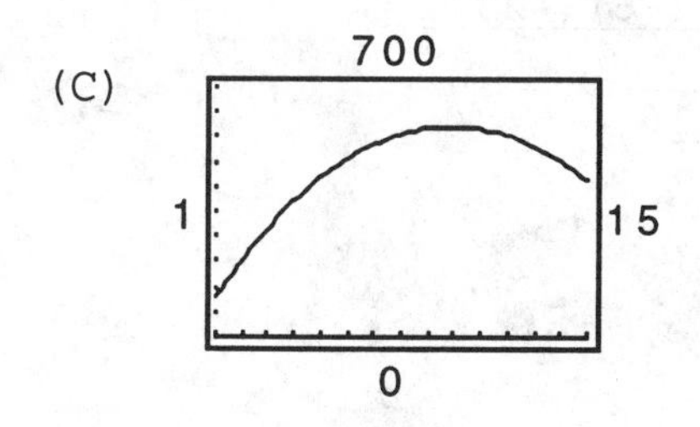

(D) 9.917 million chips (9,917,000 chips)
590.042 million dollars (\$590,042,000)

(E) $p(9.917) = 119 - 6(9.917) \approx \$59$

49. (A)

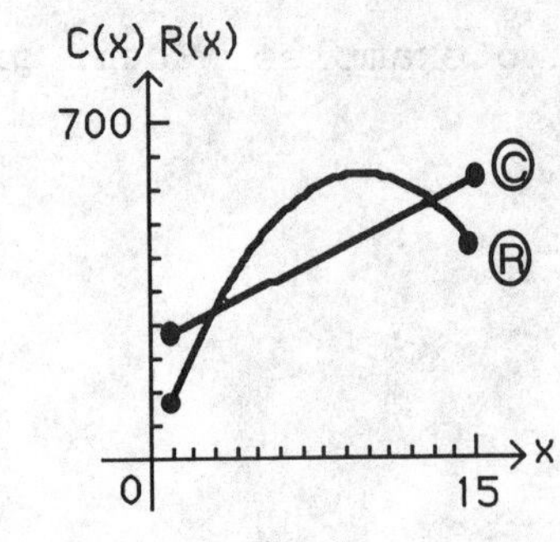

(B) $R(x) = C(x)$

$$x(119 - 6x) = 234 + 23x$$
$$-6x^2 + 96x = 234$$
$$x^2 - 16x = -39$$
$$x^2 - 16x + 39 = 0$$
$$(x - 13)(x - 3) = 0$$
$$x = 13,\ 3$$

Break-even at 3 million (3,000,000) and 13 million (13,000,000) chip production levels.

(C)

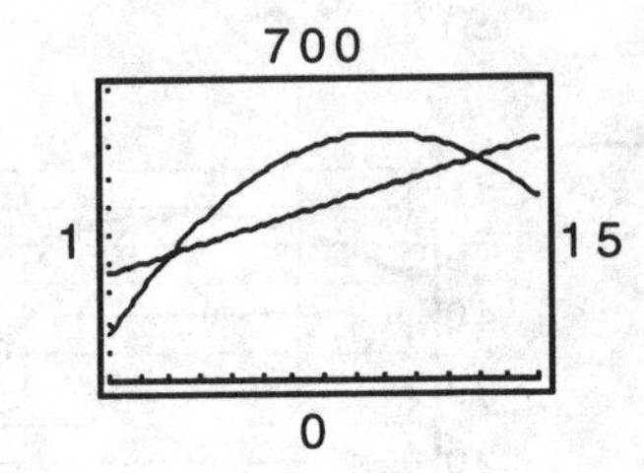

(D) Break-even at 3 million and 13 million chip production levels

(E) Loss: $1 \le x < 3$ or $13 < x \le 15$ Profit: $3 < x < 13$

51. Revenue Function: $R(x) = x(119 - 6x)$
Cost Function: $C(x) = 234 + 23x$

(A) Profit Function:

$$P(x) = R(x) - C(x) = x(119 - 6x) - (234 + 23x) = -6x^2 + 96x - 234$$

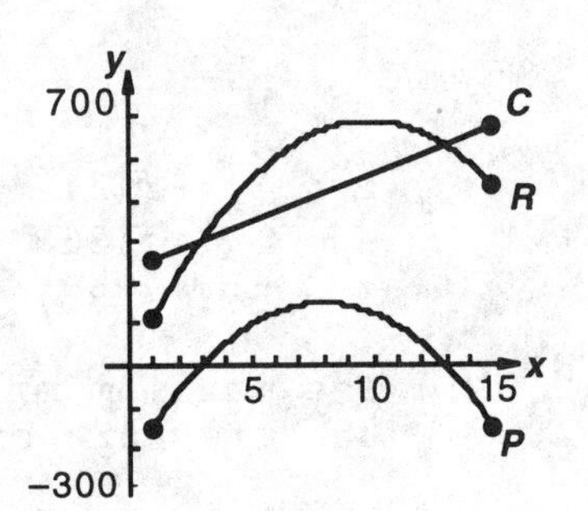

(B)

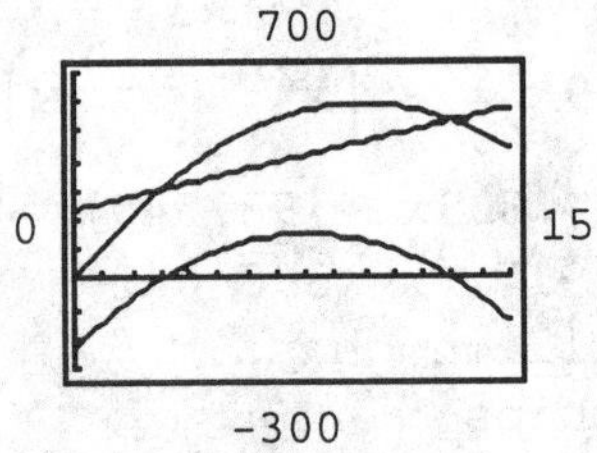

(C) The x coordinates of the intersection points of R and C are the same as the x intercepts of P.

(D) x intercepts of P: $-6x^2 + 96x - 234 = 0$

$$x = \frac{-96 \pm \sqrt{(96)^2 - 4(-6)(-234)}}{-12}$$
$$= \frac{-96 \pm \sqrt{9216 - 5616}}{-12}$$
$$= \frac{-96 \pm \sqrt{3600}}{-12}$$
$$= \frac{-96 \pm 60}{-12} = 13 \text{ or } 3$$

x intercepts of P and break-even points:
3 million (3,000,000) and 13 million (13,000,000) chips

(E) See the graphs in part (A) and (B). The x intercepts of P and the break-even points are 3 million and 13 million.

(F) The maximum profit and the maximum revenue do not occur at the same output level and they are not equal. The profit function involves both revenue and cost; the revenue function does not involve the production costs.

(G) $P(x) = -6x^2 + 96x - 234 = -6(x^2 - 16x + 64) + 384 - 234$
$= -6(x - 8)^2 + 150$
The maximum profit is 150 million dollars (\$150,000,000); it occurs at an output level of 8 million (8,000,000) chips. From Problem 47(D), the maximum revenue is 590.042 million dollars (590,042,000); the maximum profit is much smaller than the maximum revenue.

(H) See the graphs in parts (A) and (B).

53. (A) Solve: $f(x) = 1{,}000(0.04 - x^2) = 20$

$$40 - 1000x^2 = 20$$
$$1000x^2 = 20$$
$$x^2 = 0.02$$
$$x = 0.14 \text{ or } -0.14$$

Since we are measuring distance, we take the positive solution: $x = 0.14$ cm

(B) $x = 0.14$ cm

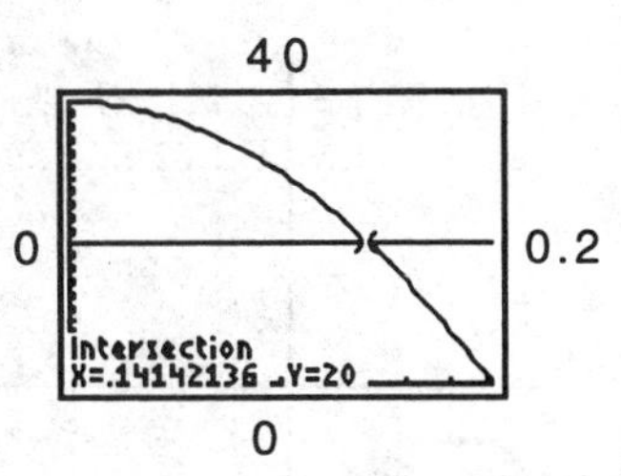

CHAPTER 1 REVIEW

1.

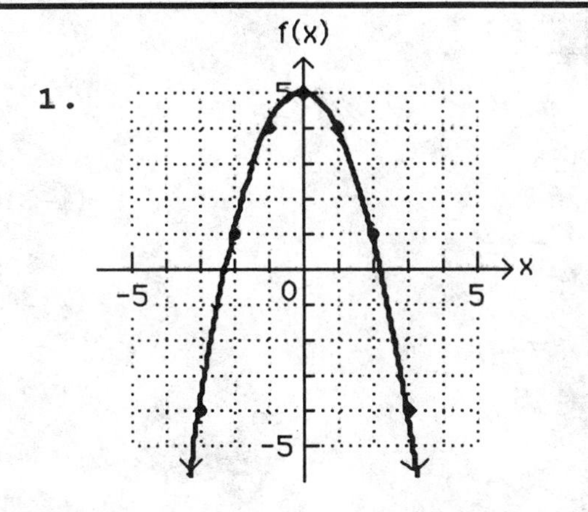

(1-1)

2. (A) Not a function; fails vertical line test (B) A function
(C) A function (D) Not a function; fails vertical line test (1-1)

3. $f(x) = 2x - 1$, $g(x) = x^2 - 2x$

(A) $f(-2) + g(-1) = 2(-2) - 1 + (-1)^2 - 2(-1) = -2$

(B) $f(0) \cdot g(4) = (2 \cdot 0 - 1)(4^2 - 2 \cdot 4) = -8$

(C) $\dfrac{g(2)}{f(3)} = \dfrac{2^2 - 2 \cdot 2}{2 \cdot 3 - 1} = 0$

(D) $\dfrac{f(3)}{g(2)}$ not defined because $g(2) = 0$ (1-1)

4. (A) $y = 4$ (B) $x = 0$ (C) $y = 1$ (D) $x = -1$ or 1
(E) $y = -2$ (F) $x = -5$ or 5 (1-1)

5. (A)

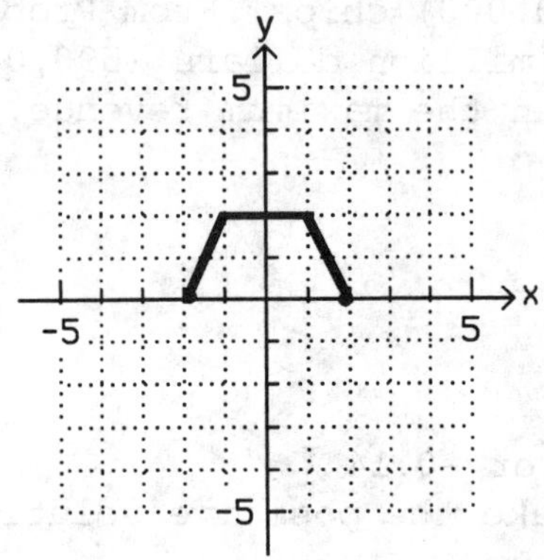

(B)

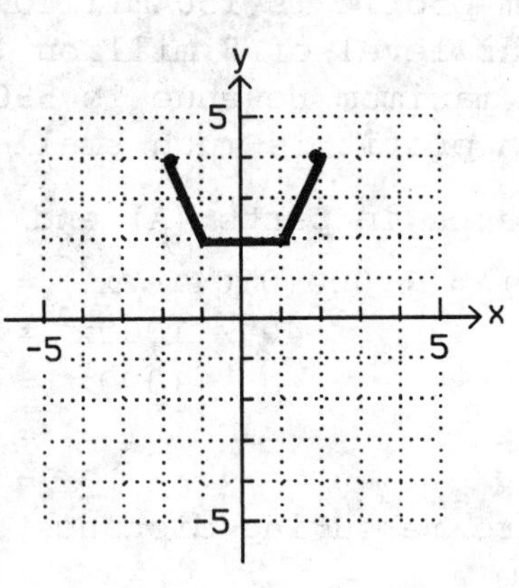

(C)

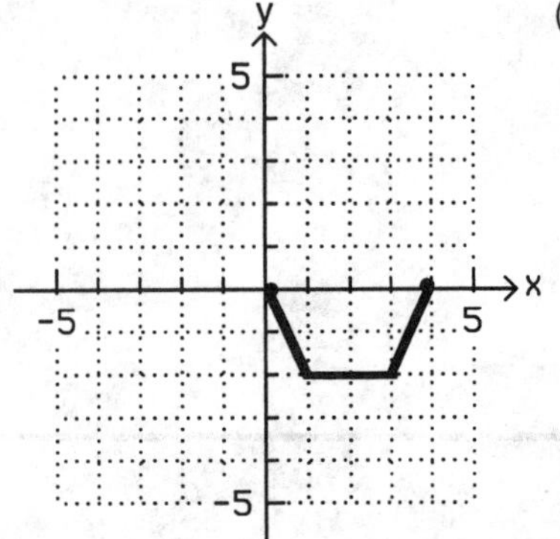

(D)

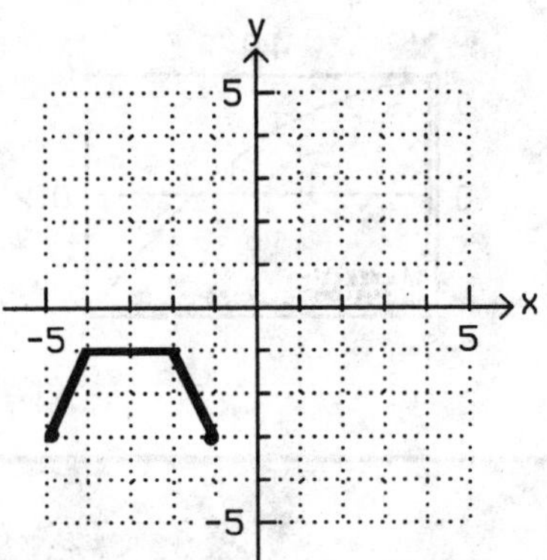

(1-2)

6. (A) (n) (B) (g) (C) (m); slope is zero
(D) (f); slope is not defined (1-3)

7. $y = -\frac{2}{3}x + 6$ (1-3)

8. vertical line: $x = -6$; horizontal line: $y = 5$ (1-3)

9. x intercept: $2x = 18$, $x = 9$;
y intercept: $-3y = 18$, $x = -6$;
slope-intercept form: $y = \frac{2}{3}x - 6$; slope $= \frac{2}{3}$
graph:

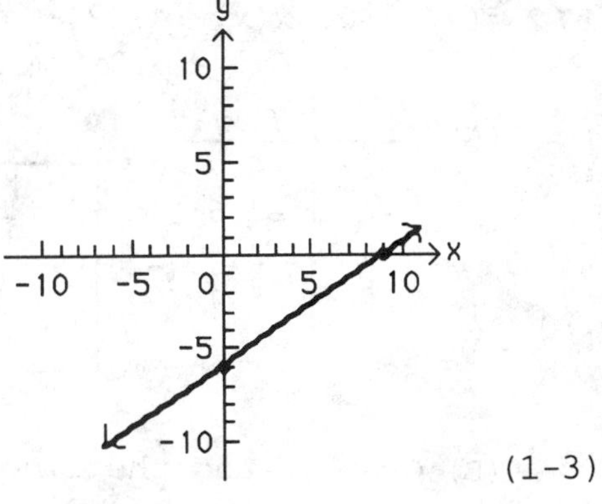

(1-3)

10. (b), (c), (d), (f) (1-4)

11. (A) g (B) m (C) n (D) f (1-2, 1-4)

12. $y = f(x) = (x + 2)^2 - 4$

(A) x intercepts: $(x + 2)^2 - 4 = 0$
$(x + 2)^2 = 4$
$x + 2 = -2$ or 2
$x = -4, 0$

y intercept: 0

(B) Vertex: $(-2, -4)$ (C) Minimum: -4 (D) Range: $y \geq -4$ or $[-4, \infty)$

(E) Increasing interval $[-2, \infty)$ (F) Decreasing interval $(-\infty, -2]$

(1-4)

13. Linear function: (a), (c), (e), (f); Constant function: (d) (1-3)

14. (A) $x^2 - x - 6 = 0$ at $x = -2, 3$
Domain: all real numbers except $x = -2, 3$

(B) $5 - x > 0$ for $x < 5$
Domain: $x < 5$ or $(-\infty, 5)$ (1-1)

15. Function g multiplies a domain element by 2 and then subtracts three times the square root of the domain element from the result. (1-1)

16. The graph of $x = -3$ is a vertical line 3 units to the *left* of the y axis; $y = 2$ is a horizontal line 2 units *above* the x axis.

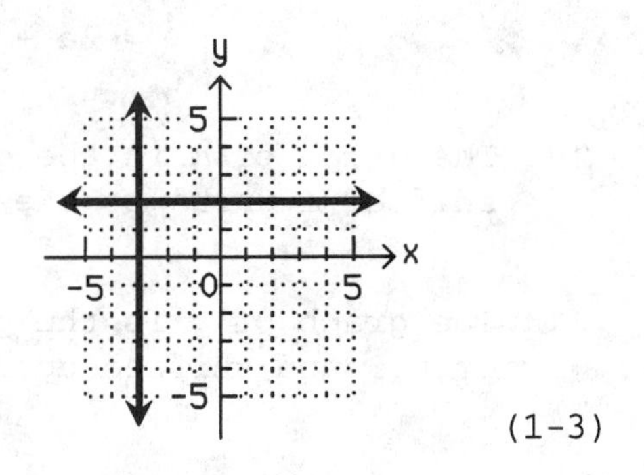

(1-3)

17. $f(x) = 0.4x(x + 4)(2 - x) = 0.4x(8 - 2x - x^2)$
$= -0.4x^3 - 0.8x^2 + 3.2x$

(A)

x	$f(x)$
-3	-6
-1	-3.6
1	2
3	-8.4

(B) Graph f and the lines $y = 3$, $y = 2$, $y = 1$ in the same coordinate system.

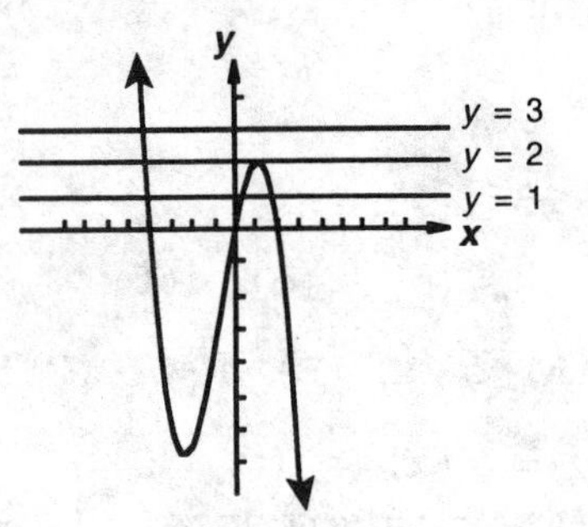

It is now easy to see that $f(x) = 3$ has one solution, $f(x) = 2$ has two solutions, and $f(x) = 1$ has three solutions.

(C) $f(x) = 3$: $x = -4.28$, $f(x) = 2$: $x = -4.19, 1$
$f(x) = 1$: $x = -4.10, 0.38, 1.75$ (1-1)

18. $f(x) = 3 - 2x$

$$\frac{f(2 + h) - f(2)}{h} = \frac{3 - 2(2 + h) - (3 - 2 \cdot 2)}{h} = \frac{3 - 4 - 2h - 3 + 4}{h} = \frac{-2h}{h} = -2$$

(1-1)

19. $f(x) = x^2 - 3x + 1$

$$\frac{f(a + h) - f(a)}{h} = \frac{(a + h)^2 - 3(a + h) + 1 - (a^2 - 3a + 1)}{h} = \frac{a^2 + 2ah + h^2 - 3a - 3h + 1 - a^2 + 3a - 1}{h} = \frac{2ah + h^2 - 3h}{h} = \frac{(2a + h - 3)h}{h} = 2a + h - 3$$

(1-1)

20. The graph of m is the graph of $y = |x|$ reflected on the x axis and shifted 4 units to the right. (1-2)

21. The graph of g is the graph of $y = x^3$ vertically contracted by a factor of 0.3 and shifted up 3 units. (1-2)

22. The graph of $y = x^2$ is vertically expanded by a factor of 2, reflected in the x axis and shifted to the left 3 units. Equation: $y = -2(x + 3)^2$ (1-2)

23. Equation: $f(x) = 2\sqrt{x + 3} - 1$

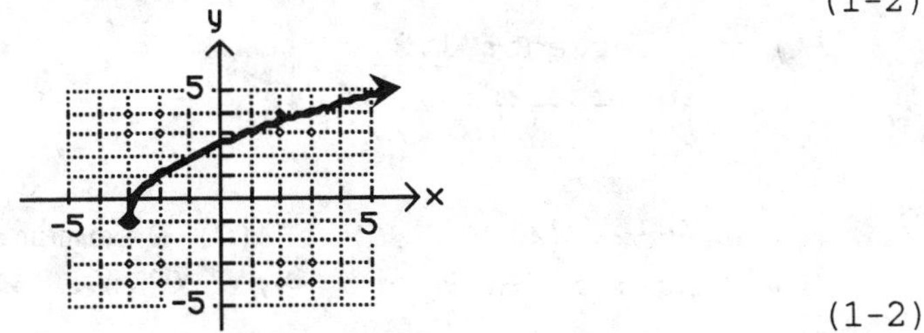

(1-2)

24. $f(x) = \begin{cases} -x - 2 & \text{for } x < 0 \\ 0.2x^2 & \text{for } x > 0 \end{cases}$

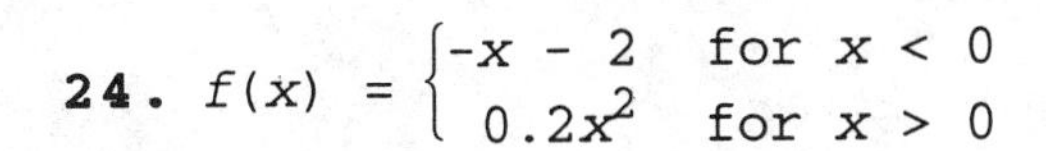

(A)

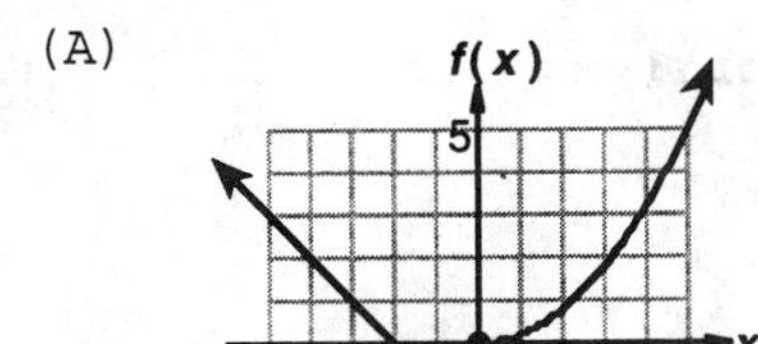

(B) Discontinuous at $x = 0$.

(C)

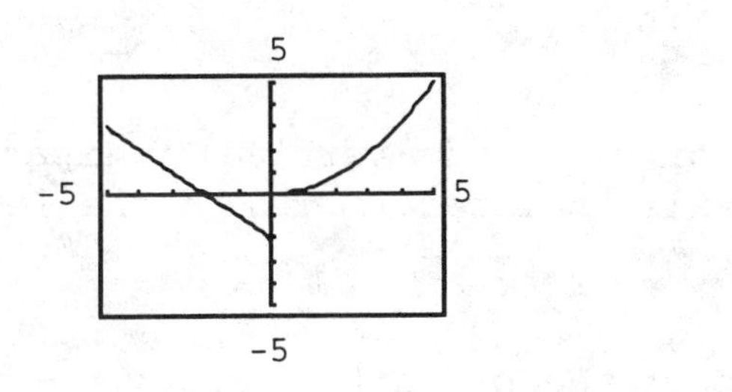

(1-2)

25. Use the point-slope form:

(A) $y - 2 = -\frac{2}{3}[x - (-3)]$

$y - 2 = -\frac{2}{3}(x + 3)$

$y = -\frac{2}{3}x$

(B) $y - 3 = 0(x - 3)$

$y = 3$

(1-3)

26. (A) Slope: $\frac{-1 - 5}{1 - (-3)} = -\frac{3}{2}$

$y - 5 = -\frac{3}{2}(x + 3)$

$3x + 2y = 1$

(B) Slope: $\frac{5 - 5}{4 - (-1)} = 0$

$y - 5 = 0(x - 1)$

$y = 5$

(C) Slope: $\frac{-2 - 7}{-2 - (-2)}$ not defined since $2 - (-2) = 0$

$x = -2$

(1-3)

27. $y = -(x - 4)^2 + 3$ (1-2, 1-4)

28. $f(x) = -0.4x^2 + 3.2x - 1.2 = -0.4(x^2 - 8x + 16) + 7.6$
$= -0.4(x - 4)^2 + 7.6$

(A) y intercept: 1.2
x intercepts: $-0.4(x - 4)^2 + 7.6 = 0$
$(x - 4)^2 = 19$
$x = 4 + \sqrt{19} \approx 8.4,\ 4 - \sqrt{19} \approx -0.4$

(B) Vertex: (4.0, 7.6) (C) Maximum: 7.6
(D) Range: $x \leq 7.6$ or $(-\infty, 7.6]$ (1-4)

29.

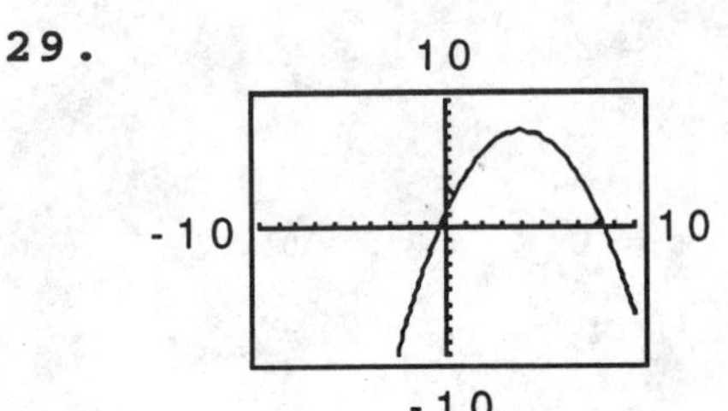

(A) y intercept: 1.2
x intercepts: -0.4, 8.4
(B) Vertex: (4.0, 7.6)
(C) Maximum: 7.6
(D) Range: $x \leq 7.6$ or $(-\infty, 7.6]$ (1-4)

30. The graph of $y = \sqrt[3]{x}$ is vertically expanded by a factor of 2, reflected in the x axis, shifted 1 unit to the left and 1 unit down.

Equation: $y = -2\sqrt[3]{x + 1} - 1$ (1-2)

31. The graphs of the pairs $\{y = 2x,\ y = -\frac{1}{2}x\}$ and $\{y = \frac{2}{3}x + 2,\ y = -\frac{3}{2}x + 2\}$ are shown below:

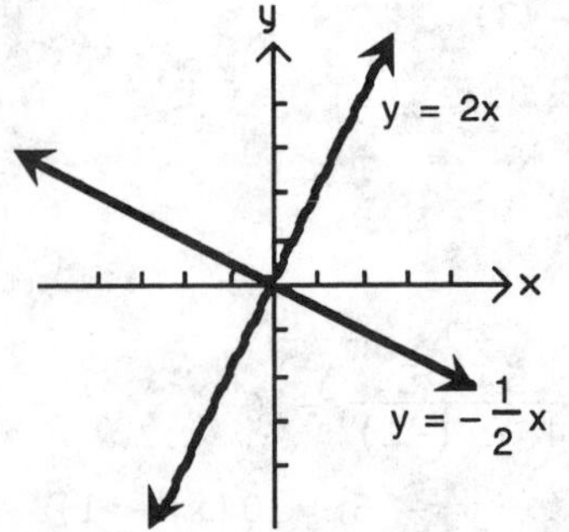

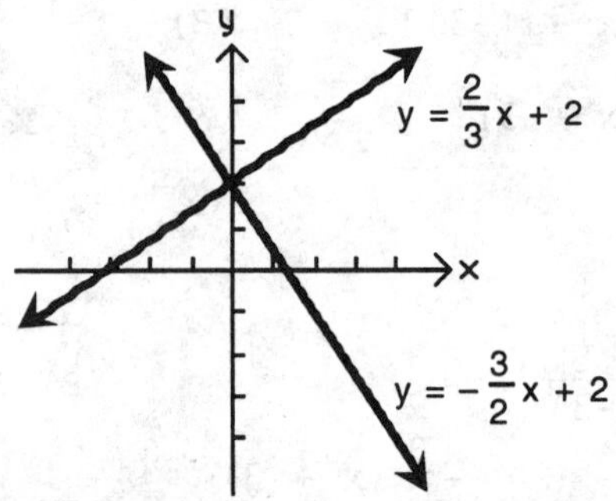

In each case, the graphs appear to be perpendicular to each other. It can be shown that two slant lines are perpendicular if and only if their slopes are negative reciprocals. (1-3)

32. (A) $f(x) = \sqrt{x}$

$$\frac{f(x + h) - f(x)}{h} = \frac{\sqrt{x + h} - \sqrt{x}}{h} = \frac{\sqrt{x + h} - \sqrt{x}}{h} \cdot \frac{\sqrt{x + h} + \sqrt{x}}{\sqrt{x + h} + \sqrt{x}}$$ rationalize the numerator

$$= \frac{x + h - x}{h[\sqrt{x + h} + \sqrt{x}]} = \frac{1}{\sqrt{x + h} + \sqrt{x}}$$

(B) $f(x) = \frac{1}{x}$

$$\frac{f(x+h) - f(x)}{h} = \frac{\frac{1}{x+h} - \frac{1}{x}}{h} = \frac{\frac{x - (x+h)}{x(x+h)}}{h}$$

$$= \frac{-h}{hx(x+h)} = \frac{-1}{x(x+h)} \qquad (1\text{-}2)$$

33. $G(x) = 0.3x^2 + 1.2x - 6.9 = 0.3(x^2 + 4x + 4) - 8.1$
$= 0.3(x + 2)^2 - 8.1$

(A) y intercept: -6.9
x intercepts: $0.3(x + 2)^2 - 8.1 = 0$
$(x + 2)^2 = 27$
$x = -2 + \sqrt{27} \approx 3.2, \; -2 - \sqrt{27} \approx -7.2$

(B) Vertex: $(-2, -8.1)$ (C) Minimum: -8.1

(D) Range: $x \geq -8.1$ or $[-8.1, \infty)$

(E) Decreasing: $(-\infty, -2]$; Increasing: $[-2, \infty)$ (1-4)

34.

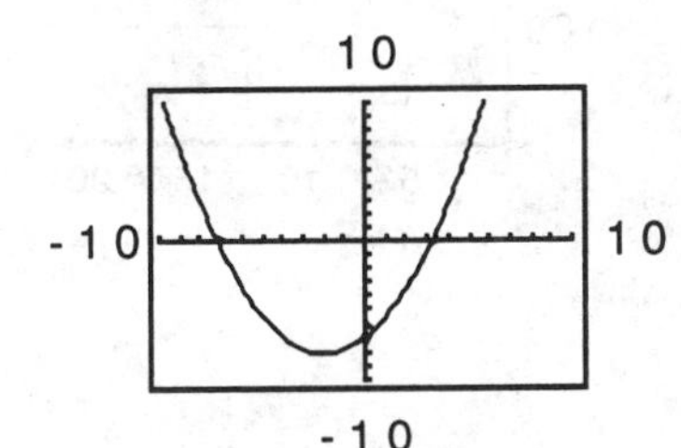

(A) y intercept: -6.9
x intercept: $-7.2, 3.2$

(B) Vertex: $(-2, -8.1)$

(C) Minimum: -8.1

(D) Range: $x \geq -8.1$ or $[-8.1, \infty)$

(E) Decreasing: $(-\infty, -2]$
Increasing: $[-2, \infty)$ (1-4)

35. (A) $V(0) = 12{,}000$, $V(8) = 2{,}000$

Slope: $\frac{2{,}000 - 12{,}000}{8 - 0} = \frac{-10{,}000}{8} = -1{,}250$

V intercept: 12,000

Equation: $V(t) = -1{,}250t + 12{,}000$

(B) $V(5) = -1{,}250(5) + 12{,}000 = \$5{,}750$

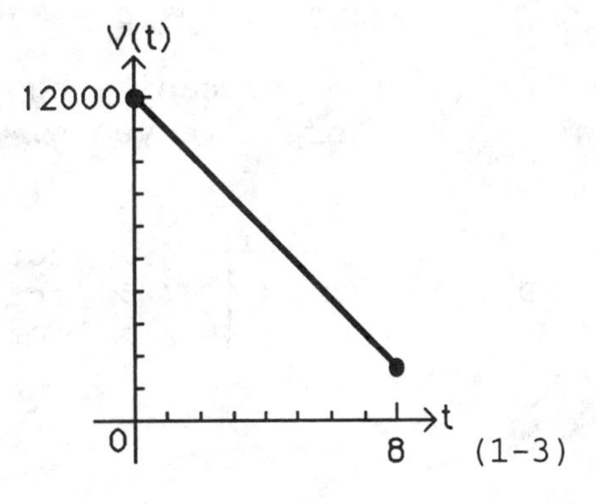

(1-3)

36. (A)

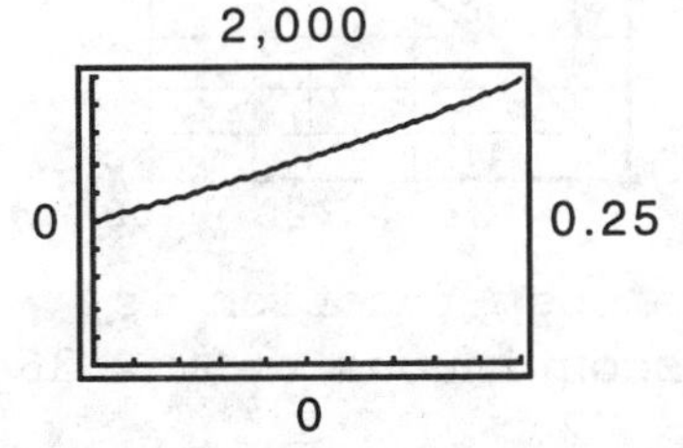

(B) $r = 0.1447$ or 14.7% compounded annually

Alternative algebraic solution:

$$1000(1 + r)^3 = 1500$$
$$(1 + r)^3 = 1.5$$
$$1 + r = \sqrt[3]{1.5} \approx 1.1447$$
$$r = 0.1447$$

(1-1, 1-2)

37. (A) $R(130) = 208$, $R(50) = 80$

Slope: $\dfrac{208 - 80}{130 - 50} = \dfrac{128}{80} = 1.6$

Equation: $R - 80 = 1.6(C - 50)$ or $R = 1.6C$

(B) $R(120) = 1.6(120) = \$192$

(C) $176 = 1.6C$; $C = \$110$

(D) 1.6; The slope gives the change in retail price per unit change in the cost. (1-3)

38. $f(x) = 303.4 - 3.46x$

(A)

x	0	5	10	15	20
Consumption	309	276	271	255	233
$f(x)$	303	286	269	252	234

(B)

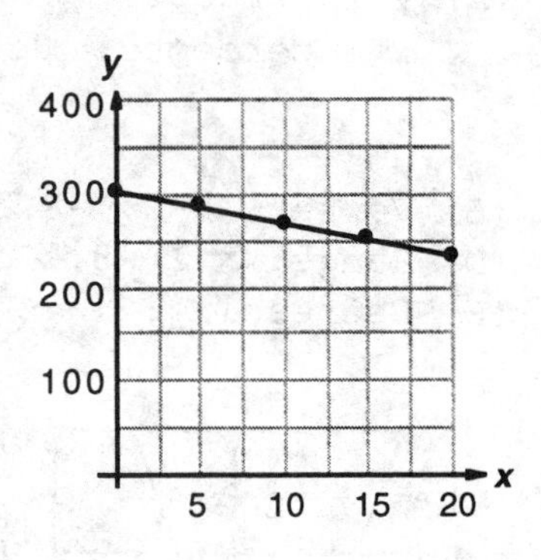

(C) The year 1995 corresponds to $x = 25$

$f(25) = 303.4 - 3.46(25) = 216.9$ or 217

The year 2000 corresponds to $x = 30$

$f(30) = 303.4 - 3.46(30) = 199.6$ or 200

(D) The per capita egg consumption is dropping approximately 17 eggs every five years. (1-3)

39. (A) $C(x) = \begin{cases} 0.49x & \text{for } 0 \le x < 36 \\ 0.44x & \text{for } 36 \le x < 72 \\ 0.39x & \text{for } x \ge 72 \end{cases}$

(B)

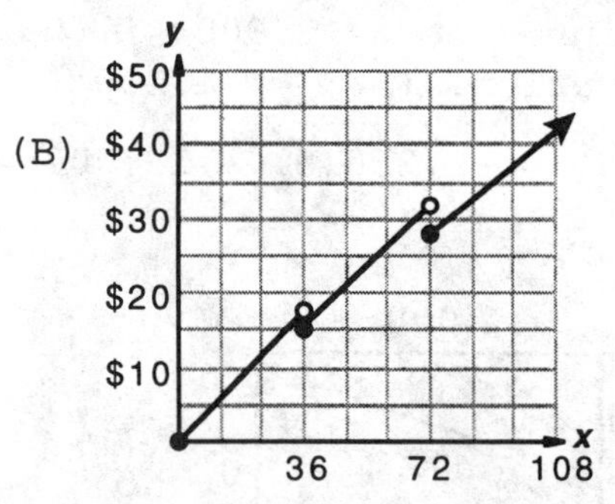

The cost function is discontinuous at $x = 36$ and $x = 72$. (1-2)

40. (A) Let x = number of video tapes produced.

$C(x) = 84,000 + 15x$

$R(x) = 50x$

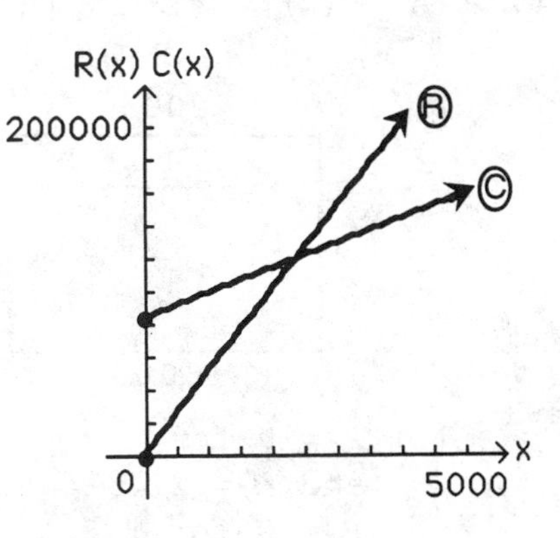

(B) $R(x) = C(x)$

$50x = 84,000 + 15x$

$35x = 84,000$

$x = 2,400$ units

$R < C$ for $0 \le x < 2,400$; $R > C$ for $x > 2,400$

(C) $R = C$ at $x = 2,400$ units

$R < C$ for $0 \le x < 2,400$; $R > C$ for $x > 2,400$ (1-3)

41. $p(x) = 50 - 1.25x$ Price-demand function

$C(x) = 160 + 10x$ Cost function

$R(x) = xp(x)$

$= x(50 - 1.25x)$ Revenue function

(A)

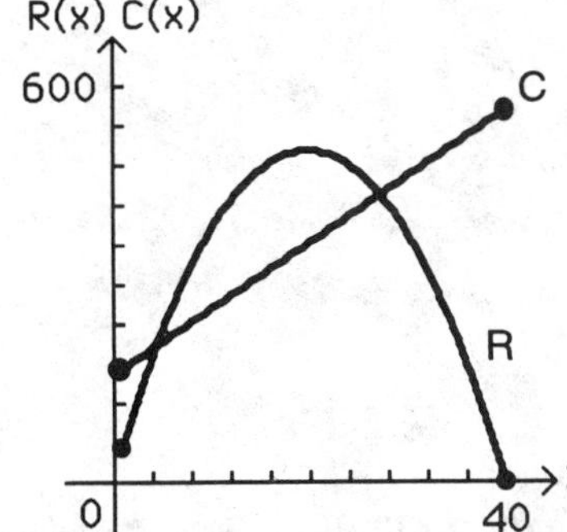

(B) $R = C$

$$x(50 - 1.25x) = 160 + 10x$$
$$-1.25x^2 + 50x = 160 + 10x$$
$$-1.25x^2 + 40x = 160$$
$$-1.25(x^2 - 32x + 256) = 160 - 320$$
$$-1.25(x - 16)^2 = -160$$
$$(x - 16)^2 = 128$$
$$x = 16 + \sqrt{128} \approx 27.314,\ 16 - \sqrt{128} \approx 4.686$$

$R = C$ at $x = 4.686$ thousand units (4,686 units) and $x = 27.314$ thousand units (27,314 units)

$R < C$ for $1 \le x < 4.686$ or $27.314 < x \le 40$

$R > C$ for $4.686 < x < 27.314$

(C) Max Rev: $50x - 1.25x^2 = R$

$$-1.25(x^2 - 40x + 400) + 500 = R$$
$$-1.25(x - 20)^2 + 500 = R$$

Vertex at (20, 500)

Max. Rev. = 500 thousand ($500,000) occurs when <u>output</u> is 20 thousand (20,000 units)

<u>Wholesale price</u> at this output: $p(x) = 50 - 1.25x$

$$p(20) = 50 - 1.25(20)$$
$$= \$25$$

(1-3, 1-4)

42. (A) $P(x) = R(x) - C(x) = x(50 - 1.25x) - (160 + 10x)$
$= -1.25x^2 + 40x - 160$

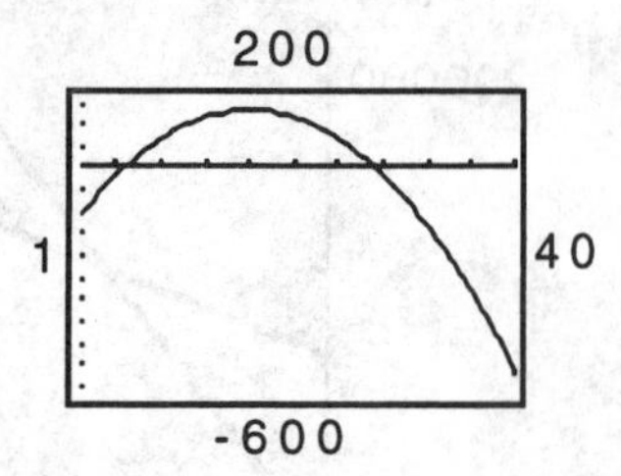

(B) $P = 0$ for $x = 4.686$ thousand uits (4,686 units) and $x = 27.314$ thousand units (27,314 units)
$P < 0$ for $1 \le x < 4.686$ or $27.314 < x \le 40$
$P > 0$ for $4.686 < x < 27.314$

(C) Maximum profit is 160 thousand dollars ($160,000), and this occurs at $x = 16$ thousand units (16,000 units). The wholesale price at this output is $p(16) = 50 - 1.25(16) = \$30$, which is $5 greater than the $25 found in 38(C). (1-4)

43. (A) The area enclosed by the pens is given by
$A = (2y)x$
Now, $3x + 4y = 840$
so $y = 210 - \frac{3}{4}x$
Thus $A(x) = 2\left(210 - \frac{3}{4}x\right)x$
$= 420x - \frac{3}{2}x^2$

(B) Clearly x and y must be nonnegative; the fact that $y \ge 0$ implies
$210 - \frac{3}{4}x \ge 0$
and $210 \ge \frac{3}{4}x$
$840 \ge 3x$
$280 \ge x$
Thus, domain A: $0 \le x \le 280$

(C)

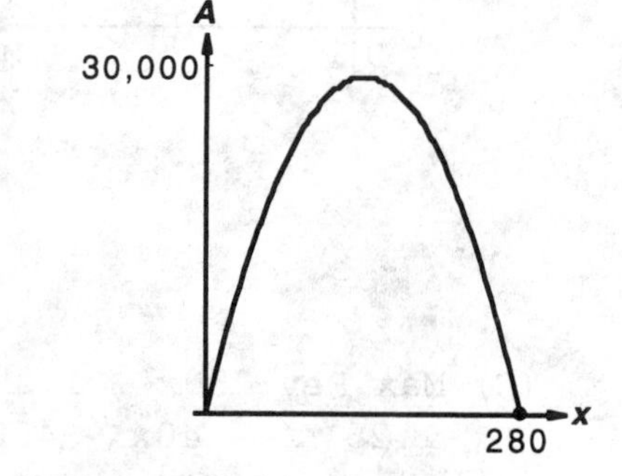

(D) Graph $A(x) = 420x - \frac{3}{2}x^2$ and $y = 25{,}000$ together.

There are two values of x that will produce storage areas with a combined area of 25,000 square feet, one near $x = 90$ and the other near $x = 190$.

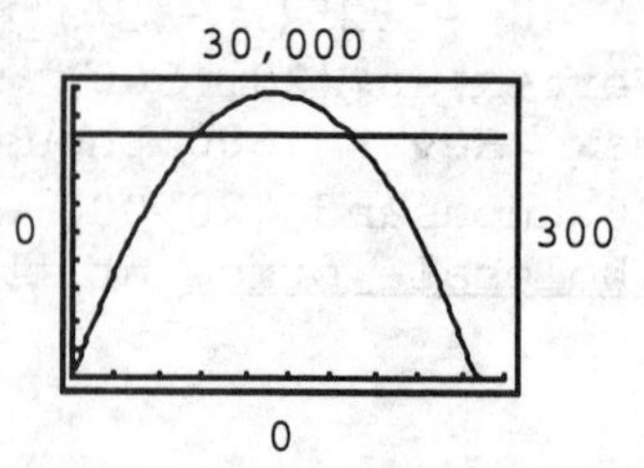

(E) $x = 86$, $x = 194$

(F) $A(x) = 420x - \frac{3}{2}x^2 = -\frac{3}{2}(x^2 - 280x)$

Completing the square, we have

$A(x) = -\frac{3}{2}(x^2 - 280x + 19,600 - 19,600)$

$= -\frac{3}{2}[(x - 140)^2 - 19,600]$

$= -\frac{3}{2}(x - 140)^2 + 29,400$

The dimensions that will produce the maximum combined area are: $x = 140$ ft, $y = 105$ ft. The maximum area is 29,400 sq. ft. (1-4)

44. (A) We are given $P(0) = 20$ and $m = 15$. Thus, $P(x) = 15x + 20$

(B) 1 PM is 5 hours after 8 AM
$P(5) = 15(5) + 20 = 95$

(C)

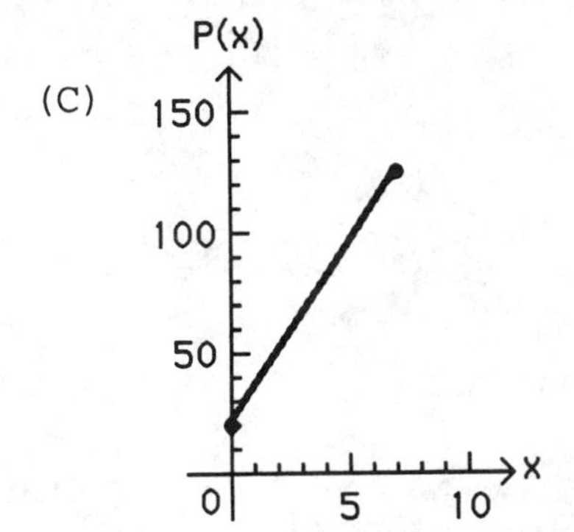

(D) Slope = 15

(1-4)

45. $\frac{\Delta s}{s} = k$. For $k = \frac{1}{30}$, $\frac{\Delta s}{s} = \frac{1}{30}$ or $\Delta s = \frac{1}{30}s$

(A) When $s = 30$, $\Delta s = \frac{1}{30}(30) = 1$ pound.

When $s = 90$, $\Delta s = \frac{1}{30}(90) = 3$ pounds.

(B) $\Delta s = \frac{1}{30}s$

Slope $m = \frac{1}{30}$

y intercept $b = 0$

(C) Slope $m = \frac{1}{30}$

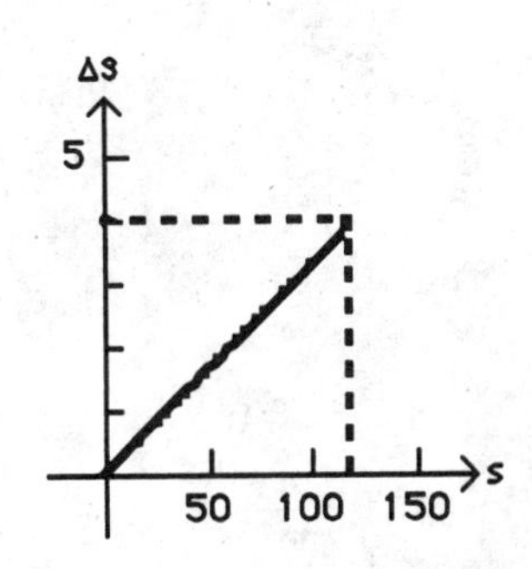

(1-4)

2 ADDITIONAL ELEMENTARY FUNCTIONS

EXERCISE 2-1

Things to remember:

1. POLYNOMIAL FUNCTION

   A POLYNOMIAL FUNCTION is a function of the form
   $$f(x) = a_nx^n + a_{n-1}x^{n-1} + \dots + a_1x + a_0$$
   for n a nonnegative integer, called the DEGREE of the polynomial. The coefficients $a_0, a_1, \dots, a_n$ are real numbers with $a_n \neq 0$. The DOMAIN of a polynomial function is the set of all real numbers.

2. TURNING POINT

   A TURNING POINT on a graph is a point that separates an increasing portion from a decreasing portion, or vice versa. The graph of a polynomial function of degree $n \geq 1$ can have at most $n - 1$ turning points and can cross the x axis at most n times.

3. LOCATING THE ZEROS OF A POLYNOMIAL

   If r is a zero of the polynomial
   $$P(x) = x_n + a_{n-1}x^{n-1} + a_{n-2}x^{n-2} + \dots + a_1x + a_0$$
   then
   $$|r| < 1 + \max\{|a_{n-1}|, |a_{n-2}|, \dots, |a_1|, |a_0|\}$$

4. A RATIONAL FUNCTION is any function of the form
   $$f(x) = \frac{n(x)}{d(x)} \qquad d(x) \neq 0$$
   where $n(x)$ and $d(x)$ are polynomials. The DOMAIN is the set of all real numbers such that $d(x) \neq 0$. We assume $n(x)/d(x)$ is reduced to lowest terms.

5. ASYMPTOTES OF RATIONAL FUNCTIONS

   Given the rational function
   $$f(x) = \frac{n(x)}{d(x)}$$
   where $n(x)$ and $d(x)$ are polynomials without common factors.

   (a) If a is a real number such that $d(a) = 0$, then the line $x = a$ is a VERTICAL ASYMPTOTE of the graph of $y = f(x)$.

   (b) HORIZONTAL ASYMPTOTES, if any exists, can be found by dividing each term of the numerator $n(x)$ and denominator $d(x)$ by the highest power of x that appears in the numerator and denominator.

1. (A) 2 (B) 1 (C) 2 (D) 0 (E) 1 (F) 1

3. (A) 5 (B) 4 (C) 5 (D) 1 (E) 1 (F) 1

5. (A) 6 (B) 5 (C) 6 (D) 0 (E) 1 (F) 1

7. (A) 3 (B) 4 (C) negative **9.** (A) 4 (B) 5 (C) negative

11. (A) 0 (B) 1 (C) negative **13.** (A) 5 (B) 6 (C) positive

15. $f(x) = \dfrac{x + 2}{x - 2}$

(A) *Intercepts:*

x intercepts: $f(x) = 0$ only if $x + 2 = 0$ or $x = -2$.
The x intercept is -2.

y intercept: $f(0) = \dfrac{0 + 2}{0 - 2} = -1$
The y intercept is -1.

(B) *Domain:* The denominator is 0 at $x = 2$. Thus, the domain is the set of all real numbers except 2.

(C) *Asymptotes:*

Vertical asymptotes: $f(x) = \dfrac{x + 2}{x - 2}$
The denominator is 0 at $x = 2$. Therefore, the line $x = 2$ is a vertical asymptote.

Horizontal asymptotes: $f(x) = \dfrac{x + 2}{x - 2} = \dfrac{1 + \frac{2}{x}}{1 - \frac{2}{x}}$

As x increases or decreases without bound, the numerator tends to 1 and the denominator tends to 1. Therefore, the line $y = 1$ is a horizontal asymptote.

(D)

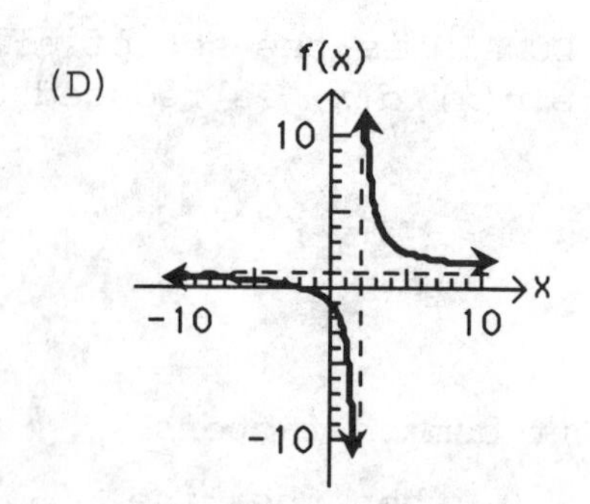

(E)

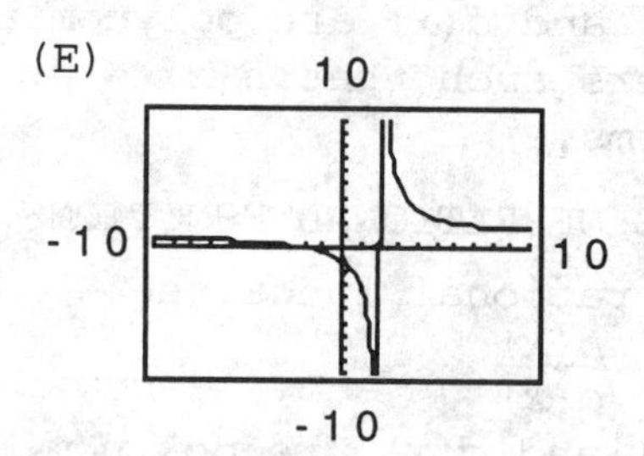

17. $f(x) = \dfrac{3x}{x + 2}$

(A) *Intercepts:*

x intercepts: $f(x) = 0$ only if $3x = 0$ or $x = 0$.
The x intercept is 0.

y intercept: $f(0) = \dfrac{3 \cdot 0}{0 + 2} = 0$
The y intercept is 0.

(B) *Domain:* The denominator is 0 at $x = -2$. Thus, the domain is the set of all real numbers except -2.

(C) *Asymptotes:*

Vertical asymptotes: $f(x) = \frac{3x}{x + 2}$

The denominator is 0 at $x = -2$. Therefore, the line $x = -2$ is a vertical asymptote.

Horizontal asymptotes: $f(x) = \frac{3x}{x + 2} = \frac{3}{1 + \frac{2}{x}}$

As x increases or decreases without bound, the numerator is 3 and the denominator tends to 1. Therefore, the line $y = 3$ is a horizontal asymptote.

(D)

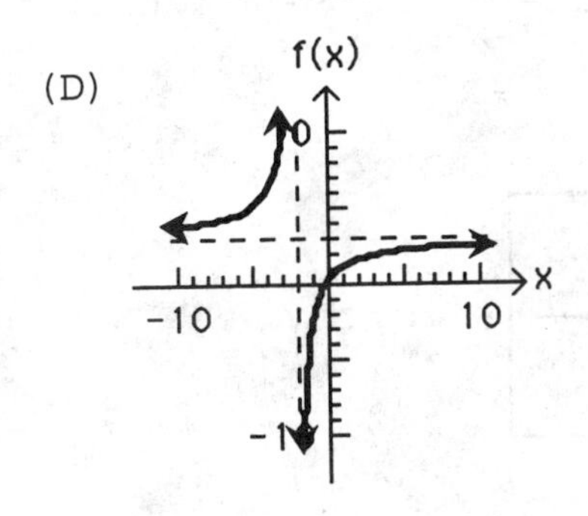

(E)

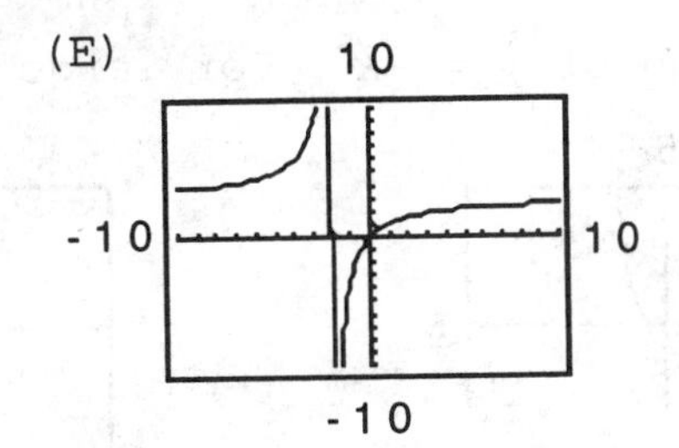

19. $f(x) = \frac{4 - 2x}{x - 4}$

(A) *Intercepts:*

x intercepts: $f(x) = 0$ only if $4 - 2x = 0$ or $x = 2$.
The x intercept is 2.

y intercept: $f(0) = \frac{4 - 2 \cdot 0}{0 - 4} = -1$
The y intercept is -1.

(B) *Domain:* The denominator is 0 at $x = 4$. Thus, the domain is the set of all real numbers except 4.

(C) *Asymptotes:*

Vertical asymptotes: $f(x) = \frac{4 - 2x}{x - 4}$

The denominator is 0 at $x = 4$. Therefore, the line $x = 4$ is a vertical asymptote.

Horizontal asymptotes: $f(x) = \frac{4 - 2x}{x - 4} = \frac{\frac{4}{x} - 2}{1 - \frac{4}{x}}$

As x increases or decreases without bound, the numerator tends to -2 and the denominator tends to 1. Therefore, the line $y = -2$ is a horizontal asymptote.

(D)

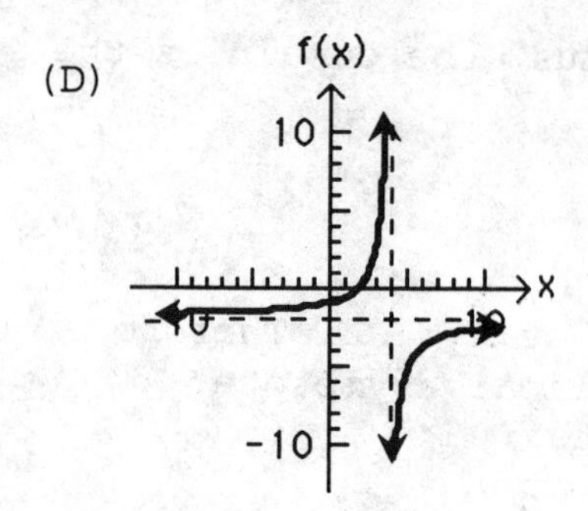

(E)

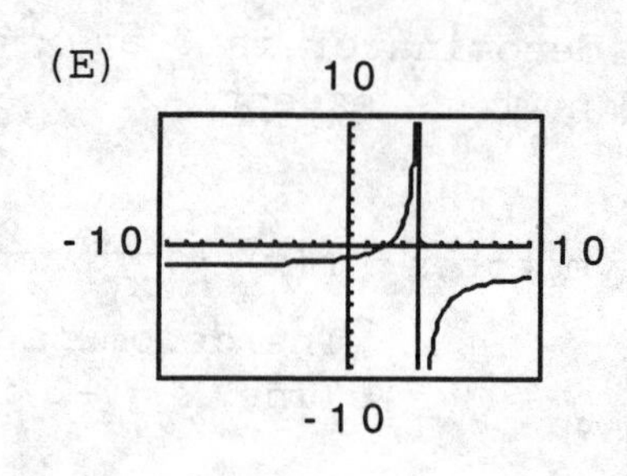

21. The graph of $f(x) = 2x^4 - 5x^2 + x + 2 = 2x^4\left(1 - \frac{5}{2x^2} + \frac{1}{2x^3} + \frac{1}{x^4}\right)$ will "look like" the graph of $y = 2x^4$. For large x, $f(x) \approx 2x^4$.

23. The graph of $f(x) = -x^5 + 4x^3 - 4x + 1 = -x^5\left(1 - \frac{4}{x^2} + \frac{4}{x^4} - \frac{1}{x^5}\right)$ will "look like" the graph of $y = -x^5$. For large x, $f(x) \approx -x^5$.

25. (A)

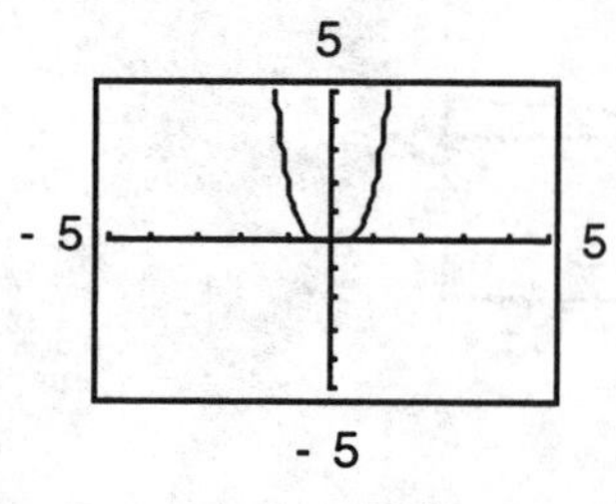

$y = 2x^4$

5 | -5 | 5 | -5

$y = 2x^4 - 5x^2 + x + 2$

(B)

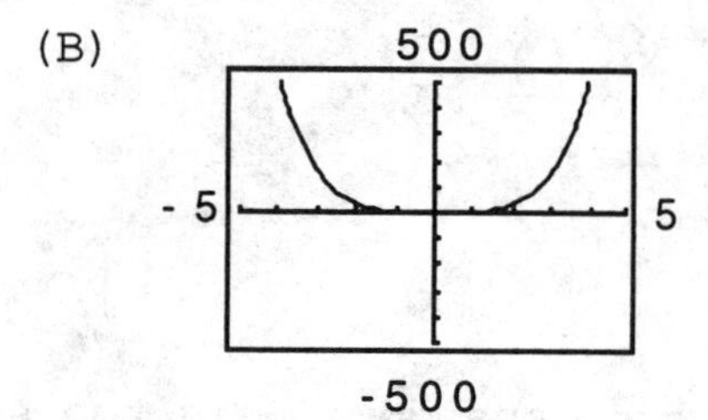

$y = 2x^4$

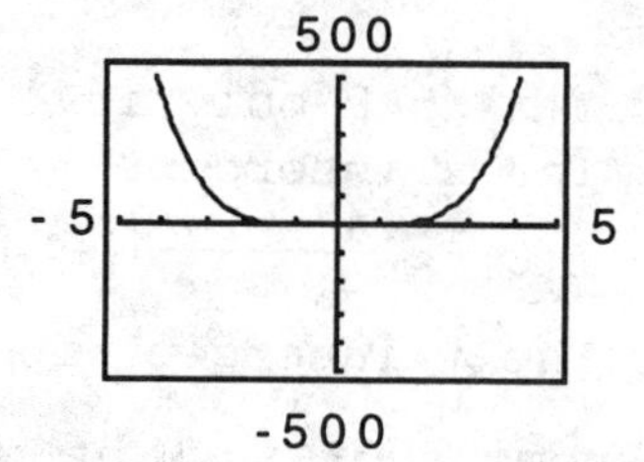

$y = 2x^4 - 5x^2 + x + 2$

27. (A)

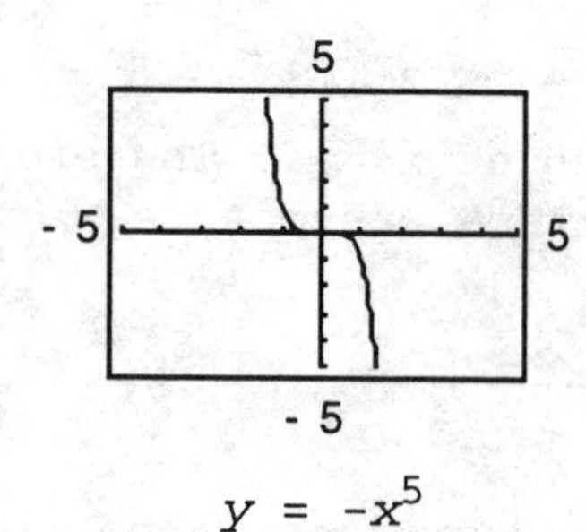

$y = -x^5$

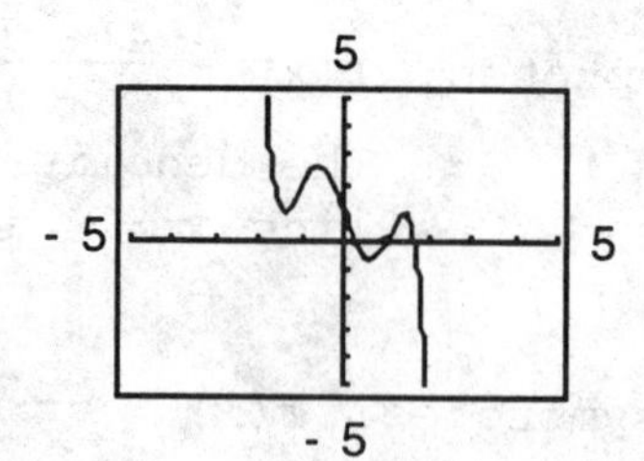

$y = -x^5 + 4x^3 - 4x + 1$

(B)

500

-5 5

-500

$y = -x^5$

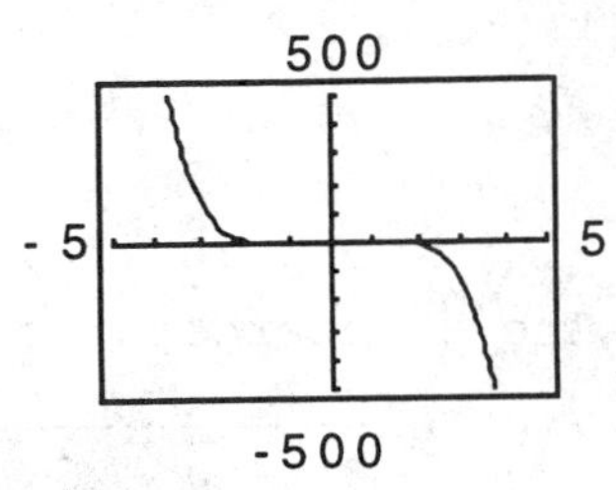

$y = -x^5 + 4x^3 - 4x + 1$

29. $P(x) = 2x^3 - x^2 - 8x - 7$

Let $Q(x) = \frac{1}{2}P(x) = x^3 - \frac{1}{2}x^2 - 4x - \frac{7}{2}$; $P(x)$ and $Q(x)$ have the same zeros. If x is a zero of $Q(x)$, then, by 3,

$$|x| < 1 + \max\left\{\left|-\frac{1}{2}\right|, |-4|, \left|-\frac{7}{2}\right|\right\} = 1 + 4 = 5$$

Graph $P(x)$ on $[-5, 5]$

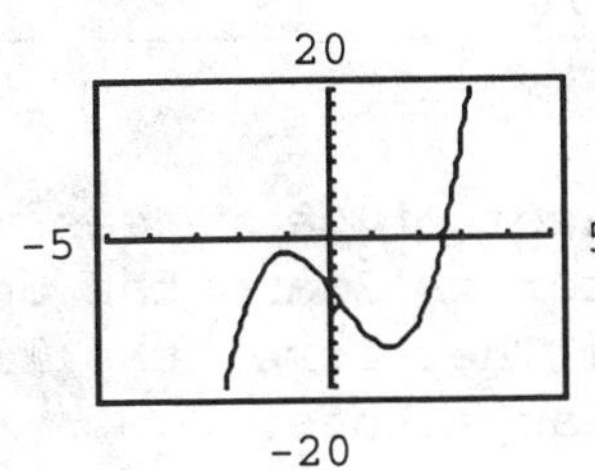

$P(x) = 0$ at $x \approx 2.58$.

31. $P(x) = x^4 + 2x^3 - 3x^2 - 2x + 3$

By 3, if x is a zero of $P(x)$, then

$|x| < 1 + \max\{|2|, |-3|, |-2|, |3|\} = 1 + 3 = 4$

Graph $P(x)$ on $[-4, 4]$

$P(x) = 0$ at $x \approx -2.68, -1.16$.

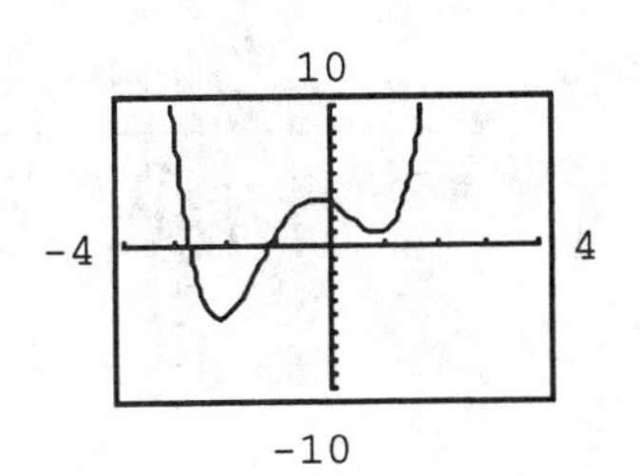

33. $P(x) = x^5 - 12x^4 + 8x^3 + 11$

By 3, if x is a zero of $P(x)$, then

$|x| < 1 + \max\{|-12|, |8|, |11|\} = 1 + 12 = 13$

Graph $P(x)$ on $[-13, 13]$

$P(x) = 0$ at $x \approx -0.84, 1.26, 11.29$.

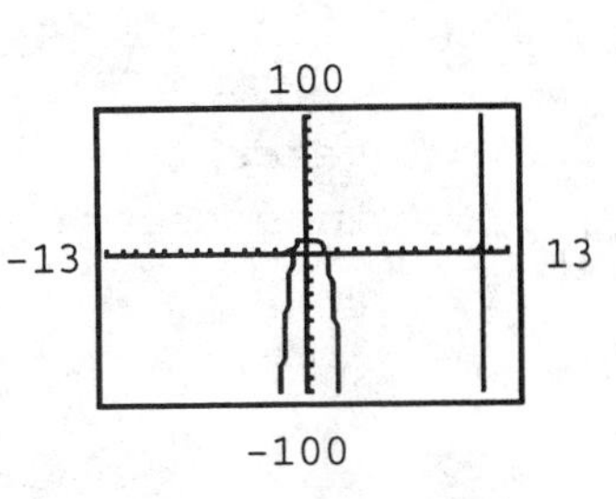

35. The linear regression model for the data set consisting of two points on the line $y = 0.5x + 3$ is simply the line $y = 0.5x + 3$. In general, the linear regression model for two points is the line that passes through the two points.

37. $f(x) = \dfrac{2x^2}{x^2 - x - 6}$

(A) *Intercepts:*

x intercepts: $f(x) = 0$ only if $2x^2 = 0$ or $x = 0$.
The x intercept is 0.

y intercept: $f(0) = \dfrac{2 \cdot 0^2}{0^2 - 0 - 6} = 0$
The y intercept is 0.

(B) *Asymptotes:*

Vertical asymptotes: $f(x) = \dfrac{2x^2}{x^2 - x - 6} = \dfrac{2x^2}{(x - 3)(x + 2)}$
The denominator is 0 at $x = -2$ and $x = 3$. Thus, the lines $x = -2$ and $x = 3$ are vertical asymptotes.

Horizontal asymptotes: $f(x) = \dfrac{2x^2}{x^2 - x - 6} = \dfrac{2}{1 - \frac{1}{x} - \frac{6}{x^2}}$

As x increases or decreases without bound, the numerator is 2 and the denominator tends to 1. Therefore, the line $y = 2$ is a horizontal asymptote.

(C)

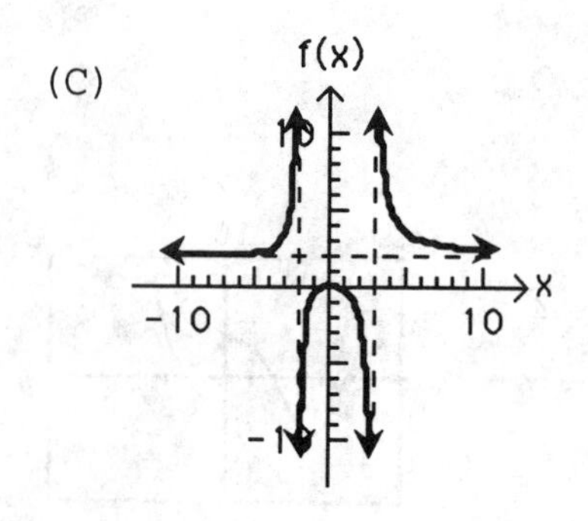

(D)

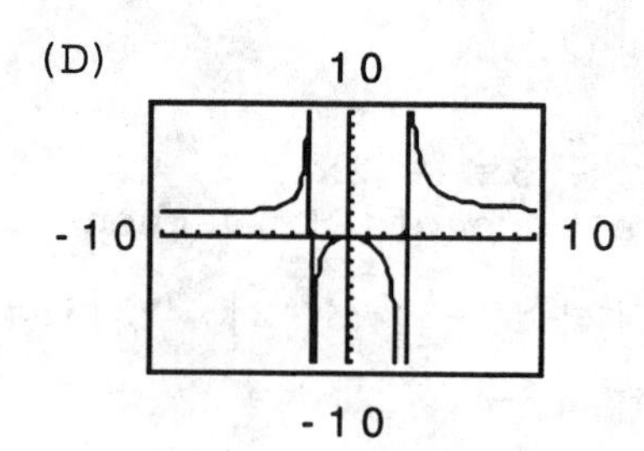

39. $f(x) = \dfrac{6 - 2x^2}{x^2 - 9}$

(A) *Intercepts:*

x intercepts: $f(x) = 0$ only if
$$6 - 2x^2 = 0$$
$$2x^2 = 6$$
$$x^2 = 3$$
$$x = \pm\sqrt{3}$$
The x intercepts are $\pm\sqrt{3}$.

y intercept: $f(0) = \dfrac{6 - 2 \cdot 0^2}{0^2 - 9} = -\dfrac{2}{3}$

The y intercept is $-\frac{2}{3}$.

(B) *Asymptotes:*

Vertical asymptotes: $f(x) = \frac{6 - 2x^2}{x^2 - 9} = \frac{6 - 2x^2}{(x - 3)(x + 3)}$

The denominator is 0 at $x = -3$ and $x = 3$. Thus, the lines $x = -3$ and $x = 3$ are vertical asymptotes.

Horizontal asymptotes: $f(x) = \frac{6 - 2x^2}{x^2 - 9} = \frac{\frac{6}{x^2} - 2}{1 - \frac{9}{x^2}}$

As x increases or decreases without bound, the numerator tends to -2 and the denominator tends to 1. Therefore, the line $y = -2$ is a horizontal asymptote.

(C)

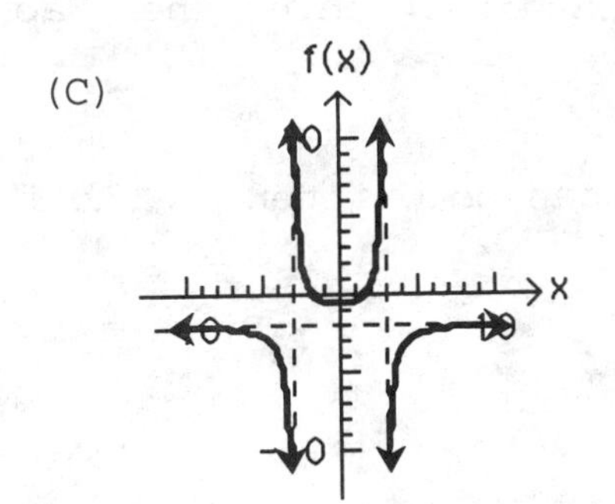

(D)

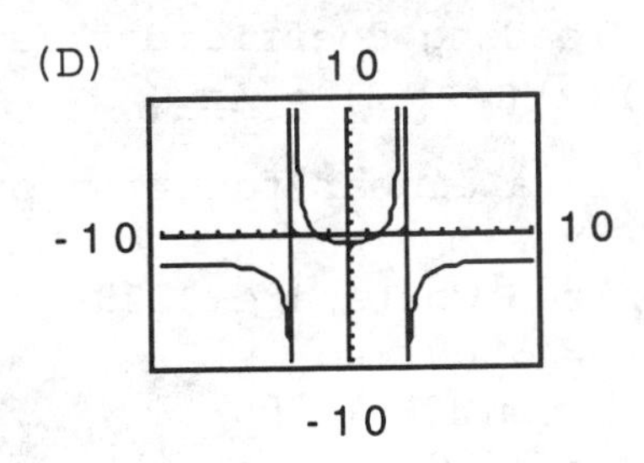

41. $f(x) = \frac{-4x}{x^2 + x - 6}$

(A) *Intercepts:*

x intercepts: $f(x) = 0$ only if $-4x = 0$ or $x = 0$.
The x intercept is 0.

y intercept: $f(0) = \frac{-4 \cdot 0}{0^2 + 0 - 6} = 0$
The y intercept is 0.

(B) *Asymptotes:*

Vertical asymptotes: $f(x) = \frac{-4x}{x^2 + x - 6} = \frac{-4x}{(x + 3)(x - 2)}$

The denominator is 0 at $x = -3$ and $x = 2$. Thus, the lines $x = -3$ and $x = 2$ are vertical asymptotes.

Horizontal asymptotes: $f(x) = \frac{-4x}{x^2 + x - 6} = \frac{-\frac{4}{x}}{1 + \frac{1}{x} - \frac{6}{x^2}}$

As x increases or decreases without bound, the numerator tends to 0 and the denominator tends to 1. Therefore, the line $y = 0$ (the x axis) is a horizontal asymptote.

(C)

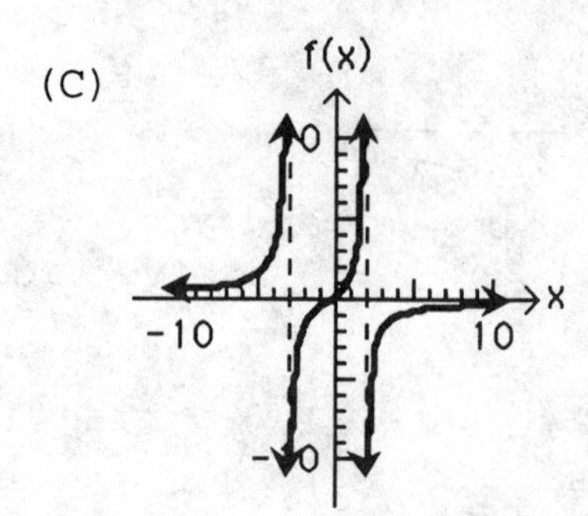

(D)

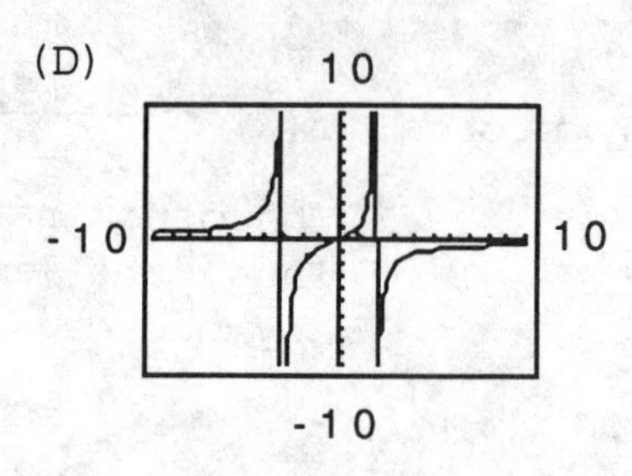

43. The graph has 1 turning point which implies degree $n = 2$.
The x intercepts are $x = -1$ and $x = 2$.
Thus, $f(x) = (x + 1)(x - 2) = x^2 - x - 2$.

45. The graph has 2 turning points which implies degree $n = 3$. The x intercepts are $x = -2$, $x = 0$, and $x = 2$. The direction of the graph indicates that leading coefficient is negative
$f(x) = -(x + 2)(x)(x - 2) = 4x - x^3$.

47. (A) Since $C(x)$ is a linear function of x, it can be written in the form

$$C(x) = mx + b$$

Since the fixed costs are \$200, $b = 200$.
Also, $C(20) = 3800$, so

$$3800 = m(20) + 200$$
$$20m = 3600$$
$$m = 180$$

Therefore, $C(x) = 180x + 200$

(B) $\overline{C}(x) = \dfrac{C(x)}{x} = \dfrac{180x + 200}{x}$

(C)

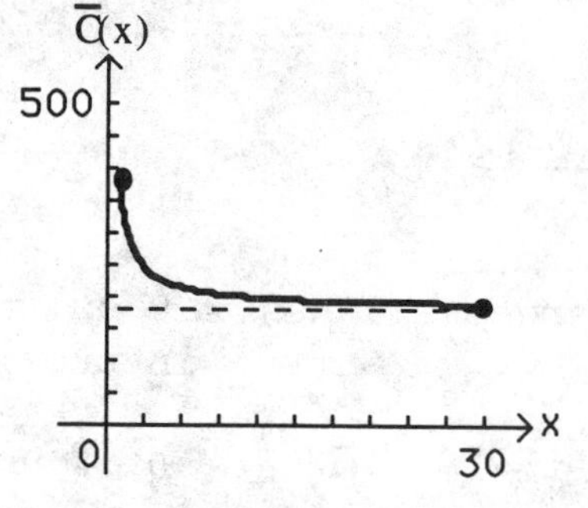

(D) $\overline{C}(x) = \dfrac{180x + 200}{x} = \dfrac{180 + \frac{200}{x}}{1}$

As x increases, the numerator tends to 180 and the denominator is 1.
Therefore, $\overline{C}(x)$ tends to 180 or \$180 per board.

49. (A) $\overline{C}(n) = \dfrac{2500 + 175n + 25n^2}{n}$

(B)

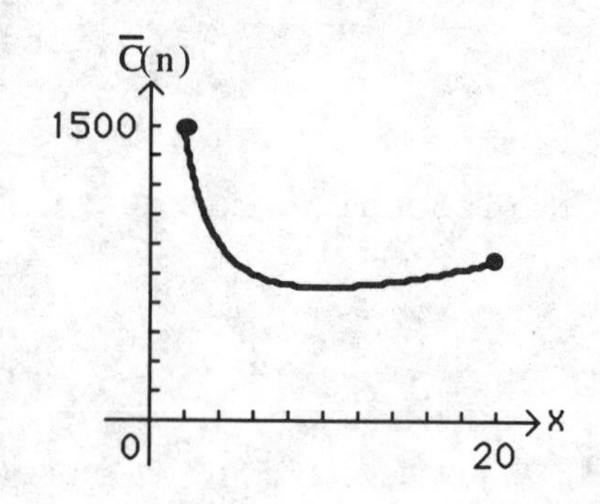

(C) Using the graph, we calculate

$$C(8) = \frac{2500 + 175(8) + 25(8)^2}{8} = 687.50$$

$$C(9) = \frac{2500 + 175(9) + 25(9)^2}{9} = 677.78$$

$$C(10) = \frac{2500 + 175(10) + 25(10)^2}{10} = 675.00$$

$$C(11) = \frac{2500 + 175(11) + 25(11)^2}{11} = 677.27$$

$$C(12) = \frac{2500 + 175(12) + 25(12)^2}{12} = 683.33$$

Thus, it appears that the average cost per year is a minimum at $n = 10$ years; at 10 years, the average minimum cost is \$675.00 per year.

(D) 10 years; \$675.00 per year

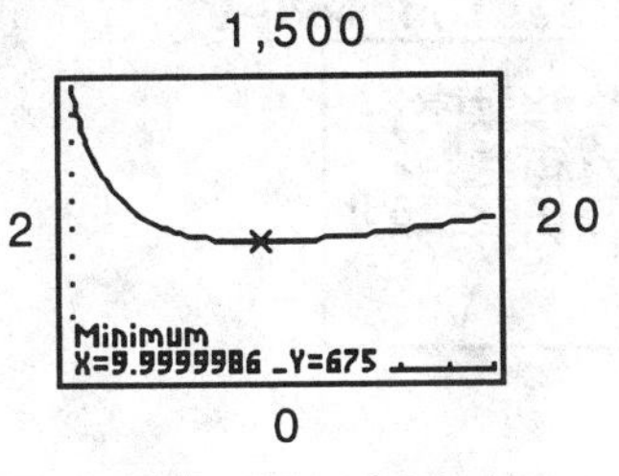

51. (A) $\overline{C}(x) = \dfrac{0.00048(x - 500)^3 + 60,000}{x}$

(B)

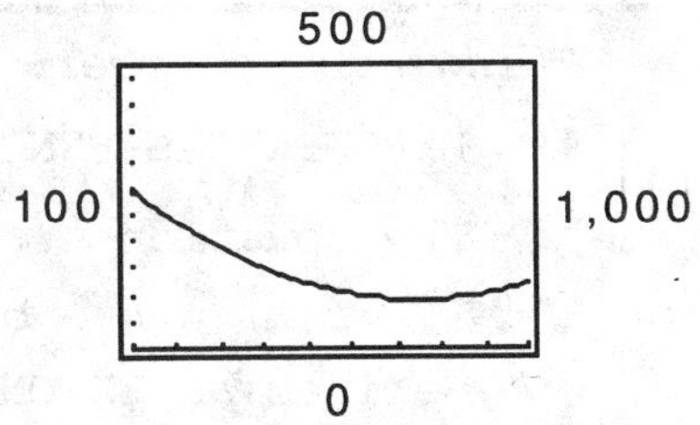

(C) The caseload which yields the minimum average cost per case is 750 cases per month. At 750 cases per month, the average cost per case is \$90.

53. (A) Linear regression model for Table 1.

```
LinReg
 y=ax+b
 a=-.0506666667
 b=19.96666667
```

Quadratic regression model for Table 2.

```
QuadReg
 y=ax²+bx+c
 a=2.4444444E-4
 b=-.0065555556
 c=2.086111111
```

(B) Graphing the two models, we have

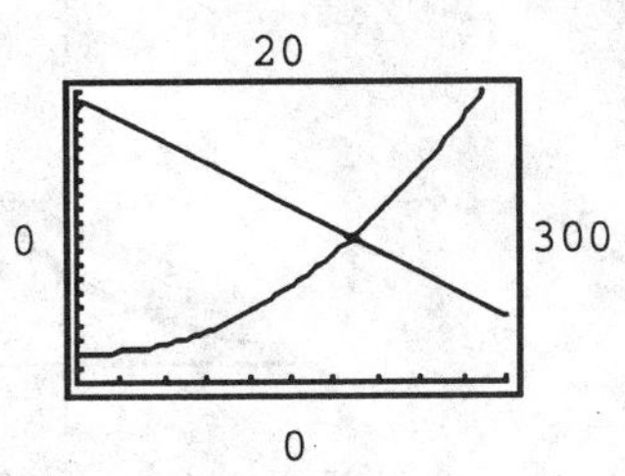

The equilibrium price is $\overline{p}$ = \$10.09 and the equilibrium quantity is $\overline{x}$ = 195.

55. (A) Cubic regression model:

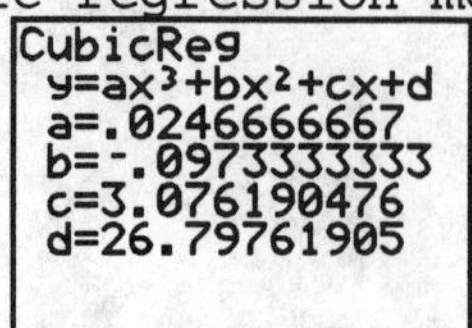

(B) 1995: $y(35) \approx \$1,072.8$ billion

57. (A) 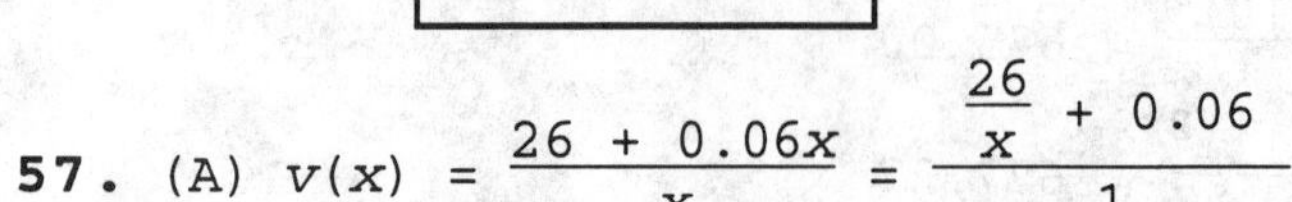

$$v(x) = \frac{26 + 0.06x}{x} = \frac{\frac{26}{x} + 0.06}{1}$$

As x increases, the numerator tends to 0.06 and the denominator is 1. Therefore, $v(x)$ approaches 0.06 centimeters per second as x increases.

(B)

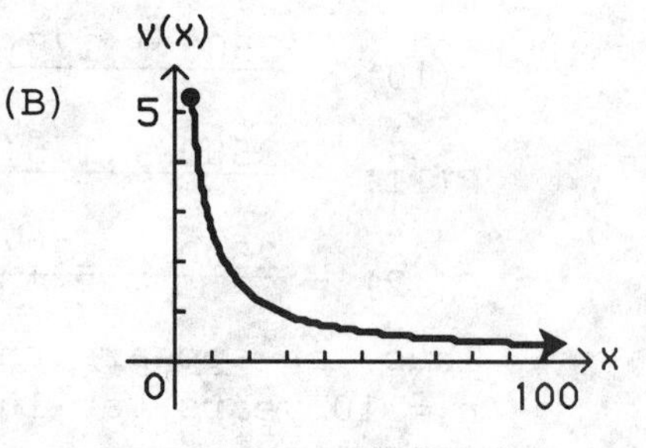

59. (A) Cubic regression model

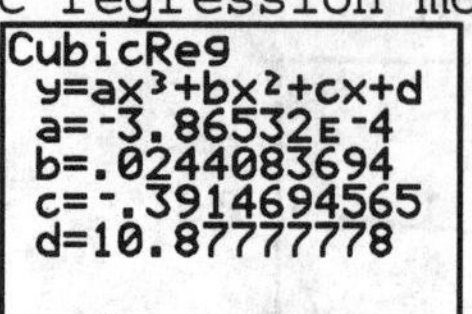

(B) 1995: $y(45) \approx 7.5$ marriages per 1,000 population.

EXERCISE 2-2

Things to remember:

1. EXPONENTIAL FUNCTION

   The equation

   $$f(x) = b^x,\ b > 0,\ b \neq 1$$

   defines an EXPONENTIAL FUNCTION for each different constant b, called the BASE. The DOMAIN of f is all real numbers, and the RANGE of f is the set of positive real numbers.

2. BASIC PROPERTIES OF THE GRAPH OF $f(x) = b^x$, $b > 0$, $b \neq 1$

   a. All graphs pass through (0,1); $b^0 = 1$ for any base b.

   b. All graphs are continuous curves; there are no holes or jumps.

   c. The x-axis is a horizontal asymptote.

   d. If $b > 1$, then b^x increases as x increases.

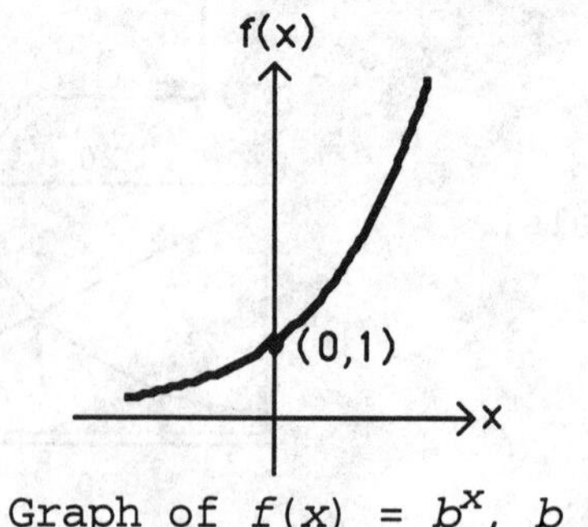

Graph of $f(x) = b^x$, $b > 1$

e. If $0 < b < 1$, then b^x decreases as x increases.

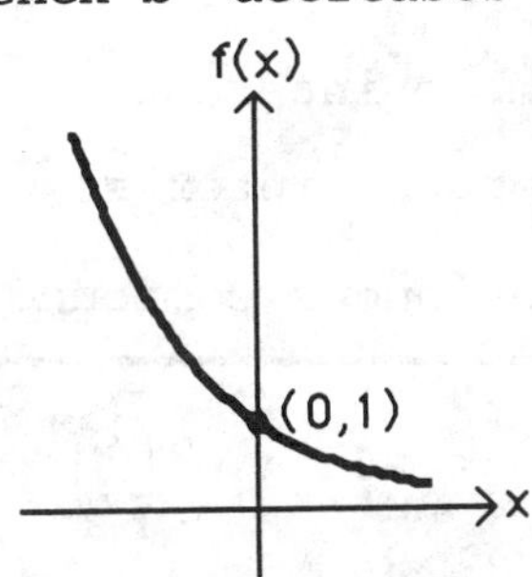

Graph of $f(x) = b^x$, $0 < b < 1$

3. EXPONENTIAL FUNCTION PROPERTIES

For $a, b > 0$, $a \neq 1$, $b \neq 1$, and x, y real numbers:

a. EXPONENT LAWS

(i) $a^x a^y = a^{x+y}$ (iv) $(ab)^x = a^x b^x$

(ii) $\dfrac{a^x}{a^y} = a^{x-y}$ (v) $\left(\dfrac{a}{b}\right)^x = \dfrac{a^x}{b^x}$

(iii) $(a^x)^y = a^{xy}$

b. $a^x = a^y$ if and only if $x = y$.

c. For $x \neq 0$, $a^x = b^x$ if and only if $a = b$.

4. EXPONENTIAL FUNCTION WITH BASE $e = 2.71828\ldots$

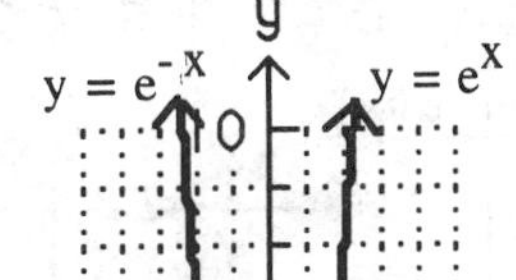

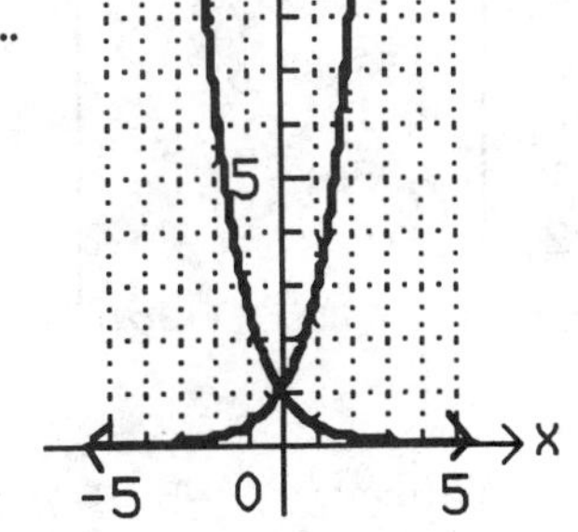

Exponential functions with base e and base $1/e$ are respectively defined by $y = e^x$ and $y = e^{-x}$.

Domain: $(-\infty, \infty)$

Range: $(0, \infty)$

5. COMPOUND INTEREST

If a principal P (present value) is invested at an annual rate r (expressed as a decimal) compounded m times per year, then the amount A (future value) in the account at the end of t years is given by:

$$A = P\left(1 + \frac{r}{m}\right)^{mt}$$

6. CONTINUOUS COMPOUND INTEREST FORMULA

If a principal P (present value) is invested at an annual rate r (expressed as a decimal) compounded continuously, then the amount A (future value) in the account at the end of t years is given by

$$A = Pe^{rt}$$

7. INTEREST FORMULAS

(a) $A = P(1 + rt)$ Simple interest

(b) $A = P\left(1 + \frac{r}{m}\right)^{mt}$ Compound interest

(c) $A = Pe^{rt}$ Continuous compound interest

---

1. $y = 5^x$, $-2 \le x \le 2$

x	y
-2	$\frac{1}{25}$
-1	$\frac{1}{5}$
0	1
1	5
2	25

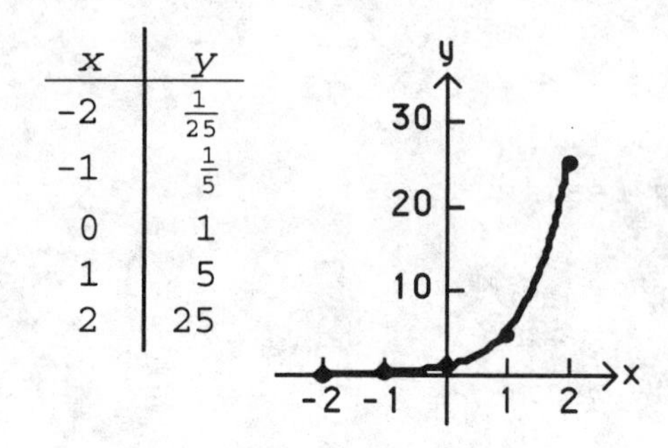

3. $y = \left(\frac{1}{5}\right)^x = 5^{-x}$, $-2 \le x \le 2$

x	y
-2	25
-1	5
0	1
1	$\frac{1}{5}$
2	$\frac{1}{25}$

5. $f(x) = -5^x$, $-2 \le x \le 2$

x	$f(x)$
-2	$-\frac{1}{25}$
-1	$-\frac{1}{5}$
0	-1
1	-5
2	-25

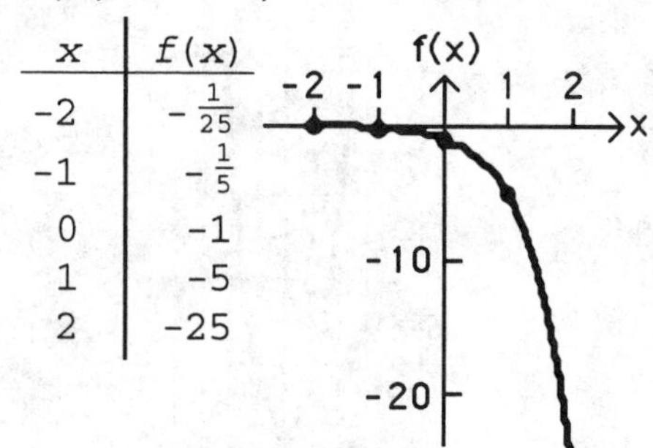

7. $y = -e^{-x}$, $-3 \le x \le 3$

x	y
-3	≈ -20
-2	≈ -7.4
-1	≈ -2.7
0	-1
1	≈ -0.4
2	≈ -0.1
3	≈ -0.05

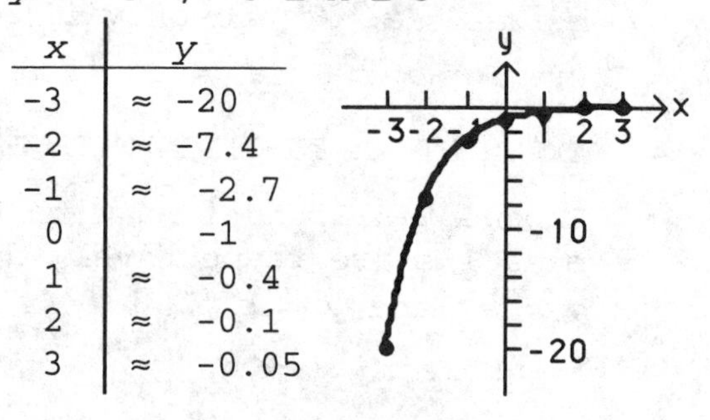

9. $y = 100e^{0.1x}$, $-5 \le x \le 5$

x	y
-5	≈ 60
-3	≈ 74
-1	≈ 90
0	100
1	≈ 111
3	≈ 135
5	≈ 165

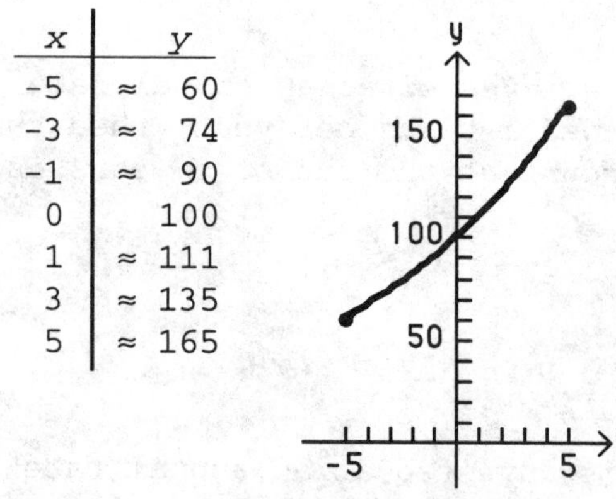

11. $g(t) = 10e^{-0.2t}$, $-5 \le t \le 5$

g	$g(t)$
-5	≈ 27.2
-3	≈ 18.2
-1	≈ 12.2
0	10
1	≈ 8.2
3	≈ 5.5
5	≈ 3.7

13. $(4^{3x})^{2y} = 4^{6xy}$ [see 3a(iii)]

15. $\frac{e^{x-3}}{e^{x-4}} = e^{(x-3)-(x-4)} = e^{x-3-x+4} = e$ [See 3a(ii)]

17. $(2e^{1.2t})^3 = 2^3 e^{3(1.2t)} = 8e^{3.6t}$ [see 3a (iv)]

19. $g(x) = -f(x)$; the graph of g is the graph of f reflected in the x axis.

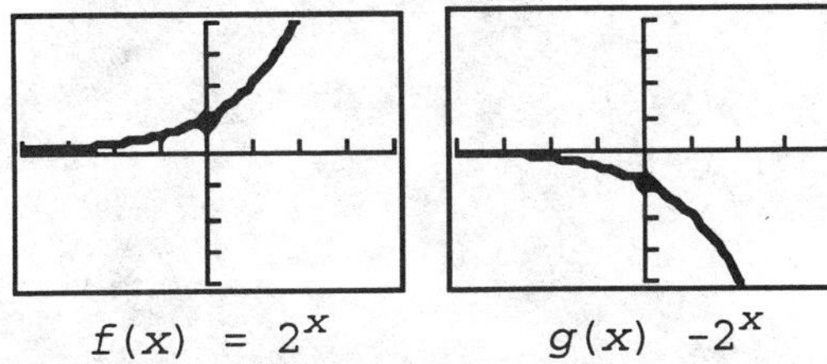

$f(x) = 2^x$ $g(x)$ -2^x

21. $g(x) = f(x + 1)$; the graph of g is the graph of f shifted one unit to the left.

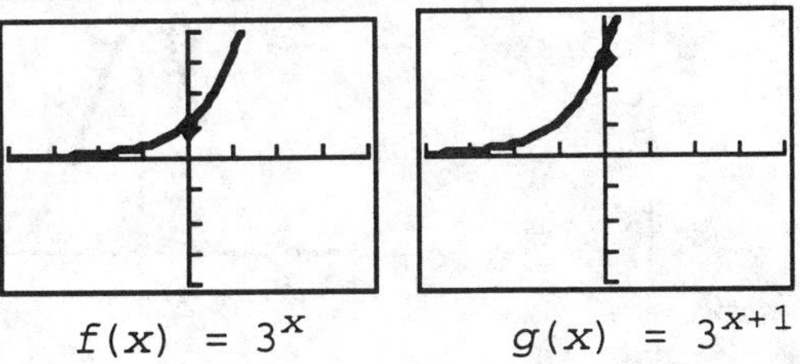

$f(x) = 3^x$ $g(x) = 3^{x+1}$

23. $g(x) = f(x) + 1$; the graph of g is the graph of f shifted one unit up.

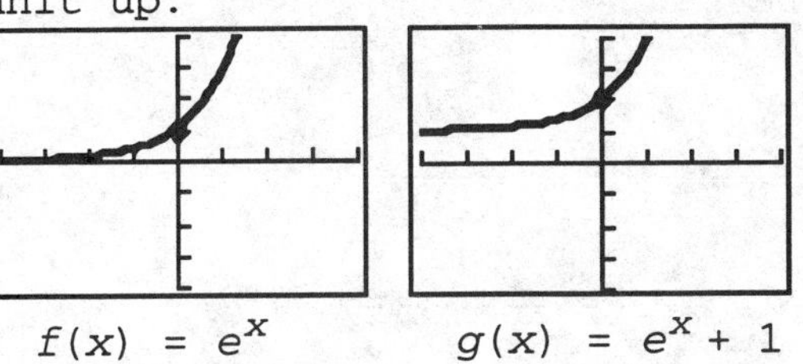

$f(x) = e^x$ $g(x) = e^x + 1$

25. $g(x) = 2f(x + 2)$; the graph of g is the graph of f vertically expanded by a factor of 2 and shifted to the left 2 units.

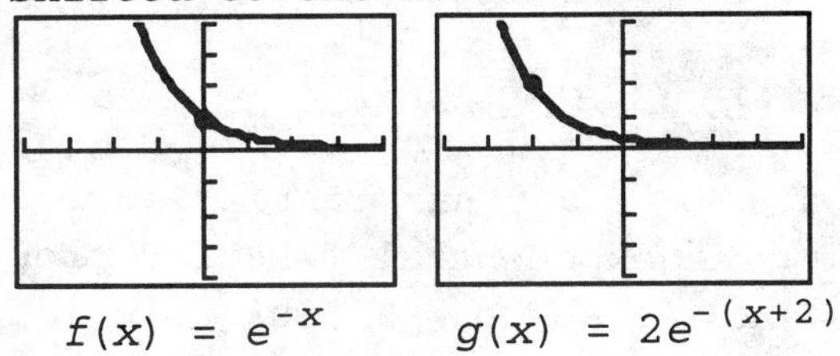

$f(x) = e^{-x}$ $g(x) = 2e^{-(x+2)}$

27. $f(t) = 2^{t/10}$, $-30 \le t \le 30$

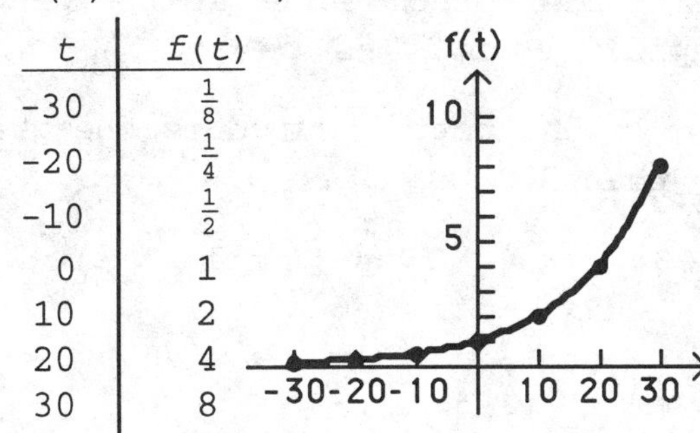

t	$f(t)$
-30	$\frac{1}{8}$
-20	$\frac{1}{4}$
-10	$\frac{1}{2}$
0	1
10	2
20	4
30	8

29. $y = -3 + e^{1+x}$, $-4 \le x \le 2$

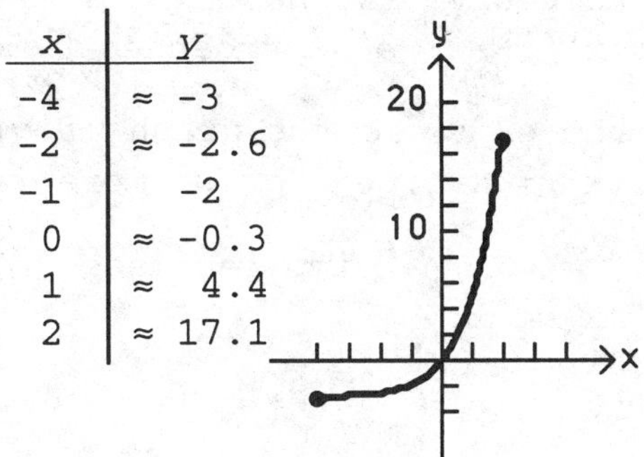

x	y
-4	$\approx$ -3
-2	$\approx$ -2.6
-1	-2
0	$\approx$ -0.3
1	$\approx$ 4.4
2	$\approx$ 17.1

31. $y = e^{|x|}$, $-3 \le x \le 3$

x	y
-3	$\approx$ 20.1
-1	$\approx$ 2.7
0	1
1	$\approx$ 2.7
3	$\approx$ 20.1

33. $C(x) = \dfrac{e^x + e^{-x}}{2}$, $-5 \le x \le 5$

x	$C(x)$
-5	$\approx$ 74
-3	$\approx$ 10
0	1
3	$\approx$ 10
5	$\approx$ 74

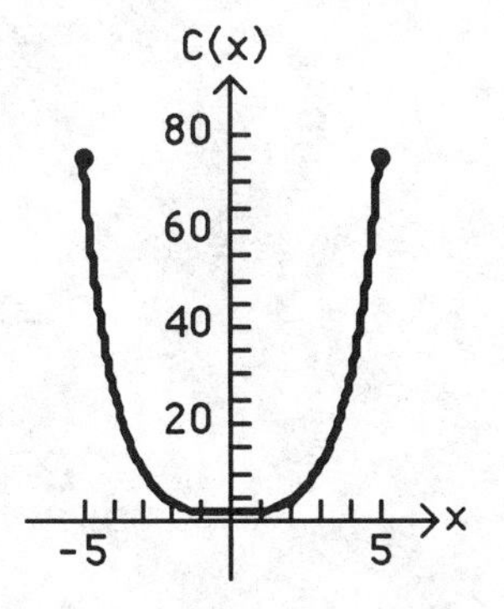

35. $y = e^{-x^2}$, $-3 \le x \le 3$

x	y
-3	0.0001
-2	0.0183
-1	0.3679
0	1
1	0.3679
2	0.0183
3	0.0001

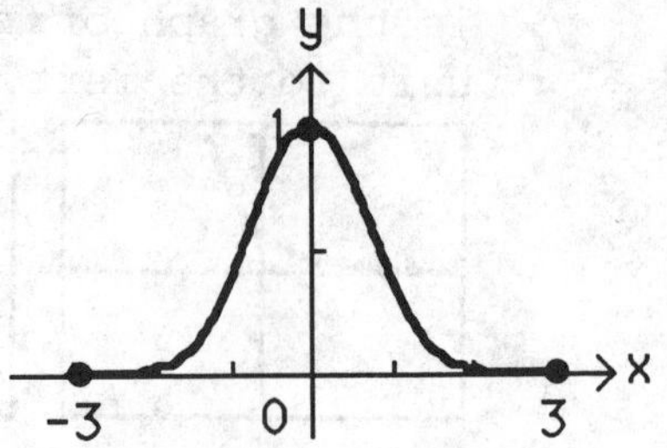

37. Solve

$$\begin{aligned} a^2 &= a^{-2} \\ a^2 &= \frac{1}{a^2} \\ a^4 &= 1 \\ a^4 - 1 &= 0 \\ (a^2 - 1)(a^2 + 1) &= 0 \end{aligned}$$

$a^2 - 1 = 0$ implies $a = 1, -1$

$a^2 + 1 = 0$ has no real solutions

The exponential function property: $a^x = a^y$ if and only if $x = y$ assumes $a > 0$ and $a \neq 1$. Our solutions are $a = 1, -1$; $1^x = 1^y$ for all real numbers x, y, $(-1)^x = (-1)^y$ for all even integers.

39. The top curve is the graph of $f(x) = 2^x$, the bottom curve is the graph of $g(x) = e^x$; e^x approaches 0 more rapidly than 2^x as $x \to -\infty$.

41. The top curve is the graph of $g(x) = e^{-x}$, the bottom curve is the graph of $f(x) = 2^{-x}$; e^{-x} grows more rapidly than 2^{-x} as $x \to -\infty$.

43. $10^{2-3x} = 10^{5x-6}$ implies (see 3b)

$$\begin{aligned} 2 - 3x &= 5x - 6 \\ -8x &= -8 \\ x &= 1 \end{aligned}$$

45. $4^{5x-x^2} = 4^{-6}$ implies

$$\begin{aligned} 5x - x^2 &= -6 \\ \text{or} \quad -x^2 + 5x + 6 &= 0 \\ x^2 - 5x - 6 &= 0 \\ (x - 6)(x + 1) &= 0 \\ x &= 6, -1 \end{aligned}$$

47. $5^3 = (x + 2)^3$ implies (by property 3c)

$5 = x + 2$

Thus, $x = 3$.

49.

$$\begin{aligned} (x - 3)e^x &= 0 \\ x - 3 &= 0 \quad (\text{since } e^x \neq 0) \\ x &= 3 \end{aligned}$$

51.

$$\begin{aligned} 3xe^{-x} + x^2e^{-x} &= 0 \\ e^{-x}(3x + x^2) &= 0 \\ 3x + x^2 &= 0 \quad (\text{since } e^{-x} \neq 0) \\ x(3 + x) &= 0 \\ x &= 0, -3 \end{aligned}$$

53. $h(x) = x2^x,\ -5 \le x \le 0$

x	$h(x)$
-5	$-\frac{5}{32}$
-4	$-\frac{1}{4}$
-3	$-\frac{3}{8}$
-2	$-\frac{1}{2}$
-1	$-\frac{1}{2}$
0	0

55. $N = \dfrac{100}{1 + e^{-t}},\ 0 \le t \le 5$

t	N
0	50
1	≈ 73.1
2	≈ 88.1
3	≈ 95.3
5	≈ 99.3

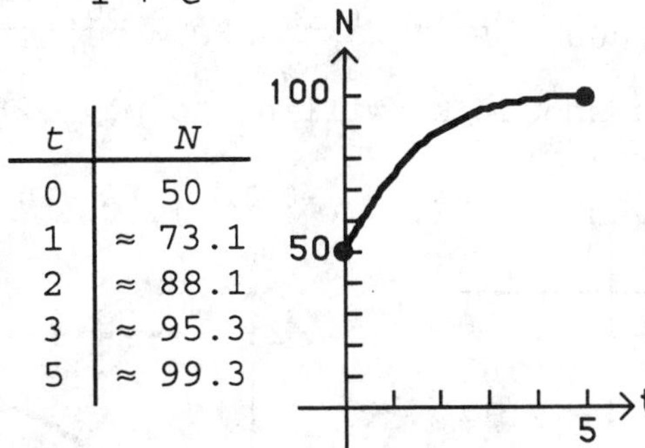

57. $f(x) = 4^x - 7$
Solve $4^x - 7 = 0$
$x \approx 1.40$

59. $f(x) = 2 + 3x + 10^x$
Solve $2 + 3x + 10^x = 0$
$x \approx -0.73$

61. Using 4, $A = P\left(1 + \frac{r}{m}\right)^{mt}$, we have:

(A) $P = 2{,}500,\ r = 0.07,\ m = 4,\ t = \frac{3}{4}$

$A = 2{,}500\left(1 + \frac{0.07}{4}\right)^{4\cdot 3/4} = 2{,}500(1 + 0.0175)^3 = 2{,}633.56$

Thus, $A = \$2{,}633.56$.

(B) $A = 2{,}500\left(1 + \frac{0.07}{4}\right)^{4\cdot 15} = 2{,}500(1 + 0.0175)^{60} = 7079.54$

Thus, $A = \$7{,}079.54$.

63. Using 6 with $P = 7{,}500$ and $r = 0.0835$, we have:
$A = 7{,}500e^{0.0835t}$

(A) $A = 7{,}500e^{(0.0835)5.5} = 7{,}500e^{0.45925} \approx 11{,}871.65$
Thus, there will be \$11,871.65 in the account after 5.5 years.

(B) $A = 7{,}500e^{(0.0835)12} = 7{,}500e^{1.002} \approx 20{,}427.93$
Thus, there will be \$20,427.93 in the account after 12 years.

65. Using $A = P\left(1 + \frac{r}{m}\right)^{mt}$, we have:
$A = 15{,}000,\ r = 0.0975,\ m = 52,\ t = 5$
Thus, $15{,}000 = P\left(1 + \frac{0.0975}{52}\right)^{52\cdot 5} = P(1 + 0.001875)^{260} \approx P(1.6275)$ and

$P = \frac{15{,}000}{1.6275} \approx 9{,}217$. Therefore, $P \approx \$9{,}217$.

67. Alamo Savings:
From Section 2-1, $A = P\left(1 + \frac{r}{m}\right)^{mt}$, where P is the principal, r is the annual rate, and m is the number of compounding periods per year. Thus:
$A = 10{,}000\left(1 + \frac{0.0825}{4}\right)^4 = 10{,}000(1.020625)^4 \approx \$10{,}850.88$

Lamar Savings:
$A = 10{,}000e^{0.0805} \approx \$10{,}838.29$

69. In $A = Pe^{rt}$, we are given $A = 50{,}000$, $r = 0.1$, and $t = 5.5$. Thus:

$50{,}000 = Pe^{(0.1)5.5}$ or $P = \dfrac{50{,}000}{e^{0.55}} \approx 28{,}847.49$

You should be willing to pay $28,847.49 for the note.

71. Given $N = 2(1 - e^{-0.037t})$, $0 \le t \le 50$

t	N
0	0
10	≈ 0.62
30	≈ 1.34
50	≈ 1.69

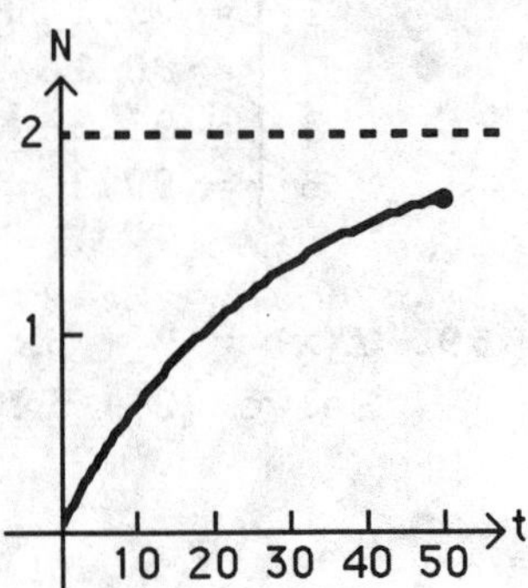

N approaches 2 as t increases without bound.

73. (A) Exponential regression model

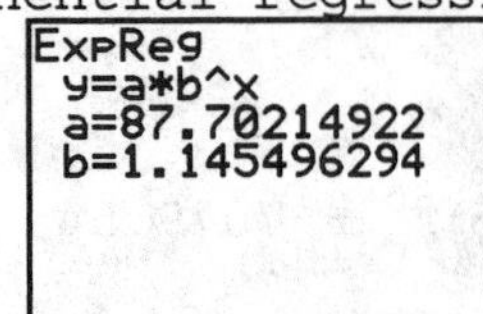

1995: $y(15) \approx \$673{,}000$
2010: $y(30) \approx \$5{,}162{,}000$

(B) The model predicts an average salary of $673,000 in 1995. This is greater than the actual average salary of $637,000. If a new model is calculated using this additional information, then the estimated salary will be less than the value $5,162,000 calculated in part (A).

75. Given $I = I_0e^{-0.23d}$

(A) $I = I_0e^{-0.23(10)} = I_0e^{-2.3} \approx I_0(0.10)$

Thus, about 10% of the surface light will reach a depth of 10 feet.

(B) $I = I_0e^{-0.23(20)} = I_0e^{-4.6} \approx I_0(0.010)$

Thus, about 1% of the surface light will reach a depth of 20 feet.

77. (A) Using $\underline{6}$ with $N_0 = 40{,}000$ and $r = 0.21$, we have $N = 40{,}000e^{0.21t}$.

(B) At the end of the year 2,000, $t = 8$ years and

$$N(8) = 40{,}000e^{0.21(8)} = 40{,}000e^{1.6} \approx 215{,}000$$

At the end of the year 2005, $t = 13$ years and

$$N(13) = 40{,}000e^{0.21(13)} = 40{,}000e^{2.73} \approx 613{,}000$$

(C)

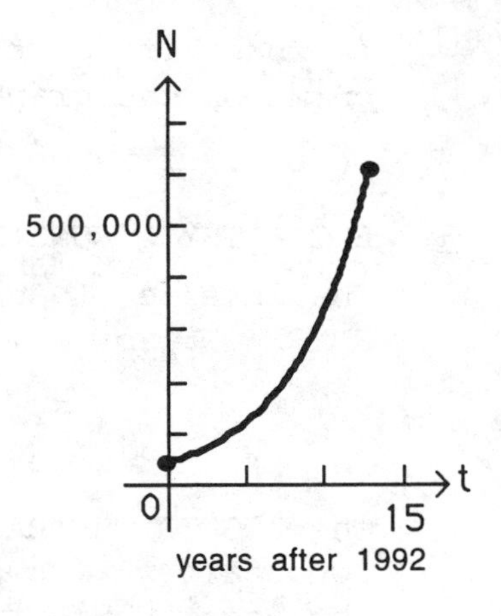

79. (A) Using $\underline{6}$ with $P_0 = 5.7$ and $r = 0.0114$, we have

$$P = 5.7e^{0.0114t}$$

(B) In the year 2010, $t = 15$ and

$$P = 5.7e^{0.0114(15)} = 5.7e^{0.171} \approx 6.8 \text{ billion}$$

In the year 2030, $t = 35$ and

$$P = 5.7e^{0.0114(35)} = 5.7e^{1.490} \approx 8.5 \text{ billion}$$

(C)

81. (A) Exponential regression model

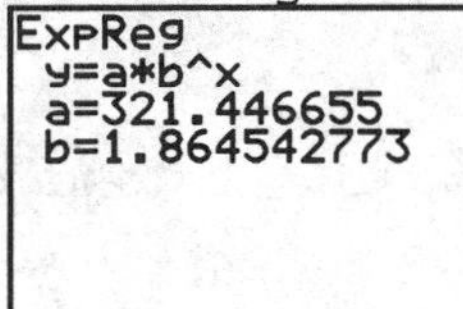

1995: $y(5) \approx 7{,}244{,}000$
1996: $y(6) \approx 13{,}507{,}000$

(B) 2017: $y(17) \approx 12.8$ billion
At this rate, the number of internet hosts will exceed the population of the world in less than 9 years.

EXERCISE 2-3

Things to remember:

1. ONE-TO-ONE FUNCTIONS

   A function f is said to be ONE-TO-ONE if each range value corresponds to exactly one domain value.

2. INVERSE OF A FUNCTION

   If f is a one-to-one function, then the INVERSE of f is the function formed by interchanging the independent and dependent variables for

f. Thus, if (a, b) is a point on the graph of f, then (b, a) is a point on the graph of the inverse of f.

Note: If f is not one-to-one, then f DOES NOT HAVE AN INVERSE.

3. LOGARITHMIC FUNCTIONS

The inverse of an exponential function is called a LOGARITHMIC FUNCTION. For $b > 0$ and $b \neq 1$,

Logarithmic form		Exponential form
$y = \log_b x$	is equivalent to	$x = b^y$

The LOG TO THE BASE b OF x is the exponent to which b must be raised to obtain x. [Remember: A logarithm is an exponent.] The DOMAIN of the logarithmic function is the range of the corresponding exponential function, and the RANGE of the logarithmic function is the domain of the corresponding exponential function. Typical graphs of an exponential function and its inverse, a logarithmic function, for $b > 1$, are shown in the figure below:

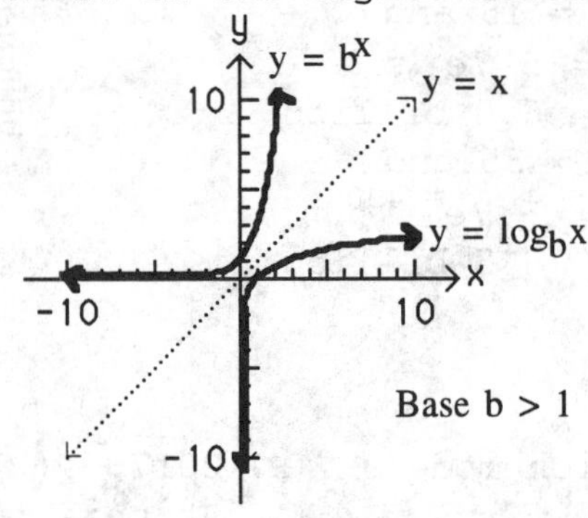

4. PROPERTIES OF LOGARITHMIC FUNCTIONS

If b, M, and N are positive real numbers, $b \neq 1$, and p and x are real numbers, then:

a. $\log_b 1 = 0$

b. $\log_b b = 1$

c. $\log_b b^x = x$

d. $b^{\log_b x} = x, \ x > 0$

e. $\log_b MN = \log_b M + \log_b N$

f. $\log_b \frac{M}{N} = \log_b M - \log_b N$

g. $\log_b M^p = p \log_b M$

h. $\log_b M = \log_b N$ if and only if $M = N$

5. LOGARITHMIC NOTATION; LOGARITHMIC-EXPONENTIAL RELATIONSHIPS

Common logarithm $\log x = \log_{10} x$

Natural logarithm $\ln x = \log_e x$

$\log x = y$ is equivalent to $x = 10^y$

$\ln x = y$ is equivalent to $x = e^y$

---

1. $27 = 3^3$ (using 3) **3.** $1 = 10^0$ **5.** $8 = 4^{3/2}$

7. $\log_7 49 = 2$

9. $\log_4 8 = \frac{3}{2}$

11. $\log_b A = u$

13. $\log_{10} 1 = y$ is equivalent to $10^y = 1$; $y = 0$.

15. $\log_e e = y$ is equivalent to $e^y = e$; $y = 1$.

17. $\log_{0.2} 0.2 = y$ is equivalent to $(0.2)^y = 0.2$; $y = 1$.

19. $\log_{10} 10^3 = 3$ (using 2a)

21. $\log_2 2^{-3} = -3$

23. $\log_{10} 1{,}000 = \log_{10} 10^3 = 3$

25. $\log_b \frac{P}{Q} = \log_b P - \log_b Q$ (using 4f)

27. $\log_b L^5 = 5 \log_b L$ (using 4g)

29. $\log_b \frac{p}{qrs} = \log_b p - \log_b qrs$ (using 4f)
$= \log_b p - (\log_b q + \log_b r + \log_b s)$ (using 4e)
$= \log_b p - \log_b q - \log_b r - \log_b s$

31. $\log_3 x = 2$
$x = 3^2$ (using 3)
$x = 9$

33. $\log_7 49 = y$
$\log_7 7^2 = y$
$2 = y$
Thus, $y = 2$.

35. $\log_b 10^{-4} = -4$
$10^{-4} = b^{-4}$
This equality implies $b = 10$ (since the exponents are the same).

37. $\log_4 x = \frac{1}{2}$
$x = 4^{1/2}$
$x = 2$

39. $\log_{1/3} 9 = y$
$9 = \left(\frac{1}{3}\right)^y$
$3^2 = (3^{-1})^y$
$3^2 = 3^{-y}$
This inequality implies that $2 = -y$ or $y = -2$.

41. $\log_b 1{,}000 = \frac{3}{2}$
$\log_b 10^3 = \frac{3}{2}$
$3 \log_b 10 = \frac{3}{2}$
$\log_b 10 = \frac{1}{2}$
$10 = b^{1/2}$
Square both sides:
$100 = b$, i.e., $b = 100$.

43. $\log_b \frac{x^5}{y^3}$
$= \log_b x^5 - \log_b y^3$
$= 5 \log_b x - 3 \log_b y$

45. $\log_b \sqrt[3]{N} = \log_b N^{1/3}$
$= \frac{1}{3} \log_b N$

47. $\log_b (x^2 \sqrt[3]{y}) = \log_b x^2 + \log_b y^{1/3} = 2 \log_b x + \frac{1}{3} \log_b y$

49. $\log_b (50 \cdot 2^{-0.2t}) = \log_b 50 + \log_b 2^{-0.2t} = \log_b 50 - 0.2t \log_b 2$

51. $\log_b P(1 + r)^t = \log_b P + \log_b (1 + r)^t = \log_b P + t\log_b (1 + r)$

53. $\log_e 100e^{-0.01t} = \log_e 100 + \log_e e^{-0.01t}$
$= \log_e 100 - 0.01t\log_e e = \log_e 100 - 0.01t$

55. $\log_b x = \frac{2}{3}\log_b 8 + \frac{1}{2}\log_b 9 - \log_b 6 = \log_b 8^{2/3} + \log_b 9^{1/2} - \log_b 6$
$= \log_b 4 + \log_b 3 - \log_b 6 = \log_b \frac{4\cdot 3}{6}$
$\log_b x = \log_b 2$
$x = 2$ (using 2e)

57. $\log_b x = \frac{3}{2}\log_b 4 - \frac{2}{3}\log_b 8 + 2\log_b 2 = \log_b 4^{3/2} - \log_b 8^{2/3} + \log_b 2^2$
$= \log_b 8 - \log_b 4 + \log_b 4 = \log_b 8$
$\log_b x = \log_b 8$
$x = 8$ (using 2e)

59. $\log_b x + \log_b (x - 4) = \log_b 21$
$\log_b x(x - 4) = \log_b 21$

Therefore, $x(x - 4) = 21$
$x^2 - 4x - 21 = 0$
$(x - 7)(x + 3) = 0$
Thus, $x = 7$.
[Note: $x = -3$ is not a solution since $\log_b(-3)$ is not defined.]

61. $\log_{10}(x - 1) - \log_{10}(x + 1) = 1$
$\log_{10}\left(\frac{x - 1}{x + 1}\right) = 1$

Therefore, $\frac{x - 1}{x + 1} = 10^1 = 10$
$x - 1 = 10(x + 1)$
$x - 1 = 10x + 10$
$-9x = 11$
$x = -\frac{11}{9}$

There is *no solution*, since
$\log_{10}\left(-\frac{11}{9} - 1\right) = \log_{10}\left(-\frac{20}{9}\right)$
is not defined. Similarly,
$\log_{10}\left(-\frac{11}{9} + 1\right) = \log_{10}\left(-\frac{2}{9}\right)$
is not defined.

63. $y = \log_2(x - 2)$
$x - 2 = 2^y$
$x = 2^y + 2$

x	y
$\frac{9}{4}$	-2
$\frac{5}{2}$	-1
3	0
4	1
6	2
18	4

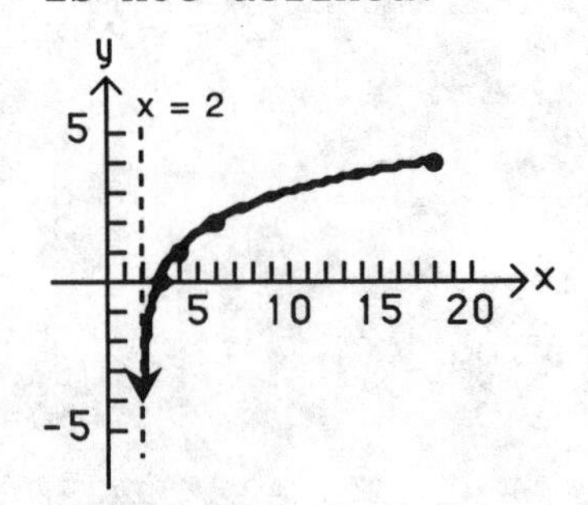

65. The graph of $y = \log_2(x - 2)$ is the graph of $y = \log_2 x$ shifted to the right 2 units.

67. Since logarithmic functions are defined only for positive "inputs", we must have $x + 1 > 0$ or $x > -1$; domain: $(-1, \infty)$. The range of $y = 1 + \ln(x + 1)$ is the set of all real numbers.

69. (A) 3.54743
(B) -2.16032
(C) 5.62629
(D) -3.19704

71. (A) $\log x = 1.1285$
$x = 13.4431$
(B) $\log x = -2.0497$
$x = 0.0089$
(C) $\ln x = 2.7763$
$x = 16.0595$
(D) $\ln x = -1.8879$
$x = 0.1514$

73. $10^x = 12$ (Take common logarithms of both sides)
$\log 10^x = \log 12 \approx 1.0792$
$x \approx 1.0792$ ($\log 10^x = x \log 10 = x$; $\log 10 = 1$)

75. $e^x = 4.304$ (Take natural logarithms of both sides)
$\ln e^x = \ln 4.304 \approx 1.4595$
$x \approx 1.4595$ ($\ln e^x = x \ln e = x$; $\ln e = 1$)

77. $1.03^x = 2.475$ (Take either common or natural logarithms of both sides; we use common logarithms)
$\log(1.03)^x = \log 2.475$
$$x = \frac{\log 2.475}{\log 1.03} \approx 30.6589$$

79. $1.005^{12t} = 3$ (Take either common or natural logarithms of both sides; here we'll use natural logarithms.)
$\ln 1.005^{12t} = \ln 3$
$$12t = \frac{\ln 3}{\ln 1.005} \approx 220.2713$$
$t = 18.3559$

81. $y = \ln x$, $x > 0$

x	y
0.5	≈ -0.69
1	0
2	≈ 0.69
4	≈ 1.39
5	≈ 1.61

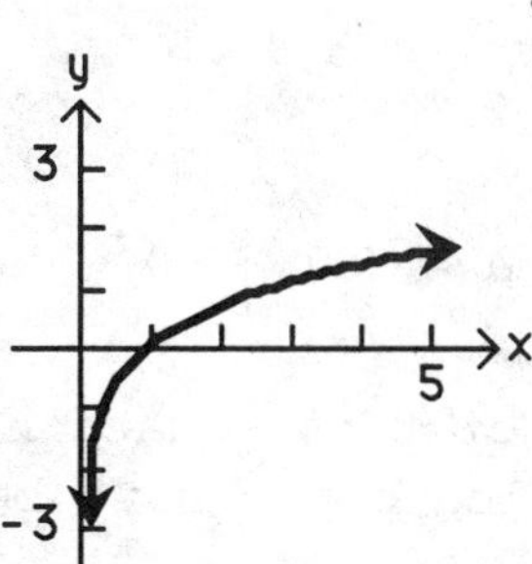

increasing $(0, \infty)$

83. $y = |\ln x|$, $x > 0$

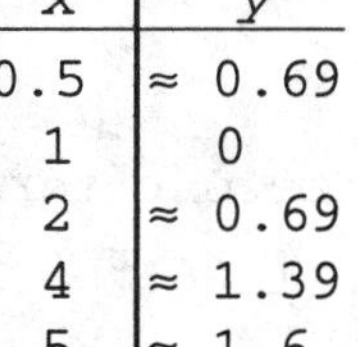

x	y
0.5	≈ 0.69
1	0
2	≈ 0.69
4	≈ 1.39
5	≈ 1.6

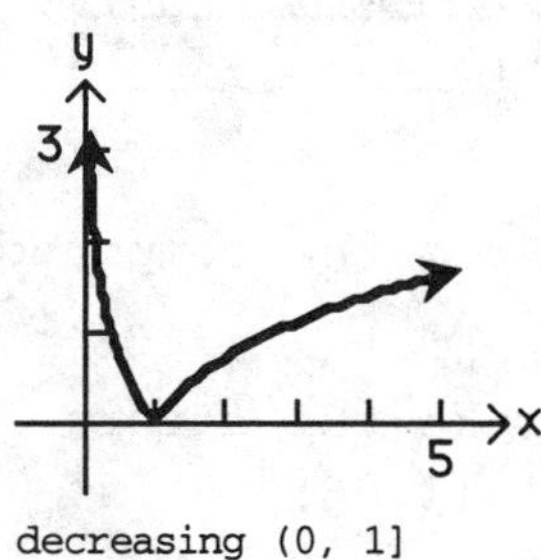

decreasing $(0, 1]$
increasing $[1, \infty)$

85. $y = 2\ln(x + 2),\ x > -2$

x	y
-1.5	$\approx$ -1.39
-1	0
0	$\approx$ 1.39
1	$\approx$ 2.2
5	$\approx$ 3.89
10	$\approx$ 4.61

increasing $(-2, \infty)$

87. $y = 4\ln x - 3,\ x > 0$

x	y
0.5	$\approx$ -5.77
1	-3
5	$\approx$ 3.44
10	$\approx$ 6.21

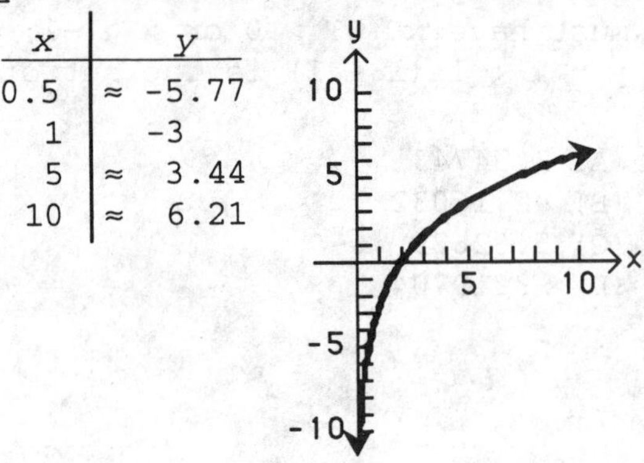

increasing $(0, \infty)$

89. The calculator interprets $\log \frac{13}{7}$ as $\frac{\log 13}{7}$ not as $\log\left(\frac{13}{7}\right)$. To find $\log\left(\frac{13}{7}\right)$, calculate $\frac{13}{7}$ and take the common logarithm of the result:

$$\log\left(\frac{13}{7}\right) = \log(1.8571\ldots) \approx 0.2688453123$$

or calculate log 13 - log 7 to get the same result.

91. For any number b, $b > 0$, $b \neq 1$, $\log_b 1 = y$ is equivalent to $b^y = 1$ which implies $y = 0$. Thus, $\log_b 1 = 0$ for any permissible base b.

93. $\log_{10} y - \log_{10} c = 0.8x$

$\log_{10} \frac{y}{c} = 0.8x$

Therefore, $\frac{y}{c} = 10^{0.8x}$ (using 1)

and $y = c \cdot 10^{0.8x}$.

95.

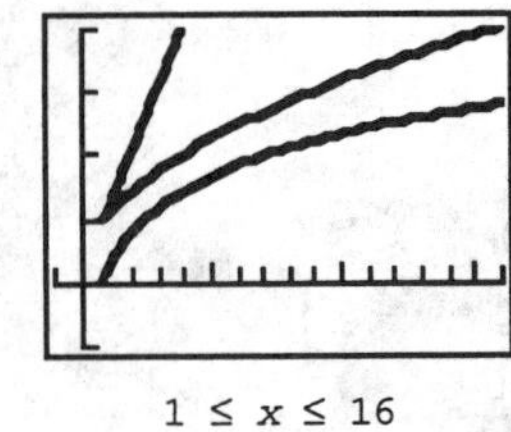

$1 \le x \le 16$

A function f is "larger than" a function g on an interval $[a, b]$ if $f(x) > g(x)$ for $a \le x \le b$. $r(x) > q(x) > p(x)$ for $1 \le x \le 16$, that is $x > \sqrt{x} > \ln x$ for $1 < x \le 16$

97. From the compound interest formula $A = P(1 + r)^t$, we have:

$2P = P(1 + .06)^t$ or $(1.06)^t = 2$

Take the natural log of both sides of this equation:

$\ln(1.06)^t = \ln 2$ [Note: The common log could have been used instead of the natural log.]

$t\ln(1.06) = \ln 2$

$$t = \frac{\ln 2}{\ln(1.06)} \approx \frac{.69315}{.05827} = 11.90 \approx 12 \text{ years}$$

99. (A) $A = P\left(1 + \frac{r}{m}\right)^{mt}$, $r = 0.06$, $m = 4$, $P = 1000$, $A = 1800$.

$$1800 = 1000\left(1 + \frac{0.06}{4}\right)^{4t} = 1000(1.015)^{4t}$$

$$(1.015)^{4t} = \frac{1800}{1000} = 1.8$$

$$4t\ln(1.015) = \ln(1.8)$$

$$t = \frac{\ln(1.8)}{4\ln(1.015)} \approx 9.87$$

\$1000 at 6% compounded quarterly will grow to \$1800 in 9.87 years.

(B) $A = Pe^{rt}$, $r = 0.06$, $P = 1000$, $A = 1800$

$$1000e^{0.06t} = 1800$$

$$e^{0.06t} = 1.8$$

$$0.06t = \ln 1.8$$

$$t = \frac{\ln 1.8}{0.06} \approx 9.80$$

\$1000 at 6% compounded continuously will grow to \$1800 in 9.80 years.

101. $A = Pe^{rt}$, $P = 10{,}000$, $A = 20{,}000$, $t = 8$

$$20{,}000 = 10{,}000e^{8r}$$

$$e^{8r} = 2$$

$$8r = \ln 2$$

$$r = \frac{\ln 2}{8} \approx 0.08664$$

\$10,000 invested at an annual interest rate of 8.664% compounded continuously will yield \$20,000 after 8 years.

103. (A) Logarithmic regression model, Table 1:

```
LnReg
 y=a+blnx
 a=256.4659159
 b=-24.03812068
```

To estimate the demand at a price level of \$50, we solve the equation

$a + b\ln x = 50$

for x. The result is $x \approx 5.373$ screwdrivers per month.

(B) Logarithmic regression model, Table 2:

```
LnReg
 y=a+blnx
 a=-127.8085281
 b=20.01315349
```

To estimate the supply at a price level of \$50, we solve

$a + b\ln x = 50$

for x. The result is $x \approx 7{,}220$ screwdrivers per month.

(C) The condition is not stable, the price is likely to decrease since the demand at a price level of \$50 is much lower than the supply at this level.

105. $I = I_0 10^{N/10}$

Take the common log of both sides of this equation. Then:

$$\log I = \log(I_0 10^{N/10}) = \log I_0 + \log 10^{N/10}$$

$$= \log I_0 + \frac{N}{10}\log 10 = \log I_0 + \frac{N}{10} \text{ (since } \log 10 = 1)$$

So, $\frac{N}{10} = \log I - \log I_0 = \log\left(\frac{I}{I_0}\right)$ and $N = 10 \log\left(\frac{I}{I_0}\right)$.

107. (A) Logarithmic regression model, Table 3:

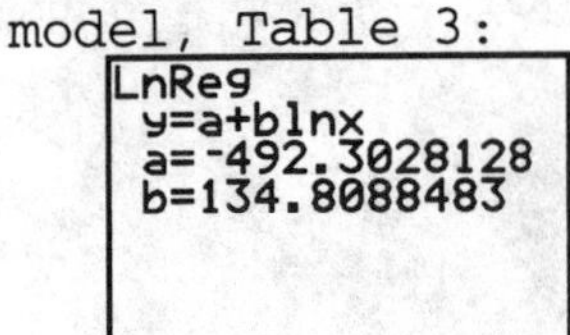

1996: $y(96) \approx 123.0$ bushels per acre
2010: $y(110) \approx 141.4$ bushels per acre

(B) The actual yield is 4.1 bushels per acre greater than the estimated yield. If a new model is calculated using the 1996 value, then the estimate for 2010 will increase.

109. Assuming that the world population is currently 5.8 billion and that it will grow at the rate of 1.14% compounded continuously, the population will be

$$P = 5.8e^{0.0114t}$$

after t years.

Given that there are 1.68×10^{14} = 168,000 billion square yards of land, we solve

$$168{,}000 = 5.8e^{0.0114t}$$

for t:

$$e^{0.0114t} \approx 28{,}966$$
$$0.0114t = \ln(28{,}966) \approx 10.2739$$
$$t \approx 901$$

It will take approximately 901 years.

CHAPTER 2 REVIEW

1. $u = e^v$
$v = \ln u$ (2-3)

2. $x = 10^y$
$y = \log x$ (2-3)

3. $\ln M = N$
$M = e^N$ (2-3)

4. $\log u = v$
$u = 10^v$ (2-3)

5. $\frac{5^{x+4}}{5^{4-x}} = 5^{x+4-(4-x)} = 5^{2x}$ (2-2)

6. $\left(\frac{e^u}{e^{-u}}\right)^u = (e^{u+u})^u = (e^{2u})^u = e^{2u^2}$ (2-2)

7. $\log_3 x = 2$
$x = 3^2$
$x = 9$ (2-3)

8. $\log_x 36 = 2$
$x^2 = 36$
$x = 6$ (2-3)

9. $\log_2 16 = x$
$2^x = 16$
$x = 4$ (2-3)

10. $10^x = 143.7$
$x = \log 143.7$
$x \approx 2.157$ (2-3)

11. $e^x = 503{,}000$

$x = \ln 503{,}000 \approx 13.128$ (2-3)

12. $\log x = 3.105$

$x = 10^{3.105} \approx 1273.503$ (2-3)

13. $\ln x = -1.147$

$x = e^{-1.147} \approx 0.318$ (2-3)

14. (A) 3 (B) 2 (C) 3 (D) 1 (E) 1 (F) 1 (2-1)

15. (A) 4 (B) 3 (C) 4 (D) 0 (E) 1 (F) 1 (2-1)

16. (A) 2 (B) 3 (C) positive (2-1)

17. (A) 3 (B) 4 (C) negative (2-1)

18. $f(x) = \dfrac{x + 4}{x - 2}$

(A) *Intercepts:*

x intercepts: $f(x) = 0$ only if $x + 4 = 0$ or $x = -4$.
The x intercept is -4.

y intercepts: $f(0) = \dfrac{0 + 4}{0 - 2} = -2$
The y intercept is -2.

(B) *Domain:* The denominator is 0 at $x = 2$. Thus, the domain is the set of all real numbers except 2.

(C) *Asymptotes:*

Vertical asymptotes: $f(x) = \dfrac{x + 4}{x - 2}$
The denominator is 0 at $x = 2$. Therefore, the line $x = 2$ is a vertical asymptote.

Horizontal asymptotes: $f(x) = \dfrac{x + 4}{x - 2} = \dfrac{1 + \frac{4}{x}}{1 - \frac{2}{x}}$

As x increases or decreases without bound, the numerator tends to 1 and the denominator tends to 1. Therefore, the line $y = 1$ is a horizontal asymptote.

(D)

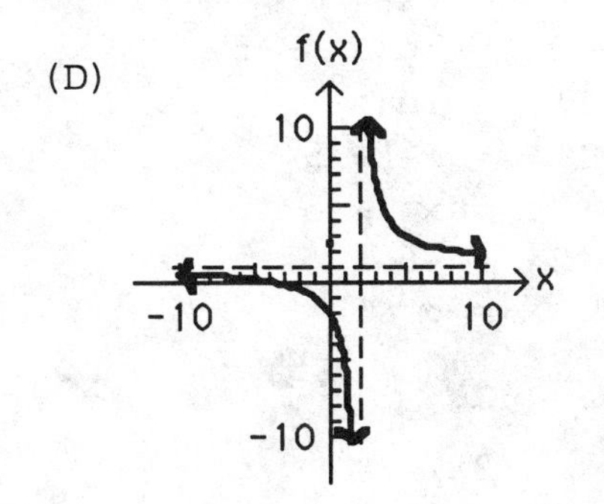

(E)

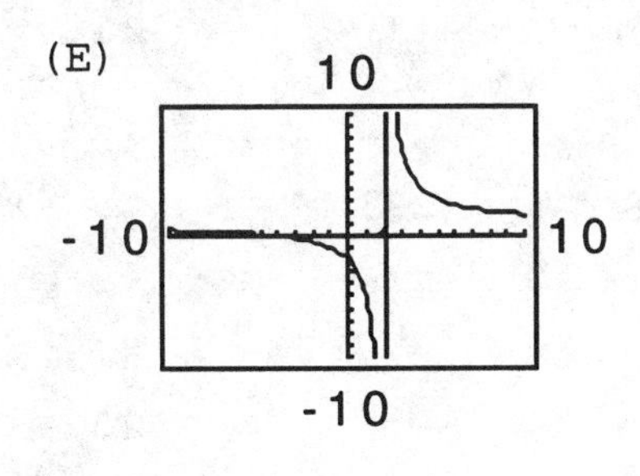

(2-1)

19. $f(x) = \frac{3x - 4}{2 + x}$

(A) *Intercepts:*

x intercepts: $f(x) = 0$ only if $3x - 4 = 0$ or $x = \frac{4}{3}$.

The x intercept is $\frac{4}{3}$.

y intercepts: $f(0) = \frac{3 \cdot 0 - 4}{2 + 0} = -2$

The y intercept is -2.

(B) *Domain:* The denominator is 0 at $x = -2$. Thus, the domain is the set of all real numbers except -2.

(C) *Asymptotes:*

Vertical asymptotes: $f(x) = \frac{3x - 4}{x + 2}$

The denominator is 0 at $x = -2$. Therefore, the line $x = -2$ is a vertical asymptote.

Horizontal asymptotes: $f(x) = \frac{3x - 4}{x + 2} = \frac{3 - \frac{4}{x}}{1 + \frac{2}{x}}$

As x increases or decreases without bound, the numerator tends to 3 and the denominator tends to 1. Therefore, the line $y = 3$ is a horizontal asymptote.

(D)

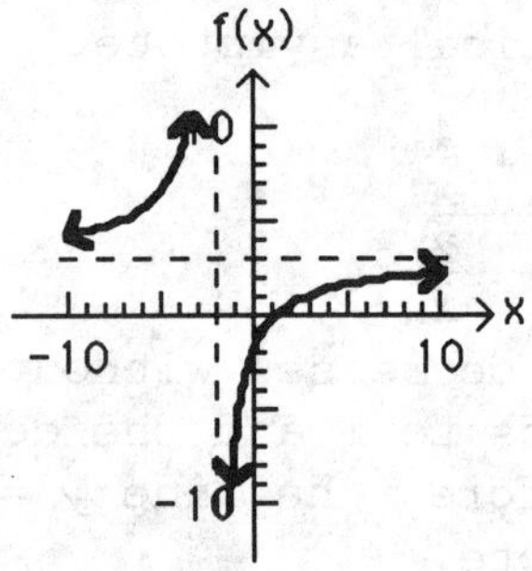
f(x)
10
-10
10
-10
x

(E)

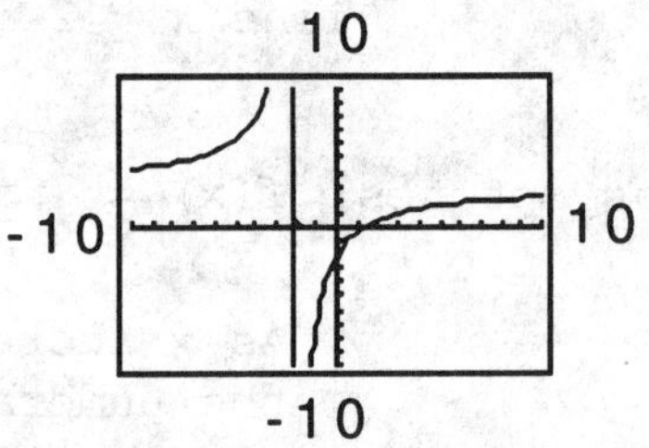
10
-10
10
-10

(2-1)

20.
$$\begin{aligned} \log(x + 5) &= \log(2x - 3) \\ x + 5 &= 2x - 3 \\ -x &= -8 \\ x &= 8 \end{aligned}$$
(2-3)

21.
$$\begin{aligned} 2\ln(x - 1) &= \ln(x^2 - 5) \\ \ln(x - 1)^2 &= \ln(x^2 - 5) \\ (x - 1)^2 &= x^2 - 5 \\ x^2 - 2x + 1 &= x^2 - 5 \\ -2x &= -6 \\ x &= 3 \end{aligned}$$
(2-3)

22.
$$\begin{aligned} 9^{x-1} &= 3^{1+x} \\ (3^2)^{x-1} &= 3^{1+x} \\ 3^{2x-2} &= 3^{1+x} \\ 2x - 2 &= 1 + x \\ x &= 3 \end{aligned}$$
(2-2)

23.
$$\begin{aligned} e^{2x} &= e^{x^2-3} \\ 2x &= x^2 - 3 \\ x^2 - 2x - 3 &= 0 \\ (x - 3)(x + 1) &= 0 \\ x &= 3, -1 \end{aligned}$$
(2-2)

24. $2x^2e^x = 3xe^x$

$2x^2 = 3x$ (divide both sides by e^x)

$2x^2 - 3x = 0$

$x(2x - 3) = 0$

$x = 0, \frac{3}{2}$ (2-2)

25. $\log_{1/3}9 = x$

$\left(\frac{1}{3}\right)^x = 9$

$\frac{1}{3^x} = 9$

$3^x = \frac{1}{9}$

$x = -2$

(2-3)

26. $\log_x 8 = -3$

$x^{-3} = 8$

$\frac{1}{x^3} = 8$

$x^3 = \frac{1}{8}$

$x = \frac{1}{2}$ (2-3)

27. $\log_9 x = \frac{3}{2}$

$9^{3/2} = x$

$x = 27$ (2-3)

28. $x = 3(e^{1.49}) \approx 13.3113$ (2-3)

29. $x = 230(10^{-0.161}) \approx 158.7552$ (2-3)

30. $\log x = -2.0144$

$x \approx 0.0097$ (2-3)

31. $\ln x = 0.3618$

$x \approx 1.4359$ (2-3)

32. $35 = 7(3^x)$

$3^x = 5$

$\ln 3^x = \ln 5$

$x \ln 3 = \ln 5$

$x = \frac{\ln 5}{\ln 3} \approx 1.4650$ (2-3)

33. $0.01 = e^{-0.05x}$

$\ln(0.01) = \ln(e^{-0.05x}) = -0.05x$

Thus, $x = \frac{\ln(0.01)}{-0.05} \approx 92.1034$ (2-3)

34. $8,000 = 4,000(1.08)^x$

$(1.08)^x = 2$

$\ln(1.08)^x = \ln 2$

$x \ln 1.08 = \ln 2$

$x = \frac{\ln 2}{\ln 1.08} \approx 9.0065$

(2-3)

35. $5^{2x-3} = 7.08$

$\ln(5^{2x-3}) = \ln 7.08$

$(2x - 3)\ln 5 = \ln 7.08$

$2x \ln 5 - 3 \ln 5 = \ln 7.08$

$x = \frac{\ln 7.08 + 3 \ln 5}{2 \ln 5}$

$= \frac{\ln 7.08 + \ln 5^3}{2 \ln 5}$

$= \frac{\ln[7.08(125)]}{2 \ln 5} \approx 2.1081$ (2-3)

36. $x = \log_2 7 = \frac{\log 7}{\log 2} \approx 2.8074$

or $x = \log_2 7 = \frac{\ln 7}{\ln 2} \approx 2.8074$

(2-3)

37. $x = \log_{0.2}5.321 = \frac{\log 5.321}{\log 0.2} \approx -1.0387$

or $x = \log_{0.2}5.321 = \frac{\ln 5.321}{\ln 0.2} \approx -1.0387$

(2-3)

38. The graph of $f(x) = x^4 - 4x^2 + 1 = x^4\left(1 - \frac{4}{x^2} + \frac{1}{x^4}\right)$ will "look like" the graph of $y = x^4$; for large x, $f(x) \approx x^4$. (2-1)

39. (A)

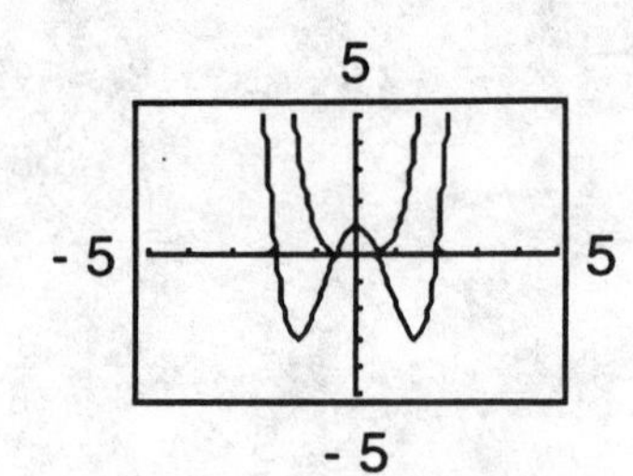

(B)

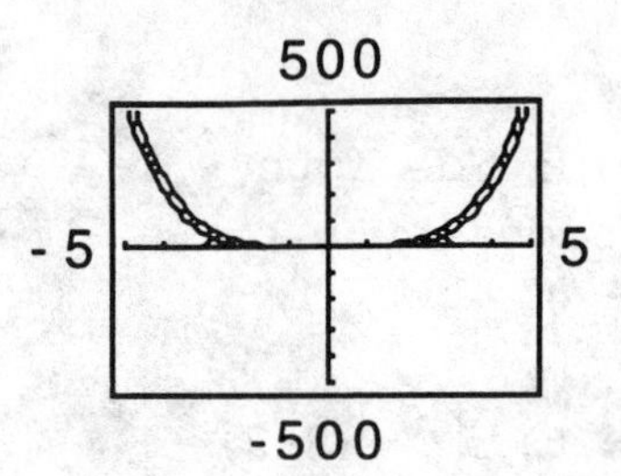

(2-1)

40. $p(x) = 2x^4 - 11x^3 - 15x^2 - 14x - 16$

Let $q(x) = \frac{1}{2}p(x) = x^4 - \frac{11}{2}x^3 - \frac{15}{2}x^2 - 7x - 8$; $p(x)$ and $q(x)$ have the same zeros. If x is a zero of $q(x)$, then

$$|x| < 1 + \max\left\{\left|-\frac{11}{2}\right|, \left|-\frac{15}{2}\right|, |-7|, |-8|\right\} = 1 + 8 = 9$$

Graph $p(x)$ on [-9, 9].

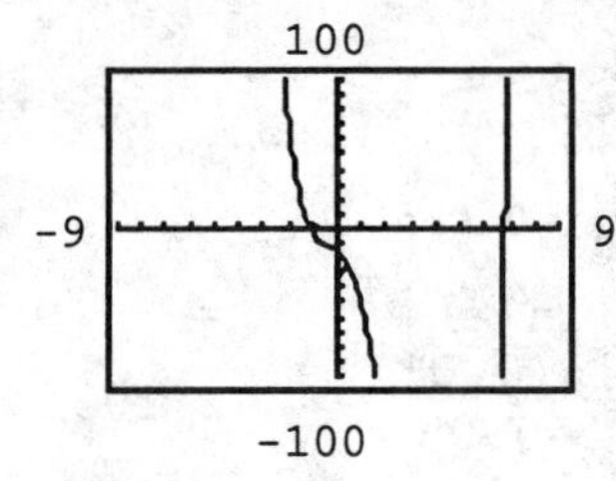

$p(x) = 0$ at $x \approx -1.14, 6.78$

(2-1)

41. $f(x) = e^x - 1$, $g(x) = \ln(x + 2)$

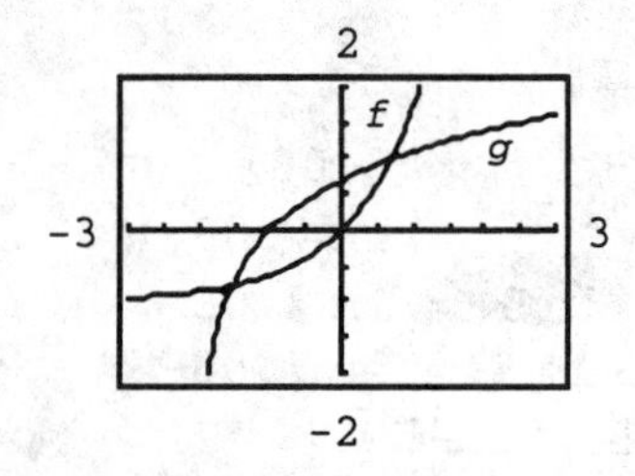

Points of intersection:
(-1.54, -0.79), (0.69, 0.99)

(2-2, 2-3)

42. $e^x(e^{-x} + 1) - (e^x + 1)(e^{-x} - 1) = 1 + e^x - (1 - e^x + e^{-x} - 1)$
$= 1 + e^x + e^x - e^{-x}$
$= 1 + 2e^x - e^{-x}$ (2-2)

43. $(e^x - e^{-x})^2 - (e^x + e^{-x})(e^x - e^{-x})$
$= (e^x)^2 - 2(e^x)(e^{-x}) + (e^{-x})^2 - [(e^x)^2 - (e^{-x})^2]$
$= e^{2x} - 2 + e^{-2x} - [e^{2x} - e^{-2x}]$
$= 2e^{-2x} - 2$ (2-2)

44. $y = 2^{x-1}$, $-2 \le x \le 4$

x	y
-2	$\frac{1}{8}$
-1	$\frac{1}{4}$
0	$\frac{1}{2}$
1	1
2	2
4	8

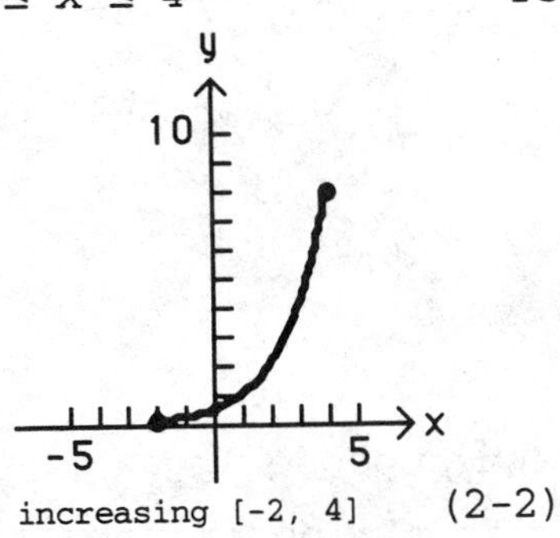

increasing [-2, 4] (2-2)

45. $f(t) = 10e^{-0.08t}$, $t \ge 0$

t	$f(t)$
0	10
10	≈ 4.5
20	≈ 2
30	≈ 0.9
40	≈ 0.4

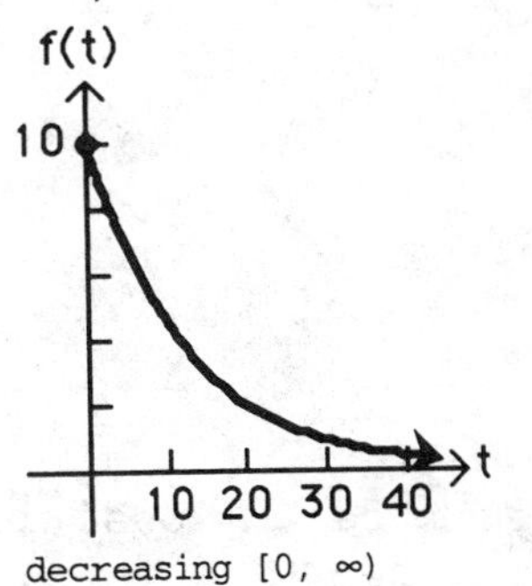

decreasing [0, ∞) (2-2)

46. $y = \ln(x + 1)$, $-1 < x \le 10$

x	y
-0.5	≈ -0.7
0	0
4	≈ 1.6
8	≈ 2.2
10	≈ 2.4

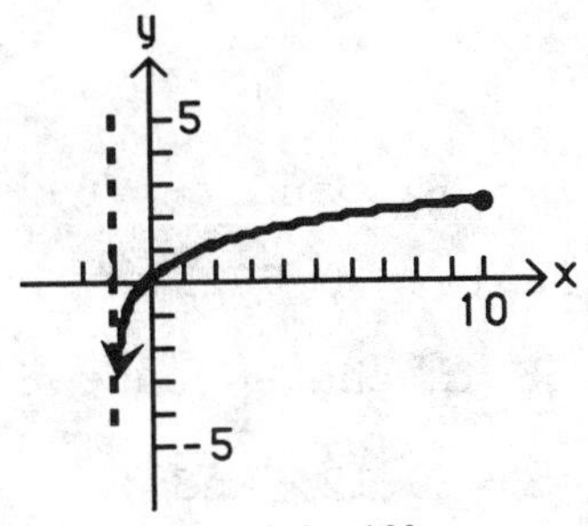

increasing (-1, 10] (2-2)

47. $\log 10^{\pi} = \pi \log 10 = \pi$ (see logarithm properties 4.b & g, Section 2-3)

$10^{\log\sqrt{2}} = y$ is equivalent to $\log y = \log\sqrt{2}$

which implies $y = \sqrt{2}$

Similarly, $\ln e^{\pi} = \pi \ln e = \pi$ (Section 2-3, 4.b & g) and $e^{\ln\sqrt{2}} = y$ implies $\ln y = \ln\sqrt{2}$ and $y = \sqrt{2}$. (2-3)

48.
$$\log x - \log 3 = \log 4 - \log(x + 4)$$
$$\log\frac{x}{3} = \log\frac{4}{x + 4}$$
$$\frac{x}{3} = \frac{4}{x + 4}$$
$$x(x + 4) = 12$$
$$x^2 + 4x - 12 = 0$$
$$(x + 6)(x - 2) = 0$$
$$x = -6,\ 2$$

Since log(-6) and log(-2) are not defined, -6 is not a solution. Therefore, the solution is $x = 2$. (2-3)

49. $\ln(2x - 2) - \ln(x - 1) = \ln x$
$$\ln\left(\frac{2x - 2}{x - 1}\right) = \ln x$$
$$\ln\left[\frac{2(x - 1)}{x - 1}\right] = \ln x$$
$$\ln 2 = \ln x$$
$$x = 2$$
(2-3)

50. $\ln(x + 3) - \ln x = 2\ln 2$
$$\ln\left(\frac{x + 3}{x}\right) = \ln(2^2)$$
$$\frac{x + 3}{x} = 4$$
$$x + 3 = 4x$$
$$3x = 3$$
$$x = 1$$
(2-3)

51.
$$\log 3x^2 = 2 + \log 9x$$
$$\log 3x^2 - \log 9x = 2$$
$$\log\left(\frac{3x^2}{9x}\right) = 2$$
$$\log\left(\frac{x}{3}\right) = 2$$
$$\frac{x}{3} = 10^2 = 100$$
$$x = 300$$
(2-3)

52.
$$\ln y = -5t + \ln c$$
$$\ln y - \ln c = -5t$$
$$\ln \frac{y}{c} = -5t$$
$$\frac{y}{c} = e^{-5t}$$
$$y = ce^{-5t}$$
(2-3)

53. Let x be *any* positive real number and suppose $\log_1 x = y$. Then $1^y = x$. But, $1^y = 1$, so $x = 1$, i.e., $x = 1$ for all positive real numbers x. This is clearly impossible. (2-3)

54. $A = P\left(1 + \frac{r}{m}\right)^{mt}$.

We let $P = 5{,}000$, $r = 0.12$, $m = 52$, and $t = 6$. Then we have:

$A = 5{,}000\left(1 + \frac{0.12}{52}\right)^{52(6)} \approx 5{,}000(1 + 0.0023)^{312} \approx 10{,}263.65$

Thus, there will be $10,263.65 in the account 6 years from now. (2-2)

55. $A = Pe^{rt}$. We let $P = 5{,}000$, $r = 0.12$, and $t = 6$. Then:

$A = 5{,}000e^{(0.12)6} \approx 10{,}272.17$

Thus, there will be $10,272.17 in the account 6 years from now. (2-2)

56. The compound interest formula for money invested at 15% compounded annually is:

$A = P(1 + 0.15)^t$

To find the tripling time, we set $A = 3P$ and solve for t:

$$3P = P(1.15)^t$$
$$(1.15)^t = 3$$
$$\ln(1.15)^t = \ln 3$$
$$t \ln 1.15 = \ln 3$$
$$t = \frac{\ln 3}{\ln 1.15} \approx 7.86$$

Thus, the tripling time (to the nearest year) is 8 years. (2-2)

57. The compound interest formula for money invested at 10% compounded continuously is:

$A = Pe^{0.1t}$

To find the doubling time, we set $A = 2P$ and solve for t:

$$2P = Pe^{0.1t}$$
$$e^{0.1t} = 2$$
$$0.1t = \ln 2$$
$$t = \frac{\ln 2}{0.1} \approx 6.93 \text{ years}$$
(2-3)

58. (A) Since $C(x)$ is a linear function of x, it can be written in the form

$$C(x) = mx + b$$

Since the fixed costs are \$300, $b = 300$.

Also, $C(100) = 4300$, so

$$4300 = 100m + 300$$
$$100m = 4000$$
$$m = 40$$

Therefore,

$$C(x) = 40x + 300$$

and $\overline{C}(x) = \dfrac{40x + 300}{x}$

(B)

(C) $\overline{C}(x) = \dfrac{40x + 300}{x} = \dfrac{40 + \frac{300}{x}}{1}$, $5 \le x \le 200$

As x increases, the numerator tends to 40 and the denominator is 1. Therefore, $\overline{C}(x)$ approaches 40; The line $y = 40$ is a horizontal asymptote.

(D) $\overline{C}(x)$ approaches \$40 per pair as production increases. (2-1)

59. (A) $\overline{C}(x) = \dfrac{C(x)}{x} = \dfrac{20x^3 - 360x^2 + 2{,}300x - 1{,}000}{x}$

(B)

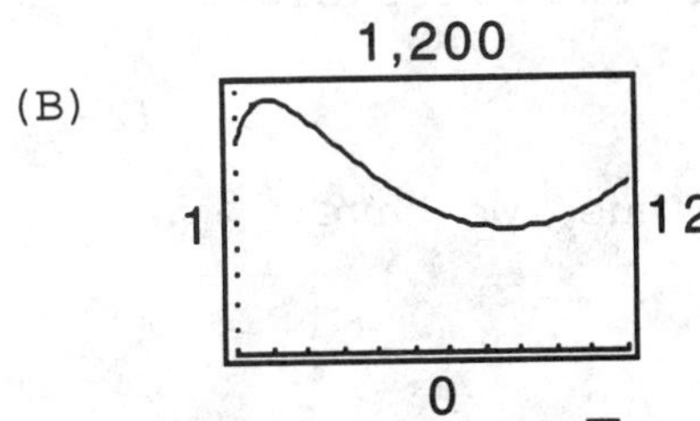

(C) From the graph, $\overline{C}(x)$ has a minimum at $x \approx 8.667$. Thus, the minimum average cost occurs when 8.667 thousand (8,667) cases are handled per year.

$$\overline{C}(8.667) = \frac{20(8.667)^3 - 360(8.667)^2 + 2300(8.667) - 1000}{8.667}$$
$$\approx \$567 \text{ per case}$$

(2-1)

60. (A) Quadratic regression model, Table 1:

To estimate the demand at a price level of \$180, we solve the equation

$$ax^2 + bx + c = 180$$

for x. The result is

$$x \approx 2{,}833 \text{ sets.}$$

(B) Linear regression model,
Table 2:

```
LinReg
 y=ax+b
 a=.0387421907
 b=-7.364689544
```

To estimate the supply at a price level of $180, we solve the equation

$$ax + b = 180$$

for x. The result is $x \approx 4{,}836$ sets.

(C) The condition is not stable; the price is likely to decrease since the supply at the price level of $180 exceeds the demand at this level.

(D) Equilibrium price: $131.59
Equilibrium quantity: 3,587 cookware sets (2-1)

61. (A) $N(0) = 1$

$$N\left(\frac{1}{2}\right) = 2$$

$$N(1) = 4 = 2^2$$

$$N\left(\frac{3}{2}\right) = 8 = 2^3$$

$$N(2) = 16 = 2^4$$

$\vdots$

Thus, we conclude that
$N(t) = 2^{2t}$ or $N = 4^t$.

(B) We need to solve:

$$2^{2t} = 10^9$$

$$\log 2^{2t} = \log 10^9 = 9$$

$$2t \log 2 = 9$$

$$t = \frac{9}{2 \log 2} \approx 14.95$$

Thus, the mouse will die in 15 days. (2-2, 2-3)

62. Given $I = I_0e^{-kd}$. When $d = 73.6$, $I = \frac{1}{2}I_0$. Thus, we have:

$$\frac{1}{2}I_0 = I_0e^{-k(73.6)}$$

$$e^{-k(73.6)} = \frac{1}{2}$$

$$-k(73.6) = \ln\frac{1}{2}$$

$$k = \frac{\ln(0.5)}{-73.6} \approx 0.00942$$

Thus, $k \approx 0.00942$.

To find the depth at which 1% of the surface light remains, we set $I = 0.01I_0$ and solve

$$0.01I_0 = I_0e^{-0.00942d}$$

for d:

$$0.01 = e^{-0.00942d}$$

$$-0.00942d = \ln 0.01$$

$$d = \frac{\ln 0.01}{-0.00942} \approx 488.87$$

Thus, 1% of the surface light remains at approximately 489 feet.

(2-2, 2-3)

63. (A) Logarithmic regression model

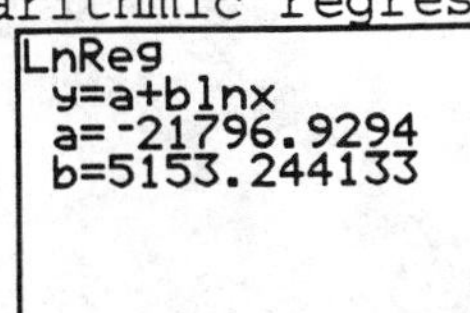

Total consumption 1996:
$y(96) \approx 1{,}724$ million bushels
Total consumption 2010:
$y(110) \approx 2{,}426$ million bushels

(B) The actual consumption was less than the consumption estimated by the model. If a new model using this additional information is computed, then the estimate for 2010 will decrease. (2-3)

64. Using the model $P = P_0(1 + r)^t$, we must solve $2P_0 = P_0(1 + 0.03)^t$ for t:

$$\begin{aligned} 2 &= (1.03)^t \\ \ln(1.03)^t &= \ln 2 \\ t\ln(1.03) &= \ln 2 \\ t &= \frac{\ln 2}{\ln 1.03} \approx 23.4 \end{aligned}$$

Thus, at a 3% growth rate, the population will double in approximately 23.4 years. (2-2, 2-3)

65. Using the continuous compounding model, we have:

$$\begin{aligned} 2P_0 &= P_0 e^{0.03t} \\ 2 &= e^{0.03t} \\ 0.03t &= \ln 2 \\ t &= \frac{\ln 2}{0.03} \approx 23.1 \end{aligned}$$

Thus, the model predicts that the population will double in approximately 23.1 years. (2-2, 2-3)

66. (A) Exponential regression model

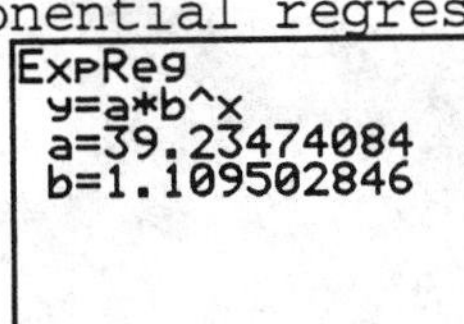

Total expenditures 1996:
$y(16) \approx \$207$ billion
Total expenditures 2010:
$y(30) \approx \$886$ billion

(B) To find when the total expenditures will reach \$500 billion, solve the equation $ab^x = 500$ for x. The result is $x \approx 24.5$ years; that is, at mid-year in 2004. (2-2)

3 MATHEMATICS OF FINANCE

EXERCISE 3-1

Things to remember:

1. SIMPLE INTEREST

$$I = Prt$$

where P = Principal
r = Annual simple interest rate expressed as a decimal
t = Time in years

2. AMOUNT—SIMPLE INTEREST

$$A = P + Prt = P(1 + rt)$$

where P = Principal or *present value*
r = Annual simple interest rate expressed as a decimal
t = Time in years
A = Amount or *future value*

---

1. $9.5\% = 0.095$; 60 days $= \frac{60}{360} = \frac{1}{6}$ year

3. $0.18 = 18\%$; 5 months $= \frac{5}{12}$ year

5. $P = \$500$, $r = 8\% = 0.08$, $t = 6$ months $= \frac{1}{2}$ year

$I = Prt$ (using 1)

$= 500(0.08)\left(\frac{1}{2}\right) = \$20$

7. $I = \$80$, $P = \$500$, $t = 2$ years

$I = Prt$

$r = \frac{I}{Pt} = \frac{80}{500(2)} = 0.08$ or 8%

9. $P = \$100$, $r = 8\% = 0.08$, $t = 18$ months $= 1.5$ years

$A = P(1 + rt) = 100(1 + 0.08 \cdot 1.5) = \$112$

11. $A = \$1000$, $r = 10\% = 0.1$, $t = 15$ months $= \frac{15}{12}$ years

$A = P(1 + rt)$

$$P = \frac{A}{1 + rt} = \frac{1000}{1 + (0.1)\left(\frac{15}{12}\right)} = \$888.89$$

13. $I = Prt$

Divide both sides by Pt.

$$\frac{I}{Pt} = \frac{Prt}{Pt}$$

$$\frac{I}{Pt} = r \quad \text{or} \quad r = \frac{I}{Pt}$$

15. $A = P + Prt = P(1 + rt)$

Divide both sides by $(1 + rt)$.

$$\frac{A}{1 + rt} = \frac{P(1 + rt)}{1 + rt}$$

$$\frac{A}{1 + rt} = P \quad \text{or} \quad P = \frac{A}{1 + rt}$$

17. Each of the graphs is a straight line; the y intercept in each case is 1000 and their slopes are 40, 80, and 120.

19. $P = \$3000$, $r = 14\% = 0.14$, $t = 4$ months $= \frac{1}{3}$ year

$$I = Prt = 3000(0.14)\left(\frac{1}{3}\right) = \$140$$

21. $P = \$554$, $r = 20\% = 0.2$, $t = 1$ month $= \frac{1}{12}$ year

$$I = Prt = 554(0.2)\left(\frac{1}{12}\right) = \$9.23$$

23. $P = \$7250$, $r = 9\% = 0.09$, $t = 8$ months $= \frac{2}{3}$ year

$$A = 7250\left[1 + 0.09\left(\frac{2}{3}\right)\right] = 7250[1.06] = \$7685.00$$

25. $P = \$4000$, $A = \$4270$, $t = 8$ months $= \frac{2}{3}$ year

The interest on the loan is $I = A - P = \$270$. From Problem 9,

$$r = \frac{I}{Pt} = \frac{270}{4000\left(\frac{2}{3}\right)} = 0.10125. \quad \text{Thus, } r = 10.125\%.$$

27. $P = \$1000$, $I = \$30$, $t = 60$ days $= \frac{1}{6}$ year

$$r = \frac{I}{Pt} = \frac{30}{1000\left(\frac{1}{6}\right)} = 0.18. \quad \text{Thus, } r = 18\%.$$

29. $P = \$1500$. The amount of interest paid is $I = (0.5)(3)(120) = \$180$. Thus, the total amount repaid is $\$1500 + \$180 = \$1680$. To find the annual interest rate, we let $t = 120$ days $= \frac{1}{3}$ year. Then

$$r = \frac{I}{Pt} = \frac{180}{1500\left(\frac{1}{3}\right)} = 0.36. \quad \text{Thus, } r = 36\%.$$

31. $P = \$9776.94$, $A = \$10{,}000$, $t = 13$ weeks $= \frac{1}{4}$ year

The interest is $I = A - P = \$223.06$.

$$r = \frac{I}{Pt} = \frac{223.06}{9776.94\left(\frac{1}{4}\right)} = 0.09126. \quad \text{Thus, } r = 9.126\%.$$

33. $A = \$10{,}000$, $r = 12.63\% = 0.1263$, $t = 13$ weeks $= \frac{1}{4}$ year.

From Problem 15, $P = \frac{A}{1 + rt} = \frac{10{,}000}{1 + (0.1263)\frac{1}{4}} = \frac{10{,}000}{1.03158} = \$9693.91.$

35. Principal plus interest on the original note:

$$A = P(1 + rt) = \$5500\left[1 + 0.12\left(\frac{90}{360}\right)\right]$$
$$= \$5665$$

The third party pays \$5,540 and will receive \$5665 in 60 days. We want to find r given that $A = 5665$, $P = 5{,}540$ and $t = \frac{60}{360} = \frac{1}{6}$

$$A = P + Prt$$
$$r = \frac{A - P}{Pt}$$
$$r = \frac{5665 - 5540}{5540\left(\frac{1}{6}\right)} = 0.13538 \text{ or } 13.538\%$$

37. The principal P is the cost of the stock plus the broker's commission. The cost of the stock is $500(14.20) = \$7100$ and the commission on this is $62 + (0.003)7100 = \$83.30$. Thus, $P = \$7183.30$. The investor sells the stock for $500(16.84) = \$8420$, and the commission on this amount is $62 + (0.003)8420 = \$87.26$. Thus, the investor has $8420 - 87.26 = \$8332.74$ after selling the stock. We can now conclude that the investor has earned $8332.74 - 7183.30 = \$1149.44$.

Now, $P = \$7183.30$, $I = \$1149.44$, $t = 39$ weeks $= \frac{3}{4}$ year. Therefore,

$$r = \frac{I}{Pt} = \frac{1149.44}{7183.30\left(\frac{3}{4}\right)} = 0.21335 \quad \text{or} \quad r = 21.335\%.$$

39. The principal P is the cost of the stock plus the broker's commission. The cost of the stock is: $2000(23.75) = \$47{,}500$, and the commission on this is: $84 + 0.002(47{,}500) = \$179$. Thus $P = \$47{,}679$. The investor sells this stock for $2000(26.15) = \$52{,}300$, and the commission on this amount is: $134 + 0.001(52{,}300) = \$186.30$. Thus, the investor has $52{,}300 - 186.30 = \$52{,}113.70$ after selling the stock. We can now conclude that the investor has earned $52{,}113.70 - 47{,}679 = \$4{,}434.70$.

Now, $P = \$47{,}679$, $I = \$4434.70$, $t = 300$ days $= \frac{300}{360} = \frac{5}{6}$ year.

Therefore, $r = \frac{I}{Pt} = \frac{4434.70}{(47{,}679)\left(\frac{5}{6}\right)} \approx 0.11161$ or $r = 11.161\%$

EXERCISE 3-2

Things to remember:

1. AMOUNT—COMPOUND INTEREST

$$A = P(1 + i)^n, \text{ where } i = \frac{r}{m} \text{ and}$$

r = Annual (quoted) rate
m = Number of compounding periods per year
n = Total number of compounding periods
$i = \frac{r}{m}$ = Rate per compounding period
P = Principal (present value)
A = Amount (future value) at the end of n periods

2. EFFECTIVE RATE

If principal P is invested at the (nominal) rate r, compounded m times per year, then the effecitve rate, r_e, is given by

$$r_e = \left(1 + \frac{r}{m}\right)^m - 1.$$

1. $P = \$100$, $i = 0.01$, $n = 12$
Using 1,
$$A = P(1 + i)^n = 100(1 + 0.01)^{12} = 100(1.01)^{12} = \$112.68$$

3. $P = \$800$, $i = 0.06$, $n = 25$
Using 1,
$$A = 800(1 + 0.06)^{25} = 800(1.06)^{25} = \$3433.50$$

5. $A = \$10{,}000$, $i = 0.03$, $n = 48$
Using 1,
$A = P(1 + i)^n$
$$P = \frac{A}{(1 + i)^n} = \frac{10{,}000}{(1 + 0.03)^{48}} = \frac{10{,}000}{(1.03)^{48}} = \$2419.99$$

7. $A = \$18{,}000$, $i = 0.01$, $n = 90$
Refer to Problem 5:
$$P = \frac{A}{(1 + i)^n} = \frac{18{,}000}{(1 + 0.01)^{90}} = \frac{18{,}000}{(1.01)^{90}} = \$7351.04$$

9. $r = 9\%$, $m = 12$. Thus, $i = \frac{r}{m} = \frac{0.09}{12} = 0.0075$ or 0.75% per month.

11. $r = 7\%$, $m = 4$. Thus, $i = \frac{r}{m} = \frac{0.07}{4} = 0.0175$ or 1.75% per quarter.

13. $i = 0.8\%$ per month ($m = 12$). Thus, $r = i \cdot m = (0.008)12 = 0.096$ or 9.6% compounded monthly.

15. $i = 4.5\%$ per half year ($m = 2$). Thus, $r = i \cdot m = (0.045)2 = 0.09$ or 9% compounded semiannually.

17. $P = \$100$, $r = 6\% = 0.06$

(A) $m = 1$, $i = 0.06$, $n = 4$

$A = (1 + i)^n$
$= 100(1 + 0.06)^4$
$= 100(1.06)^4 = \$126.25$

Interest $= 126.25 - 100 = \$26.25$

(B) $m = 4$, $i = \frac{0.06}{4} = 0.015$

$n = 4(4) = 16$
$A = 100(1 + 0.015)^{16}$
$= 100(1.015)^{16} = \$126.90$

Interest $= 126.90 - 100 = \$26.90$

(C) $m = 12$, $i = \frac{0.06}{12} = 0.005$, $n = 4(12) = 48$

$A = 100(1 + 0.005)^{48} = 100(1.005)^{48} = \$127.05$

Interest $= 127.05 - 100 = \$27.05$

19. $P = \$5000$, $r = 18\%$, $m = 12$

(A) $n = 2(12) = 24$

$i = \frac{0.18}{12} = 0.015$

$A = 5000(1 + 0.015)^{24}$
$= 5000(1.015)^{24} = \$7147.51$

(B) $n = 4(12) = 48$

$i = \frac{0.18}{12} = 0.015$

$A = 5000(1 + 0.015)^{48}$
$= 5000(1.015)^{48} = \$10,217.39$

21. Each of the graphs is increasing, curves upward and has y intercept 1000. The greater the interest rate, the greater the increase. The amounts at the end of 8 years are:

At 4%: $A = 1000\left(1 + \frac{0.04}{12}\right)^{96} = \$1376.40$

At 8%: $A = 1000\left(1 + \frac{0.08}{12}\right)^{96} = \$1892.46$

At 12%: $A = 1000\left(1 + \frac{0.12}{12}\right)^{96} = \$2599.27$

23. $P = 1,000$, $r = 9.75\% = 0.0975 = i$ since the interest is compounded annually.

1st year: $A = P(1 + i)^n = 1000(1 + 0.0975)^1 = \$1,097.50$
Interest \$97.50

2nd year: $A = 1000(1 + 0.0975)^2 = \$1,204.51$
Interest: $1,204.51 - 1,097.50 = \$107.01$

3rd year: $A = 1000(1 + 0.0975)^3 = \$1,321.95$
Interest: $1,321.95 - 1,204.51 = \$117.44$

and so on. The results are:

Period	Interest	Amount
0		\$1,000.00
1	\$97.50	\$1,097.50
2	\$107.01	\$1,204.51
3	\$117.44	\$1,321.95
4	\$128.89	\$1,450.84
5	\$141.46	\$1,592.29
6	\$155.25	\$1,747.54

25. $A = \$10,000$, $r = 8\% = 0.08$, $i = \frac{0.08}{2} = 0.04$

(A) $n = 2(5) = 10$

$$A = P(1 + i)^n$$
$$10,000 = P(1 + 0.04)^{10}$$
$$= P(1.04)^{10}$$
$$P = \frac{10,000}{(1.04)^{10}} = \$6755.64$$

(B) $n = 2(10) = 20$

$$P = \frac{A}{(1 + i)^n} = \frac{10,000}{(1 + 0.04)^{20}}$$
$$= \frac{10,000}{(1.04)^{20}}$$
$$= \$4563.87$$

27. Use the formula for r_e in 2.

(A) $r = 10\% = 0.1$, $m = 4$

$$r_e = \left(1 + \frac{0.1}{4}\right)^4 - 1 = 0.1038 \text{ or } 10.38\%$$

(B) $r = 12\% = 0.12$, $m = 12$

$$r_e = \left(1 + \frac{0.12}{12}\right)^{12} - 1 = 0.1268 \text{ or } 12.68\%$$

29. We have $P = \$4000$, $A = \$9000$, $r = 15\% = 0.15$, $m = 12$, and $i = \frac{0.15}{12} = 0.0125$. Since $A = P(1 + i)^n$, we have:

$9000 = 4000(1 + 0.0125)^n$ or $(1.0125)^n = 2.25$

Method 1: Use Table II. Look down the $(1 + i)^n$ column on the page that has $i = 0.0125$. Find the value of n in this column that is closest to and greater than 2.25. In this case, $n = 66$ months or 5 years and 6 months.

Method 2: Use logarithms and a calculator.

$$\ln(1.0125)^n = \ln 2.25$$
$$n \ln 1.0125 = \ln 2.25$$
$$n = \frac{\ln 2.25}{\ln 1.0125} \approx \frac{0.8109}{0.01242} \approx 65.29$$

Thus, $n = 66$ months or 5 years and 6 months.

31. $A = 2P$, $i = 0.06$

$$A = P(1 + i)^n$$
$$2P = P(1 + 0.06)^n$$
$$(1.06)^n = 2$$
$$\ln(1.06)^n = \ln 2$$
$$n \ln(1.06) = \ln 2$$
$$n = \frac{\ln 2}{\ln 1.06} \approx \frac{0.6931}{0.0583} \approx 11.9 \approx 12$$

33. We have $A = P(1 + i)^n$. To find the doubling time, set $A = 2P$. This yields:

$2P = P(1 + i)^n$ or $(1 + i)^n = 2$

Taking the natural logarithm of both sides, we obtain:

$$\ln(1 + i)^n = \ln 2$$
$$n \ln(1 + i) = \ln 2$$

and

$$n = \frac{\ln 2}{\ln(1 + i)}$$

(A) $r = 10\% = 0.1$, $m = 4$. Thus,

$$i = \frac{0.1}{4} = 0.025 \text{ and } n = \frac{\ln 2}{\ln(1.025)} \approx 28.07 \text{ quarters or } 7\frac{1}{4} \text{ years.}$$

(B) $r = 12\% = 0.12$, $m = 4$. Thus,

$$i = \frac{0.12}{4} = 0.03 \text{ and } n = \frac{\ln 2}{\ln(1.03)} \approx 23.44 \text{ quarters.}$$

That is, 24 quarters or 6 years.

35. $P = \$5000$, $r = 9\% = 0.09$, $m = 4$, $i = \frac{0.09}{4} = 0.0225$, $n = 17(4) = 68$

Thus,
$$\begin{aligned} A &= P(1 + i)^n \\ &= 5000(1 + 0.0225)^{68} \\ &= 5000(1.0225)^{68} \\ &= \$22{,}702.60 \end{aligned}$$

37. $P = \$110{,}000$, $r = 6\%$ or 0.06, $m = 1$, $i = 0.06$, $n = 10$

Thus,
$$\begin{aligned} A &= P(1 + i)^n \\ &= 110{,}000(1 + 0.06)^{10} \\ &= 110{,}000(1.06)^{10} \\ &\approx \$196{,}993.25 \end{aligned}$$

39. $A = \$20$, $r = 7\% = 0.07$, $m = 1$, $i = 0.07$, $n = 5$

$$A = P(1 + i)^n$$

$$P = \frac{A}{(1 + i)^n} = \frac{20}{(1.07)^5} \approx \$14.26 \text{ per square foot per month}$$

41. From Problem 33, the doubling time is:

$$n = \frac{\ln 2}{\ln(1 + i)}$$

Here $r = i = 0.04$. Thus,

$$n = \frac{\ln 2}{\ln(1.04)} \approx 17.67 \text{ or } 18 \text{ years}$$

43. The effective rate, r_e, of $r = 9\% = 0.09$ compounded monthly is:

$$r_e = \left(1 + \frac{0.09}{12}\right)^{12} - 1 = .0938 \text{ or } 9.38\%$$

The effective rate of 9.3% compounded annually is 9.3%. Thus, 9% compounded monthly is better than 9.3% compounded annually.

45. (A) The Declaration of Independence was signed in 1776, 222 years ago.

We have $P = \$100$, $r = 0.03$, $n = 4(222) = 888$ and $i = \frac{0.03}{4} = 0.0075$.

Thus,
$$\begin{aligned} A &= P(1 + i)^n \\ &= 100(1 + 0.0075)^{888} \\ &= 100(761.3926) \\ &= \$76{,}139.26 \end{aligned}$$

(B) Monthly compounding:

$$n = 12(222) = 2664; \; i = \frac{0.03}{12} = 0.0025$$

$$\begin{aligned} A &= 100(1 + 0.0025)^{2664} \\ &= \$77{,}409.05 \end{aligned}$$

Daily compounding:

$n = 365(222) = 81{,}030;\ i = \dfrac{0.03}{365} = 0.000082191$

$A = 100(1 + 0.000082)^{81{,}030}$
$= \$78{,}033.73$

Continuous compounding

$A = Pe^{rt}$
$= 100e^{0.03(222)}$
$= \$78{,}055.09$

(C)

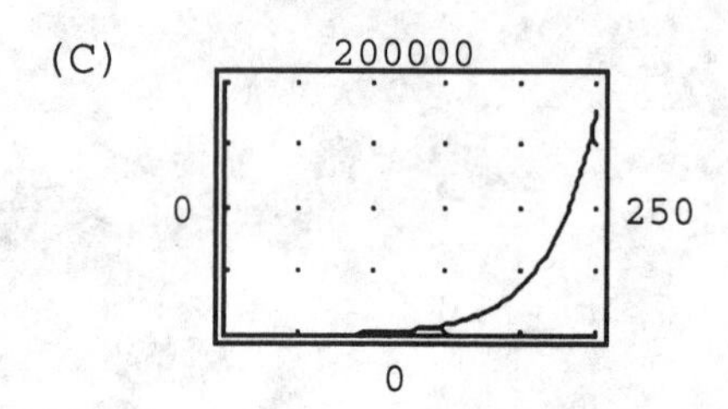

47. $P = \$7000,\ A = \$9000,\ r = 9\% = 0.09,\ m = 12,\ i = \dfrac{0.09}{12} = 0.0075$

Since $A = P(1 + i)^n$, we have:

$9000 = 7000(1 + 0.0075)^n$ or $(1.0075)^n = \dfrac{9}{7}$

Therefore, $\ln(1.0075)^n = \ln\left(\dfrac{9}{7}\right)$

$$n \ln(1.0075) = \ln\left(\frac{9}{7}\right)$$

$$n = \frac{\ln\left(\frac{9}{7}\right)}{\ln(1.0075)} \approx \frac{0.2513}{0.0075} \approx 33.6$$

Thus, it will take 34 months or 2 years and 10 months.

49. $P = \$20{,}000,\ r = 8\% = 0.08,\ m = 365,\ i = \dfrac{0.08}{365} \approx 0.0002192,$

$n = (365)35 = 12{,}775$

Since $A = P(1 + i)^n$, we have:

$$A = 20{,}000(1.0002192)^{12{,}775} \approx \$328{,}791.70$$

51. From Problem 33, the doubling time is:

$n = \dfrac{\ln 2}{\ln(1 + i)}$

(A) $r = 14\% = 0.14,\ m = 365,\ i = \dfrac{0.14}{365} \approx 0.0003836$

Thus, $n = \dfrac{\ln 2}{\ln(1.0003836)} \approx 1807.48$ days or 4.952 years.

(B) $r = 15\% = 0.15,\ m = 1,\ i = 0.15$

Thus, $n = \dfrac{\ln 2}{\ln(1.15)} \approx 4.959$ years.

53. If an investment of P dollars doubles in n years at an interest rate r compounded annually, then r satisfies the equation:

$$2P = P(1 + r)^n$$

so $1 + r = 2^{1/n}$

and $r = 2^{1/n} - 1$

This is the exact value of r.

The approximate value of r given by

$$r = \frac{72}{n}$$

is called the Rule of 72.

Setting $n = 6, 7, 8, 9, 10, 11, 12$ in these formulas gives:

Years	Exact rate	Rule of 72
6	12.2	12.0
7	10.4	10.3
8	9.1	9.0
9	8.0	8.0
10	7.2	7.2
11	6.5	6.5
12	5.9	6.0

55. \$2,400 at 13% compounded quarterly:

Value after n quarters: $A_1 = 2400\left(1 + \frac{0.13}{4}\right)^n = 2400(1.0325)^n$

\$3,000 at 9% compounded quarterly:

Value after n quarters: $A_2 = 3000\left(1 + \frac{0.09}{4}\right)^n = 3000(1.0225)^n$

Graph A_1 and A_2:

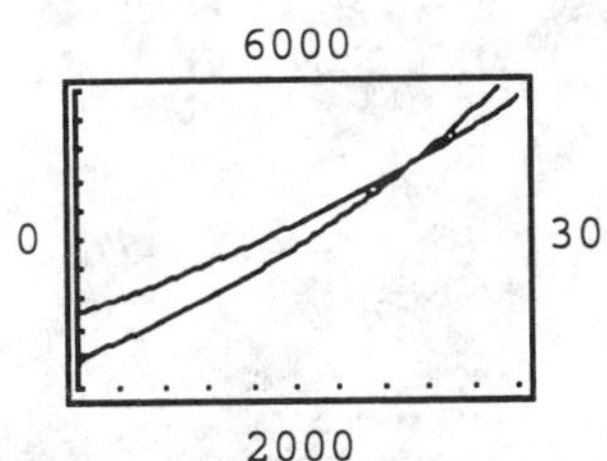

The graphs intersect at (22.96, 5000)
$A_2(n) > A_1(n)$ for $n < 22.96$, $A_1(n) > A_2(n)$ for $n > 22.96$.

Thus it will take 23 quarters for the \$2,400 investment to be worth more than the \$3,000 investment.

57. The value of P dollars at 10% simple interest after t years is given by

$$A_s = P(1 + 0.1t)$$

The value of P dollars at 7% interest compounded annually after t years is given by

$$A_c = P(1.07)^t$$

Let $P = \$1$ and graph
$A_s = 1 + 0.1t$, $A_c = (1.07)^t$.

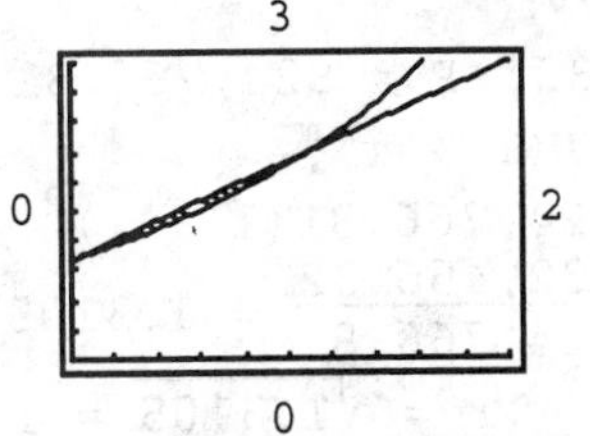

The graphs intersect at the point where $t = 10.89 \approx 11$ years.

For investments of less than 11 years, simple interest at 10% is better; for investments of more than 11 years, interest compounded annually at 7% is better.

59. The relationship between the effective rate and the annual nominal rate is:

$$r_e = \left(1 + \frac{r}{m}\right)^m - 1$$

In this case, $r_e = 0.074$ and $m = 365$. Thus, we must solve

$$0.074 = \left(1 + \frac{r}{365}\right)^{365} - 1 \text{ for } r$$

$$\left(1 + \frac{r}{365}\right)^{365} = 1.074$$

$$1 + \frac{r}{365} = (1.074)^{1/365}$$

$$r = 365[(1.074)^{1/365} - 1] \approx 0.0714 \text{ or } r = 7.14\%$$

61. $A = \$30,000$, $r = 10\% = 0.1$, $m = 1$, $i = 0.1$, $n = 17$

From $A = P(1 + i)^n$, we have:

$$P = \frac{A}{(1 + i)^n} = \frac{30,000}{(1.1)^{17}} \approx \$5935.34$$

63. $A = \$30,000$, $P = \$6844.79$, $r = i$, $n = 17$

Using $A = P(1 + r)^n$, we have:

$$30,000 = 6844.79(1 + r)^{17}$$

$$(1 + r)^{17} = \frac{30,000}{6844.79} \approx 4.3829$$

Therefore,

$1 + r \approx (4.3829)^{1/17}$ and $r \approx (4.3829)^{1/17} - 1 \approx 0.0908$ or $r = 9.08\%$

65. From 2, $r_e = \left(1 + \frac{r}{m}\right)^m - 1$.

(A) $r = 8.28\% = 0.0828$, $m = 12$

$$r_e = \left(1 + \frac{0.0828}{12}\right)^{12} - 1 \approx 0.0860 \quad \text{or} \quad 8.60\%$$

(B) $r = 8.25\% = 0.0825$, $m = 365$

$$r_e = \left(1 + \frac{0.0825}{365}\right)^{365} - 1 \approx 0.0860 \quad \text{or} \quad 8.60\%$$

(C) $r = 8.25\% = 0.0825$, $m = 12$

$$r_e = \left(1 + \frac{0.0825}{12}\right)^{12} - 1 \approx 0.0857 \quad \text{or} \quad 8.57\%$$

67. $A = \$32,456.32$, $P = \$24,766.81$, $m = 1$, $n = 2$

$$A = P(1 + r)^n$$

$$32,456.32 = 24,766.81(1 + r)^2$$

$$(1 + r)^2 = \frac{32,456.32}{24,766.81} = 1.3105$$

Therefore, $1 + r = \sqrt{1.3105} \approx 1.1448$ and $r \approx 0.1448$ or 14.48%

69. The effective rate for 8% compounded quarterly is

$$r_e = \left(1 + \frac{0.08}{4}\right)^4 - 1 = (1.02)^4 - 1 = 0.0824 \text{ or } 8.24\%$$

To find the annual nominal rate compounded monthly which has the effective rate of 8.24%, we solve

$$0.0824 = \left(1 + \frac{r}{12}\right)^{12} - 1 \text{ for } r.$$

$$\left(1 + \frac{r}{12}\right)^{12} = 1.0824$$

$$1 + \frac{r}{12} = (1.0824)^{1/12}$$

$$\frac{r}{12} = (1.0824)^{1/12} - 1$$

$$r = 12[(1.0824)^{1/12} - 1] \approx 0.0795 \text{ or } 7.95\%$$

EXERCISE 3-3

Things to remember:

1. FUTURE VALUE OF AN ORDINARY ANNUITY

$$FV = PMT\frac{(1 + i)^n - 1}{i} = PMTs_{\overline{n}|i}$$

where PMT = Periodic payment
i = Rate per period
n = Number of payments (periods)
FV = Future value (amount)

(Payments are made at the end of each period.)

2. SINKING FUND PAYMENT

$$PMT = FV\frac{i}{(1 + i)^n - 1}$$

where PMT = Sinking fund payment
FV = Value of annuity after n payments (future value)
n = Number of payments (periods)
i = Rate per period

(Payments are made at the end of each period.)

3. APPROXIMATING INTEREST RATES

Algebra can be used to solve the future value formula in 1 for PMT or n, or the payment formula in 2 for FV or n. Graphical techniques or equation solvers can be used to approximate i to as many places as desired in each of these formulas.

1. $n = 20,\ i = 0.03,\ PMT = \$500$

$$FV = PMT\,\frac{(1 + i)^n - 1}{i}$$
$$= PMTs_{\overline{n}|i} \quad \text{(using } \underline{1}\text{)}$$
$$= 500\,\frac{(1 + 0.03)^{20} - 1}{0.03} = 500s_{\overline{20}|0.03}$$
$$= 500(26.87037449) = \$13{,}435.19$$

3. $n = 40,\ i = 0.02,\ PMT = \$1000$

$$FV = 1000\,\frac{(1 + 0.02)^{40} - 1}{0.02}$$
$$= 1000s_{\overline{40}|0.02}$$
$$= 1000(60.40198318)$$
$$= \$60{,}401.98$$

5. $FV = \$3000,\ n = 20,\ i = 0.02$

$$3000 = PMT\,\frac{(1 + 0.02)^{20} - 1}{0.02}$$
$$= PMTs_{\overline{20}|0.02} \quad \text{(using } \underline{1} \text{ or } \underline{2}\text{)}$$
$$= PMT(24.29736980)$$
$$PMT = \frac{3000}{24.29736980} = \$123.47$$

7. $FV = \$5000,\ n = 15,\ i = 0.01$

$$PMT = FV\,\frac{i}{(1 + i)^n - 1}$$
$$= \frac{FV}{s_{\overline{n}|i}} \quad \text{(using } \underline{2}\text{)}$$
$$= 5000\,\frac{0.01}{(1 + 0.01)^{15} - 1}$$
$$= \frac{5000}{16.09689554} = \$310.62$$

9. $FV = \$4000,\ i = 0.02,\ PMT = 200,\ n = ?$

$$FV = PMT\,\frac{(1 + i)^n - 1}{i}$$
$$\frac{FVi}{PMT} = (1 + i)^n - 1$$
$$(1 + i)^n = \frac{FVi}{PMT} + 1$$
$$\ln(1 + i)^n = \ln\left[\frac{FVi}{PMT} + 1\right]$$
$$n \ln(1 + i) = \ln\left[\frac{FVi}{PMT} + 1\right]$$
$$n = \frac{\ln\left[\frac{FVi}{PMT} + 1\right]}{\ln(1 + i)} = \frac{\ln\left[\frac{4000(0.02)}{200} + 1\right]}{\ln(1.02)}$$
$$= \frac{\ln(1.4)}{\ln(1.02)} \approx \frac{0.3365}{0.01980} = 16.99 \quad \text{or} \quad 17 \text{ periods}$$

11. $FV = \$7{,}600;\ PMT = \$500;\ n = 10,\ i = ?$

$$FV = PMT\,\frac{(1 + i)^n - 1}{i}$$

Substituting the given values into this formula gives

$$7600 = 500\,\frac{(1 + i)^{10} - 1}{i}$$

and $\dfrac{(1 + i)^{10} - 1}{i} = 15.2$

Graph $Y_1 = \frac{(1 + x)^{10} - 1}{x}$,

$Y_2 = 15.2$.

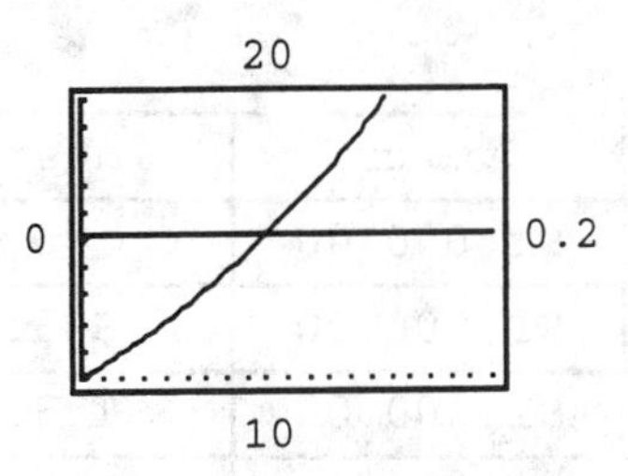

The graphs intersect at the point where $x \approx 0.09$. Thus, $i = 0.09$.

13. $PMT = \$500$, $n = 10(4) = 40$, $i = \frac{0.08}{4} = 0.02$

$$FV = 500\,\frac{(1 + 0.02)^{40} - 1}{0.02} = 500s_{\overline{40}|0.02}$$

$$= 500(60.40198318) = \$30{,}200.99$$

Total deposits = 500(40) = \$20,000.

Interest = FV - 20,000 = 30,200.99 - 20,000 = \$10,200.99.

15. $PMT = \$300$, $i = \frac{0.06}{12} = 0.005$, $n = 5(12) = 60$

$$FV = 300\,\frac{(1 + 0.005)^{60} - 1}{0.005} = 300s_{\overline{60}|0.005} \quad \text{(using }\underline{1}\text{)}$$

$$= 300(69.77003051) = \$20{,}931.01$$

After five years, \$20,931.01 will be in the account.

17. $FV = \$25{,}000$, $i = \frac{0.09}{12} = 0.0075$, $n = 12(5) = 60$

$$PMT = \frac{FV}{s_{\overline{60}|0.0075}} = \frac{25{,}000}{75.42413693} \quad \text{(using the table)}$$

$$= \$331.46 \text{ per month}$$

19. $FV = \$100{,}000$, $i = \frac{0.12}{12} = 0.01$, $n = 8(12) = 96$

$$PMT = \frac{FV}{s_{\overline{96}|0.01}} = \frac{100{,}000}{159.92729236} = \$625.28 \text{ per month}$$

21. $PMT = \$1{,}000$, $i = \frac{0.0832}{1} = 0.0832$, $n = 5$

$$FV = PMT\,\frac{(1 + i)^n - 1}{i} = 1000\,\frac{(1.0832)^n - 1}{0.0832}$$

$n = 1$: $FV = \$1{,}000$

$n = 2$: $FV = 1000\,\frac{(1.0832)^2 - 1}{0.0832} = \$2{,}083.20$

Interest: 2083.20 - 2000 = \$83.20

$n = 3$: $FV = 1000\,\frac{(1.0832)^3 - 1}{0.0832} = \$3{,}256.52$

Interest: 3256.52 - 2083.20 = \$173.32

and so on.

Balance Sheet

Period	Amount	Interest	Balance
1	$1,000.00	$0.00	$1,000.00
2	$1,000.00	$83.20	$2,083.20
3	$1,000.00	$173.32	$3,256.52
4	$1,000.00	$270.94	$4,527.46
5	$1,000.00	$376.69	$5,904.15

23. $FV = PMT \frac{(1 + i)^n - 1}{i} = 100 \frac{(1 + 0.0075)^{12} - 1}{0.0075}$ (after one year)

$= 100 \frac{(1.0075)^{12} - 1}{0.0075}$ $\left[\underline{\text{Note}}\text{: } PMT = \$100,\ i = \frac{0.09}{12} = 0.0075,\ n = 12\right]$

$= \underline{\underline{\$1250.76}}$ (1)

Total deposits in one year = 12(100) = $1200.

Interest earned in first year = FV - 1200 = 1250.76 - 1200 = $50.76.

At the end of the second year:

$FV = 100 \frac{(1 + 0.0075)^{24} - 1}{0.0075}$ [Note: $n = 24$]

$= 100 \frac{(1.0075)^{24} - 1}{0.0075} = \underline{\underline{\$2618.85}}$ (2)

Total deposits plus interest in the second year = (2) - (1)
= 2618.85 - 1250.76
= $\underline{\underline{\$1368.09}}$ (3)

Interest earned in the second year = (3) - 1200
= 1368.09 - 1200
= $168.09

At the end of the third year,

$FV = 100 \frac{(1 + 0.0075)^{36} - 1}{0.0075}$ [Note: $n = 36$]

$= 100 \frac{(1.0075)^{36} - 1}{0.0075}$

$= \underline{\underline{\$4115.27}}$ (4)

Total deposits plus interest in the third year = (4) - (2)
= 4115.27 - 2618.85
= $\underline{\underline{\$1496.42}}$ (5)

Interest earned in the third year = (5) - 1200
= 1496.42 - 1200
= $296.42

Thus,

Year	Interest earned
1	$ 50.76
2	$168.09
3	$296.42

25. (A) $PMT = \$2000$, $n = 8$, $i = 9\% = 0.09$

$$FV = 2000\,\frac{(1 + 0.09)^8 - 1}{0.09} = \frac{2000(0.99256)}{0.09} \approx 22{,}056.95$$

Thus, Jane will have \$22,056.95 in her account on her 31st birthday. On her 65th birthday, she will have:

$A = 22{,}056.95(1.09)^{34} \approx \$413{,}092$

(B) $PMT = \$2000$, $n = 34$, $i = 9\% = 0.09$

$$FV = 2000\,\frac{(1 + 0.09)^{34} - 1}{0.09} \approx \frac{2000(17.7284)}{0.09} \approx \$393{,}965$$

27. $FV = \$10{,}000$, $n = 48$, $i = \frac{8\%}{12} = \frac{0.08}{12} \approx 0.006667$

From 2, $PMT = \dfrac{10{,}000(0.006667)}{(1 + 0.006667)^{48} - 1} = \dfrac{66.67}{0.3757} \approx \$177.46$

The total of the monthly deposits for 4 years is $48 \times 177.46 = 8518.08$. Thus, the interest earned is $10{,}000 - 8518.08 = \$1481.92$.

29. $PMT = \$150$, $FV = \$7000$, $i = \frac{8.5\%}{12} = \frac{0.085}{12} \approx 0.00708$. From Problem 9:

$$n = \frac{\ln\left[\frac{FVi}{PMT} + 1\right]}{\ln(1 + i)} = \frac{\ln\left[\frac{7000(0.00708)}{150} + 1\right]}{\ln(1.00708)} \approx \frac{0.28548}{0.007055} \approx 40.46$$

Thus, $n = 41$ months or 3 years and 5 months.

31. This problem was done with a graphics calculator.

Start with the equation $\dfrac{(1 + i)^n - 1}{i} - \dfrac{FV}{PMT} = 0$

where $FV = 6300$, $PMT = 1000$, $n = 5$ and $i = \frac{r}{1} = r$

where r is the nominal annual rate with these values, the equation is:

$$\frac{(1 + r)^5 - 1}{r} - \frac{6300}{1000} = 0$$

or $\quad (1 + r)^5 - 1 - 6.3r = 0$

Set $y = (1 + r)^5 - 1 - 6.3r$ and use your calculator to find the zero r of the function y, where $0 < r < 1$. The result is $r = 0.1158$ or 11.58% to two decimal places.

33. Start with the equation $\dfrac{(1 + i)^n - 1}{i} - \dfrac{FV}{PMT} = 0$

where $FV = 620$, $PMT = 50$, $n = 12$ and $i = \frac{r}{12}$,

where r is the annual nominal rate. With these values, the equation becomes

$$\frac{(1 + i)^{12} - 1}{i} - \frac{620}{50} = 0$$

or $(1 + i)^{12} - 1 - 12.4i = 0$

Set $y = (1 + i)^{12} - 1 - 12.4i$ and use your calculator to find the zero i of the function y, where $0 < i < 1$. The result is $i = 0.005941$. Thus $r = 12(0.005941) = 0.0713$ or $r = 7.13\%$ to two decimal places.

35. Annuity: $PMT = 500$, $i = \frac{0.06}{4} = 0.015$

$$Y_1 = 500\frac{[(1 + 0.015)^{4x} - 1]}{0.015}$$

Simple interest: $P = 5000$, $r = 0.04$

$$Y_2 = 5000(1 + 0.04x)$$

The graphs of Y_1 and Y_2 are shown in the figure:

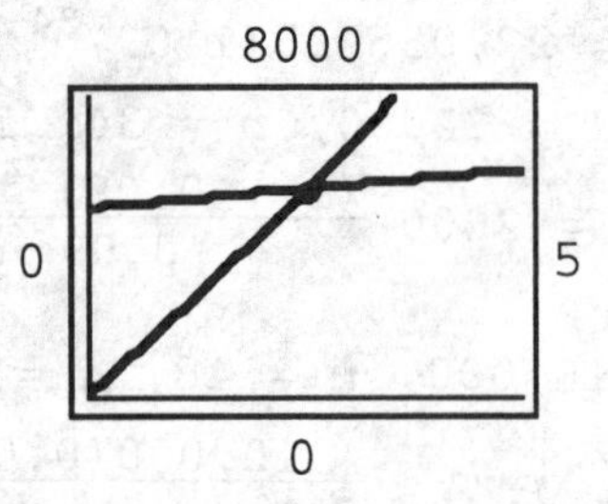

intersection: $x = 2.57$; $y = 5514$

The annuity will be worth more after 2.57 years, or 11 quarterly payments.

EXERCISE 3-4

Things to remember:

1. PRESENT VALUE OF AN ORDINARY ANNUITY

$$PV = PMT\frac{1 - (1 + i)^{-n}}{i} = PMTa_{\overline{n}|i}$$

where PMT = Periodic payment
i = Rate per period
n = Number of periods
PV = Present value of all payments

(Payments are made at the end of each period.)

2. AMORTIZATION FORMULA

$$PMT = PV\frac{i}{1 - (1 + i)^{-n}} = PV\frac{1}{a_{\overline{n}|i}}$$

where PV = Amount of loan (present value)
i = Rate per period
n = Number of payments (periods)
PMT = Periodic payment

(Payments are made at the end of each period.)

1. $PV = 200\frac{1 - (1 + 0.04)^{-30}}{0.04}$
$= PMTa_{\overline{30}|0.04}$
$= 200(17.29203330)$ (using the table)
$= \$3458.41$

3. $PV = 250\frac{1 - (1 + 0.025)^{-25}}{0.025}$
$= 250a_{\overline{25}|0.025}$
$= 250(18.42437642)$
$= \$4606.09$

5. $PMT = 6000 \dfrac{0.01}{1 - (1 + 0.01)^{-36}}$

$= \dfrac{PV}{a_{\overline{36}|0.01}}$

$= \dfrac{6000}{30.10750504} = \$199.29$

7. $PMT = 40{,}000 \dfrac{0.0075}{1 - (1 + 0.0075)^{-96}}$

$= \dfrac{40{,}000}{a_{\overline{96}|0.0075}}$

$= \dfrac{40{,}000}{68.25843856} = \$586.01$

9. $PV = \$5000$, $i = 0.01$, $PMT = 200$

We have, $PV = PMT \dfrac{1 - (1 + i)^{-n}}{i}$

$$5000 = 200 \frac{1 - (1 + 0.01)^{-n}}{0.01}$$
$$= 20{,}000[1 - (1.01)^{-n}]$$
$$\frac{1}{4} = 1 - (1.01)^{-n}$$
$$(1.01)^{-n} = \frac{3}{4} = 0.75$$
$$\ln(1.01)^{-n} = \ln(0.75)$$
$$-n \ln(1.01) = \ln(0.75)$$
$$n = \frac{-\ln(0.75)}{\ln(1.01)} \approx 29$$

11. $PV = \$9{,}000$, $PMT = \$600$, $n = 20$, $i = ?$

$PV = PMT \dfrac{1 - (1 + i)^{-n}}{i}$

Substituting the given values into this formula gives

$$9000 = 600 \frac{1 - (1 + i)^{-20}}{i}$$
$$15i = 1 - (1 + i)^{-20} = 1 - \frac{1}{(1 + i)^{20}}$$
$$15i + \frac{1}{(1 + i)^{20}} = 1$$

Graph $Y_1 = 15x + \dfrac{1}{(1 + x)^{20}}$, $Y_2 = 1$.

The curves intersect at $x = 0$ and $x \approx 0.029$. Thus $i = 0.029$.

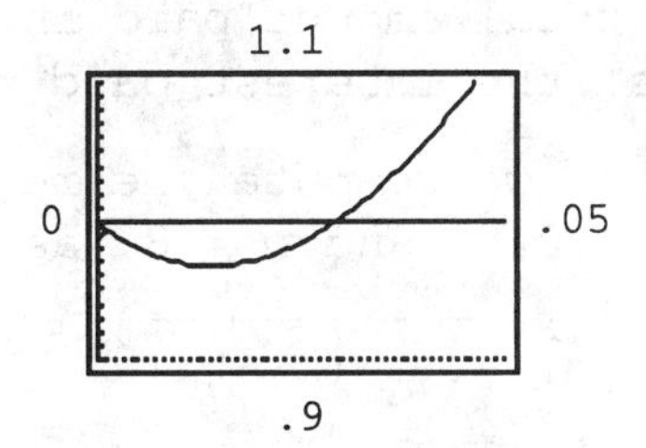

13. $PMT = \$4000$, $n = 10(4) = 40$

$i = \dfrac{0.08}{4} = 0.02$

PV = Present value

$= PMT \dfrac{1 - (1 + i)^{-n}}{i}$

$= PMTa_{\overline{n}|i}$

$= 4000a_{\overline{40}|0.02}$

$= 4000(27.35547924)$

$= \$109{,}421.92$

15. This is a present value problem.

$PMT = \$350,\ n = 4(12) = 48,\ i = \frac{0.09}{12} = 0.0075$

Hence, $PV = PMTa_{\overline{n}|i} = 350a_{\overline{48}|0.0075}$

$= 350(40.18478189) = \$14,064.67$

They should deposit $14,064.67. The child will receive 350(48) = $16,800.00.

17. (A) $PV = \$600,\ n = 18,\ i = 0.01$

Monthly payment $= PMT = PV\frac{i}{1 - (1 + i)^{-n}}$

$= \frac{PV}{a_{\overline{n}|i}} = \frac{600}{a_{\overline{18}|0.01}} = \frac{600}{16.39826858}$

$= \$36.59$ per month

The amount paid in 18 payments = 36.59(18) = $658.62.
Thus, the interest paid = 658.62 - 600 = $58.62.

(B) $PMT = \frac{600}{a_{\overline{18}|0.015}}\quad (i = 0.015)$

$= \frac{600}{15.67256089} = \$38.28$ per month

For 18 payments, the total amount = 38.28(18) = $689.04.
Thus, the interest paid = 689.04 - 600 = $89.04.

19. Amortized amount = 16,000 - (16,000)(0.25) = $12,000
Thus, $PV = \$12,000,\ n = 6(12) = 72,\ i = 0.015$

PMT = monthly payment $= \frac{PV}{a_{\overline{n}|i}} = \frac{12,000}{a_{\overline{72}|0.015}} = \frac{12,000}{43.84466677} = \$273.69$ per month

The total amount paid in 72 months = 273.69(72) = $19,705.68.
Thus, the interest paid = 19,705.68 - 12,000 = $7705.68.

21. First, we compute the required quarterly payment for $PV = \$5000$, $i = 0.045$, and $n = 8$, as follows:

$PMT = PV\frac{i}{1 - (1 + i)^{-n}} = 5000\frac{0.045}{1 - (1 + 0.045)^{-8}} = \frac{225}{1 - (1.045)^{-8}}$

$= \$758.05$ per quarter

The amortization schedule is as follows:

Payment number	Payment	Interest	Unpaid balance reduction	Unpaid balance
0				$5000.00
1	$758.05	$225.00	$533.05	4466.95
2	758.05	201.01	557.04	3909.91
3	758.05	175.95	582.10	3327.81
4	758.05	149.75	608.30	2719.51
5	758.05	122.38	635.67	2083.84
6	758.05	93.77	664.28	1419.56
7	758.05	63.88	694.17	725.39
8	758.03	32.64	725.39	0.00
Totals	$6064.38	$1064.38	$5000.00	

23. First, we compute the required monthly payment for $PV = \$6000$, $i = \frac{12}{12(100)} = 0.01$, $n = 3(12) = 36$.

$$PMT = PV \frac{i}{1 - (1 + i)^{-n}} = 6000 \frac{0.01}{1 - (1 + 0.01)^{-36}} = \frac{60}{1 - (1.01)^{-36}}$$
$$= \$199.29$$

Now, compute the unpaid balance after 12 payments by considering 24 unpaid payments: $PMT = \$199.29$, $i = 0.01$, and $n = 24$.

$$PV = PMT \frac{1 - (1 + i)^{-n}}{i} = 199.29 \frac{1 - (1 + 0.01)^{-24}}{0.01}$$
$$= 19{,}929[1 - (1.01)^{-24}] = \$4233.59$$

Thus, the amount of the loan paid in 12 months is 6000 - 4233.59 = \$1766.41, and the amount of total payment made during 12 months is 12(199.29) = \$2391.48. The interest paid during the first 12 months (first year) is:

2391.48 - 1766.41 = \$625.07

Similarly, the unpaid balance after two years can be computed by considering 12 unpaid payments: $PMT = \$199.29$, $i = 0.01$, and $n = 12$.

$$PV = 199.29 \frac{1 - (1 + 0.01)^{-12}}{0.01} = 19{,}929[1 - (1.01)^{-12}] = \$2243.02$$

Thus, the amount of the loan paid during 24 months is 6000 - 2243.02 = \$3756.98, and the amount of the loan paid during the second year is 3756.98 - 1766.41 = \$1990.57. The amount of total payment during the second year is 12(199.29) = \$2391.48. The interest paid during the second year is:

2391.48 - 1990.57 = \$400.91

The total amount paid in 36 months is 199.29(36) = \$7174.44. Thus, the total interest paid is 7174.44 - 6000 = \$1174.44 and the interest paid during the third year is 1174.44 - (625.07 + 400.91) = 1174.44 - 1025.98 = \$148.46.

25. PMT = monthly payment = \$525, $n = 30(12) = 360$, $i = \frac{0.098}{12} \approx 0.0081667$.

Thus, the present value of all payments is:

$$PV = PMT \frac{1 - (1 + i)^{-n}}{i} \approx 525 \frac{1 - (1 + 0.0081667)^{-360}}{0.0081667} \approx \$60{,}846.38$$

Hence, selling price = loan + down payment
= 60,846.38 + 25,000
= \$85,846.38

The total amount paid in 30 years (360 months) = 525(360) = \$189,000.
The interest paid is: 189,000 - 60,846.38 = \$128,153.62

27. $P = \$6000$, $n = 2(12) = 24$, $i = \frac{0.035}{12} \approx 0.0029167$

The total amount owed at the end of the two years is:

$$A = P(1 + i)^n = 6000(1 + 0.0029167)^{24} = 6000(1.0029167)^{24} \approx 6434.39$$

Now, the monthly payment is:

$$PMT = PV\frac{i}{1 - (1 + i)^{-n}}$$

where $n = 4(12) = 48$, $PV = \$6434.39$, $i = \frac{0.035}{12} \approx 0.0029167$. Thus,

$$PMT = 6434.39\frac{0.0029167}{1 - (1 + 0.0029167)^{-48}} = \$143.85 \text{ per month}$$

The total amount paid in 48 payments is 143.85(48) = $6904.80. Thus, the interest paid is 6904.80 - 6000 = $904.80.

29. First, compute the monthly payment: $PV = \$75,000$, $i = \frac{0.132}{12} = 0.011$, $n = 30(12) = 360$.

$$\text{Monthly payment} = PV\frac{i}{1 - (1 + i)^{-n}} = 75,000\frac{0.011}{1 - (1 + 0.011)^{-360}}$$

$$= 75,000\frac{0.011}{1 - (1.011)^{-360}} = \$841.39$$

(A) Now, to compute the balance after 10 years (with balance of loan to be paid in 20 years), use $PMT = \$841.39$, $i = 0.011$, and $n = 20(12) = 240$.

$$\text{Balance after 10 years} = PMT\frac{1 - (1 + i)^{-n}}{i}$$

$$= 841.39\frac{1 - (1 + 0.011)^{-240}}{0.011}$$

$$= 841.39\frac{1 - (1.011)^{-240}}{0.011} = \$70,952.33$$

(B) Similarly, the balance of the loan after 20 years (with remainder of loan to be paid in 10 years) is:

$$841.39\frac{1 - (1 + 0.011)^{-120}}{0.011} \quad [\underline{\text{Note}}: n = 12(10) = 120]$$

$$= 841.39\frac{1 - (1.011)^{-120}}{0.011} = \$55,909.02$$

(C) The balance of the loan after 25 years (with remainder of loan to be paid in 5 years) is:

$$841.39\frac{1 - (1 + 0.011)^{-60}}{0.011} \quad [\underline{\text{Note}}: n = 12(5) = 60]$$

$$= 841.39\frac{1 - (1.011)^{-60}}{0.011} = \$36,813.32$$

31. (A) $PV = \$30,000$, $i = \frac{0.15}{12} = 0.0125$, $n = 20(12) = 240$.

$$\text{Monthly payment } PMT = PV\frac{i}{1 - (1 + i)^{-n}}$$

$$= 30,000\frac{0.0125}{1 - (1 + 0.0125)^{-240}}$$

$$= 30,000\frac{0.0125}{1 - (1.0125)^{-240}} = \$395.04$$

The total amount paid in 240 payments is:
395.04(240) = $94,809.60
Thus, the interest paid is:
$94,809.60 - $30,000 = $64,809.60

(B) New payment = PMT = \$395.04 + \$100.00 = \$495.04. PV = \$30,000, $i = 0.0125$.

$$PMT = PV \frac{i}{1 - (1 + i)^{-n}}$$

$$495.04 = 30{,}000 \frac{0.0125}{1 - (1 + 0.0125)^{-n}} = \frac{375}{1 - (1.0125)^{-n}}$$

Therefore,

$$1 - (1.0125)^{-n} = \frac{375}{495.04} = 0.7575$$

$$(1.0125)^{-n} = 1 - 0.7575 = 0.2425$$

$$\ln(1.0125)^{-n} = \ln(0.2425)$$

$$-n \ln(1.0125) = \ln(0.2425)$$

$$= \frac{-\ln(0.2425)}{\ln(1.0125)} \approx 114.047 \approx 114 \text{ months or } 9.5 \text{ years}$$

The total amount paid in 114 payments of \$495.04 is:
495.04(114) = \$56,434.56
Thus, the interest paid is:
\$56,434.56 - \$30,000 = \$26,434.56
The savings on interest is:
\$64,809.60 - \$26,434.56 = \$38,375.04

33. $PV = \$500{,}000$, $i = \frac{0.075}{12} = 0.00625$

$$500{,}000 = PMT \frac{1 - (1.00625)^{-n}}{0.00625}$$

$$\frac{500{,}000(0.00625)}{PMT} = 1 - (1.00625)^{-n}$$

$$(1.00625)^{-n} = 1 - \frac{500{,}000(0.00625)}{PMT}$$

(A) $PMT = \$5{,}000$

$$(1.00625)^{-n} = 1 - \frac{500{,}000(0.00625)}{5000}$$

$$(1.00625)^{-n} = 0.375$$

$$-n = \frac{\ln(0.375)}{\ln(1.00625)} \approx -157$$

Thus, 157 withdrawals.

(B) $PMT = \$4{,}000$

$$(1.00625)^{-n} = 1 - \frac{500{,}000(0.00625)}{4000}$$

$$(1.00625)^{-n} = 0.21875$$

$$-n = \frac{\ln(0.21875)}{\ln(1.00625)} \approx -243$$

Thus, 243 withdrawals.

(C) The interest per month on \$500,000 at 7.5% compounded monthly is greater than \$3,000. For example, the interest in the first month is

$$500{,}000\left(\frac{0.075}{12}\right) = \$3{,}125$$

Thus, the owner can withdraw \$3,000 per month forever.

35. (A) First, calculate the future value of the ordinary annuity:

$PMT = \$100,\ i = \dfrac{0.0744}{12} = 0.0062,\ n = 30(12) = 360$

$$FV = PMT\frac{(1+i)^n - 1}{i} = 100\frac{(1.0062)^{360} - 1}{0.0062} = \$133{,}137$$

The interest earned during the 30 year period is:

$133{,}137 - 360(100) = 133{,}137 - 36{,}000 = \$97{,}137$

Next, using \$133,137 as the present value, determine the monthly payment:

$PV = \$133{,}137,\ i = 0.0062,\ n = 15(12) = 180$

$$PMT = PV\frac{i}{1-(1+i)^{-n}} = 133{,}137\frac{0.0062}{1-(1.0062)^{-180}} = 1229.66$$

The monthly withdrawals are \$1229.66. Interest earned during the 15 year period is:

$(1229.66)(180) - 133{,}137 = \$88{,}201.80$

Total interest: $97{,}137 + 88{,}201.80 = \$185{,}338.80$

(B) First find the present value of the ordinary annuity:

$PMT = \$2{,}000,\ i = 0.0062,\ n = 180$

$$PV = 2000\frac{1-(1.0062)^{-180}}{0.0062} = \$216{,}542.54$$

Next, using \$216,542.54, as the future value of an ordinary annuity, calculate the monthly payment.

$FV = \$216{,}542.54,\ i = 0.0062,\ n = 360$

$$PMT = 216{,}542.54\frac{0.0062}{(1.0062)^{360} - 1} = 162.65$$

The monthly deposit is \$162.65.

37. $PV = (\$79{,}000)(0.80) = \$63{,}200,\ i = \dfrac{0.12}{12} = 0.01,\ n = 12(30) = 360.$

Monthly payment $PMT = PV\dfrac{i}{1-(1+i)^{-n}} = 63{,}200\dfrac{0.01}{1-(1+0.01)^{-360}}$

$$= \frac{632}{1-(1.01)^{-360}} = \$650.08$$

Next, we find the present value of a \$650.08 per month, 18-year annuity.

$PMT = \$650.08,\ i = 0.01,$ and $n = 12(18) = 216.$

$$PV = PMT\frac{1-(1+i)^{-n}}{i} = 650.08\frac{1-(1.01)^{-216}}{0.01}$$

$$= \frac{650.08(0.8834309)}{0.01} = \$57{,}430.08$$

Finally,

Equity = (current market value) - (unpaid loan balance)

= \$100,000 - \$57,430.08 = \$42,569.92

The couple can borrow (\$42,569.92)(0.70) = \$29,799.

Problems 39 thru 43 start from the equation

$$(*)\quad \frac{1-(1+i)^{-n}}{i} - \frac{PV}{PMT} = 0$$

A graphics calculator was used to solve these problems.

39. The graphs are decreasing, curve downward, and have x intercept 30. The unpaid balances are always in the ratio 2:3:4. The monthly payments and total interest in each case are:

(use $PMT = PV\dfrac{i}{1-(1+i)^{-n}}$ where $i = \dfrac{0.09}{12} = 0.0075$, $n = 360$)

<u>$50,000 mortgage</u>

$$PMT = 50{,}000\frac{0.0075}{1-(1.0075)^{-360}}$$
$$= 50{,}000(0.0080462)$$
$$= \$402.31 \text{ per month}$$

Total interest paid = 360(402.31) - 50,000 = $94,831.60

<u>$75,000 mortgage</u>

$$PMT = 75{,}000\frac{0.0075}{1-(1.0075)^{-360}}$$
$$= 75{,}000(0.0080462)$$
$$= \$603.47 \text{ per month}$$

Total interest paid = 360(603.47) - 75,000 = $142,249.20

<u>$100,000 mortgage</u>

$$PMT = 100{,}000\frac{0.0075}{1-(1.0075)^{-360}}$$
$$= 100{,}000(0.0080462)$$
$$= \$804.62 \text{ per month}$$

Total interest paid = 360(804.62) - 100,000 = $189,663.20

41. $PV = 1000$, $PMT = 90$, $n = 12$, $i = \dfrac{r}{12}$ where r is the annual nominal rate. With these values, the equation (*) becomes

$$\frac{1-(1+i)^{-12}}{i} - \frac{1000}{90} = 0$$

or $\quad 1 - (1+i)^{-12} - 11.11i = 0$

Put $y = 1 - (1+i)^{-12} - 11.11i$ and use your calculator to find the zero i of y, where $0 < i < 1$. The result is $i \approx 0.01204$ and $r = 12(0.01204) = 0.14448$. Thus, $r = 14.45\%$ (two decimal places).

43. $PV = 90{,}000$, $PMT = 1200$, $n = 12(10) = 120$, $i = \dfrac{r}{12}$ where r is the annual nominal rate. With these values, the equation (*) becomes

$$\frac{1-(1+i)^{-120}}{i} - \frac{90{,}000}{1200} = 0$$

or $\quad 1 - (1+i)^{-120} - 75i = 0$

This equation can be written

$$(1 + i)^{120} - (1 - 75i)^{-1} = 0$$

Put $y = (1 + i)^{120} - (1 - 75i)^{-1}$ and use your calculator to find the zero i of y where $0 < i < 1$. The result is $i \approx 0.00851$ and $r = 12(0.00851) = 0.10212$. Thus $r = 10.21\%$ (two decimal places).

CHAPTER 3 REVIEW

1. $A = 100\left(1 + 0.09 \cdot \frac{1}{2}\right)$

$= 100(1.045) = \$104.50$ (3-1)

2. $808 = P\left(1 + 0.12 \cdot \frac{1}{12}\right)$

$P = \frac{808}{1.01} = \$800$ (3-1)

3. $212 = 200(1 + 0.08 \cdot t)$

$1 + 0.08t = \frac{212}{200}$

$0.08t = \frac{212}{200} - 1 = \frac{12}{200}$

$t = \frac{0.06}{0.08} = 0.75$ yr. or 9 mos.

(3-1)

4. $4120 = 4000\left(1 + r \cdot \frac{1}{2}\right)$

$1 + \frac{r}{2} = \frac{4120}{4000}$

$\frac{r}{2} = \frac{4120}{4000} - 1 = \frac{120}{4000} = 0.03$

$r = 0.06$ or 6%

(3-1)

5. $A = 1200(1 + 0.005)^{30}$

$= 1200(1.005)^{30} = \$1393.68$

(3-2)

6. $P = \frac{5000}{(1 + 0.0075)^{60}} = \frac{5000}{(1.0075)^{60}}$

$= \$3193.50$

(3-2)

7. $FV = 1000s_{\overline{60}|0.005}$

$= 1000 \cdot 69.77003051$

$= \$69,770.03$ (3-3)

8. $PMT = \frac{FV}{s_{\overline{n}|i}} = \frac{8000}{s_{\overline{48}|0.015}}$

$= \frac{8000}{69.56321929} = \$115.00$ (3-3)

9. $PV = PMTa_{\overline{n}|i} = 2500a_{\overline{16}|0.02}$

$= 2500 \cdot 13.57770931$

$= \$33,944.27$ (3-4)

10. $PMT = \frac{PV}{a_{\overline{n}|i}} = \frac{8000}{a_{\overline{60}|0.0075}}$

$= \frac{8000}{48.17337352} = \$166.07$ (3-4)

11. (A) $2500 = 1000(1.06)^n$

$(1.06)^n = 2.5$

$\ln(1.06)^n = \ln 2.5$

$n \ln 1.06 = \ln 2.5$

$n = \frac{\ln 2.5}{\ln 1.06} = 15.73 \approx 16$

(B) We find the intersection of

$Y_1 = 1000(1.06)^{\wedge}X$ and $Y_2 = 2500$

The graphs are shown in the figure at the right.

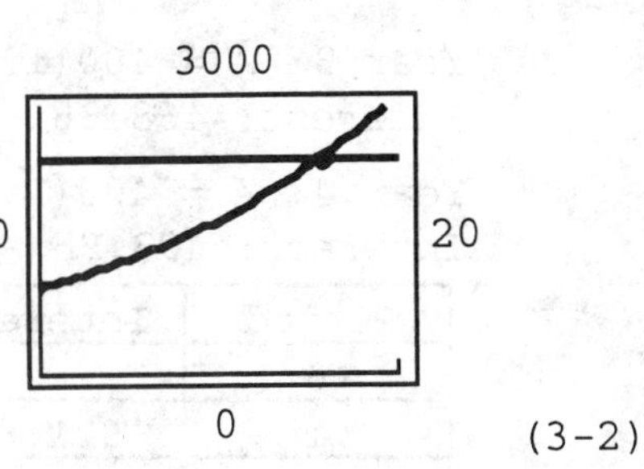

intersection: $x = 15.73$; $y = 2500$ (3-2)

12. (A)

$$5000 = 100\frac{(1.01)^n - 1}{0.01}$$
$$5000 = 10{,}000[(1.01)^n - 1]$$
$$0.5 = (1.01)^n - 1$$
$$(1.01)^n = 1.5$$
$$n \ln(1.01) = \ln(1.5)$$
$$n = \frac{\ln(1.5)}{\ln(1.01)} = 40.75 \approx 41$$

(B) We find the intersection of

$Y_1 = 100\frac{(1.01)^x - 1}{0.01} =$

$10{,}000[(1.01)^{\wedge}X - 1]$ and $Y_2 = 5000$

The graphs are shown in the figure at the right.

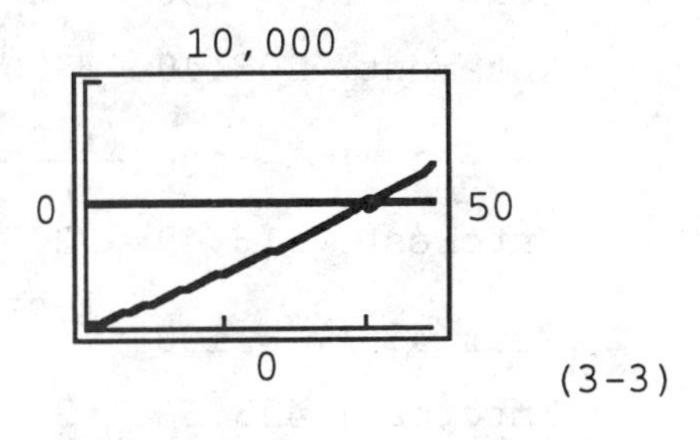

intersection: $x = 40.75$; $y = 5000$ (3-3)

13. $P = \$3000$, $r = 0.14$, $t = \frac{10}{12}$

$A = 3000\left(1 + 0.14 \cdot \frac{10}{12}\right)$ [using $A = P(1 + rt)$]

$= \$3350$

Interest $= 3350 - 3000 = \$350$ (3-1)

14. $P = \$6000$, $r = 9\% = 0.09$, $m = 12$, $i = \frac{0.09}{12} = 0.0075$, $n = 12(17) = 204$

$A = P(1 + i)^n = 6000(1 + 0.0075)^{204} = 6000(1.0075)^{204} \approx \$27{,}551.32$ (3-2)

15. $A = \$25{,}000$, $r = 10\% = 0.10$, $m = 2$, $i = \frac{0.10}{2} = 0.05$, $n = 2(10) = 20$

$P = \frac{A}{(1 + i)^n} = \frac{25{,}000}{(1 + 0.05)^{20}} = \frac{25{,}000}{(1.05)^{20}} \approx \$9422.24$ (3-2)

16. (A) $P = \$400$, $i = 0.054$

$A = P(1 + i)^n$

Year 1: $A = 400(1.054)^1 = \$421.60$

Interest: \$21.60

Year 2: $A = 400(1.054)^2 = \$444.37$

Interest: $444.37 - 421.60 = \$22.77$

Year 3: $A = 400(1.054)^3 = \$468.36$
Interest: $468.36 - 444.37 = \$23.99$

Year 4: $A = 400(1.054)^4 = \$493.65$
Interest: $493.95 - 468.36 = \$25.29$

Period	Interest	Amount
0		$400.00
1	$21.60	$421.60
2	$22.77	$444.37
3	$24.00	$468.36
4	$25.29	$493.65

(B) $PMT = \$100$, $i = 0.054$, $n = 4$

$$FV = PMT\,\frac{(1 + i)^n - 1}{i}$$

Year 1: $FV = \$100$

Year 2: $FV = 100\,\dfrac{(1.054)^2 - 1}{0.054} = \$205.40$
Interest: $205.40 - 200.00 = \$5.40$

Year 3: $FV = 100\,\dfrac{(1.054)^3 - 1}{0.054} = \$316.49$
Interest: $316.49 - 205.40 = \$11.09$

Year 4: $FV = 100\,\dfrac{(1.054)^4 - 1}{0.054} = \$433.58$
Interest: $433.58 - 316.49 = \$17.09$

Period	Interest	Payment	Balance
1		$100.00	$100.00
2	$5.40	$100.00	$205.40
3	$11.09	$100.00	$316.49
4	$17.09	$100.00	$433.58

(3-2, 3-3)

17. The value of \$1 at 13% simple interest after t years is:
$A_s = 1(1 + 0.13t) = 1 + 0.13t$
The value of \$1 at 9% interest compounded annually for t years is:
$A_c = 1(1 + 0.09)^t = (1.09)^t$

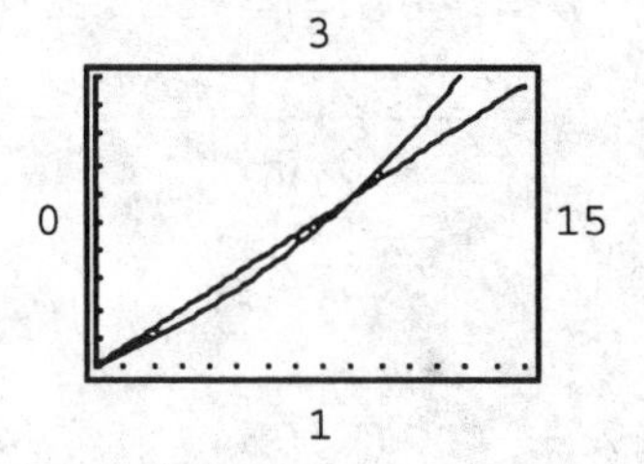

Graph $Y_1 = 1 + 0.13x$, $Y_2 = (1.09)^x$
The graphs intersect at the point where $x \approx 9$. For investments lasting less than 9 years, simple interest at 13% is better; for investments lasting more than 9 years, interest compounded annually at 9% is better. (3-2)

18. $P = \$10{,}000$, $r = 7\% = 0.07$, $m = 365$, $i = \dfrac{0.07}{365} = 0.0001918$, and

$n = 40(365) = 14{,}600$

$$A = P(1 + i)^n = 10{,}000(1 + 0.0001918)^{14{,}600}$$
$$= 10{,}000(1.0001918)^{14{,}600}$$
$$= \$164{,}402 \qquad (3\text{-}2)$$

19. The effective rate for 9% compounded quarterly is:

$$r_e = \left(1 + \frac{r}{m}\right)^m - 1,\ r = 0.09,\ m = 4$$
$$= \left(1 + \frac{0.09}{4}\right)^4 - 1 = (1.0225)^4 - 1 \approx 0.0931 \quad \text{or} \quad 9.31\%$$

The effective rate for 9.25% compounded annually is 9.25%. Thus, 9% compounded quarterly is the better investment. (3-2)

20. $PMT = \$200$, $r = 9\% = 0.09$, $m = 12$, $i = \dfrac{0.09}{12} = 0.0075$, $n = 12(8) = 96$

$$FV = PMT\,\frac{(1 + i)^n - 1}{i}$$
$$= 200\,\frac{(1 + 0.0075)^{96} - 1}{0.0075} = 200\,\frac{(1.0075)^{96} - 1}{0.0075} \approx \$27{,}971.23$$

The total amount invested with 96 payments of \$200 is: $96(200) = \$19{,}200$
Thus, the interest earned with this annuity is:
$I = \$27{,}971.23 - \$19{,}200 = \$8771.23$ (3-3)

21. $P = \$635$, $r = 22\% = 0.22$, $t = \dfrac{1}{12}$, $I = Prt = 635(0.22)\dfrac{1}{12} = \$11.64$ (3-1)

22. $P = \$8000$, $r = 5\% = 0.05$, $m = 1$, $i = \dfrac{0.05}{1} = 0.05$, $n = 5$

$$A = P(1 + i)^n = 8000(1 + 0.05)^5 = 8000(1.05)^5 \approx \$10{,}210.25 \qquad (3\text{-}2)$$

23. $A = \$8000$, $r = 5\% = 0.05$, $m = 1$, $i = \dfrac{0.05}{1} = 0.05$, $n = 5$

$$P = \frac{A}{(1 + i)^n} = \frac{8000}{(1 + 0.05)^5} = \frac{8000}{(1.05)^5} \approx \$6268.21 \qquad (3\text{-}2)$$

24. The interest paid was $\$2812.50 - \$2500 = \$312.50$. $P = \$2500$, $t = \dfrac{10}{12} = \dfrac{5}{6}$

Solving $I = Prt$ for r, we have:

$$r = \frac{I}{Pt} = \frac{312.50}{2500\left(\frac{5}{6}\right)} = 0.15 \quad \text{or} \quad 15\% \qquad (3\text{-}1)$$

25. The present value, PV, of an annuity of \$200 per month for 48 months at 14% interest compounded monthly is given by:

$$PV = PMT\frac{1-(1+i)^{-n}}{i} \quad \text{where } PMT = \$200,\ i = \frac{0.14}{12} = 0.0116667 \text{ and } n = 48$$

$$= 200\frac{1-(1+0.0116667)^{-48}}{0.0116667}$$

$$= 17{,}142.857[1-(1.0116667)^{-48}] = \$7318.91$$

With the \$3000 down payment, the selling price of the car is \$10,318.91.
The total amount paid is: \$3000 + 48(\$200) = \$12,600
Thus, the interest paid is: I = \$12,600 - \$10,318.91 = \$2281.09 (3-4)

26. $P = \$2500,\ r = 9\% = 0.09,\ m = 4,\ i = \frac{0.09}{4} = 0.0225,\ A = \$3000$

$$A = P(1+i)^n$$
$$3000 = 2500(1+0.0225)^n$$
$$(1.0225)^n = \frac{3000}{2500} = 1.2$$
$$\ln(1.0225)^n = \ln 1.2$$
$$n \ln 1.0225 = \ln 1.2$$
$$n = \frac{\ln 1.2}{\ln 1.0225} \approx 8.19$$

Thus, it will take 9 quarters, or 2 years and 3 months. (3-2)

27. (A) $r = 12\% = 0.12,\ m = 12,\ i = \frac{0.12}{12} = 0.01$

If we invest P dollars, then we want to know how long it will take to have $2P$ dollars:

$$A = P(1+i)^n$$
$$2P = P(1+0.01)^n$$
$$(1.01)^n = 2$$
$$\ln(1.01)^n = \ln 2$$
$$n \ln 1.01 = \ln 2$$
$$n = \frac{\ln 2}{\ln 1.01} \approx 69.66$$

Thus, it will take 70 months, or 5 years and 10 months, for an investment to double at 12% interest compounded monthly.

(B) $r = 18\% = 0.18,\ m = 12,\ i = \frac{0.18}{12} = 0.015$

$$2P = P(1+0.015)^n$$
$$(1.015)^n = 2$$
$$\ln(1.015)^n = \ln 2$$
$$n = \frac{\ln 2}{\ln 1.015} \approx 46.56$$

Thus, it will take 47 months, or 3 years and 11 months, for an investment to double at 18% compounded monthly. (3-2)

28. (A) $PMT = \$2000$, $m = 1$, $r = i = 7\% = 0.07$, $n = 45$

$$FV = PMT\,\frac{(1+i)^n - 1}{i}$$

$$= 2000\,\frac{(1+0.07)^{45} - 1}{0.07} = 2000\,\frac{(1.07)^{45} - 1}{0.07} \approx \$571{,}499$$

(B) $PMT = \$2000$, $m = 1$, $r = i = 11\% = 0.11$, $n = 45$

$$FV = PMT\,\frac{(1+i)^n - 1}{i}$$

$$= 2000\,\frac{(1+0.11)^{45} - 1}{0.11} = 2000\,\frac{(1.11)^{45} - 1}{0.11} \approx \$1{,}973{,}277$$

(3-3)

29. $A = \$17{,}388.17$, $P = \$12{,}903.28$, $m = 1$, $r = i$, $n = 3$

$$A = P(1+i)^n$$

$$17{,}388.17 = 12{,}903.28(1+i)^3$$

$$(1+i)^3 = \frac{17{,}388.17}{12{,}903.28} \approx 1.3475775$$

$$3\ln(1+i) = \ln(1.3475775)$$

$$\ln(1+i) \approx \frac{0.2983085}{3} \approx 0.0994362$$

$$1 + i = e^{0.0994362} \approx 1.1045$$

$$i = 0.1045 \text{ or } 10.45\%$$

(3-2)

30. $P = \$1500$, $I = \$100$,

$$t = \frac{120}{360} = \frac{1}{3} \text{ year}$$

From Problem 24,

$$r = \frac{I}{Pt} = \frac{100}{1500\left(\frac{1}{3}\right)} = 0.20 \text{ or } 20\%$$

(3-1)

31. $PMT = \$1{,}500$, $i = \frac{0.08}{4} = 0.02$, $n = 2(4) = 8$

We want to find the present value, PV, of this annuity.

$$PV = PMT\,\frac{1 - (1+i)^{-n}}{i} = 1500\,\frac{1 - (1.02)^{-8}}{0.02}$$

$$= \$10{,}988.22$$

The committee should deposit \$10,988.22. (3-4)

32. (A) The present value of an annuity which provides for quarterly withdrawals of \$5000 for 10 years at 12% interest compounded quarterly is given by:

$$PV = PMT\,\frac{1 - (1+i)^{-n}}{i} \quad \text{with } PMT = \$5000,\ i = \frac{0.12}{4} = 0.03, \text{ and } n = 10(4) = 40$$

$$= 5000\,\frac{1 - (1+0.03)^{-40}}{0.03}$$

$$= 166{,}666.67[1 - (1.03)^{-40}] = \$115{,}573.86$$

This amount will have to be in the account when he retires.

(B) To determine the quarterly deposit to accumulate the amount in part (A), we use the formula:

$PMT = FV \dfrac{i}{(1 + i)^n - 1}$ where $FV = \$115{,}573.86$, $i = 0.03$, and $n = 4(20) = 80$

$= 115{,}573.86 \dfrac{0.03}{(1 + 0.03)^{80} - 1}$

$= \dfrac{3467.22}{(1.03)^{80} - 1} = \$359.64$ quarterly payment

(C) The amount collected during the 10-year period is:
($5000)40 = $200,000
The amount deposited during the 20-year period is:
($359.64)80 = $28,771.20
Thus, the interest earned during the 30-year period is:
$200,000 - $28,771.20 = $171,228.80 (3-4)

33. The amount of the loan is $\$3000\left(\frac{2}{3}\right) = \$2000$. The monthly interest rate is $i = 1.5\% = 0.015$ and $n = 2(12) = 24$.

$PMT = PV \dfrac{i}{1 - (1 + i)^{-n}} = 2000 \dfrac{0.015}{1 - (1 + 0.015)^{-24}} = \dfrac{30}{1 - (1.015)^{-24}}$

$= \$99.85$ per month

The amount paid in 24 payments is
99.85(24) = $2396.40. So, the interest paid is 2396.40 - 2000 = $396.40. (3-4)

34. $FV = \$50{,}000$, $r = 9\% = 0.09$, $m = 12$, $i = \dfrac{0.09}{12} = 0.0075$, $n = 12(6) = 72$

$PMT = FV \dfrac{i}{(1 + i)^n - 1} = \dfrac{FV}{s_{\overline{n}|i}} = \dfrac{50{,}000}{s_{\overline{72}|0.0075}}$

$= \dfrac{50{,}000}{95.007028}$ (from Table II)

$= \$526.28$ per month (3-3)

35. To determine how long it will take money to double, we need to solve the equation $2P = P(1 + i)^n$ for n. From this equation, we obtain:

$$(1 + i)^n = 2$$
$$\ln(1 + i)^n = \ln 2$$
$$n \ln(1 + i) = \ln 2$$
$$n = \frac{\ln 2}{\ln(1 + i)}$$

(A) $i = \dfrac{0.10}{365} = 0.000274$

Thus, $n = \dfrac{\ln 2}{\ln(1.000274)} \approx 2530.08$ days or 6.93 years.

(B) $i = 0.10$

Thus, $n = \dfrac{\ln 2}{\ln(1.1)} \approx 7.27$ years. (3-2)

36. First, we must calculate the future value of \$8000 at 5.5% interest compounded monthly for 2.5 years.

$A = P(1 + i)^n$ where $P = \$8000$, $i = \dfrac{0.055}{12}$ and $n = 30$

$$= 8000\left(1 + \frac{0.055}{12}\right)^{30} = \$9176.33$$

Now, we calculate the monthly payment to amortize this debt at 5.5% interest compounded monthly over 5 years.

$$PMT = PV \frac{i}{1 + (1 + i)^{-n}} \quad \text{where } PV = \$9176.33,\ i = \frac{0.055}{12} \approx 0.0045833 \text{ and } n = 12(5) = 60$$

$$= 9176.33 \frac{0.0045833}{1 - (1 + 0.0045833)^{-60}} = \frac{42.058179}{1 - (1.0045833)^{-60}} \approx \$175.28$$

The total amount paid on the loan is:
$175.28(60) = $10,516.80
Thus, the interest paid is:
$I = \$10,516.80 - \$8000 = \$2516.80$ (3-4)

37. Use $FV = PMT\dfrac{(1 + i)^n - 1}{i}$ where $PMT = 1200$ and $i = \dfrac{0.06}{12} = 0.005$. The graphs of

$$Y_1 = 1200\frac{(1.005)^x - 1}{0.005} = 240,000[(1.005)^x - 1]$$

$$Y_2 = 100,000$$

are shown in the figure at the right.

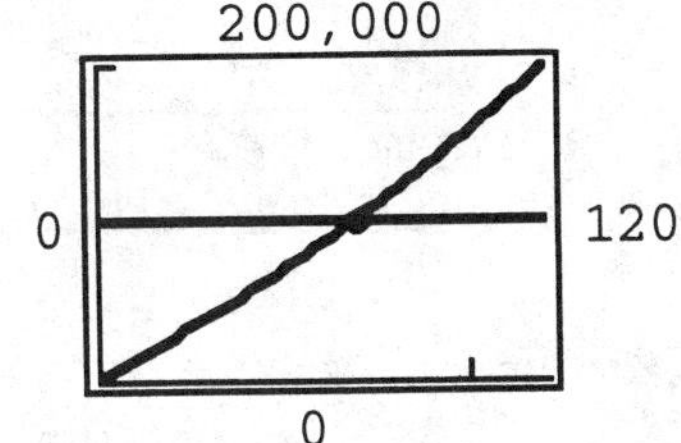

intersection: $x \approx 70$; $y = 100,000$

The fund will be worth \$100,000 after 70 payments, that is, after 5 years, 10 months. (3-3)

38. We first find the monthly payment:

$$PMT = PV\frac{i}{1 - (1 + i)^{-n}} \qquad PV = \$50,000,\ i = \frac{0.09}{12} = 0.0075,\ n = 12(20) = 240$$

$$= 50,000\frac{0.0075}{1 - (1.0075)^{-240}}$$

$= \$449.86$ per month

The present value of the \$449.86 per month, 20 year annuity at 9%, after x years, is given by

$$y = 449.86\frac{1 - (1.0075)^{-12(20-x)}}{0.0075}$$

$$= 59,981.33\left[1 - (1.0075)^{-12(20-x)}\right]$$

The graphs of

$$Y_1 = 59{,}981.33\left[1 - (1.0075)^{-12(20-x)}\right]$$
$$Y_2 = 10{,}000$$

are shown in the figure at the right.

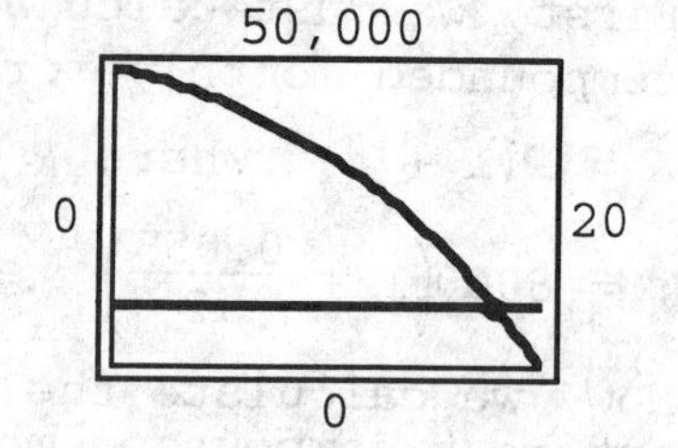

intersection: $x \approx 18$; $y = 10{,}000$

The unpaid balance will be below $10,000 after 18 years. (3-4)

39. $P = \$100$, $I = \$0.08$, $t = \frac{1}{360}$

From Problem 23, $r = \frac{I}{Pt} = \frac{0.08}{100\left(\frac{1}{360}\right)} = 0.288$ or 28.8% (3-1)

40. $PV = \$1000$, $i = 0.025$, $n = 4$

The quarterly payment is:

$$PMT = PV\frac{i}{1 - (1 + i)^{-n}}$$
$$= 1000\frac{0.025}{1 - (1 + 0.025)^{-4}} = \frac{25}{1 - (1.025)^{-4}} \approx \$265.82$$

Payment number	Payment	Interest	Unpaid balance reduction	Unpaid balance
0				$1000.00
1	$265.82	$25.00	$240.82	759.18
2	265.82	18.98	246.84	512.34
3	265.82	12.81	253.01	259.33
4	265.81	6.48	259.33	0.00
Totals	$1063.27	$63.27	$1000.00	

(3-4)

41. $PMT = \$200$, $FV = \$2500$, $i = \frac{0.0798}{12} = 0.00665$

$$FV = PMT\frac{(1 + i)^n - 1}{i}$$
$$2500 = 200\frac{(1 + 0.00665)^n - 1}{0.00665} = 30{,}075.188[(1.00665)^n - 1]$$
$$(1.00665)^n - 1 = \frac{2500}{30{,}075.188} \approx 0.083125$$
$$(1.0065)^n = 1.083125$$
$$n \ln 1.0065 = \ln 1.083125$$
$$n = \frac{\ln 1.083125}{\ln 1.0065} \approx 12.32 \text{ months}$$

Thus, it will take 13 months, or 1 year and 1 month. (3-2)

42. $FV = \$850{,}000$, $r = 8.76\% = 0.0876$, $m = 2$, $i = \frac{0.0876}{2} = 0.0438$,

$n = 2(6) = 12$

$$PMT = FV \frac{i}{(1 + i)^n - 1}$$

$$= 850{,}000 \frac{0.0438}{(1 + 0.0438)^{12} - 1} = \frac{37{,}230}{(1.0438)^{12} - 1} \approx \$55{,}347.48$$

The total amount invested is:

$12(55{,}347.48) = \$664{,}169.76$

Thus, the interest earned with this annuity is:

$I = \$850{,}000 - \$664{,}169.76 = \$185{,}830.24$ (3-3)

43. The effective rate for Security S & L is:

$$r_e = \left(1 + \frac{r}{m}\right)^m - 1 \text{ where } r = 9.38\% = 0.0938 \text{ and } m = 12$$

$$= \left(1 + \frac{0.0938}{12}\right)^{12} - 1 \approx 0.09794 \text{ or } 9.794\%$$

The effective rate for West Lake S & L is:

$$r_e = \left(1 + \frac{r}{m}\right)^m - 1 \text{ where } r = 9.35\% = 0.0935 \text{ and } m = 365$$

$$= \left(1 + \frac{0.0935}{365}\right)^{365} - 1 \approx 0.09799 \text{ or } 9.8\%$$

Thus, West Lake S & L is a better investment. (3-2)

44. $A = \$5000$, $P = \$4899.08$, $t = \frac{13}{52} = 0.25$

The interest earned is $I = \$5000.00 - \$4899.08 = \$100.92$. Thus:

$$r = \frac{I}{Pt} = \frac{100.92}{(4899.08)(0.25)} \approx 0.0824 \text{ or } 8.24\%$$ (3-1)

45. Using the sinking fund formula

$$PMT = FV \frac{i}{(1 + i)^n - 1}$$

with $PMT = \$200$, $FV = \$10{,}000$, and $i = \frac{0.09}{12} = 0.0075$, we have:

$$200 = 10{,}000 \frac{0.0075}{(1 + 0.0075)^n - 1} = \frac{75}{(1.0075)^n - 1}$$

Therefore,

$$(1.0075)^n - 1 = \frac{75}{200} = 0.375$$

$$(1.0075)^n = 0.375 + 1 = 1.375$$

$$\ln(1.0075)^n = \ln 1.375$$

$$n = \frac{\ln 1.375}{1.0075} \approx 42.62$$

The couple will have to make 43 deposits. (3-3)

46. $PV = \$80{,}000$, $i = \frac{0.15}{12} = 0.0125$, $n = 8(12) = 96$

(A) $PMT = PV \frac{i}{1 - (1 + i)^{-n}}$

$$= 80{,}000 \frac{0.0125}{1 - (1 + 0.0125)^{-96}} = \frac{1000}{1 - (1.0125)^{-96}}$$

$$= \$1435.63 \text{ monthly payment}$$

(B) Now use $PMT = \$1435.63$, $i = 0.0125$, and $n = 96 - 12 = 84$ to calculate the unpaid balance.

$PV = PMT \frac{1 - (1 + i)^{-n}}{i}$

$$= 1435.63 \frac{1 - (1 + 0.0125)^{-84}}{0.0125} = 114{,}850.40[1 - (1.0125)^{-84}]$$

$= \$74{,}397.48$ unpaid balance after the first year

(C) Amount of loan paid during the first year:
$80,000 - $74,397.48 = $5602.52
Amount of payments during the first year:
12($1435.63) = $17,227.56
Thus, the interest paid during the first year is:
$17,227.56 - $5602.52 = $11,625.04 (3-4)

47. Certificate of Deposit: $10,000 at 8.75% compounded monthly for 288 months

$A = P(1 + i)^n$ $\quad P = 10{,}000$

$= 10{,}000(1.00729)^{288}$ $\quad i = \frac{0.875}{12} = 0.00729$

$= \$81{,}041.86$ $\quad n = 288$

Reduce the principal:

Step 1: Find the monthly payment.

$PMT = PV \frac{i}{1 - (1 + i)^{-n}}$ $\quad PV = \$60{,}000$

$= 60{,}000 \frac{0.00854}{1 - (1.00854)^{-360}}$ $\quad i = \frac{0.1025}{12} = 0.00854$

$= 60{,}000(0.008961)$ $\quad n = 12(30) = 360$

$= \$537.66$ per month

Step 2: Find the unpaid balance after 72 payments, that is, find the present value of a $537.66 per month annuity at 10.25% for 288 payments.

$PV = PMT \frac{1 - (1 + i)^{-n}}{i}$ $\quad PMT = \$537.66$

$= 537.66(106.9659599)$ $\quad i = \frac{0.1025}{12} = 0.00854167$

$= \$57{,}511.32$ $\quad n = 288$

Step 3: Reduce the principal by $10,000 and determine the length of time required to pay off the loan.

$$47{,}511.32 = 537.66\frac{1 - (1.00854167)^{-n}}{0.00854167}$$

$$1 - (1.00854167)^{-n} = 0.7548005$$

$$(1.00854167)^{-n} = 0.2451995$$

$$n = -\frac{\ln(0.2451995)}{\ln(1.00854167)} \approx 165.3$$

The loan will be paid off after 166 more payments. Thus, by reducing the principal after 72 payments, the entire loan will be paid after 72 + 166 = 238 payments.

Step 4: Calculate the future value of a $537.66 per month annuity at 8.75% for 360 - 238 = 122 months.

$$FV = PMT\frac{(1 + i)^n - 1}{i} \qquad PMT = \$537.66$$

$$= 537.66\frac{(1.00729167)^{122} - 1}{0.00729167} \qquad i = \frac{0.0875}{12} = 0.00729167$$

$$= 537.66(195.6030015) \qquad n = 122$$

$$= 105{,}155.54$$

Conclusion: Use the $10,000 to reduce the principal and invest the monthly payment. (3-2, 3-3, 3-4)

48. We find the monthly payment and the total interest for each of the options. The monthly payment is given by:

$$PMT = PV\frac{i}{1 - (1 + i)^{-n}} \qquad PV = \$75{,}000, \quad n = 12(30) = 360$$

12% mortgage: $i = \frac{0.12}{12} = 0.01$

$$PMT = 75{,}000\frac{0.01}{1 - (1.01)^{-360}}$$

$$= 75{,}000(0.010286)$$

$$= 771.46 \text{ per month}$$

$$\text{Total interest paid} = 360(771.46) - 75{,}000$$

$$= \$202{,}725.60$$

11.25% mortgage: $i = \frac{0.1125}{12} = 0.009375$

$$PMT = 75{,}000\frac{0.009375}{1 - (1.009375)^{-360}}$$

$$= 75{,}000(0.0097126)$$

$$= \$728.45 \text{ per month}$$

$$\text{Total interest paid} = 360(728.47) - 75{,}000$$

$$= \$187{,}242.00$$

The lower rate would save over $15,000 in interest. (3-4)

49. $A = \$5000,\ r = i = 9.5\% = 0.095,\ n = 5$

$$P = \frac{A}{(1 + i)^n} = \frac{5000}{(1 + 0.095)^5} = \frac{5000}{(1.095)^5} \approx \$3176.14 \qquad (3\text{-}2)$$

50. $P = \$4476.20,\ A = \$10,000,\ m = 1,\ r = i,\ n = 10$

$$\begin{aligned} A &= P(1 + i)^n \\ 10,000 &= 4476.20(1 + i)^{10} \\ (1 + i)^{10} &= \frac{10,000}{4476.20} \approx 2.23404 \\ 10 \ln(1 + i) &= \ln(2.23404) \\ \ln(1 + i) &= \frac{\ln(2.23404)}{10} \approx 0.0803811 \\ 1 + i &= e^{0.0803811} \approx 1.0837 \\ i &= 0.0837 \text{ or } 8.37\% \end{aligned} \qquad (3\text{-}2)$$

51. $A = \$5000,\ r = 10.76\% = 0.1076,\ t = \frac{26}{52} = 0.5$

$$P = \frac{A}{1 + rt} = \frac{5000}{1 + (0.1076)(0.5)} = \$4744.73 \qquad (3\text{-}1)$$

52. We first compute the monthly payment using $PV = \$10,000$, $i = \frac{0.12}{12} = 0.01$, and $n = 5(12) = 60$.

$$\begin{aligned} PMT &= PV \frac{i}{1 - (1 + i)^{-n}} \\ &= 10,000 \frac{0.01}{1 - (1 + 0.01)^{-60}} = \frac{100}{1 - (1.01)^{-60}} = \$222.44 \text{ per month} \end{aligned}$$

Now, we calculate the unpaid balance after 24 payments by using $PMT = \$222.44$, $i = 0.01$, and $n = 60 - 24 = 36$.

$$\begin{aligned} PV &= PMT \frac{1 - (1 + i)^{-n}}{i} \\ &= 222.44 \frac{1 - (1 + 0.01)^{-36}}{0.01} = 22,244[1 - (1.01)^{-36}] = \$6697.11 \end{aligned}$$

Thus, the unpaid balance after 2 years is $6697.11. (3-4)

53. $r = 9\% = 0.09,\ m = 12$

$$\begin{aligned} r_e &= \left(1 + \frac{r}{m}\right)^m - 1 = \left(1 + \frac{0.09}{12}\right)^{12} - 1 \\ &= (1.0075)^{12} - 1 \approx 0.0938 \text{ or } 9.38\% \end{aligned} \qquad (3\text{-}2)$$

54. (A) We first calculate the future value of an annuity of \$2000 at 8% compounded annually for 9 years.

$FV = PMT \frac{(1 + i)^n - 1}{i}$ where $PMT = \$2000$, $i = 0.08$, and $n = 9$

$= 2000 \frac{(1 + 0.08)^9 - 1}{0.08} = 25{,}000[(1.08)^9 - 1] \approx \$24{,}975.12$

Now, we calculate the future value of this amount at 8% compounded annually for 36 years.

$A = P(1 + i)^n$, where $P = \$24{,}975.12$, $i = 0.08$, and $n = 36$

$= 24{,}975.12(1 + 0.08)^{36} = 24{,}975.12(1.08)^{36} \approx \$398{,}807$

(B) This is the future value of a \$2000 annuity at 8% compounded annually

$FV = PMT \frac{(1 + i)^n - 1}{i}$ where $PMT = \$2000$, $i = 0.08$, and $n = 36$

$= 2000 \frac{(1 + 0.08)^{36} - 1}{0.08} = 25{,}000[(1.08)^{36} - 1] \approx \$374{,}204$ (3-3)

55. The amount of the loan is (\$100,000)(0.8) = \$80,000 and

$PMT = PV \frac{i}{1 - (1 + i)^{-n}}.$

(A) First, let $i = \frac{0.1075}{12} = 0.0089583$, $n = 12(30) = 360$. Then,

$PMT = 80{,}000 \frac{0.0089583}{1 - (1 + 0.0089583)^{-360}} = \frac{716.66667}{0.9596687}$

$\approx \$746.79$ monthly payment for 30 years.

Next, let $i = \frac{0.1075}{12} = 0.0089583$, $n = 12(15) = 180$. Then

$PMT = 80{,}000 \frac{0.0089583}{1 - (1 + 0.0089593)^{-180}} = \frac{716.66667}{0.7991735}$

$\approx \$896.76$ monthly payment for 15 years.

(B) To find the unpaid balance after 10 years, we use

$PV = PMT \frac{1 - (1 + i)^{-n}}{i}$

First, for the 30-year mortgage:

$PMT = \$746.79$, $i = \frac{0.1075}{12} = 0.0089583$, $n = 12(20) = 240$

$PV = 746.79 \frac{1 - (1 + 0.0089583)^{-240}}{0.0089583} = 83{,}362.915[1 - (1.0089583)^{-240}]$

$= \$73{,}558.78$ unpaid balance for the 30-year mortgage

Next, for the 15-year mortgage:

$PMT = \$896.76$, $i = 0.0089583$, $n = 5(12) = 60$

$PV = 896.76 \frac{1 - (1.0089583)^{-60}}{0.0089583} = 100{,}103.81[1 - (1.0089583)^{-60}]$

$\approx \$41{,}482.19$ unpaid balance for the 15-year mortgage. (3-4)

56. The amount of the mortgage is:
(\$83,000)(0.8) = \$66,400

The monthly payment is given by:

$$PMT = PV\frac{i}{1 - (1 + i)^{-n}} \quad \text{where } PV = \$66{,}400,\ i = \frac{0.1125}{12} = 0.009375, \text{ and } n = 12(30) = 360$$

$$= 66{,}400\frac{0.009375}{1 - (1 + 0.009375)^{-360}}$$

$$= \frac{622.50}{1 - (1.009375)^{-360}} \approx \$644.92$$

Next, we find the present value of a \$644.92 per month, 22-year annuity:

$$PV = PMT\frac{1 - (1 + i)^{-n}}{i} \quad \text{where } PMT = \$644.92,\ i = 0.009375, \text{ and } n = 12(22) = 264$$

$$= 644.92\frac{1 - (1 + 0.009375)^{-264}}{0.009375}$$

$$= 68{,}791.467[1 - (1.009375)^{-264}] = \$62{,}934.63$$

Finally,

Equity = (current market value) - (unpaid loan balance)
= \$95,000 - \$62,934.63 = \$32,065.37

The family can borrow up to (\$32,065.37)(0.60) = \$19,239. (3-4)

57. $PV = \$600$, $PMT = 110$, $n = 6$

Solve $600 = 110\frac{1 - (1 + i)^{-6}}{i}$ for i and multiply the result by 12 to find r.

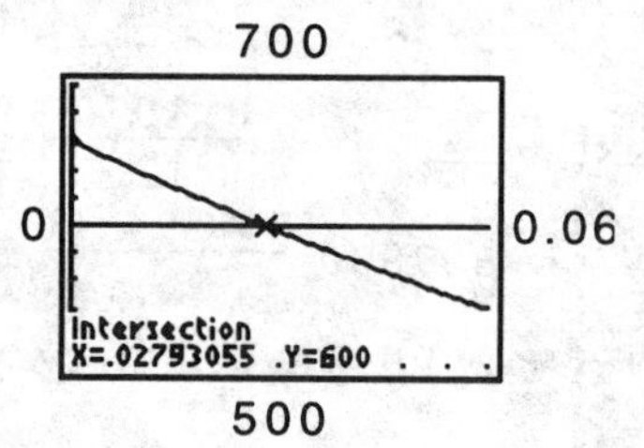

$i \approx 0.02793055$
$r \approx 0.33517$ or $r = 33.517\%$ (3-4)

58. (A) $FV = \$220{,}000$, $PMT = \$2{,}000$, $n = 25$.

Solve $220{,}000 = 2{,}000\frac{(1 + i)^{25} - 1}{i}$ for i:

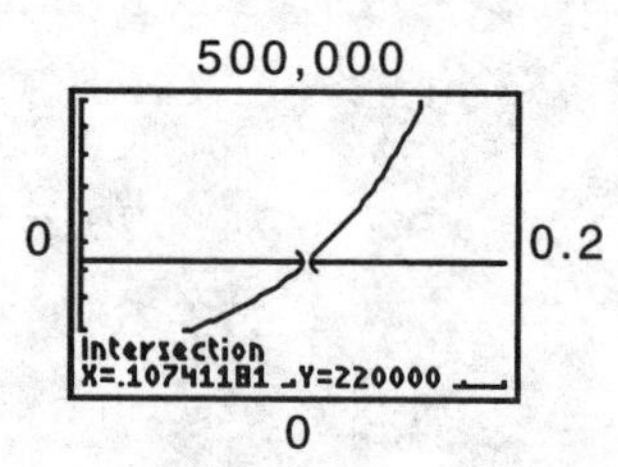

$i \approx 0.10741$ or $i = 10.741\%$

(B) Withdrawals at $30,000 per year:

Solve $220{,}000 = 30{,}000\frac{1 - (1.10741)^{-n}}{0.10741}$ for n

$n \approx 15$ years

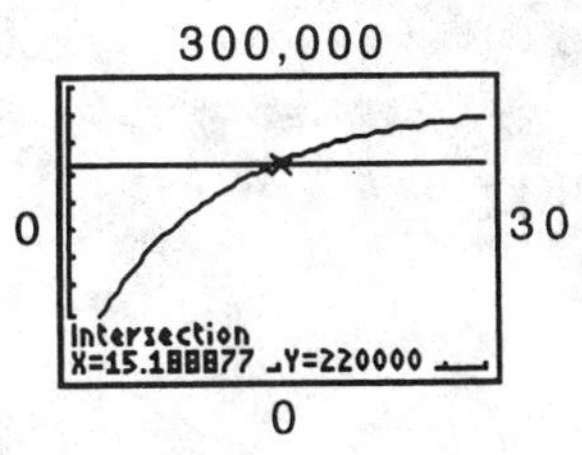

Withdrawals $24,000 per year:

Solve $220{,}000 = 24{,}000\frac{1 - (1.10741)^{-n}}{0.10741}$

$n \approx 41$ years

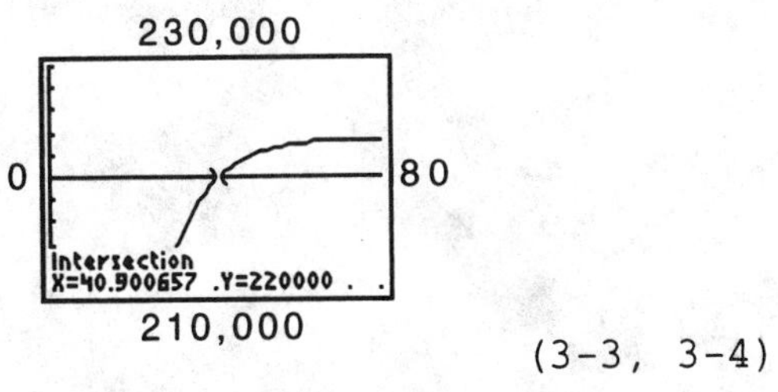

(3-3, 3-4)

4 SYSTEMS OF LINEAR EQUATIONS; MATRICES

EXERCISE 4-1

Things to remember:

1. SYSTEMS OF TWO EQUATIONS IN TWO VARIABLES

   Given the LINEAR SYSTEM
   $$ax + by = h$$
   $$cx + dy = k$$
   where a, b, c, d, h, and k are real constants, a pair of numbers $x = x_0$ and $y = y_0$ [also written as an ordered pair (x_0, y_0)] is a SOLUTION to this system if each equation is satisfied by the pair. The set of all such ordered pairs is called the SOLUTION SET for the system. To SOLVE a system is to find its solution set.

2. SYSTEMS OF LINEAR EQUATIONS: BASIC TERMS

   A system of linear equations is CONSISTENT if it has one or more solutions and INCONSISTENT if no solutions exist. Furthermore, a consistent system is said to be INDEPENDENT if it has exactly one solution (often referred to as the UNIQUE SOLUTION) and DEPENDENT if it has more than one solution.

3. The system of two linear equations in two variables
   $$ax + by = h$$
   $$cx + dy = k$$
   can be solved by:
   (a) graphing;
   (b) substitution;
   (c) elimination by addition.

4. POSSIBLE SOLUTIONS TO A LINEAR SYSTEM

   The linear system
   $$ax + by = h$$
   $$cx + dy = k$$
   must have:
   (a) exactly one solution (consistent and independent); or
   (b) no solution (inconsistent); or
   (c) infinitely many solutions (consistent and dependent).

5. Two systems of linear equations are EQUIVALENT if they have exactly the same solution set. A system of linear equations is transformed into an equivalent system if:

(a) two equations are interchanged;
(b) an equation is multiplied by a nonzero constant;
(c) a constant multiple of one equation is added to another equation.

---

1. (B); no solution

3. (A); $x = -3$, $y = 1$

5. $3x - y = 2$
$x + 2y = 10$
Point of intersection: (2, 4)
Solution: $x = 2$; $y = 4$

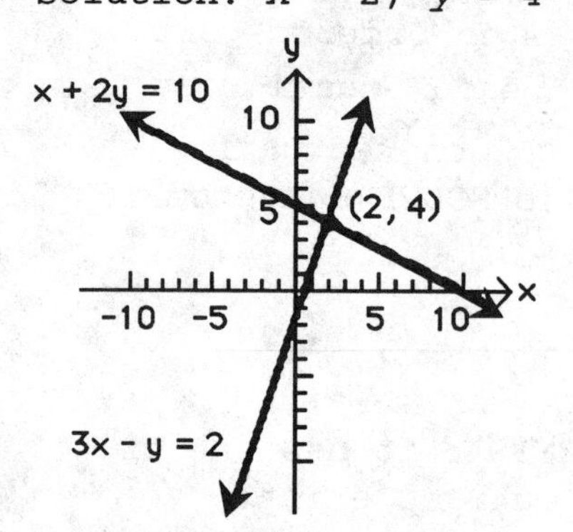

7. $m + 2n = 4$
$2m + 4n = -8$
Since the graphs of the given equations are parallel lines, there is no solution.

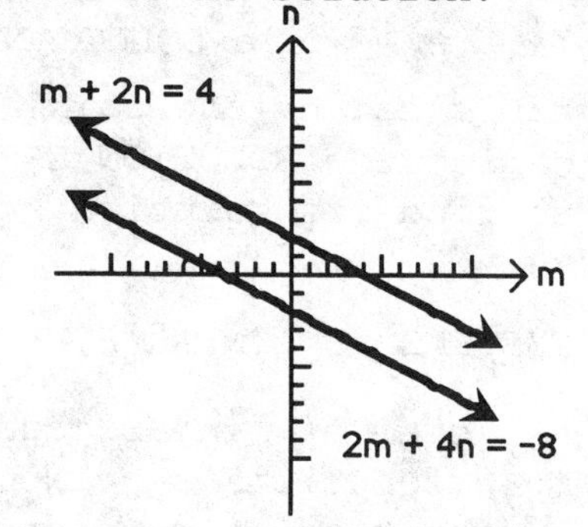

9.
$y = 2x - 3$ (1)
$x + 2y = 14$ (2)

By substituting y from (1) into (2), we get:

$$x + 2(2x - 3) = 14$$
$$x + 4x - 6 = 14$$
$$5x = 20$$
$$x = 4$$

Now, substituting $x = 4$ into (1), we have:

$y = 2(4) - 3$
$y = 5$

Solution: $x = 4$
$y = 5$

11. $2x + y = 6$ (1)
$x - y = -3$ (2)

Solve (2) for y to obtain the system:

$2x + y = 6$ (3)
$y = x + 3$ (4)

Substitute y from (4) into (3):

$$2x + x + 3 = 6$$
$$3x = 3$$
$$x = 1$$

Now, substituting $x = 1$ into (4), we get:

$y = 1 + 3$
$y = 4$

Solution: $x = 1$
$y = 4$

13. $3u - 2v = 12$ (1)
$7u + 2v = 8$ (2)

Add (1) and (2):

$10u = 20$
$u = 2$

Substituting $u = 2$ into (2), we get:

$7(2) + 2v = 8$
$2v = -6$
$v = -3$

Solution: $u = 2$
$v = -3$

15. $2m - n = 10$ (1)
$m - 2n = -4$ (2)

Multiply (1) by -2 and add to (2) to obtain:

$-3m = -24$
$m = 8$

Substituting $m = 8$ into (2), we get:

$8 - 2n = -4$
$-2n = -12$
$n = 6$

Solution: $m = 8$
$n = 6$

17. $9x - 3y = 24$ (1)
$11x + 2y = 1$ (2)

Solve (1) for y to obtain:

$y = 3x - 8$ (3)

and substitute into (2):

$11x + 2(3x - 8) = 1$
$11x + 6x - 16 = 1$
$17x = 17$
$x = 1$

Now, substitute $x = 1$ into (3):

$y = 3(1) - 8$
$y = -5$

Solution: $x = 1$
$y = -5$

19. $2x - 3y = -2$ (1)
$-4x + 6y = 7$ (2)

Multiply (1) by 2 and add to (2) to get:

$0 = 3$

This implies that the system is inconsistent, and thus there is no solution.

21. $3x + 8y = 4$ (1)
$15x + 10y = -10$ (2)

Multiply (1) by -5 and add to (2) to get:

$-30y = -30$
$y = 1$

Substituting $y = 1$ into (1), we get:

$3x + 8(1) = 4$
$3x = -4$
$x = -\frac{4}{3}$

Solution: $x = -\frac{4}{3}$; $y = 1$

23. $-6x + 10y = -30$ (1)
$3x - 5y = 15$ (2)

Multiply (2) by 2 and add to (1). This yields:

$0 = 0$

which implies that (1) and (2) are equivalent equations and there are infinitely many solutions. Geometrically, the two lines are coincident. The system is dependent.

25. $x + y = 1$ (1)

$0.3x - 0.4y = 0$ (2)

Multiply equation (2) by 10 to remove the decimals

$x + y = 1$ (1)

$3x - 4y = 0$ (3)

Multiply (1) by 4 and add to (2) to get

$7x = 4$

$x = \frac{4}{7}$

Now substitute $x = \frac{4}{7}$ in (1):

$\frac{4}{7} + y = 1$

$y = 1 - \frac{4}{7} = \frac{3}{7}$

Solution: $x = \frac{4}{7}$, $y = \frac{3}{7}$

27. $0.2x - 0.5y = 0.07$ (1)

$0.8x - 0.3y = 0.79$ (2)

Clear the decimals from (1) and (2) by multiplying each equation by 100.

$20x - 50y = 7$ (3)

$80x - 30y = 79$ (4)

Multiply (3) by -4 and add to (4) to get:

$170y = 51$

$y = \frac{51}{170}$

$y = 0.3$

Now, substitute $y = 0.3$ into (1):

$0.2x - 0.5(0.3) = 0.07$

$0.2x - 0.15 = 0.07$

$0.2x = 0.22$

$x = 1.1$

Solution: $x = 1.1$; $y = 0.3$

29. $\frac{2}{5}x + \frac{3}{2}y = 2$ (1)

$\frac{7}{3}x - \frac{5}{4}y = -5$ (2)

Multiply (1) by 10 and (2) by 12 to remove the fractions

$4x + 15y = 20$ (3)

$28x - 15y = -60$ (4)

System (3), (4) is equivalent to system (1), (2). Now add equations (3) and (4) to get

$32x = -40$

$x = -\frac{40}{32} = -\frac{5}{4}$

Now substitute $x = -\frac{5}{4}$ into either (1), (2), (3), or (4) — (3) is probably the easiest

$4\left(-\frac{5}{4}\right) + 15y = 20$

$-5 + 15y = 20$

$15y = 25$

$y = \frac{25}{15} = \frac{5}{3}$

Solution: $x = -\frac{5}{4}$, $y = \frac{5}{3}$

31. First solve each equation for y:

$y = \frac{3}{2}x - \frac{5}{2}$

$y = -\frac{4}{3}x + \frac{13}{3}$

The graphs of the equations are:

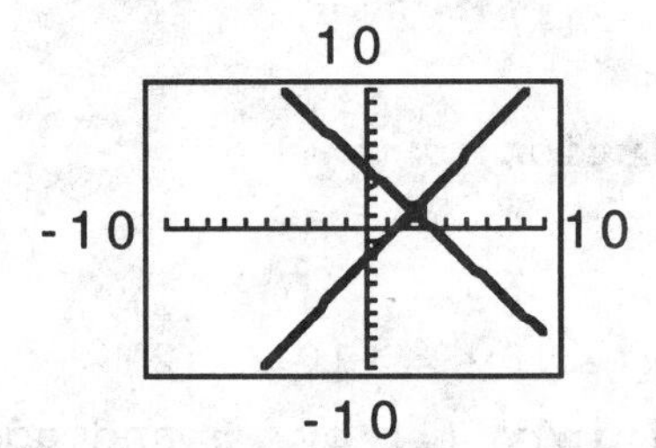

intersection: $x = 2.41$

$y = 1.12$

(2.41, 1.12)

33. Multiply each equation by 10 and then solve for y:

$$y = \frac{24}{35}x + \frac{1}{35}$$

$$y = \frac{17}{26}x - \frac{1}{13}$$

The graphs of these equations are almost indistinguishable:

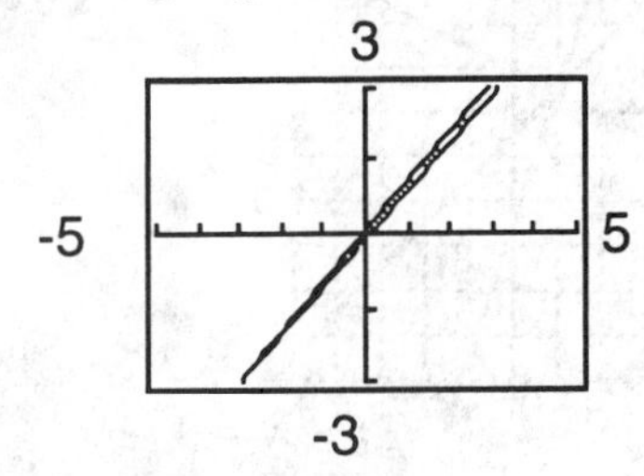

intersection: $x = -3.31$
$y = -2.24$
$(-3.31, -2.24)$

35. $x - 2y = -6$ (L_1)
$2x + y = 8$ (L_2)
$x + 2y = -2$ (L_3)

y
x − 2y = −6
x + 2y = −2
(2, 4)
(−4, 1)
x
(6, −4)
2x + y = 8

(A) L_1 and L_2 intersect:

$x - 2y = -6$ (1)
$2x + y = 8$ (2)

Multiply (2) by 2 and add to (1):
$5x = 10$
$x = 2$

Substitute $x = 2$ in (1) to get
$2 - 2y = -6$
$-2y = -8$
$y = 4$

Solution: $x = 2$, $y = 4$

(B) L_1 and L_3 intersect:

$x - 2y = -6$ (3)
$x + 2y = -2$ (4)

Add (3) and (4):
$2x = -8$
$x = -4$

Substitute $x = -4$ in (3) to get
$-4 - 2y = -6$
$-2y = -2$
$y = 1$

Solution: $x = -4$, $y = 1$

(C) L_2 and L_3 intersect:

$2x + y = 8$ (5)
$x + 2y = -2$ (6)

Multiply (6) by -2 and add to (5)
$-3y = 12$
$y = -4$

Substitute $y = -4$ in (5) to get
$2x - 4 = 8$
$2x = 12$
$x = 6$

Solution: $x = 6$, $y = -4$

37. $x + y = 1$ (L_1)

$x - 2y = -8$ (L_2)

$3x + y = -3$ (L_3)

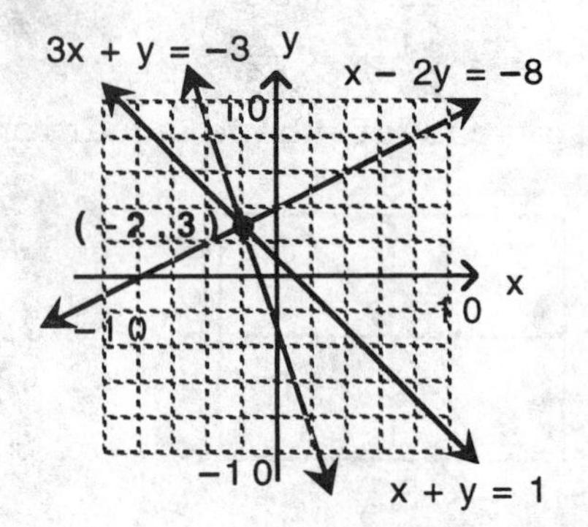

(A) L_1 and L_2 intersect

$x + y = 1$ (1)

$x - 2y = -8$ (2)

Subtract (2) from (1):

$3y = 9$

$y = 3$

Substitute $y = 3$ in (1) to get

$x + 3 = 1$

$x = -2$

Solution: $x = -2$, $y = 3$

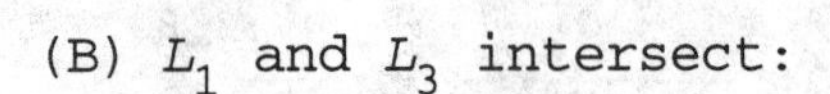
(B) L_1 and L_3 intersect:

$x + y = 1$ (3)

$3x + y = -3$ (4)

Subtract (4) from (3):

$-2x = 4$

$x = -2$

Substitute $x = -2$ in (3) to get

$-2 + y = 1$

$y = 3$

Solution: $x = -2$, $y = 3$

(C) It follows from (A) and (B), that L_2 and L_3 intersect at

$x = -2$, $y = 3$.

39. $4x - 3y = -24$ (L_1)

$2x + 3y = 12$ (L_2)

$8x - 6y = 24$ (L_3)

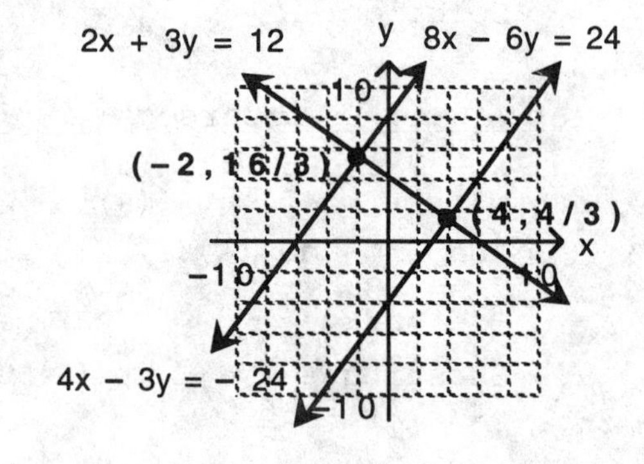

(A) L_1 and L_2 intersect:

$4x - 3y = -24$ (1)

$2x + 3y = 12$ (2)

Add (1) and (2):

$6x = -12$

$x = -2$

Substitute $x = -2$ in (2) to get

$2(-2) + 3y = 12$

$3y = 16$

$y = \frac{16}{3}$

Solution: $x = -2$, $y = \frac{16}{3}$

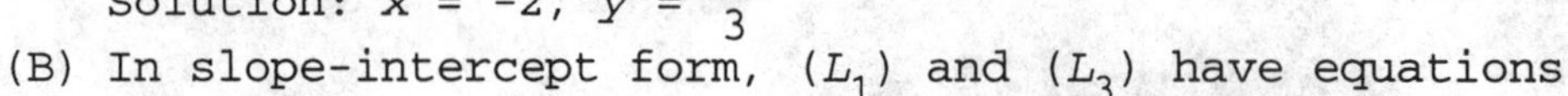
(B) In slope-intercept form, (L_1) and (L_3) have equations

$y = \frac{4}{3}x + 8$ (L_1)

$y = \frac{4}{3}x - 4$ (L_2)

Thus, (L_1) and (L_3) have the same slope and different y-intercepts; (L_1) and (L_3) are parallel; they do not intersect.

(C) L_2 and L_3 intersect:

$2x + 3y = 12$ (3)
$8x - 6y = 24$ (4)

Multiply (3) by 2 and add to (4):

$12x = 48$
$x = 4$

Substitute $x = 4$ in (3) to get

$2(4) + 3y = 12$
$3y = 4$
$y = \frac{4}{3}$

Solution: $x = 4,\ y = \frac{4}{3}$

41. (A) $5x + 4y = 4$
$11x + 9y = 4$

Multiply the top equation by 9 and the bottom equation by -4.

$45x + 36y = 36$
$-44x - 36y = -16$

Add the equations.

$x = 20$

$5(20) + 4y = 4$ Substitute $x = 20$ in the first equation.
$4y = -96$
$y = -24$

Solution: $(20, -24)$

(B) $5x + 4y = 4$
$11x + 8y = 4$

Multiply the top equation by 8 and the bottom equation by -4.

$40x + 32y = 32$
$-44x - 32y = -16$

Add the equations.

$-4x = 16$
$x = -4$

$5(-4) + 4y = 4$ Substitute $x = 4$ in the first equation.
$4y = 24$
$y = 6$

Solution: $(-4, 6)$

(C) $5x + 4y = 4$
$10x + 8y = 4$

Multiply the top equation by 8 and the bottom equation by -4.

$40x + 32y = 32$
$-40x - 32y = -16$

Add the equations.

$0 = -16$ This system has no solutions.

43. $p = 0.7q + 3$ Supply equation
$p = -1.7q + 15$ Demand equation

(A) $p = \$4$

Supply: $4 = 0.7q + 3$

$0.7q = 1$

$q = \frac{1}{0.7} \approx 1.429$

Thus, the supply will be 143 T-shirts.

Demand: $4 = -1.7q + 15$

$1.7q = 11$

$q = \frac{11}{1.7} \approx 6.471$

Thus, the demand will be 647 T-shirts. At this price level, the demand exceeds the supply; the price will rise.

(B) $p = \$9$

Supply: $9 = 0.7q + 3$

$0.7q = 6$

$q = \frac{6}{0.7} \approx 8.571$

Thus, the supply will be 857 T-shirts.

Demand: $9 = -1.7q + 15$

$1.7q = 6$

$q = \frac{6}{1.7} \approx 3.529$

Thus, the demand will be 353 T-shirts. At this price level, the supply exceeds the demand; the price will fall.

(C) Solve the pair of equations to find the equilibrium price and the equilibrium quantity.

$0.7q + 3 = -1.7q + 15$

$2.4q = 12$

$q = \frac{12}{2.4} = 5$

The equilibrium quantity is 500 T-shirts. Substitute $q = 5$ in either of the two equations to find p.

$p = (0.7)5 + 3 = 3.5 + 3$

$p = 6.50$

The equilibrium price is $6.50.

(D)

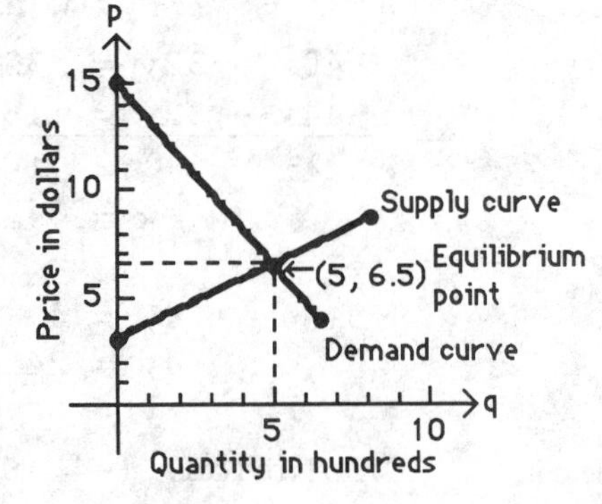

45. (A) $p = aq + b$, where a and b are to be determined. Now $q = 450$ when $p = 0.6$, and $q = 600$ when $p = 0.75$. This leads to the pair of equations

$450a + b = 0.6$ (1)

$600a + b = 0.75$ (2)

Subtracting (1) from (2), we have

$150a = 0.15$

$a = 0.001$

Substituting $a = 0.001$ in (2), we get

$600(0.001) + b = 0.75$

$b = 0.15$

Thus $p = 0.001q + 0.15$ <u>Supply equation</u>

(B) $p = aq + b$; $q = 570$ when $p = 0.6$ and $q = 495$ when $p = 0.75$. This leads to the pair of equations

$570a + b = 0.6$ (3)

$495a + b = 0.75$ (4)

Subtracting (3) from (4), we have

$-75a = 0.15$

$a = -0.002$

Substituting $a = -0.002$ in (3), we get

$570(-0.002) + b = 0.6$

$b = 1.74$

Thus $p = -0.002q + 1.74$ <u>Demand equation</u>

(C) Equilibrium occurs when supply equals demand. Equating the supply and demand equations, yields

$0.001q + 0.15 = -0.002q + 1.74$

$0.003q = 1.59$

$q = 530$ <u>Equilibrium quantity</u>

Substituting $q = 530$ into the supply equation (or into the demand equation), we get

$p = 0.001(530) + 0.15$

$p = 0.68$ or \$0.68 <u>Equilibrium price</u>

(D)

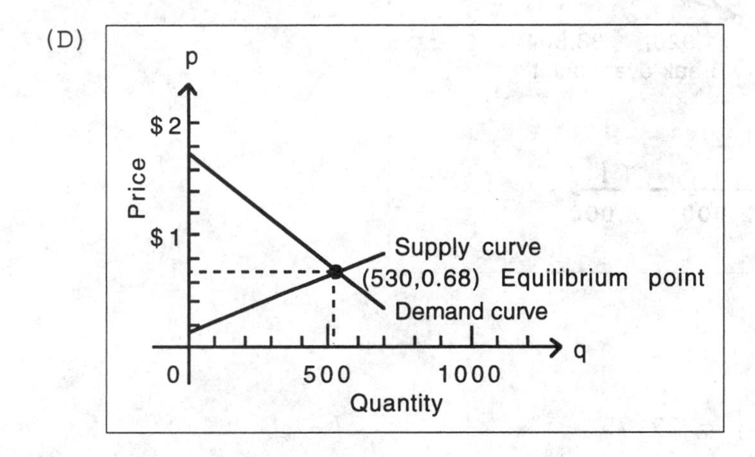

47. (A) The company breaks even when:

$$\text{Cost} = \text{Revenue}$$
$$48{,}000 + 1400x = 1800x$$
$$48{,}000 = 1800x - 1400x$$
or
$$400x = 48{,}000$$
$$x = \frac{48{,}000}{400}$$
$$x = 120$$

Thus, 120 units must be manufactured and sold to break even.

$$\text{Cost} = 48{,}000 + 1400(120)$$
$$= \$216{,}000 = \text{Revenue}$$

(B)

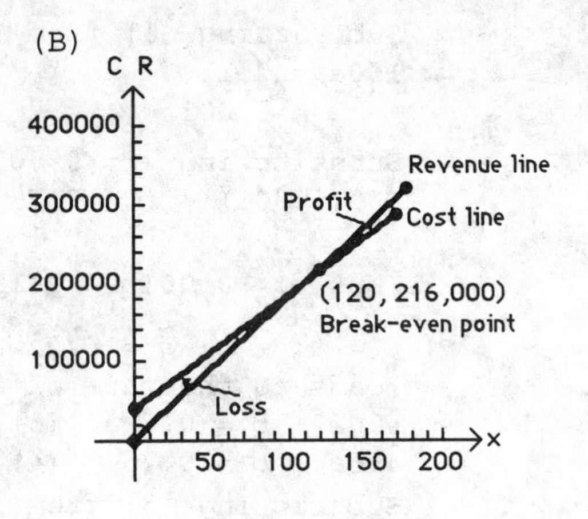

49. Let x = number of tapes marketed per month

(A) Revenue: $R = 19.95x$

Cost: $C = 7.45x + 24{,}000$

At the break-even point Revenue = Cost, that is

$19.95x = 7.45x + 24{,}000$

$12.50x = 24{,}000$

$x = 1920$

Thus, 1920 tapes must be sold per month to break even.
Cost = Revenue = $38,304 at the break-even point.

(B)

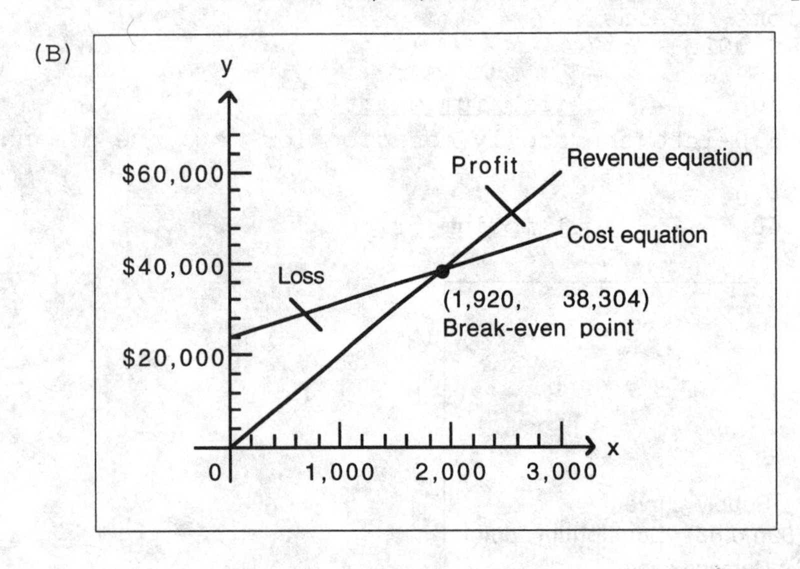

51. Let x = base price
y = surcharge

5 pound package: $x + 4y = 27.75$

20 pound package: $x + 19y = 64.50$

Solve the two equations:

$$x + 4y = 27.75 \qquad (1)$$
$$x + 19y = 64.50 \qquad (2)$$

Multiply (1) by (-1) and add to (2):

$$15y = 36.75$$
$$y = 2.45$$

Substitute $y = 2.45$ into (1):

$$x + 4(2.45) = 27.75$$
$$x = 27.75 - 9.80 = 17.95$$

Thus, the base price is \$17.95; the surcharge is \$2.45 per pound.

53. Let x = number of pounds of robust blend
y = number of pounds of mild blend

Total amount of Columbian beans: $132 \times 50 = 6,600$ lbs.

Total amount of Brazilian beans: $132 \times 40 = 5,280$ lbs.

Columbian beans needed: $\frac{12}{16}x + \frac{6}{16}y$ or $\frac{3}{4}x + \frac{3}{8}y$

Brazilian beans needed: $\frac{4}{16}x + \frac{10}{16}y$ or $\frac{1}{4}x + \frac{5}{8}y$

Thus, we need to solve:

$$\frac{3}{4}x + \frac{3}{8}y = 6,600 \quad (1)$$
$$\frac{1}{4}x + \frac{5}{8}y = 5,280 \quad (2)$$

Multiply (2) by -3 and add to (1):

$$-\frac{12}{8}y = -9,240$$
$$y = 6,160$$

Substitute $y = 6,160$ into (1):

$$\frac{3}{4}x + \frac{3}{8}(6,160) = 6,600$$
$$\frac{3}{4}x = 6,600 - 2,310 = 4,290$$
$$x = 5,720$$

Therefore, the manufacturer should produce
5,720 pounds of the robust blend and
6,160 pounds of the mild blend

55. Let x = amount of mix A, and
y = amount of mix B.

We want to solve the following system of equations:

$$0.1x + 0.2y = 20 \quad (1)$$
$$0.06x + 0.02y = 6 \quad (2)$$

Clear the decimals from (1) and (2) by multiplying both sides of (1) by 10 and both sides of (2) by 100.

$$x + 2y = 200 \quad (3)$$
$$6x + 2y = 600 \quad (4)$$

Multiply (3) by -1 and add to (4):

$5x = 400$

$x = 80$

Now substitute $x = 80$ into (3):

$80 + 2y = 200$

$2y = 120$

$y = 60$

Solution: x = mix A = 80 grams; y = mix B = 60 grams

57. $p = -\frac{1}{5}d + 70$ [Approach equation]

$p = -\frac{4}{3}d + 230$ [Avoidance equation]

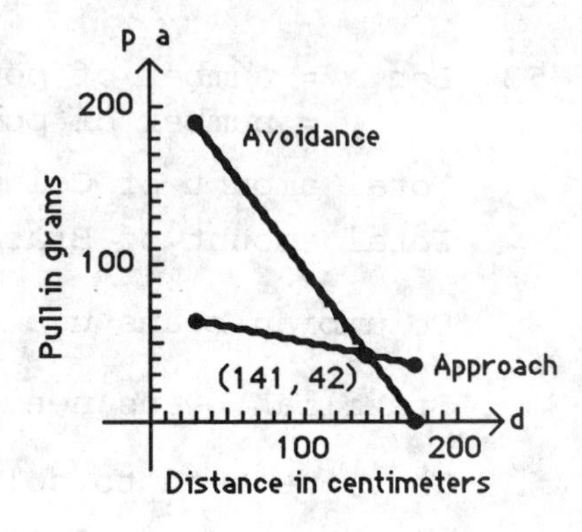

(A) The figure shows the graphs of the two equations.

(B) Setting the two equations equal to each other, we have

$$-\frac{1}{5}d + 70 = -\frac{4}{3}d + 230$$

$$-\frac{1}{5}d + \frac{4}{3}d = 230 - 70$$

$$\frac{17}{15}d = 160$$

$$d = 141 \text{ cm (approx.)}$$

(C) The rat would be very confused (!); it would vacillate.

EXERCISE 4-2

Things to remember:

1. MATRICES

   A MATRIX is a rectangular array of numbers written within brackets. Each number in a matrix is called an ELEMENT. If a matrix has m rows and n columns, it is called an $m \times n$ MATRIX; $m \times n$ is the SIZE; m and n are the DIMENSIONS. A matrix with n rows and n columns is a SQUARE MATRIX OF ORDER n. A matrix with only one column is a COLUMN MATRIX; a matrix with only one row is a ROW MATRIX. The element in the ith row and jth column of a matrix A is denoted a_{ij}. The PRINCIPAL DIAGONAL of a matrix A consists of the elements a_{11}, a_{22}, a_{33}, ...

2. A system of linear equations is transformed into an equivalent system if:

   (a) two equations ar interchanged;

   (b) an equation is multiplied by a nonzero constant;

   (c) a constant multiple of one equation is added to another equation.

3. Associated with the linear system

$$\begin{aligned} a_1x_1 + b_1x_2 &= k_1 \\ a_2x_1 + b_2x_2 &= k_2 \end{aligned} \qquad \text{(I)}$$

is the AUGMENTED MATRIX of the system

$$\left[\begin{array}{cc|c} a_1 & b_1 & k_1 \\ a_2 & b_2 & k_2 \end{array}\right]. \qquad \text{(II)}$$

4. An augmented matrix is transformed into a row-equivalent matrix if:

(a) two rows are interchanged ($R_i \leftrightarrow R_j$);

(b) a row is multiplied by a nonzero constant ($kR_i \to R_i$);

(c) a constant multiple of one row is added to another row ($kR_j + R_i \to R_i$).

(Note: The arrow $\to$ means "replaces.")

5. Given the system of linear equations (I) and its associated augmented matrix (II). If (II) is row equivalent to a matrix of the form:

(1) $\left[\begin{array}{cc|c} 1 & 0 & m \\ 0 & 1 & n \end{array}\right]$, then (I) has a unique solution; (consistent and independent);

(2) $\left[\begin{array}{cc|c} 1 & m & n \\ 0 & 0 & 0 \end{array}\right]$, then (I) has infinitely many solutions (consistent and dependent);

(3) $\left[\begin{array}{cc|c} 1 & m & n \\ 0 & 0 & p \end{array}\right]$, $p \neq 0$, then (I) has no solution (inconsistent).

---

1. A is 2×3; C is 1×3 **3.** C **5.** B

7. $a_{12} = -4$, $a_{23} = -5$ **9.** -1, 8, 0

11. (A) $E = \begin{bmatrix} 1 & -2 & 3 & 9 \\ -5 & 0 & 7 & -8 \end{bmatrix}$; size 2×4

(B) $F = \begin{bmatrix} 4 & -6 \\ 2 & 3 \\ -5 & 7 \end{bmatrix}$; size 3×2. F would have to have one more column to be square.

(C) $e_{23} = 7$; $f_{12} = -6$

13. Interchange row 1 and row 2.

$$\left[\begin{array}{cc|c} 4 & -6 & -8 \\ 1 & -3 & 2 \end{array}\right]$$

15. Multiply row 1 by -4.

$$\left[\begin{array}{cc|c} -4 & 12 & -8 \\ 4 & -6 & -8 \end{array}\right]$$

17. Multiply row 2 by 2.

$$\left[\begin{array}{cc|c} 1 & -3 & 2 \\ 8 & -12 & -16 \end{array}\right]$$

19. Replace row 2 by the sum of row 2 and -4 times row 1.

$$\left[\begin{array}{cc|c} 1 & -3 & 2 \\ 0 & 6 & -16 \end{array}\right]$$

21. Replace row 2 by the sum of row 2 and -2 times row 1.

$$\left[\begin{array}{cc|c} 1 & -3 & 2 \\ 2 & 0 & -12 \end{array}\right]$$

23. Replace row 2 by the sum of row 2 and -1 times row 1.

$$\left[\begin{array}{cc|c} 1 & -3 & 2 \\ 3 & -3 & -10 \end{array}\right]$$

25. $$\left[\begin{array}{cc|c} -1 & 2 & -3 \\ 6 & -3 & 12 \end{array}\right] \underset{\frac{1}{3}R_2 \to R_2}{\sim} \left[\begin{array}{cc|c} -1 & 2 & -3 \\ 2 & -1 & 4 \end{array}\right]$$

27. $$\left[\begin{array}{cc|c} -1 & 2 & -3 \\ 6 & -3 & 12 \end{array}\right] \underset{6R_1 + R_2 \to R_2}{\sim} \left[\begin{array}{cc|c} -1 & 2 & 3 \\ 0 & 9 & -6 \end{array}\right]$$

29. $$\left[\begin{array}{cc|c} -1 & 2 & -3 \\ 6 & -3 & 12 \end{array}\right] \underset{\frac{1}{3}R_2 + R_1 \to R_1}{\sim} \left[\begin{array}{cc|c} 1 & 1 & 1 \\ 6 & -3 & 12 \end{array}\right]$$

31. System

$$x_1 + x_2 = 5$$
$$x_1 - x_2 = 1$$

Augmented matrix

$$\left[\begin{array}{cc|c} 1 & 1 & 5 \\ 1 & -1 & 1 \end{array}\right]$$

Graphs:

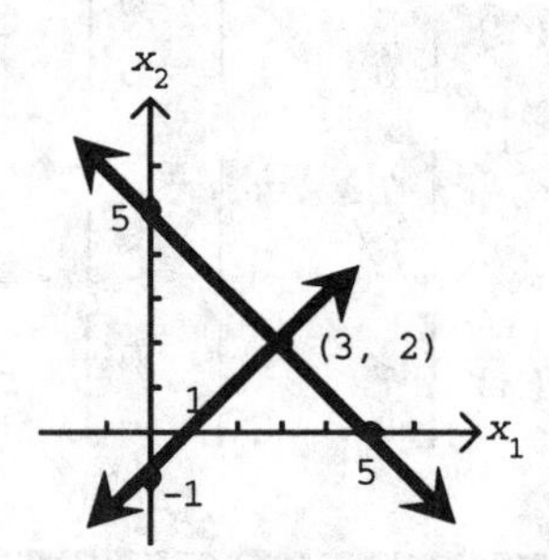

$$\left[\begin{array}{cc|c} 1 & 1 & 5 \\ 1 & -1 & 1 \end{array}\right] (-1)R_1 + R_2 \to R_2 \left[\begin{array}{cc|c} 1 & 1 & 5 \\ 0 & -2 & -4 \end{array}\right]$$

$$x_1 + x_2 = 5$$
$$-2x_2 = -4$$

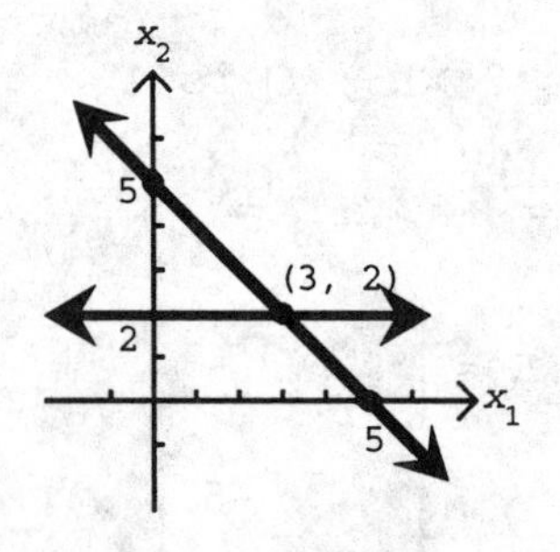

$$\begin{bmatrix} 1 & 1 & | & 5 \\ 0 & -2 & | & -4 \end{bmatrix} -\frac{1}{2}R_2 \to R_2 \begin{bmatrix} 1 & 1 & | & 5 \\ 0 & 1 & | & 2 \end{bmatrix}$$

$x_1 + x_2 = 5$
$x_2 = 2$

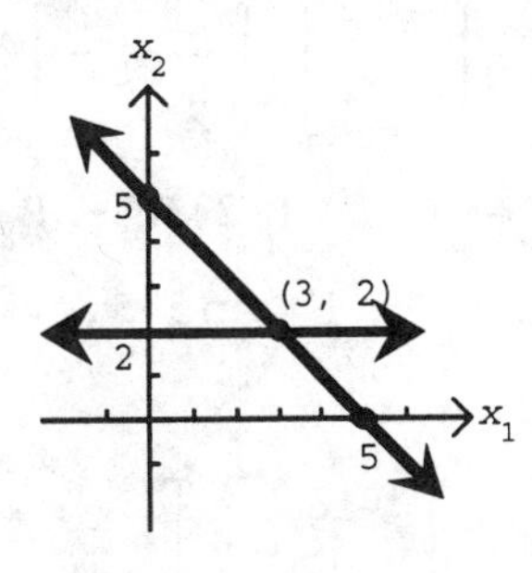

$$\begin{bmatrix} 1 & 1 & | & 5 \\ 0 & 1 & | & 2 \end{bmatrix} (-1)R_2 + R_1 \to R_1 \begin{bmatrix} 1 & 0 & | & 3 \\ 0 & 1 & | & 2 \end{bmatrix}$$

$x_1 = 3$
$x_2 = 2$

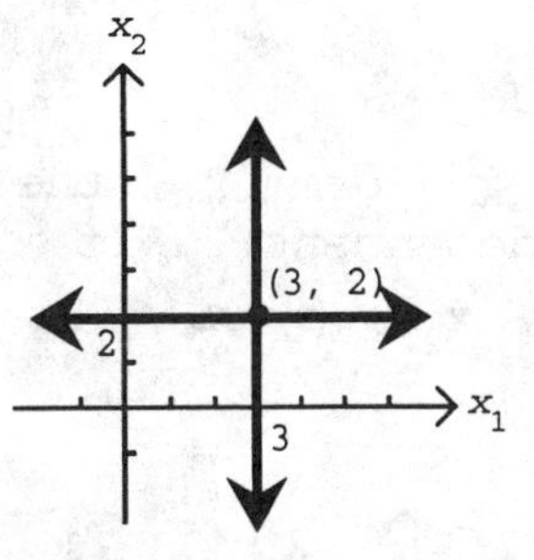

Solution: $x_1 = 3$, $x_2 = 2$. Each pair of lines has the same intersection point.

33. $\begin{bmatrix} 1 & -2 & | & 1 \\ 2 & -1 & | & 5 \end{bmatrix} \sim \begin{bmatrix} 1 & -2 & | & 1 \\ 0 & 3 & | & 3 \end{bmatrix} \sim \begin{bmatrix} 1 & -2 & | & 1 \\ 0 & 1 & | & 1 \end{bmatrix} \sim \begin{bmatrix} 1 & 0 & | & 3 \\ 0 & 1 & | & 1 \end{bmatrix}$ Thus, $x_1 = 3$ and $x_2 = 1$.

$(-2)R_1 + R_2 \to R_2$ $\quad \frac{1}{3}R_2 \to R_2$ $\quad 2R_2 + R_1 \to R_1$

35. $\begin{bmatrix} 1 & -4 & | & -2 \\ -2 & 1 & | & -3 \end{bmatrix} \sim \begin{bmatrix} 1 & -4 & | & -2 \\ 0 & -7 & | & -7 \end{bmatrix} \sim \begin{bmatrix} 1 & -4 & | & -2 \\ 0 & 1 & | & 1 \end{bmatrix} \sim \begin{bmatrix} 1 & 0 & | & 2 \\ 0 & 1 & | & 1 \end{bmatrix}$ Thus, $x_1 = 2$ and $x_2 = 1$.

$2R_1 + R_2 \to R_2$ $\quad \left(-\frac{1}{7}\right)R_2 \to R_2$ $\quad 4R_2 + R_1 \to R_1$

37. $\begin{bmatrix} 3 & -1 & | & 2 \\ 1 & 2 & | & 10 \end{bmatrix} \sim \begin{bmatrix} 1 & 2 & | & 10 \\ 3 & -1 & | & 2 \end{bmatrix} \sim \begin{bmatrix} 1 & 2 & | & 10 \\ 0 & -7 & | & -28 \end{bmatrix} \sim \begin{bmatrix} 1 & 2 & | & 10 \\ 0 & 1 & | & 4 \end{bmatrix} \sim \begin{bmatrix} 1 & 0 & | & 2 \\ 0 & 1 & | & 4 \end{bmatrix}$

$R_1 \leftrightarrow R_2$ $\quad (-3)R_1 + R_2 \to R_2$ $\quad \left(-\frac{1}{7}\right)R_2 \to R_2$ $\quad (-2)R_2 + R_1 \to R_1$

Thus, $x_1 = 2$ and $x_2 = 4$.

39. $\begin{bmatrix} 1 & 2 & | & 4 \\ 2 & 4 & | & -8 \end{bmatrix} \sim \begin{bmatrix} 1 & 2 & | & 4 \\ 0 & 0 & | & -16 \end{bmatrix}$ From 4, Form (3), the system is inconsistent; there is no solution.

$(-2)R_1 + R_2 \to R_2$

41. $\left[\begin{array}{rr|r} 2 & 1 & 6 \\ 1 & -1 & -3 \end{array}\right] \sim \left[\begin{array}{rr|r} 1 & -1 & -3 \\ 2 & 1 & 6 \end{array}\right] \sim \left[\begin{array}{rr|r} 1 & -1 & -3 \\ 0 & 3 & 12 \end{array}\right] \sim \left[\begin{array}{rr|r} 1 & -1 & -3 \\ 0 & 1 & 4 \end{array}\right] \sim \left[\begin{array}{rr|r} 1 & 0 & 1 \\ 0 & 1 & 4 \end{array}\right]$

$R_1 \leftrightarrow R_2 \qquad (-2)R_1 + R_2 \to R_2 \qquad \frac{1}{3}R_2 \to R_2 \qquad R_2 + R_1 \to R_1$

Thus, $x_1 = 1$,and $x_2 = 4$.

43. $\left[\begin{array}{rr|r} 3 & -6 & -9 \\ -2 & 4 & 6 \end{array}\right] \sim \left[\begin{array}{rr|r} 1 & -2 & -3 \\ -2 & 4 & 6 \end{array}\right] \sim \left[\begin{array}{rr|r} 1 & -2 & -3 \\ 0 & 0 & 0 \end{array}\right]$

$\frac{1}{3}R_1 \to R_1 \qquad 2R_1 + R_2 \to R_2$

From 4, Form (2), the system has infinitely many solutions (consistent and dependent). If $x_2 = s$, then $x_1 - 2s = -3$ or $x_1 = 2s - 3$.
Thus, $x_2 = s$, $x_1 = 2s - 3$, for any real number s, are the solutions.

45. $\left[\begin{array}{rr|r} 4 & -2 & 2 \\ -6 & 3 & -3 \end{array}\right] \sim \left[\begin{array}{rr|r} 1 & -\frac{1}{2} & \frac{1}{2} \\ -6 & 3 & -3 \end{array}\right] \sim \left[\begin{array}{rr|r} 1 & -\frac{1}{2} & \frac{1}{2} \\ 0 & 0 & 0 \end{array}\right]$

$\frac{1}{4}R_1 \to R_1 \qquad 6R_1 + R_2 \to R_2$

Thus, the system has infinitely many solutions (consistent and dependent). Let $x_2 = s$. Then

$x_1 - \frac{1}{2}s = \frac{1}{2}$ or $x_1 = \frac{1}{2}s + \frac{1}{2}$.

The set of solutions is $x_2 = s$, $x_1 = \frac{1}{2}s + \frac{1}{2}$ for any real number s.

47. $\left[\begin{array}{rr|r} 2 & 1 & 1 \\ 4 & -1 & -7 \end{array}\right] \sim \left[\begin{array}{rr|r} 1 & \frac{1}{2} & \frac{1}{2} \\ 4 & -1 & -7 \end{array}\right] \sim \left[\begin{array}{rr|r} 1 & \frac{1}{2} & \frac{1}{2} \\ 0 & -3 & -9 \end{array}\right] \sim \left[\begin{array}{rr|r} 1 & \frac{1}{2} & \frac{1}{2} \\ 0 & 1 & 3 \end{array}\right] \sim \left[\begin{array}{rr|r} 1 & 0 & -1 \\ 0 & 1 & 3 \end{array}\right]$

$\frac{1}{2}R_1 \to R_1 \qquad (-4)R_1 + R_2 \to R_2 \qquad \left(-\frac{1}{3}\right)R_2 \to R_2 \qquad \left(-\frac{1}{2}\right)R_2 + R_1 \to R_1$

Thus, $x_1 = -1$ and $x_2 = 3$.

49. $\left[\begin{array}{rr|r} 4 & -6 & 8 \\ -6 & 9 & -10 \end{array}\right] \sim \left[\begin{array}{rr|r} 1 & -\frac{3}{2} & 2 \\ -6 & 9 & -10 \end{array}\right] \sim \left[\begin{array}{rr|r} 1 & -\frac{3}{2} & 2 \\ 0 & 0 & 2 \end{array}\right]$

$\frac{1}{4}R_1 \to R_1 \qquad 6R_1 + R_2 \to R_2$

The second row of the final augmented matrix corresponds to the equation
$0x_1 + 0x_2 = 2$
which has no solution. Thus, the system has no solution; it is inconsistent.

51. $\left[\begin{array}{cc|c} -4 & 6 & -8 \\ 6 & -9 & 12 \end{array}\right] \sim \left[\begin{array}{cc|c} 1 & -\frac{3}{2} & 2 \\ 6 & -9 & 12 \end{array}\right] \sim \left[\begin{array}{cc|c} 1 & -\frac{3}{2} & 2 \\ 0 & 0 & 0 \end{array}\right]$

$\left(-\frac{1}{4}\right)R_1 \to R_1$ $\quad (-6)R_1 + R_2 \to R_2$

The system has infinitely many solutions (consistent and dependent). If $x_2 = t$, then

$x_1 - \frac{3}{2}t = 2$ or $x_1 = \frac{3}{2}t + 2$

Thus, the set of solutions is

$x_2 = t,\ x_1 = \frac{3}{2}t + 2$

for any real number t.

53. $\left[\begin{array}{cc|c} 3 & -1 & 7 \\ 2 & 3 & 1 \end{array}\right] \sim \left[\begin{array}{cc|c} 1 & -\frac{1}{3} & \frac{7}{3} \\ 2 & 3 & 1 \end{array}\right] \sim \left[\begin{array}{cc|c} 1 & -\frac{1}{3} & \frac{7}{3} \\ 0 & \frac{11}{3} & -\frac{11}{3} \end{array}\right] \sim \left[\begin{array}{cc|c} 1 & -\frac{1}{3} & \frac{7}{3} \\ 0 & 1 & -1 \end{array}\right] \sim \left[\begin{array}{cc|c} 1 & 0 & 2 \\ 0 & 1 & -1 \end{array}\right]$

$\frac{1}{3}R_1 \to R_1$ $\quad (-2)R_1 + R_2 \to R_2$ $\quad \frac{3}{11}R_2 \to R_2$ $\quad \frac{1}{3}R_2 + R_1 \to R_1$

Thus, $x_1 = 2$ and $x_2 = -1$.

55. $\left[\begin{array}{cc|c} 3 & 2 & 4 \\ 2 & -1 & 5 \end{array}\right] \sim \left[\begin{array}{cc|c} 1 & \frac{2}{3} & \frac{4}{3} \\ 2 & -1 & 5 \end{array}\right] \sim \left[\begin{array}{cc|c} 1 & \frac{2}{3} & \frac{4}{3} \\ 0 & -\frac{7}{3} & \frac{7}{3} \end{array}\right] \sim \left[\begin{array}{cc|c} 1 & \frac{2}{3} & \frac{4}{3} \\ 0 & 1 & -1 \end{array}\right] \sim \left[\begin{array}{cc|c} 1 & 0 & 2 \\ 0 & 1 & -1 \end{array}\right]$

$\frac{1}{3}R_1 \to R_1$ $\quad (-2)R_1 + R_2 \to R_2$ $\quad \left(-\frac{3}{7}\right)R_2 \to R_2$ $\quad \left(-\frac{2}{3}\right)R_2 + R_1 \to R_1$

Thus, $x_1 = 2$ and $x_2 = -1$.

57. $\left[\begin{array}{cc|c} 0.2 & -0.5 & 0.07 \\ 0.8 & -0.3 & 0.79 \end{array}\right] \sim \left[\begin{array}{cc|c} 1 & -2.5 & 0.35 \\ 0.8 & -0.3 & 0.79 \end{array}\right] \sim \left[\begin{array}{cc|c} 1 & -2.5 & 0.35 \\ 0 & 1.7 & 0.51 \end{array}\right]$

$\frac{1}{0.2}R_1 \to R_1$ $\quad (-0.8)R_1 + R_2 \to R_2$ $\quad \frac{1}{1.7}R_2 \to R_2$

$\sim \left[\begin{array}{cc|c} 1 & -2.5 & 0.35 \\ 0 & 1 & 0.3 \end{array}\right] \sim \left[\begin{array}{cc|c} 1 & 0 & 1.1 \\ 0 & 1 & 0.3 \end{array}\right]$ Thus, $x_1 = 1.1$ and $x_2 = 0.3$.

$2.5R_2 + R_1 \to R_1$

59. $0.8x_1 + 2.88x_2 = 4$

$1.25x_1 + 4.34x_2 = 5$

$\left[\begin{array}{cc|c} 0.8 & 2.88 & 4 \\ 1.25 & 4.34 & 5 \end{array}\right] (1.25)R_1 \to R_1$

$\left[\begin{array}{cc|c} 1 & 3.6 & 5 \\ 1.25 & 4.34 & 5 \end{array}\right] (-1.25)R_1 + R_2 \to R_2$

$\left[\begin{array}{cc|c} 1 & 3.6 & 5 \\ 0 & -0.16 & -1.25 \end{array}\right] (-6.25)R_2 \to R_2$

$$\left[\begin{array}{cc|c} 1 & 3.6 & 5 \\ 0 & 1 & 7.8125 \end{array}\right] (-3.6)R_2 + R_1 \to R_1$$

$$\left[\begin{array}{cc|c} 1 & 0 & -23.125 \\ 0 & 1 & 7.8125 \end{array}\right]$$

Solution: $x_1 = -23.125$, $x_2 = 7.8125$

61. $4.8x_1 - 40.32x_2 = 295.2$
$-3.75x_1 + 28.7x_2 = -211.2$

$$\left[\begin{array}{cc|c} 4.8 & -40.32 & 295.2 \\ -3.75 & 28.7 & -211.2 \end{array}\right] (0.20833)R_1 \to R_1$$

$$\left[\begin{array}{cc|c} 1 & -8.4 & 61.5 \\ -3.75 & 28.7 & -211.2 \end{array}\right] (3.75)R_1 + R_2 \to R_2$$

$$\left[\begin{array}{cc|c} 1 & -8.4 & 61.5 \\ 0 & -2.8 & 19.425 \end{array}\right] (-0.35714)R_2 \to R_2$$

$$\left[\begin{array}{cc|c} 1 & -8.4 & 61.5 \\ 0 & 1 & -6.9375 \end{array}\right] (8.4)R_2 + R_1 \to R_1$$

$$\left[\begin{array}{cc|c} 1 & 0 & 3.225 \\ 0 & 1 & -6.9375 \end{array}\right]$$

Solution: $x_1 = 3.225$, $x_2 = -6.9375$

EXERCISE 4-3

Things to remember:

1. A matrix is in REDUCED FORM if

   (a) each row consisting entirely of zeros is below any row having at least one nonzero element;

   (b) the left-most nonzero element in each row is 1;

   (c) all other elements in the column containing the left-most 1 of a given row are zeros;

   (d) the left-most 1 in any row is to the right of the left-most 1 in any row above.

2. GAUSS-JORDAN ELIMINATION

   Step 1. Choose the leftmost nonzero column and use appropriate row operations to get a 1 at the top.

   Step 2. Use multiples of the row containing the 1 from step 1 to get zeros in all remaining places in the column containing this 1.

   Step 3. Repeat step 1 with the SUBMATRIX formed by (mentally) deleting the row used in step 2 and all rows above this row.

Step 4. Repeat step 2 with the ENTIRE MATRIX, including the mentally deleted rows. Continue this process until it is impossible to go further.

[*Note:* If at any point in this process we obtain a row with all zeros to the left of the vertical line and a nonzero number to the right, we can stop, since we will have a contradiction: $0 = n$, $n \neq 0$. We can then conclude that the system has no solution.]

---

1. $\left[\begin{array}{cc|c} 1 & 0 & 2 \\ 0 & 1 & -1 \end{array}\right]$
is in reduced form

3. $\left[\begin{array}{cccc} 1 & 0 & 2 & 3 \\ 0 & 0 & 0 & 0 \\ 0 & 1 & -1 & 4 \end{array}\right]$
is not in reduced form: condition (a) is violated; the second row should be at the bottom.
$R_2 \leftrightarrow R_3$

5. $\left[\begin{array}{ccc|c} 0 & 1 & 0 & 2 \\ 0 & 0 & 3 & -1 \\ 0 & 0 & 0 & 0 \end{array}\right]$
is not in reduced form: The first non-zero element in the second row is not 1; condition (b) is violated
$\frac{1}{3}R_2 \to R_2$

7. $\left[\begin{array}{ccc|c} 1 & 1 & 0 & 1 \\ 0 & 0 & 1 & 1 \\ 0 & 0 & 0 & 0 \end{array}\right]$ is in reduced form.

9. $\left[\begin{array}{cccc|c} 1 & 0 & -2 & 0 & 1 \\ 0 & 0 & 1 & 1 & 0 \end{array}\right]$ not in reduced form: The column containing the left-most 1 in row 2 has a nonzero element; condition (c) is violated
$2R_2 + R_1 \to R_1$

11. $x_1 = -2$
$x_2 = 3$
$x_3 = 0$

13. $x_1 - 2x_3 = 3$ (1)
$x_2 + x_3 = -5$ (2)
Let $x_3 = t$. From (2), $x_2 = -5 - t$. From (1), $x_1 = 3 + 2t$. Thus, the solution is
$x_1 = 2t + 3$
$x_2 = -t - 5$
$x_3 = t$
t any real number.

15. $x_1 = 0$
$x_2 = 0$
$0 = 1$
Inconsistent; no solution.

17. $x_1 - 2x_2 \quad - 3x_4 = -5$

$\quad x_3 + 3x_4 = 2$

Let $x_2 = s$ and $x_4 = t$. Then

$x_1 = 2s + 3t - 5$

$x_2 = s$

$x_3 = -3t + 2$

$x_4 = t$

s and t any real numbers.

19. $\left[\begin{array}{cc|c} 1 & 2 & -1 \\ 0 & 1 & 3 \end{array}\right] \sim \left[\begin{array}{cc|c} 1 & 0 & -7 \\ 0 & 1 & 3 \end{array}\right]$

$(-2)R_2 + R_1 \to R_1$

21. $\left[\begin{array}{ccc|c} 1 & 0 & -3 & 1 \\ 0 & 1 & 2 & 0 \\ 0 & 0 & 3 & -6 \end{array}\right] \sim \left[\begin{array}{ccc|c} 1 & 0 & -3 & 1 \\ 0 & 1 & 2 & 0 \\ 0 & 0 & 1 & -2 \end{array}\right] \sim \left[\begin{array}{ccc|c} 1 & 0 & 0 & -5 \\ 0 & 1 & 0 & 4 \\ 0 & 0 & 1 & -2 \end{array}\right]$

$\frac{1}{3}R_3 \to R_3 \qquad 3R_3 + R_1 \to R_1$

$(-2)R_3 + R_2 \to R_2$

23. $\left[\begin{array}{ccc|c} 1 & 2 & -2 & -1 \\ 0 & 3 & -6 & 1 \\ 0 & -1 & 2 & -\frac{1}{3} \end{array}\right] \sim \left[\begin{array}{ccc|c} 1 & 2 & -2 & -1 \\ 0 & 1 & -2 & \frac{1}{3} \\ 0 & -1 & 2 & -\frac{1}{3} \end{array}\right] \sim \left[\begin{array}{ccc|c} 1 & 2 & -2 & -1 \\ 0 & 1 & -2 & \frac{1}{3} \\ 0 & 0 & 0 & 0 \end{array}\right] \sim \left[\begin{array}{ccc|c} 1 & 0 & 2 & -\frac{5}{3} \\ 0 & 1 & -2 & \frac{1}{3} \\ 0 & 0 & 0 & 0 \end{array}\right]$

$\frac{1}{3}R_2 \to R_2 \qquad R_2 + R_3 \to R_3 \qquad (-2)R_2 + R_1 \to R_1$

25. The corresponding augmented matrix is:

$\left[\begin{array}{ccc|c} 2 & 4 & -10 & -2 \\ 3 & 9 & -21 & 0 \\ 1 & 5 & -12 & 1 \end{array}\right] \sim \left[\begin{array}{ccc|c} 1 & 2 & -5 & -1 \\ 3 & 9 & -21 & 0 \\ 1 & 5 & -12 & 1 \end{array}\right] \sim \left[\begin{array}{ccc|c} 1 & 2 & -5 & -1 \\ 0 & 3 & -6 & 3 \\ 0 & 3 & -7 & 2 \end{array}\right] \sim \left[\begin{array}{ccc|c} 1 & 2 & -5 & -1 \\ 0 & 1 & -2 & 1 \\ 0 & 3 & -7 & 2 \end{array}\right]$

$\frac{1}{2}R_1 \to R_1 \qquad (-3)R_1 + R_2 \to R_2 \qquad \frac{1}{3}R_2 \to R_2 \qquad (-3)R_2 + R_3 \to R_3$

$(-1)R_1 + R_3 + \to R_3 \qquad (-2)R_2 + R_1 \to R_1$

$\sim \left[\begin{array}{ccc|c} 1 & 0 & -1 & -3 \\ 0 & 1 & -2 & 1 \\ 0 & 0 & -1 & -1 \end{array}\right] \sim \left[\begin{array}{ccc|c} 1 & 0 & -1 & -3 \\ 0 & 1 & -2 & 1 \\ 0 & 0 & 1 & 1 \end{array}\right] \sim \left[\begin{array}{ccc|c} 1 & 0 & 0 & -2 \\ 0 & 1 & 0 & 3 \\ 0 & 0 & 1 & 1 \end{array}\right]$

$(-1)R_3 \to R_3 \qquad 2R_3 + R_2 + \to R_2$

$R_3 + R_1 \to R_1$

Thus, $x_1 = -2$; $x_2 = 3$; $x_3 = 1$.

27. The corresponding augmented matrix is:

$\left[\begin{array}{ccc|c} 3 & 8 & -1 & -18 \\ 2 & 1 & 5 & 8 \\ 2 & 4 & 2 & -4 \end{array}\right] \sim \left[\begin{array}{ccc|c} 2 & 4 & 2 & -4 \\ 2 & 1 & 5 & 8 \\ 3 & 8 & -1 & -18 \end{array}\right] \sim \left[\begin{array}{ccc|c} 1 & 2 & 1 & -2 \\ 2 & 1 & 5 & 8 \\ 3 & 8 & -1 & -18 \end{array}\right]$

$R_1 \leftrightarrow R_3 \qquad \frac{1}{2}R_1 \to R_1 \qquad (-2)R_1 + R_2 \to R_2$

$(-3)R_1 + R_3 \to R_3$

$$\sim \left[\begin{array}{ccc|c} 1 & 2 & 1 & -2 \\ 0 & -3 & 3 & 12 \\ 0 & 2 & -4 & -12 \end{array}\right] \sim \left[\begin{array}{ccc|c} 1 & 2 & 1 & -2 \\ 0 & 1 & -1 & -4 \\ 0 & 2 & -4 & -12 \end{array}\right] \sim \left[\begin{array}{ccc|c} 1 & 0 & 3 & 6 \\ 0 & 1 & -1 & -4 \\ 0 & 0 & -2 & -4 \end{array}\right]$$

$\left(-\frac{1}{3}\right)R_2 \rightarrow R_2$ $\quad (-2)R_2 + R_3 \rightarrow R_3$, $(-2)R_2 + R_1 \rightarrow R_1$ $\quad \left(-\frac{1}{2}\right)R_3 \rightarrow R_3$

$$\sim \left[\begin{array}{ccc|c} 1 & 0 & 3 & 6 \\ 0 & 1 & -1 & -4 \\ 0 & 0 & 1 & 2 \end{array}\right] \sim \left[\begin{array}{ccc|c} 1 & 0 & 0 & 0 \\ 0 & 1 & 0 & -2 \\ 0 & 0 & 1 & 2 \end{array}\right]$$

$(-3)R_3 + R_1 \rightarrow R_1$, $R_3 + R_2 \rightarrow R_2$

Thus, $x_1 = 0$, $x_2 = -2$, $x_3 = 2$.

29. $$\left[\begin{array}{ccc|c} 2 & -1 & -3 & 8 \\ 1 & -2 & 0 & 7 \end{array}\right] \sim \left[\begin{array}{ccc|c} 1 & -2 & 0 & 7 \\ 2 & -1 & -3 & 8 \end{array}\right] \sim \left[\begin{array}{ccc|c} 1 & -2 & 0 & 7 \\ 0 & 3 & -3 & -6 \end{array}\right] \sim \left[\begin{array}{ccc|c} 1 & -2 & 0 & 7 \\ 0 & 1 & -1 & -2 \end{array}\right]$$

$R_1 \leftrightarrow R_2$ $\quad (-2)R_1 + R_2 \rightarrow R_2$ $\quad \frac{1}{3}R_2 \rightarrow R_2$ $\quad 2R_2 + R_1 \rightarrow R_1$

$$\sim \left[\begin{array}{ccc|c} 1 & 0 & -2 & 3 \\ 0 & 1 & -1 & -2 \end{array}\right]$$

Thus, $x_1 - 2x_3 = 3$ (1)
$x_2 - x_3 = -2$ (2)

Let $x_3 = t$, where t is any real number. Then:
$x_1 = 2t + 3$
$x_2 = t - 2$
$x_3 = t$

31. $$\left[\begin{array}{cc|c} 2 & -1 & 0 \\ 3 & 2 & 7 \\ 1 & -1 & -1 \end{array}\right] \sim \left[\begin{array}{cc|c} 1 & -1 & -1 \\ 3 & 2 & 7 \\ 2 & -1 & 0 \end{array}\right] \sim \left[\begin{array}{cc|c} 1 & -1 & -1 \\ 0 & 5 & 10 \\ 0 & 1 & 2 \end{array}\right] \sim \left[\begin{array}{cc|c} 1 & -1 & -1 \\ 0 & 1 & 2 \\ 0 & 5 & 10 \end{array}\right]$$

$R_1 \leftrightarrow R_3$ $\quad (-3)R_1 + R_2 \rightarrow R_2$, $(-2)R_1 + R_3 \rightarrow R_3$ $\quad R_2 \leftrightarrow R_3$ $\quad R_2 + R_1 \rightarrow R_1$, $(-5)R_2 + R_3 \rightarrow R_3$

$$\sim \left[\begin{array}{cc|c} 1 & 0 & 1 \\ 0 & 1 & 2 \\ 0 & 0 & 0 \end{array}\right]$$

Thus, $x_1 = 1$
$x_2 = 2$

33. $$\left[\begin{array}{ccc|c} 3 & -4 & -1 & 1 \\ 2 & -3 & 1 & 1 \\ 1 & -2 & 3 & 2 \end{array}\right] \sim \left[\begin{array}{ccc|c} 1 & -2 & 3 & 2 \\ 2 & -3 & 1 & 1 \\ 3 & -4 & -1 & 1 \end{array}\right] \sim \left[\begin{array}{ccc|c} 1 & -2 & 3 & 2 \\ 0 & 1 & -5 & -3 \\ 0 & 2 & -10 & -5 \end{array}\right] \sim \left[\begin{array}{ccc|c} 1 & -2 & 3 & 2 \\ 0 & 1 & -5 & -3 \\ 0 & 0 & 0 & 1 \end{array}\right]$$

$R_1 \leftrightarrow R_3$ $\quad (-2)R_1 + R_2 \rightarrow R_2$, $(-3)R_1 + R_3 \rightarrow R_3$ $\quad (-2)R_2 + R_3 \rightarrow R_3$

From the last row, we conclude that there is no solution; the system is inconsistent.

35. $\left[\begin{array}{ccc|c} 3 & -2 & 1 & -7 \\ 2 & 1 & -4 & 0 \\ 1 & 1 & -3 & 1 \end{array}\right] \sim \left[\begin{array}{ccc|c} 1 & 1 & -3 & 1 \\ 2 & 1 & -4 & 0 \\ 3 & -2 & 1 & -7 \end{array}\right] \sim \left[\begin{array}{ccc|c} 1 & 1 & -3 & 1 \\ 0 & -1 & 2 & -2 \\ 0 & -5 & 10 & -10 \end{array}\right]$

$R_1 \leftrightarrow R_3$ $\quad$ $(-2)R_1 + R_2 \to R_2$, $(-3)R_1 + R_3 \to R_3$ $\quad$ $(-1)R_2 \to R_2$

$\left[\begin{array}{ccc|c} 1 & 1 & -3 & 1 \\ 0 & 1 & -2 & 2 \\ 0 & -5 & 10 & -10 \end{array}\right] \sim \left[\begin{array}{ccc|c} 1 & 0 & -1 & -1 \\ 0 & 1 & -2 & 2 \\ 0 & 0 & 0 & 0 \end{array}\right]$

$(-1)R_2 + R_1 \to R_1$

$5R_2 + R_3 \to R_3$

From this matrix, $x_1 - x_3 = -1$ and $x_2 - 2x_3 = 2$. Let $x_3 = t$ be any real number, then $x_1 = t - 1$, $x_2 = 2t + 2$, and $x_3 = t$.

37. $\left[\begin{array}{ccc|c} 2 & 4 & -2 & 2 \\ -3 & -6 & 3 & -3 \end{array}\right] \sim \left[\begin{array}{ccc|c} 1 & 2 & -1 & 1 \\ -3 & -6 & 3 & -3 \end{array}\right] \sim \left[\begin{array}{ccc|c} 1 & 2 & -1 & 1 \\ 0 & 0 & 0 & 0 \end{array}\right]$

$\frac{1}{2}R_1 \to R_1$ $\quad$ $3R_1 + R_2 \to R_2$

From this matrix, $x_1 + 2x_2 - x_3 = 1$. Let $x_2 = s$ and $x_3 = t$. Then $x_1 = -2s + t + 1$, $x_2 = s$, and $x_3 = t$, s and t any real numbers.

39. $\left[\begin{array}{ccc|c} 4 & -1 & 2 & 3 \\ -4 & 1 & -3 & -10 \\ 8 & -2 & 9 & -1 \end{array}\right] \frac{1}{4}R_1 \to R_1 \sim \left[\begin{array}{ccc|c} 1 & -\frac{1}{4} & \frac{1}{2} & \frac{3}{4} \\ -4 & 1 & -3 & -10 \\ 8 & -2 & 9 & -1 \end{array}\right] \begin{array}{l} 4R_1 + R_2 \to R_2 \\ (-8)R_1 + R_3 \to R_3 \end{array}$

$\sim \left[\begin{array}{ccc|c} 1 & -\frac{1}{4} & \frac{1}{2} & \frac{3}{4} \\ 0 & 0 & -1 & -7 \\ 0 & 0 & 5 & -7 \end{array}\right] (-1)R_2 \to R_2 \sim \left[\begin{array}{ccc|c} 1 & -\frac{1}{4} & \frac{1}{2} & \frac{3}{4} \\ 0 & 0 & 1 & 7 \\ 0 & 0 & 5 & -7 \end{array}\right] (-5)R_2 + R_3 \to R_3$

$\sim \left[\begin{array}{ccc|c} 1 & -\frac{1}{4} & \frac{1}{2} & \frac{3}{4} \\ 0 & 0 & 1 & 7 \\ 0 & 0 & 0 & -42 \end{array}\right]$

No solution.

41. $\left[\begin{array}{ccc|c} 2 & -5 & -3 & 7 \\ -4 & 10 & 2 & 6 \\ 6 & -15 & -1 & -19 \end{array}\right] \frac{1}{2}R_1 \to R_1 \sim \left[\begin{array}{ccc|c} 1 & -\frac{5}{2} & -\frac{3}{2} & \frac{7}{2} \\ -4 & 10 & 2 & 6 \\ 6 & -15 & -1 & -19 \end{array}\right] \begin{array}{l} 4R_1 + R_2 \to R_2 \\ (-6)R_1 + R_3 \to R_3 \end{array}$

$\sim \left[\begin{array}{ccc|c} 1 & -\frac{5}{2} & -\frac{3}{2} & \frac{7}{2} \\ 0 & 0 & -4 & 20 \\ 0 & 0 & 8 & -40 \end{array}\right] \left(-\frac{1}{4}\right)R_2 \to R_2 \sim \left[\begin{array}{ccc|c} 1 & -\frac{5}{2} & -\frac{3}{2} & \frac{7}{2} \\ 0 & 0 & 1 & -5 \\ 0 & 0 & 8 & -40 \end{array}\right] (-8)R_2 + R_3 \to R_3$

$$\sim \left[\begin{array}{ccc|c} 1 & -\frac{5}{2} & -\frac{3}{2} & \frac{7}{2} \\ 0 & 0 & 1 & -5 \\ 0 & 0 & 0 & 0 \end{array}\right] \frac{3}{2}R_2 + R_1 \to R_1 \sim \left[\begin{array}{ccc|c} 1 & -\frac{5}{2} & 0 & -4 \\ 0 & 0 & 1 & -5 \\ 0 & 0 & 0 & 0 \end{array}\right]$$

The system of equations is:

$$x_1 - \frac{5}{2}x_2 = -4$$
$$x_3 = -5$$

and

$$x_1 = 2.5x_2 - 4$$
$$x_3 = -5$$

Let $x_2 = t$. Then for any real number t,

$$x_1 = 2.5t - 4$$
$$x_2 = t$$
$$x_3 = -5$$

is a solution.

43. $\left[\begin{array}{ccc|c} 5 & -3 & 2 & 13 \\ 2 & -1 & -3 & 1 \\ 4 & -2 & 4 & 12 \end{array}\right]$ $(-1)R_3 + R_1 \to R_1$ (to simplify the arithmetic)

$$\sim \left[\begin{array}{ccc|c} 1 & -1 & -2 & 1 \\ 2 & -1 & -3 & 1 \\ 4 & -2 & 4 & 12 \end{array}\right] \begin{array}{l} (-2)R_1 + R_2 \to R_2 \\ (-4)R_1 + R_3 \to R_3 \end{array} \sim \left[\begin{array}{ccc|c} 1 & -1 & -2 & 1 \\ 0 & 1 & 1 & -1 \\ 0 & 2 & 12 & 8 \end{array}\right] \begin{array}{l} R_2 + R_1 \to R_1 \\ (-2)R_2 + R_3 \to R_3 \end{array}$$

$$\sim \left[\begin{array}{ccc|c} 1 & 0 & -1 & 0 \\ 0 & 1 & 1 & -1 \\ 0 & 0 & 10 & 10 \end{array}\right] \frac{1}{10}R_3 \to R_3 \sim \left[\begin{array}{ccc|c} 1 & 0 & -1 & 0 \\ 0 & 1 & 1 & -1 \\ 0 & 0 & 1 & 1 \end{array}\right] \begin{array}{l} R_3 + R_1 \to R_1 \\ (-1)R_3 + R_2 \to R_2 \end{array}$$

$$\sim \left[\begin{array}{ccc|c} 1 & 0 & 0 & 1 \\ 0 & 1 & 0 & -2 \\ 0 & 0 & 1 & 1 \end{array}\right]$$

The system of equations is:

$$x_1 = 1$$
$$x_2 = -2$$
$$x_3 = 1$$

Solution: $x_1 = 1$, $x_2 = -2$, $x_3 = 1$

45. (A) The system is dependent with two parameters and an infinite number of solutions.

(B) The system is dependent with one parameter and an infinite number of solutions.

(C) The system is independent with a unique solution.

(D) Impossible.

47. $x_1 - x_2 = 4$
$3x_1 + kx_2 = 7$
If $k = -3$, the system is
$x_1 - x_2 = 4$
$3x_1 - 3x_2 = 7$

$$\left[\begin{array}{cc|c} 1 & -1 & 4 \\ 3 & -3 & 7 \end{array}\right] \sim \left[\begin{array}{cc|c} 1 & -1 & 4 \\ 0 & 0 & -5 \end{array}\right]$$

$-3R_1 + R_2 \to R_2$

The system has no solutions. If $k \neq -3$, the system has a unique solution.

49. $x_1 + kx_2 = 3$
$2x_1 + 6x_2 = 6$
If $k = 3$, the system is
$x_1 + 3x_2 = 3$
$2x_1 + 6x_2 = 6$

$$\left[\begin{array}{cc|c} 1 & 3 & 3 \\ 3 & 6 & 6 \end{array}\right] \sim \left[\begin{array}{cc|c} 1 & 3 & 3 \\ 0 & 0 & 0 \end{array}\right]$$

$-2R_1 + R_2 \to R_2$

The system has infinitely many solutions. If $k \neq 3$, the system has a unique solution.

51.
$$\left[\begin{array}{cccc|c} 1 & 2 & -4 & -1 & 7 \\ 2 & 5 & -9 & -4 & 16 \\ 1 & 5 & -7 & -7 & 13 \end{array}\right] \sim \left[\begin{array}{cccc|c} 1 & 2 & -4 & -1 & 7 \\ 0 & 1 & -1 & -2 & 2 \\ 0 & 3 & -3 & -6 & 6 \end{array}\right] \sim \left[\begin{array}{cccc|c} 1 & 0 & -2 & 3 & 3 \\ 0 & 1 & -1 & -2 & 2 \\ 0 & 0 & 0 & 0 & 0 \end{array}\right]$$

$(-2)R_1 + R_2 \to R_2$ $\quad$ $(-2)R_2 + R_1 \to R_1$
$(-1)R_1 + R_3 \to R_3$ $\quad$ $(-3)R_2 + R_3 \to R_3$

Thus, $x_1 - 2x_3 + 3x_4 = 3$ and $x_2 - x_3 - 2x_4 = 2$. Let $x_3 = s$ and $x_4 = t$. Then $x_1 = 2s - 3t + 3$, $x_2 = s + 2t + 2$, $x_3 = s$, $x_4 = t$, where s, t are any real numbers.

53.
$$\left[\begin{array}{cccc|c} 1 & -1 & 3 & -2 & 1 \\ -2 & 4 & -3 & 1 & 0.5 \\ 3 & -1 & 10 & -4 & 2.9 \\ 4 & -3 & 8 & -2 & 0.6 \end{array}\right] \begin{array}{l} 2R_1 + R_2 \to R_2 \\ (-3)R_1 + R_3 \to R_3 \\ (-4)R_1 + R_4 \to R_4 \end{array} \sim \left[\begin{array}{cccc|c} 1 & -1 & 3 & -2 & 1 \\ 0 & 2 & 3 & -3 & 2.5 \\ 0 & 2 & 1 & 2 & -0.1 \\ 0 & 1 & -4 & 6 & -3.4 \end{array}\right] R_2 \leftrightarrow R_4$$

$$\sim \left[\begin{array}{cccc|c} 1 & -1 & 3 & -2 & 1 \\ 0 & 1 & -4 & 6 & -3.4 \\ 0 & 2 & 1 & 2 & -0.1 \\ 0 & 2 & 3 & -3 & 2.5 \end{array}\right] \begin{array}{l} R_2 + R_1 \to R_1 \\ (-2)R_2 + R_3 \to R_3 \\ (-2)R_2 + R_4 \to R_4 \end{array}$$

$$\sim \left[\begin{array}{cccc|c} 1 & 0 & -1 & 4 & -2.4 \\ 0 & 1 & -4 & 6 & -3.4 \\ 0 & 0 & 9 & -10 & 6.7 \\ 0 & 0 & 11 & -15 & 9.3 \end{array}\right] (-1)R_4 + R_3 \to R_3 \text{ (to simplify arithmetic)}$$

$$\sim \left[\begin{array}{cccc|c} 1 & 0 & -1 & 4 & -2.4 \\ 0 & 1 & -4 & 6 & -3.4 \\ 0 & 0 & -2 & 5 & -2.6 \\ 0 & 0 & 11 & -15 & 9.3 \end{array}\right] \left(-\frac{1}{2}\right)R_3 \to R_3$$

$$\sim \left[\begin{array}{cccc|c} 1 & 0 & -1 & 4 & -2.4 \\ 0 & 1 & -4 & 6 & -3.4 \\ 0 & 0 & 1 & -2.5 & 1.3 \\ 0 & 0 & 11 & -15 & 9.3 \end{array}\right] \begin{array}{r} R_3 + R_1 \to R_1 \\ 4R_3 + R_2 \to R_2 \\ (-11)R_3 + R_4 \to R_4 \end{array}$$

$$\sim \left[\begin{array}{cccc|c} 1 & 0 & 0 & 1.5 & -1.1 \\ 0 & 1 & 0 & -4 & 1.8 \\ 0 & 0 & 1 & -2.5 & 1.3 \\ 0 & 0 & 0 & 12.5 & -5 \end{array}\right] \frac{1}{12.5}R_4 \to R_4$$

$$\sim \left[\begin{array}{cccc|c} 1 & 0 & 0 & 1.5 & -1.1 \\ 0 & 1 & 0 & -4 & 1.8 \\ 0 & 0 & 1 & -2.5 & 1.3 \\ 0 & 0 & 0 & 1 & -0.4 \end{array}\right] \begin{array}{r} (-1.5)R_4 + R_1 \to R_1 \\ 4R_4 + R_2 \to R_2 \\ 2.5R_4 + R_3 \to R_3 \end{array}$$

$$\sim \left[\begin{array}{cccc|c} 1 & 0 & 0 & 0 & -0.5 \\ 0 & 1 & 0 & 0 & 0.2 \\ 0 & 0 & 1 & 0 & 0.3 \\ 0 & 0 & 0 & 1 & -0.4 \end{array}\right]$$

The system of equations is:

$$\begin{aligned} x_1 &= -0.5 \\ x_2 &= 0.2 \\ x_3 &= 0.3 \\ x_4 &= -0.4 \end{aligned}$$

Solution: $x_1 = -0.5$, $x_2 = 0.2$, $x_3 = 0.3$, $x_2 = -0.4$

55. $$\left[\begin{array}{ccccc|c} 1 & -2 & 1 & 1 & 2 & 2 \\ -2 & 4 & 2 & 2 & -2 & 0 \\ 3 & -6 & 1 & 1 & 5 & 4 \\ -1 & 2 & 3 & 1 & 1 & 3 \end{array}\right] \begin{array}{r} 2R_1 + R_2 \to R_2 \\ (-3)R_1 + R_3 \to R_3 \\ R_1 + R_4 \to R_4 \end{array}$$

$$\sim \left[\begin{array}{ccccc|c} 1 & -2 & 1 & 1 & 2 & 2 \\ 0 & 0 & 4 & 4 & 2 & 4 \\ 0 & 0 & 2 & -2 & -1 & -2 \\ 0 & 0 & 4 & 2 & 3 & 5 \end{array}\right] \frac{1}{4}R_2 \to R_2$$

$$\sim \left[\begin{array}{ccccc|c} 1 & -2 & 1 & 1 & 2 & 2 \\ 0 & 0 & 1 & 1 & \frac{1}{2} & 1 \\ 0 & 0 & -2 & -2 & -1 & -2 \\ 0 & 0 & 4 & 2 & 3 & 5 \end{array}\right] \begin{array}{r} (-1)R_2 + R_1 \to R_1 \\ 2R_2 + R_3 \to R_3 \\ (-4)R_2 + R_4 \to R_4 \end{array}$$

$$\sim \left[\begin{array}{ccccc|c} 1 & -2 & 1 & 1 & \frac{3}{2} & 1 \\ 0 & 0 & 1 & 1 & \frac{1}{2} & 1 \\ 0 & 0 & 0 & 0 & 0 & 0 \\ 0 & 0 & 0 & -2 & 1 & 1 \end{array}\right] R_3 \leftrightarrow R_4 \sim \left[\begin{array}{ccccc|c} 1 & -2 & 0 & 0 & \frac{3}{2} & 1 \\ 0 & 0 & 1 & 1 & \frac{1}{2} & 1 \\ 0 & 0 & 0 & -2 & 1 & 1 \\ 0 & 0 & 0 & 0 & 0 & 0 \end{array}\right] \left(-\frac{1}{2}\right)R_3 \to R_3$$

$$\sim \left[\begin{array}{ccccc|c} 1 & -2 & 0 & 0 & \frac{3}{2} & 1 \\ 0 & 0 & 1 & 1 & \frac{1}{2} & 1 \\ 0 & 0 & 0 & 1 & -\frac{1}{2} & -\frac{1}{2} \\ 0 & 0 & 0 & 0 & 0 & 0 \end{array}\right] (-1)R_3 + R_2 \to R_2 \sim \left[\begin{array}{ccccc|c} 1 & -2 & 0 & 0 & \frac{3}{2} & 1 \\ 0 & 0 & 1 & 0 & 1 & \frac{3}{2} \\ 0 & 0 & 0 & 1 & -\frac{1}{2} & -\frac{1}{2} \\ 0 & 0 & 0 & 0 & 0 & 0 \end{array}\right]$$

The system of equations is:

$$x_1 - 2x_2 + \frac{3}{2}x_5 = 1$$
$$x_3 + x_5 = \frac{3}{2}$$
$$x_4 - \frac{1}{2}x_5 = -\frac{1}{2}$$

or $x_1 = 2x_2 - \frac{3}{2}x_5 + 1$

$x_3 = -x_5 + \frac{3}{2}$

$x_4 = \frac{1}{2}x_5 - \frac{1}{2}$

Let $x_2 = s$ and $x_5 = t$. Then $x_1 = 2s - \frac{3}{2}t + 1$, $x_2 = s$, $x_3 = -t + \frac{3}{2}$, $x_4 = \frac{1}{2}t - \frac{1}{2}$, $x_5 = t$ for any real numbers s and t.

57. Let x_1 = Number of one-person boats
x_2 = Number of two-person boats
x_3 = Number of three-person boats.

(A) The mathematical model is:

$$0.5x_1 + x_2 + 1.5x_3 = 380$$
$$0.6x_1 + 0.9x_2 + 1.2x_3 = 330$$
$$0.2x_1 + 0.3x_2 + 0.5x_3 = 120$$

$$\left[\begin{array}{ccc|c} 0.5 & 1 & 1.5 & 380 \\ 0.6 & 0.9 & 1.2 & 330 \\ 0.2 & 0.3 & 0.5 & 120 \end{array}\right] \sim \left[\begin{array}{ccc|c} 1 & 2 & 3 & 760 \\ 0.6 & 0.9 & 1.2 & 330 \\ 0.2 & 0.3 & 0.5 & 120 \end{array}\right] \sim \left[\begin{array}{ccc|c} 1 & 2 & 3 & 760 \\ 0 & -0.3 & -0.6 & -126 \\ 0 & -0.1 & -0.1 & -32 \end{array}\right]$$

$2R_1 \to R_1$ $\quad$ $(-0.6)R_1 + R_2 \to R_2$, $(-0.2)R_1 + R_3 \to R_3$ $\quad$ $\left(-\frac{1}{0.3}\right)R_2 \to R_2$

$$\sim \left[\begin{array}{ccc|c} 1 & 2 & 3 & 760 \\ 0 & 1 & 2 & 420 \\ 0 & -0.1 & -0.1 & -32 \end{array}\right] \sim \left[\begin{array}{ccc|c} 1 & 0 & -1 & -80 \\ 0 & 1 & 2 & 420 \\ 0 & 0 & 0.1 & 10 \end{array}\right] \sim \left[\begin{array}{ccc|c} 1 & 0 & -1 & -80 \\ 0 & 1 & 2 & 420 \\ 0 & 0 & 1 & 100 \end{array}\right]$$

$(0.1)R_2 + R_3 \to R_3$, $(-2)R_2 + R_1 \to R_1$ $\quad$ $10R_3 \to R_3$ $\quad$ $R_3 + R_1 \to R_1$, $(-2)R_3 + R_2 \to R_2$

$$\sim \left[\begin{array}{ccc|c} 1 & 0 & 0 & 20 \\ 0 & 1 & 0 & 220 \\ 0 & 0 & 1 & 100 \end{array}\right]$$

Thus, $x_1 = 20$, $x_2 = 220$, and $x_3 = 100$, or 20 one-person boats, 220 two-person boats, and 100 four-person boats.

(B) The mathematical model is:

$$\begin{aligned} 0.5x_1 + x_2 + 1.5x_3 &= 380 \\ 0.6x_1 + 0.9x_2 + 1.2x_3 &= 330 \end{aligned}$$

$$\left[\begin{array}{ccc|c} 0.5 & 1 & 1.5 & 380 \\ 0.6 & 0.9 & 1.2 & 330 \end{array}\right] \sim \left[\begin{array}{ccc|c} 1 & 2 & 3 & 760 \\ 0.6 & 0.9 & 1.2 & 330 \end{array}\right] \sim \left[\begin{array}{ccc|c} 1 & 2 & 3 & 760 \\ 0 & -0.3 & -0.6 & -126 \end{array}\right]$$

$2R_1 \to R_1$ $\qquad$ $(-0.6)R_1 + R_2 \to R_2$ $\qquad$ $\left(-\frac{1}{0.3}\right)R_2 \to R_2$

$$\sim \left[\begin{array}{ccc|c} 1 & 2 & 3 & 760 \\ 0 & 1 & 2 & 420 \end{array}\right] \sim \left[\begin{array}{ccc|c} 1 & 0 & -1 & -80 \\ 0 & 1 & 2 & 420 \end{array}\right]$$

$(-2)R_2 + R_1 \to R_1$

Thus, $x_1 - x_3 = -80$ (1)

$x_2 + 2x_3 = 420$ (2)

Let $x_3 = t$ (t any real number)

Then, $x_2 = 420 - 2t$ [from (2)]

$x_1 = t - 80$ [from (1)]

In order to keep x_1 and x_2 positive, $t \le 210$ and $t \ge 80$.

Thus, $x_1 = t - 80$ (one-person boats)

$x_2 = 420 - 2t$ (two-person boats)

$x_3 = t$ (four-person boats)

where $80 \le t \le 210$ and t is an integer.

(C) The mathematical model is:

$$\begin{aligned} 0.5x_1 + x_2 &= 380 \\ 0.6x_1 + 0.9x_2 &= 330 \\ 0.2x_1 + 0.3x_2 &= 120 \end{aligned}$$

$$\left[\begin{array}{cc|c} 0.5 & 1 & 380 \\ 0.6 & 0.9 & 330 \\ 0.2 & 0.3 & 120 \end{array}\right] \sim \left[\begin{array}{cc|c} 1 & 2 & 760 \\ 0.6 & 0.9 & 330 \\ 0.2 & 0.3 & 120 \end{array}\right] \sim \left[\begin{array}{cc|c} 1 & 2 & 760 \\ 0 & -0.3 & -126 \\ 0 & -0.1 & -32 \end{array}\right] \sim \left[\begin{array}{cc|c} 1 & 2 & 760 \\ 0 & 1 & 420 \\ 0 & -0.1 & -32 \end{array}\right]$$

$2R_1 \to R_1$ $\qquad$ $(-0.6R_1) + R_2 \to R_2$ $\qquad$ $\left(-\frac{1}{0.3}\right)R_2 \to R_2$ $\qquad$ $0.1R_2 + R_3 \to R_3$

$(-0.2R_1) + R_3 \to R_3$

$$\sim \left[\begin{array}{cc|c} 1 & 2 & 760 \\ 0 & 1 & 420 \\ 0 & 0 & 10 \end{array}\right]$$

From this matrix, we conclude that there is no solution; there is no production schedule that will use all the labor-hours in all departments.

59. Let x_1 = number of 6000 gallon tank cars
x_2 = number of 8000 gallon tank cars
x_3 = number of 18,000 gallon tank cars

Then, the mathematical model is:

$$x_1 + x_2 + x_3 = 24$$
$$6000x_1 + 8000x_2 + 18,000x_3 = 250,000$$

Dividing the second equation by 2000, we get the system

$$x_1 + x_2 + x_3 = 24$$
$$3x_1 + 4x_2 + 9x_3 = 125$$

The augmented matrix corresponding to this system is:

$$\left[\begin{array}{ccc|c} 1 & 1 & 1 & 24 \\ 3 & 4 & 9 & 125 \end{array}\right] \sim \left[\begin{array}{ccc|c} 1 & 1 & 1 & 24 \\ 0 & 1 & 6 & 53 \end{array}\right] \sim \left[\begin{array}{ccc|c} 1 & 0 & -5 & -29 \\ 0 & 1 & 6 & 53 \end{array}\right]$$

$(-3)R_1 + R_2 \to R_2$ $\quad$ $(-1)R_2 + R_1 \to R_1$

Thus $x_1 - 5x_3 = -29$
$x_2 + 6x_3 = 53$

Let $x_3 = t$. Then $x_1 = 5t - 29$ and $x_2 = 53 - 6t$

Thus, $(5t - 29)$ 6000-gallon tank cars, $(53 - 6t)$ 8000-gallon tank cars and (t) 18000-gallon tank cars should be purchased. Also, since t, $5t - 29$ and $53 - 6t$ must each be non-negative integers, it follows that $t = 6, 7$ or 8.

61. Let x_1 = federal income tax
x_2 = state income tax
x_3 = local income tax

Then, the mathematical model is:

$$x_1 = 0.50[7,650,000 - (x_2 + x_3)]$$
$$x_2 = 0.20[7,650,000 - (x_1 + x_3)]$$
$$x_3 = 0.10[7,650,000 - (x_1 + x_2)]$$

and

$$x_1 + 0.5x_2 + 0.5x_3 = 3,825,000$$
$$0.2x_1 + x_2 + 0.2x_3 = 1,530,000$$
$$0.1x_1 + 0.1x_2 + x_3 = 765,000$$

The corresponding augmented matrix is:

$$\left[\begin{array}{ccc|c} 1 & 0.5 & 0.5 & 3,825,000 \\ 0.2 & 1 & 0.2 & 1,530,000 \\ 0.1 & 0.1 & 1 & 765,000 \end{array}\right] \begin{array}{l} (-0.2)R_1 + R_2 \to R_2 \\ (-0.1)R_1 + R_3 \to R_3 \end{array}$$

$$\sim \left[\begin{array}{ccc|c} 1 & 0.5 & 0.5 & 3,825,000 \\ 0 & 0.9 & 0.1 & 765,000 \\ 0 & 0.05 & 0.95 & 382,500 \end{array}\right] \quad 20R_3 \to R_3 \text{ (simplify arithmetic)}$$

$$\sim \left[\begin{array}{ccc|c} 1 & 0.5 & 0.5 & 3,825,000 \\ 0 & 0.9 & 0.1 & 765,000 \\ 0 & 1 & 19 & 7,650,000 \end{array}\right] \quad R_2 \leftrightarrow R_3$$

$$\sim \left[\begin{array}{ccc|c} 1 & 0.5 & 0.5 & 3,825,000 \\ 0 & 1 & 19 & 7,650,000 \\ 0 & 0.9 & 0.1 & 765,000 \end{array}\right] \quad \begin{array}{l} (-0.5)R_2 + R_1 \to R_1 \\ (-0.9)R_2 + R_3 \to R_3 \end{array}$$

$$\sim \left[\begin{array}{ccc|c} 1 & 0 & -9 & 0 \\ 0 & 1 & 19 & 7,650,000 \\ 0 & 0 & -17 & -6,120,000 \end{array}\right] \quad \left(-\frac{1}{17}\right)R_3 \to R_3$$

$$\sim \left[\begin{array}{ccc|c} 1 & 0 & -9 & 0 \\ 0 & 1 & 19 & 7,650,000 \\ 0 & 0 & 1 & 360,000 \end{array}\right] \quad \begin{array}{l} 9R_3 + R_1 \to R_1 \\ (-19)R_3 + R_2 \to R_2 \end{array}$$

$$\sim \left[\begin{array}{ccc|c} 1 & 0 & 0 & 3,240,000 \\ 0 & 1 & 0 & 810,000 \\ 0 & 0 & 1 & 360,000 \end{array}\right]$$

Thus, x_1 = \$3,240,000, x_2 = \$810,000, x_3 = \$360,000. The total tax liability is $x_1 + x_2 + x_3$ = \$4,410,000 which is 57.65% of the taxable income $\left(\frac{4,410,000}{7,650,000} = 0.5765 \text{ or } 57.65\%\right)$.

63. Let x_1 = Taxable income of company A
x_2 = Taxable income of company B
x_3 = Taxable income of company C
x_4 = Taxable income of company D

The taxable income of each company is given by the system of equations:

$$\begin{aligned} x_1 &= 0.71(3.2) + 0.08x_2 + 0.03x_3 + 0.07x_4 \\ x_2 &= 0.12x_1 + 0.81(2.6) + 0.11x_3 + 0.13x_4 \\ x_3 &= 0.11x_1 + 0.09x_2 + 0.72(3.8) + 0.08x_4 \\ x_4 &= 0.06x_1 + 0.02x_2 + 0.14x_3 + 0.78(4.4) \end{aligned}$$

which is the same as:

$$\begin{aligned} x_1 - 0.08x_2 - 0.03x_3 - 0.07x_4 &= 2.272 \\ -0.12x_1 + x_2 - 0.11x_3 - 0.13x_4 &= 2.106 \\ -0.11x_1 - 0.09x_2 + x_3 - 0.08x_4 &= 2.736 \\ -0.06x_1 - 0.02x_2 - 0.14x_3 + x_4 &= 3.432 \end{aligned}$$

This is the mathematical model.

$$\begin{bmatrix} 1 & -0.08 & -0.03 & -0.07 \\ -0.12 & 1 & -0.11 & -0.13 \\ -0.11 & -0.09 & 1 & -0.08 \\ -0.06 & -0.02 & -0.14 & 1 \end{bmatrix}\begin{bmatrix} x_1 \\ x_2 \\ x_3 \\ x_4 \end{bmatrix} = \begin{bmatrix} 2.272 \\ 2.106 \\ 2.736 \\ 3.432 \end{bmatrix}$$

Therefore,

$$\begin{bmatrix} x_1 \\ x_2 \\ x_3 \\ x_4 \end{bmatrix} = \begin{bmatrix} 1 & -0.08 & -0.03 & -0.07 \\ -0.12 & 1 & -0.11 & -0.13 \\ -0.11 & -0.09 & 1 & -0.08 \\ -0.06 & -0.02 & -0.14 & 1 \end{bmatrix}^{-1}\begin{bmatrix} 2.272 \\ 2.106 \\ 2.736 \\ 3.432 \end{bmatrix} \text{ and } \begin{bmatrix} x_1 \\ x_2 \\ x_3 \\ x_4 \end{bmatrix} = \begin{bmatrix} 2.950 \\ 3.413 \\ 3.703 \\ 4.196 \end{bmatrix}$$

The taxable incomes are: Company A - \$2,950,000, Company B - \$3,413,000, Company C - \$3,703,000, Company D - \$4,196,000

65. Let x_1 = number of ounces of food A,
x_2 = number of ounces of food B,
and x_3 = number of ounces of food C.

(A) The mathematical model is:

$$30x_1 + 10x_2 + 20x_3 = 340$$
$$10x_1 + 10x_2 + 20x_3 = 180$$
$$10x_1 + 30x_2 + 20x_3 = 220$$

$$\left[\begin{array}{ccc|c} 30 & 10 & 20 & 340 \\ 10 & 10 & 20 & 180 \\ 10 & 30 & 20 & 220 \end{array}\right] \sim \left[\begin{array}{ccc|c} 10 & 10 & 20 & 180 \\ 30 & 10 & 20 & 340 \\ 10 & 30 & 20 & 220 \end{array}\right] \sim \left[\begin{array}{ccc|c} 1 & 1 & 2 & 18 \\ 3 & 1 & 2 & 34 \\ 1 & 3 & 2 & 22 \end{array}\right]$$

$R_1 \leftrightarrow R_2$ $\quad \frac{1}{10}R_1 \to R_1$, $\frac{1}{10}R_2 \to R_2$, $\frac{1}{10}R_3 \to R_3$ $\quad (-3)R_1 + R_2 \to R_2$, $(-1)R_1 + R_3 \to R_3$

$$\sim \left[\begin{array}{ccc|c} 1 & 1 & 2 & 18 \\ 0 & -2 & -4 & -20 \\ 0 & 2 & 0 & 4 \end{array}\right] \sim \left[\begin{array}{ccc|c} 1 & 1 & 2 & 18 \\ 0 & 1 & 2 & 10 \\ 0 & 2 & 0 & 4 \end{array}\right] \sim \left[\begin{array}{ccc|c} 1 & 0 & 0 & 8 \\ 0 & 1 & 2 & 10 \\ 0 & 0 & -4 & -16 \end{array}\right]$$

$-\frac{1}{2}R_2 \to R_2$ $\quad (-1)R_2 + R_1 \to R_1$, $(-2)R_2 + R_3 \to R_3$ $\quad -\frac{1}{4}R_3 \to R_3$

$$\sim \left[\begin{array}{ccc|c} 1 & 0 & 0 & 8 \\ 0 & 1 & 2 & 10 \\ 0 & 0 & 1 & 4 \end{array}\right] \sim \left[\begin{array}{ccc|c} 1 & 0 & 0 & 8 \\ 0 & 1 & 0 & 2 \\ 0 & 0 & 1 & 4 \end{array}\right]$$

$(-2)R_3 + R_2 \to R_2$

Thus, $x_1 = 8$, $x_2 = 2$, and $x_3 = 4$ or 8 ounces of food A, 2 ounces of food B, and 4 ounces of food C.

(B) The mathematical model is:

$$30x_1 + 10x_2 = 340$$
$$10x_1 + 10x_2 = 180$$
$$10x_1 + 30x_2 = 220$$

$$\begin{bmatrix} 30 & 10 & | & 340 \\ 10 & 10 & | & 180 \\ 10 & 30 & | & 220 \end{bmatrix} \sim \begin{bmatrix} 3 & 1 & | & 34 \\ 1 & 1 & | & 18 \\ 1 & 3 & | & 22 \end{bmatrix} \sim \begin{bmatrix} 1 & 1 & | & 18 \\ 3 & 1 & | & 34 \\ 1 & 3 & | & 22 \end{bmatrix} \sim \begin{bmatrix} 1 & 1 & | & 18 \\ 0 & -2 & | & -20 \\ 0 & 2 & | & 4 \end{bmatrix} \sim \begin{bmatrix} 1 & 1 & | & 18 \\ 0 & 1 & | & 10 \\ 0 & 2 & | & 4 \end{bmatrix}$$

$\frac{1}{10}R_1 \to R_1$, $\frac{1}{10}R_2 \to R_2$, $\frac{1}{10}R_3 \to R_3$ $\quad R_1 \leftrightarrow R_2 \quad (-3)R_1 + R_2 \to R_2$, $(-1)R_1 + R_3 \to R_3 \quad \left(-\frac{1}{2}\right)R_2 \to R_2 \quad (-2)R_2 + R_3 \to R_3$

$$\sim \begin{bmatrix} 1 & 1 & | & 18 \\ 0 & 1 & | & 10 \\ 0 & 0 & | & -16 \end{bmatrix}$$

From this matrix, we conclude that there is no solution.

(C) The mathematical model is:

$30x_1 + 10x_2 + 20x_3 = 340$
$10x_1 + 10x_2 + 20x_3 = 180$

$$\begin{bmatrix} 30 & 10 & 20 & | & 340 \\ 10 & 10 & 20 & | & 180 \end{bmatrix} \sim \begin{bmatrix} 10 & 10 & 20 & | & 180 \\ 30 & 10 & 20 & | & 340 \end{bmatrix} \sim \begin{bmatrix} 1 & 1 & 2 & | & 18 \\ 3 & 1 & 2 & | & 34 \end{bmatrix} \sim \begin{bmatrix} 1 & 1 & 2 & | & 18 \\ 0 & -2 & -4 & | & -20 \end{bmatrix}$$

$R_1 \leftrightarrow R_2 \quad \frac{1}{10}R_1 \to R_1$, $\frac{1}{10}R_2 \to R_2 \quad (-3)R_1 + R_2 \to R_2 \quad \left(-\frac{1}{2}\right)R_2 \to R_2$

$$\sim \begin{bmatrix} 1 & 1 & 2 & | & 18 \\ 0 & 1 & 2 & | & 10 \end{bmatrix} \sim \begin{bmatrix} 1 & 0 & 0 & | & 8 \\ 0 & 1 & 2 & | & 10 \end{bmatrix}$$

$(-1)R_2 + R_1 \to R_1$

Thus, $x_1 = 8$
$x_2 + 2x_3 = 10$

Let $x_3 = t$ (t any real number). Then, $x_2 = 10 - 2t$, $0 \le t \le 5$, for x_2 to be positive.

The solution is: $x_1 = 8$ ounces of food A; $x_2 = 10 - 2t$ ounces of food B; $x_3 = t$ ounces of food C, $0 \le t \le 5$.

67. Let x_1 = number of barrels of mix A,
x_2 = number of barrels of mix B,
x_3 = number of barrels of mix C,
and x_4 = number of barrels of mix D,

The mathematical model is:

$30x_1 + 30x_2 + 30x_3 + 60x_4 = 900 \quad (1)$
$50x_1 + 75x_2 + 25x_3 + 25x_4 = 750 \quad (2)$
$30x_1 + 20x_2 + 20x_3 + 50x_4 = 700 \quad (3)$

Divide each side of equation (1) by 30, each side of equation (2) by 25, and each side of equation (3) by 10. This yields the system of linear equations:

$x_1 + x_2 + x_3 + 2x_4 = 30$
$2x_1 + 3x_2 + x_3 + x_4 = 30$
$3x_1 + 2x_2 + 2x_3 + 5x_4 = 70$

$$\left[\begin{array}{cccc|c} 1 & 1 & 1 & 2 & 30 \\ 2 & 3 & 1 & 1 & 30 \\ 3 & 2 & 2 & 5 & 70 \end{array}\right] \sim \left[\begin{array}{cccc|c} 1 & 1 & 1 & 2 & 30 \\ 0 & 1 & -1 & -3 & -30 \\ 0 & -1 & -1 & -1 & -20 \end{array}\right] \sim \left[\begin{array}{cccc|c} 1 & 0 & 2 & 5 & 60 \\ 0 & 1 & -1 & -3 & -30 \\ 0 & 0 & -2 & -4 & -50 \end{array}\right]$$

$(-2)R_1 + R_2 \rightarrow R_2$ $\quad$ $R_3 + R_2 \rightarrow R_3$ $\quad$ $\left(-\frac{1}{2}\right)R_3 \rightarrow R_3$

$(-3)R_1 + R_3 \rightarrow R_3$ $\quad$ $(-1)R_2 + R_1 \rightarrow R_1$

$$\sim \left[\begin{array}{cccc|c} 1 & 0 & 2 & 5 & 60 \\ 0 & 1 & -1 & -3 & -30 \\ 0 & 0 & 1 & 2 & 25 \end{array}\right] \sim \left[\begin{array}{cccc|c} 1 & 0 & 0 & 1 & 10 \\ 0 & 1 & 0 & -1 & -5 \\ 0 & 0 & 1 & 2 & 25 \end{array}\right]$$

$R_3 + R_2 \rightarrow R_2$

$(-2)R_3 + R_1 \rightarrow R_1$

Thus,
$$\begin{aligned} x_1 \quad + x_4 &= 10 \\ x_2 \quad - x_4 &= -5 \\ x_3 + 2x_4 &= 25 \end{aligned}$$

Let $x_4 = t$ = number of barrels of mix D. Then $x_1 = 10 - t$ = number of barrels of mix A, $x_2 = t - 5$ = number of barrels of mix B, and $x_3 = 25 - 2t$ = number of barrels of mix C. Since the number of barrels of each mix must be nonnegative, $5 \leq t \leq 10$. Also, t is an integer.

69. Let x_1 = number of hours for Company A,
and x_2 = number of hours for Company B.
The mathematical model is:
$$\begin{aligned} 30x_1 + 20x_2 &= 600 \\ 10x_1 + 20x_2 &= 400 \end{aligned}$$

Divide each side of each equation by 10. This yields the system of linear equations:

$$\begin{aligned} 3x_1 + 2x_2 &= 60 \\ x_1 + 2x_2 &= 40 \end{aligned}$$

$$\left[\begin{array}{cc|c} 3 & 2 & 60 \\ 1 & 2 & 40 \end{array}\right] \sim \left[\begin{array}{cc|c} 1 & 2 & 40 \\ 3 & 2 & 60 \end{array}\right] \sim \left[\begin{array}{cc|c} 1 & 2 & 40 \\ 0 & -4 & -60 \end{array}\right] \sim \left[\begin{array}{cc|c} 1 & 2 & 40 \\ 0 & 1 & 15 \end{array}\right] \sim \left[\begin{array}{cc|c} 1 & 0 & 10 \\ 0 & 1 & 15 \end{array}\right]$$

$R_1 \leftrightarrow R_2$ $\quad$ $(-3)R_1 + R_2 \rightarrow R_2$ $\quad$ $\left(-\frac{1}{4}\right)R_2 \rightarrow R_2$ $\quad$ $(-2)R_2 + R_1 \rightarrow R_1$

Thus, $x_1 = 10$ and $x_2 = 15$, or 10 hours for Company A and 15 hours for Company B.

71. (A) 6th and Washington Ave.: $x_1 + x_2 = 1200$
6th and Lincoln Ave.: $x_2 + x_3 = 1000$
5th and Lincoln Ave.: $x_3 + x_4 = 1300$

(B) The system of equations is:

$$\begin{aligned} x_1 \qquad\qquad + x_4 &= 1500 \\ x_1 + x_2 \qquad\qquad &= 1200 \\ x_2 + x_3 \qquad &= 1000 \\ x_3 + x_4 &= 1300 \end{aligned}$$

$$\left[\begin{array}{cccc|c} 1 & 0 & 0 & 1 & 1500 \\ 1 & 1 & 0 & 0 & 1200 \\ 0 & 1 & 1 & 0 & 1000 \\ 0 & 0 & 1 & 1 & 1300 \end{array}\right] \sim \left[\begin{array}{cccc|c} 1 & 0 & 0 & 1 & 1500 \\ 0 & 1 & 0 & -1 & -300 \\ 0 & 1 & 1 & 0 & 1000 \\ 0 & 0 & 1 & 1 & 1300 \end{array}\right] \sim \left[\begin{array}{cccc|c} 1 & 0 & 0 & 1 & 1500 \\ 0 & 1 & 0 & -1 & -300 \\ 0 & 0 & 1 & 1 & 1300 \\ 0 & 0 & 1 & 1 & 1300 \end{array}\right]$$

$(-1)R_1 + R_2 \rightarrow R_2$ $\qquad$ $(-1)R_2 + R_3 \rightarrow R_3$ $\qquad$ $(-1)R_3 + R_4 \rightarrow R_4$

$$\sim \left[\begin{array}{cccc|c} 1 & 0 & 0 & 1 & 1500 \\ 0 & 1 & 0 & -1 & -300 \\ 0 & 0 & 1 & 1 & 1300 \\ 0 & 0 & 0 & 0 & 0 \end{array}\right]$$

Thus

$$\begin{aligned} x_1 \qquad + x_4 &= 1500 \\ x_2 \qquad - x_4 &= -300 \\ x_3 + x_4 &= 1300 \end{aligned}$$

Let $x_4 = t$. Then $x_1 = 1500 - t$, $x_2 = t - 300$ and $x_3 = 1300 - t$. Since x_1, x_2, x_3, and x_4 must be nonnegative integers, we have $300 \le t \le 1300$.

(C) The flow from Washington Ave. to Lincoln Ave. on 5th Street is given by $x_4 = t$. As shown in part (B), $300 \le t \le 1300$, that is, the maximum number of vehicles is 1300 and the minimum number is 300.

(D) If $x_4 = t = 1000$, then
Washington Ave.: $x_1 = 1500 - 1000 = 500$
6th St.: $x_2 = 1000 - 300 = 700$
Lincoln Ave.: $x_3 = 1300 - 1000 = 300$

EXERCISE 4-4

Things to remember:

1. A matrix with m rows and n columns is said to have SIZE or DIMENSION $m \times n$. If a matrix has the same number of rows and columns, then it is called a SQUARE MATRIX. A matrix with only one column is a COLUMN MATRIX, and a matrix with only one row is a ROW MATRIX.

2. Two matrices are EQUAL if they have the same dimension and their corresponding elements are equal.

3. The SUM of two matrices of the same dimension, $m \times n$, is an $m \times n$ matrix whose elements are the sum of the corresponding elements of the two given matrices. Addition is not defined for matrices with different dimensions. Matrix addition is commutative: $A + B = B + A$, and assocative: $(A + B) + C = A + (B + C)$.

4. A matrix with all elements equal to zero is called a ZERO MATRIX.

5. The NEGATIVE OF A MATRIX M, denoted by $-M$, is the matrix whose elements are the negatives of the elements of M.

6. If A and B are matrices of the same dimension, then subtraction is defined by $A - B = A + (-B)$. Thus, to subtract B from A, simply subtract corresponding elements.

7. If M is a matrix and k is a number, then kM is the matrix formed by multiplying each element of M by k.

8. PRODUCT OF A ROW MATRIX AND A COLUMN MATRIX

The product of a $1 \times n$ row matrix and an $n \times 1$ column matrix is the 1×1 matrix given by

$$\underset{1 \times n}{\begin{bmatrix} a_1 & a_2 & \cdots & a_n \end{bmatrix}} \underset{n \times 1}{\begin{bmatrix} b_1 \\ b_2 \\ \vdots \\ b_n \end{bmatrix}} = [a_1b_1 + a_2b_2 + \cdots + a_nb_n]$$

Note that the number of elements in the row matrix and the number of elements in the column matrix must be the same for the product to be defined.

9. Let A be an $m \times p$ matrix and B be a $p \times n$ matrix. The MATRIX PRODUCT of A and B, denoted AB, is the $m \times n$ matrix whose element in the ith row and the jth column is the real number obtained from the product of the ith row of A and the jth column of B. If the number of columns in A does not equal the number of rows in B, then the matrix product AB is not defined.

NOTE: Matrix multiplication is *not* commutative. That is AB does not always equal BA, even when both multiplications are defined.

---

1. $\begin{bmatrix} 2 & -1 \\ 3 & 0 \end{bmatrix} + \begin{bmatrix} -3 & 1 \\ 2 & -3 \end{bmatrix} = \begin{bmatrix} 2 + (-3) & -1 + 1 \\ 3 + 2 & 0 + (-3) \end{bmatrix} = \begin{bmatrix} -1 & 0 \\ 5 & -3 \end{bmatrix}$

3. Addition not defined; the matrices have different dimensions.

5. $\begin{bmatrix} 4 & -5 \\ 1 & 0 \\ 1 & -3 \end{bmatrix} - \begin{bmatrix} -1 & 2 \\ 6 & -2 \\ 1 & -7 \end{bmatrix} = \begin{bmatrix} 4 - (-1) & -5 - 2 \\ 1 - 6 & 0 - (-2) \\ 1 - 1 & -3 - (-7) \end{bmatrix} = \begin{bmatrix} 5 & -7 \\ -5 & 2 \\ 0 & 4 \end{bmatrix}$

7. $5\begin{bmatrix} 1 & -2 & 0 & 4 \\ -3 & 2 & -1 & 6 \end{bmatrix} = \begin{bmatrix} 5(1) & 5(-2) & 5(0) & 5(4) \\ 5(-3) & 5(2) & 5(-1) & 5(6) \end{bmatrix} = \begin{bmatrix} 5 & -10 & 0 & 20 \\ -15 & 10 & -5 & 30 \end{bmatrix}$

9. $\begin{bmatrix} 3 & 4 \\ -1 & -2 \end{bmatrix}\begin{bmatrix} -1 \\ 2 \end{bmatrix} = \begin{bmatrix} [3 \quad 4]\begin{bmatrix} -1 \\ 2 \end{bmatrix} \\ [-1 \quad -2]\begin{bmatrix} -1 \\ 2 \end{bmatrix} \end{bmatrix} = \begin{bmatrix} -3 + 8 \\ 1 - 4 \end{bmatrix} = \begin{bmatrix} 5 \\ -3 \end{bmatrix}$

11. $\begin{bmatrix} 2 & -3 \\ 1 & 2 \end{bmatrix}\begin{bmatrix} 1 & -1 \\ 0 & -2 \end{bmatrix} = \begin{bmatrix} [2 \quad -3]\begin{bmatrix} 1 \\ 0 \end{bmatrix} & [2 \quad -3]\begin{bmatrix} -1 \\ -2 \end{bmatrix} \\ [1 \quad 2]\begin{bmatrix} 1 \\ 0 \end{bmatrix} & [1 \quad 2]\begin{bmatrix} -1 \\ -2 \end{bmatrix} \end{bmatrix} = \begin{bmatrix} 2 + 0 & -2 + 6 \\ 1 + 0 & -1 - 4 \end{bmatrix} = \begin{bmatrix} 2 & 4 \\ 1 & -5 \end{bmatrix}$

13. $\begin{bmatrix} 1 & -1 \\ 0 & -2 \end{bmatrix}\begin{bmatrix} 2 & -3 \\ 1 & 2 \end{bmatrix} = \begin{bmatrix} [1 \quad -1]\begin{bmatrix} 2 \\ 1 \end{bmatrix} & [1 \quad -1]\begin{bmatrix} -3 \\ 2 \end{bmatrix} \\ [0 \quad -2]\begin{bmatrix} 2 \\ 1 \end{bmatrix} & [0 \quad -2]\begin{bmatrix} -3 \\ 2 \end{bmatrix} \end{bmatrix} = \begin{bmatrix} 2 - 1 & -3 - 2 \\ 0 - 2 & 0 - 4 \end{bmatrix} = \begin{bmatrix} 1 & -5 \\ -2 & -4 \end{bmatrix}$

15. $[5 \quad -2]\begin{bmatrix} -3 \\ -4 \end{bmatrix} = [-15 + 8] = [-7]$

17. $\begin{bmatrix} -3 \\ -4 \end{bmatrix}[5 \quad -2] = \begin{bmatrix} (-3)(5) & (-3)(-2) \\ (-4)(5) & (-4)(-2) \end{bmatrix} = \begin{bmatrix} -15 & 6 \\ -20 & 8 \end{bmatrix}$

19. $[3 \quad -2 \quad -4]\begin{bmatrix} 1 \\ 2 \\ -3 \end{bmatrix} = [(3 - 4 + 12)] = [11]$

21. $\begin{bmatrix} 1 \\ 2 \\ -3 \end{bmatrix}[3 \quad -2 \quad -4] = \begin{bmatrix} (1)(3) & (1)(-2) & (1)(-4) \\ (2)(3) & (2)(-2) & (2)(-4) \\ (-3)(3) & (-3)(-2) & (-3)(-4) \end{bmatrix} = \begin{bmatrix} 3 & -2 & -4 \\ 6 & -4 & -8 \\ -9 & 6 & 12 \end{bmatrix}$

23. $AC = \begin{bmatrix} 2 & -1 & 3 \\ 0 & 4 & -2 \end{bmatrix}\begin{bmatrix} -1 & 0 & 2 \\ 4 & -3 & 1 \\ -2 & 3 & 5 \end{bmatrix} = \begin{bmatrix} -12 & 12 & 18 \\ 20 & -18 & -6 \end{bmatrix}$

25. AB is not defined; the number of columns of A (3) does not equal the number of rows of B (2).

27. $B^2 = BB = \begin{bmatrix} -3 & 1 \\ 2 & 5 \end{bmatrix}\begin{bmatrix} -3 & 1 \\ 2 & 5 \end{bmatrix} = \begin{bmatrix} 11 & 2 \\ 4 & 27 \end{bmatrix}$

29. $B + AD = \begin{bmatrix} -3 & 1 \\ 2 & 5 \end{bmatrix} + \begin{bmatrix} 2 & -1 & 3 \\ 0 & 4 & -2 \end{bmatrix}\begin{bmatrix} 3 & -2 \\ 0 & -1 \\ 1 & 2 \end{bmatrix}$

$= \begin{bmatrix} -3 & 1 \\ 2 & 5 \end{bmatrix} + \begin{bmatrix} 9 & 3 \\ -2 & -8 \end{bmatrix} = \begin{bmatrix} 6 & 4 \\ 0 & -3 \end{bmatrix}$

31. $0.1DB = 0.1\begin{bmatrix} 3 & -2 \\ 0 & -1 \\ 1 & 2 \end{bmatrix}\begin{bmatrix} -3 & 1 \\ 2 & 5 \end{bmatrix}$

$= 0.1\begin{bmatrix} -13 & -7 \\ -2 & -5 \\ 1 & 11 \end{bmatrix} = \begin{bmatrix} -1.3 & -0.7 \\ -0.2 & -0.5 \\ 0.1 & 1.1 \end{bmatrix}$

33. $3BA + 4AC = 3\begin{bmatrix} -3 & 1 \\ 2 & 5 \end{bmatrix}\begin{bmatrix} 2 & -1 & 3 \\ 0 & 4 & -2 \end{bmatrix} + 4\begin{bmatrix} 2 & -1 & 3 \\ 0 & 4 & -2 \end{bmatrix}\begin{bmatrix} -1 & 0 & 2 \\ 4 & -3 & 1 \\ -2 & 3 & 5 \end{bmatrix}$

$= 3\begin{bmatrix} -6 & 7 & -11 \\ 4 & 18 & -4 \end{bmatrix} + 4\begin{bmatrix} -12 & 12 & 18 \\ 20 & -18 & -6 \end{bmatrix}$

$= \begin{bmatrix} -18 & 21 & -33 \\ 12 & 54 & -12 \end{bmatrix} + \begin{bmatrix} -48 & 48 & 72 \\ 80 & -72 & -24 \end{bmatrix}$

$= \begin{bmatrix} -66 & 69 & 39 \\ 92 & -18 & -36 \end{bmatrix}$

35. $(-2)BA + 6CD$ is not defined; $-2BA$ is 2×3, $6CD$ is 3×2

37. $ACD = A(CD) = \begin{bmatrix} 2 & -1 & 3 \\ 0 & 4 & -2 \end{bmatrix}\left(\begin{bmatrix} -1 & 0 & 2 \\ 4 & -3 & 1 \\ -2 & 3 & 5 \end{bmatrix}\begin{bmatrix} 3 & -2 \\ 0 & -1 \\ 1 & 2 \end{bmatrix}\right)$

$= \begin{bmatrix} 2 & -1 & 3 \\ 0 & 4 & -2 \end{bmatrix}\begin{bmatrix} -1 & 6 \\ 13 & -3 \\ -1 & 11 \end{bmatrix}$

$= \begin{bmatrix} -18 & 48 \\ 54 & -34 \end{bmatrix}$

39. $DBA = D(BA) = \begin{bmatrix} 3 & -2 \\ 0 & -1 \\ 1 & 2 \end{bmatrix}\left(\begin{bmatrix} -3 & 1 \\ 2 & 5 \end{bmatrix}\begin{bmatrix} 2 & -1 & 3 \\ 0 & 4 & -2 \end{bmatrix}\right)$

$= \begin{bmatrix} 3 & -2 \\ 0 & -1 \\ 1 & 2 \end{bmatrix}\begin{bmatrix} -6 & 7 & -11 \\ 4 & 18 & -4 \end{bmatrix}$

$= \begin{bmatrix} -26 & -15 & -25 \\ -4 & -18 & 4 \\ 2 & 43 & -19 \end{bmatrix}$

41. $A = [0.3 \quad 0.7]$, $B = \begin{bmatrix} 0.4 & 0.6 \\ 0.2 & 0.8 \end{bmatrix}$

$B^2 = \begin{bmatrix} 0.28 & 0.72 \\ 0.24 & 0.76 \end{bmatrix}$, $B^3 = \begin{bmatrix} 0.256 & 0.744 \\ 0.248 & 0.752 \end{bmatrix}\ldots,$

$B^8 = \begin{bmatrix} 0.250 & 0.74999 \\ 0.24999 & 0.75000 \end{bmatrix}$, $B^n \to \begin{bmatrix} 0.25 & 0.75 \\ 0.25 & 0.75 \end{bmatrix}$

$AB = [0.26 \quad 0.74]$, $AB^2 = [0.252 \quad 0.748]$,

$AB^3 = [0.2504 \quad 0.7496]\ldots,$

$AB^{10} = [0.2500\ldots \quad 0.7499\ldots]$, $AB^n \to [0.25 \quad 0.75]$

43. $\begin{bmatrix} a & b \\ c & d \end{bmatrix} + \begin{bmatrix} 2 & -3 \\ 0 & 1 \end{bmatrix} = \begin{bmatrix} a+2 & b-3 \\ c+0 & d+1 \end{bmatrix} = \begin{bmatrix} 1 & -2 \\ 3 & -4 \end{bmatrix}$

Thus, $a + 2 = 1$, $a = -1$
$b - 3 = -2$, $b = 1$
$c + 0 = 3$, $c = 3$
$d + 1 = -4$, $d = -5$

45. $\begin{bmatrix} 2x & 4 \\ -3 & 5x \end{bmatrix} + \begin{bmatrix} 3y & -2 \\ -2 & -y \end{bmatrix} = \begin{bmatrix} -5 & 2 \\ -5 & 13 \end{bmatrix}$

$\begin{bmatrix} 2x + 3y & 4 - 2 \\ -3 - 2 & 5x - y \end{bmatrix} = \begin{bmatrix} -5 & 2 \\ -5 & 13 \end{bmatrix}$

$\begin{bmatrix} 2x + 3y & 2 \\ -5 & 5x - y \end{bmatrix} = \begin{bmatrix} -5 & 2 \\ -5 & 13 \end{bmatrix}$

Thus, $2x + 3y = -5$ (1)
$5x - y = 13$ (2)
Solve the above system for x, y.

From (2), $y = 5x - 13$. Substitute $y = 5x - 13$ in (1):

$2x + 3(5x - 13) = -5$
$2x + 15x - 39 = -5$
$17x = -5 + 39$
$17x = 34$
$x = 2$

Substitute $x = 2$ in (1):

$2(2) + 3y = -5$
$4 + 3y = -5$
$3y = -9$
$y = -3$

Thus, the solution is $x = 2$ and $y = -3$.

47. $\begin{bmatrix} x & -1 \\ 1 & 0 \end{bmatrix}\begin{bmatrix} 2 & 1 \\ 4 & 1 \end{bmatrix} = \begin{bmatrix} y & y \\ 2 & 1 \end{bmatrix}$

$\begin{bmatrix} 2x - 4 & x - 1 \\ 2 & 1 \end{bmatrix} = \begin{bmatrix} y & y \\ 2 & 1 \end{bmatrix}$

implies $2x - 4 = y$
$x - 1 = y$

Thus, $2x - 4 = x - 1$
and $x = 3$, $y = 2$

49. $\begin{bmatrix} 1 & -2 \\ 2 & -3 \end{bmatrix}\begin{bmatrix} a & b \\ c & d \end{bmatrix} = \begin{bmatrix} 1 & 0 \\ 3 & 2 \end{bmatrix}$

$\begin{bmatrix} a - 2c & b - 2d \\ 2a - 3c & 2b - 3d \end{bmatrix} = \begin{bmatrix} 1 & 0 \\ 3 & 2 \end{bmatrix}$

implies $a - 2c = 1$ $\quad b - 2d = 0$
$2a - 3c = 3$ $\quad 2b - 3d = 2$

The augmented matrix for the first system is:

$\left[\begin{array}{cc|c} 1 & -2 & 1 \\ 2 & -3 & 3 \end{array}\right] (-2)R_1 + R_2 \rightarrow R_2 \sim \left[\begin{array}{cc|c} 1 & -2 & 1 \\ 0 & 1 & 1 \end{array}\right] 2R_2 + R_1 \rightarrow R_1$

$\sim \left[\begin{array}{cc|c} 1 & 0 & 3 \\ 0 & 1 & 1 \end{array}\right]$ Thus, $a = 3$, $c = 1$.

For the second system, substitute $b = 2d$ from the first equation into the second equation:

$2(2d) - 3d = 2$
$d = 2$
$b = 4$

Solution: $a = 3$, $b = 4$, $c = 1$, $d = 2$

51. Let $A = \begin{bmatrix} a_1 & 0 \\ 0 & a_2 \end{bmatrix}$ and $B = \begin{bmatrix} b_1 & 0 \\ 0 & b_2 \end{bmatrix}$

(A) Always true:

$$A + B = \begin{bmatrix} a_1 & 0 \\ 0 & a_2 \end{bmatrix} + \begin{bmatrix} b_1 & 0 \\ 0 & b_2 \end{bmatrix} = \begin{bmatrix} a_1 + b_1 & 0 \\ 0 & a_2 + b_2 \end{bmatrix}$$

(B) Always true: matrix addition is commutative, $A + B = B + A$ for *any* pair of matrices of the same size.

(C) Always true:

$$AB = \begin{bmatrix} a_1 & 0 \\ 0 & a_2 \end{bmatrix}\begin{bmatrix} b_1 & 0 \\ 0 & b_2 \end{bmatrix} = \begin{bmatrix} a_1b_1 & 0 \\ 0 & a_2b_2 \end{bmatrix}$$

(D) Always true:

$$BA = \begin{bmatrix} b_1 & 0 \\ 0 & b_2 \end{bmatrix}\begin{bmatrix} a_1 & 0 \\ 0 & a_2 \end{bmatrix} = \begin{bmatrix} b_1a_1 & 0 \\ 0 & b_2a_2 \end{bmatrix} = \begin{bmatrix} a_1b_1 & 0 \\ 0 & a_2b_2 \end{bmatrix} = AB$$

53. $A + B = \begin{bmatrix} \$30 & \$25 \\ \$60 & \$80 \end{bmatrix} + \begin{bmatrix} \$36 & \$27 \\ \$54 & \$74 \end{bmatrix} = \begin{bmatrix} \$66 & \$52 \\ \$114 & \$154 \end{bmatrix}$

$$\frac{1}{2}(A + B) = \frac{1}{2}\begin{bmatrix} \$66 & \$52 \\ \$114 & \$154 \end{bmatrix} = \begin{array}{c} \text{Guitar} \quad \text{Banjo} \\ \begin{bmatrix} \$33 & \$26 \\ \$57 & \$77 \end{bmatrix} \end{array} \begin{array}{l} \text{Materials} \\ \text{Labor} \end{array}$$

55. The dealer is increasing the retail prices by 10%. Thus, the new retail price matrix M' (to the nearest dollar) is given by

$$M' = M + 0.1M = 1.1M = 1.1\begin{bmatrix} 10900 & 683 & 253 & 195 \\ 13000 & 738 & 382 & 206 \\ 16300 & 867 & 537 & 225 \end{bmatrix} = \begin{bmatrix} 11990 & 751 & 278 & 214 \\ 14300 & 812 & 420 & 227 \\ 17930 & 954 & 591 & 248 \end{bmatrix}$$

The new dealer invoice matrix N' is given by

$$N' = N + 0.15N = 1.15N = 1.15\begin{bmatrix} 9400 & 582 & 195 & 160 \\ 11500 & 621 & 295 & 171 \\ 14100 & 737 & 420 & 184 \end{bmatrix}$$

$$= \begin{bmatrix} 10810 & 669 & 224 & 184 \\ 13225 & 714 & 339 & 197 \\ 16215 & 848 & 483 & 212 \end{bmatrix}$$

The new markup is:

$$M' - N' = \begin{bmatrix} 11990 & 751 & 278 & 214 \\ 14300 & 812 & 420 & 227 \\ 17930 & 954 & 591 & 248 \end{bmatrix} - \begin{bmatrix} 10810 & 669 & 224 & 184 \\ 13225 & 714 & 339 & 197 \\ 16215 & 848 & 483 & 212 \end{bmatrix}$$

$$= \begin{array}{l} \text{Model A} \\ \text{Model B} \\ \text{Model C} \end{array} \begin{array}{c} \text{Basic Car} \quad \text{Air/Cond.} \quad \text{AM/FM} \quad \text{Cruise} \\ \begin{bmatrix} \$1180 & \$82 & \$54 & \$30 \\ \$1075 & \$98 & \$81 & \$30 \\ \$1715 & \$106 & \$108 & \$36 \end{bmatrix} \end{array}$$

57. (A) $[0.6 \quad 0.6 \quad 0.2]\begin{bmatrix} 8 \\ 10 \\ 5 \end{bmatrix} = [4.8 + 6.0 + 1.0] = [11.8]$

Thus, the labor cost per boat for one-person boats at plant I is \$11.80.

(B) $[1.5 \quad 1.2 \quad 0.4]\begin{bmatrix} 9 \\ 12 \\ 6 \end{bmatrix} = [13.5 + 14.4 + 2.4] = [30.3]$

Thus, the labor cost per boat for four-person boats at plant II is \$30.30.

(C) MN gives the labor costs per boat at each plant; NM is not defined.

(D) $MN = \begin{bmatrix} 0.6 & 0.6 & 0.2 \\ 1.0 & 0.9 & 0.3 \\ 1.5 & 1.2 & 0.4 \end{bmatrix} \begin{bmatrix} 8 & 9 \\ 10 & 12 \\ 5 & 6 \end{bmatrix}$

$$\begin{array}{c} \\ = \end{array} \begin{matrix} \text{Plant I} & \text{Plant II} \end{matrix}$$
$$= \begin{bmatrix} \$11.80 & \$13.80 \\ \$18.50 & \$21.60 \\ \$26.00 & \$30.30 \end{bmatrix} \begin{matrix} \text{One-person boat} \\ \text{Two-person boat} \\ \text{Four-person boat} \end{matrix}$$

This matrix represents the labor cost per boat for each kind of boat at each plant. For example, \$21.60 represents labor costs per boat for two-person boats at plant II.

59. $A = \begin{bmatrix} 0 & 1 & 0 & 1 & 0 \\ 0 & 0 & 1 & 0 & 0 \\ 1 & 0 & 0 & 0 & 1 \\ 0 & 0 & 1 & 0 & 0 \\ 0 & 0 & 0 & 1 & 0 \end{bmatrix}$

(A) $A^2 = \begin{bmatrix} 0 & 1 & 0 & 1 & 0 \\ 0 & 0 & 1 & 0 & 0 \\ 1 & 0 & 0 & 0 & 1 \\ 0 & 0 & 1 & 0 & 0 \\ 0 & 0 & 0 & 1 & 0 \end{bmatrix} \begin{bmatrix} 0 & 1 & 0 & 1 & 0 \\ 0 & 0 & 1 & 0 & 0 \\ 1 & 0 & 0 & 0 & 1 \\ 0 & 0 & 1 & 0 & 0 \\ 0 & 0 & 0 & 1 & 0 \end{bmatrix} = \begin{bmatrix} 0 & 0 & 2 & 0 & 0 \\ 1 & 0 & 0 & 0 & 1 \\ 0 & 1 & 0 & 2 & 0 \\ 1 & 0 & 0 & 0 & 1 \\ 0 & 0 & 1 & 0 & 0 \end{bmatrix}$

The 1 in row two, column one indicates that there is one way to travel from Baltimore to Atlanta with one intermediate connection, namely Baltimore-to-Chicago-to-Atlanta.

The 2 in row one, column three indicates that there are two ways to travel from Atlanta to Chicago with one intermediate connection, namely Atlanta-to-Baltimore-to-Chicago, and Atlanta-to-Denver-to-Chicago.

In general, the element b_{ij}, $i \neq j$, in A^2 indicates the number of different ways to travel from the ith city to the jth city with one intermediate connection.

(B) $A^3 = AA^2 = \begin{bmatrix} 0 & 1 & 0 & 1 & 0 \\ 0 & 0 & 1 & 0 & 0 \\ 1 & 0 & 0 & 0 & 1 \\ 0 & 0 & 1 & 0 & 0 \\ 0 & 0 & 0 & 1 & 0 \end{bmatrix} \begin{bmatrix} 0 & 0 & 2 & 0 & 0 \\ 1 & 0 & 0 & 0 & 1 \\ 0 & 1 & 0 & 2 & 0 \\ 1 & 0 & 0 & 0 & 1 \\ 0 & 0 & 1 & 0 & 0 \end{bmatrix} = \begin{bmatrix} 2 & 0 & 0 & 0 & 2 \\ 0 & 1 & 0 & 2 & 0 \\ 0 & 0 & 3 & 0 & 0 \\ 0 & 1 & 0 & 2 & 0 \\ 1 & 0 & 0 & 0 & 1 \end{bmatrix}$

The 1 in row four, column 2 indicates that there is one way to travel from Denver to Baltimore with two intermediate connections.
The 2 in row one, column five indicates that there are two ways to travel from Atlanta to El Paso with two intermediate connections.
In general, the element c_{ij}, $i \neq j$, in A^3 indicates the number of ways to travel from the ith city to the jth city with two intermediate connections.

(C) From parts (A) and (B)

$$A + A^2 = \begin{bmatrix} 0 & 1 & 0 & 1 & 0 \\ 0 & 0 & 1 & 0 & 0 \\ 1 & 0 & 0 & 0 & 1 \\ 0 & 0 & 1 & 0 & 0 \\ 0 & 0 & 0 & 1 & 0 \end{bmatrix} + \begin{bmatrix} 0 & 0 & 2 & 0 & 0 \\ 1 & 0 & 0 & 0 & 1 \\ 0 & 1 & 0 & 2 & 0 \\ 1 & 0 & 0 & 0 & 1 \\ 0 & 0 & 1 & 0 & 0 \end{bmatrix} = \begin{bmatrix} 0 & 1 & 2 & 1 & 0 \\ 1 & 0 & 1 & 0 & 1 \\ 1 & 1 & 0 & 2 & 1 \\ 1 & 0 & 1 & 0 & 1 \\ 0 & 0 & 1 & 1 & 0 \end{bmatrix}$$

$$A + A^2 + A^3 = \begin{bmatrix} 0 & 1 & 2 & 1 & 0 \\ 1 & 0 & 1 & 0 & 1 \\ 1 & 1 & 0 & 2 & 1 \\ 1 & 0 & 1 & 0 & 1 \\ 0 & 0 & 1 & 1 & 0 \end{bmatrix} + \begin{bmatrix} 2 & 0 & 0 & 0 & 2 \\ 0 & 1 & 0 & 2 & 0 \\ 0 & 0 & 3 & 0 & 0 \\ 0 & 1 & 0 & 2 & 0 \\ 1 & 0 & 0 & 0 & 1 \end{bmatrix} = \begin{bmatrix} 2 & 1 & 2 & 1 & 2 \\ 1 & 1 & 1 & 2 & 1 \\ 1 & 1 & 3 & 2 & 1 \\ 1 & 1 & 1 & 2 & 1 \\ 1 & 0 & 1 & 1 & 1 \end{bmatrix}$$

$$A^4 = AA^3 = \begin{bmatrix} 0 & 1 & 0 & 1 & 0 \\ 0 & 0 & 1 & 0 & 0 \\ 1 & 0 & 0 & 0 & 1 \\ 0 & 0 & 1 & 0 & 0 \\ 0 & 0 & 0 & 1 & 0 \end{bmatrix} \begin{bmatrix} 2 & 0 & 0 & 0 & 2 \\ 0 & 1 & 0 & 2 & 0 \\ 0 & 0 & 3 & 0 & 0 \\ 0 & 1 & 0 & 2 & 0 \\ 1 & 0 & 0 & 0 & 1 \end{bmatrix} = \begin{bmatrix} 0 & 2 & 0 & 4 & 0 \\ 0 & 0 & 3 & 0 & 0 \\ 3 & 0 & 0 & 0 & 3 \\ 0 & 0 & 3 & 0 & 0 \\ 0 & 1 & 0 & 2 & 0 \end{bmatrix}$$

$$A + A^2 + A^3 + A^4 = \begin{bmatrix} 2 & 1 & 2 & 1 & 2 \\ 1 & 1 & 1 & 2 & 1 \\ 1 & 1 & 3 & 2 & 1 \\ 1 & 1 & 1 & 2 & 1 \\ 1 & 0 & 1 & 1 & 1 \end{bmatrix} + \begin{bmatrix} 0 & 2 & 0 & 4 & 0 \\ 0 & 0 & 3 & 0 & 0 \\ 3 & 0 & 0 & 0 & 3 \\ 0 & 0 & 3 & 0 & 0 \\ 0 & 1 & 0 & 2 & 0 \end{bmatrix}$$

$$= \begin{bmatrix} 2 & 3 & 2 & 5 & 2 \\ 1 & 1 & 4 & 2 & 1 \\ 4 & 1 & 3 & 2 & 4 \\ 1 & 1 & 4 & 2 & 1 \\ 1 & 1 & 1 & 3 & 1 \end{bmatrix}$$

It is possible to travel from any origin to any destination with at most 3 intermediate connections.

61. (A) $[4 \quad 2]\begin{bmatrix} 15 \\ 5 \end{bmatrix} = 70$

There are 70 g of protein in Mix X.

(B) $[3 \quad 1]\begin{bmatrix} 5 \\ 15 \end{bmatrix} = 30$

There are 30 g of fat in Mix Z.

(C) *MN* gives the amount (in grams) of protein, carbohydrates and fat in 20 ounces of each mix. The product *NM* is not defined.

(D) $MN = \begin{bmatrix} 4 & 2 \\ 20 & 16 \\ 3 & 1 \end{bmatrix} \begin{bmatrix} 15 & 10 & 5 \\ 5 & 10 & 15 \end{bmatrix} = \begin{bmatrix} 70 & 60 & 50 \\ 380 & 360 & 340 \\ 50 & 40 & 30 \end{bmatrix}$

	Mix X	Mix Y	Mix Z
Protein	70	60	50
Carbohydrate	380	360	340
Fat	50	40	30

63. (A) $[1000 \quad 500 \quad 5000]\begin{bmatrix} 0.40 \\ 1.00 \\ 0.35 \end{bmatrix} = 2{,}650$; Total amount spent in Berkeley $= \$2{,}650$.

(B) $[2000 \quad 800 \quad 8000]\begin{bmatrix} 0.40 \\ 1.00 \\ 0.35 \end{bmatrix} = 4{,}400$; Total amount spent in Oakland $= \$4{,}400$.

(C) NM gives the total cost per town.

(D) $NM = \begin{bmatrix} 1000 & 500 & 5000 \\ 2000 & 800 & 8000 \end{bmatrix}\begin{bmatrix} 0.40 \\ 1.00 \\ 0.35 \end{bmatrix} = \begin{bmatrix} 2{,}650 \\ 4{,}400 \end{bmatrix}$ Cost/Town: Berkeley, Oakland

(E) $[1 \quad 1]N = [1 \quad 1]\begin{bmatrix} 1000 & 500 & 5000 \\ 2000 & 800 & 8000 \end{bmatrix}$

$= [3{,}000 \quad 1300 \quad 13{,}000]$ (Telephone Calls, House Calls, Letters)

(F) $N\begin{bmatrix} 1 \\ 1 \\ 1 \end{bmatrix} = \begin{bmatrix} 1{,}000 & 500 & 5{,}000 \\ 2{,}000 & 800 & 8{,}000 \end{bmatrix}\begin{bmatrix} 1 \\ 1 \\ 1 \end{bmatrix} = \begin{bmatrix} 6{,}500 \\ 10{,}800 \end{bmatrix}$ Total contacts: Berkeley, Oakland

65. (A)

	a	b	c	d	e	f
a	0	0	1	1	1	0
b	1	0	0	1	1	0
c	0	1	0	1	0	0
d	0	0	0	0	0	1
e	0	0	1	1	0	1
f	1	1	1	0	0	0

(B) $B = A + A^2 =$

$$\begin{bmatrix} 0 & 0 & 1 & 1 & 1 & 0 \\ 1 & 0 & 0 & 1 & 1 & 0 \\ 0 & 1 & 0 & 1 & 0 & 0 \\ 0 & 0 & 0 & 0 & 0 & 1 \\ 0 & 0 & 1 & 1 & 0 & 1 \\ 1 & 1 & 1 & 0 & 0 & 0 \end{bmatrix} + \begin{bmatrix} 0 & 1 & 1 & 2 & 0 & 2 \\ 0 & 0 & 2 & 2 & 1 & 2 \\ 1 & 0 & 0 & 1 & 1 & 1 \\ 1 & 1 & 1 & 0 & 0 & 0 \\ 1 & 2 & 1 & 1 & 0 & 1 \\ 1 & 1 & 1 & 3 & 2 & 0 \end{bmatrix} = \begin{bmatrix} 0 & 1 & 2 & 3 & 1 & 2 \\ 1 & 0 & 2 & 3 & 2 & 2 \\ 1 & 1 & 0 & 2 & 1 & 1 \\ 1 & 1 & 1 & 0 & 0 & 1 \\ 1 & 2 & 2 & 2 & 0 & 2 \\ 2 & 2 & 2 & 3 & 2 & 0 \end{bmatrix} = B$$

(C) Let $C = \begin{bmatrix} 1 \\ 1 \\ 1 \\ 1 \\ 1 \\ 1 \end{bmatrix}$ and calculate BC: $BC = \begin{bmatrix} 9 \\ 10 \\ 6 \\ 4 \\ 9 \\ 11 \end{bmatrix}$

(D) The ranking from strongest to weakest is:
Frank, Bart, Aaron and Elvis (tie), Charles, Dan.

The entries in A are the first stage dominances. The entries in A^2 are the second stage dominances. By summing the rows of $A + A^2$, we calculate the sum of the first and second stage dominances.

Things to remember:

1. The IDENTITY element for multiplication for the set of square matrices of order n (dimension $n \times n$) is the square matrix I of order n which has 1's on the principal diagonal (upper left corner to lower right corner) and 0's elsewhere. The identity matrices of order 2 and 3, respectively, are

$$I = \begin{bmatrix} 1 & 0 \\ 0 & 1 \end{bmatrix} \text{ and } I = \begin{bmatrix} 1 & 0 & 0 \\ 0 & 1 & 0 \\ 0 & 0 & 1 \end{bmatrix}.$$

2. If M is any square matrix of order n and I is the identity matrix of order n, then

$$IM = MI = M.$$

3. INVERSE OF A SQUARE MATRIX

Let M be a square matrix of order n and I be the identity matrix of order n. If there exists a matrix M^{-1} such that

$$MM^{-1} = M^{-1}M = I$$

then M^{-1} is called the MULTIPLICATIVE INVERSE OF M or, more simply, the INVERSE OF M. M^{-1} is read "M inverse."

4. If the augmented matrix $[M \mid I]$ is transformed by row operations into $[I \mid B]$, then the resulting matrix B is M^{-1}. However, if all zeros are obtained in one or more rows to the left of the vertical line during the row transformation procedure, then M^{-1} does not exist.

---

1. $\begin{bmatrix} 1 & 0 \\ 0 & 1 \end{bmatrix}\begin{bmatrix} 2 & -3 \\ 4 & 5 \end{bmatrix} = \begin{bmatrix} 1\cdot2 + 0\cdot4 & 1(-3) + 0\cdot5 \\ 0\cdot2 + 1\cdot4 & 0(-3) + 1\cdot5 \end{bmatrix} = \begin{bmatrix} 2 & -3 \\ 4 & 5 \end{bmatrix}$

3. $\begin{bmatrix} 2 & -3 \\ 4 & 5 \end{bmatrix}\begin{bmatrix} 1 & 0 \\ 0 & 1 \end{bmatrix} = \begin{bmatrix} 2\cdot1 + (-3)0 & 2\cdot0 + (-3)1 \\ 4\cdot1 + 5\cdot0 & 4\cdot0 + 5\cdot1 \end{bmatrix} = \begin{bmatrix} 2 & -3 \\ 4 & 5 \end{bmatrix}$

5. $\begin{bmatrix} 1 & 0 & 0 \\ 0 & 1 & 0 \\ 0 & 0 & 1 \end{bmatrix}\begin{bmatrix} -2 & 1 & 3 \\ 2 & 4 & -2 \\ 5 & 1 & 0 \end{bmatrix}$

$$= \begin{bmatrix} 1(-2) + 0\cdot2 + 0\cdot5 & 1\cdot1 + 0\cdot4 + 0\cdot1 & 1\cdot3 + 0(-2) + 0\cdot0 \\ 0(-2) + 1\cdot2 + 0\cdot5 & 0\cdot1 + 1\cdot4 + 0\cdot1 & 0\cdot3 + 1(-2) + 0\cdot0 \\ 0(-2) + 0\cdot2 + 1\cdot5 & 0\cdot1 + 0\cdot4 + 1\cdot1 & 0\cdot3 + 0(-2) + 1\cdot0 \end{bmatrix} = \begin{bmatrix} -2 & 1 & 3 \\ 2 & 4 & -2 \\ 5 & 1 & 0 \end{bmatrix}$$

7. $\begin{bmatrix} -2 & 1 & 3 \\ 2 & 4 & -2 \\ 5 & 1 & 0 \end{bmatrix}\begin{bmatrix} 1 & 0 & 0 \\ 0 & 1 & 0 \\ 0 & 0 & 1 \end{bmatrix}$

$$= \begin{bmatrix} (-2)\cdot 1 + 1\cdot 0 + 3\cdot 0 & (-2)0 + 1\cdot 1 + 3\cdot 0 & (-2)0 + 1\cdot 0 + 3\cdot 1 \\ 2\cdot 1 + 4\cdot 0 + (-2)0 & 2\cdot 0 + 4\cdot 1 + (-2)0 & 2\cdot 0 + 4\cdot 0 + (-2)1 \\ 5\cdot 1 + 1\cdot 0 + 0\cdot 0 & 5\cdot 0 + 1\cdot 1 + 0\cdot 0 & 5\cdot 0 + 1\cdot 0 + 0\cdot 1 \end{bmatrix}$$

$$= \begin{bmatrix} -2 & 1 & 3 \\ 2 & 4 & -2 \\ 5 & 1 & 0 \end{bmatrix}$$

9. $\begin{bmatrix} 3 & -4 \\ -2 & 3 \end{bmatrix}\begin{bmatrix} 3 & 4 \\ 2 & 3 \end{bmatrix} = \begin{bmatrix} 1 & 0 \\ 0 & 1 \end{bmatrix}$ Yes

11. $\begin{bmatrix} 2 & 2 \\ -1 & -1 \end{bmatrix}\begin{bmatrix} 1 & 1 \\ -1 & -1 \end{bmatrix} = \begin{bmatrix} 0 & 0 \\ 0 & 0 \end{bmatrix}$ No

13. $\begin{bmatrix} -5 & 2 \\ -8 & 3 \end{bmatrix}\begin{bmatrix} 3 & -2 \\ 8 & -5 \end{bmatrix} = \begin{bmatrix} 1 & 0 \\ 0 & 1 \end{bmatrix}$ Yes

15. No. The second matrix has a column of zeros; it does not have an inverse.

17. $\begin{bmatrix} 1 & -1 & 1 \\ 0 & 2 & -1 \\ 2 & 3 & 0 \end{bmatrix}\begin{bmatrix} 3 & 3 & -1 \\ -2 & -2 & 1 \\ -4 & -5 & 2 \end{bmatrix} = \begin{bmatrix} 1 & 0 & 0 \\ 0 & 1 & 0 \\ 0 & 0 & 1 \end{bmatrix}$ Yes

19. $\left[\begin{array}{cc|cc} -1 & 0 & 1 & 0 \\ -3 & 1 & 0 & 1 \end{array}\right] \sim \left[\begin{array}{cc|cc} 1 & 0 & -1 & 0 \\ -3 & 1 & 0 & 1 \end{array}\right] \sim \left[\begin{array}{cc|cc} 1 & 0 & -1 & 0 \\ 0 & 1 & -3 & 1 \end{array}\right]$

$(-1)R_1 \to R_1$ $\qquad 3R_1 + R_2 \to R_2$

Thus, $M^{-1} = \begin{bmatrix} -1 & 0 \\ -3 & 1 \end{bmatrix}$

<u>Check</u>:

$$M\cdot M^{-1} = \begin{bmatrix} -1 & 0 \\ -3 & 1 \end{bmatrix}\begin{bmatrix} -1 & 0 \\ -3 & 1 \end{bmatrix} = \begin{bmatrix} (-1)(-1) + 0(-3) & (-1)0 + 0\cdot 1 \\ (-3)(-1) + 1(-3) & (-3)0 + 1\cdot 1 \end{bmatrix} = \begin{bmatrix} 1 & 0 \\ 0 & 1 \end{bmatrix}$$

21. $\left[\begin{array}{cc|cc} 1 & 2 & 1 & 0 \\ 1 & 3 & 0 & 1 \end{array}\right] \sim \left[\begin{array}{cc|cc} 1 & 2 & 1 & 0 \\ 0 & 1 & -1 & 1 \end{array}\right] \sim \left[\begin{array}{cc|cc} 1 & 0 & 3 & -2 \\ 0 & 1 & -1 & 1 \end{array}\right]$

$(-1)R_1 + R_2 \to R_2$ $\qquad (-2)R_2 + R_1 \to R_1$

Thus, $M^{-1} = \begin{bmatrix} 3 & -2 \\ -1 & 1 \end{bmatrix}$.

<u>Check</u>:

$$M\cdot M^{-1} = \begin{bmatrix} 1 & 2 \\ 1 & 3 \end{bmatrix}\begin{bmatrix} 3 & -2 \\ -1 & 1 \end{bmatrix} = \begin{bmatrix} 1\cdot 3 + 2(-1) & 1(-2) + 2\cdot 1 \\ 1\cdot 3 + 3(-1) & 1(-2) + 3\cdot 1 \end{bmatrix} = \begin{bmatrix} 1 & 0 \\ 0 & 1 \end{bmatrix}$$

23. $\left[\begin{array}{cc|cc} 1 & 3 & 1 & 0 \\ 2 & 7 & 0 & 1 \end{array}\right] \sim \left[\begin{array}{cc|cc} 1 & 3 & 1 & 0 \\ 0 & 1 & -2 & 1 \end{array}\right] \sim \left[\begin{array}{cc|cc} 1 & 0 & 7 & -3 \\ 0 & 1 & -2 & 1 \end{array}\right]$

$(-2)R_1 + R_2 \rightarrow R_2$ $\quad$ $(-3)R_2 + R_1 \rightarrow R_1$

Thus, $M^{-1} = \begin{bmatrix} 7 & -3 \\ -2 & 1 \end{bmatrix}$.

Check:

$$\begin{bmatrix} 1 & 3 \\ 2 & 7 \end{bmatrix}\begin{bmatrix} 7 & -3 \\ -2 & 1 \end{bmatrix} = \begin{bmatrix} 1\cdot7 + 3(-2) & 1(-3) + 3\cdot1 \\ 2\cdot7 + 7(-2) & 2(-3) + 7\cdot1 \end{bmatrix} = \begin{bmatrix} 1 & 0 \\ 0 & 1 \end{bmatrix}$$

25. $\left[\begin{array}{ccc|ccc} 1 & -3 & 0 & 1 & 0 & 0 \\ 0 & 1 & 1 & 0 & 1 & 0 \\ 2 & -1 & 4 & 0 & 0 & 1 \end{array}\right]$ $(-2)R_1 + R_3 \rightarrow R_3$

$\sim \left[\begin{array}{ccc|ccc} 1 & -3 & 0 & 1 & 0 & 0 \\ 0 & 1 & 1 & 0 & 1 & 0 \\ 0 & 5 & 4 & -2 & 0 & 1 \end{array}\right]$ $\begin{array}{r} 3R_2 + R_1 \rightarrow R_1 \\ (-5)R_2 + R_3 \rightarrow R_3 \end{array}$

$\sim \left[\begin{array}{ccc|ccc} 1 & 0 & 3 & 1 & 3 & 0 \\ 0 & 1 & 1 & 0 & 1 & 0 \\ 0 & 0 & -1 & -2 & -5 & 1 \end{array}\right]$ $(-1)R_3 \rightarrow R_3$

$\sim \left[\begin{array}{ccc|ccc} 1 & 0 & 3 & 1 & 3 & 0 \\ 0 & 1 & 1 & 0 & 1 & 0 \\ 0 & 0 & 1 & 2 & 5 & -1 \end{array}\right]$ $\begin{array}{r} (-3)R_3 + R_1 \rightarrow R_1 \\ (-1)R_3 + R_2 \rightarrow R_2 \end{array}$

$\sim \left[\begin{array}{ccc|ccc} 1 & 0 & 0 & -5 & -12 & 3 \\ 0 & 1 & 0 & -2 & -4 & 1 \\ 0 & 0 & 1 & 2 & 5 & -1 \end{array}\right]$ $\quad M^{-1} = \begin{bmatrix} -5 & -12 & 3 \\ -2 & -4 & 1 \\ 2 & 5 & -1 \end{bmatrix}$

$$M \cdot M^{-1} = \begin{bmatrix} 1 & -3 & 0 \\ 0 & 1 & 1 \\ 2 & -1 & 4 \end{bmatrix}\begin{bmatrix} -5 & -12 & 3 \\ -2 & -4 & 1 \\ 2 & 5 & -1 \end{bmatrix} = \begin{bmatrix} 1 & 0 & 0 \\ 0 & 1 & 0 \\ 0 & 0 & 1 \end{bmatrix}$$

27. $\left[\begin{array}{ccc|ccc} 1 & 1 & 0 & 1 & 0 & 0 \\ 2 & 3 & -1 & 0 & 1 & 0 \\ 1 & 0 & 2 & 0 & 0 & 1 \end{array}\right]$ $\begin{array}{r} (-2)R_1 + R_2 \rightarrow R_2 \\ (-1)R_1 + R_3 \rightarrow R_3 \end{array}$

$\sim \left[\begin{array}{ccc|ccc} 1 & 1 & 0 & 1 & 0 & 0 \\ 0 & 1 & -1 & -2 & 1 & 0 \\ 0 & -1 & 2 & -1 & 0 & 1 \end{array}\right]$ $\begin{array}{r} (-1)R_2 + R_1 \rightarrow R_1 \\ R_2 + R_3 \rightarrow R_3 \end{array}$

$\sim \left[\begin{array}{ccc|ccc} 1 & 0 & 1 & 3 & -1 & 0 \\ 0 & 1 & -1 & -2 & 1 & 0 \\ 0 & 0 & 1 & -3 & 1 & 1 \end{array}\right]$ $\begin{array}{r} (-1)R_3 + R_1 \rightarrow R_1 \\ R_3 + R_2 \rightarrow R_2 \end{array}$

$$\sim \left[\begin{array}{ccc|ccc} 1 & 0 & 0 & 6 & -2 & -1 \\ 0 & 1 & 0 & -5 & 2 & 1 \\ 0 & 0 & 1 & -3 & 1 & 1 \end{array}\right] \quad M^{-1} = \begin{bmatrix} 6 & -2 & -1 \\ -5 & 2 & 1 \\ -3 & 1 & 1 \end{bmatrix}$$

$$M \cdot M^{-1} = \begin{bmatrix} 1 & 1 & 0 \\ 2 & 3 & -1 \\ 1 & 0 & 2 \end{bmatrix} \begin{bmatrix} 6 & -2 & -1 \\ -5 & 2 & 1 \\ -3 & 1 & 1 \end{bmatrix} = \begin{bmatrix} 1 & 0 & 0 \\ 0 & 1 & 0 \\ 0 & 0 & 1 \end{bmatrix}$$

29. $\left[\begin{array}{cc|cc} 4 & 3 & 1 & 0 \\ -3 & -2 & 0 & 1 \end{array}\right] R_2 + R_1 \to R_1$

$\sim \left[\begin{array}{cc|cc} 1 & 1 & 1 & 1 \\ -3 & -2 & 0 & 1 \end{array}\right] 3R_1 + R_2 \to R_2$

$\sim \left[\begin{array}{cc|cc} 1 & 1 & 1 & 1 \\ 0 & 1 & 3 & 4 \end{array}\right] (-1)R_2 + R_1 \to R_1$

$\sim \left[\begin{array}{cc|cc} 1 & 0 & -2 & -3 \\ 0 & 1 & 3 & 4 \end{array}\right] (-1)R_2 + R_1 \to R_1$

$M^{-1} = \begin{bmatrix} -2 & -3 \\ 3 & 4 \end{bmatrix}$

31. $\left[\begin{array}{cc|cc} 2 & 6 & 1 & 0 \\ 3 & 9 & 0 & 1 \end{array}\right] \frac{1}{2}R_1 \to R_1$

$\sim \left[\begin{array}{cc|cc} 1 & 3 & \frac{1}{2} & 0 \\ 3 & 9 & 0 & 1 \end{array}\right] (-3)R_1 + R_2 \to R_2$

$\sim \left[\begin{array}{cc|cc} 1 & 3 & \frac{1}{2} & 0 \\ 0 & 0 & -\frac{3}{2} & 1 \end{array}\right]$ The inverse does not exist.

35. $\left[\begin{array}{cc|cc} 2 & 1 & 1 & 0 \\ 4 & 3 & 0 & 1 \end{array}\right] \frac{1}{2}R_1 \to R_1$

$\sim \left[\begin{array}{cc|cc} 1 & \frac{1}{2} & \frac{1}{2} & 0 \\ 4 & 3 & 0 & 1 \end{array}\right] (-4)R_1 + R_2 \to R_2$

$\sim \left[\begin{array}{cc|cc} 1 & \frac{1}{2} & \frac{1}{2} & 0 \\ 0 & 1 & -2 & 1 \end{array}\right] \left(-\frac{1}{2}\right)R_2 + R_1 \to R_1$

$\sim \left[\begin{array}{cc|cc} 1 & 0 & \frac{3}{2} & -\frac{1}{2} \\ 0 & 1 & -2 & 1 \end{array}\right]$

$M^{-1} = \begin{bmatrix} \frac{3}{2} & -\frac{1}{2} \\ -2 & 1 \end{bmatrix} = \begin{bmatrix} 1.5 & -0.5 \\ -2 & 1 \end{bmatrix}$

37. $\left[\begin{array}{ccc|ccc} -5 & -2 & -2 & 1 & 0 & 0 \\ 2 & 1 & 0 & 0 & 1 & 0 \\ 1 & 0 & 1 & 0 & 0 & 1 \end{array}\right] \sim \left[\begin{array}{ccc|ccc} 1 & 0 & 1 & 0 & 0 & 1 \\ 2 & 1 & 0 & 0 & 1 & 0 \\ -5 & -2 & -2 & 1 & 0 & 0 \end{array}\right]$

$R_1 \leftrightarrow R_3$ $\qquad$ $(-2)R_1 + R_2 \rightarrow R_2$, $5R_1 + R_3 \rightarrow R_3$

$\sim \left[\begin{array}{ccc|ccc} 1 & 0 & 1 & 0 & 0 & 1 \\ 0 & 1 & -2 & 0 & 1 & -2 \\ 0 & -2 & 3 & 1 & 0 & 5 \end{array}\right] \sim \left[\begin{array}{ccc|ccc} 1 & 0 & 1 & 0 & 0 & 1 \\ 0 & 1 & -2 & 0 & 1 & -2 \\ 0 & 0 & -1 & 1 & 2 & 1 \end{array}\right]$

$2R_2 + R_3 \rightarrow R_3$ $\qquad$ $(-1)R_3 \rightarrow R_3$

$\sim \left[\begin{array}{ccc|ccc} 1 & 0 & 1 & 0 & 0 & 1 \\ 0 & 1 & -2 & 0 & 1 & -2 \\ 0 & 0 & 1 & -1 & -2 & -1 \end{array}\right] \sim \left[\begin{array}{ccc|ccc} 1 & 0 & 0 & 1 & 2 & 2 \\ 0 & 1 & 0 & -2 & -3 & -4 \\ 0 & 0 & 1 & -1 & -2 & -1 \end{array}\right]$

$2R_3 + R_2 \rightarrow R_2$

$(-1)R_3 + R_1 \rightarrow R_1$

Thus, the inverse is $\begin{bmatrix} 1 & 2 & 2 \\ -2 & -3 & -4 \\ -1 & -2 & -1 \end{bmatrix}$.

39. $\left[\begin{array}{ccc|ccc} 2 & 1 & 1 & 1 & 0 & 0 \\ 1 & 1 & 0 & 0 & 1 & 0 \\ -1 & -1 & 0 & 0 & 0 & 1 \end{array}\right] \sim \left[\begin{array}{ccc|ccc} 1 & 1 & 0 & 0 & 1 & 0 \\ 2 & 1 & 1 & 1 & 0 & 0 \\ -1 & -1 & 0 & 0 & 0 & 1 \end{array}\right] \sim \left[\begin{array}{ccc|ccc} 1 & 1 & 0 & 0 & 1 & 0 \\ 0 & -1 & 1 & 1 & -2 & 0 \\ 0 & 0 & 0 & 0 & 1 & 1 \end{array}\right]$

$R_1 \leftrightarrow R_2$ $\qquad$ $(-2)R_1 + R_2 \rightarrow R_2$, $R_1 + R_3 \rightarrow R_3$

From this matrix, we conclude that the inverse does not exist.

41. $\left[\begin{array}{ccc|ccc} -1 & -2 & 2 & 1 & 0 & 0 \\ 4 & 3 & 0 & 0 & 1 & 0 \\ 4 & 0 & 4 & 0 & 0 & 1 \end{array}\right]$ $(-1)R_1 \rightarrow R_1$

$\sim \left[\begin{array}{ccc|ccc} 1 & 2 & -2 & -1 & 0 & 0 \\ 4 & 3 & 0 & 0 & 1 & 0 \\ 4 & 0 & 4 & 0 & 0 & 1 \end{array}\right]$ $(-4)R_1 + R_2 \rightarrow R_2$, $(-4)R_1 + R_3 \rightarrow R_3$

$\sim \left[\begin{array}{ccc|ccc} 1 & 2 & -2 & -1 & 0 & 0 \\ 0 & -5 & 8 & 4 & 1 & 0 \\ 0 & -8 & 12 & 4 & 0 & 1 \end{array}\right]$ $\left(-\frac{1}{5}\right)R_2 \rightarrow R_2$

$\sim \left[\begin{array}{ccc|ccc} 1 & 2 & -2 & -1 & 0 & 0 \\ 0 & 1 & -\frac{8}{5} & -\frac{4}{5} & -\frac{1}{5} & 0 \\ 0 & -8 & 12 & 4 & 0 & 1 \end{array}\right]$ $(-2)R_2 + R_1 \rightarrow R_1$, $8R_2 + R_3 \rightarrow R_3$

$$\sim \left[\begin{array}{ccc|ccc} 1 & 0 & \frac{6}{5} & \frac{3}{5} & \frac{2}{5} & 0 \\ 0 & 1 & -\frac{8}{5} & -\frac{4}{5} & -\frac{1}{5} & 0 \\ 0 & 0 & -\frac{4}{5} & -\frac{12}{5} & -\frac{8}{5} & 1 \end{array}\right] \left(-\frac{5}{4}\right)R_3 \to R_3$$

$$\sim \left[\begin{array}{ccc|ccc} 1 & 0 & \frac{6}{5} & \frac{3}{5} & \frac{2}{5} & 0 \\ 0 & 1 & -\frac{8}{5} & -\frac{4}{5} & -\frac{1}{5} & 0 \\ 0 & 0 & 1 & 3 & 2 & -\frac{5}{4} \end{array}\right] \begin{array}{l} \left(-\frac{6}{5}\right)R_3 + R_1 \to R_1 \\ \left(\frac{8}{5}\right)R_3 + R_2 \to R_2 \end{array}$$

$$\sim \left[\begin{array}{ccc|ccc} 1 & 0 & 0 & -3 & -2 & \frac{3}{2} \\ 0 & 1 & 0 & 4 & 3 & -2 \\ 0 & 0 & 1 & 3 & 2 & -\frac{5}{4} \end{array}\right]$$

$$M^{-1} = \begin{bmatrix} -3 & -2 & \frac{3}{2} \\ 4 & 3 & -2 \\ 3 & 2 & -\frac{5}{4} \end{bmatrix} = \begin{bmatrix} -3 & -2 & 1.5 \\ 4 & 3 & -2 \\ 3 & 2 & -1.25 \end{bmatrix}$$

43. $$\left[\begin{array}{ccc|ccc} 2 & -1 & -2 & 1 & 0 & 0 \\ -4 & 2 & 8 & 0 & 1 & 0 \\ 6 & -2 & -1 & 0 & 0 & 1 \end{array}\right] \frac{1}{2}R_1 \to R_1$$

$$\sim \left[\begin{array}{ccc|ccc} 1 & -\frac{1}{2} & -1 & \frac{1}{2} & 0 & 0 \\ -4 & 2 & 8 & 0 & 1 & 0 \\ 6 & -2 & -1 & 0 & 0 & 1 \end{array}\right] \begin{array}{l} 4R_1 + R_2 \to R_2 \\ (-6)R_1 + R_3 \to R_3 \end{array}$$

$$\sim \left[\begin{array}{ccc|ccc} 1 & -\frac{1}{2} & -1 & \frac{1}{2} & 0 & 0 \\ 0 & 0 & 4 & 2 & 1 & 0 \\ 0 & 1 & 5 & -3 & 0 & 1 \end{array}\right] R_2 \leftrightarrow R_3$$

$$\sim \left[\begin{array}{ccc|ccc} 1 & -\frac{1}{2} & -1 & \frac{1}{2} & 0 & 0 \\ 0 & 1 & 5 & -3 & 0 & 1 \\ 0 & 0 & 4 & 2 & 1 & 0 \end{array}\right] \begin{array}{l} \frac{1}{2}R_2 + R_1 \to R_1 \\ \frac{1}{4}R_3 \to R_3 \end{array}$$

$$\sim \left[\begin{array}{ccc|ccc} 1 & 0 & \frac{3}{2} & -1 & 0 & \frac{1}{2} \\ 0 & 1 & 5 & -3 & 0 & 1 \\ 0 & 0 & 1 & \frac{1}{2} & \frac{1}{4} & 0 \end{array}\right] \begin{array}{l} \left(-\frac{3}{2}\right)R_3 + R_1 \to R_1 \\ (-5)R_3 + R_2 \to R_2 \end{array}$$

$$\sim \left[\begin{array}{ccc|ccc} 1 & 0 & 0 & -\frac{7}{4} & -\frac{3}{8} & \frac{1}{2} \\ 0 & 1 & 0 & -\frac{11}{2} & -\frac{5}{4} & 1 \\ 0 & 0 & 1 & \frac{1}{2} & \frac{1}{4} & 0 \end{array}\right]$$

$$M^{-1} = \begin{bmatrix} -\frac{7}{4} & -\frac{3}{8} & \frac{1}{2} \\ -\frac{11}{2} & -\frac{5}{4} & 1 \\ \frac{1}{2} & \frac{1}{4} & 0 \end{bmatrix} = \begin{bmatrix} -1.75 & -0.375 & 0.5 \\ -5.5 & -1.25 & 1 \\ 0.5 & 0.25 & 0 \end{bmatrix}$$

45. $A = \begin{bmatrix} 4 & 3 \\ 3 & 2 \end{bmatrix}$; $\left[\begin{array}{cc|cc} 4 & 3 & 1 & 0 \\ 3 & 2 & 0 & 1 \end{array}\right]$ $(-1)R_2 + R_1 \rightarrow R_1$

$\sim \left[\begin{array}{cc|cc} 1 & 1 & 1 & -1 \\ 3 & 2 & 0 & 1 \end{array}\right]$ $(-3)R_1 + R_2 \rightarrow R_2$ $\sim \left[\begin{array}{cc|cc} 1 & 1 & 1 & -1 \\ 0 & -1 & -3 & 4 \end{array}\right]$ $(-1)R_2 \rightarrow R_2$

$\sim \left[\begin{array}{cc|cc} 1 & 1 & 1 & -1 \\ 0 & 1 & 3 & -4 \end{array}\right]$ $(-1)R_2 + R_1 \rightarrow R_1$ $\sim \left[\begin{array}{cc|cc} 1 & 0 & -2 & 3 \\ 0 & 1 & 3 & -4 \end{array}\right]$

$A^{-1} = \begin{bmatrix} -2 & 3 \\ 3 & -4 \end{bmatrix}$; $\left[\begin{array}{cc|cc} -2 & 3 & 1 & 0 \\ 3 & -4 & 0 & 1 \end{array}\right]$ $R_2 + R_1 \rightarrow R_1$

$\sim \left[\begin{array}{cc|cc} 1 & -1 & 1 & 1 \\ 3 & -4 & 0 & 1 \end{array}\right]$ $(-3)R_1 + R_2 \rightarrow R_2$ $\sim \left[\begin{array}{cc|cc} 1 & -1 & 1 & 1 \\ 0 & -1 & -3 & -2 \end{array}\right]$ $(-1)R_2 \rightarrow R_2$

$\sim \left[\begin{array}{cc|cc} 1 & -1 & 1 & 1 \\ 0 & 1 & 3 & 2 \end{array}\right]$ $R_2 + R_1 \rightarrow R_1$ $\sim \left[\begin{array}{cc|cc} 1 & 0 & 4 & 3 \\ 0 & 1 & 3 & 2 \end{array}\right]$ Thus, $(A^{-1})^{-1} = \begin{bmatrix} 4 & 3 \\ 3 & 2 \end{bmatrix} = A$

45. $\left[\begin{array}{cc|cc} a & 0 & 1 & 0 \\ 0 & d & 0 & 1 \end{array}\right]$ $\frac{1}{a}R_1 \rightarrow R_1$, provided $a \neq 0$
$\frac{1}{d}R_2 \rightarrow R_2$, provided $d \neq 0$

$\left[\begin{array}{cc|cc} 1 & 0 & \frac{1}{a} & 0 \\ 0 & 1 & 0 & \frac{1}{d} \end{array}\right]$

M^{-1} exists and equals $\begin{bmatrix} \frac{1}{a} & 0 \\ 0 & \frac{1}{d} \end{bmatrix}$ if and only if $a \neq 0$, $d \neq 0$. In general, the inverse of a diagonal matrix exists if and only if each of the diagonal elements is non-zero.

47. $A = \begin{bmatrix} 3 & 2 \\ -4 & -3 \end{bmatrix}$

A^{-1}: $\left[\begin{array}{cc|cc} 3 & 2 & 1 & 0 \\ -4 & -3 & 0 & 1 \end{array}\right] \sim \left[\begin{array}{cc|cc} -1 & -1 & 1 & 1 \\ -4 & -3 & 0 & 1 \end{array}\right] \sim \left[\begin{array}{cc|cc} 1 & 1 & -1 & -1 \\ -4 & -3 & 0 & 1 \end{array}\right]$

$R_1 + R_2 \rightarrow R_1$ $\quad (-1)R_1 \rightarrow R_1$ $\quad 4R_1 + R_2 \rightarrow R_2$

$\sim \left[\begin{array}{cc|cc} 1 & 1 & -1 & -1 \\ 0 & 1 & -4 & -3 \end{array}\right] \sim \left[\begin{array}{cc|cc} 1 & 0 & 3 & 2 \\ 0 & 1 & -4 & -3 \end{array}\right]$

$(-1)R_2 + R_1 \rightarrow R_1$

$A^{-1} = \begin{bmatrix} 3 & 2 \\ -4 & -3 \end{bmatrix} = A$; $A^2 = AA = AA^{-1} = \begin{bmatrix} 1 & 0 \\ 0 & 1 \end{bmatrix} = I$

49. $A = \begin{bmatrix} 4 & 3 \\ -5 & -4 \end{bmatrix}$

A^{-1}: $\left[\begin{array}{cc|cc} 4 & 3 & 1 & 0 \\ -5 & -4 & 0 & 1 \end{array}\right] \sim \left[\begin{array}{cc|cc} -1 & -1 & 1 & 0 \\ -5 & -4 & 0 & 1 \end{array}\right]$

$R_2 + R_1 \rightarrow R_1$ $\quad (-1)R_1 \rightarrow R_1$

$$\sim \left[\begin{array}{rr|rr} 1 & 1 & -1 & -1 \\ -5 & -4 & 0 & 1 \end{array}\right] \sim \left[\begin{array}{rr|rr} 1 & 1 & -1 & -1 \\ 0 & 1 & -5 & -4 \end{array}\right]$$

$5R_1 + R_2 \to R_2$ $\qquad (-1)R_2 + R_1 \to R_1$

$$\sim \left[\begin{array}{rr|rr} 1 & 0 & 4 & 3 \\ 0 & 1 & -5 & -4 \end{array}\right]$$

$$A^{-1} = \begin{bmatrix} 4 & 3 \\ -5 & -4 \end{bmatrix} = A;\ A^2 = AA = AA^{-1} = I$$

51. $A = \begin{bmatrix} 6 & 2 & 0 & 4 \\ 5 & 3 & 2 & 1 \\ 0 & -1 & 1 & -2 \\ 2 & -3 & 1 & 0 \end{bmatrix}$; $A^{-1} = \begin{bmatrix} 0.5 & -0.3 & 0.85 & -0.25 \\ 0 & 0.1 & 0.05 & -0.25 \\ -1 & 0.9 & -1.55 & 0.75 \\ -0.5 & 0.4 & -1.3 & 0.5 \end{bmatrix}$

53. $A = \begin{bmatrix} 3 & 2 & 3 & 4 & 4 \\ 5 & 4 & 3 & 2 & 1 \\ -1 & -1 & 2 & -2 & 3 \\ 3 & -3 & 1 & 0 & 1 \\ 1 & 1 & 2 & 0 & 2 \end{bmatrix}$;

$$A^{-1} = \begin{bmatrix} 1.75 & 5.25 & 8.75 & -1 & -18.75 \\ 1.25 & 3.75 & 6.25 & -1 & -13.25 \\ -4.75 & -13.25 & -22.75 & 3 & 48.75 \\ -1.375 & -4.625 & -7.875 & 1 & 16.375 \\ 3.25 & 8.75 & 15.25 & -2 & -32.25 \end{bmatrix}$$

55. $A = \begin{bmatrix} 1 & 2 \\ 1 & 3 \end{bmatrix}$

Assign the numbers 1—26 to the letters of the alphabet, in order, and let 27 correspond to a blank space. Then the message "THE SUN ALSO RISES" corresponds to the sequence

20 8 5 27 19 21 14 27 1 12 19 15 27 18 9 19 5 19

To encode this message, divide the numbers into groups of two and use the groups as columns of a matrix B with two rows

$$B = \begin{bmatrix} 20 & 5 & 19 & 14 & 1 & 19 & 27 & 9 & 5 \\ 8 & 27 & 21 & 27 & 12 & 15 & 18 & 19 & 19 \end{bmatrix}$$

Now

$$AB = \begin{bmatrix} 1 & 2 \\ 1 & 3 \end{bmatrix}\begin{bmatrix} 20 & 5 & 19 & 14 & 1 & 19 & 27 & 9 & 5 \\ 8 & 27 & 21 & 27 & 12 & 15 & 18 & 19 & 19 \end{bmatrix}$$

$$= \begin{bmatrix} 36 & 59 & 61 & 68 & 25 & 49 & 63 & 47 & 43 \\ 44 & 86 & 82 & 95 & 37 & 64 & 81 & 66 & 62 \end{bmatrix}$$

The coded message is:

36 44 59 86 61 82 68 95 25 37 49 64 63 81 47 66 43 62.

57. First, we must find the inverse of $A = \begin{bmatrix} 1 & 2 \\ 1 & 3 \end{bmatrix}$

$$\left[\begin{array}{cc|cc} 1 & 2 & 1 & 0 \\ 1 & 3 & 0 & 1 \end{array}\right] \sim \left[\begin{array}{cc|cc} 1 & 2 & 1 & 0 \\ 0 & 1 & -1 & 1 \end{array}\right] \sim \left[\begin{array}{cc|cc} 1 & 0 & 3 & -2 \\ 0 & 1 & -1 & 1 \end{array}\right]$$

$(-1)R_1 + R_2 \rightarrow R_2$ $\quad$ $(-2)R_2 + R_1 \rightarrow R_1$

Thus, $A^{-1} = \begin{bmatrix} 3 & -2 \\ -1 & 1 \end{bmatrix}$

Now $\begin{bmatrix} 3 & -2 \\ -1 & 1 \end{bmatrix}\begin{bmatrix} 37 & 24 & 73 & 49 & 62 & 36 & 59 & 41 & 22 \\ 52 & 29 & 96 & 69 & 89 & 44 & 86 & 50 & 26 \end{bmatrix}$

$= \begin{bmatrix} 7 & 14 & 27 & 9 & 8 & 20 & 5 & 23 & 14 \\ 15 & 5 & 23 & 20 & 27 & 8 & 27 & 9 & 4 \end{bmatrix}$

Thus, the decoded message is

7 15 14 5 27 23 9 20 8 27 20 8 5 27 23 9 14 4

which corresponds to

GONE WITH THE WIND

59. "THE BEST YEARS OF OUR LIVES" corresponds to the sequence

20 8 5 27 2 5 19 20 27 25 5 1 18 19 27 15 6 27 15
21 18 27 12 9 22 5 19

We divide the numbers in the sequence into groups of 5 and use these groups as the columns of a matrix with 5 rows, adding 3 blanks at the end to make the columns come out even. Then we multiply this matrix on the left by the given matrix B.

$$\begin{bmatrix} 1 & 0 & 1 & 0 & 1 \\ 0 & 1 & 1 & 0 & 3 \\ 2 & 1 & 1 & 1 & 1 \\ 0 & 0 & 1 & 0 & 2 \\ 1 & 1 & 1 & 2 & 1 \end{bmatrix}\begin{bmatrix} 20 & 5 & 5 & 15 & 18 & 5 \\ 8 & 19 & 1 & 6 & 27 & 19 \\ 5 & 20 & 18 & 27 & 12 & 27 \\ 27 & 27 & 19 & 15 & 9 & 27 \\ 2 & 25 & 27 & 21 & 22 & 27 \end{bmatrix}$$

$$= \begin{bmatrix} 27 & 50 & 50 & 63 & 52 & 59 \\ 19 & 114 & 100 & 96 & 105 & 127 \\ 82 & 101 & 75 & 99 & 106 & 110 \\ 9 & 70 & 72 & 69 & 56 & 81 \\ 89 & 123 & 89 & 99 & 97 & 132 \end{bmatrix}$$

The encoded message is:

27 19 82 9 89 50 114 101 70 123 50 100 75 72 89 63 96
99 69 99 52 105 106 56 97 59 127 110 81 132.

61. First, we must find the inverse of B:

$$B^{-1} = \begin{bmatrix} -2 & -1 & 2 & 2 & -1 \\ 3 & 2 & -2 & -4 & 1 \\ 6 & 2 & -4 & -5 & 2 \\ -2 & -1 & 1 & 2 & -7 \\ 3 & -1 & 2 & 3 & -1 \end{bmatrix}$$

Now $B^{-1}\begin{bmatrix} 32 & 24 & 54 & 56 & 48 & 62 \\ 34 & 21 & 71 & 92 & 66 & 135 \\ 87 & 67 & 112 & 109 & 98 & 124 \\ 19 & 11 & 43 & 55 & 41 & 81 \\ 94 & 69 & 112 & 109 & 89 & 143 \end{bmatrix} = \begin{bmatrix} 20 & 18 & 19 & 15 & 27 & 8 \\ 8 & 5 & 20 & 23 & 5 & 27 \\ 5 & 1 & 27 & 27 & 1 & 27 \\ 27 & 20 & 19 & 15 & 18 & 27 \\ 7 & 5 & 8 & 14 & 20 & 27 \end{bmatrix}$

Thus, the decoded message is

20 8 5 27 7 18 5 1 20 5 19 20 27 19 8 15 23 27 15
14 27 5 1 18 20 8

which corresponds to

THE GREATEST SHOW ON EARTH

EXERCISE 4-6

Things to remember:

1. BASIC PROPERTIES OF MATRICES

Assuming all products and sums are defined for the indicated matrices A, B, C, I, and O, then

ADDITION PROPERTIES

ASSOCIATIVE: $(A + B) + C = A + (B + C)$
COMMUTATIVE: $A + B = B + A$
ADDITIVE IDENTITY: $A + 0 = 0 + A = A$
ADDITIVE INVERSE: $A + (-A) = (-A) + A = 0$

MULTIPLICATION PROPERTIES

ASSOCIATIVE PROPERTY: $A(BC) = (AB)C$
MULTIPLICATIVE IDENTITY: $AI = IA = A$
MULTIPLICATIVE INVERSE: If A is a square matrix and A^{-1} exists, then $AA^{-1} = A^{-1}A = I$.

COMBINED PROPERTIES

LEFT DISTRIBUTIVE: $A(B + C) = AB + AC$
RIGHT DISTRIBUTIVE: $(B + C)A = BA + CA$

EQUALITY

ADDITION: If $A = B$ then $A + C = B + C$.

LEFT MULTIPLICATION: If $A = B$, then $CA = CB$.

RIGHT MULTIPLICATION: If $A = B$, then $AC = BC$.

2. USING INVERSE METHODS TO SOLVE SYSTEMS OF EQUATIONS

If the number of equations in a system equals the number of variables and the coefficient matrix has an inverse, then the system will always have a unique solution that can be found by using the inverse of the coefficient matrix to solve the corresponding matrix equation.

Matrix Equation	Solution
$AX = B$	$X = A^{-1}B$

---

1. $\begin{bmatrix} 3 & 1 \\ 2 & -1 \end{bmatrix}\begin{bmatrix} x_1 \\ x_2 \end{bmatrix} = \begin{bmatrix} 5 \\ -4 \end{bmatrix}$

$\begin{bmatrix} 3x_1 + x_2 \\ 2x_1 - x_2 \end{bmatrix} = \begin{bmatrix} 5 \\ -4 \end{bmatrix}$

Thus, $3x_1 + x_2 = 5$
$2x_1 - x_2 = -4$

3. $\begin{bmatrix} -3 & 1 & 0 \\ 2 & 0 & 1 \\ -1 & 3 & -2 \end{bmatrix}\begin{bmatrix} x_1 \\ x_2 \\ x_3 \end{bmatrix} = \begin{bmatrix} 3 \\ -4 \\ 2 \end{bmatrix}$

$\begin{bmatrix} -3x_1 + x_2 \\ 2x_1 + x_3 \\ -x_1 + 3x_2 - 2x_3 \end{bmatrix} = \begin{bmatrix} 3 \\ -4 \\ 2 \end{bmatrix}$

Thus, $-3x_1 + x_2 = 3$
$2x_1 + x_3 = -4$
$-x_1 + 3x_2 - 2x_3 = 2$

5. $3x_1 - 4x_2 = 1$
$2x_1 + x_2 = 5$

$\begin{bmatrix} 3x_1 - 4x_2 \\ 2x_1 + x_2 \end{bmatrix} = \begin{bmatrix} 1 \\ 5 \end{bmatrix}$ and $\begin{bmatrix} 3 & -4 \\ 2 & 1 \end{bmatrix}\begin{bmatrix} x_1 \\ x_2 \end{bmatrix} = \begin{bmatrix} 1 \\ 5 \end{bmatrix}$

7. $x_1 - 3x_2 + 2x_3 = -3$
$-2x_1 + 3x_2 = 1$
$x_1 + x_2 + 4x_3 = -2$

$\begin{bmatrix} x_1 - 3x_2 + 2x_3 \\ -2x_1 + 3x_2 \\ x_1 + x_2 + 4x_3 \end{bmatrix} = \begin{bmatrix} -3 \\ 1 \\ -2 \end{bmatrix}$ and $\begin{bmatrix} 1 & -3 & 2 \\ -2 & 3 & 0 \\ 1 & 1 & 4 \end{bmatrix}\begin{bmatrix} x_1 \\ x_2 \\ x_3 \end{bmatrix} = \begin{bmatrix} -3 \\ 1 \\ -2 \end{bmatrix}$

9. $\begin{bmatrix} x_1 \\ x_2 \end{bmatrix} = \begin{bmatrix} 3 & -2 \\ 1 & 4 \end{bmatrix}\begin{bmatrix} -2 \\ 1 \end{bmatrix} = \begin{bmatrix} 3(-2) + (-2)1 \\ 1(-2) + 4\cdot 1 \end{bmatrix} = \begin{bmatrix} -8 \\ 2 \end{bmatrix}$ Thus, $x_1 = -8$ and $x_2 = 2$

11. $\begin{bmatrix} x_1 \\ x_2 \end{bmatrix} = \begin{bmatrix} -2 & 3 \\ 2 & -1 \end{bmatrix}\begin{bmatrix} 3 \\ 2 \end{bmatrix} = \begin{bmatrix} (-2)3 + 3\cdot 2 \\ 2\cdot 3 + (-1)2 \end{bmatrix} = \begin{bmatrix} 0 \\ 4 \end{bmatrix}$ Thus, $x_1 = 0$ and $x_2 = 4$

13. $\begin{bmatrix} 1 & -1 \\ 1 & -2 \end{bmatrix}\begin{bmatrix} x_1 \\ x_2 \end{bmatrix} = \begin{bmatrix} 5 \\ 7 \end{bmatrix}$

If $A = \begin{bmatrix} 1 & -1 \\ 1 & -2 \end{bmatrix}$ has an inverse, then $\begin{bmatrix} x_1 \\ x_2 \end{bmatrix} = A^{-1}\begin{bmatrix} 5 \\ 7 \end{bmatrix}$.

$$\left[\begin{array}{cc|cc} 1 & -1 & 1 & 0 \\ 1 & -2 & 0 & 1 \end{array}\right] \sim \left[\begin{array}{cc|cc} 1 & -1 & 1 & 0 \\ 0 & -1 & -1 & 1 \end{array}\right] \sim \left[\begin{array}{cc|cc} 1 & -1 & 1 & 0 \\ 0 & 1 & 1 & -1 \end{array}\right]$$

$(-1)R_1 \to R_2$ $\quad$ $(-1)R_2 \to R_2$ $\quad$ $R_2 + R_1 \to R_1$

$$\sim \left[\begin{array}{cc|cc} 1 & 0 & 2 & -1 \\ 0 & 1 & 1 & -1 \end{array}\right]; \; A^{-1} = \begin{bmatrix} 2 & -1 \\ 1 & -1 \end{bmatrix} \quad \begin{bmatrix} x_1 \\ x_2 \end{bmatrix} = \begin{bmatrix} 2 & -1 \\ 1 & -1 \end{bmatrix}\begin{bmatrix} 5 \\ 7 \end{bmatrix} = \begin{bmatrix} 3 \\ -2 \end{bmatrix}$$

Therefore, $x_1 = 3$, $x_2 = -2$.

15. $\begin{bmatrix} 1 & 1 \\ 2 & -3 \end{bmatrix}\begin{bmatrix} x_1 \\ x_2 \end{bmatrix} = \begin{bmatrix} 15 \\ 10 \end{bmatrix}$

If $A = \begin{bmatrix} 1 & 1 \\ 2 & -3 \end{bmatrix}$ has an inverse, then $\begin{bmatrix} x_1 \\ x_2 \end{bmatrix} = A^{-1}\begin{bmatrix} 15 \\ 10 \end{bmatrix}$

$$\left[\begin{array}{cc|cc} 1 & 1 & 1 & 0 \\ 2 & -3 & 0 & 1 \end{array}\right] \sim \left[\begin{array}{cc|cc} 1 & 1 & 1 & 0 \\ 0 & -5 & -2 & 1 \end{array}\right] \sim \left[\begin{array}{cc|cc} 1 & 1 & 1 & 0 \\ 0 & 1 & \frac{2}{5} & -\frac{1}{5} \end{array}\right]$$

$(-2)R_1 + R_2 \to R_2$ $\quad$ $\left(-\frac{1}{5}\right)R_2 \to R_2$ $\quad$ $(-1)R_2 + R_1 \to R_1$

$$\sim \left[\begin{array}{cc|cc} 1 & 0 & \frac{3}{5} & \frac{1}{5} \\ 0 & 1 & \frac{2}{5} & -\frac{1}{5} \end{array}\right] \; A^{-1} = \begin{bmatrix} \frac{3}{5} & \frac{1}{5} \\ \frac{2}{5} & -\frac{1}{5} \end{bmatrix}; \; \begin{bmatrix} x_1 \\ x_2 \end{bmatrix} = \begin{bmatrix} \frac{3}{5} & \frac{1}{5} \\ \frac{2}{5} & -\frac{1}{5} \end{bmatrix}\begin{bmatrix} 15 \\ 10 \end{bmatrix} = \begin{bmatrix} 11 \\ -4 \end{bmatrix}$$

Therefore, $x_1 = 11$, $x_2 = -4$.

17. The matrix equation for the given system is:

$$\begin{bmatrix} 1 & 2 \\ 1 & 3 \end{bmatrix}\begin{bmatrix} x_1 \\ x_2 \end{bmatrix} = \begin{bmatrix} k_1 \\ k_2 \end{bmatrix}$$

From Exercise 4-5, Problem 21, $\begin{bmatrix} 1 & 2 \\ 1 & 3 \end{bmatrix}^{-1} = \begin{bmatrix} 3 & -2 \\ -1 & 1 \end{bmatrix}$

Thus, $\begin{bmatrix} x_1 \\ x_2 \end{bmatrix} = \begin{bmatrix} 3 & -2 \\ -1 & 1 \end{bmatrix}\begin{bmatrix} k_1 \\ k_2 \end{bmatrix}$

(A) $\begin{bmatrix} x_1 \\ x_2 \end{bmatrix} = \begin{bmatrix} 3 & -2 \\ -1 & 1 \end{bmatrix}\begin{bmatrix} 1 \\ 3 \end{bmatrix} = \begin{bmatrix} -3 \\ 2 \end{bmatrix}$ Thus, $x_1 = -3$ and $x_2 = 2$

(B) $\begin{bmatrix} x_1 \\ x_2 \end{bmatrix} = \begin{bmatrix} 3 & -2 \\ -1 & 1 \end{bmatrix}\begin{bmatrix} 3 \\ 5 \end{bmatrix} = \begin{bmatrix} -1 \\ 2 \end{bmatrix}$ Thus, $x_1 = -1$ and $x_2 = 2$

(C) $\begin{bmatrix} x_1 \\ x_2 \end{bmatrix} = \begin{bmatrix} 3 & -2 \\ -1 & 1 \end{bmatrix}\begin{bmatrix} -2 \\ 1 \end{bmatrix} = \begin{bmatrix} -8 \\ 3 \end{bmatrix}$ Thus, $x_1 = -8$ and $x_2 = 3$

19. The matrix equation for the given system is:

$$\begin{bmatrix} 1 & 3 \\ 2 & 7 \end{bmatrix}\begin{bmatrix} x_1 \\ x_2 \end{bmatrix} = \begin{bmatrix} k_1 \\ k_2 \end{bmatrix}$$

From Exercise 4-5, Problem 23, $\begin{bmatrix} 1 & 3 \\ 2 & 7 \end{bmatrix}^{-1} = \begin{bmatrix} 7 & -3 \\ -2 & 1 \end{bmatrix}$

Thus, $\begin{bmatrix} x_1 \\ x_2 \end{bmatrix} = \begin{bmatrix} 7 & -3 \\ -2 & 1 \end{bmatrix}\begin{bmatrix} k_1 \\ k_2 \end{bmatrix}$

(A) $\begin{bmatrix} x_1 \\ x_2 \end{bmatrix} = \begin{bmatrix} 7 & -3 \\ -2 & 1 \end{bmatrix}\begin{bmatrix} 2 \\ -1 \end{bmatrix} = \begin{bmatrix} 17 \\ -5 \end{bmatrix}$ Thus, $x_1 = 17$ and $x_2 = -5$

(B) $\begin{bmatrix} x_1 \\ x_2 \end{bmatrix} = \begin{bmatrix} 7 & -3 \\ -2 & 1 \end{bmatrix}\begin{bmatrix} 1 \\ 0 \end{bmatrix} = \begin{bmatrix} 7 \\ -2 \end{bmatrix}$ Thus, $x_1 = 7$ and $x_2 = -2$

(C) $\begin{bmatrix} x_1 \\ x_2 \end{bmatrix} = \begin{bmatrix} 7 & -3 \\ -2 & 1 \end{bmatrix}\begin{bmatrix} 3 \\ -1 \end{bmatrix} = \begin{bmatrix} 24 \\ -7 \end{bmatrix}$ Thus, $x_1 = 24$ and $x_2 = -7$

21. The matrix equation for the given system is:

$$\begin{bmatrix} 1 & -3 & 0 \\ 0 & 1 & 1 \\ 2 & -1 & 4 \end{bmatrix}\begin{bmatrix} x_1 \\ x_2 \\ x_3 \end{bmatrix} = \begin{bmatrix} k_1 \\ k_2 \\ k_3 \end{bmatrix}$$

From Exercise 4-5, Problem 25, $\begin{bmatrix} 1 & -3 & 0 \\ 0 & 1 & 1 \\ 2 & -1 & 4 \end{bmatrix}^{-1} = \begin{bmatrix} -5 & -12 & 3 \\ -2 & -4 & 1 \\ 2 & 5 & -1 \end{bmatrix}$

Thus,

$$\begin{bmatrix} x_1 \\ x_2 \\ x_3 \end{bmatrix} = \begin{bmatrix} -5 & -12 & 3 \\ -2 & -4 & 1 \\ 2 & 5 & -1 \end{bmatrix}\begin{bmatrix} k_1 \\ k_2 \\ k_3 \end{bmatrix}$$

(A) $\begin{bmatrix} x_1 \\ x_2 \\ x_3 \end{bmatrix} = \begin{bmatrix} -5 & -12 & 3 \\ -2 & -4 & 1 \\ 2 & 5 & -1 \end{bmatrix}\begin{bmatrix} 1 \\ 0 \\ 2 \end{bmatrix} = \begin{bmatrix} 1 \\ 0 \\ 0 \end{bmatrix}$; $x_1 = 1$, $x_2 = 0$, $x_3 = 0$

(B) $\begin{bmatrix} x_1 \\ x_2 \\ x_3 \end{bmatrix} = \begin{bmatrix} -5 & -12 & 3 \\ -2 & -4 & 1 \\ 2 & 5 & -1 \end{bmatrix} \begin{bmatrix} -1 \\ 1 \\ 0 \end{bmatrix} = \begin{bmatrix} -7 \\ -2 \\ 3 \end{bmatrix}$; $x_1 = -7$, $x_2 = -2$, $x_3 = 3$

(C) $\begin{bmatrix} x_1 \\ x_2 \\ x_3 \end{bmatrix} = \begin{bmatrix} -5 & -12 & 3 \\ -2 & -4 & 1 \\ 2 & 5 & -1 \end{bmatrix} \begin{bmatrix} 2 \\ -2 \\ 1 \end{bmatrix} = \begin{bmatrix} 17 \\ 5 \\ -7 \end{bmatrix}$; $x_1 = 17$, $x_2 = 5$, $x_3 = -7$

23. The matrix equation for the given system is:

$$\begin{bmatrix} 1 & 1 & 0 \\ 2 & 3 & -1 \\ 1 & 0 & 2 \end{bmatrix} \begin{bmatrix} x_1 \\ x_2 \\ x_3 \end{bmatrix} = \begin{bmatrix} k_1 \\ k_2 \\ k_3 \end{bmatrix}$$

From Exercise 4-5, Problem 27, $\begin{bmatrix} 1 & 1 & 0 \\ 2 & 3 & -1 \\ 1 & 0 & 2 \end{bmatrix}^{-1} = \begin{bmatrix} 6 & -2 & -1 \\ -5 & 2 & 1 \\ -3 & 1 & 1 \end{bmatrix}$

Thus,

$$\begin{bmatrix} x_1 \\ x_2 \\ x_3 \end{bmatrix} = \begin{bmatrix} 6 & -2 & -1 \\ -5 & 2 & 1 \\ -3 & 1 & 1 \end{bmatrix} \begin{bmatrix} k_1 \\ k_2 \\ k_3 \end{bmatrix}$$

(A) $\begin{bmatrix} x_1 \\ x_2 \\ x_3 \end{bmatrix} = \begin{bmatrix} 6 & -2 & -1 \\ -5 & 2 & 1 \\ -3 & 1 & 1 \end{bmatrix} \begin{bmatrix} 2 \\ 0 \\ 4 \end{bmatrix} = \begin{bmatrix} 8 \\ -6 \\ -2 \end{bmatrix}$; $x_1 = 8$, $x_2 = -6$, $x_3 = -2$

(B) $\begin{bmatrix} x_1 \\ x_2 \\ x_3 \end{bmatrix} = \begin{bmatrix} 6 & -2 & -1 \\ -5 & 2 & 1 \\ -3 & 1 & 1 \end{bmatrix} \begin{bmatrix} 0 \\ 4 \\ -2 \end{bmatrix} = \begin{bmatrix} -6 \\ 6 \\ 2 \end{bmatrix}$; $x_1 = -6$, $x_2 = 6$, $x_3 = 2$

(C) $\begin{bmatrix} x_1 \\ x_2 \\ x_3 \end{bmatrix} = \begin{bmatrix} 6 & -2 & -1 \\ -5 & 2 & 1 \\ -3 & 1 & 1 \end{bmatrix} \begin{bmatrix} 4 \\ 2 \\ 0 \end{bmatrix} = \begin{bmatrix} 20 \\ -16 \\ -10 \end{bmatrix}$; $x_1 = 20$, $x_2 = -16$, $x_3 = -10$

25. $-2x_1 + 4x_2 = 5$
$6x_1 - 12x_2 = 15$

The second equation is a multiple (-3) of the first. Therefore, the system has infinitely many solutions. The solutions are:

$x_1 = 2t + \frac{5}{2}$, $x_2 = t$, t any real number

27. $x_1 - 3x_2 - 2x_3 = -1$
$-2x_1 + 6x_2 + 4x_3 = 3$

The system is not "square"--2 equations in 3 unknowns. The matrix of coefficients is 2×3; it does not have an inverse.

Solve the system by Gauss-Jordan elimination:

$$\left(\begin{array}{rrr|r} 1 & -3 & -2 & -1 \\ -2 & 6 & 4 & 3 \end{array}\right) \sim \left(\begin{array}{rrr|r} 1 & -3 & -2 & -1 \\ 0 & 0 & 0 & 1 \end{array}\right)$$

$2R_1 + R_2 \to R_2$

The system has no solution.

29. e_1: $x_1 - 2x_2 + 3x_3 = 1$
e_2: $2x_1 - 3x_2 - 2x_3 = 3$
e_3: $x_1 - x_2 - 5x_3 = 2$

Note that $e_3 = (-1)e_1 + e_2$. This implies that the system has infinitely many solutions; the coefficient matrix does not have an inverse. Solve the system by Gauss-Jordan elimination:

$$\left[\begin{array}{rrr|r} 1 & -2 & 3 & 1 \\ 2 & -3 & -2 & 3 \\ 1 & -1 & -5 & 2 \end{array}\right] \sim \left[\begin{array}{rrr|r} 1 & -2 & 3 & 1 \\ 0 & 1 & -8 & 1 \\ 0 & 1 & -8 & 1 \end{array}\right] \sim \left[\begin{array}{rrr|r} 1 & -2 & 3 & 1 \\ 0 & 1 & -8 & 1 \\ 0 & 0 & 0 & 0 \end{array}\right]$$

$(-2)R_1 + R_2 \to R_2$ $\quad (-1)R_2 + R_3 \to R_3$ $\quad 2R_2 + R_1 \to R_1$
$(-1)R_1 + R_3 \to R_3$

$$\sim \left[\begin{array}{rrr|r} 1 & 0 & -13 & 3 \\ 0 & 1 & -8 & 1 \\ 0 & 0 & 0 & 0 \end{array}\right]$$

Solutions: $x_1 = 3 + 13t$, $x_2 = 1 + 8t$, $x_3 = t$, t any real number.

31. $AX - BX = C$
$(A - B)X = C$
$X = (A - B)^{-1}C$

33. $AX + X = C$
$(A + I)X = C$, where I is the identity matrix of order n
$X = (A + I)^{-1}C$

35. $AX - C = D - BX$
$AX + BX = C + D$
$(A + B)X = C + D$
$X = (A + B)^{-1}(C + D)$

37. The matrix equation for the given system is:

$$\begin{bmatrix} 1 & 2.001 \\ 1 & 2 \end{bmatrix} \begin{bmatrix} x_1 \\ x_2 \end{bmatrix} = \begin{bmatrix} k_1 \\ k_2 \end{bmatrix}$$

First we compute the inverse of $\begin{bmatrix} 1 & 2.001 \\ 1 & 2 \end{bmatrix}$

$$\left[\begin{array}{rr|rr} 1 & 2.001 & 1 & 0 \\ 1 & 2 & 0 & 1 \end{array}\right] (-1)R_1 + R_2 \to R_2$$

$$\sim \left[\begin{array}{rr|rr} 1 & 2.001 & 1 & 0 \\ 0 & -0.001 & -1 & 1 \end{array}\right] (-1000)R_2 \to R_2$$

$$\sim \left[\begin{array}{cc|cc} 1 & 2.001 & 1 & 0 \\ 0 & 1 & 1000 & -1000 \end{array}\right] (-2.001)R_2 + R_1 \to R_1$$

$$\left[\begin{array}{cc|cc} 1 & 0 & -2000 & 2001 \\ 0 & 1 & 1000 & -1000 \end{array}\right]$$

Thus, $\begin{bmatrix} 1 & 2.001 \\ 1 & 2 \end{bmatrix}^{-1} = \begin{bmatrix} -2000 & 2001 \\ 1000 & -1000 \end{bmatrix}$ and $\begin{bmatrix} x_1 \\ x_2 \end{bmatrix} = \begin{bmatrix} -2000 & 2001 \\ 1000 & -1000 \end{bmatrix}\begin{bmatrix} k_1 \\ k_2 \end{bmatrix}$

(A) $\begin{bmatrix} x_1 \\ x_2 \end{bmatrix} = \begin{bmatrix} -2000 & 2001 \\ 1000 & -1000 \end{bmatrix}\begin{bmatrix} 1 \\ 1 \end{bmatrix} = \begin{bmatrix} 1 \\ 0 \end{bmatrix}$; $x_1 = 1$, $x_2 = 0$

(B) $\begin{bmatrix} x_1 \\ x_2 \end{bmatrix} = \begin{bmatrix} -2000 & 2001 \\ 1000 & -1000 \end{bmatrix}\begin{bmatrix} 1 \\ 0 \end{bmatrix} = \begin{bmatrix} -2000 \\ 1000 \end{bmatrix}$; $x_1 = -2{,}000$, $x_2 = 1{,}000$

(C) $\begin{bmatrix} x_1 \\ x_2 \end{bmatrix} = \begin{bmatrix} -2000 & 2001 \\ 1000 & -1000 \end{bmatrix}\begin{bmatrix} 0 \\ 1 \end{bmatrix} = \begin{bmatrix} 2001 \\ -1000 \end{bmatrix}$; $x_1 = 2{,}001$, $x_2 = -1{,}000$

39. The matrix equation for the given system is:

$$\begin{bmatrix} 1 & 8 & 7 \\ 6 & 6 & 8 \\ 3 & 4 & 6 \end{bmatrix}\begin{bmatrix} x_1 \\ x_2 \\ x_3 \end{bmatrix} = \begin{bmatrix} 135 \\ 155 \\ 75 \end{bmatrix}$$

Thus, $\begin{bmatrix} x_1 \\ x_2 \\ x_3 \end{bmatrix} = \begin{bmatrix} 1 & 8 & 7 \\ 6 & 6 & 8 \\ 3 & 4 & 6 \end{bmatrix}^{-1}\begin{bmatrix} 135 \\ 155 \\ 75 \end{bmatrix} = \begin{bmatrix} -0.08 & 0.4 & -0.44 \\ 0.24 & 0.3 & -0.68 \\ -0.12 & -0.4 & 0.84 \end{bmatrix}\begin{bmatrix} 135 \\ 155 \\ 75 \end{bmatrix}$

$= \begin{bmatrix} 18.2 \\ 27.9 \\ -15.2 \end{bmatrix}$ and $x_1 = 18.2$, $x_2 = 27.9$, $x_3 = -15.2$

41. The matrix equation for the given system is:

$$\begin{bmatrix} 6 & 9 & 7 & 5 \\ 6 & 4 & 7 & 3 \\ 4 & 5 & 3 & 2 \\ 4 & 3 & 8 & 2 \end{bmatrix}\begin{bmatrix} x_1 \\ x_2 \\ x_3 \\ x_4 \end{bmatrix} = \begin{bmatrix} 250 \\ 195 \\ 145 \\ 125 \end{bmatrix}$$

Thus

$$\begin{bmatrix} x_1 \\ x_2 \\ x_3 \\ x_4 \end{bmatrix} = \begin{bmatrix} 6 & 9 & 7 & 5 \\ 6 & 4 & 7 & 3 \\ 4 & 5 & 3 & 2 \\ 4 & 3 & 8 & 2 \end{bmatrix}^{-1} \begin{bmatrix} 250 \\ 195 \\ 145 \\ 125 \end{bmatrix} = \begin{bmatrix} -0.25 & 0.37 & 0.28 & -0.21 \\ 0 & -0.4 & 0.4 & 0.2 \\ 0 & -0.16 & -0.04 & 0.28 \\ 0.5 & 0.5 & -1 & -0.5 \end{bmatrix} \begin{bmatrix} 250 \\ 195 \\ 145 \\ 125 \end{bmatrix}$$

$$= \begin{bmatrix} 24 \\ 5 \\ -2 \\ 15 \end{bmatrix} \text{ and } \begin{aligned} x_1 &= 24 \\ x_2 &= 5 \\ x_3 &= -2 \\ x_4 &= 15 \end{aligned}$$

43. Let x_1 = Number of \$4 tickets sold
x_2 = Number of \$8 tickets sold

The mathematical model is:

$$\begin{aligned} x_1 + x_2 &= 10{,}000 \\ 4x_1 + 8x_2 &= k \quad , \quad k = 56{,}000,\ 60{,}000,\ 68{,}000 \end{aligned}$$

The corresponding matrix equation is:

$$\begin{bmatrix} 1 & 1 \\ 4 & 8 \end{bmatrix} \begin{bmatrix} x_1 \\ x_2 \end{bmatrix} = \begin{bmatrix} 10{,}000 \\ k \end{bmatrix}$$

Compute the inverse of the coefficient matrix A.

$$\left[\begin{array}{cc|cc} 1 & 1 & 1 & 0 \\ 4 & 8 & 0 & 1 \end{array}\right] \sim \left[\begin{array}{cc|cc} 1 & 1 & 1 & 0 \\ 0 & 4 & -4 & 1 \end{array}\right] \sim \left[\begin{array}{cc|cc} 1 & 1 & 1 & 0 \\ 0 & 1 & -1 & \frac{1}{4} \end{array}\right]$$

$(-4)R_1 + R_2 \to R_2$ $\quad \frac{1}{4}R_2 \to R_2$ $\quad (-1)R_2 + R_1 \to R_1$

$$\sim \left[\begin{array}{cc|cc} 1 & 0 & 2 & -\frac{1}{4} \\ 0 & 1 & -1 & \frac{1}{4} \end{array}\right] \quad \text{Thus, } A^{-1} = \begin{bmatrix} 2 & -\frac{1}{4} \\ -1 & \frac{1}{4} \end{bmatrix}$$

Concert 1: Return \$56,000

$$\begin{bmatrix} x_1 \\ x_2 \end{bmatrix} = \begin{bmatrix} 2 & -\frac{1}{4} \\ -1 & \frac{1}{4} \end{bmatrix} \begin{bmatrix} 10{,}000 \\ 56{,}000 \end{bmatrix} = \begin{bmatrix} 6{,}000 \\ 4{,}000 \end{bmatrix}$$

Thus, 6,000 \$4 tickets and 4,000 \$8 tickets must be sold.

Concert 2: Return \$60,000

$$\begin{bmatrix} x_1 \\ x_2 \end{bmatrix} = \begin{bmatrix} 2 & -\frac{1}{4} \\ -1 & \frac{1}{4} \end{bmatrix} \begin{bmatrix} 10{,}000 \\ 60{,}000 \end{bmatrix} = \begin{bmatrix} 5{,}000 \\ 5{,}000 \end{bmatrix}$$

Thus, 5,000 \$4 tickets and 5,000 \$8 tickets must be sold.

Concert 3: Return \$68,000

$$\begin{bmatrix} x_1 \\ x_2 \end{bmatrix} = \begin{bmatrix} 2 & -\frac{1}{4} \\ -1 & \frac{1}{4} \end{bmatrix} \begin{bmatrix} 10{,}000 \\ 68{,}000 \end{bmatrix} = \begin{bmatrix} 3{,}000 \\ 7{,}000 \end{bmatrix}$$

Thus, 3,000 \$4 tickets and 7,000 \$8 tickets must be sold.

(B) \$9,000 Return?

$$\begin{bmatrix} x_1 \\ x_2 \end{bmatrix} \stackrel{?}{=} \begin{bmatrix} 2 & -\frac{1}{4} \\ -1 & \frac{1}{4} \end{bmatrix} \begin{bmatrix} 10,000 \\ 9,000 \end{bmatrix} = \begin{bmatrix} 17,750 \\ -7,750 \end{bmatrix}$$

This is not possible; x_2 cannot be negative.

\$3,000 Return?

$$\begin{bmatrix} x_1 \\ x_2 \end{bmatrix} \stackrel{?}{=} \begin{bmatrix} 2 & -\frac{1}{4} \\ -1 & \frac{1}{4} \end{bmatrix} \begin{bmatrix} 10,000 \\ 3,000 \end{bmatrix} = \begin{bmatrix} 19,250 \\ -9,250 \end{bmatrix}$$

This is not possible; x_2 cannot be negative.

(C) Fix a return k. Then

$$\begin{bmatrix} x_1 \\ x_2 \end{bmatrix} = \begin{bmatrix} 2 & -\frac{1}{4} \\ -1 & \frac{1}{4} \end{bmatrix} \begin{bmatrix} 10,000 \\ k \end{bmatrix} = \begin{bmatrix} 20,000 - \frac{k}{4} \\ -10,000 + \frac{k}{4} \end{bmatrix}$$

Thus, $x_1 = 20,000 - \frac{k}{4}$ and $x_2 = -10,000 + \frac{k}{4}$.

Since $x_1 \geq 0$, $20,000 - \frac{k}{4} \geq 0$

$$-\frac{k}{4} \geq -20,000$$

$$k \leq 80,000$$

Since $x_2 \geq 0$, $-10,000 + \frac{k}{4} \geq 0$

$$\frac{k}{4} \geq 10,000$$

$$k \geq 40,000$$

Thus, $40,000 \leq k \leq 80,000$; any number between (and including) \$40,000 and \$80,000 is a possible return.

45. Let x_1 = number of hours at Plant A
and x_2 = number of hours at Plant B

Then, the mathematical model is:

$10x_1 + 8x_2 = k_1$ (number of car frames)

$5x_1 + 8x_2 = k_2$ (number of truck frames)

The corresponding matrix equation is:

$$\begin{bmatrix} 10 & 8 \\ 5 & 8 \end{bmatrix} \begin{bmatrix} x_1 \\ x_2 \end{bmatrix} = \begin{bmatrix} k_1 \\ k_2 \end{bmatrix}$$

First we compute the inverse of $\begin{bmatrix} 10 & 8 \\ 5 & 8 \end{bmatrix}$

$$\left[\begin{array}{cc|cc} 10 & 8 & 1 & 0 \\ 5 & 8 & 0 & 1 \end{array}\right] \sim \left[\begin{array}{cc|cc} 1 & \frac{4}{5} & \frac{1}{10} & 0 \\ 5 & 8 & 0 & 1 \end{array}\right] \sim \left[\begin{array}{cc|cc} 1 & \frac{4}{5} & \frac{1}{10} & 0 \\ 0 & 4 & -\frac{1}{2} & 1 \end{array}\right]$$

$\frac{1}{10}R_1 \rightarrow R_1$ $(-5)R_1 + R_2 \rightarrow R_2$ $\frac{1}{4}R_2 \rightarrow R_2$

$$\sim \left[\begin{array}{cc|cc} 1 & \frac{4}{5} & \frac{1}{10} & 0 \\ 0 & 1 & -\frac{1}{8} & \frac{1}{4} \end{array}\right] \sim \left[\begin{array}{cc|cc} 1 & 0 & \frac{1}{5} & -\frac{1}{5} \\ 0 & 1 & -\frac{1}{8} & \frac{1}{4} \end{array}\right]$$

$\left(-\frac{4}{5}\right)R_2 + R_1 \to R_1$

Thus $\begin{bmatrix} 10 & 8 \\ 5 & 8 \end{bmatrix}^{-1} = \begin{bmatrix} \frac{1}{5} & -\frac{1}{5} \\ -\frac{1}{8} & \frac{1}{4} \end{bmatrix}$ and $\begin{bmatrix} x_1 \\ x_2 \end{bmatrix} = \begin{bmatrix} \frac{1}{5} & -\frac{1}{5} \\ -\frac{1}{8} & \frac{1}{4} \end{bmatrix}\begin{bmatrix} k_1 \\ k_2 \end{bmatrix}$

Now, for order 1:

$\begin{bmatrix} x_1 \\ x_2 \end{bmatrix} = \begin{bmatrix} \frac{1}{5} & -\frac{1}{5} \\ -\frac{1}{8} & \frac{1}{4} \end{bmatrix}\begin{bmatrix} 3000 \\ 1600 \end{bmatrix} = \begin{bmatrix} 280 \\ 25 \end{bmatrix}$ and x_1 = 280 hours at Plant A, x_2 = 25 hours at Plant B

For order 2:

$\begin{bmatrix} x_1 \\ x_2 \end{bmatrix} = \begin{bmatrix} \frac{1}{5} & -\frac{1}{5} \\ -\frac{1}{8} & \frac{1}{4} \end{bmatrix}\begin{bmatrix} 2800 \\ 2000 \end{bmatrix} = \begin{bmatrix} 160 \\ 150 \end{bmatrix}$ and x_1 = 160 hours at Plant A, x_2 = 150 hours at Plant B

For order 3:

$\begin{bmatrix} x_1 \\ x_2 \end{bmatrix} = \begin{bmatrix} \frac{1}{5} & -\frac{1}{5} \\ -\frac{1}{8} & \frac{1}{4} \end{bmatrix}\begin{bmatrix} 2600 \\ 2200 \end{bmatrix} = \begin{bmatrix} 80 \\ 225 \end{bmatrix}$ and x_1 = 80 hours at Plant A, x_2 = 225 hours at Plant B

47. Let x_1 = President's bonus
x_2 = Executive Vice President's bonus
x_3 = Associate Vice President's bonus
x_4 = Assistant Vice President's bonus

Then, the mathematical model is:

$x_1 = 0.03(2{,}000{,}000 - x_2 - x_3 - x_4)$
$x_2 = 0.025(2{,}000{,}000 - x_1 - x_3 - x_4)$
$x_3 = 0.02(2{,}000{,}000 - x_1 - x_2 - x_4)$
$x_4 = 0.015(2{,}000{,}000 - x_1 - x_2 - x_3)$

or

$$\begin{aligned} x_1 + 0.03x_2 + 0.03x_3 + 0.03x_4 &= 60{,}000 \\ 0.025x_1 + x_2 + 0.025x_3 + 0.025x_4 &= 50{,}000 \\ 0.02x_1 + 0.02x_2 + x_3 + 0.02x_4 &= 40{,}000 \\ 0.015x_1 + 0.015x_2 + 0.015x_3 + x_4 &= 30{,}000 \end{aligned}$$

and $\begin{bmatrix} 1 & 0.03 & 0.03 & 0.03 \\ 0.025 & 1 & 0.025 & 0.025 \\ 0.02 & 0.02 & 1 & 0.02 \\ 0.015 & 0.015 & 0.015 & 1 \end{bmatrix}\begin{bmatrix} x_1 \\ x_2 \\ x_3 \\ x_4 \end{bmatrix} = \begin{bmatrix} 60{,}000 \\ 50{,}000 \\ 40{,}000 \\ 30{,}000 \end{bmatrix}$

Thus $\begin{bmatrix} x_1 \\ x_2 \\ x_3 \\ x_4 \end{bmatrix} = \begin{bmatrix} 1 & 0.03 & 0.03 & 0.03 \\ 0.025 & 1 & 0.025 & 0.025 \\ 0.02 & 0.02 & 1 & 0.02 \\ 0.015 & 0.015 & 0.015 & 1 \end{bmatrix}^{-1}\begin{bmatrix} 60{,}000 \\ 50{,}000 \\ 40{,}000 \\ 30{,}000 \end{bmatrix} \approx \begin{bmatrix} 56{,}600 \\ 47{,}000 \\ 37{,}400 \\ 27{,}900 \end{bmatrix}$

or x_1 = \$56,600, x_2 = \$47,000, x_3 = \$37,400, x_4 = \$27,900 to the nearest hundred dollars.

49. Let x_1 = Number of ounces of mix A

x_2 = Number of ounces of mix B

The mathematical model is:

$$0.2x_1 + 0.1x_2 = k_1 \quad \text{(protein)}$$
$$0.02x_1 + 0.06x_2 = k_2 \quad \text{(fat)}$$

The corresponding matrix equation is:

$$\begin{bmatrix} 0.2 & 0.1 \\ 0.02 & 0.06 \end{bmatrix}\begin{bmatrix} x_1 \\ x_2 \end{bmatrix} = \begin{bmatrix} k_1 \\ k_2 \end{bmatrix}$$

Next, compute the inverse of the coefficient matrix A:

$$\left[\begin{array}{cc|cc} 0.2 & 0.1 & 1 & 0 \\ 0.02 & 0.06 & 0 & 1 \end{array}\right] \sim \left[\begin{array}{cc|cc} 1 & 0.5 & 5 & 0 \\ 0.02 & 0.06 & 0 & 1 \end{array}\right]$$

$5R_1 \to R_1$ $\qquad$ $(-0.02)R_1 + R_2 \to R_2$

$$\sim \left[\begin{array}{cc|cc} 1 & 0.5 & 5 & 0 \\ 0 & 0.05 & -0.1 & 1 \end{array}\right] \sim \left[\begin{array}{cc|cc} 1 & 0.5 & 5 & 0 \\ 0 & 1 & -2 & 20 \end{array}\right]$$

$20R_2 \to R_2$ $\qquad$ $(-0.5)R_2 + R_1 \to R_1$

$$\sim \left[\begin{array}{cc|cc} 1 & 0 & 6 & -10 \\ 0 & 1 & -2 & 20 \end{array}\right] \quad \text{Thus, } A^{-1} = \begin{bmatrix} 6 & -10 \\ -2 & 20 \end{bmatrix}$$

(A) Diet 1: Protein - 20 oz, Fat - 6 oz

$$\begin{bmatrix} x_1 \\ x_2 \end{bmatrix} = \begin{bmatrix} 6 & -10 \\ -2 & 20 \end{bmatrix}\begin{bmatrix} 20 \\ 6 \end{bmatrix} = \begin{bmatrix} 60 \\ 80 \end{bmatrix}$$

Thus, 60 ounces of mix A, 80 ounces of mix B.

Diet 2: Protein - 10 oz, Fat - 4 oz

$$\begin{bmatrix} x_1 \\ x_2 \end{bmatrix} = \begin{bmatrix} 6 & -10 \\ -2 & 20 \end{bmatrix}\begin{bmatrix} 10 \\ 4 \end{bmatrix} = \begin{bmatrix} 20 \\ 60 \end{bmatrix}$$

Thus, 20 ounces of mix A, 60 ounces of mix B.

Diet 3: Protein - 10 oz, Fat - 6 oz

$$\begin{bmatrix} x_1 \\ x_2 \end{bmatrix} = \begin{bmatrix} 6 & -10 \\ -2 & 20 \end{bmatrix}\begin{bmatrix} 10 \\ 6 \end{bmatrix} = \begin{bmatrix} 0 \\ 100 \end{bmatrix}$$

Thus, 0 ounces of mix A, 100 ounces of mix B.

(B) Protein - 20 oz, Fat - 14 oz ?

$$\begin{bmatrix} x_1 \\ x_2 \end{bmatrix} = \begin{bmatrix} 6 & -10 \\ -2 & 20 \end{bmatrix}\begin{bmatrix} 20 \\ 14 \end{bmatrix} = \begin{bmatrix} -20 \\ 240 \end{bmatrix}, \quad x_1 = -20, \; x_2 = 20$$

This is not possible; x_1 and x_2 must both be nonnegative.

Protein - 20 oz, Fat - 1 oz ?

$$\begin{bmatrix} x_1 \\ x_2 \end{bmatrix} = \begin{bmatrix} 6 & -10 \\ -2 & 20 \end{bmatrix}\begin{bmatrix} 20 \\ 1 \end{bmatrix} = \begin{bmatrix} 110 \\ -20 \end{bmatrix}, \quad x_1 = 110, \; x_2 = -20$$

This is not possible; x_1 and x_2 must both be nonnegative.

EXERCISE 4-7

Things to remember:

1. Given two industries C_1 and C_2, with

$$M = \begin{matrix} & C_1 & C_2 \\ C_1 & a_{11} & a_{12} \\ C_2 & a_{21} & a_{22} \end{matrix}, \quad X = \begin{bmatrix} x_1 \\ x_2 \end{bmatrix}, \quad D = \begin{bmatrix} d_1 \\ d_2 \end{bmatrix},$$

Technology Matrix, Output Matrix, Final Demand Matrix

where a_{ij} is the input required from C_i to produce a dollar's worth of output for C_j. The solution to the input-output matrix equation $X = MX + D$ is

$$X = (I - M)^{-1}D,$$

where I is the identity matrix, assuming $I - M$ has an inverse.

1. 40¢ from A and 20¢ from E are required to produce a dollar's worth of output for A.

3. $I - M = \begin{bmatrix} 1 & 0 \\ 0 & 1 \end{bmatrix} - \begin{bmatrix} 0.4 & 0.2 \\ 0.2 & 0.1 \end{bmatrix} = \begin{bmatrix} 0.6 & -0.2 \\ -0.2 & 0.9 \end{bmatrix}$

Converting the decimals to fractions to calculate the inverse, we have:

$$\left[\begin{array}{cc|cc} \frac{3}{5} & -\frac{1}{5} & 1 & 0 \\ -\frac{1}{5} & \frac{9}{10} & 0 & 1 \end{array}\right] \sim \left[\begin{array}{cc|cc} 1 & -\frac{1}{3} & \frac{5}{3} & 0 \\ -\frac{1}{5} & \frac{9}{10} & 0 & 1 \end{array}\right] \sim \left[\begin{array}{cc|cc} 1 & -\frac{1}{3} & \frac{5}{3} & 0 \\ 0 & \frac{5}{6} & \frac{1}{3} & 1 \end{array}\right] \sim \left[\begin{array}{cc|cc} 1 & -\frac{1}{3} & \frac{5}{3} & 0 \\ 0 & 1 & \frac{2}{5} & \frac{6}{5} \end{array}\right]$$

$\frac{5}{3}R_1 \to R_1$ $\quad \frac{1}{5}R_1 + R_2 \to R_2$ $\quad \frac{6}{5}R_2 \to R_2$ $\quad \frac{1}{3}R_2 + R_1 \to R_1$

$$\sim \left[\begin{array}{cc|cc} 1 & 0 & \frac{9}{5} & \frac{2}{5} \\ 0 & 1 & \frac{2}{5} & \frac{6}{5} \end{array}\right]$$

Thus, $I - M = \begin{bmatrix} 0.6 & -0.2 \\ -0.2 & 0.9 \end{bmatrix}$ and $(I - M)^{-1} = \begin{bmatrix} 1.8 & 0.4 \\ 0.4 & 1.2 \end{bmatrix}$

5. $X = (I - M)^{-1}D_2 = \begin{bmatrix} 1.8 & 0.4 \\ 0.4 & 1.2 \end{bmatrix}\begin{bmatrix} 8 \\ 5 \end{bmatrix}$ Thus, $\begin{bmatrix} x_1 \\ x_2 \end{bmatrix} = \begin{bmatrix} 16.4 \\ 9.2 \end{bmatrix}$ and $x_1 = 16.4$, $x_2 = 9.2$.

7. 20¢ from A, 10¢ from B, and 10¢ from E are required to produce a dollar's worth of output for B.

9. $\begin{bmatrix} 1 & 0 & 0 \\ 0 & 1 & 0 \\ 0 & 0 & 1 \end{bmatrix} - \begin{bmatrix} 0.3 & 0.2 & 0.2 \\ 0.1 & 0.1 & 0.1 \\ 0.2 & 0.1 & 0.1 \end{bmatrix} = \begin{bmatrix} 0.7 & -0.2 & -0.2 \\ -0.1 & 0.9 & -0.1 \\ -0.2 & -0.1 & 0.9 \end{bmatrix}$

11. $X = (I - M)^{-1}D_1$

Therefore, $\begin{bmatrix} x_1 \\ x_2 \\ x_3 \end{bmatrix} = \begin{bmatrix} 1.6 & 0.4 & 0.4 \\ 0.22 & 1.18 & 0.18 \\ 0.38 & 0.22 & 1.22 \end{bmatrix}\begin{bmatrix} 5 \\ 10 \\ 15 \end{bmatrix} = \begin{bmatrix} (1.6)5 + (0.4)10 + (0.4)15 \\ (0.22)5 + (1.18)10 + (0.18)15 \\ (0.38)5 + (0.22)10 + (1.22)15 \end{bmatrix}$

$= \begin{bmatrix} 8 + 4 + 6 \\ 1.1 + 11.8 + 2.7 \\ 1.9 + 2.2 + 18.3 \end{bmatrix} = \begin{bmatrix} 18 \\ 15.6 \\ 22.4 \end{bmatrix}$

Thus, agriculture, \$18 billion; building, \$15.6 billion; and energy, \$22.4 billion.

13. $I - M = \begin{bmatrix} 1 & 0 \\ 0 & 1 \end{bmatrix} - \begin{bmatrix} 0.2 & 0.2 \\ 0.3 & 0.3 \end{bmatrix} = \begin{bmatrix} 0.8 & -0.2 \\ -0.3 & 0.7 \end{bmatrix} = \begin{bmatrix} \frac{4}{5} & -\frac{1}{5} \\ -\frac{3}{10} & \frac{7}{10} \end{bmatrix}$,

converting the decimals to fractions.

$\left[\begin{array}{cc|cc} \frac{4}{5} & -\frac{1}{5} & 1 & 0 \\ -\frac{3}{10} & \frac{7}{10} & 0 & 1 \end{array}\right] \sim \left[\begin{array}{cc|cc} 1 & -\frac{1}{4} & \frac{5}{4} & 0 \\ -\frac{3}{10} & \frac{7}{10} & 0 & 1 \end{array}\right] \sim \left[\begin{array}{cc|cc} 1 & -\frac{1}{4} & \frac{5}{4} & 0 \\ 0 & \frac{5}{8} & \frac{3}{8} & 1 \end{array}\right] \sim \left[\begin{array}{cc|cc} 1 & -\frac{1}{4} & \frac{5}{4} & 0 \\ 0 & 1 & \frac{3}{5} & \frac{8}{5} \end{array}\right]$

$\frac{5}{4}R_1 \to R_1$ $\quad \frac{3}{10}R_1 + R_2 \to R_2$ $\quad \frac{8}{5}R_2 \to R_2$ $\quad \frac{1}{4}R_2 + R_1 \to R_1$

$\sim \left[\begin{array}{cc|cc} 1 & 0 & \frac{7}{5} & \frac{2}{5} \\ 0 & 1 & \frac{3}{5} & \frac{8}{5} \end{array}\right]$ Thus, $(I - M)^{-1} = \begin{bmatrix} 1.4 & 0.4 \\ 0.6 & 1.6 \end{bmatrix}$.

Now, $X = (I - M)^{-1}D = \begin{bmatrix} 1.4 & 0.4 \\ 0.6 & 1.6 \end{bmatrix}\begin{bmatrix} 10 \\ 25 \end{bmatrix} = \begin{bmatrix} 24 \\ 46 \end{bmatrix}$.

15. $I - M = \begin{bmatrix} 1 & 0 & 0 \\ 0 & 1 & 0 \\ 0 & 0 & 1 \end{bmatrix} - \begin{bmatrix} 0.3 & 0.1 & 0.3 \\ 0.2 & 0.1 & 0.2 \\ 0.1 & 0.1 & 0.1 \end{bmatrix} = \begin{bmatrix} 0.7 & -0.1 & -0.3 \\ -0.2 & 0.9 & -0.2 \\ -0.1 & -0.1 & 0.9 \end{bmatrix}$

$\left[\begin{array}{ccc|ccc} 0.7 & -0.1 & -0.3 & 1 & 0 & 0 \\ -0.2 & 0.9 & -0.2 & 0 & 1 & 0 \\ -0.1 & -0.1 & 0.9 & 0 & 0 & 1 \end{array}\right] \sim \left[\begin{array}{ccc|ccc} 7 & -1 & -3 & 10 & 0 & 0 \\ -2 & 9 & -2 & 0 & 10 & 0 \\ 1 & 1 & -9 & 0 & 0 & -10 \end{array}\right]$

$10R_1 \to R_1$
$10R_2 \to R_2$
$-10R_3 \to R_3$

$R_1 \leftrightarrow R_3$

$$\sim\left[\begin{array}{ccc|ccc} 1 & 1 & -9 & 0 & 0 & -10 \\ -2 & 9 & -2 & 0 & 10 & 0 \\ 7 & -1 & -3 & 10 & 0 & 0 \end{array}\right] \sim \left[\begin{array}{ccc|ccc} 1 & 1 & -9 & 0 & 0 & -10 \\ 0 & 11 & -20 & 0 & 10 & -20 \\ 0 & -8 & 60 & 10 & 0 & 70 \end{array}\right]$$

$2R_1 + R_2 \to R_2$ $\quad$ $\frac{1}{11}R_2 \to R_2$

$(-7)R_1 + R_3 \to R_3$

$$\sim\left[\begin{array}{ccc|ccc} 1 & 1 & -9 & 0 & 0 & -10 \\ 0 & 1 & -1.82 & 0 & 0.91 & -1.82 \\ 0 & -8 & 60 & 10 & 0 & 70 \end{array}\right] \sim \left[\begin{array}{ccc|ccc} 1 & 0 & -7.18 & 0 & -0.91 & -8.18 \\ 0 & 1 & -1.82 & 0 & 0.91 & -1.82 \\ 0 & 0 & 45.44 & 10 & 7.28 & 55.44 \end{array}\right]$$

$8R_2 + R_3 \to R_3$ $\quad$ $\frac{1}{45.44}R_3 \to R_3$

$$\sim\left[\begin{array}{ccc|ccc} 1 & 0 & -7.18 & 0 & -0.91 & -8.18 \\ 0 & 1 & -1.82 & 0 & 0.91 & -1.82 \\ 0 & 0 & 1 & 0.22 & 0.16 & 1.22 \end{array}\right] \sim \left[\begin{array}{ccc|ccc} 1 & 0 & 0 & 1.58 & 0.24 & 0.58 \\ 0 & 1 & 0 & 0.4 & 1.2 & 0.4 \\ 0 & 0 & 1 & 0.22 & 0.16 & 1.22 \end{array}\right]$$

$1.82R_3 + R_2 \to R_2$

$7.18R_3 + R_1 \to R_1$

Thus, $(I - M)^{-1} = \begin{bmatrix} 1.58 & 0.24 & 0.58 \\ 0.4 & 1.2 & 0.4 \\ 0.22 & 0.16 & 1.22 \end{bmatrix}$,

and $X = (I - M)^{-1}D = \begin{bmatrix} 1.58 & 0.24 & 0.58 \\ 0.4 & 1.2 & 0.4 \\ 0.22 & 0.16 & 1.22 \end{bmatrix}\begin{bmatrix} 20 \\ 5 \\ 10 \end{bmatrix}$

$$= \begin{bmatrix} (1.58)20 + (0.24)5 + (0.58)10 \\ (0.4)20 + (1.2)5 + (0.4)10 \\ (0.22)20 + (0.16)5 + (1.22)10 \end{bmatrix} = \begin{bmatrix} 38.6 \\ 18 \\ 17.4 \end{bmatrix}.$$

17. (A) The technology matrix $M = \begin{bmatrix} 0.3 & 0.25 \\ 0.1 & 0.25 \end{bmatrix}$ and the final demand matrix

$D = \begin{bmatrix} 40 \\ 40 \end{bmatrix}$. The input-output matrix equation is $X = MX + D$ or

$X = \begin{bmatrix} 0.3 & 0.25 \\ 0.1 & 0.25 \end{bmatrix}X + \begin{bmatrix} 40 \\ 40 \end{bmatrix}$, where $X = \begin{bmatrix} x_1 \\ x_2 \end{bmatrix}$.

The solution is $X = (I - M)^{-1}D$, provided $I - M$ has an inverse. Now,

$I - M = \begin{bmatrix} 1 & 0 \\ 0 & 1 \end{bmatrix} - \begin{bmatrix} 0.3 & 0.25 \\ 0.1 & 0.25 \end{bmatrix} = \begin{bmatrix} 0.7 & -0.25 \\ -0.1 & 0.75 \end{bmatrix}$

$(I - M)^{-1}$: $\left[\begin{array}{cc|cc} 0.7 & -0.25 & 1 & 0 \\ -0.1 & 0.75 & 0 & 1 \end{array}\right]$ $-10R_2 \to R_2$

$$\sim\left[\begin{array}{cc|cc} 0.7 & -0.25 & 1 & 0 \\ 1 & -7.5 & 0 & -10 \end{array}\right] \quad R_1 \leftrightarrow R_2$$

$$\sim \left[\begin{array}{cc|cc} 1 & -7.5 & 0 & -10 \\ 0.7 & -0.25 & 1 & 0 \end{array}\right] \quad (-0.7)R_1 + R_2 \rightarrow R_2$$

$$\sim \left[\begin{array}{cc|cc} 1 & -7.5 & 0 & -10 \\ 0 & 5 & 1 & 7 \end{array}\right] \quad (0.2)R_2 \rightarrow R_2$$

$$\sim \left[\begin{array}{cc|cc} 1 & -7.5 & 0 & -10 \\ 0 & 1 & 0.2 & 1.4 \end{array}\right] \quad (7.5)R_2 + R_1 \rightarrow R_1$$

$$\sim \left[\begin{array}{cc|cc} 1 & 0 & 1.5 & 0.5 \\ 0 & 1 & 0.2 & 1.4 \end{array}\right]$$

Thus, $(I - M)^{-1} = \begin{bmatrix} 1.5 & 0.5 \\ 0.2 & 1.4 \end{bmatrix}$ and $X = \begin{bmatrix} 1.5 & 0.5 \\ 0.2 & 1.4 \end{bmatrix}\begin{bmatrix} 40 \\ 40 \end{bmatrix} = \begin{bmatrix} 80 \\ 64 \end{bmatrix}$

Thus, the output for each sector is:
Agriculture: \$80 million; Manufacturing: \$64 million

(B) If the agricultural output is increased by \$20 million and the manufacturing output remains at \$64 million, then the final demand D is given by

$$D = (I - M)X = \begin{bmatrix} 0.7 & -0.25 \\ -0.1 & 0.75 \end{bmatrix}\begin{bmatrix} 100 \\ 64 \end{bmatrix} = \begin{bmatrix} 54 \\ 38 \end{bmatrix}$$

The final demand for agriculture increases to \$54 million and the final demand for manufacturing decreases to \$38 million.

19. Let x_1 = total output of energy
x_2 = total output of mining

Then the final demand matrix $D = \begin{pmatrix} 0.4x_1 \\ 0.4x_2 \end{pmatrix}$ and the input-output matrix equation is:

$$\begin{bmatrix} x_1 \\ x_2 \end{bmatrix} = \begin{bmatrix} 0.2 & 0.3 \\ 0.4 & 0.3 \end{bmatrix}\begin{bmatrix} x_1 \\ x_2 \end{bmatrix} + \begin{bmatrix} 0.4x_1 \\ 0.4x_2 \end{bmatrix}$$

This yields the dependent system of equations

$$0.6x_1 = 0.2x_1 + 0.3x_2$$
$$0.6x_2 = 0.4x_1 + 0.3x_2$$

which is equivalent to

$$0.4x_1 - 0.3x_2 = 0$$

or $\quad x_1 = \frac{3}{4}x_2$

Thus, the total ouput of the energy sector should be 75% of the total output of the mining sector.

21. Each element of a technology matrix represents the input needed from C_i to produce \$1 dollar's worth of output for C_j. Hence, each element must be a number between 0 and 1, inclusive.

23. The technology matrix $M = \begin{bmatrix} 0.1 & 0.2 \\ 0.2 & 0.4 \end{bmatrix}$ and the final demand matrix $D = \begin{bmatrix} 20 \\ 10 \end{bmatrix}$.

The input-output matrix equation is $X = MX + D$ or

$X = \begin{bmatrix} 0.1 & 0.2 \\ 0.2 & 0.4 \end{bmatrix} X + \begin{bmatrix} 20 \\ 10 \end{bmatrix}$ where $X = \begin{bmatrix} x_1 \\ x_2 \end{bmatrix}$.

The solution is $X = (I - M)^{-1}D$, provided $(I - M)$ has an inverse. Now,

$$I - M = \begin{bmatrix} 1 & 0 \\ 0 & 1 \end{bmatrix} - \begin{bmatrix} 0.1 & 0.2 \\ 0.2 & 0.4 \end{bmatrix} = \begin{bmatrix} 0.9 & -0.2 \\ -0.2 & 0.6 \end{bmatrix} = \begin{bmatrix} \frac{9}{10} & -\frac{1}{5} \\ -\frac{1}{5} & \frac{3}{5} \end{bmatrix}$$

$$\left[\begin{array}{cc|cc} \frac{9}{10} & -\frac{1}{5} & 1 & 0 \\ -\frac{1}{5} & \frac{3}{5} & 0 & 1 \end{array}\right] \sim \left[\begin{array}{cc|cc} 1 & -\frac{2}{9} & \frac{10}{9} & 0 \\ -\frac{1}{5} & \frac{3}{5} & 0 & 1 \end{array}\right] \sim \left[\begin{array}{cc|cc} 1 & -\frac{2}{9} & \frac{10}{9} & 0 \\ 0 & \frac{5}{9} & \frac{2}{9} & 1 \end{array}\right] \sim \left[\begin{array}{cc|cc} 1 & -\frac{2}{9} & \frac{10}{9} & 0 \\ 0 & 1 & \frac{2}{5} & \frac{9}{5} \end{array}\right]$$

$\frac{10}{9}R_1 \to R_1$ $\quad$ $\frac{1}{5}R_1 + R_2 \to R_2$ $\quad$ $\frac{9}{5}R_2 \to R_2$ $\quad$ $\frac{2}{9}R_2 + R_1 \to R_1$

$\sim \left[\begin{array}{cc|cc} 1 & 0 & \frac{6}{5} & \frac{2}{5} \\ 0 & 1 & \frac{2}{5} & \frac{9}{5} \end{array}\right]$ Thus, $(I - M)^{-1} = \begin{bmatrix} \frac{6}{5} & \frac{2}{5} \\ \frac{2}{5} & \frac{9}{5} \end{bmatrix} = \begin{bmatrix} 1.2 & 0.4 \\ 0.4 & 1.8 \end{bmatrix}$, and

$X = \begin{bmatrix} 1.2 & 0.4 \\ 0.4 & 1.8 \end{bmatrix} \begin{bmatrix} 20 \\ 10 \end{bmatrix} = \begin{bmatrix} 28 \\ 26 \end{bmatrix}$.

Therefore, the output for each sector is: coal, \$28 billion; steel, \$26 billion.

25. The technology matrix $M = \begin{bmatrix} 0.20 & 0.40 \\ 0.15 & 0.30 \end{bmatrix} = \begin{bmatrix} \frac{1}{5} & \frac{2}{5} \\ \frac{3}{20} & \frac{3}{10} \end{bmatrix}$ and the final demand

matrix $D = \begin{bmatrix} 60 \\ 80 \end{bmatrix}$. The input-output matrix equation is $X = MX + D$ or

$X = \begin{bmatrix} \frac{1}{5} & \frac{2}{5} \\ \frac{3}{20} & \frac{3}{10} \end{bmatrix} X + \begin{bmatrix} 60 \\ 80 \end{bmatrix}$ where $X = \begin{bmatrix} x_1 \\ x_2 \end{bmatrix}$. The solution is $X = (I - M)^{-1}D$,

provided $(I - M)$ has an inverse. Now

$$I - M = \begin{bmatrix} 1 & 0 \\ 0 & 1 \end{bmatrix} - \begin{bmatrix} \frac{1}{5} & \frac{2}{5} \\ \frac{3}{20} & \frac{3}{10} \end{bmatrix} = \begin{bmatrix} \frac{4}{5} & -\frac{2}{5} \\ -\frac{3}{20} & \frac{7}{10} \end{bmatrix}.$$

$$\left[\begin{array}{cc|cc} \frac{4}{5} & -\frac{2}{5} & 1 & 0 \\ -\frac{3}{20} & \frac{7}{10} & 0 & 1 \end{array}\right] \sim \left[\begin{array}{cc|cc} 1 & -\frac{1}{2} & \frac{5}{4} & 0 \\ -\frac{3}{20} & \frac{7}{10} & 0 & 1 \end{array}\right] \sim \left[\begin{array}{cc|cc} 1 & -\frac{1}{2} & \frac{5}{4} & 0 \\ 0 & \frac{5}{8} & \frac{3}{16} & 1 \end{array}\right]$$

$\frac{5}{4}R_1 \to R_1$ $\quad$ $\frac{3}{20}R_1 + R_2 \to R_2$ $\quad$ $\frac{8}{5}R_2 \to R_2$

$$\sim \left[\begin{array}{cc|cc} 1 & -\frac{1}{2} & \frac{5}{4} & 0 \\ 0 & 1 & \frac{3}{10} & \frac{8}{5} \end{array}\right] \sim \left[\begin{array}{cc|cc} 1 & 0 & \frac{7}{5} & \frac{4}{5} \\ 0 & 1 & \frac{3}{10} & \frac{8}{5} \end{array}\right]$$

$\frac{1}{2}R_2 + R_1 \to R_1$

Thus $(I - M)^{-1} = \begin{bmatrix} \frac{7}{5} & \frac{4}{5} \\ \frac{3}{10} & \frac{8}{5} \end{bmatrix}$ and $X = \begin{bmatrix} \frac{7}{5} & \frac{4}{5} \\ \frac{3}{10} & \frac{8}{5} \end{bmatrix}\begin{bmatrix} 60 \\ 80 \end{bmatrix} = \begin{bmatrix} 148 \\ 146 \end{bmatrix}$

Therefore, the output for each sector is:
Agriculture—\$148 million; Tourism—\$146 million

27. The technology matrix $M = \begin{bmatrix} 0.2 & 0.4 & 0.3 \\ 0.2 & 0.1 & 0.1 \\ 0.2 & 0.1 & 0.1 \end{bmatrix}$ and the final demand matrix $D = \begin{bmatrix} 10 \\ 15 \\ 20 \end{bmatrix}$.

The input-output matrix equation is $X = MX + D$ or

$$X = \begin{bmatrix} 0.2 & 0.4 & 0.3 \\ 0.2 & 0.1 & 0.1 \\ 0.2 & 0.1 & 0.1 \end{bmatrix}X + \begin{bmatrix} 10 \\ 15 \\ 20 \end{bmatrix}.$$

The solution is $X = (I - M)^{-1}D$, provided $I - M$ has an inverse. Now,

$$I - M = \begin{bmatrix} 1 & 0 & 0 \\ 0 & 1 & 0 \\ 0 & 0 & 1 \end{bmatrix} - \begin{bmatrix} 0.2 & 0.4 & 0.3 \\ 0.2 & 0.1 & 0.1 \\ 0.2 & 0.1 & 0.1 \end{bmatrix} = \begin{bmatrix} 0.8 & -0.4 & -0.3 \\ -0.2 & 0.9 & -0.1 \\ -0.2 & -0.1 & 0.9 \end{bmatrix}.$$

$$\left[\begin{array}{ccc|ccc} 0.8 & -0.4 & -0.3 & 1 & 0 & 0 \\ -0.2 & 0.9 & -0.1 & 0 & 1 & 0 \\ -0.2 & -0.1 & 0.9 & 0 & 0 & 1 \end{array}\right] \sim \left[\begin{array}{ccc|ccc} 8 & -4 & -3 & 10 & 0 & 0 \\ -2 & 9 & -1 & 0 & 10 & 0 \\ -2 & -1 & 9 & 0 & 0 & 10 \end{array}\right]$$

$10R_1 \to R_1$, $10R_2 \to R_2$, $10R_3 \to R_3$; $\left(-\frac{1}{2}\right)R_2 \to R_2$

$$\sim \left[\begin{array}{ccc|ccc} 8 & -4 & -3 & 10 & 0 & 0 \\ 1 & -\frac{9}{2} & \frac{1}{2} & 0 & -5 & 0 \\ -2 & -1 & 9 & 0 & 0 & 10 \end{array}\right] \sim \left[\begin{array}{ccc|ccc} 1 & -\frac{9}{2} & \frac{1}{2} & 0 & -5 & 0 \\ 8 & -4 & -3 & 10 & 0 & 0 \\ -2 & -1 & 9 & 0 & 0 & 10 \end{array}\right]$$

$R_1 \leftrightarrow R_2$; $(-8)R_1 + R_2 \to R_2$, $2R_1 + R_3 \to R_3$

$$\left[\begin{array}{ccc|ccc} 1 & -\frac{9}{2} & \frac{1}{2} & 0 & -5 & 0 \\ 0 & 32 & -7 & 10 & 40 & 0 \\ 0 & -10 & 10 & 0 & -10 & 10 \end{array}\right] \sim \left[\begin{array}{ccc|ccc} 1 & -\frac{9}{2} & \frac{1}{2} & 0 & -5 & 0 \\ 0 & 32 & -7 & 10 & 40 & 0 \\ 0 & 1 & -1 & 0 & 1 & -1 \end{array}\right]$$

$\left(-\frac{1}{10}\right)R_3 \to R_3$; $R_2 \leftrightarrow R_3$

$$\left[\begin{array}{ccc|ccc} 1 & -\frac{9}{2} & \frac{1}{2} & 0 & -5 & 0 \\ 0 & 1 & -1 & 0 & 1 & -1 \\ 0 & 32 & -7 & 10 & 40 & 0 \end{array}\right] \sim \left[\begin{array}{ccc|ccc} 1 & 0 & -4 & 0 & -\frac{1}{2} & -\frac{9}{2} \\ 0 & 1 & -1 & 0 & 1 & -1 \\ 0 & 0 & 25 & 10 & 8 & 32 \end{array}\right]$$

$(-32)R_2 + R_3 \to R_3$, $\frac{9}{2}R_2 + R_1 \to R_1$; $\frac{1}{25}R_3 \to R_3$

$$\sim \left[\begin{array}{ccc|ccc} 1 & 0 & -4 & 0 & -\frac{1}{2} & -\frac{9}{2} \\ 0 & 1 & -1 & 0 & 1 & -1 \\ 0 & 0 & 1 & 0.4 & 0.32 & 1.28 \end{array}\right] \sim \left[\begin{array}{ccc|ccc} 1 & 0 & 0 & 1.6 & 0.78 & 0.62 \\ 0 & 1 & 0 & 0.4 & 1.32 & 0.28 \\ 0 & 0 & 1 & 0.4 & 0.32 & 1.28 \end{array}\right]$$

$R_3 + R_2 \to R_2$
$4R_3 + R_1 \to R_1$

Thus, $(I - M)^{-1} = \begin{bmatrix} 1.6 & 0.78 & 0.62 \\ 0.4 & 1.32 & 0.28 \\ 0.4 & 0.32 & 1.28 \end{bmatrix}$, and

$$X = (I - M)^{-1}D = \begin{bmatrix} 1.6 & 0.78 & 0.62 \\ 0.4 & 1.32 & 0.28 \\ 0.4 & 0.32 & 1.28 \end{bmatrix}\begin{bmatrix} 10 \\ 15 \\ 20 \end{bmatrix}$$

$$= \begin{bmatrix} (1.6)10 + (0.78)15 + (0.62)20 \\ (0.4)10 + (1.32)15 + (0.28)20 \\ (0.4)10 + (0.32)15 + (1.28)20 \end{bmatrix} = \begin{bmatrix} 40.1 \\ 29.4 \\ 34.4 \end{bmatrix}.$$

Therefore, agriculture, \$40.1 billion; manufacturing, \$29.4 billion; and energy, \$34.4 billion.

29. The technology matrix is $M = \begin{bmatrix} 0.05 & 0.17 & 0.23 & 0.09 \\ 0.07 & 0.12 & 0.15 & 0.19 \\ 0.25 & 0.08 & 0.03 & 0.32 \\ 0.11 & 0.19 & 0.28 & 0.16 \end{bmatrix}$.

The input-output matrix equation is $X = MX + D$

where $X = \begin{bmatrix} A \\ E \\ L \\ M \end{bmatrix}$ and D is the final demand matrix. Thus, $X = (I - M)^{-1}D$,

where $I - M = \begin{bmatrix} 0.95 & -0.17 & -0.23 & -0.09 \\ -0.07 & 0.88 & -0.15 & -0.19 \\ -0.25 & -0.08 & 0.97 & -0.32 \\ -0.11 & -0.19 & -0.28 & 0.84 \end{bmatrix}$

Now, $X = \begin{bmatrix} 1.25 & 0.37 & 0.47 & 0.40 \\ 0.26 & 1.33 & 0.41 & 0.48 \\ 0.47 & 0.36 & 1.39 & 0.66 \\ 0.38 & 0.47 & 0.62 & 1.57 \end{bmatrix} D$

Year 1: $D = \begin{bmatrix} 23 \\ 41 \\ 18 \\ 31 \end{bmatrix}$ and $(I - M)^{-1}D \approx \begin{bmatrix} 65 \\ 83 \\ 71 \\ 88 \end{bmatrix}$

Agriculture: \$65 billion; Energy: \$83 billion; Labor: \$71 billion; Manufacturing: \$88 billion

Year 2: $D = \begin{bmatrix} 32 \\ 48 \\ 21 \\ 33 \end{bmatrix}$ and $(I - M)^{-1}D \approx \begin{bmatrix} 81 \\ 97 \\ 83 \\ 99 \end{bmatrix}$

Agriculture: \$81 billion; Energy: \$97 billion; Labor: \$83 billion; Manufacturing: \$99 billion

Year 3: $D = \begin{bmatrix} 55 \\ 62 \\ 25 \\ 35 \end{bmatrix}$ and $(I - M)^{-1}D \approx \begin{bmatrix} 117 \\ 124 \\ 106 \\ 120 \end{bmatrix}$

Agriculture: $117 billion; Energy: $124 billion; Labor: $106 billion; Manufacturing: $120 billion

CHAPTER 4 REVIEW

1. $y = 2x - 4$ (1)

$y = \frac{1}{2}x + 2$ (2)

The point of intersection is the solution. This is $x = 4$, $y = 4$.

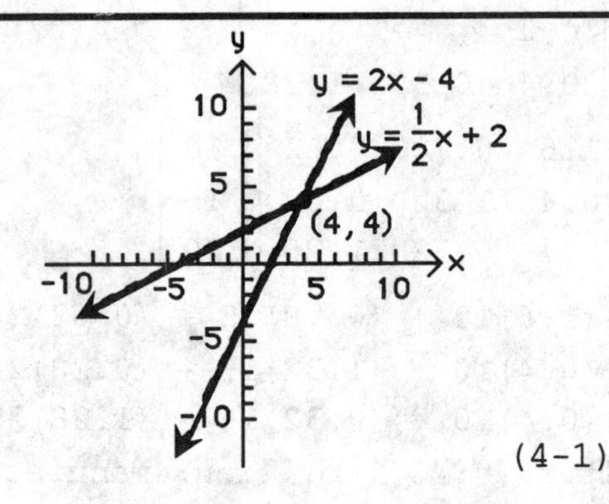

(4-1)

2. Substitute equation (1) into (2):

$2x - 4 = \frac{1}{2}x + 2$

$\frac{3}{2}x = 6$

$x = 4$

Substitute $x = 4$ into (1):

$y = 2 \cdot 4 - 4 = 4$

Solution:

$x = 4$, $y = 4$ (4-1)

3. (A) $\left[\begin{array}{cc|c} 0 & 1 & 2 \\ 1 & 0 & 3 \end{array}\right]$ is not in reduced form; the left-most 1 in the second row is not to the right of the left-most 1 in the first row. [condition (d)]

$R_1 \leftrightarrow R_2$

(B) $\left[\begin{array}{cc|c} 1 & 0 & 2 \\ 0 & 3 & 3 \end{array}\right]$ is not in reduced form; the left-most nonzero element in row 2 is not 1. [condition (b)]

$\frac{1}{3}R_2 \to R_2$

(C) $\left[\begin{array}{ccc|c} 1 & 0 & 1 & 2 \\ 0 & 1 & 1 & 3 \end{array}\right]$ is in reduced form.

(D) $\left[\begin{array}{ccc|c} 1 & 1 & 0 & 2 \\ 0 & 1 & 1 & 3 \end{array}\right]$ is not in reduced form; the left-most 1 in the second row is not the only non-zero element in its column. [condition (c)]

$(-1)R_2 + R_1 \to R_1$ (4-3)

4. $A = \begin{bmatrix} 5 & 3 & -1 & 0 & 2 \\ -4 & 8 & 1 & 3 & 0 \end{bmatrix}$, $B = \begin{bmatrix} -3 & 2 \\ 0 & 4 \\ -1 & 7 \end{bmatrix}$

(A) A is 2×5, B is 3×2

(B) $a_{24} = 3$, $a_{15} = 2$, $b_{31} = -1$, $b_{22} = 4$

(C) AB is not defined; the number of columns of $A \neq$ the number of rows of B.

BA is defined. (4-2, 4-4)

5. $\begin{bmatrix} 1 & -2 \\ 1 & -3 \end{bmatrix}\begin{bmatrix} x_1 \\ x_2 \end{bmatrix} = \begin{bmatrix} 4 \\ 2 \end{bmatrix}$

$$\left[\begin{array}{cc|c} 1 & -2 & 4 \\ 1 & -3 & 2 \end{array}\right] \sim \left[\begin{array}{cc|c} 1 & -2 & 4 \\ 0 & -1 & -2 \end{array}\right] \sim \left[\begin{array}{cc|c} 1 & -2 & 4 \\ 0 & 1 & 2 \end{array}\right]$$

$(-1)R_1 + R_2 \to R_2 \quad (-1)R_2 \to R_2 \quad (2)R_2 + R_1 \to R_1$

$\sim \left[\begin{array}{cc|c} 1 & 0 & 8 \\ 0 & 1 & 2 \end{array}\right]$ Therefore, $x_1 = 8$, $x_2 = 2$.

or calculate the inverse of the coefficient matrix A:

$A^{-1} = \begin{bmatrix} 3 & -2 \\ 1 & -1 \end{bmatrix}$ and $\begin{bmatrix} x_1 \\ x_2 \end{bmatrix} = \begin{bmatrix} 3 & -2 \\ 1 & -1 \end{bmatrix}\begin{bmatrix} 4 \\ 2 \end{bmatrix} = \begin{bmatrix} 8 \\ 2 \end{bmatrix}$; $x_1 = 8$, $x_2 = 2$. (4-2, 4-6)

6. $A + B = \begin{bmatrix} 1+2 & 2+1 \\ 3+1 & 1+1 \end{bmatrix} = \begin{bmatrix} 3 & 3 \\ 4 & 2 \end{bmatrix}$ (4-4)

7. $B + D = \begin{bmatrix} 2 & 1 \\ 1 & 1 \end{bmatrix} + \begin{bmatrix} 1 \\ 2 \end{bmatrix}$

The matrices B and D cannot be added because their dimensions are different. (4-4)

8. $A - 2B = \begin{bmatrix} 1 & 2 \\ 3 & 1 \end{bmatrix} - 2\begin{bmatrix} 2 & 1 \\ 1 & 1 \end{bmatrix} = \begin{bmatrix} 1 & 2 \\ 3 & 1 \end{bmatrix} + \begin{bmatrix} -4 & -2 \\ -2 & -2 \end{bmatrix} = \begin{bmatrix} -3 & 0 \\ 1 & -1 \end{bmatrix}$ (4-4)

9. $AB = \begin{bmatrix} 1 & 2 \\ 3 & 1 \end{bmatrix}\begin{bmatrix} 2 & 1 \\ 1 & 1 \end{bmatrix} = \begin{bmatrix} [1 \ \ 2]\begin{bmatrix} 2 \\ 1 \end{bmatrix} & [1 \ \ 2]\begin{bmatrix} 1 \\ 1 \end{bmatrix} \\ [3 \ \ 1]\begin{bmatrix} 2 \\ 1 \end{bmatrix} & [3 \ \ 1]\begin{bmatrix} 1 \\ 1 \end{bmatrix} \end{bmatrix} = \begin{bmatrix} 4 & 3 \\ 7 & 4 \end{bmatrix}$ (4-4)

10. AC is *not defined* because the dimension of A is 2×2 and the dimension of C is 1×2. So, the number of columns in A is not equal to the number of rows in C. (4-4)

11. $AD = \begin{bmatrix} 1 & 2 \\ 3 & 1 \end{bmatrix}\begin{bmatrix} 1 \\ 2 \end{bmatrix} = \begin{bmatrix} [1 \ \ 2]\begin{bmatrix} 1 \\ 2 \end{bmatrix} \\ [3 \ \ 1]\begin{bmatrix} 1 \\ 2 \end{bmatrix} \end{bmatrix} = \begin{bmatrix} 5 \\ 5 \end{bmatrix}$ (4-4)

12. $DC = \begin{bmatrix} 1 \\ 2 \end{bmatrix}[2 \ \ 3] = \begin{bmatrix} (1)\cdot(2) & (1)\cdot(3) \\ (2)\cdot(2) & (2)\cdot(3) \end{bmatrix} = \begin{bmatrix} 2 & 3 \\ 4 & 6 \end{bmatrix}$ (4-4)

13. $CD = [2 \ \ 3]\begin{bmatrix} 1 \\ 2 \end{bmatrix} = [2 + 6] = [8]$ (4-4)

14. $C + D = [2 \quad 3] + \begin{bmatrix} 1 \\ 2 \end{bmatrix}$

Not defined because the dimensions of C and D are different. (4-4)

15. $\left[\begin{array}{cc|cc} 4 & 3 & 1 & 0 \\ 3 & 2 & 0 & 1 \end{array}\right]$ $(-1)R_2 + R_1 \to R_1$

$\sim \left[\begin{array}{cc|cc} 1 & 1 & 1 & -1 \\ 3 & 2 & 0 & 1 \end{array}\right]$ $(-3)R_1 + R_2 \to R_2$

$\sim \left[\begin{array}{cc|cc} 1 & 1 & 1 & -1 \\ 0 & -1 & -3 & 4 \end{array}\right]$ $(-1)R_2 \to R_2$

$\sim \left[\begin{array}{cc|cc} 1 & 1 & 1 & -1 \\ 0 & 1 & 3 & -4 \end{array}\right]$ $(-1)R_2 + R_1 \to R_1$

$\left[\begin{array}{cc|cc} 1 & 0 & -2 & 3 \\ 0 & 1 & 3 & -4 \end{array}\right]$

Thus, $A^{-1} = \begin{bmatrix} -2 & 3 \\ 3 & -4 \end{bmatrix}$ and $A^{-1}A = \begin{bmatrix} -2 & 3 \\ 3 & -4 \end{bmatrix}\begin{bmatrix} 4 & 3 \\ 3 & 2 \end{bmatrix} = \begin{bmatrix} 1 & 0 \\ 0 & 1 \end{bmatrix}$ (4-5)

16. (1) $4x_1 + 3x_2 = 3$
(2) $3x_1 + 2x_2 = 5$

Multiply (1) by 2 and (2) by -3.

$8x_1 + 6x_2 = 6$
$-9x_1 - 6x_2 = -15$

Add the two equations.

$-x_1 = -9$
$x_1 = 9$

Substitute $x_1 = 9$ into either (1) or (2); we choose (2).

$3(9) + 2x_2 = 5$
$27 + 2x_2 = 5$
$2x_2 = -22$
$x_2 = -11$

Solution: $x_1 = 9$, $x_2 = -11$ (4-1)

17. The augmented matrix of the system is:

$\left[\begin{array}{cc|c} 4 & 3 & 3 \\ 3 & 2 & 5 \end{array}\right]$ $(-1)R_2 + R_1 \to R_1$

$\sim \left[\begin{array}{cc|c} 1 & 1 & -2 \\ 3 & 2 & 5 \end{array}\right]$ $(-3)R_1 + R_2 \to R_2$ $\sim \left[\begin{array}{cc|c} 1 & 1 & -2 \\ 0 & -1 & 11 \end{array}\right]$ $(-1)R_2 \to R_2$

$\sim \left[\begin{array}{cc|c} 1 & 1 & -2 \\ 0 & 1 & -11 \end{array}\right]$ $(-1)R_2 + R_1 \to R_1$

$\sim \left[\begin{array}{cc|c} 1 & 0 & 9 \\ 0 & 1 & -11 \end{array}\right]$ The system of equations is:

$x_1 = 9$
$x_2 = -11$

Solution: $x_1 = 9$, $x_2 = -11$ (4-2)

18. The system of equations in matrix form is:

$$\begin{bmatrix} 4 & 3 \\ 3 & 2 \end{bmatrix}\begin{bmatrix} x_1 \\ x_2 \end{bmatrix} = \begin{bmatrix} 3 \\ 5 \end{bmatrix}$$

Thus, $\begin{bmatrix} x_1 \\ x_2 \end{bmatrix} = \begin{bmatrix} 4 & 3 \\ 3 & 2 \end{bmatrix}^{-1}\begin{bmatrix} 3 \\ 5 \end{bmatrix} = \begin{bmatrix} -2 & 3 \\ 3 & -4 \end{bmatrix}\begin{bmatrix} 3 \\ 5 \end{bmatrix}$ (by Problem 15) $= \begin{bmatrix} 9 \\ -11 \end{bmatrix}$

Solution: $x_1 = 9$, $x_2 = -11$.

Replacing the constants 3, 5 by 7, 10, respectively:

$$\begin{bmatrix} x_1 \\ x_2 \end{bmatrix} = \begin{bmatrix} -2 & 3 \\ 3 & -4 \end{bmatrix}\begin{bmatrix} 7 \\ 10 \end{bmatrix} = \begin{bmatrix} 16 \\ -19 \end{bmatrix}$$

Solution: $x_1 = 16$, $x_2 = -19$

Replacing the constants 3, 5 by 4, 2, respectively:

$$\begin{bmatrix} x_1 \\ x_2 \end{bmatrix} = \begin{bmatrix} -2 & 3 \\ 3 & -4 \end{bmatrix}\begin{bmatrix} 4 \\ 2 \end{bmatrix} = \begin{bmatrix} -2 \\ 4 \end{bmatrix}$$

Solution: $x_1 = -2$, $x_2 = 4$ (4-6)

19. $A + D = \begin{bmatrix} 2 & -2 \\ 1 & 0 \\ 3 & 2 \end{bmatrix} + \begin{bmatrix} 3 & -2 & 1 \\ -1 & 1 & 2 \end{bmatrix}$ Not defined, because the dimensions of A and D are different. (4-4)

20. $E + DA = \begin{bmatrix} 3 & -4 \\ -1 & 0 \end{bmatrix} + \begin{bmatrix} 3 & -2 & 1 \\ -1 & 1 & 2 \end{bmatrix}\begin{bmatrix} 2 & -2 \\ 1 & 0 \\ 3 & 2 \end{bmatrix} = \begin{bmatrix} 3 & -4 \\ -1 & 0 \end{bmatrix} + \begin{bmatrix} 7 & -4 \\ 5 & 6 \end{bmatrix} = \begin{bmatrix} 10 & -8 \\ 4 & 6 \end{bmatrix}$

(4-4)

21. From Problem 20, $DA = \begin{bmatrix} 7 & -4 \\ 5 & 6 \end{bmatrix}$. Thus,

$DA - 3E = \begin{bmatrix} 7 & -4 \\ 5 & 6 \end{bmatrix} - 3\begin{bmatrix} 3 & -4 \\ -1 & 0 \end{bmatrix} = \begin{bmatrix} 7 & -4 \\ 5 & 6 \end{bmatrix} + \begin{bmatrix} -9 & 12 \\ 3 & 0 \end{bmatrix} = \begin{bmatrix} -2 & 8 \\ 8 & 6 \end{bmatrix}$ (4-4)

22. $BC = \begin{bmatrix} -1 \\ 2 \\ 3 \end{bmatrix}[2 \quad 1 \quad 3] = \begin{bmatrix} -2 & -1 & -3 \\ 4 & 2 & 6 \\ 6 & 3 & 9 \end{bmatrix}$ (4-4)

23. $CB = [2 \quad 1 \quad 3]\begin{bmatrix} -1 \\ 2 \\ 3 \end{bmatrix} = [-2 + 2 + 9] = [9]$ (a 1×1 matrix) (4-4)

24. $AD - BC$

$$AD = \begin{bmatrix} 2 & -2 \\ 1 & 0 \\ 3 & 2 \end{bmatrix}\begin{bmatrix} 3 & -2 & 1 \\ -1 & 1 & 2 \end{bmatrix} = \begin{bmatrix} [2 \quad -2]\begin{bmatrix} 3 \\ -1 \end{bmatrix} & [2 \quad -2]\begin{bmatrix} -2 \\ 1 \end{bmatrix} & [2 \quad -2]\begin{bmatrix} 1 \\ 2 \end{bmatrix} \\ [1 \quad 0]\begin{bmatrix} 3 \\ -1 \end{bmatrix} & [1 \quad 0]\begin{bmatrix} -2 \\ 1 \end{bmatrix} & [1 \quad 0]\begin{bmatrix} 1 \\ 2 \end{bmatrix} \\ [3 \quad 2]\begin{bmatrix} 3 \\ -1 \end{bmatrix} & [3 \quad 2]\begin{bmatrix} -2 \\ 1 \end{bmatrix} & [3 \quad 2]\begin{bmatrix} 1 \\ 2 \end{bmatrix} \end{bmatrix}$$

$$= \begin{bmatrix} 8 & -6 & -2 \\ 3 & -2 & 1 \\ 7 & -4 & 7 \end{bmatrix}$$

$$BC = \begin{bmatrix} -1 \\ 2 \\ 3 \end{bmatrix} [2 \quad 1 \quad 3] = \begin{bmatrix} -2 & -1 & -3 \\ 4 & 2 & 6 \\ 6 & 3 & 9 \end{bmatrix}$$

$$AD - BC = \begin{bmatrix} 8 & -6 & -2 \\ 3 & -2 & 1 \\ 7 & -4 & 7 \end{bmatrix} - \begin{bmatrix} -2 & -1 & -3 \\ 4 & 2 & 6 \\ 6 & 3 & 9 \end{bmatrix} = \begin{bmatrix} 8-(-2) & -6-(-1) & -2-(-3) \\ 3-4 & -2-2 & 1-6 \\ 7-6 & -4-3 & 7-9 \end{bmatrix} = \begin{bmatrix} 10 & -5 & 1 \\ -1 & -4 & -5 \\ 1 & -7 & -2 \end{bmatrix}$$

(4-4)

25.
$$\left[\begin{array}{ccc|ccc} 1 & 2 & 3 & 1 & 0 & 0 \\ 2 & 3 & 4 & 0 & 1 & 0 \\ 1 & 2 & 1 & 0 & 0 & 1 \end{array}\right] \sim \left[\begin{array}{ccc|ccc} 1 & 2 & 3 & 1 & 0 & 0 \\ 0 & -1 & -2 & -2 & 1 & 0 \\ 0 & 0 & -2 & -1 & 0 & 1 \end{array}\right] \sim \left[\begin{array}{ccc|ccc} 1 & 2 & 3 & 1 & 0 & 0 \\ 0 & 1 & 2 & 2 & -1 & 0 \\ 0 & 0 & -2 & -1 & 0 & 1 \end{array}\right]$$

$(-2)R_1 + R_2 \to R_2$ $\quad$ $(-1)R_1 \to R_2$ $\quad$ $(-2)R_2 + R_1 \to R_1$

$(-1)R_1 + R_3 \to R_3$ $\quad$ $\left(-\frac{1}{2}\right)R_3 \to R_3$

$$\sim \left[\begin{array}{ccc|ccc} 1 & 0 & -1 & -3 & 2 & 0 \\ 0 & 1 & 2 & 2 & -1 & 0 \\ 0 & 0 & 1 & \frac{1}{2} & 0 & -\frac{1}{2} \end{array}\right] \sim \left[\begin{array}{ccc|ccc} 1 & 0 & 0 & -\frac{5}{2} & 2 & -\frac{1}{2} \\ 0 & 1 & 0 & 1 & -1 & 1 \\ 0 & 0 & 1 & \frac{1}{2} & 0 & -\frac{1}{2} \end{array}\right]; A^{-1} = \begin{bmatrix} -\frac{5}{2} & 2 & -\frac{1}{2} \\ 1 & -1 & 1 \\ \frac{1}{2} & 0 & -\frac{1}{2} \end{bmatrix}$$

$R_3 + R_1 \to R_1$

$(-2)R_3 + R_2 \to R_2$

<u>Check</u>:

$$A^{-1}A = \begin{bmatrix} -\frac{5}{2} & 2 & -\frac{1}{2} \\ 1 & -1 & 1 \\ \frac{1}{2} & 0 & -\frac{1}{2} \end{bmatrix} \begin{bmatrix} 1 & 2 & 3 \\ 2 & 3 & 4 \\ 1 & 2 & 1 \end{bmatrix} = \begin{bmatrix} -\frac{5}{2} + 4 - \frac{1}{2} & -5 + 6 - 1 & -\frac{15}{2} + 8 - \frac{1}{2} \\ 1 - 2 + 1 & 2 - 3 + 2 & 3 - 4 + 1 \\ \frac{1}{2} + 0 - \frac{1}{2} & 1 + 0 - 1 & \frac{3}{2} + 0 - \frac{1}{2} \end{bmatrix}$$

$$= \begin{bmatrix} 1 & 0 & 0 \\ 0 & 1 & 0 \\ 0 & 0 & 1 \end{bmatrix}$$

(4-5)

26. (A) The augmented matrix corresponding to the given system is:

$$\left[\begin{array}{ccc|c} 1 & 2 & 3 & 1 \\ 2 & 3 & 4 & 3 \\ 1 & 2 & 1 & 3 \end{array}\right] \sim \left[\begin{array}{ccc|c} 1 & 2 & 3 & 1 \\ 0 & -1 & -2 & 1 \\ 0 & 0 & -2 & 2 \end{array}\right] \sim \left[\begin{array}{ccc|c} 1 & 2 & 3 & 1 \\ 0 & 1 & 2 & -1 \\ 0 & 0 & -2 & 2 \end{array}\right] \sim \left[\begin{array}{ccc|c} 1 & 0 & -1 & 3 \\ 0 & 1 & 2 & -1 \\ 0 & 0 & -2 & 2 \end{array}\right]$$

$(-2)R_1 + R_2 \to R_2$ $\quad$ $(-1)R_2 \to R_2$ $\quad$ $(-2)R_2 + R_1 \to R_1$ $\quad$ $\left(-\frac{1}{2}\right)R_3 \to R_3$

$(-1)R_1 + R_3 \to R_3$

$$\sim \left[\begin{array}{ccc|c} 1 & 0 & -1 & 3 \\ 0 & 1 & 2 & -1 \\ 0 & 0 & 1 & -1 \end{array}\right] \sim \left[\begin{array}{ccc|c} 1 & 0 & 0 & 2 \\ 0 & 1 & 0 & 1 \\ 0 & 0 & 1 & -1 \end{array}\right]$$

Thus, the solution is: $x_1 = 2$, $x_2 = 1$, $x_3 = -1$.

$(-2)R_3 + R_2 \to R_2$

$R_3 + R_1 \to R_1$

(4-3)

(B) The augmented matrix corresponding to the given system is:

$$\left[\begin{array}{ccc|c} 1 & 2 & -1 & 2 \\ 2 & 3 & 1 & -3 \\ 3 & 5 & 0 & -1 \end{array}\right] \sim \left[\begin{array}{ccc|c} 1 & 2 & -1 & 2 \\ 0 & -1 & 3 & -7 \\ 0 & -1 & 3 & -7 \end{array}\right] \sim \left[\begin{array}{ccc|c} 1 & 2 & -1 & 2 \\ 0 & 1 & -3 & 7 \\ 0 & -1 & 3 & -7 \end{array}\right] \sim \left[\begin{array}{ccc|c} 1 & 0 & 5 & -12 \\ 0 & 1 & -3 & 7 \\ 0 & 0 & 0 & 0 \end{array}\right]$$

$(-2)R_1 + R_2 \to R_2$ $\quad$ $(-1)R_2 \to R_2$ $\quad$ $R_2 + R_3 \to R_3$

$(-3)R_1 + R_3 \to R_3$ $\quad$ $(-2)R_2 + R_1 \to R_1$

Thus, $x_1 \quad + 5x_3 = -12 \quad (1)$

$x_2 - 3x_3 = 7 \quad (2)$

Let $x_3 = t$ (t any real number). Then, from (1),

$x_1 = -5t - 12$

and, from (2),

$x_2 = 3t + 7$.

Thus, the solution is $x_1 = -5t - 12$, $x_2 = 3t + 7$, $x_3 = t$. (4-3)

27. (A) The matrix equation for the given system is:

$$\begin{bmatrix} 1 & 2 & 3 \\ 2 & 3 & 4 \\ 1 & 2 & 1 \end{bmatrix}\begin{bmatrix} x_1 \\ x_2 \\ x_3 \end{bmatrix} = \begin{bmatrix} 1 \\ 3 \\ 3 \end{bmatrix}$$

The inverse matrix of the coefficient matrix of the system, from Problem 25, is:

$$\begin{bmatrix} -\frac{5}{2} & 2 & -\frac{1}{2} \\ 1 & -1 & 1 \\ \frac{1}{2} & 0 & -\frac{1}{2} \end{bmatrix}$$ Thus,

$$\begin{bmatrix} x_1 \\ x_2 \\ x_3 \end{bmatrix} = \begin{bmatrix} -\frac{5}{2} & 2 & -\frac{1}{2} \\ 1 & -1 & 1 \\ \frac{1}{2} & 0 & -\frac{1}{2} \end{bmatrix}\begin{bmatrix} 1 \\ 3 \\ 3 \end{bmatrix} = \begin{bmatrix} \frac{-5 + 12 - 3}{2} \\ 1 - 3 + 3 \\ \frac{1 + 0 - 3}{2} \end{bmatrix} = \begin{bmatrix} 2 \\ 1 \\ -1 \end{bmatrix}$$

Solution: $x_1 = 2$, $x_2 = 1$, $x_3 = -1$.

(B) $$\begin{bmatrix} x_1 \\ x_2 \\ x_3 \end{bmatrix} = \begin{bmatrix} -\frac{5}{2} & 2 & -\frac{1}{2} \\ 1 & -1 & 1 \\ \frac{1}{2} & 0 & -\frac{1}{2} \end{bmatrix}\begin{bmatrix} 0 \\ 0 \\ -2 \end{bmatrix} = \begin{bmatrix} 1 \\ -2 \\ 1 \end{bmatrix}$$

Solution: $x_1 = 1$, $x_2 = -2$, $x_3 = 1$.

(C) $$\begin{bmatrix} x_1 \\ x_2 \\ x_3 \end{bmatrix} = \begin{bmatrix} -\frac{5}{2} & 2 & -\frac{1}{2} \\ 1 & -1 & 1 \\ \frac{1}{2} & 0 & -\frac{1}{2} \end{bmatrix}\begin{bmatrix} -3 \\ -4 \\ 1 \end{bmatrix} = \begin{bmatrix} -1 \\ 2 \\ -2 \end{bmatrix}$$

Solution: $x_1 = -1$, $x_2 = 2$, $x_3 = -2$. (4-6)

28. $2x_1 - 6x_2 = 4$

$-x_1 + kx_2 = -2$

If $k = 3$, then the first equation is a multiple (-2) of the second equation which implies there are infinitely many solutions. If $k \neq 3$, then the system has a unique solution. (4-3)

29. $M = \begin{bmatrix} 0.2 & 0.15 \\ 0.4 & 0.3 \end{bmatrix}$, $D = \begin{bmatrix} 30 \\ 20 \end{bmatrix}$

$$I - M = \begin{bmatrix} 1 & 0 \\ 0 & 1 \end{bmatrix} - \begin{bmatrix} 0.2 & 0.15 \\ 0.4 & 0.3 \end{bmatrix} = \begin{bmatrix} 0.8 & -0.15 \\ -0.4 & 0.7 \end{bmatrix} = \begin{bmatrix} \frac{4}{5} & -\frac{3}{20} \\ -\frac{2}{5} & \frac{7}{10} \end{bmatrix}$$

Now,

$$\left[\begin{array}{cc|cc} \frac{4}{5} & -\frac{3}{20} & 1 & 0 \\ -\frac{2}{5} & \frac{7}{10} & 0 & 1 \end{array}\right] \sim \left[\begin{array}{cc|cc} 1 & -\frac{3}{16} & \frac{5}{4} & 0 \\ -\frac{2}{5} & \frac{7}{10} & 0 & 1 \end{array}\right] \sim \left[\begin{array}{cc|cc} 1 & -\frac{3}{16} & \frac{5}{4} & 0 \\ 0 & \frac{5}{8} & \frac{1}{2} & 1 \end{array}\right] \sim \left[\begin{array}{cc|cc} 1 & -\frac{3}{16} & \frac{5}{4} & 0 \\ 0 & 1 & \frac{4}{5} & \frac{8}{5} \end{array}\right]$$

$\frac{5}{4}R_1 \to R_1$ $\qquad \frac{2}{5}R_1 + R_2 \to R_2$ $\qquad \frac{8}{5}R_2 \to R_2$ $\qquad \frac{3}{16}R_2 + R_1 \to R_1$

$$\sim \left[\begin{array}{cc|cc} 1 & 0 & \frac{7}{5} & \frac{3}{10} \\ 0 & 1 & \frac{4}{5} & \frac{8}{5} \end{array}\right]$$ Thus, $(I - M)^{-1} = \begin{bmatrix} \frac{7}{5} & \frac{3}{10} \\ \frac{4}{5} & \frac{8}{5} \end{bmatrix} = \begin{bmatrix} 1.4 & 0.3 \\ 0.8 & 1.6 \end{bmatrix}$

The output matrix $X = (I - M)^{-1}D = \begin{bmatrix} 1.4 & 0.3 \\ 0.8 & 1.6 \end{bmatrix}\begin{bmatrix} 30 \\ 20 \end{bmatrix} = \begin{bmatrix} 48 \\ 56 \end{bmatrix}$ (4-7)

30. The graphs of the two equations are:

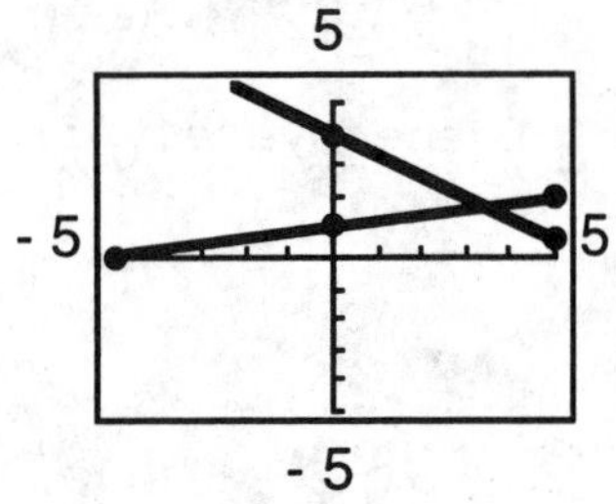

5

-5 5

-5

$x \approx 3.46$, $y \approx 1.69$ (4-1)

31. $\left[\begin{array}{ccc|ccc} 4 & 5 & 6 & 1 & 0 & 0 \\ 4 & 5 & -4 & 0 & 1 & 0 \\ 1 & 1 & 1 & 0 & 0 & 1 \end{array}\right]$ $R_1 \leftrightarrow R_3$

$\sim \left[\begin{array}{ccc|ccc} 1 & 1 & 1 & 0 & 0 & 1 \\ 4 & 5 & -4 & 0 & 1 & 0 \\ 4 & 5 & 6 & 1 & 0 & 0 \end{array}\right]$ $\begin{array}{l} (-4)R_1 + R_2 \to R_2 \\ (-4)R_1 + R_3 \to R_3 \end{array}$

$\sim \left[\begin{array}{ccc|ccc} 1 & 1 & 1 & 0 & 0 & 1 \\ 0 & 1 & -8 & 0 & 1 & -4 \\ 0 & 1 & 2 & 1 & 0 & -4 \end{array}\right]$ $\begin{array}{l} (-1)R_2 + R_1 \to R_1 \\ (-1)R_2 + R_3 \to R_3 \end{array}$

$\sim \left[\begin{array}{ccc|ccc} 1 & 0 & 9 & 0 & -1 & 5 \\ 0 & 1 & -8 & 0 & 1 & -4 \\ 0 & 0 & 10 & 1 & -1 & 0 \end{array}\right]$ $\frac{1}{10}R_3 \to R_3$

$$\sim\left[\begin{array}{ccc|ccc} 1 & 0 & 9 & 0 & -1 & 5 \\ 0 & 1 & -8 & 0 & 1 & -4 \\ 0 & 0 & 1 & \frac{1}{10} & -\frac{1}{10} & 0 \end{array}\right] \begin{array}{l} (-9)R_3 + R_1 \rightarrow R_1 \\ 8R_3 + R_2 \rightarrow R_2 \end{array}$$

$$\sim\left[\begin{array}{ccc|ccc} 1 & 0 & 0 & -\frac{9}{10} & -\frac{1}{10} & 5 \\ 0 & 1 & 0 & \frac{8}{10} & \frac{2}{10} & -4 \\ 0 & 0 & 1 & \frac{1}{10} & -\frac{1}{10} & 0 \end{array}\right]$$

Thus, $A^{-1} = \begin{bmatrix} -0.9 & -0.1 & 5 \\ 0.8 & 0.2 & -4 \\ 0.1 & -0.1 & 0 \end{bmatrix}$;

$$A^{-1}A = \begin{bmatrix} -0.9 & -0.1 & 5 \\ 0.8 & 0.2 & -4 \\ 0.1 & -0.1 & 0 \end{bmatrix}\begin{bmatrix} 4 & 5 & 6 \\ 4 & 5 & -4 \\ 1 & 1 & 1 \end{bmatrix} = \begin{bmatrix} 1 & 0 & 0 \\ 0 & 1 & 0 \\ 0 & 0 & 1 \end{bmatrix} \qquad (4\text{-}5)$$

32. The given system is equivalent to:

$$\begin{aligned} 4x_1 + 5x_2 + 6x_3 &= 36{,}000 \\ 4x_1 + 5x_2 - 4x_3 &= 12{,}000 \\ x_1 + x_2 + x_3 &= 7{,}000 \end{aligned}$$

In matrix form, this system is:

$$\begin{bmatrix} 4 & 5 & 6 \\ 4 & 5 & -4 \\ 1 & 1 & 1 \end{bmatrix}\begin{bmatrix} x_1 \\ x_2 \\ x_3 \end{bmatrix} = \begin{bmatrix} 36{,}000 \\ 12{,}000 \\ 7{,}000 \end{bmatrix}$$

Thus,

$$\begin{bmatrix} x_1 \\ x_2 \\ x_3 \end{bmatrix} = \begin{bmatrix} 4 & 5 & 6 \\ 4 & 5 & -4 \\ 1 & 1 & 1 \end{bmatrix}^{-1}\begin{bmatrix} 36{,}000 \\ 12{,}000 \\ 7{,}000 \end{bmatrix} = \begin{bmatrix} -0.9 & -0.1 & 5 \\ 0.8 & 0.2 & -4 \\ 0.1 & -0.1 & 0 \end{bmatrix}\begin{bmatrix} 36{,}000 \\ 12{,}000 \\ 7{,}000 \end{bmatrix} = \begin{bmatrix} 1{,}400 \\ 3{,}200 \\ 2{,}400 \end{bmatrix}$$

Solution: $x_1 = 1{,}400$, $x_2 = 3{,}200$, $x_3 = 2{,}400$ (4-6)

33. First, multiply the first two equations of the system by 100. Then the augmented matrix of the resulting system is:

$$\left[\begin{array}{ccc|c} 4 & 5 & 6 & 36{,}000 \\ 4 & 5 & -4 & 12{,}000 \\ 1 & 1 & 1 & 7{,}000 \end{array}\right] R_1 \leftrightarrow R_3$$

$$\sim\left[\begin{array}{ccc|c} 1 & 1 & 1 & 7{,}000 \\ 4 & 5 & -4 & 12{,}000 \\ 4 & 5 & 6 & 36{,}000 \end{array}\right] \begin{array}{l} (-4)R_1 + R_2 \rightarrow R_2 \\ (-4)R_1 + R_3 \rightarrow R_3 \end{array}$$

$$\sim\left[\begin{array}{ccc|c} 1 & 1 & 1 & 7{,}000 \\ 0 & 1 & -8 & -16{,}000 \\ 0 & 1 & 2 & 8{,}000 \end{array}\right] \begin{array}{l} (-1)R_2 + R_1 \rightarrow R_1 \\ (-1)R_2 + R_3 \rightarrow R_3 \end{array}$$

$$\sim\left[\begin{array}{ccc|c} 1 & 0 & 9 & 23{,}000 \\ 0 & 1 & -8 & -16{,}000 \\ 0 & 0 & 10 & 24{,}000 \end{array}\right] \frac{1}{10}R_3 \rightarrow R_3$$

$$\sim\left[\begin{array}{ccc|c} 1 & 0 & 9 & 23{,}000 \\ 0 & 1 & -8 & -16{,}000 \\ 0 & 0 & 1 & 2{,}400 \end{array}\right] \begin{array}{r} (-9)R_3 + R_1 \to R_1 \\ 8R_3 + R_2 \to R_2 \end{array}$$

$$\left[\begin{array}{ccc|c} 1 & 0 & 0 & 1{,}400 \\ 0 & 1 & 0 & 3{,}200 \\ 0 & 0 & 1 & 2{,}400 \end{array}\right] \begin{array}{l} x_1 = 1{,}400 \\ x_2 = 3{,}200 \\ x_3 = 2{,}400 \end{array}$$

Solution: $x_1 = 1{,}400$, $x_2 = 3{,}200$, $x_3 = 2{,}400$ (4-3)

34. $M = \begin{bmatrix} 0.2 & 0 & 0.4 \\ 0.1 & 0.3 & 0.1 \\ 0 & 0.4 & 0.2 \end{bmatrix} = \begin{bmatrix} \frac{1}{5} & 0 & \frac{2}{5} \\ \frac{1}{10} & \frac{3}{10} & \frac{1}{10} \\ 0 & \frac{2}{5} & \frac{1}{5} \end{bmatrix}$ and $D = \begin{bmatrix} 40 \\ 20 \\ 30 \end{bmatrix}$

$$I - M = \begin{bmatrix} 1 & 0 & 0 \\ 0 & 1 & 0 \\ 0 & 0 & 1 \end{bmatrix} - \begin{bmatrix} \frac{1}{5} & 0 & \frac{2}{5} \\ \frac{1}{10} & \frac{3}{10} & \frac{1}{10} \\ 0 & \frac{2}{5} & \frac{1}{5} \end{bmatrix} = \begin{bmatrix} \frac{4}{5} & 0 & -\frac{2}{5} \\ -\frac{1}{10} & \frac{7}{10} & -\frac{1}{10} \\ 0 & -\frac{2}{5} & \frac{4}{5} \end{bmatrix}.$$

$$\left[\begin{array}{ccc|ccc} \frac{4}{5} & 0 & -\frac{2}{5} & 1 & 0 & 0 \\ -\frac{1}{10} & \frac{7}{10} & -\frac{1}{10} & 0 & 1 & 0 \\ 0 & -\frac{2}{5} & \frac{4}{5} & 0 & 0 & 1 \end{array}\right] \sim \left[\begin{array}{ccc|ccc} 1 & 0 & -\frac{1}{2} & \frac{5}{4} & 0 & 0 \\ -\frac{1}{10} & \frac{7}{10} & -\frac{1}{10} & 0 & 1 & 0 \\ 0 & -\frac{2}{5} & \frac{4}{5} & 0 & 0 & 1 \end{array}\right]$$

$\frac{5}{4}R_1 \to R_1$ $\qquad$ $\frac{1}{10}R_1 + R_2 \to R_2$

$$\sim \left[\begin{array}{ccc|ccc} 1 & 0 & -\frac{1}{2} & \frac{5}{4} & 0 & 0 \\ 0 & \frac{7}{10} & -\frac{3}{20} & \frac{1}{8} & 1 & 0 \\ 0 & -\frac{2}{5} & \frac{4}{5} & 0 & 0 & 1 \end{array}\right] \sim \left[\begin{array}{ccc|ccc} 1 & 0 & -\frac{1}{2} & \frac{5}{4} & 0 & 0 \\ 0 & -\frac{2}{5} & \frac{4}{5} & 0 & 0 & 1 \\ 0 & \frac{7}{10} & -\frac{3}{20} & \frac{1}{8} & 1 & 0 \end{array}\right]$$

$R_2 \leftrightarrow R_3$ $\qquad$ $\left(-\frac{5}{2}\right)R_2 \to R_2$

$$\sim \left[\begin{array}{ccc|ccc} 1 & 0 & -\frac{1}{2} & \frac{5}{4} & 0 & 0 \\ 0 & 1 & -2 & 0 & 0 & -\frac{5}{2} \\ 0 & \frac{7}{10} & -\frac{3}{20} & \frac{1}{8} & 1 & 0 \end{array}\right] \sim \left[\begin{array}{ccc|ccc} 1 & 0 & -\frac{1}{2} & \frac{5}{4} & 0 & 0 \\ 0 & 1 & -2 & 0 & 0 & -\frac{5}{2} \\ 0 & 0 & \frac{5}{4} & \frac{1}{8} & 1 & \frac{7}{4} \end{array}\right]$$

$\left(-\frac{7}{10}\right)R_2 + R_3 \to R_3$ $\qquad$ $\frac{4}{5}R_3 \to R_3$

$$\sim \left[\begin{array}{ccc|ccc} 1 & 0 & -\frac{1}{2} & \frac{5}{4} & 0 & 0 \\ 0 & 1 & -2 & 0 & 0 & -\frac{5}{2} \\ 0 & 0 & 1 & \frac{1}{10} & \frac{4}{5} & \frac{7}{5} \end{array}\right] \sim \left[\begin{array}{ccc|ccc} 1 & 0 & 0 & \frac{13}{10} & \frac{2}{5} & \frac{7}{10} \\ 0 & 1 & 0 & \frac{1}{5} & \frac{8}{5} & \frac{3}{10} \\ 0 & 0 & 1 & \frac{1}{10} & \frac{4}{5} & \frac{7}{5} \end{array}\right]$$

$2R_3 + R_2 \to R_2$, $\frac{1}{2}R_3 + R_1 \to R_1$

Thus $(I - M)^{-1} = \begin{bmatrix} \frac{13}{10} & \frac{2}{5} & \frac{7}{10} \\ \frac{1}{5} & \frac{8}{5} & \frac{3}{10} \\ \frac{1}{10} & \frac{4}{5} & \frac{7}{5} \end{bmatrix} = \begin{bmatrix} 1.3 & 0.4 & 0.7 \\ 0.2 & 1.6 & 0.3 \\ 0.1 & 0.8 & 1.4 \end{bmatrix}$ and

$$X = (I - M)^{-1}D = \begin{bmatrix} 1.3 & 0.4 & 0.7 \\ 0.2 & 1.6 & 0.3 \\ 0.1 & 0.8 & 1.4 \end{bmatrix}\begin{bmatrix} 40 \\ 20 \\ 30 \end{bmatrix} = \begin{bmatrix} 81 \\ 49 \\ 62 \end{bmatrix}$$ (4-7)

35. (A) The system has a unique solution.

(B) The system either has no solutions or infinitely many solutions. (4-6)

36. (A) The system has a unique solution.

(B) The system has **no** solutions.

(C) The system has infinitely many solutions. (4-3)

37. The third step in (A) is incorrect:

$X - MX = (I - M)X$ **not** $X(I - M)$

Each step in (B) is correct. (4-6)

38. Let x = the number of machines produced.

(A) $C(x) = 243{,}000 + 22.45x$
$R(x) = 59.95x$

(B) Set $C(x) = R(x)$:

$59.95x = 243{,}000 + 22.45x$
$37.5x = 243{,}000$
$x = 6{,}480$

If 6,480 machines are produced, $C = R = \$388{,}476$;
break-even point (6,480, 388,476).

(C) A profit occurs if $x > 6{,}480$; a loss occurs if $x < 6{,}480$

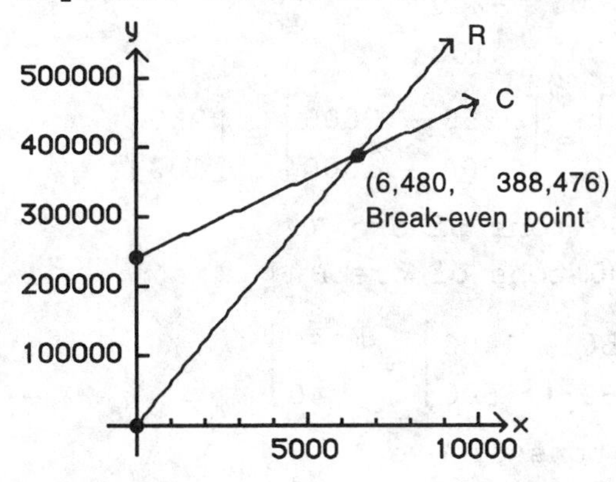

(4-1)

39. Let x_1 = number of tons of ore A
and x_2 = number of tons of ore B.

Then, we have the following system of equations:

$0.01x_1 + 0.02x_2 = 4.5$
$0.02x_1 + 0.05x_2 = 10$

Multiply each equation by 100. This yields

$x_1 + 2x_2 = 450$
$2x_1 + 5x_2 = 1000$

The augmented matrix corresponding to this system is:

$$\left[\begin{array}{cc|c} 1 & 2 & 450 \\ 2 & 5 & 1000 \end{array}\right] \sim \left[\begin{array}{cc|c} 1 & 2 & 450 \\ 0 & 1 & 100 \end{array}\right] \sim \left[\begin{array}{cc|c} 1 & 0 & 250 \\ 0 & 1 & 100 \end{array}\right]$$

$(-2)R_1 + R_2 \rightarrow R_2 \quad (-2)R_2 + R_1 \rightarrow R_1$

Thus, the solution is: x_1 = 250 tons of ore A, x_2 = 100 tons of ore B.

(4-3)

40. (A) The matrix equation for Problem 35 is:

$$\begin{bmatrix} 0.01 & 0.02 \\ 0.02 & 0.05 \end{bmatrix}\begin{bmatrix} x_1 \\ x_2 \end{bmatrix} = \begin{bmatrix} 4.5 \\ 10 \end{bmatrix}$$

First, compute the inverse of $\begin{bmatrix} 0.01 & 0.02 \\ 0.02 & 0.05 \end{bmatrix}$;

$$\left[\begin{array}{cc|cc} 0.01 & 0.02 & 1 & 0 \\ 0.02 & 0.05 & 0 & 1 \end{array}\right] \sim \left[\begin{array}{cc|cc} 1 & 2 & 100 & 0 \\ 0.02 & 0.05 & 0 & 1 \end{array}\right] \sim \left[\begin{array}{cc|cc} 1 & 2 & 100 & 0 \\ 0 & 0.01 & -2 & 1 \end{array}\right]$$

$100R_1 \rightarrow R_1 \qquad (-0.02)R_1 + R_2 \rightarrow R_2 \qquad 100R_2 \rightarrow R_2$

$$\sim \left[\begin{array}{cc|cc} 1 & 2 & 100 & 0 \\ 0 & 1 & -200 & 100 \end{array}\right] \sim \left[\begin{array}{cc|cc} 1 & 0 & 500 & -200 \\ 0 & 1 & -200 & 100 \end{array}\right]$$

$(-2)R_2 + R_1 \rightarrow R_1$

Thus, the inverse matrix is $\begin{bmatrix} 500 & -200 \\ -200 & 100 \end{bmatrix}$.

Hence, $\begin{bmatrix} x_1 \\ x_2 \end{bmatrix} = \begin{bmatrix} 500 & -200 \\ -200 & 100 \end{bmatrix}\begin{bmatrix} 4.5 \\ 10 \end{bmatrix} = \begin{bmatrix} 2250 - 2000 \\ -900 + 1000 \end{bmatrix} = \begin{bmatrix} 250 \\ 100 \end{bmatrix}$

Again the solution is: x_1 = 250 tons of ore A.
x_2 = 100 tons of ore B.

(B) $\begin{bmatrix} x_1 \\ x_2 \end{bmatrix} = \begin{bmatrix} 500 & -200 \\ -200 & 100 \end{bmatrix}\begin{bmatrix} 2.3 \\ 5 \end{bmatrix} = \begin{bmatrix} 1150 - 1000 \\ -460 + 500 \end{bmatrix} = \begin{bmatrix} 150 \\ 40 \end{bmatrix}$

Now the solution is: x_1 = 150 tons of ore A.
x_2 = 40 tons of ore B.

(4-6)

41. Let x_1 = number of model A trucks

x_2 = number of model B trucks

x_3 = number of model C trucks

Then $x_1 + x_2 + x_3 = 12$

and $18{,}000x_1 + 22{,}000x_2 + 30{,}000x_3 = 300{,}000$

or
$$\begin{aligned} x_1 + x_2 + x_3 &= 12 \\ 9x_1 + 11x_2 + 15x_3 &= 150 \end{aligned}$$

The augmented matrix corresponding to this system is $\left[\begin{array}{ccc|c} 1 & 1 & 1 & 12 \\ 9 & 11 & 15 & 150 \end{array}\right]$

Now

$$\left[\begin{array}{ccc|c} 1 & 1 & 1 & 12 \\ 9 & 11 & 15 & 150 \end{array}\right] \sim \left[\begin{array}{ccc|c} 1 & 1 & 1 & 12 \\ 0 & 2 & 6 & 42 \end{array}\right] \sim \left[\begin{array}{ccc|c} 1 & 1 & 1 & 12 \\ 0 & 1 & 3 & 21 \end{array}\right] \sim \left[\begin{array}{ccc|c} 1 & 0 & -2 & -9 \\ 0 & 1 & 3 & 21 \end{array}\right]$$

$(-9)R_1 + R_2 \to R_2$ $\quad \frac{1}{2}R_2 \to R_2$ $\quad (-1)R_2 + R_1 \to R_1$

The corresponding system of equations is

$$\begin{aligned} x_1 \quad - 2x_3 &= -9 \\ x_2 + 3x_3 &= 21 \end{aligned}$$

and the solutions are

$$\begin{aligned} x_1 &= 2t - 9 \\ x_2 &= 21 - 3t \\ x_3 &= t \end{aligned}$$

Now, since x_1, x_2, and x_3 are nonnegative integers, we must have $\frac{9}{2} \le t \le 7$ or $t = 5$, 6, or 7.

For $t = 5$: 1 model A truck, 6 model B trucks, 5 model C trucks

$t = 6$: 3 model A trucks, 3 model B trucks, 6 model C trucks

$t = 7$: 5 model A trucks, 0 model B trucks, 7 model C trucks (4-3)

42. (A) The elements of MN give the cost of materials for each alloy from each supplier. The product NM is also defined, but does not have an interpretation in the context of this problem.

(B) $MN = \begin{bmatrix} 4{,}800 & 600 & 300 \\ 6{,}000 & 1{,}400 & 700 \end{bmatrix} \begin{bmatrix} 0.75 & 0.70 \\ 6.50 & 6.70 \\ 0.40 & 0.50 \end{bmatrix}$

$$= \begin{array}{c} \text{Supplier A} \quad \text{Supplier B} \\ \begin{bmatrix} \$\ 7{,}620 & \$\ 7{,}530 \\ \$13{,}880 & \$13{,}930 \end{bmatrix} \end{array} \begin{array}{l} \text{Alloy 1} \\ \text{Alloy 2} \end{array}$$

(C) The total costs of materials from Supplier A is:

$7,620 + $13,880 = $21,500

The total costs of materials from Supplier B is:

$7,530 + $13,930 = $21,460

These values can be obtained from the matrix product

$$\begin{bmatrix} 1 & 1 \end{bmatrix} \begin{bmatrix} 7{,}620 & 7{,}530 \\ 13{,}880 & 13{,}930 \end{bmatrix}$$

Supplier B will provide the materials at lower cost. (4-4)

43. (A) $[0.25 \quad 0.20 \quad 0.05]\begin{bmatrix} 15 \\ 12 \\ 4 \end{bmatrix} = 6.35$

The labor cost for one model *B* calculator at the California plant is $6.35.

(B) The elements of *MN* give the total labor costs for each calculator at each plant. The product *NM* is also defined, but does not have an interpretation in the context of this problem.

(C) $MN = \begin{bmatrix} 0.15 & 0.10 & 0.05 \\ 0.25 & 0.20 & 0.05 \end{bmatrix}\begin{bmatrix} 15 & 12 \\ 12 & 10 \\ 4 & 4 \end{bmatrix} = \begin{bmatrix} \$3.65 & \$3.00 \\ \$6.35 & \$5.20 \end{bmatrix}$ (columns: Calif., Texas; rows: Model A, Model B) (4-4)

44. Let x_1 = amount invested at 5%
and x_2 = amount invested at 10%.

Then,

$$x_1 + x_2 = 5000$$
$$0.05x_1 + 0.1x_2 = 400$$

The augmented matrix for the system given above is:

$$\left[\begin{array}{cc|c} 1 & 1 & 5000 \\ 0.05 & 0.1 & 400 \end{array}\right] \sim \left[\begin{array}{cc|c} 1 & 1 & 5000 \\ 0 & 0.05 & 150 \end{array}\right] \sim \left[\begin{array}{cc|c} 1 & 1 & 5000 \\ 0 & 1 & 3000 \end{array}\right] \sim \left[\begin{array}{cc|c} 1 & 0 & 2000 \\ 0 & 1 & 3000 \end{array}\right]$$

$(-0.05)R_1 + R_2 \to R_2$ $\quad \frac{1}{0.05}R_2 \to R_2$ $\quad (-1)R_2 + R_1 \to R_1$

Hence, x_1 = $2000 at 5%, x_2 = $3000 at 10%. (4-3)

45. The matrix equation corresponding to the system in Problem 35 is:

$$\begin{bmatrix} 1 & 1 \\ 0.05 & 0.1 \end{bmatrix}\begin{bmatrix} x_1 \\ x_2 \end{bmatrix} = \begin{bmatrix} 5000 \\ 400 \end{bmatrix}$$

Now we compute the inverse matrix of $\begin{bmatrix} 1 & 1 \\ 0.05 & 0.1 \end{bmatrix}$.

$$\left[\begin{array}{cc|cc} 1 & 1 & 1 & 0 \\ 0.05 & 0.1 & 0 & 1 \end{array}\right] \sim \left[\begin{array}{cc|cc} 1 & 1 & 1 & 0 \\ 0 & 0.05 & -0.05 & 1 \end{array}\right] \sim \left[\begin{array}{cc|cc} 1 & 1 & 1 & 0 \\ 0 & 1 & -1 & 20 \end{array}\right] \sim \left[\begin{array}{cc|cc} 1 & 0 & 2 & -20 \\ 0 & 1 & -1 & 20 \end{array}\right]$$

$(-0.05)R_1 + R_2 \to R_2$ $\quad \frac{1}{0.05}R_2 \to R_2$ $\quad (-1)R_2 + R_1 \to R_1$

Thus, the inverse of the coefficient matrix is $\begin{bmatrix} 2 & -20 \\ -1 & 20 \end{bmatrix}$, and

$$\begin{bmatrix} x_1 \\ x_2 \end{bmatrix} = \begin{bmatrix} 2 & -20 \\ -1 & 20 \end{bmatrix}\begin{bmatrix} 5000 \\ 400 \end{bmatrix} = \begin{bmatrix} 10,000 - 8,000 \\ -5,000 + 8,000 \end{bmatrix} = \begin{bmatrix} 2000 \\ 3000 \end{bmatrix}.$$

So, x_1 = $2000 at 5%,
x_2 = $3000 at 10%.

46. From Problem 44, the system of equations is:

$$x_1 + x_2 = 5000$$
$$0.05x_1 + 0.1x_2 = k \quad \text{(annual return)}$$

Calculate the inverse of the coefficient matrix A.

$$\left[\begin{array}{cc|cc} 1 & 1 & 1 & 0 \\ 0.05 & 0.1 & 0 & 1 \end{array}\right] \sim \left[\begin{array}{cc|cc} 1 & 1 & 1 & 0 \\ 0 & 0.05 & -0.05 & 1 \end{array}\right]$$

$(-0.05)R_1 + R_2 \to R_2$ $\qquad \frac{1}{0.05}R_2 \to R_2$

$$\sim \left[\begin{array}{cc|cc} 1 & 1 & 1 & 0 \\ 0 & 1 & -1 & 20 \end{array}\right] \sim \left[\begin{array}{cc|cc} 1 & 0 & 2 & -20 \\ 0 & 1 & -1 & 20 \end{array}\right]$$

$(-1)R_2 + R_1 \to R_1$

Thus, $A^{-1} = \begin{bmatrix} 2 & -20 \\ -1 & 20 \end{bmatrix}$

$k = \$200$?

$$\begin{bmatrix} x_1 \\ x_2 \end{bmatrix} = \begin{bmatrix} 2 & -20 \\ -1 & 20 \end{bmatrix}\begin{bmatrix} 5000 \\ 200 \end{bmatrix} = \begin{bmatrix} 6000 \\ -1000 \end{bmatrix}; \; x_1 = \$6{,}000, \; x_2 = -\$1{,}000$$

This is not possible, x_2 cannot be negative.

$k = \$600$?

$$\begin{bmatrix} x_1 \\ x_2 \end{bmatrix} = \begin{bmatrix} 2 & -20 \\ -1 & 20 \end{bmatrix}\begin{bmatrix} 5000 \\ 600 \end{bmatrix} = \begin{bmatrix} -2000 \\ 7000 \end{bmatrix}; \; x_1 = -\$2{,}000, \; x_2 = \$7{,}000$$

This is not possible, x_1 cannot be negative.

Fix a return k. Then

$$\begin{bmatrix} x_1 \\ x_2 \end{bmatrix} = \begin{bmatrix} 2 & -20 \\ -1 & 20 \end{bmatrix}\begin{bmatrix} 5000 \\ k \end{bmatrix} = \begin{bmatrix} 10{,}000 - 20k \\ -5{,}000 + 20k \end{bmatrix}$$

so $x_1 = 10{,}000 - 20k$, $x_2 = -5{,}000 + 20k$

Since $x_1 \ge 0$, we have $\quad 10{,}000 - 20k \ge 0$

$$20k \le 10{,}000$$
$$k \le 500$$

Since $x_2 \ge 0$, we have $\quad -5{,}000 + 20k \ge 0$

$$20k \ge 5{,}000$$
$$k \ge 250$$

The possible annual yields must satisfy $250 \le k \le 500$. (4-6)

47. Let x_1 = number of \$8 tickets
x_2 = number of \$12 tickets
x_3 = number of \$20 tickets

Since the number of \$8 tickets must equal the number of \$20 tickets, we have

$$x_1 = x_3 \quad \text{or} \quad x_1 - x_3 = 0$$

Also, since all seats are sold

$$x_1 + x_2 + x_3 = 25{,}000$$

Finally, the return is

$$8x_1 + 12x_2 + 20x_3 = R \text{ (where } R \text{ is the return required).}$$

Thus, the system of equations is:

$$\begin{aligned} x_1 \quad\quad - x_3 &= 0 \\ x_1 + x_2 + x_3 &= 25{,}000 \\ 8x_1 + 12x_2 + 20x_3 &= R \end{aligned} \quad \text{or} \quad \begin{bmatrix} 1 & 0 & -1 \\ 1 & 1 & 1 \\ 8 & 12 & 20 \end{bmatrix}\begin{bmatrix} x_1 \\ x_2 \\ x_3 \end{bmatrix} = \begin{bmatrix} 0 \\ 25{,}000 \\ R \end{bmatrix}$$

First, we compute the inverse of the coefficient matrix

$$\left[\begin{array}{ccc|ccc} 1 & 0 & -1 & 1 & 0 & 0 \\ 1 & 1 & 1 & 0 & 1 & 0 \\ 8 & 12 & 20 & 0 & 0 & 1 \end{array}\right] \sim \left[\begin{array}{ccc|ccc} 1 & 0 & -1 & 1 & 0 & 0 \\ 0 & 1 & 2 & -1 & 1 & 0 \\ 0 & 12 & 28 & -8 & 0 & 1 \end{array}\right]$$

$(-1)R_1 + R_2 \to R_2$ $\quad$ $(-12)R_2 + R_3 \to R_3$
$(-8)R_1 + R_3 \to R_3$

$$\sim \left[\begin{array}{ccc|ccc} 1 & 0 & -1 & 1 & 0 & 0 \\ 0 & 1 & 2 & -1 & 1 & 0 \\ 0 & 0 & 4 & 4 & -12 & 1 \end{array}\right] \sim \left[\begin{array}{ccc|ccc} 1 & 0 & -1 & 1 & 0 & 0 \\ 0 & 1 & 2 & -1 & 1 & 0 \\ 0 & 0 & 1 & 1 & -3 & \frac{1}{4} \end{array}\right]$$

$\frac{1}{4}R_3 \to R_3$ $\quad$ $(-2)R_3 + R_2 \to R_2$
$R_3 + R_1 \to R_1$

$$\sim \left[\begin{array}{ccc|ccc} 1 & 0 & 0 & 2 & -3 & \frac{1}{4} \\ 0 & 1 & 0 & -3 & 7 & -\frac{1}{2} \\ 0 & 0 & 1 & 1 & -3 & \frac{1}{4} \end{array}\right]. \text{ Thus, the inverse is } \begin{bmatrix} 2 & -3 & \frac{1}{4} \\ -3 & 7 & -\frac{1}{2} \\ 1 & -3 & \frac{1}{4} \end{bmatrix}$$

Concert 1:

$$\begin{aligned} x_1 \quad\quad - x_3 &= 0 \\ x_1 + x_2 + x_3 &= 25{,}000 \\ 8x_1 + 12x_2 + 20x_3 &= 320{,}000 \end{aligned} \quad \text{or} \quad \begin{bmatrix} 1 & 0 & -1 \\ 1 & 1 & 1 \\ 8 & 12 & 20 \end{bmatrix}\begin{bmatrix} x_1 \\ x_2 \\ x_3 \end{bmatrix} = \begin{bmatrix} 0 \\ 25{,}000 \\ 320{,}000 \end{bmatrix}$$

$$\text{Thus } \begin{bmatrix} x_1 \\ x_2 \\ x_3 \end{bmatrix} = \begin{bmatrix} 2 & -3 & \frac{1}{4} \\ -3 & 7 & -\frac{1}{2} \\ 1 & -3 & \frac{1}{4} \end{bmatrix}\begin{bmatrix} 0 \\ 25{,}000 \\ 320{,}000 \end{bmatrix} = \begin{bmatrix} 5{,}000 \\ 15{,}000 \\ 5{,}000 \end{bmatrix}$$

and $x_1 = 5,000$ \$8 tickets
$x_2 = 15,000$ \$12 tickets
$x_3 = 5,000$ \$20 tickets

Concert 2:

$$\begin{aligned} x_1 \qquad\quad - x_3 &= 0 \\ x_1 + x_2 + x_3 &= 25,000 \\ 8x_1 + 12x_2 + 20x_3 &= 330,000 \end{aligned} \quad \text{or} \quad \begin{bmatrix} 1 & 0 & -1 \\ 1 & 1 & 1 \\ 8 & 12 & 20 \end{bmatrix} \begin{bmatrix} x_1 \\ x_2 \\ x_3 \end{bmatrix} = \begin{bmatrix} 0 \\ 25,000 \\ 330,000 \end{bmatrix}$$

$$\text{Thus } \begin{bmatrix} x_1 \\ x_2 \\ x_3 \end{bmatrix} = \begin{bmatrix} 2 & -3 & \frac{1}{4} \\ -3 & 7 & -\frac{1}{2} \\ 1 & -3 & \frac{1}{4} \end{bmatrix} \begin{bmatrix} 0 \\ 25,000 \\ 330,000 \end{bmatrix} = \begin{bmatrix} 7,500 \\ 10,000 \\ 7,500 \end{bmatrix}$$

and $x_1 = 7,500$ \$8 tickets
$x_2 = 10,000$ \$12 tickets
$x_3 = 7,500$ \$20 tickets

Concert 3:

$$\begin{aligned} x_1 \qquad\quad - x_3 &= 0 \\ x_1 + x_2 + x_3 &= 25,000 \\ 8x_1 + 12x_2 + 20x_3 &= 340,000 \end{aligned} \quad \text{or} \quad \begin{bmatrix} 1 & 0 & -1 \\ 1 & 1 & 1 \\ 8 & 12 & 20 \end{bmatrix} \begin{bmatrix} x_1 \\ x_2 \\ x_3 \end{bmatrix} = \begin{bmatrix} 0 \\ 25,000 \\ 340,000 \end{bmatrix}$$

$$\text{Thus } \begin{bmatrix} x_1 \\ x_2 \\ x_3 \end{bmatrix} = \begin{bmatrix} 2 & -3 & \frac{1}{4} \\ -3 & 7 & -\frac{1}{2} \\ 1 & -3 & \frac{1}{4} \end{bmatrix} \begin{bmatrix} 0 \\ 25,000 \\ 340,000 \end{bmatrix} = \begin{bmatrix} 10,000 \\ 5,000 \\ 10,000 \end{bmatrix}$$

and $x_1 = 10,000$ \$8 tickets
$x_2 = 5,000$ \$12 tickets
$x_3 = 10,000$ \$20 tickets

(4-6)

48. From Problem 47, if it is not required to have an equal number of \$8 tickets and \$12 tickets, then the new mathematical model is:

$$\begin{aligned} x_1 + x_2 + x_3 &= 25,000 \\ 8x_1 + 12x_2 + 20x_3 &= k \quad \text{(return requested)} \end{aligned}$$

The augmented matrix is:

$$\left[\begin{array}{ccc|c} 1 & 1 & 1 & 25,000 \\ 8 & 12 & 20 & k \end{array}\right]$$

$$\left[\begin{array}{ccc|c} 1 & 1 & 1 & 25,000 \\ 8 & 12 & 20 & k \end{array}\right] \sim \left[\begin{array}{ccc|c} 1 & 1 & 1 & 25,000 \\ 0 & 4 & 12 & k - 200,000 \end{array}\right]$$

$(-8)R_1 + R_2 \to R_2$ $\qquad \frac{1}{4}R_2 \to R_2$

$$\sim \left[\begin{array}{ccc|c} 1 & 1 & 1 & 25,000 \\ 0 & 1 & 3 & \frac{k}{4} - 50,000 \end{array}\right] \sim \left[\begin{array}{ccc|c} 1 & 0 & -2 & -\frac{k}{4} + 75,000 \\ 0 & 1 & 3 & \frac{k}{4} - 50,000 \end{array}\right]$$

$(-1)R_2 + R_1 \to R_1$

Concert 1: $k = \$320{,}000$; $\frac{k}{4} = 80{,}000$

$$\left[\begin{array}{ccc|c} 1 & 0 & -2 & -5{,}000 \\ 0 & 1 & 3 & 30{,}000 \end{array}\right]; \quad \begin{array}{lll} x_1 = 2t - 5{,}000 & \$8 \text{ tickets} \\ x_2 = 30{,}000 - 3t & \$12 \text{ tickets} \\ x_3 = t \quad \$20 \text{ tickets, } t \text{ an integer} \end{array}$$

Since $x_1, x_2 \geq 0$, t must satisfy $2{,}500 \leq t \leq 10{,}000$.

Concert 2: $k = \$330{,}000$; $\frac{k}{4} = 82{,}500$

$$\left[\begin{array}{ccc|c} 1 & 0 & -2 & -7{,}500 \\ 0 & 1 & 3 & 32{,}500 \end{array}\right]; \quad \begin{array}{lll} x_1 = 2t - 7{,}500 & \$8 \text{ tickets} \\ x_2 = 32{,}500 - 3t & \$12 \text{ tickets} \\ x_3 = t \quad \$20 \text{ tickets, } t \text{ an integer} \end{array}$$

Since $x_1, x_2 \geq 0$, t must satisfy $3{,}750 \leq t \leq 10{,}833$.

Concert 3: $k = \$340{,}000$; $\frac{k}{4} = 85{,}000$

$$\left[\begin{array}{ccc|c} 1 & 0 & -2 & -10{,}000 \\ 0 & 1 & 3 & 35{,}000 \end{array}\right]; \quad \begin{array}{lll} x_1 = 2t - 10{,}000 & \$8 \text{ tickets} \\ x_2 = 35{,}000 - 3t & \$12 \text{ tickets} \\ x_3 = t \quad \$20 \text{ tickets, } t \text{ an integer} \end{array}$$

Since $x_1, x_2 \geq 0$, t must satisfy $5{,}000 \leq t \leq 11{,}666$. (4-3)

49. The technology matrix is

$$M = \begin{array}{r} \\ \text{Agriculture} \\ \text{Fabrication} \end{array} \begin{array}{c} \begin{array}{cc} \text{Agriculture} & \text{Fabrication} \end{array} \\ \begin{bmatrix} 0.30 & 0.10 \\ 0.20 & 0.40 \end{bmatrix} \end{array}$$

Now $I - M = \begin{bmatrix} 1 & 0 \\ 0 & 1 \end{bmatrix} - \begin{bmatrix} 0.30 & 0.10 \\ 0.20 & 0.40 \end{bmatrix} = \begin{bmatrix} 0.70 & -0.10 \\ -0.20 & 0.60 \end{bmatrix} = \begin{bmatrix} \frac{7}{10} & -\frac{1}{10} \\ -\frac{1}{5} & \frac{3}{5} \end{bmatrix}$

Next, we calculate the inverse of $I - M$

$$\left[\begin{array}{cc|cc} \frac{7}{10} & -\frac{1}{10} & 1 & 0 \\ -\frac{1}{5} & \frac{3}{5} & 0 & 1 \end{array}\right] \sim \left[\begin{array}{cc|cc} 1 & -3 & 0 & -5 \\ \frac{7}{10} & -\frac{1}{10} & 1 & 0 \end{array}\right] \sim \left[\begin{array}{cc|cc} 1 & -3 & 0 & -5 \\ 0 & 2 & 1 & \frac{7}{2} \end{array}\right] \sim \left[\begin{array}{cc|cc} 1 & -3 & 0 & -5 \\ 0 & 1 & \frac{1}{2} & \frac{7}{4} \end{array}\right]$$

$-5R_2 \to R_2$, $R_1 \leftrightarrow R_2$ $\quad \left(-\frac{7}{10}\right)R_1 + R_2 \to R_2$ $\quad \frac{1}{2}R_2 \to R_2$ $\quad 3R_2 + R_1 \to R_1$

$$\sim \left[\begin{array}{cc|cc} 1 & 0 & \frac{3}{2} & \frac{1}{4} \\ 0 & 1 & \frac{1}{2} & \frac{7}{4} \end{array}\right] \quad \text{Thus, } (I - M)^{-1} = \begin{bmatrix} \frac{3}{2} & \frac{1}{4} \\ \frac{1}{2} & \frac{7}{4} \end{bmatrix}.$$

Let x_1 = output for agriculture and x_2 = output for fabrication.

Then the output $X = \begin{bmatrix} x_1 \\ x_2 \end{bmatrix}$ needed to satisfy a final demand $D = \begin{bmatrix} d_1 \\ d_2 \end{bmatrix}$ for agriculture and fabrication is given by $X = (I - M)^{-1}D$.

(A) Let $D = \begin{bmatrix} 50 \\ 20 \end{bmatrix}$. Then $X = \begin{bmatrix} \frac{3}{2} & \frac{1}{4} \\ \frac{1}{2} & \frac{7}{4} \end{bmatrix} \begin{bmatrix} 50 \\ 20 \end{bmatrix} = \begin{bmatrix} 75 + 5 \\ 25 + 35 \end{bmatrix} = \begin{bmatrix} 80 \\ 60 \end{bmatrix}$

Thus, the total output for agriculture is \$80 billion; the total output for fabrication is \$60 billion.

(B) Let $D = \begin{bmatrix} 80 \\ 60 \end{bmatrix}$. Then $X = \begin{bmatrix} \frac{3}{2} & \frac{1}{4} \\ \frac{1}{2} & \frac{7}{4} \end{bmatrix} \begin{bmatrix} 80 \\ 60 \end{bmatrix} = \begin{bmatrix} 120 + 15 \\ 40 + 105 \end{bmatrix} = \begin{bmatrix} 135 \\ 145 \end{bmatrix}$

Thus, the total output for agriculture is \$135 billion; the total output for fabrication is \$145 billion. (4-7)

50. First we find the inverse of $B = \begin{bmatrix} 1 & 1 & 0 \\ 1 & 0 & 1 \\ 1 & 1 & 1 \end{bmatrix}$.

$$\left[\begin{array}{ccc|ccc} 1 & 1 & 0 & 1 & 0 & 0 \\ 1 & 0 & 1 & 0 & 1 & 0 \\ 1 & 1 & 1 & 0 & 0 & 1 \end{array}\right] \sim \left[\begin{array}{ccc|ccc} 1 & 1 & 0 & 1 & 0 & 0 \\ 0 & -1 & 1 & -1 & 1 & 0 \\ 0 & 0 & 1 & -1 & 0 & 1 \end{array}\right]$$

$(-1)R_1 + R_2 \to R_2$ $\quad$ $(-1)R_2 \to R_2$

$(-1)R_1 + R_3 \to R_3$

$$\sim \left[\begin{array}{ccc|ccc} 1 & 1 & 0 & 1 & 0 & 0 \\ 0 & 1 & -1 & 1 & -1 & 0 \\ 0 & 0 & 1 & -1 & 0 & 1 \end{array}\right] \sim \left[\begin{array}{ccc|ccc} 1 & 0 & 1 & 0 & 1 & 0 \\ 0 & 1 & -1 & 1 & -1 & 0 \\ 0 & 0 & 1 & -1 & 0 & 1 \end{array}\right]$$

$(-1)R_2 + R_1 \to R_1$ $\quad$ $(-1)R_3 + R_1 \to R_1$

$R_3 + R_2 \to R_2$

$$\sim \left[\begin{array}{ccc|ccc} 1 & 0 & 0 & 1 & 1 & -1 \\ 0 & 1 & 0 & 0 & -1 & 1 \\ 0 & 0 & 1 & -1 & 0 & 1 \end{array}\right] \quad \text{Thus, } B^{-1} = \begin{bmatrix} 1 & 1 & -1 \\ 0 & -1 & 1 \\ -1 & 0 & 1 \end{bmatrix}$$

Now, $\begin{bmatrix} 1 & 1 & -1 \\ 0 & -1 & 1 \\ -1 & 0 & 1 \end{bmatrix} \begin{bmatrix} 25 & 24 & 21 & 41 & 21 & 52 \\ 8 & 25 & 41 & 30 & 32 & 52 \\ 26 & 33 & 48 & 50 & 41 & 79 \end{bmatrix} = \begin{bmatrix} 7 & 16 & 14 & 21 & 12 & 25 \\ 18 & 8 & 7 & 20 & 9 & 27 \\ 1 & 9 & 27 & 9 & 20 & 27 \end{bmatrix}$

Thus, the decoded message is

7 18 1 16 8 9 14 7 27 21 20 9 12 9 20 25 27 27

which corresponds to GRAPHING UTILITY. (4-5)

51. (A) 1st & Elm: $x_1 + x_4 = 1300$
2nd & Elm: $x_1 - x_2 = 400$
2nd & Oak: $x_2 + x_3 = 700$
1st & Oak: $x_3 - x_4 = -200$

(B) The augmented matrix for the system in part A is:

$$\left[\begin{array}{cccc|c} 1 & 0 & 0 & 1 & 1300 \\ 1 & -1 & 0 & 0 & 400 \\ 0 & 1 & 1 & 0 & 700 \\ 0 & 0 & 1 & -1 & -200 \end{array}\right] \sim \left[\begin{array}{cccc|c} 1 & 0 & 0 & 1 & 1300 \\ 0 & -1 & 0 & -1 & -900 \\ 0 & 1 & 1 & 0 & 700 \\ 0 & 0 & 1 & -1 & -200 \end{array}\right]$$

$(-1)R_1 + R_2 \to R_2$ $\qquad$ $(-1)R_2 \to R_2$

$$\sim \left[\begin{array}{cccc|c} 1 & 0 & 0 & 1 & 1300 \\ 0 & 1 & 0 & 1 & 900 \\ 0 & 1 & 1 & 0 & 700 \\ 0 & 0 & 1 & -1 & -200 \end{array}\right] \sim \left[\begin{array}{cccc|c} 1 & 0 & 0 & 1 & 1300 \\ 0 & 1 & 0 & 1 & 900 \\ 0 & 0 & 1 & -1 & -200 \\ 0 & 0 & 1 & -1 & -200 \end{array}\right]$$

$(-1)R_2 + R_3 \to R_3$ $\qquad$ $(-1)R_3 + R_4 \to R_4$

$$\sim \left[\begin{array}{cccc|c} 1 & 0 & 0 & 1 & 1300 \\ 0 & 1 & 0 & 1 & 900 \\ 0 & 0 & 1 & -1 & -200 \\ 0 & 0 & 0 & 0 & 0 \end{array}\right]$$

The corresponding system of equations is

$$\begin{aligned} x_1 \quad\quad + x_4 &= 1300 \\ x_2 \quad + x_4 &= 900 \\ x_3 - x_4 &= -200 \end{aligned}$$

Let $x_4 = t$. Then, $x_1 = 1300 - t$, $x_2 = 900 - t$, $x_3 = t - 200$, $x_4 = t$ where $200 \le t \le 900$. *($t \ge 200$ so that x_3 is non-negative; $t \le 900$ so that x_2 is non-negative)

(C) maximum: 900
minimum: 200

(D) Elm St.: $x_1 = 800$; 2nd St.: $x_2 = 400$; Oak St.: $x_3 = 300$. (4-3)

52. (A) $M = \begin{array}{c} \\ A \\ B \\ C \\ D \end{array}\!\!\begin{array}{c} \begin{array}{cccc} A & B & C & D \end{array} \\ \begin{bmatrix} 0 & 1 & 0 & 0 \\ 0 & 0 & 1 & 1 \\ 1 & 0 & 0 & 1 \\ 1 & 0 & 0 & 0 \end{bmatrix} \end{array}$

$$M + M^2 = \begin{bmatrix} 0 & 1 & 0 & 0 \\ 0 & 0 & 1 & 1 \\ 1 & 0 & 0 & 1 \\ 1 & 0 & 0 & 0 \end{bmatrix} + \begin{bmatrix} 0 & 0 & 1 & 1 \\ 2 & 0 & 0 & 1 \\ 1 & 1 & 0 & 0 \\ 0 & 1 & 0 & 0 \end{bmatrix} = \begin{array}{c} \\ A \\ B \\ C \\ D \end{array}\!\!\begin{array}{c} \begin{array}{cccc} A & B & C & D \end{array} \\ \begin{bmatrix} 0 & 1 & 1 & 1 \\ 2 & 0 & 1 & 2 \\ 2 & 1 & 0 & 1 \\ 1 & 1 & 0 & 0 \end{bmatrix} \end{array}$$

Summing the rows, we have

$A{:}3$, $B{:}5$, $C{:}4$, $D{:}2$

Thus, the ranking is: B, C, A, D.

The entries in M are the first stage dominances. The entries in M^2 are the second stage dominances. For example, the entry $0 + 2 = 2$ the (2, 1) position indicates that Team B does not beat team A, but it did beat two other teams (C and D), each of which beat team A, and so on. By summing the rows of $M + M^2$, we calculate the sum of the first and second stage dominances. (4-7)

5 LINEAR INEQUALITIES AND LINEAR PROGRAMMING

EXERCISE 5-1

Things to remember:

1. The graph of the linear inequality

   $Ax + By < C$ or $Ax + By > C$

   with $B \neq 0$ is either the upper half-plane or the lower half-plane (but not both) determined by the line $Ax + By = C$. If $B = 0$, the graph of

   $Ax < C$ or $Ax > C$

   is either the right half-plane or the left half-plane (but not both) determined by the vertical line $Ax = C$.

2. For strict inequalities ("<" or ">"), the line is not included in the graph. For weak inequalities ("≤" or "≥"), the line is included in the graph.

3. PROCEDURE FOR GRAPHING LINEAR INEQUALITIES

   (a) First graph $Ax + By = C$ as a broken line if equality is not included in the original statement or as a solid line if equality is included.

   (b) Choose a test point anywhere in the plane not on the line [the origin (0, 0) often requires the least computation] and substitute the coordinates into the inequality.

   (c) The graph of the original inequality includes the half-plane containing the test point if the inequality is satisfied by that point or the half-plane not containing the test point if the inequality is not satisfied by that point.

4. To solve a system of linear inequalities graphically, graph each inequality in the system and then take the intersection of all the graphs. The resulting graph is called the SOLUTION REGION, or FEASIBLE REGION.

5. A CORNER POINT of a solution region is a point in the solution region which is the intersection of two boundary lines.

6. The solution region of a system of linear inequalities is BOUNDED if it can be enclosed within a circle; if it cannot be enclosed within a circle, then it is UNBOUNDED.

1. $y \leq x - 1$

Graph $y = x - 1$ as a solid line.

x	y
0	-1
1	0

Test point (0, 0):

$0 \leq 0 - 1$

$0 \leq -1$

The inequality is false. Thus, the graph is below the line $y = x - 1$, including the line.

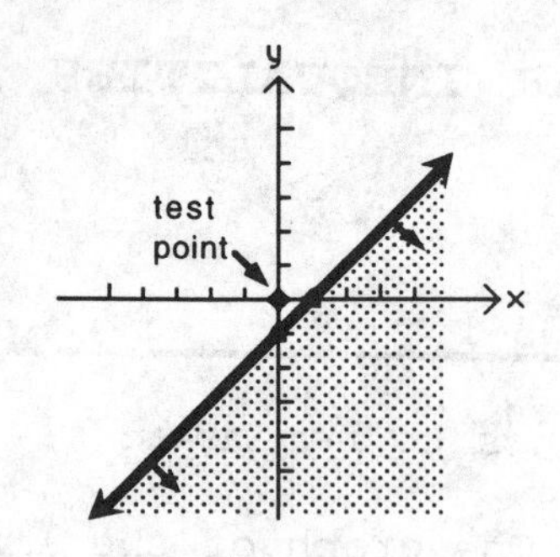

3. $3x - 2y > 6$

Graph $3x - 2y = 6$ as a broken line.

Test point (0, 0):

$3 \cdot 0 - 2 \cdot 0 > 6$

$0 > 6$

The inequality is false. Thus, the graph is below the line $3x - 2y = 6$, not including the line.

x	y
0	-3
2	0

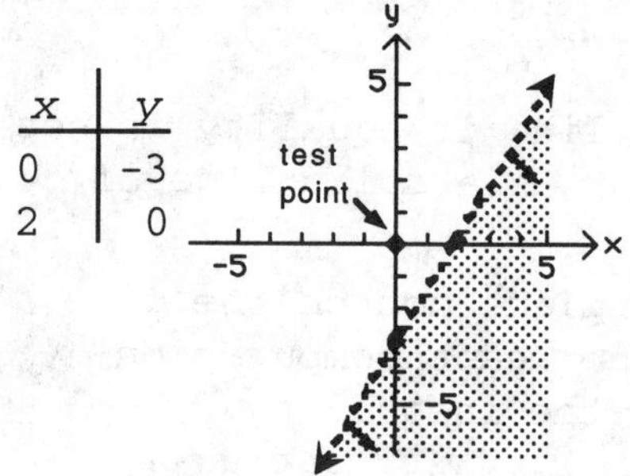

5. $x \geq -4$

Graph $x = -4$ [the vertical line through (-4, 0)] as a solid line.

Test point (0, 0):

$0 \geq -4$

The inequality is true. Thus, the graph is to the right of the line $x = -4$, including the line.

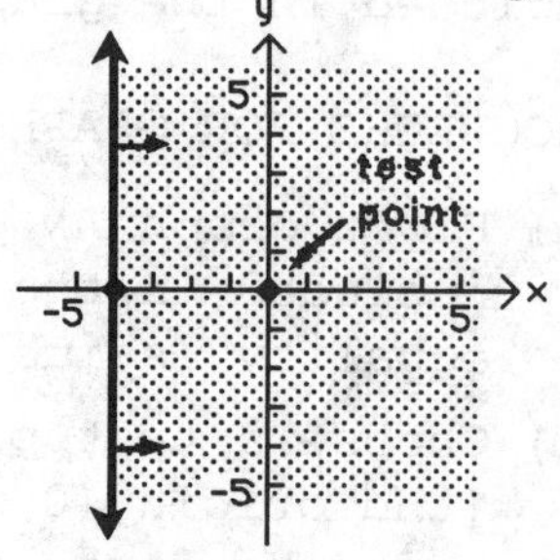

7. $6x + 4y \geq 24$

Graph the line $6x + 4y = 24$ as a solid line.

Test point (0, 0):

$6 \cdot 0 + 4 \cdot 0 \geq 24$

$0 \geq 24$

The inequality is false. Thus, the graph is the region above the line, including the line.

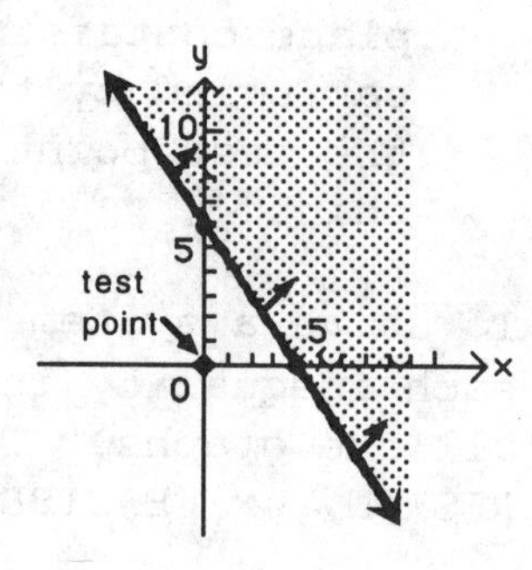

9. $5x \le -2y$ or $5x + 2y \le 0$

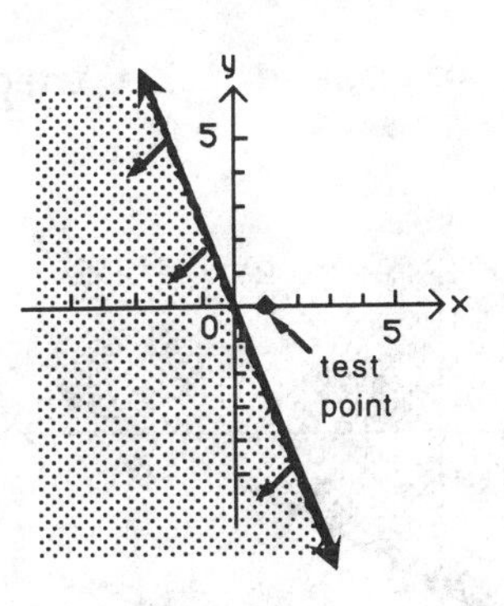

Graph the line $5x + 2y = 0$ as a solid line. Since the line passes through the origin (0, 0), we use (1, 0) as a test point:

$$5 \cdot 1 + 2 \cdot 0 \le 0$$
$$5 \le 0$$

This inequality is false. Thus, the graph is below the line $5x + 2y = 0$, including the line.

11. The graph of $x + 2y \le 8$ is the region below the line $x + 2y = 8$ [e.g., (0, 0) satisfies the inequality]. The graph of $3x - 2y \ge 0$ is the region below the line $3x - 2y = 0$ [e.g., (1, 0) satisfies the inequality]. The intersection of these two regions is region IV.

13. The graph of $x + 2y \ge 8$ is the region above the line $x + 2y = 8$ [e.g., (0, 0) does not satisfy the inequality]. The graph of $3x - 2y \ge 0$ is the region below the line $3x - 2y = 0$ [e.g., (1, 0) satisfies the inequality]. The intersection of these two regions is region I.

15. The graphs of the inequalities $3x + y \ge 6$ and $x \le 4$ are:

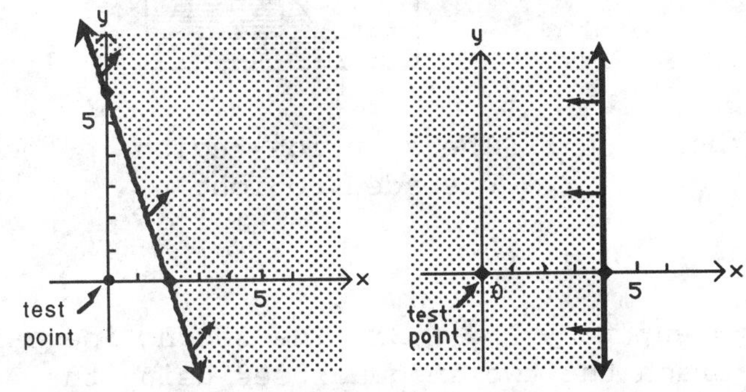

The intersection of these regions (drawn on the same coordinate plane) is shown in the graph at the right.

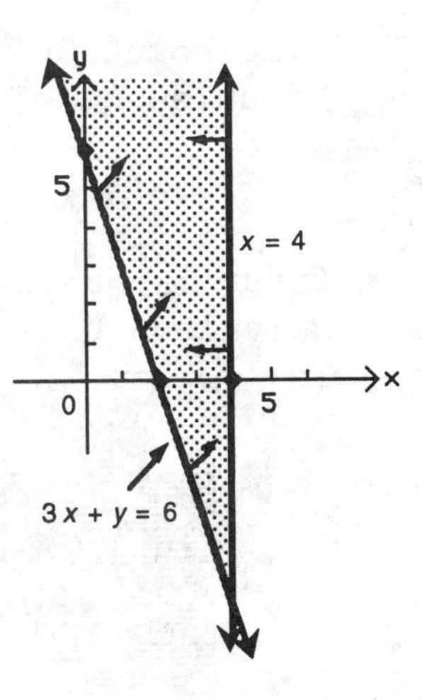

17. The graphs of the inequalities $x - 2y \le 12$ and $2x + y \ge 4$ are:

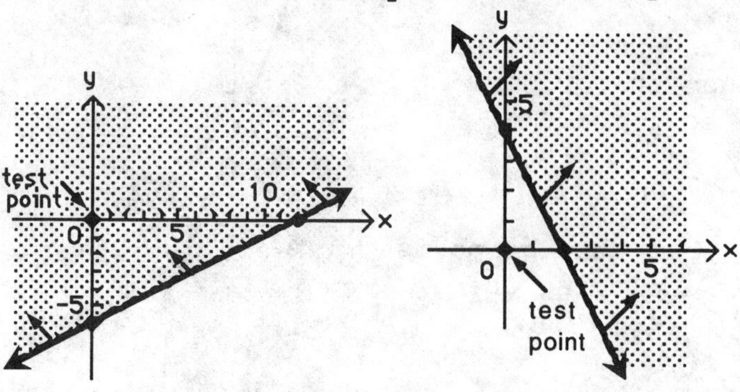

The intersection of these regions (drawn on the same coordinate plane) is shown in the graph at the right.

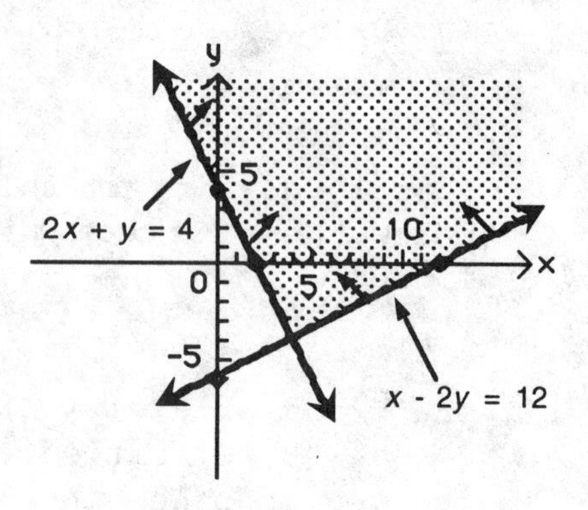

19. $x + y \le 5$
$2x - y \le 1$

The first inequality is equivalent to $y \le 5 - x$. The second inequality is equivalent to $y \ge 2x - 1$. Graph the two inequalities using the shading option.

(A)

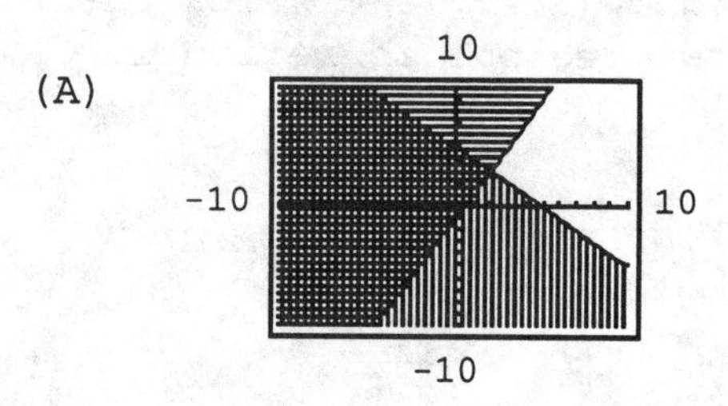

The solution region is the double-shaded region.

(B)

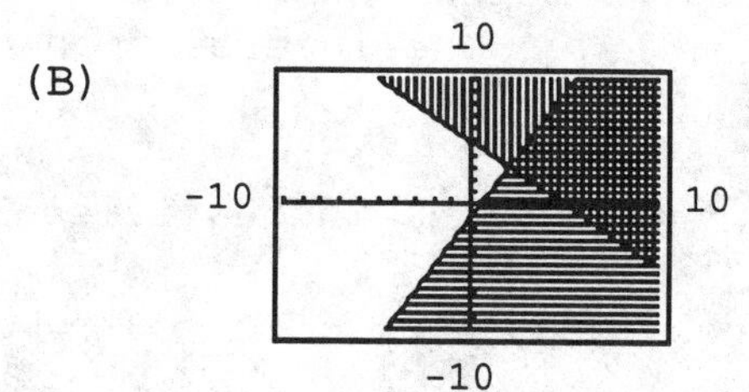

The solution region is the unshaded region.

21. $2x + y \ge 4$
$3x - y \le 7$

The first inequality is equivalent to $y \ge 4 - 2x$. The second inequality is equivalent to $y \ge 3x - 7$. Graph the two inequalities using the shading option.

(A)

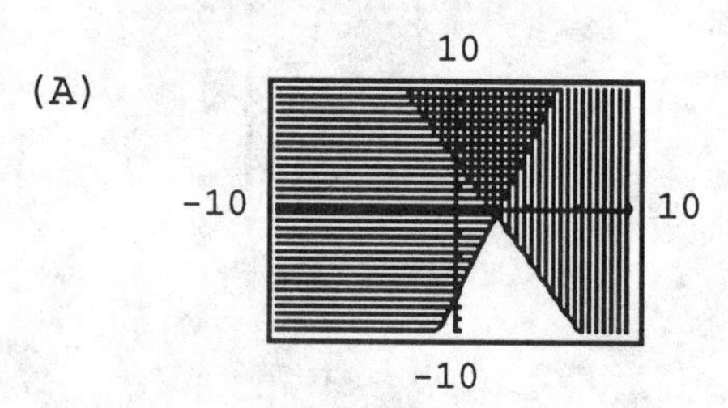

The solution region is the double-shaded region.

(B)

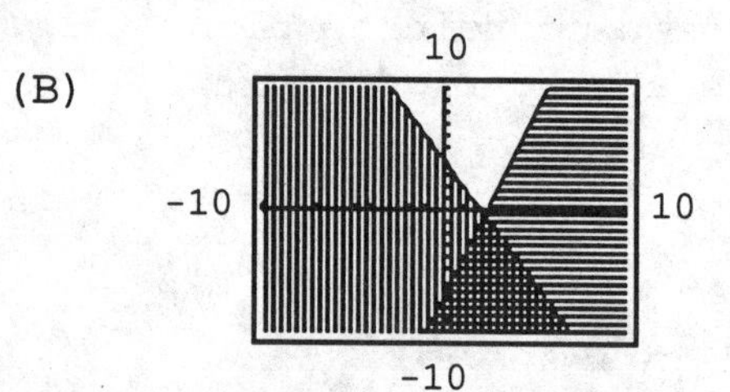

The solution region is the unshaded region.

23. The graph of $x + 3y \le 18$ is the region below the line $x + 3y = 18$ and the graph of $2x + y \ge 16$ is the region above the line $2x + y = 16$. The graph of $x \ge 0$, $y \ge 0$ is the first quadrant. The intersection of these regions is region IV. The corner points are (8, 0), (18, 0), and (6, 4).

25. The graph of $x + 3y \ge 18$ is the region above the line $x + 3y = 18$ and the graph of $2x + y \ge 16$ is the region above the line $2x + y = 16$. The graph of $x \ge 0$, $y \ge 0$ is the first quadrant. The intersection of these regions is region I. The corner points are (0, 16), (6, 4), and (18, 0).

27. The graphs of the inequalities are shown at the right. The solution region is indicated by the shaded region. The solution region is *bounded*.

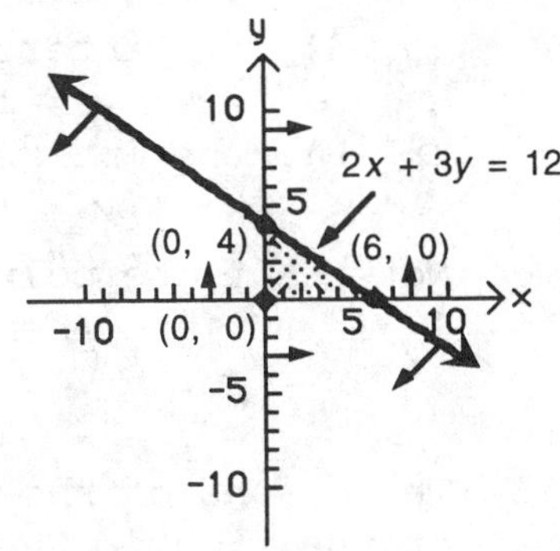

The corner points of the solution region are:

(0, 0), the intersection of $x = 0$, $y = 0$;
(0, 4), the intersection of $x = 0$, $2x + 3y = 12$;
(6, 0), the intersection of $y = 0$, $2x + 3y = 12$.

29. The graphs of the inequalities are shown at the right. The solution region is shaded. The solution region is *bounded*.

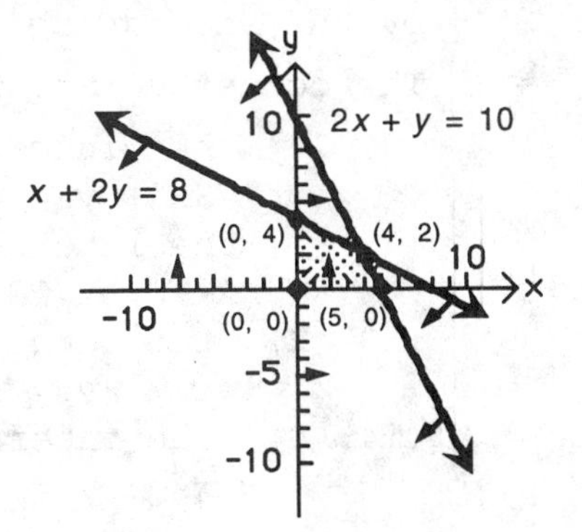

The corner points of the solution region are:

(0, 0), the intersection of $x = 0$, $y = 0$;
(0, 4), the intersection of $x = 0$, $x + 2y = 8$;
(4, 2), the intersection of $x + 2y = 8$, $2x + y = 10$;
(5, 0), the intersection of $y = 0$, $2x + y = 10$.

31. The graphs of the inequalities are shown at the right. The solution region is shaded. The solution region is *unbounded*.

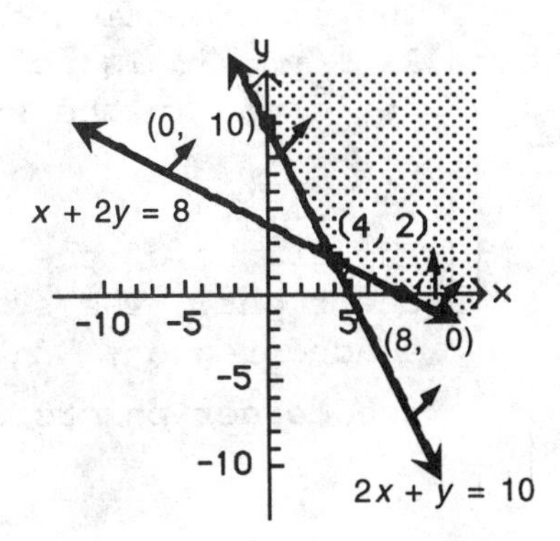

The corner points of the solution region are:

(0, 10), the intersection of $x = 0$, $2x + y = 10$;
(4, 2), the intersection of $x + 2y = 8$, $2x + y = 10$;
(8, 0), the intersection of $y = 0$, $x + 2y = 8$.

33. The graphs of the inequalities are shown at the right. The solution is indicated by the shaded region. The solution region is *bounded*.

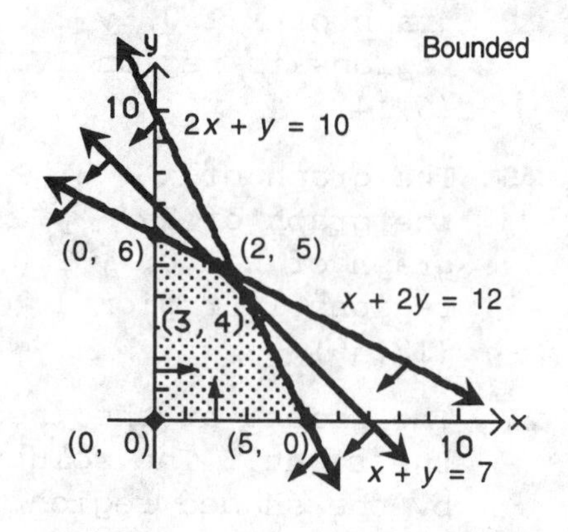

The corner points of the solution region are:

(0, 0), the intersection of $x = 0$, $y = 0$,
(0, 6), the intersection of $x = 0$, $x + 2y = 12$;
(2, 5), the intersection of $x + 2y = 12$, $x + y = 7$;
(3, 4), the intersection of $x + y = 7$, $2x + y = 10$;
(5, 0), the intersection of $y = 0$, $2x + y = 10$.

Note that the point of intersection of the lines $2x + y = 10$, $x + 2y = 12$ is not a corner point because it is not in the solution region.

35. The graphs of the inequalities are shown at the right. The solution is indicated by the shaded region, which is *unbounded*.

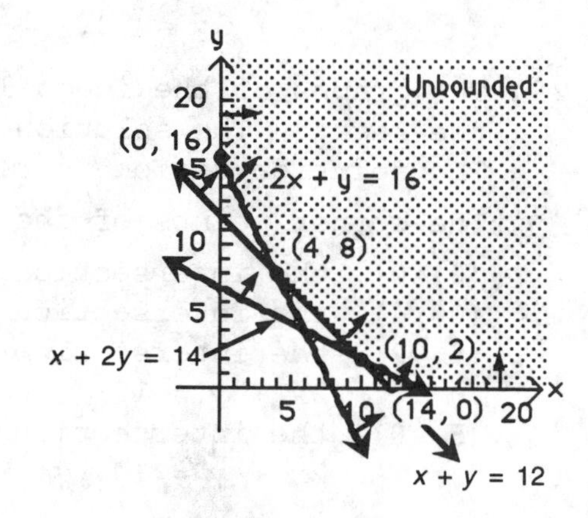

The corner points are:

(0, 16), the intersection of $x = 0$, $2x + y = 16$;
(4, 8), the intersection of $2x + y = 16$, $x + y = 12$;
(10, 2), the intersection of $x + y = 12$, $x + 2y = 14$;
(14, 0), the intersection of $y = 0$, $x + 2y = 14$.

The intersection of $x + 2y = 14$, $2x + y = 16$ is not a corner point because it is not in the solution region.

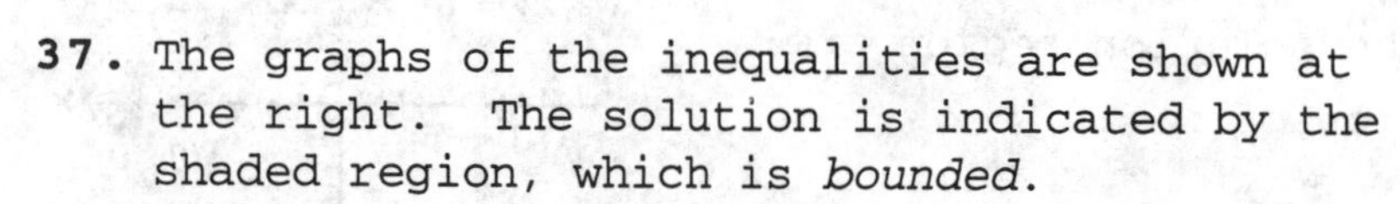

37. The graphs of the inequalities are shown at the right. The solution is indicated by the shaded region, which is *bounded*.

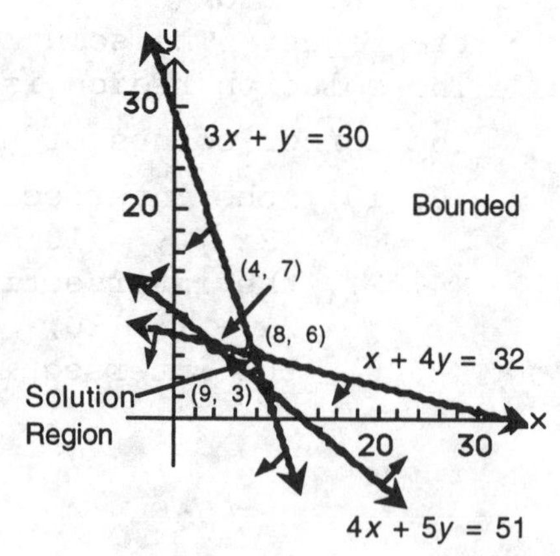

The corner points are (8, 6), (4, 7), and (9, 3).

39. The graphs of the inequalities are shown at the right. The system of inequalities does not have a solution because the intersection of the graphs is empty.

$x = 9$

$2x + y = 24$

$4x + 3y = 48$

41. The graphs of the inequalities are shown at the right. The solution is indicated by the shaded region, which is *unbounded*.

The corner points are (0, 0), (4, 4), and (8, 12).

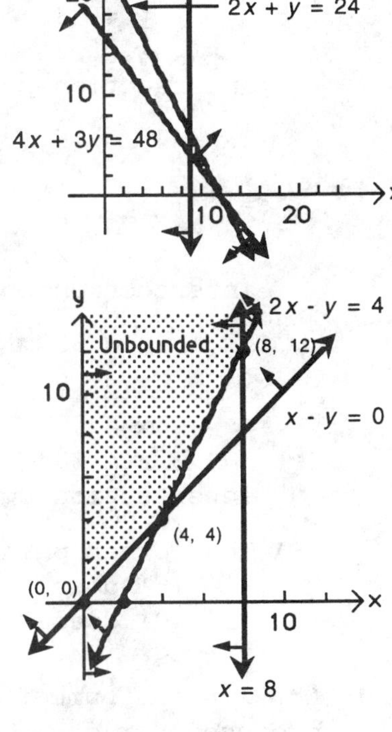

43. The graphs of the inequalities are shown at the right. The solution is indicated by the shaded region, which is *bounded*.

The corner points are (2, 1), (3, 6), (5, 4), and (5, 2).

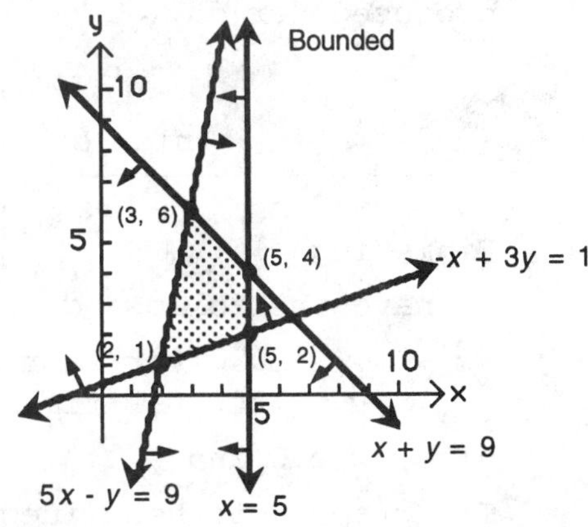

45. The graphs of the inequalities are shown at the right. The solution is indicated by the shaded region, which is *bounded*. The corner points are 1.27, 5.36), (2.14, 6.52), and (5.91, 1.88).

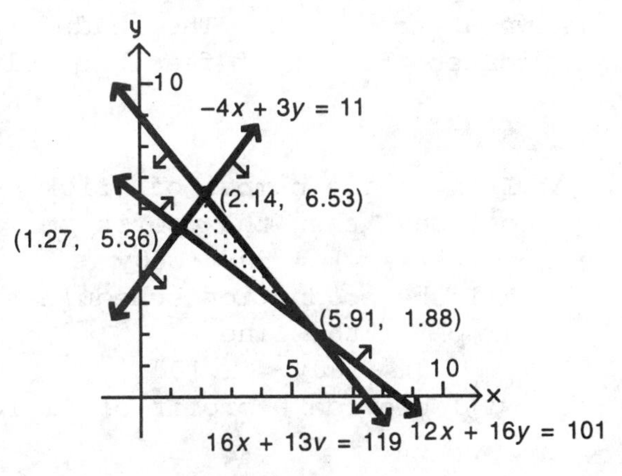

47. (A) $3x + 4y = 36$
$3x + 2y = 30$ subtract
$2y = 6$
$y = 3$
$x = 8$
intersection point: (8, 3)

$3x + 4y = 36$
$x = 0$
$4y = 36$
$y = 9$
intersection point: (0, 9)

$3x + 4y = 36$
$y = 0$
$3x = 36$
$x = 12$
intersection point: (12, 0)

$3x + 2y = 30$
$x = 0$
$2y = 30$
$y = 15$
intersection point: (0, 15)

$3x + 2y = 30$
$y = 0$
$3x = 30$
$x = 10$
intersection point: (10, 0)

$x = 0$
$y = 0$
intersection point: (0, 0)

(B) The corner points are: (8, 3), (0, 9), (10, 0), (0, 0); (0, 15) does not satisfy $3x + 4y \le 36$, (12, 0) does not satisfy $3x + 2y \le 30$.

49. Let x = the number of trick skis and y = the number of slalom skis produced per day. The information is summarized in the following table.

	Hours per ski		Maximum labor-hours per day available
	Trick ski	Slalom ski	
Fabrication	6 hrs	4 hrs	108 hrs
Finishing	1 hr	1 hr	24 hrs

We have the following inequalities:

$6x + 4y \le 108$ for fabrication
$x + y \le 24$ for finishing

Also, $x \ge 0$ and $y \ge 0$.

The graphs of these inequalities are shown at the right. The shaded region indicates the set of feasible solutions.

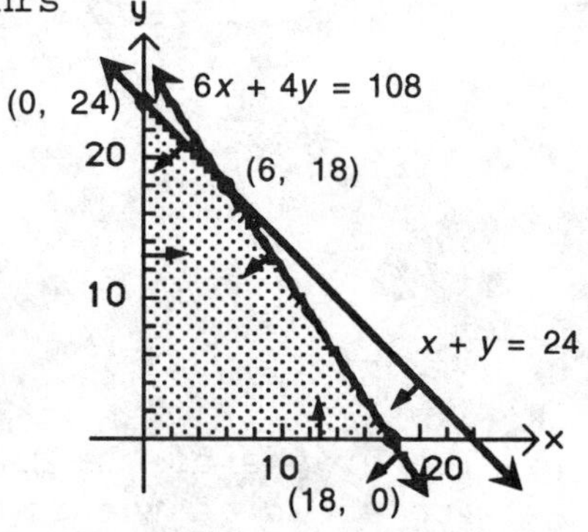

51. (A) If x is the number of trick skis and y is the number of slalom skis per day, then the profit per day is given by

$$P(x, y) = 50x + 60y$$

All the production schedules in the feasible region that lie on the graph of the line

$$50x + 60y = 1{,}100$$

will provide a profit of \$1,100.

(B) There are many possible choices. For example, producing 5 trick skis and 15 slalom skis per day will produce a profit of

$$P(5, 15) = 50(5) + 60(15) = 1{,}150$$

All the production schedules in the feasible region that lie on the graph of $50x + 60y = 1{,}150$ will provide a profit of $1,150.

53. Let x = the number of cubic yards of mix A and y = the number of cubic yards of mix B. The information is summarized in the following table:

	Amount of substance per cubic yard		
	Mix A	Mix B	Minimum monthly requirement
Phosphoric acid	20 lbs	10 lbs	460 lbs
Nitrogen	30 lbs	30 lbs	960 lbs
Potash	5 lbs	10 lbs	220 lbs

We have the following inequalities:

$20x + 10y \geq 460$
$30x + 30y \geq 960$
$5x + 10y \geq 220$

Also, $x \geq 0$ and $y \geq 0$.

The graphs of these inequalities are shown at the right. The shaded region indicates the set of feasible solutions.

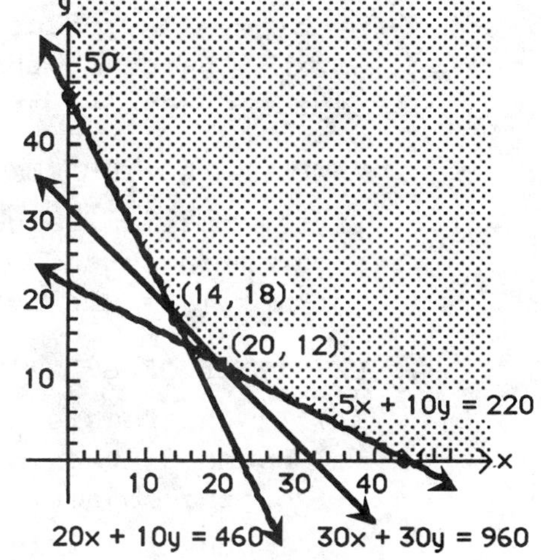

55. Let x = the number of mice used and y = the number of rats used. The information is summarized in the following table.

	Mice	Rats	Maximum time available per day
Box A	10 min	20 min	800 min
Box B	20 min	10 min	640 min

We have the following inequalities:

$10x + 20y \leq 800$ for box A
$20x + 10y \leq 640$ for box B

Also, $x \geq 0$ and $y \geq 0$.

The graphs of these inequalities are shown at the right. The shaded region indicates the set of feasible solutions.

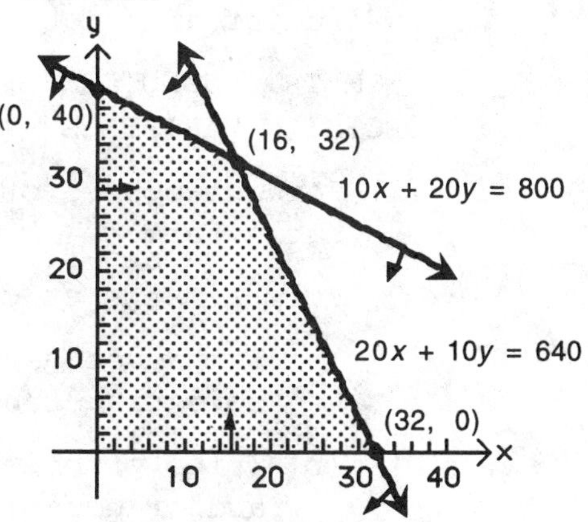

Things to remember:

1. A LINEAR PROGRAMMING PROBLEM is a problem concerned with finding the maximum or minimum value of a linear OBJECTIVE FUNCTION of the form

$$z = c_1x_1 + c_2x_2 + \cdots + c_nx_n,$$

where the DECISION VARIABLES $x_1, x_2, \ldots, x_n$ are subject to PROBLEM CONSTRAINTS in the form of linear inequalities and equations. In addition, the decision variables must satisfy the NONNEGATIVE CONSTRAINTS $x_i \geq 0$, for $i = 1, 2, \ldots, n$. The set of points satisfying both the problem constraints and the nonnegative constraints is called the FEASIBLE REGION for the problem. Any point in the feasible region that produces the optimal value of the objective function over the feasible region is called an OPTIMAL SOLUTION.

2. FUNDAMENTAL THEOREM OF LINEAR PROGRAMMING

If the optimal value of the objective function in a linear programming problem exists, then that value must occur at one (or more) of the corner points of the feasible region.

3. EXISTENCE OF SOLUTIONS

(A) If the feasible region for a linear programming problem is bounded, then both the maximum value and the minimum value of the objective function always exist.

(B) If the feasible region is unbounded, and the coefficients of the objective function are positive, then the minimum value of the objective function exists, but the maximum value does not.

(C) If the feasible region is empty (that is, there are no points that satisfy all the constraints), then both the maximum value and the minimum value of the objective function do not exist.

4. GEOMETRIC SOLUTION OF A LINEAR PROGRAMMING PROBLEM WITH TWO DECISION VARIABLES.

(1) For an applied problem, summarize relevant material in table form.

(2) Form a mathematical model for the problem:

(a) Introduce decision variables and write a linear objective function.

(b) Write problem constraints using linear inequalities and/or equations.

(c) Write nonnegative constraints.

(3) Graph the feasible region. Then, if according to 3 an optimal solution exists, find the coordinates of each corner point.

(4) Make a table listing the value of the objective function at each corner point.

(5) Determine the optimal solution(s) from the table in Step (4).

(6) For an applied problem, interpret the optimal solution(s) in terms of the original problem.

---

1. Steps (1)–(3) in 4 do not apply. Thus, we begin with Step (4).

Step (4): Evaluate the objective function at each corner point.

Corner Point	$z = x + y$
(0, 0)	0
(0, 12)	12
(7, 9)	16
(10, 0)	10

Step (5): Determine the optimal solution from Step (4).
The maximum value of z is 16 at (7, 9).

3. Steps (1)–(3) in 4 do not apply. Thus, we begin with Step (4).

Step (4): Evaluate the objective function at each corner point.

Corner Point	$z = 3x + 7y$
(0, 0)	0
(0, 12)	84
(7, 9)	84
(10, 0)	30

Step (5): Determine the optimal solution from Step (4).
The maximum value of z is 84 at (0, 12) *and* (7, 9). This is a multiple optimal solution.

5. Steps (1)–(3) in 4 do not apply. Thus, we begin with Step (4).

Step (4): Evaluate the objective function at each corner point.

Corner Point	$z = 7x + 4y$
(0, 12)	48
(0, 8)	32
(4, 3)	40
(12, 0)	84

Step (5): Determine the optimal solution from Step (4).
The minimum value of z is 32 at (0, 8).

7. Steps (1)–(3) in 4 do not apply. Thus, we begin with Step (4).

Step (4): Evaluate the objective function at each corner point.

Corner Point	$z = 3x + 8y$
(0, 12)	96
(0, 8)	64
(4, 3)	36
(12, 0)	36

Step (5): Determine the optimal solution from Step (4).
The minimum value of z is 36 at (4, 3) and (12, 0). This is a multiple optimal solution.

9. Step (3): Graph the feasible region and find the corner points.

The feasible region S is the solution set of the given inequalities. This region is indicated by the shading in the graph at the right.

The corner points are (0, 0), (0, 4), (4, 2), and (5, 0).

Since S is bounded, it follows from 3(a) that P has a maximum value.

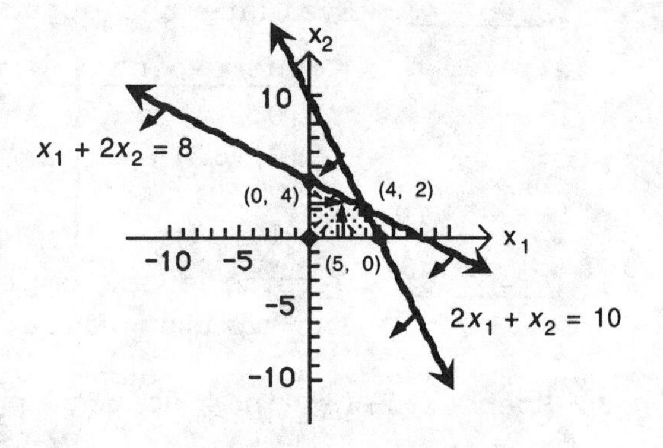

Step (4): Evaluate the objective function at each corner point.

The value of P at each corner point is given in the following table.

Corner Point	$P = 5x_1 + 5x_2$
(0, 0)	$P = 5(0) + 5(0) = 0$
(0, 4)	$P = 5(0) + 5(4) = 20$
(4, 2)	$P = 5(4) + 5(2) = 30$
(5, 0)	$P = 5(5) + 5(0) = 25$

Step (5): Determine the optimal solution.
The maximum value of P is 30 at $x_1 = 4$, $x_2 = 2$.

11. Step (3): Graph the feasible region and find the corner points.

The feasible region S is the solution set of the given inequalities. This region is indicated by the shading in the graph at the right.

The corner points are (0, 10), (4, 2), and (8, 0).

Since S is unbounded and $a = 2 > 0$, $b = 3 > 0$, it follows from 3(b) that P has a minimum value but not a maximum value.

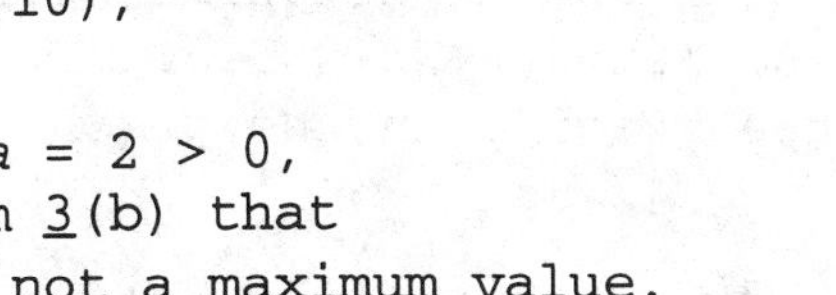

Step (4): Evaluate the objective function at each corner point.

The value of P at each corner point is given in the following table:

Corner Point	$z = 2x_1 + 3x_2$
(0, 10)	$z = 2(0) + 3(10) = 30$
(4, 2)	$z = 2(4) + 3(2) = 14$
(8, 0)	$z = 2(8) + 3(0) = 16$

Step (5): Determine the optimal solutions.
The minimum occurs at $x_1 = 4$, $x_2 = 2$, and the minimum value is $z = 14$; z does not have a maximum value.

13. Step (3): Graph the feasible region and find the corner points.

The feasible region S is the solution set of the given inequalities. This region is indicated by the shading in the graph at the right.

The corner points are (0, 0), (0, 6), (2, 5), (3, 4), and (5, 0).

Since S is bounded, it follows from 3(a) that P has a maximum value.

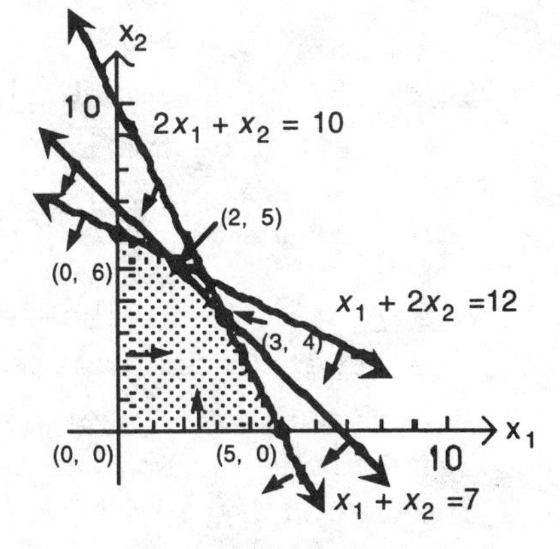

Step (4): Evaluate the objective function at each corner point.

The value of P at each corner point is:

Corner Point	$P = 30x_1 + 40x_2$
(0, 0)	$P = 30(0) + 40(0) = 0$
(0, 6)	$P = 30(0) + 40(6) = 240$
(2, 5)	$P = 30(2) + 40(5) = 260$
(3, 4)	$P = 30(3) + 40(4) = 250$
(5, 0)	$P = 30(5) + 40(0) = 150$

Step (5): Determine the optimal solution.
The maximum occurs at $x_1 = 2$, $x_2 = 5$, and the maximum value is $P = 260$.

15. Step (3): Graph the feasible region and find the corner points.

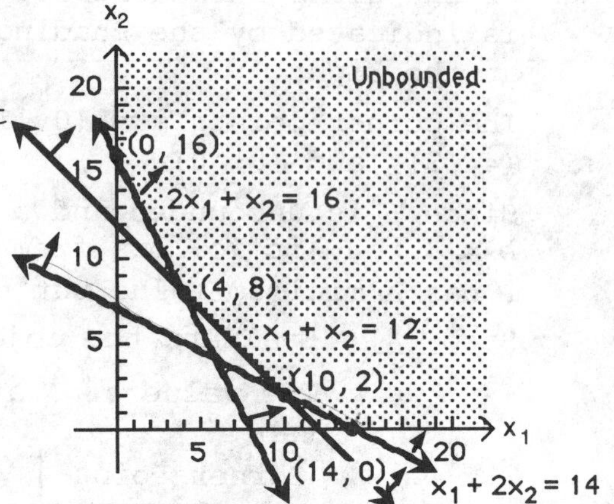

The feasible region S is the solution set of the given inequalities. This region is indicated by the shading in the graph at the right.

The corner points are (0, 16), (4, 8), (10, 2), and (14, 0).

Since S is unbounded and $a = 10 > 0$, $b = 30 > 0$, it follows from 3(b) that z has a minimum value but not a maximum value.

Step (4): Evaluate the objective function at each corner point.

The value of z at each corner point is:

Corner Point	$z = 10x_1 + 30x_2$
(0, 16)	$z = 10(0) + 30(16) = 480$
(4, 8)	$z = 10(4) + 30(8) = 280$
(10, 2)	$z = 10(10) + 30(2) = 160$
(14, 0)	$z = 10(14) + 30(0) = 140$

Step (5): Determine the optimal solution.
The minimum occurs at $x_1 = 14$, $x_2 = 0$, and the minimum value is $z = 140$; z does not have a maximum value.

17. Step (3): Graph the feasible region and find the corner points.

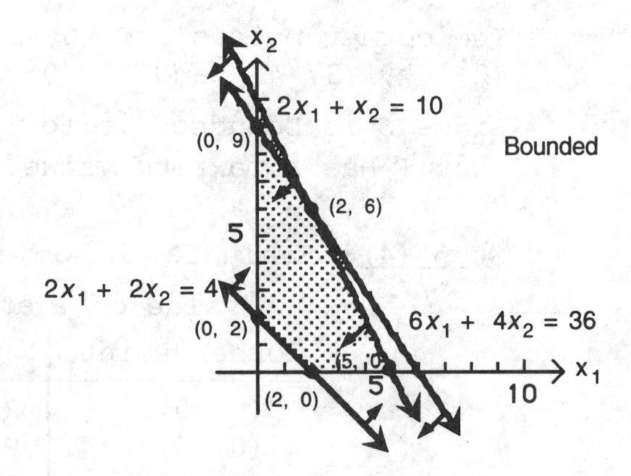

The feasible region S is the solution set of the given inequalities, and is indicated by the shading in the graph at the right.

The corner points are (0, 2), (0, 9), (2, 6), (5, 0), and (2, 0).

Since S is bounded, it follows from 3(a) that P has a maximum value and a minimum value.

Step (4): Evaluate the objective function at each corner point.

The value of P at each corner point is given in the following table:

Corner Point	$P = 30x_1 + 10x_2$
(0, 2)	$P = 30(0) + 10(2) = 20$
(0, 9)	$P = 30(0) + 10(9) = 90$
(2, 6)	$P = 30(2) + 10(6) = 120$
(5, 0)	$P = 30(5) + 10(0) = 150$
(2, 0)	$P = 30(2) + 10(0) = 60$

Step (5): Determine the optimal solutions.

The maximum occurs at $x_1 = 5$, $x_2 = 0$, and the maximum value is $P = 150$; the minimum occurs at $x_1 = 0$, $x_2 = 2$, and the minimum value is $P = 20$.

19. Step (3): Graph the feasible region and find the corner points.

The feasible region S is the solution set of the given inequalities. As indicated, the feasible region is empty. Thus, by 3(c), there are no optimal solutions.

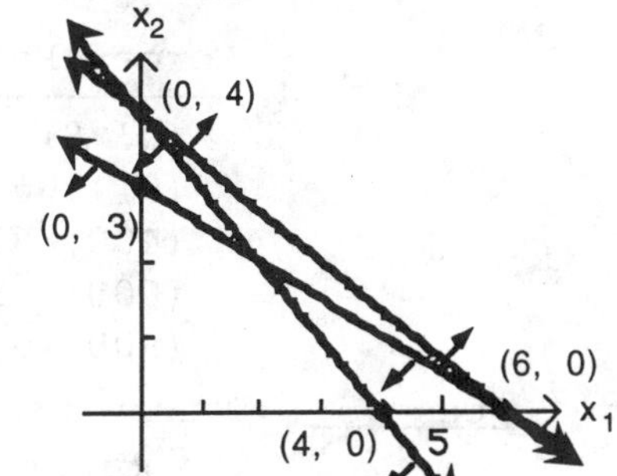

21. Step (3): Graph the feasible region and find the corner points.

The feasible region S is the solution set of the given inequalities, and is indicated by the shading in the graph at the right.

The corner points are (3, 8), (8, 10), and (12, 2).

Since S is bounded, it follows from 3(a) that P has a maximum value and a minimum value.

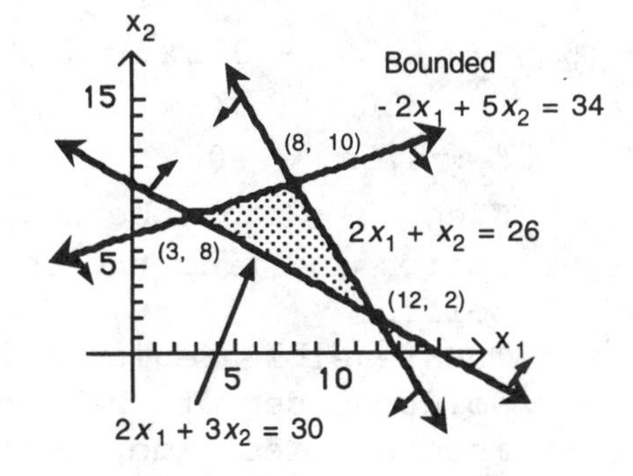

Step (4): Evaluate the objective function at each corner point.

The value of P at each corner point is:

Corner Point	$P = 20x_1 + 10x_2$
(3, 8)	$P = 20(3) + 10(8) = 140$
(8, 10)	$P = 20(8) + 10(10) = 260$
(12, 2)	$P = 20(12) + 10(2) = 260$

Step (5): Determine the optimal solutions.

The minimum occurs at $x_1 = 3$, $x_2 = 8$, and the minimum value is $P = 140$; the maximum occurs at $x_1 = 8$, $x_2 = 10$, at $x_1 = 12$, $x_2 = 2$, and at any point along the line segment joining (8, 10) and (12, 2). The maximum value is $P = 260$.

23. Step (3): Graph the feasible region and find the corner points. The feasible region S is the set of solutions of the given inequalities, and is indicated by the shading in the graph at the right.

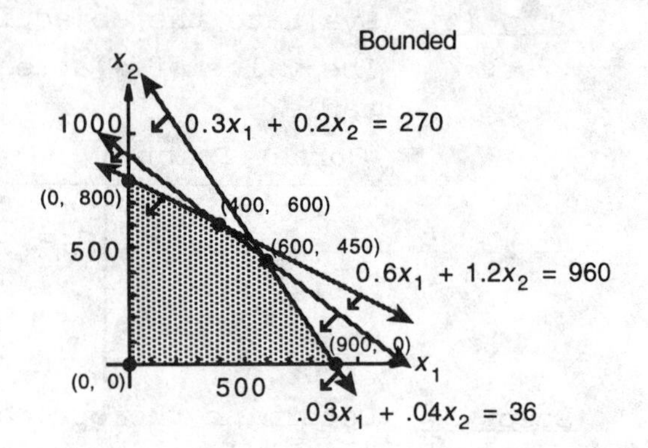

The corner points are (0, 0), (0, 800), (400, 600), (600, 450), and (900, 0). Since S is bounded, it follows from 3(a) that P has a maximum value.

Step (4): Evaluate the objective function at each corner point. The value of P at each corner point is:

Corner Point	$P = 20x_1 + 30x_2$
(0, 0)	$P = 20(0) + 30(0) = 0$
(0, 800)	$P = 20(0) + 30(800) = 24,000$
(400, 600)	$P = 20(400) + 30(600) = 26,000$
(600, 450)	$P = 20(600) + 30(450) = 25,500$
(900, 0)	$P = 20(900) + 30(0) = 18,000$

Step (5): Determine the optimal solution. The maximum occurs at $x_1 = 400$, $x_2 = 600$, and the maximum value is $P = 26,000$.

25. ℓ_1: $275x_1 + 322x_2 = 3,381$
ℓ_2: $350x_1 + 340x_2 = 3,762$
ℓ_3: $425x_1 + 306x_2 = 4,114$.

Step (3): Graph the feasible region and find the corner points.
The feasible region S is the solution set of the given inequalities, and is indicated by the shading in the graph at the right.

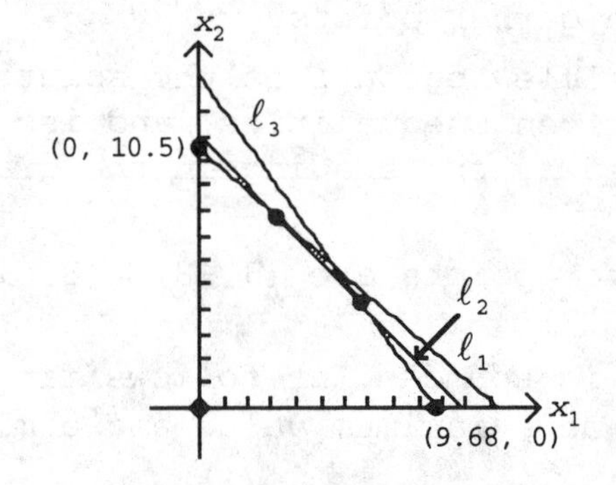

The corner points are (0, 0), (0, 10.5), (3.22, 7.75), (6.62, 4.25), (9.68, 0).

Step (4): Evaluate the objective function at each corner point. The value of P at each corner point is

Corner Point	$P = 525x_1 + 478x_2$
(0, 0)	$P = 525(0) + 478(0) = 0$
(0, 10.5)	$P = 525(0) + 478(10.5) = 5,019$
(3.22, 7.75)	$P = 525(3.22) + 478(7.75) = 5,395$
(6.62, 4.25)	$P = 525(6.62) + 478(4.25) = 5,507$
(9.68, 0)	$P = 525(9.68) + 478(0) = 5,082$

Step (5): Determine the optimal solution.

The maximum occurs at $x_1 = 6.62$, $x_2 = 4.25$, and the maximum value is $P = 5{,}507$.

27. Minimize and maximize $z = x_1 - x_2$

Subject to

$$x_1 - 2x_2 \le 0$$
$$2x_1 - x_2 \le 6$$
$$x_1, x_2 \ge 0$$

The feasible region and several values of the objective function are shown in the figure.

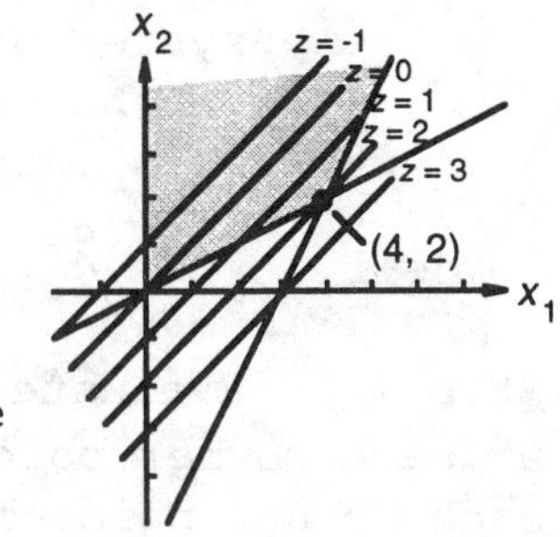

The points (0, 0) and (4, 2) are the corner points; $z = x_1 - x_2$ does not have a minimum value. Its maximum value is 2 at (4, 2).

29. The value of $P = ax_1 + bx_2$, $a > 0$, $b > 0$, at each corner point is:

Corner Point	P
O: (0, 0)	$P = a(0) + b(0) = 0$
A: (0, 5)	$P = a(0) + b(5) = 5b$
B: (4, 3)	$P = a(4) + b(3) = 4a + 3b$
C: (5, 0)	$P = a(5) + b(0) = 5a$

(A) For the maximum value of P to occur at A only, we must have $5b > 4a + 3b$ and $5b > 5a$. Solving the first inequality, we get $2b > 4a$ or $b > 2a$; from the second inequality, we get $b > a$. Therefore, we must have $b > 2a$ or $2a < b$ in order for P to have its maximum value at A only.

(B) For the maximum value of P to occur at B only, we must have $4a + 3b > 5b$ and $4a + 3b > 5a$. Solving this pair of inequalities, we get $4a > 2b$ and $3b > a$, which is the same as $\frac{a}{3} < b < 2a$.

(C) For the maximum value of P to occur at C only, we must have $5a > 4a + 3b$ and $5a > 5b$. This pair of inequalities implies that $a > 3b$ or $b < \frac{a}{3}$.

(D) For the maximum value of P to occur at both A and B, we must have $5b = 4a + 3b$ or $b = 2a$.

(E) For the maximum value of P to occur at both B and C, we must have $4a + 3b = 5a$ or $b = \frac{a}{3}$.

31. (A) Step (1): Has been done.

Step (2): Form a mathematical model for the problem.

Let x_1 = the number of trick skis

and x_2 = the number of slalom skis produced per day.

The mathematical model for this problem is:

Maximize $P = 40x_1 + 30x_2$

Subject to: $6x_1 + 4x_2 \le 108$

$x_1 + x_2 \le 24$

$x_1 \ge 0$, $x_2 \ge 0$

Step (3): Graph the feasible region and find the corner points.

The feasible region S is the solution set of the given system of inequalities, and is indicated by the shading in the graph below.

The corner points are (0, 0), (0, 24), (6, 18), and (18, 0).

Since S is bounded, P has a maximum value by 3(a).

Step (4): Evaluate the objective function at each corner point. The value of P at each corner point is:

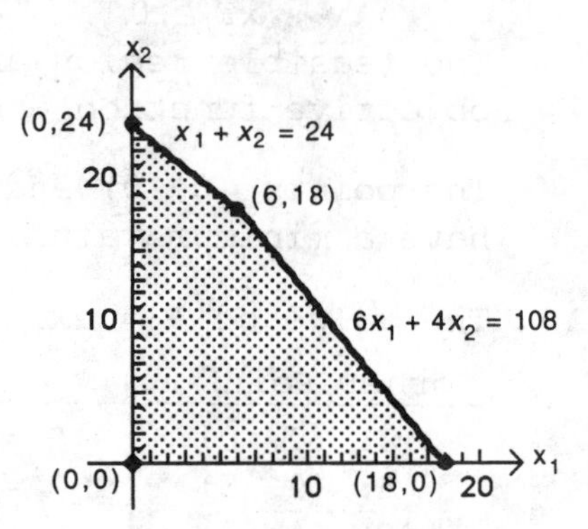

Corner Point	$P = 40x_1 + 30x_2$
(0, 0)	$P = 40(0) + 30(0) = 0$
(0, 24)	$P = 40(0) + 30(24) = 720$
(6, 18)	$P = 40(6) + 30(18) = 780$
(18, 0)	$P = 40(18) + 30(0) = 720$

Step (5): Determine the optimal solution. The maximum occurs when $x_1 = 6$ (trick skis) and $x_2 = 18$ (slalom skis) are produced. The maximum profit is $P = \$780$.

(B)

Corner Point	$P = 40x_1 + 25x_2$
(0, 0)	$P = 40(0) + 25(0) = 0$
(0, 24)	$P = 40(0) + 25(24) = 600$
(6, 18)	$P = 40(6) + 25(18) = 690$
(18, 0)	$P = 40(18) + 25(0) = 720$

The maximum profit decreases to $720 when 18 trick skis and no slalom skis are produced.

(C)

Corner Point	$P = 40x_1 + 45x_2$
(0, 0)	$P = 40(0) + 45(0) = 0$
(0, 24)	$P = 40(0) + 45(24) = 1080$
(6, 18)	$P = 40(6) + 45(18) = 1050$
(18, 0)	$P = 40(18) + 45(0) = 720$

The maximum profit increases to $1,080 when no trick skis and 24 slalom skis are produced.

33. (A) Step (1): Summarize relevant material in table form.

	Plant A	Plant B	Amount required
Tables	20	25	200
Chairs	60	50	500
Cost per day	$1000	$900	

Step (2): Form a mathematical model for the problem.

Let x_1 = the number of days to operate Plant A
and x_2 = the number of days to operate Plant B.

The mathematical model for this problem is:

Minimize $C = 1000x_1 + 900x_2$

Subject to: $20x_1 + 25x_2 \geq 200$

$60x_1 + 50x_2 \geq 500$

$x_1 \geq 0,\ x_2 \geq 0$

Step (3): Graph the feasible region and find the corner points.

The feasible region S is the solution set of the system of inequalities, and is indicated by the shading in the graph shown below.

The corner points are (0, 10), (5, 4), and (10, 0).

Since S is unbounded and $a = 1000 > 0$, $b = 900 > 0$, C has a minimum value by 3(b).

Step (4): Evaluate the objective function at each corner point.

The value of C at each corner point is:

Corner Point	$C = 1000x_1 + 900x_2$
(0, 10)	$C = 1000(0) + 900(10) = 9,000$
(5, 4)	$C = 1000(5) + 900(4) = 8,600$
(10, 0)	$C = 1000(10) + 900(0) = 10,000$

Step (5): Determine the optimal solution.

The minimum occurs when $x_1 = 5$ and $x_2 = 4$.

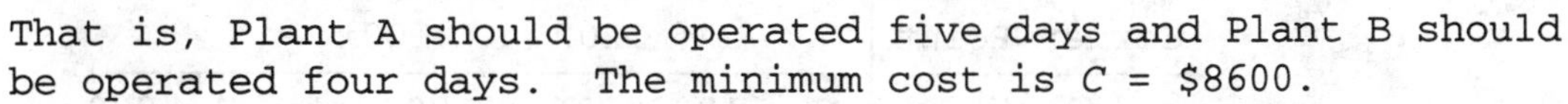

That is, Plant A should be operated five days and Plant B should be operated four days. The minimum cost is C = \$8600.

(B)

Corner Point	$C = 600x_1 + 900x_2$
(0, 10)	$C = 600(0) + 900(10) = 9000$
(5, 4)	$C = 600(5) + 900(4) = 6600$
(10, 0)	$C = 600(10) + 900(0) = 6000$

The minimum cost decreases to \$6,000 per day when Plant A is operated 10 days and Plant B is operated 0 days.

(C)

Corner Point	$C = 1000x_1 + 800x_2$
(0, 10)	$C = 600(0) + 800(10) = 8,000$
(5, 4)	$C = 1000(5) + 800(4) = 8,200$
(10, 0)	$C = 1000(10) + 800(0) = 10,000$

The minimum cost decreases to \$8,000 per day when Plant A is operated 0 days and Plant B is operated 10 days.

35. (A) Step (1): Summarize relevant material.

	Buses	Vans	Number to accommodate
Students	40	8	400
Chaperones	3	1	36
Rental cost	\$1200 per bus	\$100 per van	

Step (2): Form a mathematical model for the problem.
Let x_1 = the number of buses
and x_2 = the number of vans.
The mathematical model for this problem is:
Minimize $C = 1200x_1 + 100x_2$
Subject to: $40x_1 + 8x_2 \geq 400$
$3x_1 + x_2 \leq 36$
$x_1 \geq 0,\ x_2 \geq 0$

Step (3): Graph the feasible region and find the corner points.

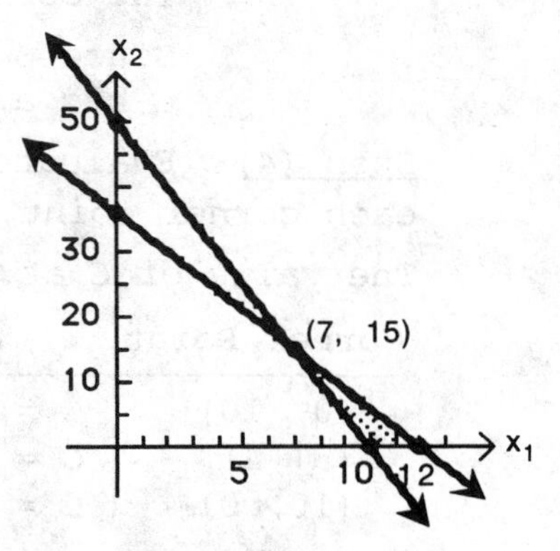

The feasible region S is the solution set of the system of inequalities, and is indicated by the shading in the graph at the right.

The corner points are (10, 0), (7, 15), and (12, 0).

Since S is bounded, C has a minimum value by 3(a).

Step (4): Evaluate the objective function at each corner point.
The value of C at each corner point is:

Corner Point	$C = 1200x_1 + 100x_2$
(10, 0)	$C = 1200(10) + 100(0) = 12{,}000$
(7, 15)	$C = 1200(7) + 100(15) = 9{,}900$
(12, 0)	$C = 1200(12) + 100(0) = 14{,}400$

Step (5): Determine the optimal solution.
The minimum occurs when $x_1 = 7$ and $x_2 = 15$. That is, the officers should rent 7 buses and 15 vans at the minimum cost of \$9900.

37. Let x_1 = amount invested in the CD
and x_2 = amount invested in the mutual fund.

(A) The mathematical model for this problem is
Maximize $P = 0.05x_1 + 0.09x_2$
Subject to $x_1 + x_2 \leq 60{,}000$
$x_2 \geq 10{,}000$
$x_1 \geq 2x_2$
$x_1,\ x_2 \geq 0$

The feasible region S is the solution set of the system of inequalities and is indicated by the shading in the graph.

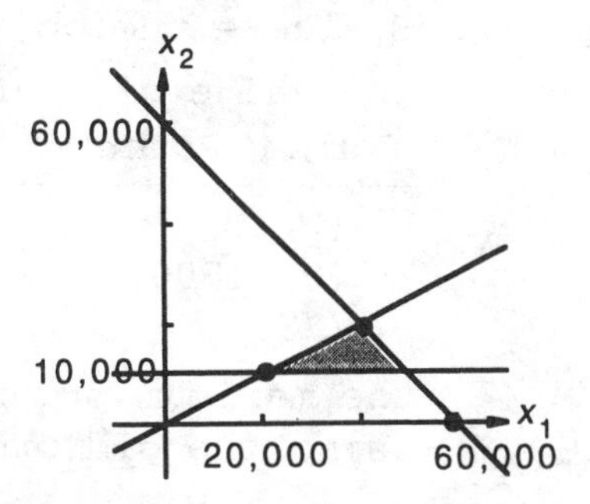

The corner points are (20,000, 10,000), (40,000, 20,000) and (50,000, 10,000).

Since S is bounded, P has a maximum value by 3a.

The value of P at each corner point is given in the table below.

Corner Point	$P = x_1 + x_2$
(20,000, 10,000)	$P = 0.05(20,000) + 0.09(10,000) = 1900$
(40,000, 20,000)	$P = 0.05(40,000) + 0.09(20,000) = 3800$
(50,000, 10,000)	$P = 0.05(50,000) + 0.09(10,000) = 3400$

Thus, the maximum return is \$3,800 when \$40,000 is invested in the CD and \$20,000 is invested in the mutual fund.

39. (A) Step (1): Summarize relevant material.

	Grams per Gallon Emitted by: Old process	New Process	Maximum allowed
Sulfur dioxide	20	5	16,000
Particulate	40	20	30,000

Step (2): Form a mathematical model for the problem.

Let x_1 = the number of gallons produced by the old process and x_2 = the number of gallons produced by the new process.

The mathematical model for this problem is:

Maximize $P = 60x_1 + 20x_2$

Subject to: $20x_1 + 5x_2 \le 16,000$

$40x_1 + 20x_2 \le 30,000$

$x_1 \ge 0,\ x_2 \ge 0$

Step (3): Graph the feasible region and find the corner points.

The feasible region S is the solution set of the given inequalities, and is indicated by the shading in the graph at the right.

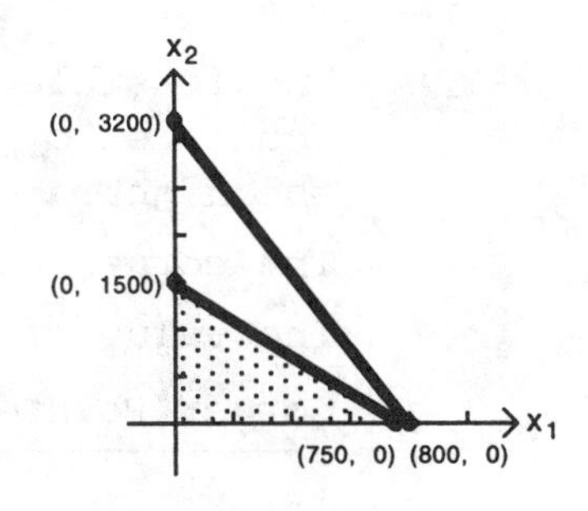

The corner points are (0, 0), (0, 1500), and (750, 0).

Since S is bounded, P has a maximum value by 3(a).

Step (4): Evaluate the objective function at each corner point.

The value of P at each corner point is:

Corner Point	$P = 60x_1 + 20x_2$
(0, 0)	$P = 60(0) + 20(0) = 0$
(0, 1500)	$P = 60(0) + 20(1500) = 30,000$
(750, 0)	$P = 60(750) + 20(0) = 45,000$

The maximum profit is \$450 when 750 gallons are produced using the old process exclusively.

(B) The mathematical model for this problem is:

Maximize $P = 60x_1 + 20x_2$

Subject to: $20x_1 + 5x_2 \le 11,500$

$40x_1 + 20x_2 \le 30,000$

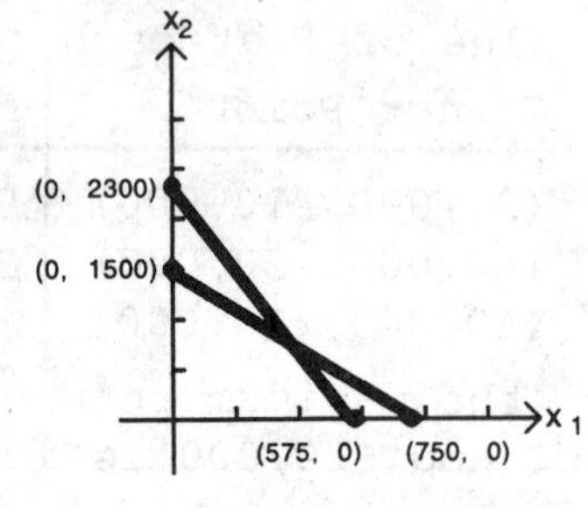

The feasible region S for this problem is indicated by the shading in the graph at the right.

The corner points are (0, 0), (0, 1500), (400, 700), and (575, 0).

The value P at each corner point is:

Corner Point	$P = 60x_1 + 20x_2$
(0, 0)	$P = 60(0) + 20(0) = 0$
(0, 1500)	$P = 60(0) + 20(1,500) = 30,000$
(400, 700)	$P = 60(400) + 20(700) = 38,000$
(575, 0)	$P = 60(575) + 20(0) = 34,500$

The maximum profit is \$380 when 400 gallons are produced using the old process and 700 gallons using the new process.

(C) The mathematical model for this problem is:

Maximize $P = 60x_1 + 20x_2$

Subject to: $20x_1 + 5x_2 \le 7,200$

$40x_1 + 20x_2 \le 30,000$

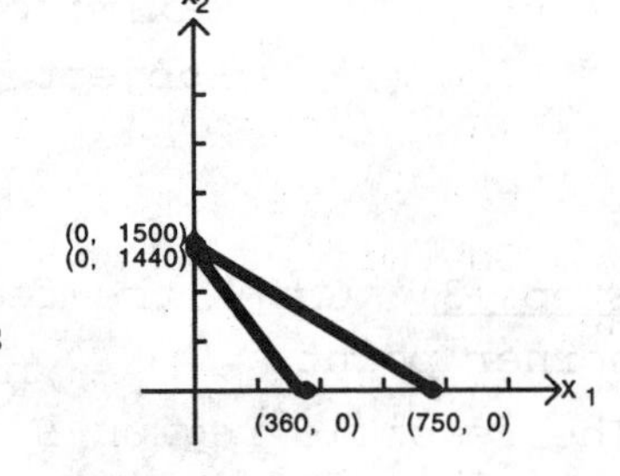

The feasible region S for this problem is indicated by the shading in the graph at the right.

The corner points are (0, 0), (0, 1440), and (360, 0).

The value of P at each corner point is:

Corner Point	$P = 60x_1 + 20x_2$
(0, 0)	$P = 60(0) + 20(0) = 0$
(0, 1440)	$P = 60(0) + 20(1,440) = 28,800$
(360, 0)	$P = 60(360) + 20(0) = 21,600$

The maximum profit is \$288 when 1,440 gallons are produced by the new process exclusively.

41. Let x_1 = the number of bags of Brand A
and x_2 = the number of bags of Brand B.

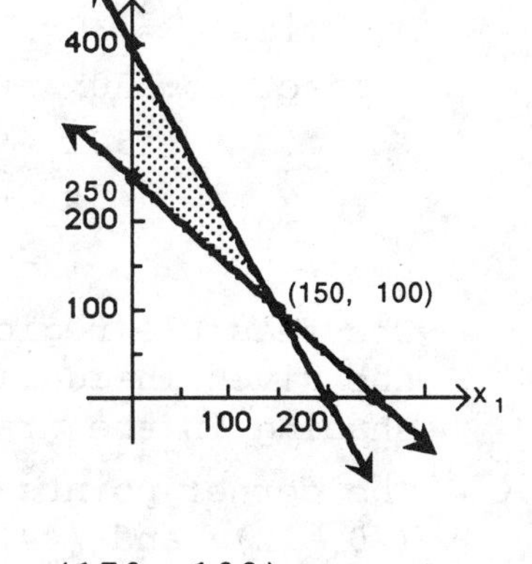

(A) The mathematical model for this problem is:

Maximize $N = 8x_1 + 3x_2$

Subject to: $4x_1 + 4x_2 \geq 1000$

$2x_1 + x_2 \leq 400$

$x_1 \geq 0,\ x_2 \geq 0$

The feasible region S is the solution set of the system of inequalities, and is indicated by the shading in the graph at the right.

The corner points are (0, 250), (0, 400), and (150, 100).

Since S is bounded, N has a maximum value by 3(a).

The value of N at each corner point is given in the table below:

Corner Point	$N = 8x_1 + 3x_2$
(0, 250)	$N = 8(0) + 3(250) = 750$
(150, 100)	$N = 8(150) + 3(100) = 1500$
(0, 400)	$N = 8(0) + 3(400) = 1200$

Thus, the maximum occurs when $x_1 = 150$ and $x_2 = 100$. That is, the grower should use 150 bags of Brand A and 100 bags of Brand B. The maximum number of pounds of nitrogen is 1500.

(B) The mathematical model for this problem is:

Minimize $N = 8x_1 + 3x_2$

Subject to: $4x_1 + 4x_2 \geq 1000$

$2x_1 + x_2 \leq 400$

$x_1 \geq 0,\ x_2 \geq 0$

The feasible region S and the corner points are the same as in part (A). Thus, the minimum occurs when $x_1 = 0$ and $x_2 = 250$. That is, the grower should use 0 bags of Brand A and 250 bags of Brand B. The minimum number of pounds of nitrogen is 750.

43.

	Amount per Cubic Yard (in pounds) Mix A	Mix B	Minimum monthly requirement
Phosphoric acid	20	10	460
Nitrogen	30	30	960
Potash	5	10	220
Cost/cubic yd.	\$30	\$35	

Let x_1 = the number of cubic yards of mix A
and x_2 = the number of cubic yards of mix B.

The mathematical model for this problem is:

Minimize $C = 30x_1 + 35x_2$

Subject to: $20x_1 + 10x_2 \geq 460$

$30x_1 + 30x_2 \geq 960$

$5x_1 + 10x_2 \geq 220$

$x_1 \geq 0,\ x_2 \geq 0$

The feasible region S is the solution set of the given inequalities and is indicated by the shading in the graph at the right.

The corner points are (0, 46), (14, 18), (20, 12), and (44, 0).

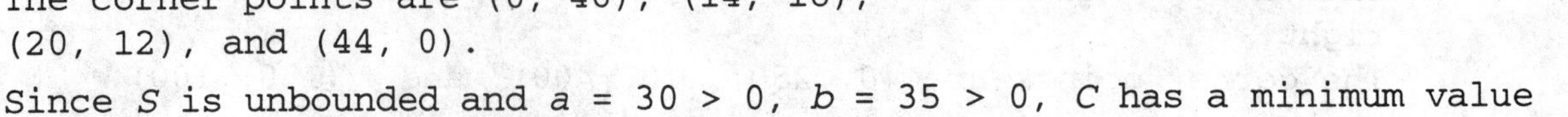

Since S is unbounded and $a = 30 > 0$, $b = 35 > 0$, C has a minimum value by 3(b).

The value of C at each corner point is:

Corner Point	$C = 30x_1 + 35x_2$
(0, 46)	$C = 30(0) + 35(46) = 1610$
(14, 18)	$C = 30(14) + 35(18) = 1050$
(20, 12)	$C = 30(20) + 35(12) = 1020$
(44, 0)	$C = 30(44) + 35(0) = 1320$

Thus, the minimum occurs when the amount of mix A used is 20 cubic yards and the amount of mix B used is 12 cubic yards. The minimum cost is $C = \$1020$.

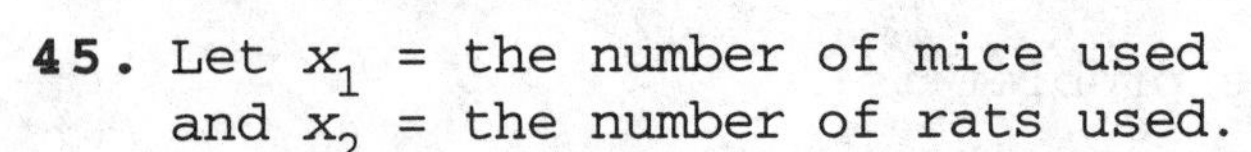

45. Let x_1 = the number of mice used and x_2 = the number of rats used.

The mathematical model for this problem is:

Maximize $P = x_1 + x_2$

Subject to: $10x_1 + 20x_2 \leq 800$

$20x_1 + 10x_2 \leq 640$

$x_1 \geq 0,\ x_2 \geq 0$

The feasible region S is the solution set of the given inequalities, and is indicated by the shading in the graph at the right.

The corner points are (0, 0), (0, 40), (16, 32), and (32, 0).

Since S is bounded, P has a maximum value by 3(a).

The value of P at each corner point is:

Corner Point	$P = x_1 + x_2$
(0, 0)	$P = 0 + 0 = 0$
(0, 40)	$P = 0 + 40 = 40$
(16, 32)	$P = 16 + 32 = 48$
(32, 0)	$P = 32 + 0 = 32$

Thus, the maximum occurs when the number of mice used is 16 and the number of rats used is 32. The maximum number of mice and rats that can be used is 48.

Things to remember:

1. STANDARD MAXIMIZATION PROBLEM IN STANDARD FORM

   A linear programming problem is said to be a STANDARD MAXIMIZATION PROBLEM IN STANDARD FORM if its mathematical model is of the form:

   Maximize $P = c_1x_1 + c_2x_2 + \cdots + c_nx_n$

   Subject to problem constraints of the form:

   $a_1x_1 + a_2x_2 + \cdots + a_nx_n \leq b, \quad b \geq 0$

   with nonnegative constraints:

   $x_1, x_2, \ldots, x_n \geq 0.$

   [Note: The coefficients of the objective function can be any real numbers.]

2. SLACK VARIABLES

   Given a linear programming problem. SLACK VARIABLES are nonnegative quantities that are introduced to convert problem constraint inequalities into equations.

3. BASIC VARIABLES AND NONBASIC VARIABLES; BASIC SOLUTIONS AND BASIC FEASIBLE SOLUTIONS

   Given a system of linear equations associated with a linear programming problem. (Such a system will always have more variables than equations.)

   The variables are divided into two (mutually exclusive) groups, called BASIC VARIABLES and NONBASIC VARIABLES, as follows: Basic variables are selected arbitrarily with the one restriction that there be as many basic variables as there are equations. The remaining variables are called nonbasic variables.

   A solution found by setting the nonbasic variables equal to zero and solving for the basic variables is called a BASIC SOLUTION. If a basic solution has no negative values, it is a BASIC FEASIBLE SOLUTION.

4. FUNDAMENTAL THEOREM OF LINEAR PROGRAMMING

   If the optimal value of the objective function in a linear programming problem exists, then that value must occur at one (or more) of the basic feasible solutions.

1. (A) Since there are 2 problem constraints, 2 slack variables are introduced.

(B) Since there are two equations (from the two problem constraints) and three decision variables, there are two basic variables and three nonbasic variables.

(C) There will be two linear equations and two variables.

3. (A) There are 5 constraint equations; the number of equations is the same as the number of slack variables.

(B) There are 4 decision variables since there are 9 variables altogether, and 5 of them are slack variables.

(C) There are 5 basic variables and 4 nonbasic variables; the number of basic variables equals the number of equations.

(D) Five linear equations with 5 variables.

5.

	Nonbasic	Basic	Feasible?
(A)	x_1, x_2	s_1, s_2	Yes, all values are nonnegative.
(B)	x_1, s_1	x_2, s_2	Yes, all values are nonnegative.
(C)	x_1, s_2	x_2, s_1	No, $s_1 = -12 < 0$.
(D)	x_2, s_1	x_1, s_2	No, $s_2 = -12 < 0$.
(E)	x_2, s_2	x_1, s_1	Yes, all values are nonnegative.
(F)	s_1, s_2	x_1, x_2	Yes, all values are nonnegative.

7.

	x_1	x_2	s_1	s_2	Feasible?
(A)	0	0	50	40	Yes, all values are nonnegative.
(B)	0	50	0	-60	No, $s_2 = -60 < 0$.
(C)	0	20	30	0	Yes, all values are nonnegative.
(D)	25	0	0	15	Yes, all values are nonnegative.
(E)	40	0	-30	0	No, $s_1 = -30 < 0$.
(F)	20	10	0	0	Yes, all values are nonnegative.

9.

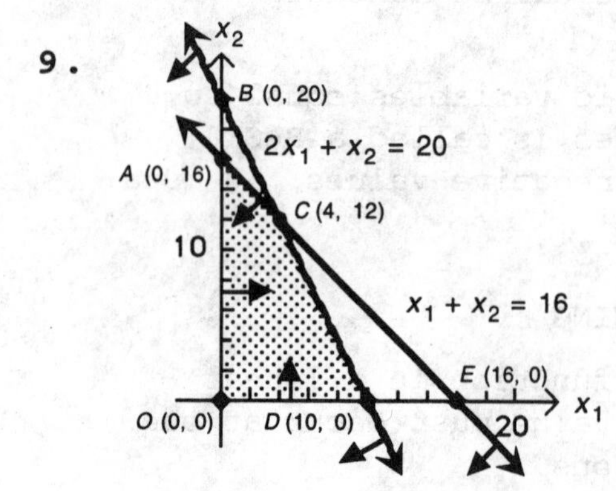

Introduce slack variables s_1 and s_2 to obtain the system of equations:

$$\begin{aligned} x_1 + x_2 + s_1 &= 16 \\ 2x_1 + x_2 + s_2 &= 20 \end{aligned}$$

x_1	x_2	s_1	s_2	Intersection Point	Feasible?
0	0	16	20	O	Yes
0	16	0	4	A	Yes
0	20	-4	0	B	No, $s_1 = -4 < 0$
16	0	0	-12	E	No, $s_2 = -12 < 0$
10	0	6	0	D	Yes
4	12	0	0	C	Yes

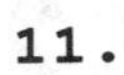

11.

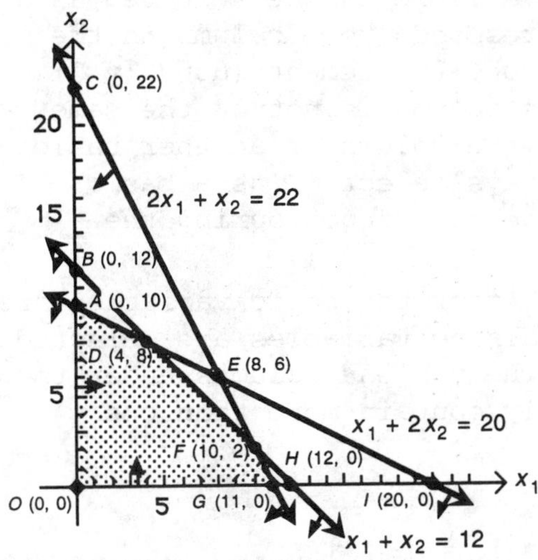

Introduce slack variables s_1, s_2, and s_3 to obtain the system of equations:

$$
\begin{aligned}
2x_1 + x_2 + s_1 \quad\quad &= 22 \\
x_1 + x_2 \quad + s_2 \quad &= 12 \\
x_1 + 2x_2 \quad\quad + s_3 &= 20
\end{aligned}
$$

x_1	x_2	s_1	s_2	s_3	Intersection Point	Feasible?
0	0	22	12	20	O	Yes
0	22	0	-10	-24	C	No
0	12	10	0	-4	B	No
0	10	12	2	0	A	Yes
11	0	0	1	9	G	Yes
12	0	-2	0	8	H	No
20	0	-18	-8	0	I	No
10	2	0	0	6	F	Yes
8	6	0	-2	0	E	No
4	8	6	0	0	D	Yes

Things to remember:

1. SELECTING BASIC AND NONBASIC VARIABLES FOR THE SIMPLEX PROCESS

   Given a simplex tableau.

   (a) Determine the number of basic and the number of nonbasic variables. These numbers do not change during the simplex process.

   (b) SELECTING BASIC VARIABLES: A variable can be selected as a basic variable only if it corresponds to a column in the tableau that has exactly one nonzero element (usually 1) and the nonzero element in the column is not in the same row as the nonzero element in the column of another basic variable. (This procedure always selects P as a basic variable, since the P column never changes during the simplex process.)

   (c) SELECTING NONBASIC VARIABLES: After the basic variables are selected in Step (b), the remaining variables are selected as the nonbasic variables. (The tableau columns under the nonbasic variables will usually contain more than one nonzero element.)

2. SELECTING THE PIVOT ELEMENT

   (a) Locate the most negative indicator in the bottom row of the tableau to the left of the P column (the negative number with the largest absolute value). The column containing this element is the PIVOT COLUMN. If there is a tie for the most negative, choose either.

   (b) Divide each POSITIVE element in the pivot column above the dashed line into the corresponding element in the last column. The PIVOT ROW is the row corresponding to the smallest quotient. If there is a tie for the smallest quotient, choose either. If the pivot column above the dashed line has no positive elements, then there is no solution and we stop.

   (c) The PIVOT (or PIVOT ELEMENT) is the element in the intersection of the pivot column and pivot row. [Note: The pivot element is always positive and is never in the bottom row.]

   [Remember: The entering variable is at the top of the pivot column and the exiting variable is at the left of the pivot row.]

3. PERFORMING THE PIVOT OPERATION

A PIVOT OPERATION or PIVOTING consists of performing row operations as follows:

(a) Multiply the pivot row by the reciprocal of the pivot element to transform the pivot element into a 1. (If the pivot element is already a 1, omit this step.)

(b) Add multiples of the pivot row to other rows in the tableau to transform all other nonzero elements in the pivot column into 0's.

[Note: Rows are not to be interchanged while performing a pivot operation. The only way the (positive) pivot element can be transformed into 1 (if it is not a 1 already) is for the pivot row to be multiplied by the reciprocal of the pivot element.]

4. SIMPLEX ALGORITHM FOR STANDARD MAXIMIZATION PROBLEMS

Problem constraints are of the ≤ form with nonnegative constants on the right hand side. The coefficients of the objective function can be any real numbers.

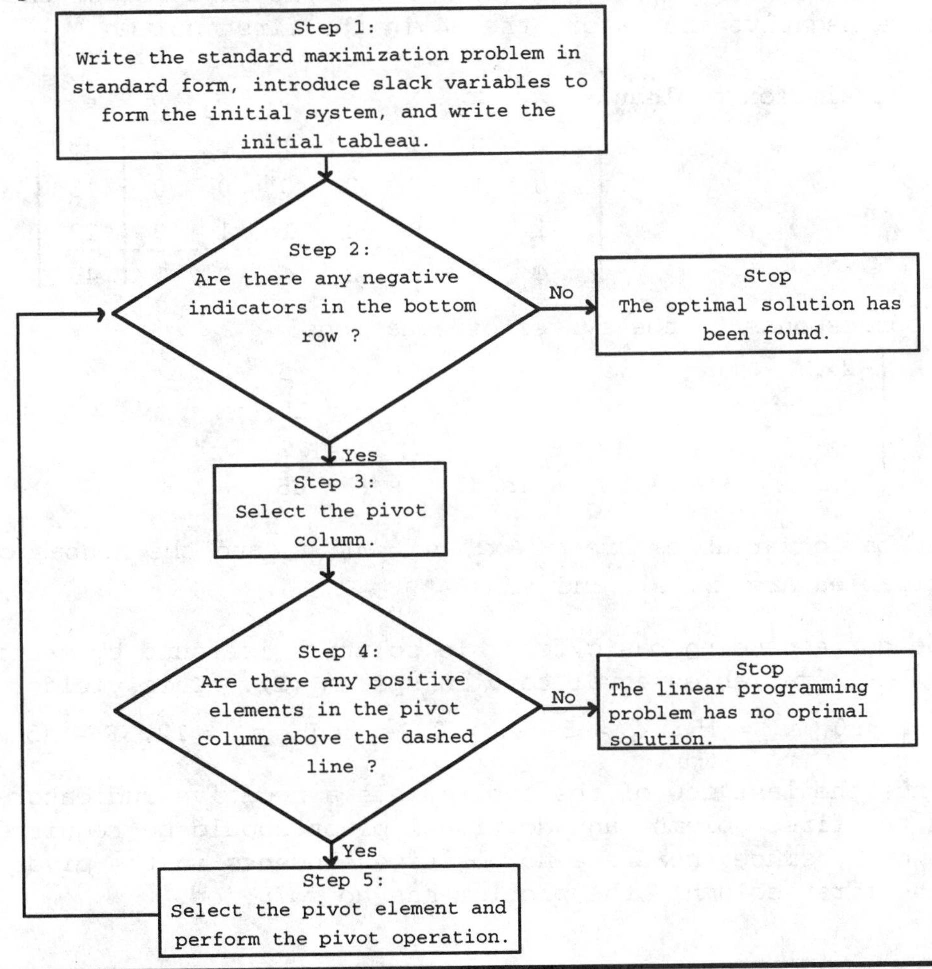

1. Given the simplex tableau:

$$\begin{array}{ccccc|c} x_1 & x_2 & s_1 & s_2 & P & \\ \hline 2 & 1 & 0 & 3 & 0 & 12 \\ 3 & 0 & 1 & -2 & 0 & 15 \\ \hline -4 & 0 & 0 & 4 & 1 & 20 \end{array}$$

which corresponds to the system of equations:

$$\text{(I)} \begin{cases} 2x_1 + x_2 + 3s_2 = 12 \\ 3x_1 + s_1 - 2s_2 = 15 \\ -4x_1 + 4s_2 + P = 20 \end{cases}$$

(A) The basic variables are x_2, s_1, and P, and the nonbasic variables are x_1 and s_2.

(B) The corresponding basic feasible solution is found by setting the nonbasic variables equal to 0 in system (I). This yields:

$x_1 = 0$, $x_2 = 12$, $s_1 = 15$, $s_2 = 0$, $P = 20$

(C) An additional pivot is required, since the last row of the tableau has a negative indicator, the -4 in the first column.

3. Given the simplex tableau:

$$\begin{array}{ccccccc|c} x_1 & x_2 & x_3 & s_1 & s_2 & s_3 & P & \\ \hline -2 & 0 & 1 & 3 & 1 & 0 & 0 & 5 \\ 0 & 1 & 0 & -2 & 0 & 0 & 0 & 15 \\ -1 & 0 & 0 & 4 & 1 & 1 & 0 & 12 \\ \hline -4 & 0 & 0 & 2 & 4 & 0 & 1 & 45 \end{array}$$

which corresponds to the system of equations:

$$\text{(I)} \begin{cases} -2x_1 + x_3 + 3s_1 + s_2 = 5 \\ x_2 - 2s_1 = 15 \\ -x_1 + 4s_1 + s_2 + s_3 = 12 \\ -4x_1 + 2s_1 + 4s_2 + P = 45 \end{cases}$$

(A) The basic variables are x_2, x_3, s_3, and P, and the nonbasic variables are x_1, s_1, and s_2.

(B) The corresponding basic feasible solution is found by setting the nonbasic variables equal to 0 in system (I). This yields:

$x_1 = 0$, $x_2 = 15$, $x_3 = 5$, $s_1 = 0$, $s_2 = 0$, $s_3 = 12$, $P = 45$

(C) Since the last row of the tableau has a negative indicator, the -4 in the first column, an additional pivot should be required. However, since there are no positive elements in the pivot column (the first column), the problem has *no solution*.

5. Given the simplex tableau:

$$\begin{array}{c} \\ \end{array}\begin{array}{ccccc|c} x_1 & x_2 & s_1 & s_2 & P & \\ \hline 1 & 4 & 1 & 0 & 0 & 4 \\ 3 & 5 & 0 & 1 & 0 & 24 \\ \hdashline -8 & -5 & 0 & 0 & 1 & 0 \end{array}$$

The most negative indicator is -8 in the first column. Thus, the first column is the pivot column. Now, $\frac{4}{1} = 4$ and $\frac{24}{3} = 8$. Thus, the first row is the pivot row and the pivot element is the element in the first row, first column. These are indicated in the following tableau.

$$\begin{array}{c|ccccc|c|l} & \text{Enter} & & & & & & \\ & x_1 & x_2 & s_1 & s_2 & P & & \\ \hline \text{Exit } s_1 & \textcircled{1} & 4 & 1 & 0 & 0 & 4 & \frac{4}{1} = 4 \text{ (minimum)} \\ s_2 & 3 & 5 & 0 & 1 & 0 & 24 & \frac{24}{3} = 8 \\ \hdashline P & -8 & -5 & 0 & 0 & 1 & 0 & \end{array}$$

$$\left[\begin{array}{ccccc|c} \textcircled{1} & 4 & 1 & 0 & 0 & 4 \\ 3 & 5 & 0 & 1 & 0 & 24 \\ \hdashline -8 & -5 & 0 & 0 & 1 & 0 \end{array}\right] \sim \left[\begin{array}{ccccc|c} 1 & 4 & 1 & 0 & 0 & 4 \\ 0 & -7 & -3 & 1 & 0 & 12 \\ \hdashline 0 & 27 & 8 & 0 & 1 & 32 \end{array}\right]$$

$(-3)R_1 + R_2 \rightarrow R_2$

$8R_1 + R_3 \rightarrow R_3$

7. Given the simplex tableau:

$$\begin{array}{cccccc|c} x_1 & x_2 & s_1 & s_2 & s_3 & P & \\ \hline 2 & 1 & 1 & 0 & 0 & 0 & 4 \\ 3 & 0 & 1 & 1 & 0 & 0 & 8 \\ 0 & 0 & 2 & 0 & 1 & 0 & 2 \\ \hdashline -4 & 0 & -3 & 0 & 0 & 1 & 5 \end{array}$$

The most negative indicator is -4. Thus, the first column is the pivot column. Now, $\frac{4}{2} = 2$, $\frac{8}{3} = 2\frac{2}{3}$. Thus, the first row is the pivot row, and the pivot element is the element in the first row, first column. These are indicated in the tableau.

$$\begin{array}{c|cccccc|c|l} & \text{Enter} & & & & & & & \\ & x_1 & x_2 & s_1 & s_2 & s_3 & P & & \\ \hline \text{Exit } x_2 & \textcircled{2} & 1 & 1 & 0 & 0 & 0 & 4 & \frac{4}{2} = 2 \text{ (minimum)} \\ s_2 & 3 & 0 & 1 & 1 & 0 & 0 & 8 & \frac{8}{3} = 2\frac{2}{3} \\ s_3 & 0 & 0 & 2 & 0 & 1 & 0 & 2 & \\ \hdashline P & -4 & 0 & -3 & 0 & 0 & 1 & 5 & \end{array}$$

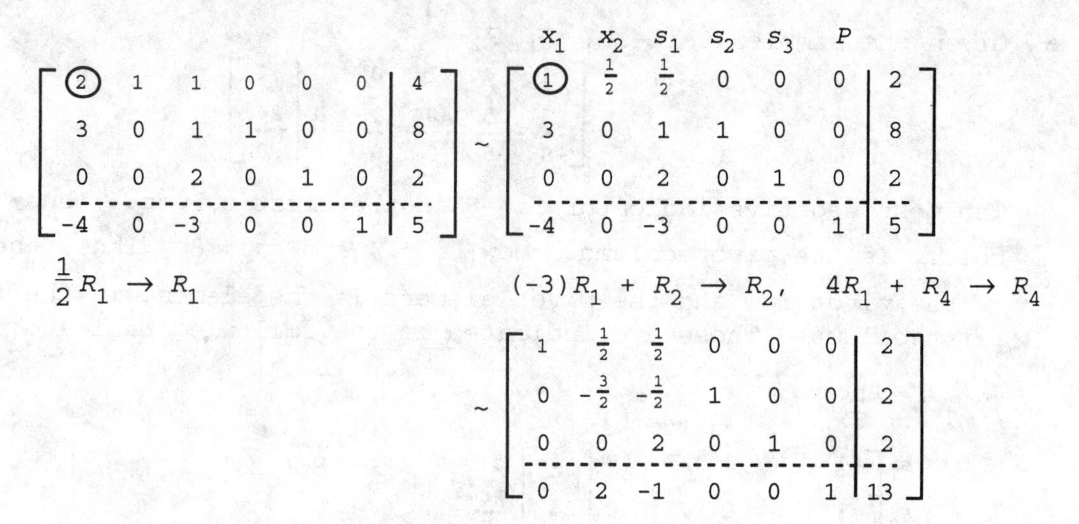

9. (A) Introduce slack variables s_1 and s_2 to obtain:

Maximize $P = 15x_1 + 10x_2$

Subject to: $2x_1 + x_2 + s_1 = 10$

$x_1 + 3x_2 + s_2 = 10$

$x_1, x_2, s_1, s_2 \geq 0$

This system can be written in initial form:

$$\begin{aligned} 2x_1 + x_2 + s_1 &= 10 \\ x_1 + 3x_2 + s_2 &= 10 \\ -15x_1 - 10x_2 + P &= 0 \\ x_1, x_2, s_1, s_2 &\geq 0 \end{aligned}$$

(B) The simplex tableau for this problem is:

$$\begin{array}{ll} & \text{Enter} \\ & \begin{array}{ccccc} x_1 & x_2 & s_1 & s_2 & P \end{array} \\ \text{Exit } s_1 & \left[\begin{array}{ccccc|c} \textcircled{2} & 1 & 1 & 0 & 0 & 10 \\ 1 & 3 & 0 & 1 & 0 & 10 \\ \hline -15 & -10 & 0 & 0 & 1 & 0 \end{array}\right] \begin{array}{l} \frac{10}{2} = 5 \text{ (minimum)} \\ \frac{10}{1} = 10 \\ \ \end{array} \\ \quad s_2 & \\ \quad P & \end{array}$$

Column 1 is the pivot column (-15 is the most negative indicator). Row 1 is the pivot row (5 is the smallest positive quotient). Thus, the pivot element is the circled 2.

(C) We use the simplex method as outlined above. The pivot elements are circled.

$$\begin{array}{c} \\ s_1 \\ s_2 \\ P \end{array} \begin{array}{c} \begin{array}{ccccc} x_1 & x_2 & s_1 & s_2 & P \end{array} \\ \left[\begin{array}{ccccc|c} \textcircled{2} & 1 & 1 & 0 & 0 & 10 \\ 1 & 3 & 0 & 1 & 0 & 10 \\ \hline -15 & -10 & 0 & 0 & 1 & 0 \end{array}\right] \end{array} \sim \left[\begin{array}{ccccc|c} \textcircled{1} & \frac{1}{2} & \frac{1}{2} & 0 & 0 & 5 \\ 1 & 3 & 0 & 1 & 0 & 10 \\ \hline -15 & -10 & 0 & 0 & 1 & 0 \end{array}\right] \sim$$

$\frac{1}{2}R_1 \to R_1$ $(-1)R_1 + R_2 \to R_2$

$15R_1 + R_3 \to R_3$

$$\sim \left[\begin{array}{ccccc|c} 1 & \frac{1}{2} & \frac{1}{2} & 0 & 0 & 5 \\ 0 & \textcircled{\frac{5}{2}} & -\frac{1}{2} & 1 & 0 & 5 \\ \hline 0 & -\frac{5}{2} & \frac{15}{2} & 0 & 1 & 75 \end{array}\right] \sim \left[\begin{array}{ccccc|c} 1 & \frac{1}{2} & \frac{1}{2} & 0 & 0 & 5 \\ 0 & \textcircled{1} & -\frac{1}{5} & \frac{2}{5} & 0 & 2 \\ \hline 0 & -\frac{5}{2} & \frac{15}{2} & 0 & 1 & 75 \end{array}\right]$$

$\frac{2}{5}R_2 \to R_2$

$\left(-\frac{1}{2}\right)R_2 + R_1 \to R_1$

$\left(\frac{5}{2}\right)R_2 + R_3 \to R_3$

$$\sim \begin{array}{c} \\ x_1 \\ x_2 \\ P \end{array} \begin{array}{c} \begin{array}{ccccc} x_1 & x_2 & s_1 & s_2 & P \end{array} \\ \left[\begin{array}{ccccc|c} 1 & 0 & \frac{3}{5} & -\frac{1}{5} & 0 & 4 \\ 0 & 1 & -\frac{1}{5} & \frac{2}{5} & 0 & 2 \\ \hline 0 & 0 & 7 & 1 & 1 & 80 \end{array}\right] \end{array}$$

All elements in the last row are nonnegative. Thus, max $P = 80$ at $x_1 = 4$, $x_2 = 2$, $s_1 = 0$, $s_2 = 0$.

11. (A) Introduce slack variables s_1 and s_2 to obtain:

Maximize $P = 30x_1 + x_2$

Subject to:
$$\begin{aligned} 2x_1 + x_2 + s_1 \quad &= 10 \\ x_1 + 3x_2 \quad + s_2 &= 10 \\ x_1, x_2, s_1, s_2 &\geq 0 \end{aligned}$$

This system can be written in the initial form:

$$\begin{aligned} 2x_1 + x_2 + s_1 \quad\quad &= 10 \\ x_1 + 3x_2 \quad + s_2 \quad &= 10 \\ -30x_1 - x_2 \quad\quad + P &= 0 \end{aligned}$$

(B) The simplex tableau for this problem is:

$$\begin{array}{cc} & \text{Enter} \\ & \begin{array}{ccccc} x_1 & x_2 & s_1 & s_2 & P \end{array} \\ \begin{array}{c} \text{Exit } s_1 \\ s_2 \\ P \end{array} & \left[\begin{array}{ccccc|c} \textcircled{2} & 1 & 1 & 0 & 0 & 10 \\ 1 & 3 & 0 & 1 & 0 & 10 \\ \hline -30 & -1 & 0 & 0 & 1 & 0 \end{array}\right] \end{array} \begin{array}{l} \frac{10}{2} = 5 \text{ (minimum)} \\ \frac{10}{1} = 10 \\ \end{array}$$

↑ pivot column

(C)

$$\begin{array}{c} \\ s_1 \\ s_2 \\ P \end{array}\begin{array}{c} \begin{array}{ccccc} x_1 & x_2 & s_1 & s_2 & P \end{array} \\ \left[\begin{array}{ccccc|c} \textcircled{2} & 1 & 1 & 0 & 0 & 10 \\ 1 & 3 & 0 & 1 & 0 & 10 \\ \hdashline -30 & -1 & 0 & 0 & 1 & 0 \end{array}\right] \end{array} \sim \left[\begin{array}{ccccc|c} \textcircled{1} & \frac{1}{2} & \frac{1}{2} & 0 & 0 & 5 \\ 1 & 3 & 0 & 1 & 0 & 10 \\ \hdashline -30 & -1 & 0 & 0 & 1 & 0 \end{array}\right]$$

$\frac{1}{2}R_1 \to R_1$ $\qquad$ $(-1)R_1 + R_2 \to R_2$

$30R_1 + R_3 \to R_3$

$$\sim \begin{array}{c} \\ x_1 \\ s_2 \\ P \end{array}\begin{array}{c} \begin{array}{ccccc} x_1 & x_2 & s_1 & s_2 & P \end{array} \\ \left[\begin{array}{ccccc|c} 1 & \frac{1}{2} & \frac{1}{2} & 0 & 0 & 5 \\ 0 & \frac{5}{2} & -\frac{1}{2} & 1 & 0 & 5 \\ \hdashline 0 & 14 & 15 & 0 & 1 & 150 \end{array}\right] \end{array}$$

All the elements in the last row are nonnegative. Thus, max $P = 150$ at $x_1 = 5$, $x_2 = 0$, $s_1 = 0$, $s_2 = 5$.

13. The simplex tableau for this problem is:

[Note: The pivot elements have been circled.]

$$\begin{array}{c} \\ s_1 \\ s_2 \\ \text{pivot row} \to s_3 \ (\uparrow \text{Exit}) \\ \\ \end{array}\begin{array}{c} \begin{array}{cccccc} x_1 & x_2 \ (\text{Enter}) & s_1 & s_2 & s_3 & P \end{array} \\ \left[\begin{array}{cccccc|c} 2 & 1 & 1 & 0 & 0 & 0 & 10 \\ 1 & 1 & 0 & 1 & 0 & 0 & 7 \\ 1 & \textcircled{2} & 0 & 0 & 1 & 0 & 12 \\ \hdashline -30 & -40 & 0 & 0 & 0 & 1 & 0 \end{array}\right] \end{array} \begin{array}{l} 10 \\ 7 \\ \frac{12}{2} = 6 \text{ (minimum)} \\ \\ \end{array}$$

$\uparrow$ pivot column $\qquad \frac{1}{2}R_3 \to R_3$

$$\sim \left[\begin{array}{cccccc|c} 2 & 1 & 1 & 0 & 0 & 0 & 10 \\ 1 & 1 & 0 & 1 & 0 & 0 & 7 \\ \frac{1}{2} & \textcircled{1} & 0 & 0 & \frac{1}{2} & 0 & 6 \\ \hdashline -30 & -40 & 0 & 0 & 0 & 1 & 0 \end{array}\right]$$

$(-1)R_3 + R_1 \to R_1$, $(-1)R_3 + R_2 \to R_2$, and $40R_3 + R_4 \to R_4$

$$\begin{array}{c} \\ \text{pivot row} \to \\ \\ \\ \end{array} \sim \left[\begin{array}{cccccc|c} \frac{3}{2} & 0 & 1 & 0 & -\frac{1}{2} & 0 & 4 \\ \textcircled{\frac{1}{2}} & 0 & 0 & 1 & -\frac{1}{2} & 0 & 1 \\ \frac{1}{2} & 1 & 0 & 0 & \frac{1}{2} & 0 & 6 \\ \hdashline -10 & 0 & 0 & 0 & 20 & 1 & 240 \end{array}\right] \begin{array}{l} \frac{4}{3/2} = \frac{8}{3} \\ \frac{1}{1/2} = 2 \text{ (minimum)} \\ \frac{6}{1/2} = 12 \\ \\ \end{array}$$

$\uparrow$ pivot column $\qquad 2R_2 \to R_2$

$$\sim \left[\begin{array}{cccccc|c} -\frac{3}{2} & 0 & 1 & 0 & -\frac{1}{2} & 0 & 4 \\ \textcircled{1} & 0 & 0 & 2 & -1 & 0 & 2 \\ \frac{1}{2} & 1 & 0 & 0 & \frac{1}{2} & 0 & 6 \\ \hdashline -10 & 0 & 0 & 0 & 20 & 1 & 240 \end{array}\right]$$

$\left(-\frac{3}{2}\right)R_2 + R_1 \to R_1$, $\left(-\frac{1}{2}\right)R_2 + R_3 \to R_3$, and $10R_2 + R_4 \to R_4$

$$\sim \begin{array}{c} \\ s_1 \\ x_1 \\ x_2 \\ \\ \end{array} \begin{array}{c} \begin{array}{cccccc} x_1 & x_2 & s_1 & s_2 & s_3 & P \end{array} \\ \left[\begin{array}{cccccc|c} 0 & 0 & 1 & -3 & 1 & 0 & 1 \\ 1 & 0 & 0 & 2 & -1 & 0 & 2 \\ 0 & 1 & 0 & -1 & 1 & 0 & 5 \\ \hdashline 0 & 0 & 0 & 20 & 10 & 1 & 260 \end{array}\right] \end{array}$$

Optimal solution: max $P = 260$ at $x_1 = 2$, $x_2 = 5$, $s_1 = 1$, $s_2 = 0$, $s_3 = 0$.

15. The simplex tableau for this problem is:

$$\begin{array}{c} \text{Exit} \\ \\ \text{pivot row} \to s_1 \\ s_2 \\ s_3 \\ P \end{array} \begin{array}{c} \begin{array}{cccccc} x_1 & \overset{\text{Enter}}{x_2} & s_1 & s_2 & s_3 & P \end{array} \\ \left[\begin{array}{cccccc|c} -2 & \textcircled{1} & 1 & 0 & 0 & 0 & 2 \\ -1 & 1 & 0 & 1 & 0 & 0 & 5 \\ 0 & 1 & 0 & 0 & 1 & 0 & 6 \\ \hdashline -2 & -3 & 0 & 0 & 0 & 1 & 0 \end{array}\right] \\ \begin{array}{c} \uparrow \\ \text{pivot column} \end{array} \end{array} \begin{array}{l} \frac{2}{1} = 2 \text{ (minimum)} \\ \frac{5}{1} = 5 \\ \frac{6}{1} = 6 \\ \\ \end{array}$$

$(-1)R_1 + R_2 \to R_2$, $(-1)R_1 + R_3 \to R_3$, and $3R_1 + R_4 \to R_4$

$$\sim \begin{array}{c} \\ \\ \text{pivot row} \to \\ \\ \end{array} \left[\begin{array}{cccccc|c} -2 & 1 & 1 & 0 & 0 & 0 & 2 \\ 1 & 0 & -1 & 1 & 0 & 0 & 3 \\ \textcircled{2} & 0 & -1 & 0 & 1 & 0 & 4 \\ \hdashline -8 & 0 & 3 & 0 & 0 & 1 & 6 \end{array}\right] \begin{array}{l} \\ \frac{3}{1} = 3 \\ \frac{4}{2} = 2 \text{ (minimum)} \\ \\ \end{array}$$

$\uparrow$ pivot column

$\frac{1}{2}R_3 \to R_3$

$$\sim \left[\begin{array}{cccccc|c} -2 & 1 & 1 & 0 & 0 & 0 & 2 \\ 1 & 0 & -1 & 1 & 0 & 0 & 3 \\ \textcircled{1} & 0 & -\frac{1}{2} & 0 & \frac{1}{2} & 0 & 2 \\ \hdashline -8 & 0 & 3 & 0 & 0 & 1 & 6 \end{array}\right] \sim \begin{array}{c} \begin{array}{cccccc} x_1 & x_2 & s_1 & s_2 & s_3 & P \end{array} \\ \left[\begin{array}{cccccc|c} 0 & 1 & 0 & 0 & 1 & 0 & 6 \\ 0 & 0 & -\frac{1}{2} & 1 & -\frac{1}{2} & 0 & 1 \\ 1 & 0 & -\frac{1}{2} & 0 & \frac{1}{2} & 0 & 2 \\ \hdashline 0 & 0 & -1 & 0 & 4 & 1 & 22 \end{array}\right] \end{array}$$

$2R_3 + R_1 \to R_1$, $(-1)R_3 + R_2 \to R_2$, and $8R_3 + R_4 \to R_4$

($\uparrow$ pivot column: s_1)

Since there are no positive elements in the pivot column (above the dashed line), we conclude that there is no solution.

17. The simplex tableau for this problem is:

$$\begin{array}{c} \\ \text{pivot row} \rightarrow s_1 \\ s_2 \\ s_3 \\ P \end{array} \begin{array}{c} \begin{matrix} x_1 & x_2 & s_1 & s_2 & s_3 & P \end{matrix} \\ \left[\begin{array}{cccccc|c} -1 & \textcircled{1} & 1 & 0 & 0 & 0 & 2 \\ -1 & 3 & 0 & 1 & 0 & 0 & 12 \\ 1 & -4 & 0 & 0 & 1 & 0 & 4 \\ \hdashline 1 & -2 & 0 & 0 & 0 & 1 & 0 \end{array}\right] \end{array} \begin{array}{l} \\ \frac{2}{1} = 2 \text{ (minimum)} \\ \frac{12}{3} = 4 \\ \\ \\ \end{array}$$

$\uparrow$ pivot column

$(-3)R_1 + R_2 \rightarrow R_2$, $4R_1 + R_3 \rightarrow R_3$, and $2R_1 + R_4 \rightarrow R_4$

$$\sim \left[\begin{array}{cccccc|c} -1 & 1 & 1 & 0 & 0 & 0 & 2 \\ \textcircled{2} & 0 & -3 & 1 & 0 & 0 & 6 \\ -3 & 0 & 4 & 0 & 1 & 0 & 12 \\ \hdashline -1 & 0 & 2 & 0 & 0 & 1 & 4 \end{array}\right] \begin{array}{l} \\ \frac{6}{2} = 3 \leftarrow \text{pivot row} \\ \\ \\ \end{array}$$

(pivot row $\rightarrow$ second row; $\uparrow$ pivot column: first column)

[Note: We only use the *positive* elements above the dashed line in the pivot column.]

$\frac{1}{2}R_2 \rightarrow R_2$

$$\sim \left[\begin{array}{cccccc|c} -1 & 1 & 1 & 0 & 0 & 0 & 2 \\ \textcircled{1} & 0 & -\frac{3}{2} & \frac{1}{2} & 0 & 0 & 3 \\ -3 & 0 & 4 & 0 & 1 & 0 & 12 \\ \hdashline -1 & 0 & 2 & 0 & 0 & 1 & 4 \end{array}\right]$$

$R_2 + R_1 \rightarrow R_1$, $3R_2 + R_3 \rightarrow R_3$, and $R_2 + R_4 \rightarrow R_4$

$$\sim \begin{array}{c} \\ x_2 \\ x_1 \\ s_3 \\ \\ \end{array} \begin{array}{c} \begin{matrix} x_1 & x_2 & s_1 & s_2 & s_3 & P \end{matrix} \\ \left[\begin{array}{cccccc|c} 0 & 1 & -\frac{1}{2} & \frac{1}{2} & 0 & 0 & 5 \\ 1 & 0 & -\frac{3}{2} & \frac{1}{2} & 0 & 0 & 3 \\ 0 & 0 & -\frac{1}{2} & \frac{3}{2} & 1 & 0 & 21 \\ \hdashline 0 & 0 & \frac{1}{2} & \frac{1}{2} & 0 & 1 & 7 \end{array}\right] \end{array}$$

Optimal solution: max $P = 7$ at $x_1 = 3$, $x_2 = 5$, $s_1 = 0$, $s_2 = 0$, $s_3 = 21$.

19. The simplex tableau for this problem is:

$$\begin{array}{c} \\ \text{pivot row} \rightarrow s_1 \\ s_2 \\ P \end{array} \begin{array}{c} \begin{matrix} x_1 & x_2 & x_3 & s_1 & s_2 & P \end{matrix} \\ \left[\begin{array}{cccccc|c} \textcircled{1} & 1 & -1 & 1 & 0 & 0 & 10 \\ 2 & 4 & 3 & 0 & 1 & 0 & 30 \\ \hdashline -5 & -2 & 1 & 0 & 0 & 1 & 0 \end{array}\right] \end{array} \begin{array}{l} \\ \frac{10}{1} = 10 \text{ (minimum)} \\ \frac{30}{2} = 15 \\ \\ \end{array}$$

$\uparrow$ pivot column

$(-2)R_1 + R_2 \rightarrow R_2$, $5R_1 + R_3 \rightarrow R_3$

$$\sim \left[\begin{array}{cccccc|c} 1 & 1 & -1 & 1 & 0 & 0 & 10 \\ 0 & 2 & \textcircled{5} & -2 & 1 & 0 & 10 \\ \hdashline 0 & 3 & -4 & 5 & 0 & 1 & 50 \end{array}\right] \sim \left[\begin{array}{cccccc|c} 1 & 1 & -1 & 1 & 0 & 0 & 10 \\ 0 & \frac{2}{5} & \textcircled{1} & -\frac{2}{5} & \frac{1}{5} & 0 & 2 \\ \hdashline 0 & 3 & -4 & 5 & 0 & 1 & 50 \end{array}\right]$$

$\frac{1}{5}R_2 \rightarrow R_2$

$R_2 + R_1 \rightarrow R_1$, $4R_2 + R_3 \rightarrow R_3$

$$\begin{array}{c} \\ x_1 \\ x_3 \\ P \end{array} \begin{array}{c} \begin{matrix} x_1 & x_2 & x_3 & s_1 & s_2 & P \end{matrix} \\ \sim \left[\begin{array}{cccccc|c} 1 & \frac{7}{5} & 0 & \frac{3}{5} & \frac{1}{5} & 0 & 12 \\ 0 & \frac{2}{5} & 1 & -\frac{2}{5} & \frac{1}{5} & 0 & 2 \\ \hdashline 0 & \frac{23}{5} & 0 & \frac{17}{5} & \frac{4}{5} & 1 & 58 \end{array}\right] \end{array}$$

Optimal solution: max $P = 58$ at $x_1 = 12$, $x_2 = 0$, $x_3 = 2$, $s_1 = 0$, $s_2 = 0$.

21. The simplex tableau for this problem is:

$$\begin{array}{c} \\ s_1 \\ \text{pivot row} \rightarrow s_2 \\ P \end{array} \begin{array}{c} \begin{matrix} x_1 & x_2 & x_3 & s_1 & s_2 & P \end{matrix} \\ \left[\begin{array}{cccccc|c} 1 & 0 & 1 & 1 & 0 & 0 & 4 \\ 0 & 1 & \textcircled{1} & 0 & 1 & 0 & 3 \\ \hdashline -2 & -3 & -4 & 0 & 0 & 1 & 0 \end{array}\right] \end{array} \begin{array}{l} \frac{4}{1} = 4 \\ \frac{3}{1} = 3 \text{ (minimum)} \\ \\ \end{array}$$

$\uparrow$ pivot column $\quad (-1)R_2 + R_1 \rightarrow R_1$, $4R_2 + R_3 \rightarrow R_3$

$$\sim \left[\begin{array}{cccccc|c} \textcircled{1} & -1 & 0 & 1 & -1 & 0 & 1 \\ 0 & 1 & 1 & 0 & 1 & 0 & 3 \\ \hdashline -2 & 1 & 0 & 0 & 4 & 1 & 12 \end{array}\right] \sim \left[\begin{array}{cccccc|c} 1 & -1 & 0 & 1 & -1 & 0 & 1 \\ 0 & \textcircled{1} & 1 & 0 & 1 & 0 & 3 \\ \hdashline 0 & -1 & 0 & 2 & 2 & 1 & 14 \end{array}\right]$$

$2R_1 + R_3 \rightarrow R_3 \qquad\qquad R_2 + R_1 \rightarrow R_1$ and $R_2 + R_3 \rightarrow R_3$

$$\begin{array}{c} \begin{matrix} x_1 & x_2 & x_3 & s_1 & s_2 & P \end{matrix} \\ \sim \left[\begin{array}{cccccc|c} 1 & 0 & 1 & 1 & 0 & 0 & 4 \\ 0 & 1 & 1 & 0 & 1 & 0 & 3 \\ \hdashline 0 & 0 & 1 & 2 & 3 & 1 & 17 \end{array}\right] \end{array}$$

Optimal solution: max $P = 17$ at $x_1 = 4$, $x_2 = 3$, $x_3 = 0$, $s_1 = 0$, $s_2 = 0$.

23. The simplex tableau for this problem is:

$$\begin{array}{c} \\ s_1 \\ \text{pivot row} \rightarrow s_2 \\ s_3 \\ \\ \end{array} \begin{array}{c} \begin{matrix} x_1 & x_2 & x_3 & s_1 & s_2 & s_3 & P \end{matrix} \\ \left[\begin{array}{ccccccc|c} 3 & 2 & 5 & 1 & 0 & 0 & 0 & 23 \\ \textcircled{2} & 1 & 1 & 0 & 1 & 0 & 0 & 8 \\ 1 & 1 & 2 & 0 & 0 & 1 & 0 & 7 \\ \hdashline -4 & -3 & -2 & 0 & 0 & 0 & 1 & 0 \end{array}\right] \end{array} \begin{array}{l} \frac{23}{3} = 7\frac{2}{3} \\ \frac{8}{2} = 4 \text{ (minimum)} \\ \frac{7}{1} = 7 \\ \\ \end{array}$$

$\uparrow$ pivot column $\quad \frac{1}{2}R_2 \rightarrow R_2$

$$\sim \left[\begin{array}{ccccccc|c} 3 & 2 & 5 & 1 & 0 & 0 & 0 & 23 \\ \textcircled{1} & \frac{1}{2} & \frac{1}{2} & 0 & \frac{1}{2} & 0 & 0 & 4 \\ 1 & 1 & 2 & 0 & 0 & 1 & 0 & 7 \\ \hdashline -4 & -3 & -2 & 0 & 0 & 0 & 1 & 0 \end{array}\right] \sim \left[\begin{array}{ccccccc|c} 0 & \frac{1}{2} & \frac{7}{2} & 1 & -\frac{3}{2} & 0 & 0 & 11 \\ 1 & \frac{1}{2} & \frac{1}{2} & 0 & \frac{1}{2} & 0 & 0 & 4 \\ 0 & \textcircled{\frac{1}{2}} & \frac{3}{2} & 0 & -\frac{1}{2} & 1 & 0 & 3 \\ \hdashline 0 & -1 & 0 & 0 & 2 & 0 & 1 & 16 \end{array}\right]$$

$(-3)R_2 + R_1 \rightarrow R_1$, $(-1)R_2 + R_3 \rightarrow R_3$, and $4R_2 + R_4 \rightarrow R_4 \qquad 2R_3 \rightarrow R_3$

$$\sim \left[\begin{array}{ccccccc|c} 0 & \frac{1}{2} & \frac{7}{2} & 1 & -\frac{3}{2} & 0 & 0 & 11 \\ 1 & \frac{1}{2} & \frac{1}{2} & 0 & \frac{1}{2} & 0 & 0 & 4 \\ 0 & \textcircled{1} & 3 & 0 & -1 & 2 & 0 & 6 \\ \hdashline 0 & -1 & 0 & 0 & 2 & 0 & 1 & 16 \end{array}\right] \sim \begin{array}{c} \\ s_1 \\ x_1 \\ x_2 \\ P \end{array} \begin{array}{c} \begin{array}{ccccccc} x_1 & x_2 & x_3 & s_1 & s_2 & s_3 & P \end{array} \\ \left[\begin{array}{ccccccc|c} 0 & 0 & 2 & 1 & -1 & -1 & 0 & 8 \\ 1 & 0 & -1 & 0 & 1 & -1 & 0 & 1 \\ 0 & 1 & 3 & 0 & -1 & 2 & 0 & 6 \\ \hdashline 0 & 0 & 3 & 0 & 1 & 2 & 1 & 22 \end{array}\right] \end{array}$$

$\left(-\frac{1}{2}\right)R_3 + R_1 \to R_1$, $\left(-\frac{1}{2}\right)R_3 + R_2 \to R_2$, and

$R_3 + R_4 \to R_4$

Optimal solution: max $P = 22$ at $x_1 = 1$, $x_2 = 6$, $x_3 = 0$, $s_1 = 8$, $s_2 = 0$, $s_3 = 0$.

25. Multiply the first problem constraint by $\frac{10}{6}$, the second by 100, and the third by 10 to clear the fractions. Then, the simplex tableau for this problem is:

$$\begin{array}{c} \\ s_1 \\ s_2 \\ s_3 \\ P \end{array} \begin{array}{c} \begin{array}{cccccc} x_1 & x_2 & s_1 & s_2 & s_3 & P \end{array} \\ \left[\begin{array}{cccccc|c} 1 & \textcircled{2} & 1 & 0 & 0 & 0 & 1{,}600 \\ 3 & 4 & 0 & 1 & 0 & 0 & 3{,}600 \\ 3 & 2 & 0 & 0 & 1 & 0 & 2{,}700 \\ \hdashline -20 & -30 & 0 & 0 & 0 & 1 & 0 \end{array}\right] \end{array} \begin{array}{l} \frac{1{,}600}{2} = 800 \\ \frac{3{,}600}{4} = 900 \\ \frac{2{,}700}{2} = 1{,}350 \end{array}$$

$\frac{1}{2}R_1 \to R_1$

$$\sim \left[\begin{array}{cccccc|c} \frac{1}{2} & \textcircled{1} & \frac{1}{2} & 0 & 0 & 0 & 800 \\ 3 & 4 & 0 & 1 & 0 & 0 & 3{,}600 \\ 3 & 2 & 0 & 0 & 1 & 0 & 2{,}700 \\ \hdashline -20 & -30 & 0 & 0 & 0 & 1 & 0 \end{array}\right]$$

$(-4)R_1 + R_2 \to R_2$, $(-2)R_1 + R_3 \to R_3$, and $30R_1 + R_4 \to R_4$

$$\sim \left[\begin{array}{cccccc|c} \frac{1}{2} & 1 & \frac{1}{2} & 0 & 0 & 0 & 800 \\ \textcircled{1} & 0 & -2 & 1 & 0 & 0 & 400 \\ 2 & 0 & -1 & 0 & 1 & 0 & 1{,}100 \\ \hdashline -5 & 0 & 15 & 0 & 0 & 1 & 24{,}000 \end{array}\right] \begin{array}{l} \frac{800}{1/2} = 1{,}600 \\ \frac{400}{1} = 400 \\ \frac{1{,}100}{2} = 550 \end{array}$$

$\left(-\frac{1}{2}\right)R_2 + R_1 \to R_1$, $(-2)R_2 + R_3 \to R_3$, and $5R_2 + R_4 \to R_4$

$$\sim \begin{array}{c} \\ x_2 \\ x_1 \\ s_3 \\ P \end{array} \begin{array}{c} \begin{array}{cccccc|c} x_1 & x_2 & s_1 & s_2 & s_3 & P & \end{array} \\ \left[\begin{array}{cccccc|c} 0 & 1 & \frac{3}{2} & -\frac{1}{2} & 0 & 0 & 600 \\ 1 & 0 & -2 & 1 & 0 & 0 & 400 \\ 0 & 0 & 3 & -2 & 1 & 0 & 300 \\ \hline 0 & 0 & 5 & 5 & 0 & 1 & 26{,}000 \end{array}\right] \end{array}$$

Optimal solution: max $P = 26{,}000$ at $x_1 = 400$, $x_2 = 600$, $s_1 = 0$, $s_2 = 0$, $s_3 = 300$.

27. The simplex tableau for this problem is:

$$\begin{array}{c} \\ s_1 \\ s_2 \\ s_3 \\ P \end{array} \begin{array}{c} \begin{array}{ccccccc|c} x_1 & x_2 & x_3 & s_1 & s_2 & s_3 & P & \end{array} \\ \left[\begin{array}{ccccccc|c} 2 & 2 & \textcircled{8} & 1 & 0 & 0 & 0 & 600 \\ 1 & 3 & 2 & 0 & 1 & 0 & 0 & 600 \\ 3 & 2 & 1 & 0 & 0 & 1 & 0 & 400 \\ \hline -1 & -2 & -3 & 0 & 0 & 0 & 1 & 0 \end{array}\right] \end{array} \begin{array}{l} \frac{600}{8} = 75 \\ \frac{600}{2} = 300 \\ \frac{400}{1} = 400 \\ \end{array}$$

$\frac{1}{8}R_1 \to R_1$

$$\sim \left[\begin{array}{ccccccc|c} \frac{1}{4} & \frac{1}{4} & \textcircled{1} & \frac{1}{8} & 0 & 0 & 0 & 75 \\ 1 & 3 & 2 & 0 & 1 & 0 & 0 & 600 \\ 3 & 2 & 1 & 0 & 0 & 1 & 0 & 400 \\ \hline -1 & -2 & -3 & 0 & 0 & 0 & 1 & 0 \end{array}\right]$$

$(-2)R_1 + R_2 \to R_2$, $(-1)R_1 + R_3 \to R_3$, and $3R_1 + R_4 \to R_4$

$$\sim \left[\begin{array}{ccccccc|c} \frac{1}{4} & \frac{1}{4} & 1 & \frac{1}{8} & 0 & 0 & 0 & 75 \\ \frac{1}{2} & \textcircled{\frac{5}{2}} & 0 & -\frac{1}{4} & 1 & 0 & 0 & 450 \\ \frac{11}{4} & \frac{7}{4} & 0 & -\frac{1}{8} & 0 & 1 & 0 & 325 \\ \hline -\frac{1}{4} & -\frac{5}{4} & 0 & \frac{3}{8} & 0 & 0 & 1 & 225 \end{array}\right] \begin{array}{l} \frac{75}{1/4} = 300 \\ \frac{450}{5/2} = 180 \\ \frac{325}{7/4} = 185.71 \\ \end{array}$$

$\frac{2}{5}R_2 \to R_2$

$$\sim \left[\begin{array}{ccccccc|c} \frac{1}{4} & \frac{1}{4} & 1 & \frac{1}{8} & 0 & 0 & 0 & 75 \\ \frac{1}{5} & \textcircled{1} & 0 & -\frac{1}{10} & \frac{2}{5} & 0 & 0 & 180 \\ \frac{11}{4} & \frac{7}{4} & 0 & -\frac{1}{8} & 0 & 1 & 0 & 325 \\ \hline -\frac{1}{4} & -\frac{5}{4} & 0 & \frac{3}{8} & 0 & 0 & 1 & 225 \end{array}\right]$$

$\left(-\frac{1}{4}\right)R_2 + R_1 \to R_1$, $\left(-\frac{7}{4}\right)R_2 + R_3 \to R_3$, and $\frac{5}{4}R_2 + R_4 \to R_4$

$$\sim \begin{array}{c} \\ x_3 \\ x_2 \\ s_3 \\ P \end{array} \begin{array}{c} \begin{array}{ccccccc|c} x_1 & x_2 & x_3 & s_1 & s_2 & s_3 & P & \end{array} \\ \left[\begin{array}{ccccccc|c} \frac{1}{5} & 0 & 1 & \frac{3}{20} & -\frac{1}{10} & 0 & 0 & 30 \\ \frac{1}{5} & 1 & 0 & -\frac{1}{10} & \frac{2}{5} & 0 & 0 & 180 \\ \frac{12}{5} & 0 & 0 & \frac{1}{20} & -\frac{7}{10} & 1 & 0 & 10 \\ \hline 0 & 0 & 0 & \frac{1}{4} & \frac{1}{2} & 0 & 1 & 450 \end{array}\right] \end{array}$$

Optimal solution: max $P = 450$ at $x_1 = 0$, $x_2 = 180$, $x_3 = 30$, $s_1 = 0$, $s_2 = 0$, $s_3 = 10$.

29. The simplex tableau for this problem is:

$$
\begin{array}{c} \\ s_1 \\ s_2 \\ s_3 \\ s_4 \\ P \end{array}
\begin{array}{c}
\begin{array}{ccccccc} x_1 & x_2 & s_1 & s_2 & s_3 & s_4 & P \end{array} \\
\left[\begin{array}{ccccccc|c}
1 & 2 & 1 & 0 & 0 & 0 & 0 & 40 \\
1 & 3 & 0 & 1 & 0 & 0 & 0 & 48 \\
1 & 4 & 0 & 0 & 1 & 0 & 0 & 60 \\
0 & \textcircled{1} & 0 & 0 & 0 & 1 & 0 & 14 \\
\hdashline
-2 & -5 & 0 & 0 & 0 & 0 & 1 & 0
\end{array}\right]
\end{array}
\begin{array}{l} \\ \frac{40}{2} = 20 \\ \frac{48}{3} = 16 \\ \frac{60}{4} = 15 \\ \frac{14}{1} = 14 \\ \\ \end{array}
$$

$(-2)R_4 + R_1 \rightarrow R_1$, $(-3)R_4 + R_2 \rightarrow R_2$, $(-4)R_4 + R_3 \rightarrow R_3$, and $5R_4 + R_5 \rightarrow R_5$

$$
\sim \left[\begin{array}{ccccccc|c}
1 & 0 & 1 & 0 & 0 & -2 & 0 & 12 \\
1 & 0 & 0 & 1 & 0 & -3 & 0 & 6 \\
\textcircled{1} & 0 & 0 & 0 & 1 & -4 & 0 & 4 \\
0 & 1 & 0 & 0 & 0 & 1 & 0 & 14 \\
\hdashline
-2 & 0 & 0 & 0 & 0 & 5 & 1 & 70
\end{array}\right]
\begin{array}{l} \frac{12}{1} = 12 \\ \frac{6}{1} = 6 \\ \frac{4}{1} = 4 \\ \\ \\ \end{array}
$$

$(-1)R_3 + R_1 \rightarrow R_1$, $(-1)R_3 + R_2 \rightarrow R_2$, and $2R_3 + R_5 \rightarrow R_5$

$$
\sim \left[\begin{array}{ccccccc|c}
0 & 0 & 1 & 0 & -1 & 2 & 0 & 8 \\
0 & 0 & 0 & 1 & -1 & \textcircled{1} & 0 & 2 \\
1 & 0 & 0 & 0 & 1 & -4 & 0 & 4 \\
0 & 1 & 0 & 0 & 0 & 1 & 0 & 14 \\
\hdashline
0 & 0 & 0 & 0 & 2 & -3 & 1 & 78
\end{array}\right]
\begin{array}{l} \frac{8}{2} = 4 \\ \frac{2}{1} = 2 \\ \\ \frac{14}{1} = 14 \\ \\ \end{array}
$$

$(-2)R_2 + R_1 \rightarrow R_1$, $4R_2 + R_3 \rightarrow R_3$, $(-1)R_2 + R_4 \rightarrow R_4$, and $3R_2 + R_5 \rightarrow R_5$

$$
\sim \left[\begin{array}{ccccccc|c}
0 & 0 & 1 & -2 & \textcircled{1} & 0 & 0 & 4 \\
0 & 0 & 0 & 1 & -1 & 1 & 0 & 2 \\
1 & 0 & 0 & 4 & -3 & 0 & 0 & 12 \\
0 & 1 & 0 & -1 & 1 & 0 & 0 & 12 \\
\hdashline
0 & 0 & 0 & 3 & -1 & 0 & 1 & 84
\end{array}\right]
\begin{array}{l} \frac{4}{1} = 4 \\ \\ \\ \frac{12}{1} = 12 \\ \\ \end{array}
$$

$R_1 + R_2 \rightarrow R_2$, $3R_1 + R_3 \rightarrow R_3$, $(-1)R_1 + R_4 \rightarrow R_4$, and $R_1 + R_5 \rightarrow R_5$

$$
\begin{array}{c}
 \\ s_3 \\ s_4 \\ \sim x_1 \\ x_2 \\ P
\end{array}
\begin{array}{c}
\begin{array}{ccccccc|c} x_1 & x_2 & s_1 & s_2 & s_3 & s_4 & P & \end{array} \\
\left[\begin{array}{ccccccc|c}
0 & 0 & 1 & -2 & 1 & 0 & 0 & 4 \\
0 & 0 & 1 & -1 & 0 & 1 & 0 & 6 \\
1 & 0 & 3 & -2 & 0 & 0 & 0 & 24 \\
0 & 1 & -1 & 1 & 0 & 0 & 0 & 8 \\
\hdashline
0 & 0 & 1 & 1 & 0 & 0 & 1 & 88
\end{array}\right]
\end{array}
$$

Optimal solution: max $P = 88$ at $x_1 = 24$, $x_2 = 8$, $s_1 = 0$, $s_2 = 0$, $s_3 = 4$, $s_4 = 6$.

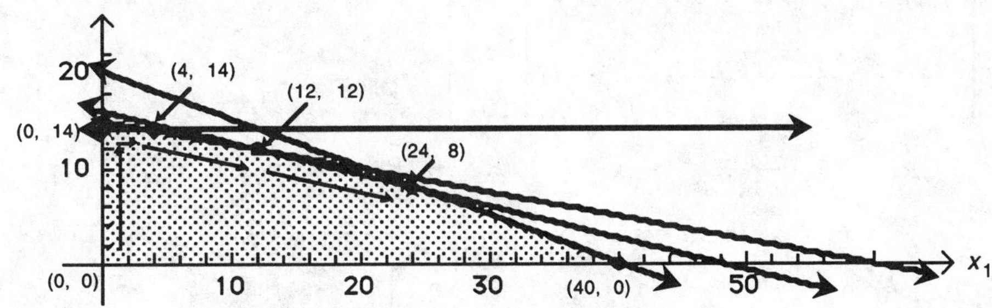

31. The simplex tableau for this problem is:

$$
\begin{array}{c}
 \\ s_1 \\ s_2 \\ s_3 \\ P
\end{array}
\begin{array}{c}
\begin{array}{cccccc|c} x_1 & x_2 & s_1 & s_2 & s_3 & P & \end{array} \\
\left[\begin{array}{cccccc|c}
2 & 1 & 1 & 0 & 0 & 0 & 16 \\
1 & 0 & 0 & 1 & 0 & 0 & 6 \\
0 & 1 & 0 & 0 & 1 & 0 & 10 \\
\hdashline
-1 & -1 & 0 & 0 & 0 & 1 & 0
\end{array}\right]
\end{array}
$$

(A) Solution using the first column as the pivot column

$$
\begin{array}{c}
\begin{array}{cccccc|c} x_1 & x_2 & s_1 & s_2 & s_3 & P & \end{array} \\
\left[\begin{array}{cccccc|c}
2 & 1 & 1 & 0 & 0 & 0 & 16 \\
\textcircled{1} & 0 & 0 & 1 & 0 & 0 & 6 \\
0 & 1 & 0 & 0 & 1 & 0 & 10 \\
\hdashline
-1 & -1 & 0 & 0 & 0 & 1 & 0
\end{array}\right]
\end{array}
\begin{array}{l}
\frac{16}{2} = 8 \\
\frac{6}{1} = 6 \\
 \\
 \\
\end{array}
$$

$(-2)R_2 + R_1 \to R_1$, $R_2 + R_4 \to R_4$

$$
\sim \left[\begin{array}{cccccc|c}
0 & \textcircled{1} & 1 & -2 & 0 & 0 & 4 \\
1 & 0 & 0 & 1 & 0 & 0 & 6 \\
0 & 1 & 0 & 0 & 1 & 0 & 10 \\
\hdashline
0 & -1 & 0 & 1 & 0 & 1 & 6
\end{array}\right]
\begin{array}{l}
\frac{4}{1} = 4 \\
 \\
\frac{10}{1} = 10 \\
 \\
\end{array}
$$

$(-1)R_1 + R_3 \to R_3$, $R_1 + R_4 \to R_4$

$$
\sim \left[\begin{array}{cccccc|c}
0 & 1 & 1 & -2 & 0 & 0 & 4 \\
1 & 0 & 0 & 1 & 0 & 0 & 6 \\
0 & 0 & -1 & \textcircled{2} & 1 & 0 & 6 \\
\hdashline
0 & 0 & 1 & -1 & 0 & 1 & 10
\end{array}\right]
\begin{array}{l}
\\
\frac{6}{1} = 6 \\
\frac{6}{2} = 3 \\
\\
\end{array}
$$

$\frac{1}{2}R_3 \rightarrow R_3$

$$
\sim \left[\begin{array}{cccccc|c}
0 & 1 & 1 & -2 & 0 & 0 & 4 \\
1 & 0 & 0 & 1 & 0 & 0 & 6 \\
0 & 0 & -\frac{1}{2} & 1 & \frac{1}{2} & 0 & 3 \\
\hdashline
0 & 0 & 1 & -1 & 0 & 1 & 10
\end{array}\right]
$$

$2R_3 + R_1 \rightarrow R_1$, $(-1)R_3 + R_2 \rightarrow R_2$, $R_3 + R_4 \rightarrow R_4$

$$
\begin{array}{c}
\\
x_2 \\
x_1 \\
s_2 \\
P
\end{array}
\begin{array}{c}
\begin{array}{cccccc}
x_1 & x_2 & s_1 & s_2 & s_3 & P
\end{array} \\
\sim \left[\begin{array}{cccccc|c}
0 & 1 & 0 & 0 & 1 & 0 & 10 \\
1 & 0 & \frac{1}{2} & 0 & -\frac{1}{2} & 0 & 3 \\
0 & 0 & -\frac{1}{2} & 1 & \frac{1}{2} & 0 & 3 \\
\hdashline
0 & 0 & \frac{1}{2} & 0 & \frac{1}{2} & 1 & 13
\end{array}\right]
\end{array}
$$

Optimal solution: max $P = 13$ at $x_1 = 3$, $x_2 = 10$, $s_1 = 0$, $s_2 = 3$, $s_3 = 0$

(B) Solution using the second column as the pivot column

$$
\begin{array}{c}
\\
s_1 \\
s_2 \\
s_3 \\
P
\end{array}
\begin{array}{c}
\begin{array}{cccccc}
x_1 & x_2 & s_1 & s_2 & s_3 & P
\end{array} \\
\left[\begin{array}{cccccc|c}
2 & 1 & 1 & 0 & 0 & 0 & 16 \\
1 & 0 & 0 & 1 & 0 & 0 & 6 \\
0 & \textcircled{1} & 0 & 0 & 1 & 0 & 10 \\
\hdashline
-1 & -1 & 0 & 0 & 0 & 1 & 0
\end{array}\right]
\end{array}
\begin{array}{l}
\\
\frac{16}{1} = 16 \\
\\
\frac{10}{1} = 10 \\
\\
\end{array}
$$

$(-1)R_3 + R_1 \rightarrow R_1$, $R_3 + R_4 \rightarrow R_4$

$$
\sim \left[\begin{array}{cccccc|c}
\textcircled{2} & 0 & 1 & 0 & -1 & 0 & 6 \\
1 & 0 & 0 & 1 & 0 & 0 & 6 \\
0 & 1 & 0 & 0 & 1 & 0 & 10 \\
\hdashline
-1 & 0 & 0 & 0 & 1 & 1 & 10
\end{array}\right]
\begin{array}{l}
\frac{6}{2} = 3 \\
\frac{6}{1} = 6 \\
\\
\\
\end{array}
$$

$\frac{1}{2}R_1 \rightarrow R_1$

$$\sim \left[\begin{array}{cccccc|c} 1 & 0 & \frac{1}{2} & 0 & -\frac{1}{2} & 0 & 3 \\ 1 & 0 & 0 & 1 & 0 & 0 & 6 \\ 0 & 1 & 0 & 0 & 1 & 0 & 10 \\ \hdashline -1 & 0 & 0 & 0 & 1 & 1 & 10 \end{array}\right]$$

$(-1)R_1 + R_2 \to R_2$, $R_1 + R_4 \to R_4$

$$\sim \begin{array}{c} \\ x_1 \\ s_2 \\ x_2 \\ P \end{array} \begin{array}{c} \begin{array}{cccccc} x_1 & x_2 & s_1 & s_2 & s_3 & P \end{array} \\ \left[\begin{array}{cccccc|c} 1 & 0 & \frac{1}{2} & 0 & -\frac{1}{2} & 0 & 3 \\ 0 & 0 & -\frac{1}{2} & 1 & \frac{1}{2} & 0 & 3 \\ 0 & \textcircled{1} & 0 & 0 & 1 & 0 & 10 \\ \hdashline 0 & 0 & \frac{1}{2} & 0 & \frac{1}{2} & 1 & 13 \end{array}\right] \end{array}$$

Optimal solution: max $P = 13$ at $x_1 = 3$, $x_2 = 10$, $s_1 = 0$, $s_2 = 3$, $s_3 = 0$

Choosing either solution produces the *same* optimal solution.

33. The simplex tableau for this problem is:

$$\begin{array}{c} \\ s_1 \\ s_2 \\ P \end{array} \begin{array}{c} \begin{array}{cccccc} x_1 & x_2 & x_3 & s_1 & s_2 & P \end{array} \\ \left[\begin{array}{cccccc|c} 1 & 1 & 2 & 1 & 0 & 0 & 20 \\ 2 & 1 & 4 & 0 & 1 & 0 & 32 \\ \hdashline -3 & -3 & -2 & 0 & 0 & 1 & 0 \end{array}\right] \end{array}$$

(A) Solution using the first column as the pivot column

$$\begin{array}{c} \begin{array}{cccccc} x_1 & x_2 & x_3 & s_1 & s_2 & P \end{array} \\ \left[\begin{array}{cccccc|c} 1 & 1 & 2 & 1 & 0 & 0 & 20 \\ \textcircled{2} & 1 & 4 & 0 & 1 & 0 & 32 \\ \hdashline -3 & -3 & -2 & 0 & 0 & 1 & 0 \end{array}\right] \end{array} \begin{array}{l} \frac{20}{1} = 20 \\ \frac{32}{2} = 16 \\ \end{array}$$

$\frac{1}{2}R_2 \to R_2$

$$\sim \left[\begin{array}{cccccc|c} 1 & 1 & 2 & 1 & 0 & 0 & 20 \\ 1 & \frac{1}{2} & 2 & 0 & \frac{1}{2} & 0 & 16 \\ \hdashline -3 & -3 & -2 & 0 & 0 & 1 & 0 \end{array}\right]$$

$(-1)R_2 + R_1 \to R_1$, $3R_2 + R_3 \to R_3$

$$\sim \left[\begin{array}{cccccc|c} 0 & \textcircled{\tfrac{1}{2}} & 0 & 1 & -\frac{1}{2} & 0 & 4 \\ 1 & \frac{1}{2} & 2 & 0 & \frac{1}{2} & 0 & 16 \\ \hdashline 0 & -\frac{3}{2} & 4 & 0 & \frac{3}{2} & 1 & 48 \end{array}\right] \begin{array}{l} \dfrac{4}{1/2} = 8 \\ \dfrac{16}{1/2} = 32 \\ \end{array}$$

$2R_1 \to R_1$

$$\sim \left[\begin{array}{cccccc|c} 0 & 1 & 0 & 2 & -1 & 0 & 8 \\ 1 & \frac{1}{2} & 2 & 0 & \frac{1}{2} & 0 & 16 \\ \hdashline 0 & -\frac{3}{2} & 4 & 0 & \frac{3}{2} & 1 & 48 \end{array}\right]$$

$\left(-\frac{1}{2}\right)R_1 + R_2 \to R_2,\ \frac{3}{2}R_1 + R_3 \to R_3$

$$\sim \begin{array}{c} \\ x_2 \\ x_1 \\ P \end{array} \begin{array}{c} \begin{array}{cccccc} x_1 & x_2 & x_3 & s_1 & s_2 & P \end{array} \\ \left[\begin{array}{cccccc|c} 0 & 1 & 0 & 2 & -1 & 0 & 8 \\ 1 & 0 & 2 & -1 & 1 & 0 & 12 \\ \hdashline 0 & 0 & 4 & 3 & 0 & 1 & 60 \end{array}\right] \end{array}$$

Optimal solution: max $P = 60$ at $x_1 = 12$, $x_2 = 8$, $x_3 = 0$, $s_1 = 0$, $s_2 = 0$

(B) Solution using the second column as the pivot column

$$\begin{array}{c} \\ s_1 \\ s_2 \\ P \end{array} \begin{array}{c} \begin{array}{cccccc} x_1 & x_2 & x_3 & s_1 & s_2 & P \end{array} \\ \left[\begin{array}{cccccc|c} 1 & \textcircled{1} & 2 & 1 & 0 & 0 & 20 \\ 2 & 1 & 4 & 0 & 1 & 0 & 32 \\ \hdashline -3 & -3 & -2 & 0 & 0 & 1 & 0 \end{array}\right] \end{array} \begin{array}{l} \dfrac{20}{1} = 20 \\ \dfrac{32}{1} = 32 \\ \end{array}$$

$(-1)R_1 + R_2 \to R_2,\ 3R_1 + R_3 \to R_3$

$$\sim \begin{array}{c} \\ x_2 \\ s_2 \\ P \end{array} \begin{array}{c} \begin{array}{cccccc} x_1 & x_2 & x_3 & s_1 & s_2 & P \end{array} \\ \left[\begin{array}{cccccc|c} 1 & 1 & 2 & 1 & 0 & 0 & 20 \\ 1 & 0 & 2 & -1 & 1 & 0 & 12 \\ \hdashline 0 & 0 & 4 & 3 & 0 & 1 & 60 \end{array}\right] \end{array}$$

Optimal solution: max $P = 60$ at $x_1 = 0$, $x_2 = 20$, $x_3 = 0$, $s_1 = 0$, $s_2 = 12$

The maximum value of P is 60. Since the optimal solution is obtained at two corner points, (12, 8, 0) and (0, 20, 0), every point on the line segment connecting these points is also an optimal solution.

35. Let x_1 = the number of A components
x_2 = the number of B components
x_3 = the number of C components

The mathematical model for this problem is:

Maximize $P = 7x_1 + 8x_2 + 10x_3$

Subject to

$$\begin{aligned} 2x_1 + 3x_2 + 2x_3 &\le 1000 \\ x_1 + x_2 + 2x_3 &\le 800 \\ x_1, x_2, x_3 &\ge 0 \end{aligned}$$

We introduce slack variables s_1, s_2 to obtain the equivalent form:

$$\begin{aligned} 2x_1 + 3x_2 + 2x_3 + s_1 \qquad &= 1000 \\ x_1 + x_2 + 2x_3 \qquad + s_2 &= 800 \\ -7x_1 - 8x_2 - 10x_3 \qquad + P &= 0 \end{aligned}$$

The simplex tableau for this problem is:

$$\begin{array}{c} \\ s_1 \\ s_2 \\ P \end{array} \begin{array}{c} \begin{array}{cccccc} x_1 & x_2 & x_3 & s_1 & s_2 & P \end{array} \\ \left[\begin{array}{cccccc|c} 2 & 3 & 2 & 1 & 0 & 0 & 1000 \\ 1 & 1 & \textcircled{2} & 0 & 1 & 0 & 800 \\ \hdashline -7 & -8 & -10 & 0 & 0 & 1 & 0 \end{array}\right] \end{array} \begin{array}{l} \frac{1000}{2} = 500 \\ \frac{800}{2} = 400 \\ \end{array}$$

$\frac{1}{2}R_2 \to R_2$

$$\sim \left[\begin{array}{cccccc|c} 2 & 3 & 2 & 1 & 0 & 0 & 1000 \\ \frac{1}{2} & \frac{1}{2} & \textcircled{1} & 0 & \frac{1}{2} & 0 & 400 \\ \hdashline -7 & -8 & -10 & 0 & 0 & 1 & 0 \end{array}\right]$$

$(-2)R_2 + R_1 \to R_1$, $10R_2 + R_3 \to R_3$

$$\sim \left[\begin{array}{cccccc|c} 1 & \textcircled{2} & 0 & 1 & -1 & 0 & 200 \\ \frac{1}{2} & \frac{1}{2} & 1 & 0 & \frac{1}{2} & 0 & 400 \\ \hdashline -2 & -3 & 0 & 0 & 5 & 1 & 4000 \end{array}\right] \begin{array}{l} \frac{200}{2} = 100 \\ \frac{400}{1/2} = 800 \\ \end{array}$$

$\frac{1}{2}R_1 \to R_1$

$$\sim \left[\begin{array}{cccccc|c} \frac{1}{2} & \textcircled{1} & 0 & \frac{1}{2} & -\frac{1}{2} & 0 & 100 \\ \frac{1}{2} & \frac{1}{2} & 1 & 0 & \frac{1}{2} & 0 & 400 \\ \hdashline -2 & -3 & 0 & 0 & 5 & 1 & 4000 \end{array}\right]$$

$\left(-\frac{1}{2}\right)R_1 + R_2 \to R_2$, $3R_1 + R_3 \to R_3$

$$\sim \left[\begin{array}{cccccc|c} \textcircled{\frac{1}{2}} & 1 & 0 & \frac{1}{2} & -\frac{1}{2} & 0 & 100 \\ \frac{1}{4} & 0 & 1 & -\frac{1}{4} & \frac{3}{4} & 0 & 350 \\ \hdashline -\frac{1}{2} & 0 & 0 & \frac{3}{2} & \frac{7}{2} & 1 & 4300 \end{array}\right] \begin{array}{l} \frac{100}{1/2} = 200 \\ \frac{350}{1/4} = 1400 \\ \end{array}$$

$2R_1 \to R_1$

$$\sim \left[\begin{array}{cccccc|c} 1 & 2 & 0 & 1 & -1 & 0 & 200 \\ \frac{1}{4} & 0 & 1 & -\frac{1}{4} & \frac{3}{4} & 0 & 350 \\ \hline -\frac{1}{2} & 0 & 0 & \frac{3}{2} & \frac{7}{2} & 1 & 4300 \end{array}\right]$$

$\left(-\frac{1}{4}\right)R_1 + R_2 \to R_2$, $\frac{1}{2}R_1 + R_3 \to R_3$

$$\sim \begin{array}{c} \\ x_1 \\ x_3 \\ P \end{array} \begin{array}{c} \begin{array}{cccccc} x_1 & x_2 & x_3 & s_1 & s_2 & P \end{array} \\ \left[\begin{array}{cccccc|c} 1 & 2 & 0 & 1 & -1 & 0 & 200 \\ 0 & -\frac{1}{2} & 1 & -\frac{1}{2} & 1 & 0 & 300 \\ \hline 0 & 1 & 0 & 2 & 3 & 1 & 4400 \end{array}\right] \end{array}$$

Optimal solution: the maximum profit is \$4400 when 200 A components, 0 B components and 300 C components are manufactured.

37. Let x_1 = the amount invested in government bonds,
x_2 = the amount invested in mutual funds,
and x_3 = the amount invested in money market funds.

The mathematical model for this problem is:

Maximize $P = .08x_1 + .13x_2 + .15x_3$

Subject to: $x_1 + x_2 + x_3 \le 100{,}000$

$x_2 + x_3 \le x_1$

$x_1, x_2, x_3 \ge 0$

We introduce slack variables s_1 and s_2 to obtain the equivalent form:

$$\begin{aligned} x_1 + x_2 + x_3 + s_1 \qquad\quad &= 100{,}000 \\ -x_1 + x_2 + x_3 \qquad + s_2 \quad &= 0 \\ -.08x_1 - .13x_2 - .15x_3 \qquad\quad + P &= 0 \end{aligned}$$

The simplex tableau for this problem is:

$$\begin{array}{c} \\ s_1 \\ s_2 \\ P \end{array} \begin{array}{c} \begin{array}{cccccc} x_1 & x_2 & x_3 & s_1 & s_2 & P \end{array} \\ \left[\begin{array}{cccccc|c} 1 & 1 & 1 & 1 & 0 & 0 & 100{,}000 \\ -1 & 1 & \textcircled{1} & 0 & 1 & 0 & 0 \\ \hline -.08 & -.13 & -.15 & 0 & 0 & 1 & 0 \end{array}\right] \end{array} \quad \frac{100{,}000}{1} = 100{,}000$$

$(-1)R_2 + R_1 \to R_1$ and $.15R_2 + R_3 \to R_3$

$$\sim \left[\begin{array}{cccccc|c} \textcircled{2} & 0 & 0 & 1 & -1 & 0 & 100{,}000 \\ -1 & 1 & 1 & 0 & 1 & 0 & 0 \\ \hline -.23 & .02 & 0 & 0 & .15 & 1 & 0 \end{array}\right] \sim \left[\begin{array}{cccccc|c} \textcircled{1} & 0 & 0 & \frac{1}{2} & -\frac{1}{2} & 0 & 50{,}000 \\ -1 & 1 & 1 & 0 & 1 & 0 & 0 \\ \hline -.23 & .02 & 0 & 0 & .15 & 1 & 0 \end{array}\right]$$

$\frac{1}{2}R_1 \to R_1$ $\qquad$ $R_1 + R_2 \to R_2$ and $.23R_1 + R_3 \to R_3$

$$\begin{array}{c} \\ x_1 \\ x_2 \\ P \end{array} \begin{bmatrix} x_1 & x_2 & x_3 & s_1 & s_2 & P & \\ 1 & 0 & 0 & \frac{1}{2} & -\frac{1}{2} & 0 & 50{,}000 \\ 0 & 1 & 1 & \frac{1}{2} & \frac{1}{2} & 0 & 50{,}000 \\ \hline 0 & .02 & 0 & .115 & .035 & 1 & 11{,}500 \end{bmatrix}$$

Optimal solution: the maximum return is \$11,500 when x_1 = \$50,000 is invested in government bonds, x_2 = \$0 is invested in mutual funds, and x_3 = \$50,000 is invested in money market funds.

39. Let x_1 = the number of daytime ads,
x_2 = the number of prime-time ads,
and x_3 = the number of late-night ads.

The mathematical model for this problem is:

Maximize $P = 14{,}000x_1 + 24{,}000x_2 + 18{,}000x_3$

Subject to:
$$1000x_1 + 2000x_2 + 1500x_3 \le 20{,}000$$
$$x_1 + x_2 + x_3 \le 15$$
$$x_1, x_2, x_3 \ge 0$$

We introduce slack variables to obtain the following initial form:

$$\begin{aligned} 1000x_1 + 2000x_2 + 1500x_3 + s_1 \quad &= 20{,}000 \\ x_1 + x_2 + x_3 \quad + s_2 &= 15 \\ -14{,}000x_1 - 24{,}000x_2 - 18{,}000x_3 \quad + P &= 0 \end{aligned}$$

The simplex tableau for this problem is:

$$\begin{array}{c} \\ s_1 \\ s_2 \\ P \end{array} \begin{bmatrix} x_1 & x_2 & x_3 & s_1 & s_2 & P & \\ 1000 & \boxed{2000} & 1500 & 1 & 0 & 0 & 20{,}000 \\ 1 & 1 & 1 & 0 & 1 & 0 & 15 \\ \hline -14{,}000 & -24{,}000 & -18{,}000 & 0 & 0 & 1 & 0 \end{bmatrix} \begin{array}{l} \\ \frac{20{,}000}{2000} = 10 \\ \frac{15}{1} = 15 \\ \\ \end{array}$$

$\frac{1}{2000}R_1 \to R_1$

$$\sim \begin{bmatrix} \frac{1}{2} & \boxed{1} & \frac{3}{4} & \frac{1}{2000} & 0 & 0 & 10 \\ 1 & 1 & 1 & 0 & 1 & 0 & 15 \\ \hline -14{,}000 & -24{,}000 & -18{,}000 & 0 & 0 & 1 & 0 \end{bmatrix}$$

$(-1)R_1 + R_2 \to R_2,\ 24{,}000R_1 + R_3 \to R_3$

$$\sim \begin{bmatrix} \frac{1}{2} & 1 & \frac{3}{4} & \frac{1}{2000} & 0 & 0 & 10 \\ \boxed{\frac{1}{2}} & 0 & \frac{1}{4} & -\frac{1}{2000} & 1 & 0 & 5 \\ \hline -2000 & 0 & 0 & 12 & 0 & 1 & 240{,}000 \end{bmatrix}$$

$2R_2 \to R_2$

$$\sim \begin{bmatrix} \frac{1}{2} & 1 & \frac{3}{4} & \frac{1}{2000} & 0 & 0 & 10 \\ \boxed{1} & 0 & \frac{1}{2} & -\frac{1}{1000} & 2 & 0 & 10 \\ \hline -2000 & 0 & 0 & 12 & 0 & 1 & 240{,}000 \end{bmatrix}$$

$\left(-\frac{1}{2}\right)R_2 + R_1 \to R_1,\ 2000R_2 + R_3 \to R_3$

$$\begin{array}{c} \\ x_2 \\ \sim\; x_1 \\ P \end{array}\left[\begin{array}{cccccc|c} x_1 & x_2 & x_3 & s_1 & s_2 & P & \\ 0 & 1 & \frac{1}{2} & \frac{1}{1000} & -1 & 0 & 5 \\ 1 & 0 & \frac{1}{2} & -\frac{1}{1000} & 2 & 0 & 10 \\ \hdashline 0 & 0 & 1000 & 10 & 4000 & 1 & 260{,}000 \end{array}\right]$$

Optimal solution: maximum number of potential customers is 260,000 when $x_1 = 10$ daytime ads, $x_2 = 5$ prime-time ads, and $x_3 = 0$ late-night ads are placed.

41. Let x_1 = the number of colonial houses,
x_2 = the number of split-level houses,
and x_3 = the number of ranch-style houses.

(A) The mathematical model for this problem is:

Maximize $P = 20{,}000x_1 + 18{,}000x_2 + 24{,}000x_3$

$$\text{Subject to: } \begin{aligned} \tfrac{1}{2}x_1 + \tfrac{1}{2}x_2 + x_3 &\le 30 \\ 60{,}000x_1 + 60{,}000x_2 + 80{,}000x_3 &\le 3{,}200{,}000 \\ 4{,}000x_1 + 3{,}000x_2 + 4{,}000x_3 &\le 180{,}000 \\ x_1, x_2, x_3 &\ge 0 \end{aligned}$$

We simplify the inequalities and then introduce slack variables to obtain the initial form:

$$\begin{aligned} \tfrac{1}{2}x_1 + \tfrac{1}{2}x_2 + x_3 + s_1 &= 30 \\ 6x_1 + 6x_2 + 8x_3 + s_2 &= 320 \\ 4x_1 + 3x_2 + 4x_3 + s_3 &= 180 \\ -20{,}000x_1 - 18{,}000x_2 - 24{,}000x_3 + P &= 0 \end{aligned}$$

[Note: This simplification will change the interpretation of the slack variables.]

The simplex tableau for this problem is:

$$\begin{array}{c} \\ s_1 \\ s_2 \\ s_3 \\ P \end{array}\left[\begin{array}{ccccccc|c} x_1 & x_2 & x_3 & s_1 & s_2 & s_3 & P & \\ \frac{1}{2} & \frac{1}{2} & \textcircled{1} & 1 & 0 & 0 & 0 & 30 \\ 6 & 6 & 8 & 0 & 1 & 0 & 0 & 320 \\ 4 & 3 & 4 & 0 & 0 & 1 & 0 & 180 \\ \hdashline -20{,}000 & -18{,}000 & -24{,}000 & 0 & 0 & 0 & 1 & 0 \end{array}\right] \begin{array}{l} \\ \frac{30}{1} = 30 \\ \frac{320}{8} = 40 \\ \frac{180}{4} = 45 \\ \\ \end{array}$$

$(-8)R_1 + R_2 \to R_2,\ (-4)R_1 + R_3 \to R_3,\ 24{,}000R_1 + R_4 \to R_4$

$$\sim \left[\begin{array}{ccccccc|c} \frac{1}{2} & \frac{1}{2} & 1 & 1 & 0 & 0 & 0 & 30 \\ 2 & 2 & 0 & -8 & 1 & 0 & 0 & 80 \\ \textcircled{2} & 1 & 0 & -4 & 0 & 1 & 0 & 60 \\ \hdashline -8000 & -6000 & 0 & 24{,}000 & 0 & 0 & 1 & 720{,}000 \end{array}\right]$$

$\frac{1}{2}R_3 \to R_3$

$$\sim\left[\begin{array}{ccccccc|c} \frac{1}{2} & \frac{1}{2} & 1 & 1 & 0 & 0 & 0 & 30 \\ 2 & 2 & 0 & -8 & 1 & 0 & 0 & 80 \\ \textcircled{1} & \frac{1}{2} & 0 & -2 & 0 & \frac{1}{2} & 0 & 30 \\ \hline -8000 & -6000 & 0 & 24,000 & 0 & 0 & 1 & 720,000 \end{array}\right]$$

$\left(-\frac{1}{2}\right)R_3 + R_1 \rightarrow R_1$, $(-2)R_3 + R_2 \rightarrow R_2$, $8000R_3 + R_4 \rightarrow R_4$

$$\sim\left[\begin{array}{ccccccc|c} 0 & \frac{1}{4} & 1 & 2 & 0 & -\frac{1}{4} & 0 & 15 \\ 0 & \textcircled{1} & 0 & -4 & 1 & -1 & 0 & 20 \\ 1 & \frac{1}{2} & 0 & -2 & 0 & \frac{1}{2} & 0 & 30 \\ \hline 0 & -2000 & 0 & 8000 & 0 & 4000 & 1 & 960,000 \end{array}\right]$$

$\left(-\frac{1}{4}\right)R_2 + R_1 \rightarrow R_1$, $\left(-\frac{1}{2}\right)R_2 + R_3 \rightarrow R_3$, $2000R_2 + R_4 \rightarrow R_4$

$$\sim\begin{array}{c} \\ x_3 \\ x_2 \\ x_1 \\ P \end{array}\begin{array}{c} \begin{array}{ccccccc} x_1 & x_2 & x_3 & s_1 & s_2 & s_3 & P \end{array} \\ \left[\begin{array}{ccccccc|c} 0 & 0 & 1 & 3 & -\frac{1}{4} & 0 & 0 & 10 \\ 0 & 1 & 0 & -4 & 1 & -1 & 0 & 20 \\ 1 & 0 & 0 & 0 & -\frac{1}{2} & 1 & 0 & 20 \\ \hline 0 & 0 & 0 & 0 & 2000 & 2000 & 1 & 1,000,000 \end{array}\right] \end{array}$$

Optimal solution: maximum profit is \$1,000,000 when $x_1 = 20$ colonial houses, $x_2 = 20$ split-level houses, and $x_3 = 10$ ranch-style houses are built.

(B) The mathematical model for this problem is:
Maximize $P = 17,000x_1 + 18,000x_2 + 24,000x_3$

Subject to:
$$\begin{aligned} \frac{1}{2}x_1 + \frac{1}{2}x_2 + x_3 &\le 30 \\ 60,000x_1 + 60,000x_2 + 80,000x_3 &\le 3,200,000 \\ 4,000x_1 + 3,000x_2 + 4,000x_3 &\le 180,000 \end{aligned}$$

Following the solution in part (A), we obtain the simplex tableau:

$$\begin{array}{c} \\ s_1 \\ s_2 \\ s_3 \\ P \end{array}\begin{array}{c} \begin{array}{ccccccc} x_1 & x_2 & x_3 & s_1 & s_2 & s_3 & P \end{array} \\ \left[\begin{array}{ccccccc|c} \frac{1}{2} & \frac{1}{2} & \textcircled{1} & 1 & 0 & 0 & 0 & 30 \\ 6 & 6 & 8 & 0 & 1 & 0 & 0 & 320 \\ 4 & 3 & 4 & 0 & 0 & 1 & 0 & 180 \\ \hline -17,000 & -18,000 & -24,000 & 0 & 0 & 0 & 1 & 0 \end{array}\right] \end{array}\begin{array}{l} \\ \frac{30}{1} = 30 \\ \frac{320}{8} = 40 \\ \frac{180}{4} = 45 \\ \\ \end{array}$$

$(-8)R_1 + R_2 \rightarrow R_2$, $(-4)R_1 + R_3 \rightarrow R_3$, $24,000R_1 + R_4 \rightarrow R_4$

$$\sim\left[\begin{array}{ccccccc|c} \frac{1}{2} & \frac{1}{2} & 1 & 1 & 0 & 0 & 0 & 30 \\ 2 & \textcircled{2} & 0 & -8 & 1 & 0 & 0 & 80 \\ 2 & 1 & 0 & -4 & 0 & 1 & 0 & 60 \\ \hdashline -5000 & -6000 & 0 & 24,000 & 0 & 0 & 1 & 720,000 \end{array}\right] \begin{array}{l} \frac{30}{1/2} = 60 \\ \frac{80}{2} = 40 \\ \frac{60}{1} = 60 \\ \\ \end{array}$$

$\frac{1}{2}R_2 \to R_2$

$$\sim\left[\begin{array}{ccccccc|c} \frac{1}{2} & \frac{1}{2} & 1 & 1 & 0 & 0 & 0 & 30 \\ 1 & \textcircled{1} & 0 & -4 & \frac{1}{2} & 0 & 0 & 40 \\ 2 & 1 & 0 & -4 & 0 & 1 & 0 & 60 \\ \hdashline -5000 & -6000 & 0 & 24,000 & 0 & 0 & 1 & 720,000 \end{array}\right]$$

$\left(-\frac{1}{2}\right)R_2 + R_1 \to R_1$, $(-1)R_2 + R_3 \to R_3$, $6,000R_2 + R_4 \to R_4$

$$\sim\begin{array}{c} \\ x_3 \\ x_2 \\ s_3 \\ P \end{array}\begin{array}{c} \begin{array}{ccccccc} x_1 & x_2 & x_3 & s_1 & s_2 & s_3 & P \end{array} \\ \left[\begin{array}{ccccccc|c} 0 & 0 & 1 & 3 & -\frac{1}{4} & 0 & 0 & 10 \\ 1 & 1 & 0 & -4 & \frac{1}{2} & 0 & 0 & 40 \\ 1 & 0 & 0 & 0 & -\frac{1}{2} & 1 & 0 & 20 \\ \hdashline 1000 & 0 & 0 & 0 & 3000 & 0 & 1 & 960,000 \end{array}\right] \end{array}$$

Optimal solution: maximum profit is \$960,000 when $x_1 = 0$ colonial houses, $x_2 = 40$ split level houses and $x_3 = 10$ ranch houses are built. In this case, $s_3 = 20$ (thousand) labor hours are not used.

(C) The mathematical model for this problem is:
Maximize $P = 25,000x_1 + 18,000x_2 + 24,000x_3$

Subject to:
$$\begin{aligned} \frac{1}{2}x_1 + \frac{1}{2}x_2 + x_3 &\le 30 \\ 60,000x_1 + 60,000x_2 + 80,000x_3 &\le 3,200,000 \\ 4,000x_1 + 3,000x_2 + 4,000x_3 &\le 180,000 \end{aligned}$$

Following the solutions in parts (A) and (B), we obtain the simplex tableau:

$$\begin{array}{c} \\ s_1 \\ s_2 \\ s_3 \\ P \end{array}\begin{array}{c} \begin{array}{ccccccc} x_1 & x_2 & x_3 & s_1 & s_2 & s_3 & P \end{array} \\ \left[\begin{array}{ccccccc|c} \frac{1}{2} & \frac{1}{2} & 1 & 1 & 0 & 0 & 0 & 30 \\ 6 & 6 & 8 & 0 & 1 & 0 & 0 & 320 \\ \textcircled{4} & 3 & 4 & 0 & 0 & 1 & 0 & 180 \\ \hdashline -25,000 & -18,000 & -24,000 & 0 & 0 & 0 & 1 & 0 \end{array}\right] \end{array} \begin{array}{l} \\ \frac{30}{1/2} = 60 \\ \frac{320}{6} = 53.33 \\ \frac{180}{4} = 45 \\ \\ \end{array}$$

$\frac{1}{4}R_3 \to R_3$

$$\sim \left[\begin{array}{ccccccc|c} \frac{1}{2} & \frac{1}{2} & 1 & 1 & 0 & 0 & 0 & 30 \\ 6 & 6 & 8 & 0 & 1 & 0 & 0 & 320 \\ \textcircled{1} & \frac{3}{4} & 1 & 0 & 0 & \frac{1}{4} & 0 & 45 \\ \hline -25,000 & -18,000 & -24,000 & 0 & 0 & 0 & 1 & 0 \end{array}\right]$$

$\left(-\frac{1}{2}\right)R_3 + R_1 \to R_1$, $-6R_3 + R_2 \to R_2$, $25,000R_3 + R_4 \to R_4$

$$\sim \begin{array}{c} \\ s_1 \\ s_2 \\ x_1 \\ P \end{array} \begin{array}{c} \begin{array}{ccccccc} x_1 & x_2 & x_3 & s_1 & s_2 & s_3 & P \end{array} \\ \left[\begin{array}{ccccccc|c} 0 & \frac{1}{8} & \frac{1}{2} & 1 & 0 & -\frac{1}{8} & 0 & 7.5 \\ 0 & \frac{3}{2} & 2 & 0 & 1 & \frac{3}{2} & 0 & 50 \\ 1 & \frac{3}{4} & 1 & 0 & 0 & \frac{1}{4} & 0 & 45 \\ \hline 0 & 750 & 1000 & 0 & 0 & 6250 & 1 & 1,125,000 \end{array}\right] \end{array}$$

Optimal solution: maximum profit is $1,125,000 when $x_1 = 45$ colonial houses, $x_2 = 0$ split level houses and $x_3 = 0$ ranch houses are built. In this case, $s_1 = 7.5$ acres of land, and $s_2 = 50(10,000) =$ $500,000 of capital are not used.

43. Let x_1 = the number of boxes of Assortment I,
x_2 = the number of boxes of Assortment II,
and x_3 = the number of boxes of Assortment III.

(A) The profit per box of Assortment I is:
$9.40 - [4(0.20) + 4(0.25) + 12(0.30)] = \$4.00$
The profit per box of Assortment II is:
$7.60 - [12(0.20) + 4(0.25) + 4(0.30)] = \$3.00$
The profit per box of Assortment III is:
$11.00 - [8(0.20) + 8(0.25) + 8(0.30)] = \$5.00$
The mathematical model for this problem is:
Maximize $P = 4x_1 + 3x_2 + 5x_3$
Subject to:
$$\begin{aligned} 4x_1 + 12x_2 + 8x_3 &\le 4800 \\ 4x_1 + 4x_2 + 8x_3 &\le 4000 \\ 12x_1 + 4x_2 + 8x_3 &\le 5600 \\ x_1, x_2, x_3 &\ge 0 \end{aligned}$$
We introduce slack variables to obtain the initial form:
$$\begin{aligned} 4x_1 + 12x_2 + 8x_3 + s_1 \qquad\qquad &= 4800 \\ 4x_1 + 4x_2 + 8x_3 \qquad + s_2 \qquad &= 4000 \\ 12x_1 + 4x_2 + 8x_3 \qquad\qquad + s_3 &= 5600 \\ -4x_1 - 3x_2 - 5x_3 \qquad\qquad + P &= 0 \end{aligned}$$

$$
\begin{array}{c}
\\ s_1 \\ s_2 \\ s_3 \\ P
\end{array}
\begin{array}{ccccccc|c}
x_1 & x_2 & x_3 & s_1 & s_2 & s_3 & P & \\
4 & 12 & 8 & 1 & 0 & 0 & 0 & 4800 \\
4 & 4 & \textcircled{8} & 0 & 1 & 0 & 0 & 4000 \\
12 & 4 & 8 & 0 & 0 & 1 & 0 & 5600 \\
\hline
-4 & -3 & -5 & 0 & 0 & 0 & 1 & 0
\end{array}
\quad
\begin{array}{l}
\\ \frac{4800}{8} = 600 \\ \frac{4000}{8} = 500 \\ \frac{5600}{8} = 700 \\ \\
\end{array}
$$

$\frac{1}{8}R_2 \to R_2$

$$
\sim \left[\begin{array}{ccccccc|c}
4 & 12 & 8 & 1 & 0 & 0 & 0 & 4800 \\
\frac{1}{2} & \frac{1}{2} & \textcircled{1} & 0 & \frac{1}{8} & 0 & 0 & 500 \\
12 & 4 & 8 & 0 & 0 & 1 & 0 & 5600 \\
\hline
-4 & -3 & -5 & 0 & 0 & 0 & 1 & 0
\end{array}\right]
\sim \left[\begin{array}{ccccccc|c}
0 & 8 & 0 & 1 & -1 & 0 & 0 & 800 \\
\frac{1}{2} & \frac{1}{2} & 1 & 0 & \frac{1}{8} & 0 & 0 & 500 \\
\textcircled{8} & 0 & 0 & 0 & -1 & 1 & 0 & 1600 \\
\hline
-\frac{3}{2} & -\frac{1}{2} & 0 & 0 & \frac{5}{8} & 0 & 1 & 2500
\end{array}\right]
$$

$(-8)R_2 + R_1 \to R_1$, $(-8)R_2 + R_3 \to R_3$, $\quad \frac{1}{8}R_3 \to R_3$
$5R_2 + R_4 \to R_4$

$$
\sim \left[\begin{array}{ccccccc|c}
0 & 8 & 0 & 1 & -1 & 0 & 0 & 800 \\
\frac{1}{2} & \frac{1}{2} & 1 & 0 & \frac{1}{8} & 0 & 0 & 500 \\
\textcircled{1} & 0 & 0 & 0 & -\frac{1}{8} & \frac{1}{8} & 0 & 200 \\
\hline
-\frac{3}{2} & -\frac{1}{2} & 0 & 0 & \frac{5}{8} & 0 & 1 & 2500
\end{array}\right]
\sim \left[\begin{array}{ccccccc|c}
0 & \textcircled{8} & 0 & 1 & -1 & 0 & 0 & 800 \\
0 & \frac{1}{2} & 1 & 0 & \frac{3}{16} & -\frac{1}{16} & 0 & 400 \\
1 & 0 & 0 & 0 & -\frac{1}{8} & \frac{1}{8} & 0 & 200 \\
\hline
0 & -\frac{1}{2} & 0 & 0 & \frac{7}{16} & \frac{3}{16} & 1 & 2800
\end{array}\right]
$$

$\left(-\frac{1}{2}\right)R_3 + R_2 \to R_2$, $\frac{3}{2}R_3 + R_4 \to R_4 \quad \frac{1}{8}R_1 \to R_1$

$$
\sim \left[\begin{array}{ccccccc|c}
0 & \textcircled{1} & 0 & \frac{1}{8} & -\frac{1}{8} & 0 & 0 & 100 \\
0 & \frac{1}{2} & 1 & 0 & \frac{3}{16} & -\frac{1}{16} & 0 & 400 \\
1 & 0 & 0 & 0 & -\frac{1}{8} & \frac{1}{8} & 0 & 200 \\
\hline
0 & -\frac{1}{2} & 0 & 0 & \frac{7}{16} & \frac{3}{16} & 1 & 2800
\end{array}\right]
\sim
\begin{array}{ccccccc|c}
x_1 & x_2 & x_3 & s_1 & s_2 & s_3 & P & \\
0 & 1 & 0 & \frac{1}{8} & -\frac{1}{8} & 0 & 0 & 100 \\
0 & 0 & 1 & -\frac{1}{16} & \frac{1}{4} & -\frac{1}{16} & 0 & 350 \\
1 & 0 & 0 & 0 & -\frac{1}{8} & \frac{1}{8} & 0 & 200 \\
\hline
0 & 0 & 0 & \frac{1}{16} & \frac{3}{8} & \frac{3}{16} & 1 & 2850
\end{array}
$$

$\left(-\frac{1}{2}\right)R_1 + R_2 \to R_2$, $\frac{1}{2}R_1 + R_4 \to R_4$

Optimal solution: maximum profit is \$2850 when 200 boxes of Assortment I, 100 boxes of Assortment II, and 350 boxes of Assortment III are made.

(B) The mathematical model for this problem is:
Maximize $P = 4x_1 + 3x_2 + 5x_3$
Subject to:

$$
\begin{aligned}
4x_1 + 12x_2 + 8x_3 &\le 4800 \\
4x_1 + 4x_2 + 8x_3 &\le 5000 \\
12x_1 + 4x_2 + 8x_3 &\le 5600
\end{aligned}
$$

Following the solution in part (A), we obtain the simplex tableau:

$$\begin{array}{c} \\ s_1 \\ s_2 \\ s_3 \\ P \end{array} \begin{array}{c} \begin{array}{ccccccc} x_1 & x_2 & x_3 & s_1 & s_2 & s_3 & P \end{array} \\ \left[\begin{array}{ccccccc|c} 4 & 12 & \boxed{8} & 1 & 0 & 0 & 0 & 4800 \\ 4 & 4 & 8 & 0 & 1 & 0 & 0 & 5000 \\ 12 & 4 & 8 & 0 & 0 & 1 & 0 & 5600 \\ \hdashline -4 & -3 & -5 & 0 & 0 & 0 & 1 & 0 \end{array}\right] \end{array} \begin{array}{l} \frac{4800}{8} = 600 \\ \frac{5000}{8} = 625 \\ \frac{5600}{8} = 700 \end{array}$$

$\frac{1}{8}R_1 \rightarrow R_1$

$$\sim \left[\begin{array}{ccccccc|c} \frac{1}{2} & \frac{3}{2} & \boxed{1} & \frac{1}{8} & 0 & 0 & 0 & 600 \\ 4 & 4 & 8 & 0 & 1 & 0 & 0 & 5000 \\ 12 & 4 & 8 & 0 & 0 & 1 & 0 & 5600 \\ \hdashline -4 & -3 & -5 & 0 & 0 & 0 & 1 & 0 \end{array}\right]$$

$(-8)R_1 + R_2 \rightarrow R_2$, $(-8)R_1 + R_3 \rightarrow R_3$, $5R_1 + R_4 \rightarrow R_4$

$$\sim \left[\begin{array}{ccccccc|c} \frac{1}{2} & \frac{3}{2} & 1 & \frac{1}{8} & 0 & 0 & 0 & 600 \\ 0 & -8 & 0 & -1 & 1 & 0 & 0 & 200 \\ \boxed{8} & -8 & 0 & -1 & 0 & 1 & 0 & 800 \\ \hdashline -\frac{3}{2} & \frac{9}{2} & 0 & \frac{5}{8} & 0 & 0 & 1 & 3000 \end{array}\right] \begin{array}{l} \frac{600}{1/2} = 1200 \\ \\ \frac{800}{8} = 100 \\ \\ \end{array}$$

$\frac{1}{8}R_3 \rightarrow R_3$

$$\sim \left[\begin{array}{ccccccc|c} \frac{1}{2} & \frac{3}{2} & 1 & \frac{1}{8} & 0 & 0 & 0 & 600 \\ 0 & -8 & 0 & -1 & 1 & 0 & 0 & 200 \\ \boxed{1} & -1 & 0 & -\frac{1}{8} & 0 & \frac{1}{8} & 0 & 100 \\ \hdashline -\frac{3}{2} & \frac{9}{2} & 0 & \frac{5}{8} & 0 & 0 & 1 & 3000 \end{array}\right]$$

$\left(-\frac{1}{2}\right)R_3 + R_1 \rightarrow R_1$, $\left(\frac{3}{2}\right)R_3 + R_4 \rightarrow R_4$

$$\sim \begin{array}{c} \\ x_3 \\ s_2 \\ x_1 \\ P \end{array} \begin{array}{c} \begin{array}{ccccccc} x_1 & x_2 & x_3 & s_1 & s_2 & s_3 & P \end{array} \\ \left[\begin{array}{ccccccc|c} 0 & 2 & 1 & \frac{3}{16} & 0 & -\frac{1}{16} & 0 & 550 \\ 0 & -8 & 0 & -1 & 1 & 0 & 0 & 200 \\ 1 & -1 & 0 & -\frac{1}{8} & 0 & \frac{1}{8} & 0 & 100 \\ \hdashline 0 & 3 & 0 & \frac{7}{16} & 0 & \frac{3}{16} & 1 & 3150 \end{array}\right] \end{array}$$

Optimal solution: maximum profit is \$3,150 when $x_1 = 100$ boxes of assortment I, $x_2 = 0$ boxes of assortment II, and $x_3 = 550$ boxes of assortment III are produced. In this case, $s_2 = 200$ fruit-filled candies are not used.

(C) The mathematical model for this problem is:

Maximize $P = 4x_1 + 3x_2 + 5x_3$

Subject to:
$$\begin{aligned} 4x_1 + 12x_2 + 8x_3 &\le 6000 \\ 4x_1 + 4x_2 + 8x_3 &\le 6000 \\ 12x_1 + 4x_2 + 8x_3 &\le 5600 \end{aligned}$$

Following the solutions in parts (A) and (B), we obtain the simplex tableau:

$$\begin{array}{c} \\ s_1 \\ s_2 \\ s_3 \\ P \end{array} \begin{array}{c} \begin{array}{ccccccc} x_1 & x_2 & x_3 & s_1 & s_2 & s_3 & P \end{array} \\ \left[\begin{array}{ccccccc|c} 4 & 12 & 8 & 1 & 0 & 0 & 0 & 6000 \\ 4 & 4 & 8 & 0 & 1 & 0 & 0 & 6000 \\ 12 & 4 & \textcircled{8} & 0 & 0 & 1 & 0 & 5600 \\ \hline -4 & -3 & -5 & 0 & 0 & 0 & 1 & 0 \end{array}\right] \end{array} \begin{array}{l} \frac{6000}{8} = 750 \\ \frac{6000}{8} = 750 \\ \frac{5600}{8} = 700 \\ \\ \end{array}$$

$\frac{1}{8}R_3 \to R_3$

$$\sim \left[\begin{array}{ccccccc|c} 4 & 12 & 8 & 1 & 0 & 0 & 0 & 6000 \\ 4 & 4 & 8 & 0 & 1 & 0 & 0 & 6000 \\ \frac{3}{2} & \frac{1}{2} & 1 & 0 & 0 & \frac{1}{8} & 0 & 700 \\ \hline -4 & -3 & -5 & 0 & 0 & 0 & 1 & 0 \end{array}\right]$$

$(-8)R_3 + R_1 \to R_1$, $(-8)R_3 + R_2 \to R_2$, $5R_3 + R_4 \to R_4$

$$\sim \left[\begin{array}{ccccccc|c} -8 & \textcircled{8} & 0 & 1 & 0 & -1 & 0 & 400 \\ -8 & 0 & 0 & 0 & 1 & -1 & 0 & 400 \\ \frac{3}{2} & \frac{1}{2} & 1 & 0 & 0 & \frac{1}{8} & 0 & 700 \\ \hline \frac{7}{2} & -\frac{1}{2} & 0 & 0 & 0 & \frac{5}{8} & 1 & 3500 \end{array}\right] \begin{array}{l} \frac{400}{8} = 50 \\ \\ \frac{700}{1/2} = 1400 \\ \\ \end{array}$$

$\frac{1}{8}R_1 \to R_1$

$$\sim \left[\begin{array}{ccccccc|c} -1 & \textcircled{1} & 0 & \frac{1}{8} & 0 & -\frac{1}{8} & 0 & 50 \\ -8 & 0 & 0 & 0 & 1 & -1 & 0 & 400 \\ \frac{3}{2} & \frac{1}{2} & 1 & 0 & 0 & \frac{1}{8} & 0 & 700 \\ \hline \frac{7}{2} & -\frac{1}{2} & 0 & 0 & 0 & \frac{5}{8} & 1 & 3500 \end{array}\right]$$

$\left(-\frac{1}{2}\right)R_1 + R_3 \to R_3$, $\frac{1}{2}R_1 + R_4 \to R_4$

$$\sim \begin{array}{c} \\ x_2 \\ s_2 \\ x_3 \\ P \end{array} \begin{array}{c} \begin{array}{ccccccc} x_1 & x_2 & x_3 & s_1 & s_2 & s_3 & P \end{array} \\ \left[\begin{array}{ccccccc|c} -1 & 1 & 0 & \frac{1}{8} & 0 & -\frac{1}{8} & 0 & 50 \\ -8 & 0 & 0 & 0 & 1 & -1 & 0 & 400 \\ 2 & 0 & 1 & -\frac{1}{16} & 0 & \frac{3}{16} & 0 & 675 \\ \hdashline 3 & 0 & 0 & \frac{1}{16} & 0 & \frac{9}{16} & 1 & 3525 \end{array}\right] \end{array}$$

Optimal solution: maximum profit is \$3,525 when $x_1 = 0$ boxes of assortment I, $x_2 = 50$ boxes of assortment II, and $x_3 = 675$ boxes of assortment III are produced. In this case, $s_2 = 400$ fruit-filled candies are not used.

45. Let x_1 = the number of grams of food A,
x_2 = the number of grams of food B,
and x_3 = the number of grams of food C.

The mathematical model for this problem is:

Maximize $P = 3x_1 + 4x_2 + 5x_3$

Subject to:
$$x_1 + 3x_2 + 2x_3 \le 30$$
$$2x_1 + x_2 + 2x_3 \le 24$$
$$x_1, x_2, x_3 \ge 0$$

We introduce slack variables s_1 and s_2 to obtain the initial form:

$$\begin{aligned} x_1 + 3x_2 + 2x_3 + s_1 \qquad\quad &= 30 \\ 2x_1 + x_2 + 2x_3 \qquad + s_2 \quad &= 24 \\ -3x_1 - 4x_2 - 5x_3 \qquad\qquad + P &= 0 \end{aligned}$$

The simplex tableau for this problem is:

$$\begin{array}{c} \\ s_1 \\ s_2 \\ P \end{array} \begin{array}{c} \begin{array}{cccccc} x_1 & x_2 & x_3 & s_1 & s_2 & P \end{array} \\ \left[\begin{array}{cccccc|c} 1 & 3 & 2 & 1 & 0 & 0 & 30 \\ 2 & 1 & \textcircled{2} & 0 & 1 & 0 & 24 \\ \hdashline -3 & -4 & -5 & 0 & 0 & 1 & 0 \end{array}\right] \end{array} \begin{array}{l} \frac{30}{2} = 15 \\ \frac{24}{2} = 12 \\ \end{array}$$

$\frac{1}{2}R_2 \to R_2$

$$\sim \left[\begin{array}{cccccc|c} 1 & 3 & 2 & 1 & 0 & 0 & 30 \\ 1 & \frac{1}{2} & \textcircled{1} & 0 & \frac{1}{2} & 0 & 12 \\ \hdashline -3 & -4 & -5 & 0 & 0 & 1 & 0 \end{array}\right] \sim \left[\begin{array}{cccccc|c} -1 & \textcircled{2} & 0 & 1 & -1 & 0 & 6 \\ 1 & \frac{1}{2} & 1 & 0 & \frac{1}{2} & 0 & 12 \\ \hdashline 2 & -\frac{3}{2} & 0 & 0 & \frac{5}{2} & 1 & 60 \end{array}\right] \begin{array}{l} \frac{6}{2} = 3 \\ \frac{12}{1/2} = 24 \\ \end{array}$$

$(-2)R_2 + R_1 \to R_1$, $5R_2 + R_3 \to R_3$ $\qquad$ $\frac{1}{2}R_1 \to R_1$

$$\sim \left[\begin{array}{cccccc|c} -\frac{1}{2} & \textcircled{1} & 0 & \frac{1}{2} & -\frac{1}{2} & 0 & 3 \\ 1 & \frac{1}{2} & 1 & 0 & \frac{1}{2} & 0 & 12 \\ \hline 2 & -\frac{3}{2} & 0 & 0 & \frac{5}{2} & 1 & 60 \end{array}\right] \sim \begin{array}{c} \\ x_2 \\ x_3 \\ P \end{array} \begin{array}{c} \begin{array}{cccccc} x_1 & x_2 & x_3 & s_1 & s_2 & P \end{array} \\ \left[\begin{array}{cccccc|c} -\frac{1}{2} & 1 & 0 & \frac{1}{2} & -\frac{1}{2} & 0 & 3 \\ \frac{5}{4} & 0 & 1 & -\frac{1}{4} & \frac{3}{4} & 0 & \frac{21}{2} \\ \hline \frac{5}{4} & 0 & 0 & \frac{3}{4} & \frac{7}{4} & 1 & \frac{129}{2} \end{array}\right] \end{array}$$

$\left(-\frac{1}{2}\right)R_1 + R_2 \to R_2,\ \frac{3}{2}R_1 + R_3 \to R_3$

Optimal solution: the maximum amount of protein is 64.5 units when $x_1 = 0$ grams of food *A*, $x_2 = 3$ grams of food *B* and $x_3 = 10.5$ grams of food *C* are used.

47. Let x_1 = the number of undergraduate students,
x_2 = the number of graduate students,
and x_3 = the number of faculty members.

The mathematical model for this problem is:

Maximize $P = 18x_1 + 25x_2 + 30x_3$

Subject to:
$$\begin{aligned} x_1 + x_2 + x_3 &\le 20 \\ 100x_1 + 150x_2 + 200x_3 &\le 3200 \\ x_1, x_2, x_3 &\ge 0 \end{aligned}$$

Divide the second inequality by 50 to simplify the arithmetic. Then introduce slack variables s_1 and s_2 to obtain the initial form.

$$\begin{aligned} x_1 + x_2 + x_3 + s_1 \qquad\quad &= 20 \\ 2x_1 + 3x_2 + 4x_3 \qquad + s_2 \quad &= 64 \\ -18x_1 - 25x_2 - 30x_3 \qquad\qquad + P &= 0 \end{aligned}$$

The simplex tableau for this problem is:

$$\begin{array}{c} \\ s_1 \\ s_2 \\ P \end{array} \begin{array}{c} \begin{array}{cccccc} x_1 & x_2 & x_3 & s_1 & s_2 & P \end{array} \\ \left[\begin{array}{cccccc|c} 1 & 1 & 1 & 1 & 0 & 0 & 20 \\ 2 & 3 & \textcircled{4} & 0 & 1 & 0 & 64 \\ \hline -18 & -25 & -30 & 0 & 0 & 1 & 0 \end{array}\right] \end{array} \begin{array}{l} \frac{20}{1} = 20 \\ \frac{64}{4} = 16 \\ \\ \end{array}$$

$\frac{1}{4}R_2 \to R_2$

$$\sim \left[\begin{array}{cccccc|c} 1 & 1 & 1 & 1 & 0 & 0 & 20 \\ \frac{1}{2} & \frac{3}{4} & \textcircled{1} & 0 & \frac{1}{4} & 0 & 16 \\ \hline -18 & -25 & -30 & 0 & 0 & 1 & 0 \end{array}\right] \sim \left[\begin{array}{cccccc|c} \textcircled{\frac{1}{2}} & \frac{1}{4} & 0 & 1 & -\frac{1}{4} & 0 & 4 \\ \frac{1}{2} & \frac{3}{4} & 1 & 0 & \frac{1}{4} & 0 & 16 \\ \hline -3 & -\frac{5}{2} & 0 & 0 & \frac{15}{2} & 1 & 480 \end{array}\right] \begin{array}{l} \frac{4}{1/2} = 8 \\ \frac{16}{1/2} = 32 \\ \\ \end{array}$$

$(-1)R_2 + R_1 \to R_1,\ 30R_2 + R_3 \to R_3 \qquad 2R_1 \to R_1$

$$\sim \left[\begin{array}{cccccc|c} \textcircled{1} & \frac{1}{2} & 0 & 2 & -\frac{1}{2} & 0 & 8 \\ \frac{1}{2} & \frac{3}{4} & 1 & 0 & \frac{1}{4} & 0 & 16 \\ \hline -3 & -\frac{5}{2} & 0 & 0 & \frac{15}{2} & 1 & 480 \end{array}\right] \sim \left[\begin{array}{cccccc|c} 1 & \textcircled{\frac{1}{2}} & 0 & 2 & -\frac{1}{2} & 0 & 8 \\ 0 & \frac{1}{2} & 1 & -1 & \frac{1}{2} & 0 & 12 \\ \hline 0 & -1 & 0 & 6 & 6 & 1 & 504 \end{array}\right] \begin{array}{l} \frac{8}{1/2} = 16 \\ \frac{12}{1/2} = 24 \\ \\ \end{array}$$

$\left(-\frac{1}{2}\right)R_1 + R_2 \to R_2,\ 3R_1 + R_3 \to R_3 \qquad 2R_1 \to R_1$

$$\sim \left[\begin{array}{cccccc|c} 2 & 1 & 0 & 4 & -1 & 0 & 16 \\ 0 & \frac{1}{2} & 1 & -1 & \frac{1}{2} & 0 & 12 \\ \hline 0 & -1 & 0 & 6 & 6 & 1 & 504 \end{array}\right] \sim \begin{array}{c} \\ x_2 \\ x_3 \\ P \end{array} \begin{array}{c} \begin{array}{cccccc} x_1 & x_2 & x_3 & s_1 & s_2 & P \end{array} \\ \left[\begin{array}{cccccc|c} 2 & 1 & 0 & 4 & -1 & 0 & 16 \\ -1 & 0 & 1 & -3 & 1 & 0 & 4 \\ \hline 2 & 0 & 0 & 10 & 5 & 1 & 520 \end{array}\right] \end{array}$$

$\left(-\frac{1}{2}\right)R_1 + R_2 \to R_2$, $R_1 + R_3 \to R_3$

Optimal solution: the maximum number of interviews is 520 when $x_1 = 0$ undergraduates, $x_2 = 16$ graduate students, and $x_3 = 4$ faculty members are hired.

EXERCISE 5-5

Things to remember:

1. Given a matrix A. The transpose of A, denoted A^T, is the matrix formed by interchanging the rows and corresponding columns of A (first row with first columnn, second row with second column, and so on.)

2. FORMATION OF THE DUAL PROBLEM

   Given a minimization problem with ≥ problem constraints:

   Step 1. Use the coefficients and constants in the problem constraints and the objective function to form a matrix A with the coefficients of the objective function in the last row.

   Step 2. Interchange the rows and columns of matrix A to form the matrix A^T, the transpose of A.

   Step 3. Use the rows of A^T to form a maximization problem with ≤ problem constraints.

3. THE FUNDAMENTAL PRINCIPLE OF DUALITY

   A minimization problem has a solution if and only if its dual problem has a solution. If a solution exists, then the optimal value of the minimization problem is the same as the optimal value of the dual problem.

4. SOLUTION OF A MINIMIZATION PROBLEM

   Given a minimization problem with nonnegative coefficients in the objective function:

   (i) Write all problem constraints as ≥ inequalities. (This may introduce negative numbers on the right side of the problem constraints.)

(ii) Form the dual problem.

(iii) Write the initial system of the dual problem, using the variables from the minimization problem as the slack variables.

(iv) Use the simplex method to solve the dual problem.

(v) Read the solution of the minimization problem from the bottom row of the final simplex tableau in Step (iv).

[Note: If the dual problem has no solution, then the minimization problem has no solution.]

---

1. $A = [-5 \quad 0 \quad 3 \quad -1 \quad 8]$; $A^T = \begin{bmatrix} -5 \\ 0 \\ 3 \\ -1 \\ 8 \end{bmatrix}$

3. $A = \begin{bmatrix} 1 \\ -2 \\ 0 \\ 4 \end{bmatrix}$; $A^T = [1 \quad -2 \quad 0 \quad 4]$

5. $A = \begin{bmatrix} 2 & 1 & -6 & 0 & -1 \\ 5 & 2 & 0 & 1 & 3 \end{bmatrix}$; $A^T = \begin{bmatrix} 2 & 5 \\ 1 & 2 \\ -6 & 0 \\ 0 & 1 \\ -1 & 3 \end{bmatrix}$

7. $A = \begin{bmatrix} 1 & 2 & -1 \\ 0 & 2 & -7 \\ 8 & 0 & 1 \\ 4 & -1 & 3 \end{bmatrix}$; $A^T = \begin{bmatrix} 1 & 0 & 8 & 4 \\ 2 & 2 & 0 & -1 \\ -1 & -7 & 1 & 3 \end{bmatrix}$

9. (A) Given the minimization problem:

Minimize $C = 8x_1 + 9x_2$

Subject to: $x_1 + 3x_2 \geq 4$

$2x_1 + x_2 \geq 5$

$x_1, x_2 \geq 0$

The matrix A^T corresponding to this problem is: $A = \left[\begin{array}{cc|c} 1 & 3 & 4 \\ 2 & 1 & 5 \\ \hline 8 & 9 & 1 \end{array}\right]$

The matrix A^T corresponding to the dual problem has the rows of A as its columns. Thus:

$$A^T = \left[\begin{array}{cc|c} 1 & 2 & 8 \\ 3 & 1 & 9 \\ \hline 4 & 5 & 1 \end{array}\right]$$

The dual problem is: Maximize $P = 4y_1 + 5y_2$

Subject to: $y_1 + 2y_2 \le 8$

$3y_1 + y_2 \le 9$

$y_1, y_2 \ge 0$

(B) Letting x_1 and x_2 be slack variables, the initial system for the dual problem is:

$$\begin{aligned} y_1 + 2y_2 + x_1 \quad\quad &= 8 \\ 3y_1 + y_2 \quad + x_2 \quad &= 9 \\ -4y_1 - 5y_2 \quad\quad + P &= 0 \end{aligned}$$

(C) The simplex tableau for this problem is:

$$\begin{array}{c} \\ x_1 \\ x_2 \\ P \end{array} \begin{array}{c} \begin{array}{ccccc} y_1 & y_2 & x_1 & x_2 & P \end{array} \\ \left[\begin{array}{ccccc|c} 1 & 2 & 1 & 0 & 0 & 8 \\ 3 & 1 & 0 & 1 & 0 & 9 \\ \hline -4 & -5 & 0 & 0 & 1 & 0 \end{array}\right] \end{array}$$

11. From the final simplex tableau,

$$\begin{array}{c} \\ y_2 \\ y_1 \\ P \end{array} \begin{array}{c} \begin{array}{ccccc} y_1 & y_2 & x_1 & x_2 & P \end{array} \\ \left[\begin{array}{ccccc|c} 0 & 1 & 5 & -2 & 0 & 5 \\ 1 & 0 & -7 & 3 & 0 & 3 \\ \hline 0 & 0 & 1 & 2 & 1 & 121 \end{array}\right] \end{array}$$

(A) the optimal solution of the dual problem is:
maximum value of $P = 121$ at $y_1 = 3$ and $y_2 = 5$;

(B) the optimal solution of the minimization problem is:
minimum value of $C = 121$ at $x_1 = 1$, $x_2 = 2$.

13. (A) The matrix corresponding to the given problem is: $A = \left[\begin{array}{cc|c} 4 & 1 & 13 \\ 3 & 1 & 12 \\ \hline 9 & 2 & 1 \end{array}\right]$

The matrix A^T corresponding to the dual problem has the rows of A as its columns, that is:

$$A^T = \left[\begin{array}{cc|c} 4 & 3 & 9 \\ 1 & 1 & 2 \\ \hline 13 & 12 & 1 \end{array}\right]$$

Thus, the dual problem is: Maximize $P = 13y_1 + 12y_2$

Subject to: $4y_1 + 3y_2 \le 9$

$y_1 + y_2 \le 2$

$y_1, y_2 \ge 0$

(B) We introduce slack variables x_1 and x_2 to obtain the initial system for the dual problem:

$$\begin{aligned} 4y_1 + 3y_2 + x_1 \quad &= 9 \\ y_1 + y_2 \quad + x_2 &= 2 \\ -13y_1 - 12y_2 \quad + P &= 0 \end{aligned}$$

The simplex tableau for this problem is:

$$\begin{array}{c} \\ x_1 \\ x_2 \\ P \end{array} \begin{array}{c} \begin{matrix} y_1 & y_2 & x_1 & x_2 & P & \end{matrix} \\ \left[\begin{array}{ccccc|c} 4 & 3 & 1 & 0 & 0 & 9 \\ \textcircled{1} & 1 & 0 & 1 & 0 & 2 \\ \hdashline -13 & -12 & 0 & 0 & 1 & 0 \end{array}\right] \end{array} \begin{array}{l} \frac{9}{4} = 2.25 \\ \frac{2}{1} = 2 \\ \end{array} \sim \begin{array}{c} \\ x_1 \\ y_1 \\ P \end{array} \begin{array}{c} \begin{matrix} y_1 & y_2 & x_1 & x_2 & P & \end{matrix} \\ \left[\begin{array}{ccccc|c} 0 & -1 & 1 & -4 & 0 & 1 \\ 1 & 1 & 0 & 1 & 0 & 2 \\ \hdashline 0 & 1 & 0 & 13 & 1 & 26 \end{array}\right] \end{array}$$

$(-4)R_2 + R_1 \to R_1$ and $13R_2 + R_3 \to R_3$

Optimal solution: min $C = 26$ at $x_1 = 0$, $x_2 = 13$.

15. (A) The matrix corresponding to the given problem is: $A = \left[\begin{array}{cc|c} 2 & 3 & 15 \\ 1 & 2 & 8 \\ \hline 7 & 12 & 1 \end{array}\right]$

The matrix A^T corresponding to the dual problem has the rows of A as its columns, that is:

$$A^T = \left[\begin{array}{cc|c} 2 & 1 & 7 \\ 3 & 2 & 12 \\ \hline 15 & 8 & 1 \end{array}\right]$$

Thus, the dual problem is: Maximize $P = 15y_1 + 8y_2$

Subject to: $2y_1 + y_2 \le 7$

$3y_1 + 2y_2 \le 12$

$y_1, y_2 \ge 0$

(B) We introduce slack variables x_1 and x_2 to obtain the initial system for the dual problem:

$$\begin{aligned} 2y_1 + y_2 + x_1 \quad &= 7 \\ 3y_1 + 2y_2 \quad + x_2 &= 12 \\ -15y_1 - 8y_2 \quad + P &= 0 \end{aligned}$$

The simplex tableau for this problem is:

$$\begin{array}{c} \\ x_1 \\ x_2 \\ P \end{array} \begin{array}{c} \begin{matrix} y_1 & y_2 & x_1 & x_2 & P & \end{matrix} \\ \left[\begin{array}{ccccc|c} \textcircled{2} & 1 & 1 & 0 & 0 & 7 \\ 3 & 2 & 0 & 1 & 0 & 12 \\ \hdashline -15 & -8 & 0 & 0 & 1 & 0 \end{array}\right] \end{array} \begin{array}{l} \frac{7}{2} = 3.5 \\ \frac{12}{3} = 4 \\ \end{array} \sim \left[\begin{array}{ccccc|c} \textcircled{1} & \frac{1}{2} & \frac{1}{2} & 0 & 0 & \frac{7}{2} \\ 3 & 2 & 0 & 1 & 0 & 12 \\ \hdashline -15 & -8 & 0 & 0 & 1 & 0 \end{array}\right]$$

$\frac{1}{2}R_1 \to R_1$ $\qquad$ $(-3)R_1 + R_2 \to R_2$ and $15R_1 + R_3 \to R_3$

$$\sim\left[\begin{array}{ccccc|c} 1 & \frac{1}{2} & \frac{1}{2} & 0 & 0 & \frac{7}{2} \\ 0 & \textcircled{\frac{1}{2}} & -\frac{3}{2} & 1 & 0 & \frac{3}{2} \\ \hdashline 0 & -\frac{1}{2} & \frac{15}{2} & 0 & 1 & \frac{105}{2} \end{array}\right] \begin{array}{l} \frac{7/2}{1/2} = 7 \\ \frac{3/2}{1/2} = 3 \\ \\ \end{array} \sim \left[\begin{array}{ccccc|c} 1 & \frac{1}{2} & \frac{1}{2} & 0 & 0 & \frac{7}{2} \\ 0 & \textcircled{1} & -3 & 2 & 0 & 3 \\ \hdashline 0 & -\frac{1}{2} & \frac{15}{2} & 0 & 1 & \frac{105}{2} \end{array}\right]$$

$2R_2 \to R_2$ $\qquad$ $\left(-\frac{1}{2}\right)R_2 + R_1 \to R_1$ and $\frac{1}{2}R_2 + R_3 \to R_3$

$$\begin{array}{c} \\ y_1 \\ \sim y_2 \\ P \end{array} \begin{array}{c} \begin{array}{ccccc} y_1 & y_2 & x_1 & x_2 & P \end{array} \\ \left[\begin{array}{ccccc|c} 1 & 0 & 2 & -1 & 0 & 2 \\ 0 & 1 & -3 & 2 & 0 & 3 \\ \hdashline 0 & 0 & 6 & 1 & 1 & 54 \end{array}\right] \end{array}$$

Optimal solution: min $C = 54$ at $x_1 = 6$, $x_2 = 1$.

17. (A) The matrices corresponding to the given problem and to the dual problem are:

$$A = \left[\begin{array}{cc|c} 2 & 1 & 8 \\ -2 & 3 & 4 \\ \hline 11 & 4 & 1 \end{array}\right] \quad \text{and} \quad A^T = \left[\begin{array}{cc|c} 2 & -2 & 11 \\ 1 & 3 & 4 \\ \hline 8 & 4 & 1 \end{array}\right]$$

respectively.

Thus, the dual problem is:

Maximize $P = 8y_1 + 4y_2$

Subject to: $2y_1 - 2y_2 \le 11$

$y_1 + 3y_2 \le 4$

$y_1, y_2 \ge 0$

(B) We introduce slack variables x_1 and x_2 to obtain the initial system for the dual problem:

$$\begin{aligned} 2y_1 - 2y_2 + x_1 \qquad\qquad &= 11 \\ y_1 + 3y_2 \qquad + x_2 \qquad &= 4 \\ -8y_1 - 4y_2 \qquad\qquad + P &= 0 \end{aligned}$$

The simplex tableau for this problem is:

$$\begin{array}{c} \\ x_1 \\ \sim x_2 \\ P \end{array} \begin{array}{c} \begin{array}{ccccc} y_1 & y_2 & x_1 & x_2 & P \end{array} \\ \left[\begin{array}{ccccc|c} 2 & -2 & 1 & 0 & 0 & 11 \\ \textcircled{1} & 3 & 0 & 1 & 0 & 4 \\ \hdashline -8 & -4 & 0 & 0 & 1 & 0 \end{array}\right] \end{array} \begin{array}{l} \\ \frac{11}{2} = 5.5 \\ \frac{4}{1} = 4 \\ \\ \end{array} \sim \begin{array}{c} \\ x_1 \\ y_1 \\ P \end{array} \begin{array}{c} \begin{array}{ccccc} y_1 & y_2 & x_1 & x_2 & P \end{array} \\ \left[\begin{array}{ccccc|c} 0 & -8 & 1 & -2 & 0 & 3 \\ 1 & 3 & 0 & 1 & 0 & 4 \\ \hdashline 0 & 20 & 0 & 8 & 1 & 32 \end{array}\right] \end{array}$$

$(-2)R_2 + R_1 \to R_1$ and $8R_2 + R_3 \to R_3$

Optimal solution: min $C = 32$ at $x_1 = 0$, $x_2 = 8$.

19. (A) The matrices corresponding to the given problem and the dual problem are:

$$A = \left[\begin{array}{cc|c} -3 & 1 & 6 \\ 1 & -2 & 4 \\ \hline 7 & 9 & 1 \end{array}\right] \quad \text{and} \quad A^T = \left[\begin{array}{cc|c} -3 & 1 & 7 \\ 1 & -2 & 9 \\ \hline 6 & 4 & 1 \end{array}\right]$$

respectively. Thus, the dual problem is:

Maximize $P = 6y_1 + 4y_2$

Subject to: $-3y_1 + y_2 \le 7$

$y_1 - 2y_2 \le 9$

$y_1, y_2 \ge 0$

(B) We introduce slack variables x_1 and x_2 to obtain the initial system for the dual problem:

$$\begin{aligned} -3y_1 + y_2 + x_1 \qquad\quad &= 7 \\ y_1 - 2y_2 \qquad + x_2 \quad &= 9 \\ -6y_1 - 4y_2 \qquad\qquad + P &= 0 \end{aligned}$$

The simplex tableau for this problem is:

$$\begin{array}{c} \\ x_1 \\ x_2 \\ P \end{array} \begin{array}{c} \begin{array}{ccccc} y_1 & y_2 & x_1 & x_2 & P \end{array} \\ \left[\begin{array}{ccccc|c} -3 & 1 & 1 & 0 & 0 & 7 \\ \textcircled{1} & -2 & 0 & 1 & 0 & 9 \\ \hdashline -6 & -4 & 0 & 0 & 1 & 0 \end{array}\right] \end{array} \sim \begin{array}{c} \begin{array}{ccccc} y_1 & y_2 & x_1 & x_2 & P \end{array} \\ \left[\begin{array}{ccccc|c} 0 & -5 & 1 & 3 & 0 & 34 \\ 1 & -2 & 0 & 1 & 0 & 9 \\ \hdashline 0 & -16 & 0 & 6 & 1 & 54 \end{array}\right] \end{array}$$

$3R_2 + R_1 \to R_1$ and $6R_2 + R_3 \to R_3$

The negative elements in the second column above the dashed line indicate that the problem does not have a solution.

21. The matrices corresponding to the given problem and the dual problem are:

$$A = \left[\begin{array}{cc|c} 2 & 1 & 8 \\ 1 & 2 & 8 \\ \hline 3 & 9 & 1 \end{array}\right] \quad \text{and} \quad A^T = \left[\begin{array}{cc|c} 2 & 1 & 3 \\ 1 & 2 & 9 \\ \hline 8 & 8 & 1 \end{array}\right]$$

respectively. Thus, the dual problem is: Maximize $P = 8y_1 + 8y_2$

Subject to: $2y_1 + y_2 \le 3$

$y_1 + 2y_2 \le 9$

$y_1, y_2 \ge 0$

We introduce slack variables x_1 and x_2 to obtain the initial system:

$$\begin{aligned} 2y_1 + y_2 + x_1 \qquad\quad &= 3 \\ y_1 + 2y_2 \qquad + x_2 \quad &= 9 \\ -8y_1 - 8y_2 \qquad\qquad + P &= 0 \end{aligned}$$

The simplex tableau for this problem is:

$$\begin{array}{c} \\ x_1 \\ x_2 \\ P \end{array}\begin{array}{c} \begin{array}{ccccc} y_1 & y_2 & x_1 & x_2 & P \end{array} \\ \left[\begin{array}{ccccc|c} 2 & \textcircled{1} & 1 & 0 & 0 & 3 \\ 1 & 2 & 0 & 1 & 0 & 9 \\ \hdashline -8 & -8 & 0 & 0 & 1 & 0 \end{array}\right] \end{array} \begin{array}{l} \frac{3}{1} = 3 \\ \frac{9}{2} = 4.5 \\ \\ \end{array} \sim \begin{array}{c} \\ y_2 \\ x_2 \\ P \end{array}\begin{array}{c} \begin{array}{ccccc} y_1 & y_2 & x_1 & x_2 & P \end{array} \\ \left[\begin{array}{ccccc|c} 2 & 1 & 1 & 0 & 0 & 3 \\ -3 & 0 & -2 & 1 & 0 & 3 \\ \hdashline 8 & 0 & 8 & 0 & 1 & 24 \end{array}\right] \end{array}$$

$(-2)R_1 + R_2 \to R_2$ and $8R_1 + R_3 \to R_3$

Optimal solution: min $C = 24$ at $x_1 = 8$, $x_2 = 0$.

[Note: We could use either column 1 or column 2 as the pivot column. Column 2 involves slightly simpler calculations.]

23. The matrices corresponding to the given problem and the dual problem are:

$$A = \left[\begin{array}{cc|c} 1 & 1 & 4 \\ 1 & -2 & -8 \\ -2 & 1 & -8 \\ \hline 7 & 5 & 1 \end{array}\right] \quad \text{and} \quad A^T = \left[\begin{array}{ccc|c} 1 & 1 & -2 & 7 \\ 1 & -2 & 1 & 5 \\ \hline 4 & -8 & -8 & 1 \end{array}\right] \text{ respectively.}$$

Thus, the dual problem is: Maximize $P = 4y_1 - 8y_2 - 8y_3$

Subject to:
$$\begin{aligned} y_1 + y_2 - 2y_3 &\le 7 \\ y_1 - 2y_2 + y_3 &\le 5 \\ y_1, y_2, y_3 &\ge 0 \end{aligned}$$

We introduce slack variables x_1 and x_2 to obtain the initial system:

$$\begin{aligned} y_1 + y_2 - 2y_3 + x_1 \qquad\qquad &= 7 \\ y_1 - 2y_2 + y_3 \qquad + x_2 \qquad &= 5 \\ -4y_1 + 8y_2 + 8y_3 \qquad\qquad + P &= 0 \end{aligned}$$

The simplex tableau for this problem is:

$$\begin{array}{c} \\ x_1 \\ x_2 \\ P \end{array}\begin{array}{c} \begin{array}{cccccc} y_1 & y_2 & y_3 & x_1 & x_2 & P \end{array} \\ \left[\begin{array}{cccccc|c} 1 & 1 & -2 & 1 & 0 & 0 & 7 \\ \textcircled{1} & -2 & 1 & 0 & 1 & 0 & 5 \\ \hdashline -4 & 8 & 8 & 0 & 0 & 1 & 0 \end{array}\right] \end{array} \begin{array}{l} \frac{7}{1} = 7 \\ \frac{5}{1} = 5 \\ \\ \end{array} \sim \begin{array}{c} \\ x_1 \\ y_1 \\ P \end{array}\begin{array}{c} \begin{array}{cccccc} y_1 & y_2 & y_3 & x_1 & x_2 & P \end{array} \\ \left[\begin{array}{cccccc|c} 0 & 3 & -3 & 1 & -1 & 0 & 2 \\ 1 & -2 & 1 & 0 & 1 & 0 & 5 \\ \hdashline 0 & 0 & 12 & 0 & 4 & 1 & 20 \end{array}\right] \end{array}$$

$(-1)R_2 + R_1 \to R_1$ and $4R_2 + R_3 \to R_3$

Optimal solution: min $C = 20$ at $x_1 = 0$, $x_2 = 4$.

25. The matrices corresponding to the given problem and the dual problem are:

$$A = \left[\begin{array}{cc|c} 2 & 1 & 16 \\ 1 & 1 & 12 \\ 1 & 2 & 14 \\ \hline 10 & 30 & 1 \end{array}\right] \quad \text{and} \quad A^T = \left[\begin{array}{ccc|c} 2 & 1 & 1 & 10 \\ 1 & 1 & 2 & 30 \\ \hline 16 & 12 & 14 & 1 \end{array}\right] \text{ respectively.}$$

Thus, the dual problem is: Maximize $P = 16y_1 + 12y_2 + 14y_3$

Subject to: $2y_1 + y_2 + y_3 \le 10$

$y_1 + y_2 + 2y_3 \le 30$

$y_1, y_2, y_3 \ge 0$

We introduce slack variables x_1 and x_2 to obtain the initial system:

$$\begin{aligned} 2y_1 + y_2 + y_3 + x_1 \quad &= 10 \\ y_1 + y_2 + 2y_3 \quad + x_2 &= 30 \\ -16y_1 - 12y_2 - 14y_3 \quad + P &= 0 \end{aligned}$$

The simplex tableau for this problem is:

$$\begin{matrix} & y_1 & y_2 & y_3 & x_1 & x_2 & P & \\ x_1 & \textcircled{2} & 1 & 1 & 1 & 0 & 0 & 10 \\ x_2 & 1 & 1 & 2 & 0 & 1 & 0 & 30 \\ P & -16 & -12 & -14 & 0 & 0 & 1 & 0 \end{matrix} \quad \begin{matrix} \frac{10}{2} = 5 \\ \frac{30}{1} = 30 \end{matrix} \sim \left[\begin{array}{cccccc|c} \textcircled{1} & \frac{1}{2} & \frac{1}{2} & \frac{1}{2} & 0 & 0 & 5 \\ 1 & 1 & 2 & 0 & 1 & 0 & 30 \\ \hline -16 & -12 & -14 & 0 & 0 & 1 & 0 \end{array}\right]$$

$\frac{1}{2}R_1 \to R_1$ $\qquad$ $(-1)R_1 + R_2 \to R_2$ and $16R_1 + R_3 \to R_3$

$$\sim \left[\begin{array}{cccccc|c} 1 & \frac{1}{2} & \textcircled{\frac{1}{2}} & \frac{1}{2} & 0 & 0 & 5 \\ 0 & \frac{1}{2} & \frac{3}{2} & -\frac{1}{2} & 1 & 0 & 25 \\ \hline 0 & -4 & -6 & 8 & 0 & 1 & 80 \end{array}\right] \begin{matrix} \frac{5}{1/2} = 10 \\ \frac{25}{3/2} = 16.66 \end{matrix} \sim \left[\begin{array}{cccccc|c} 2 & 1 & \textcircled{1} & 1 & 0 & 0 & 10 \\ 0 & \frac{1}{2} & \frac{3}{2} & -\frac{1}{2} & 1 & 0 & 25 \\ \hline 0 & -4 & -6 & 8 & 0 & 1 & 80 \end{array}\right]$$

$2R_1 \to R_1$ $\qquad$ $\left(-\frac{3}{2}\right)R_1 + R_2 \to R_2$ and $6R_1 + R_3 \to R_3$

$$\begin{matrix} & y_1 & y_2 & y_3 & x_1 & x_2 & P & \\ y_3 & 2 & 1 & 1 & 1 & 0 & 0 & 10 \\ \sim x_2 & -3 & -1 & 0 & -2 & 1 & 0 & 10 \\ P & 12 & 2 & 0 & 14 & 0 & 1 & 140 \end{matrix}$$

Optimal solution: min $C = 140$ at $x_1 = 14$, $x_2 = 0$.

27. The matrices corresponding to the given problem and the dual problem are:

$$A = \left[\begin{array}{cc|c} 1 & 0 & 4 \\ 1 & 1 & 8 \\ 1 & 2 & 10 \\ \hline 5 & 7 & 1 \end{array}\right] \text{ and } A^T = \left[\begin{array}{ccc|c} 1 & 1 & 1 & 5 \\ 0 & 1 & 2 & 7 \\ \hline 4 & 8 & 10 & 1 \end{array}\right] \text{ respectively.}$$

Thus, the dual problem is: Maximize $P = 4y_1 + 8y_2 + 10y_3$

Subject to: $y_1 + y_2 + y_3 \le 5$

$y_2 + 2y_3 \le 7$

$y_1, y_2, y_3 \ge 0$

We introduce slack variables x_1 and x_2 to obtain the initial system:

$$\begin{aligned} y_1 + y_2 + y_3 + x_1 &= 5 \\ y_2 + 2y_3 + x_2 &= 7 \\ -4y_1 - 8y_2 - 10y_3 + P &= 0 \end{aligned}$$

The simplex tableau for this problem is:

$$\begin{matrix} & y_1 & y_2 & y_3 & x_1 & x_2 & P & \\ x_1 & 1 & 1 & 1 & 1 & 0 & 0 & 5 \\ x_2 & 0 & 1 & \textcircled{2} & 0 & 1 & 0 & 7 \\ P & -4 & -8 & -10 & 0 & 0 & 1 & 0 \end{matrix} \quad \frac{5}{1} = 5, \quad \frac{7}{2} = 3.5$$

$\frac{1}{2}R_2 \to R_2$

$$\sim \left[\begin{array}{cccccc|c} 1 & 1 & 1 & 1 & 0 & 0 & 5 \\ 0 & \frac{1}{2} & \textcircled{1} & 0 & \frac{1}{2} & 0 & \frac{7}{2} \\ \hline -4 & -8 & -10 & 0 & 0 & 1 & 0 \end{array}\right]$$

$(-1)R_2 + R_1 \to R_1$ and $10R_2 + R_3 \to R_3$

$$\sim \left[\begin{array}{cccccc|c} \textcircled{1} & \frac{1}{2} & 0 & 1 & -\frac{1}{2} & 0 & \frac{3}{2} \\ 0 & \frac{1}{2} & 1 & 0 & \frac{1}{2} & 0 & \frac{7}{2} \\ \hline -4 & -3 & 0 & 0 & 5 & 1 & 35 \end{array}\right]$$

$4R_1 + R_3 \to R_3$

$$\sim \left[\begin{array}{cccccc|c} 1 & \textcircled{\tfrac{1}{2}} & 0 & 1 & -\frac{1}{2} & 0 & \frac{3}{2} \\ 0 & \frac{1}{2} & 1 & 0 & \frac{1}{2} & 0 & \frac{7}{2} \\ \hline 0 & -1 & 0 & 4 & 3 & 1 & 41 \end{array}\right] \quad \frac{3/2}{1/2} = 3, \quad \frac{7/2}{1/2} = 7$$

$2R_1 \to R_1$

$$\sim \left[\begin{array}{cccccc|c} 2 & \textcircled{1} & 0 & 2 & -1 & 0 & 3 \\ 0 & \frac{1}{2} & 1 & 0 & \frac{1}{2} & 0 & \frac{7}{2} \\ \hline 0 & -1 & 0 & 4 & 3 & 1 & 41 \end{array}\right]$$

$\left(-\frac{1}{2}\right)R_1 + R_2 \to R_2$ and $R_3 + R_1 \to R_3$

$$\sim \begin{matrix} & y_1 & y_2 & y_3 & x_1 & x_2 & P & \\ y_2 & 2 & 1 & 0 & 2 & -1 & 0 & 3 \\ y_3 & -1 & 0 & 1 & -1 & 1 & 0 & 2 \\ P & 2 & 0 & 0 & 6 & 2 & 1 & 44 \end{matrix}$$

Optimal solution: min $C = 44$ at $x_1 = 6$, $x_2 = 2$.

29. The matrices corresponding to the given problem and the dual problem are:

$$A = \left[\begin{array}{ccc|c} 1 & 1 & 2 & 7 \\ 2 & 1 & 1 & 4 \\ \hline 10 & 7 & 12 & 1 \end{array}\right] \quad \text{and} \quad A^T = \left[\begin{array}{cc|c} 1 & 2 & 10 \\ 1 & 1 & 7 \\ 2 & 1 & 12 \\ \hline 7 & 4 & 1 \end{array}\right] \text{ respectively.}$$

Thus, the dual problem is: Maximize $P = 7y_1 + 4y_2$

Subject to:
$$\begin{aligned} y_1 + 2y_2 &\le 10 \\ y_1 + y_2 &\le 7 \\ 2y_1 + y_2 &\le 12 \\ y_1, y_2 &\ge 0 \end{aligned}$$

We introduce slack variables x_1, x_2, and x_3 to obtain the initial system:

$$\begin{aligned} y_1 + 2y_2 + x_1 &= 10 \\ y_1 + y_2 + x_2 &= 7 \\ 2y_1 + y_2 + x_3 &= 12 \\ -7y_1 - 4y_2 + P &= 0 \end{aligned}$$

The simplex tableau for this problem is:

$$\begin{array}{c} \\ x_1 \\ x_2 \\ x_3 \\ P \end{array} \begin{array}{c} \begin{array}{cccccc} y_1 & y_2 & x_1 & x_2 & x_3 & P \end{array} \\ \left[\begin{array}{cccccc|c} 1 & 2 & 1 & 0 & 0 & 0 & 10 \\ 1 & 1 & 0 & 1 & 0 & 0 & 7 \\ \textcircled{2} & 1 & 0 & 0 & 1 & 0 & 12 \\ \hline -7 & -4 & 0 & 0 & 0 & 1 & 0 \end{array}\right] \end{array} \begin{array}{l} \frac{10}{1} = 10 \\ \frac{7}{1} = 7 \\ \frac{12}{2} = 6 \end{array} \sim \left[\begin{array}{cccccc|c} 1 & 2 & 1 & 0 & 0 & 0 & 10 \\ 1 & 1 & 0 & 1 & 0 & 0 & 7 \\ \textcircled{1} & \frac{1}{2} & 0 & 0 & \frac{1}{2} & 0 & 6 \\ \hline -7 & -4 & 0 & 0 & 0 & 1 & 0 \end{array}\right]$$

$\frac{1}{2}R_3 \to R_3$

$(-1)R_3 + R_1 \to R_1$, $(-1)R_3 + R_2 \to R_2$, and $7R_3 + R_4 \to R_4$

$$\sim \left[\begin{array}{cccccc|c} 0 & \frac{3}{2} & 1 & 0 & -\frac{1}{2} & 0 & 4 \\ 0 & \textcircled{\tfrac{1}{2}} & 0 & 1 & -\frac{1}{2} & 0 & 1 \\ 1 & \frac{1}{2} & 0 & 0 & \frac{1}{2} & 0 & 6 \\ \hline 0 & -\frac{1}{2} & 0 & 0 & \frac{7}{2} & 1 & 42 \end{array}\right] \begin{array}{l} \frac{4}{3/2} = \frac{8}{3} \\ \frac{1}{1/2} = 2 \\ \frac{6}{1/2} = 12 \end{array} \sim \left[\begin{array}{cccccc|c} 0 & \frac{3}{2} & 1 & 0 & -\frac{1}{2} & 0 & 4 \\ 0 & \textcircled{1} & 0 & 2 & -1 & 0 & 2 \\ 1 & \frac{1}{2} & 0 & 0 & \frac{1}{2} & 0 & 6 \\ \hline 0 & -\frac{1}{2} & 0 & 0 & \frac{7}{2} & 1 & 42 \end{array}\right]$$

$2R_2 \to R_2$

$\left(-\frac{3}{2}\right)R_2 + R_1 \to R_1$, $\left(-\frac{1}{2}\right)R_2 + R_3 \to R_3$, and $\frac{1}{2}R_2 + R_4 \to R_4$

$$\sim \begin{array}{c} \\ x_1 \\ y_2 \\ y_1 \\ P \end{array} \begin{array}{c} \begin{array}{cccccc} y_1 & y_2 & x_1 & x_2 & x_3 & P \end{array} \\ \left[\begin{array}{cccccc|c} 0 & 0 & 1 & -3 & 1 & 0 & 1 \\ 0 & 1 & 0 & 2 & -1 & 0 & 2 \\ 1 & 0 & 0 & -1 & 1 & 0 & 5 \\ \hline 0 & 0 & 0 & 1 & 3 & 1 & 43 \end{array}\right] \end{array}$$

Optimal solution: min $C = 43$ at $x_1 = 0$, $x_2 = 1$, $x_3 = 3$.

31. The matrices corresponding to the given problem and the dual problem are:

$$A = \left[\begin{array}{ccc|c} 1 & -4 & 1 & 6 \\ -1 & 1 & -2 & 4 \\ \hline 5 & 2 & 2 & 1 \end{array}\right] \text{ and } A^T = \left[\begin{array}{cc|c} 1 & -1 & 5 \\ -4 & 1 & 2 \\ 1 & -2 & 2 \\ \hline 6 & 4 & 1 \end{array}\right] \text{ respectively.}$$

Thus, the dual problem is: Maximize $P = 6y_1 + 4y_2$

Subject to:
$$\begin{aligned} y_1 - y_2 &\le 5 \\ -4y_1 + y_2 &\le 2 \\ y_1 - 2y_2 &\le 2 \\ y_1, y_2 &\ge 0 \end{aligned}$$

We introduce slack variables x_1, x_2, and x_3 to obtain the initial system:

$$\begin{aligned} y_1 - y_2 + x_1 \qquad\qquad &= 5 \\ -4y_1 + y_2 \qquad + x_2 \qquad &= 2 \\ y_1 - 2y_2 \qquad\qquad + x_3 \quad &= 2 \\ -6y_1 - 4y_2 \qquad\qquad\qquad + P &= 0 \end{aligned}$$

The simplex tableau for this problem is:

$$\begin{array}{c} \\ x_1 \\ x_2 \\ x_3 \\ P \end{array}\begin{array}{c} \begin{array}{cccccc} y_1 & y_2 & x_1 & x_2 & x_3 & P \end{array} \\ \left[\begin{array}{cccccc|c} 1 & -1 & 1 & 0 & 0 & 0 & 5 \\ -4 & 1 & 0 & 1 & 0 & 0 & 2 \\ \textcircled{1} & -2 & 0 & 0 & 1 & 0 & 2 \\ \hline -6 & -4 & 0 & 0 & 0 & 1 & 0 \end{array}\right] \end{array} \begin{array}{l} \frac{5}{1} = 5 \\ \\ \frac{2}{1} = 2 \\ \end{array} \sim \left[\begin{array}{cccccc|c} 0 & \textcircled{1} & 1 & 0 & -1 & 0 & 3 \\ 0 & -7 & 0 & 1 & 4 & 0 & 10 \\ 1 & -2 & 0 & 0 & 1 & 0 & 2 \\ \hline 0 & -16 & 0 & 0 & 6 & 1 & 12 \end{array}\right]$$

$(-1)R_3 + R_1 \to R_1$, $4R_3 + R_2 \to R_2$, and $6R_3 + R_4 \to R_4$

$7R_1 + R_2 \to R_2$, $2R_1 + R_3 \to R_3$, and $16R_1 + R_4 \to R_4$

$$\sim \begin{array}{c} \\ y_2 \\ x_2 \\ y_1 \\ P \end{array}\begin{array}{c} \begin{array}{cccccc} y_1 & y_2 & x_1 & x_2 & x_3 & P \end{array} \\ \left[\begin{array}{cccccc|c} 0 & 1 & 1 & 0 & -1 & 0 & 3 \\ 0 & 0 & 7 & 1 & -3 & 0 & 31 \\ 1 & 0 & 2 & 0 & -1 & 0 & 8 \\ \hline 0 & 0 & 16 & 0 & -10 & 1 & 60 \end{array}\right] \end{array}$$

Since all the entries above the dashed line in the pivot column, the x_3 column, are negative, the problem does not have an optimal solution.

33. The dual problem has 2 variables and 4 problem constraints.

35. The original problem must have two problem constraints, and any number of variables.

37. No. The dual problem will not be a standard maximization problem (one of the elements in the last column will be negative.)

39. Yes. Multiply both sides of the inequality by -1.

41. The matrices corresponding to the given problem and the dual problem are:

$$A = \left[\begin{array}{ccc|c} 3 & 2 & 2 & 16 \\ 4 & 3 & 1 & 14 \\ 5 & 3 & 1 & 12 \\ \hline 16 & 8 & 4 & 1 \end{array}\right] \text{ and } A^T = \left[\begin{array}{ccc|c} 3 & 4 & 5 & 16 \\ 2 & 3 & 3 & 8 \\ 2 & 1 & 1 & 4 \\ \hline 16 & 14 & 12 & 1 \end{array}\right] \text{ respectively.}$$

Thus, the dual problem is: Maximize $P = 16y_1 + 14y_2 + 12y_3$

Subject to:
$$\begin{aligned} 3y_1 + 4y_2 + 5y_3 &\le 16 \\ 2y_1 + 3y_2 + 3y_3 &\le 8 \\ 2y_1 + y_2 + y_3 &\le 4 \\ y_1, y_2, y_3 &\ge 0 \end{aligned}$$

We introduce slack variables x_1, x_2, and x_3 to obtain the initial system:

$$\begin{aligned} 3y_1 + 4y_2 + 5y_3 + x_1 \qquad\qquad\qquad &= 16 \\ 2y_1 + 3y_2 + 3y_3 \qquad + x_2 \qquad\qquad &= 8 \\ 2y_1 + y_2 + y_3 \qquad\qquad + x_3 \qquad &= 4 \\ -16y_1 - 14y_2 - 12y_3 \qquad\qquad\qquad + P &= 0 \end{aligned}$$

The simplex tableau for this problem is:

$$\begin{array}{c} \\ x_1 \\ x_2 \\ x_3 \\ P \end{array} \begin{array}{c} \begin{array}{ccccccc} y_1 & y_2 & y_3 & x_1 & x_2 & x_3 & P \end{array} \\ \left[\begin{array}{ccccccc|c} 3 & 4 & 5 & 1 & 0 & 0 & 0 & 16 \\ 2 & 3 & 3 & 0 & 1 & 0 & 0 & 8 \\ \textcircled{2} & 1 & 1 & 0 & 0 & 1 & 0 & 4 \\ \hdashline -16 & -14 & -12 & 0 & 0 & 0 & 1 & 0 \end{array}\right] \end{array} \begin{array}{l} \frac{16}{3} = 5.33 \\ \frac{8}{2} = 4 \\ \frac{4}{2} = 2 \end{array}$$

$\frac{1}{2}R_3 \to R_3$

$$\sim \left[\begin{array}{ccccccc|c} 3 & 4 & 5 & 1 & 0 & 0 & 0 & 16 \\ 2 & 3 & 3 & 0 & 1 & 0 & 0 & 8 \\ \textcircled{1} & \frac{1}{2} & \frac{1}{2} & 0 & 0 & \frac{1}{2} & 0 & 2 \\ \hdashline -16 & -14 & -12 & 0 & 0 & 0 & 1 & 0 \end{array}\right]$$

$(-3)R_3 + R_1 \to R_1$, $(-2)R_3 + R_2 \to R_2$, and $16R_3 + R_4 \to R_4$

$$\sim \left[\begin{array}{ccccccc|c} 0 & \frac{5}{2} & \frac{7}{2} & 1 & 0 & -\frac{3}{2} & 0 & 10 \\ 0 & \textcircled{2} & 2 & 0 & 1 & -1 & 0 & 4 \\ 1 & \frac{1}{2} & \frac{1}{2} & 0 & 0 & \frac{1}{2} & 0 & 2 \\ \hdashline 0 & -6 & -4 & 0 & 0 & 8 & 1 & 32 \end{array}\right] \begin{array}{l} \frac{10}{5/2} = 4 \\ \frac{4}{2} = 2 \\ \frac{2}{1/2} = 4 \end{array}$$

$\frac{1}{2}R_2 \to R_2$

$$\sim \left[\begin{array}{ccccccc|c} 0 & \frac{5}{2} & \frac{7}{2} & 1 & 0 & -\frac{3}{2} & 0 & 10 \\ 0 & \textcircled{1} & 1 & 0 & \frac{1}{2} & -\frac{1}{2} & 0 & 2 \\ 1 & \frac{1}{2} & \frac{1}{2} & 0 & 0 & \frac{1}{2} & 0 & 2 \\ \hdashline 0 & -6 & -4 & 0 & 0 & 8 & 1 & 32 \end{array}\right]$$

$\left(-\frac{5}{2}\right)R_2 + R_1 \to R_1$, $\left(-\frac{1}{2}\right)R_2 + R_3 \to R_3$, and $6R_2 + R_4 \to R_4$

$$\sim \begin{array}{c} \\ y_3 \\ y_2 \\ y_1 \\ P \end{array} \begin{array}{c} \begin{array}{ccccccc} y_1 & y_2 & y_3 & x_1 & x_2 & x_3 & P \end{array} \\ \left[\begin{array}{ccccccc|c} 0 & 0 & 1 & 1 & -\frac{5}{4} & -\frac{1}{4} & 0 & 5 \\ 0 & 1 & 1 & 0 & \frac{1}{2} & -\frac{1}{2} & 0 & 2 \\ 1 & 0 & 0 & 0 & -\frac{1}{4} & \frac{3}{4} & 0 & 1 \\ \hdashline 0 & 0 & 2 & 0 & 3 & 5 & 1 & 44 \end{array}\right] \end{array}$$

Optimal solution: min $C = 44$ at $x_1 = 0$, $x_2 = 3$, $x_3 = 5$.

43. The first and second inequalities must be rewritten before forming the dual.

Minimize $C = 5x_1 + 4x_2 + 5x_3 + 6x_4$

Subject to:

$$\begin{aligned} -x_1 - x_2 &\geq -12 \\ -x_3 - x_4 &\geq -25 \\ x_1 + x_3 &\geq 20 \\ x_2 + x_4 &\geq 15 \\ x_1, x_2, x_3, x_4 &\geq 0 \end{aligned}$$

The matrices corresponding to the given problem and the dual problem are:

$$A = \left[\begin{array}{cccc|c} -1 & -1 & 0 & 0 & -12 \\ 0 & 0 & -1 & -1 & -25 \\ 1 & 0 & 1 & 0 & 20 \\ 0 & 1 & 0 & 1 & 15 \\ \hline 5 & 4 & 5 & 6 & 1 \end{array}\right] \text{ and } A^T = \left[\begin{array}{cccc|c} -1 & 0 & 1 & 0 & 5 \\ -1 & 0 & 0 & 1 & 4 \\ 0 & -1 & 1 & 0 & 5 \\ 0 & -1 & 0 & 1 & 6 \\ \hline -12 & -25 & 20 & 15 & 1 \end{array}\right]$$

The dual problem is: Maximize $P = -12y_1 - 25y_2 + 20y_3 + 15y_4$

Subject to:

$$\begin{aligned} -y_1 \quad + y_3 \quad &\le 5 \\ -y_1 \quad\quad + y_4 &\le 4 \\ -y_2 + y_3 \quad &\le 5 \\ -y_2 \quad + y_4 &\le 6 \\ y_1, y_2, y_3, y_4 &\ge 0 \end{aligned}$$

We introduce the slack variables x_1, x_2, x_3, and x_4 to obtain the initial system:

$$\begin{aligned} -y_1 + y_3 + x_1 &= 5 \\ -y_1 + y_4 + x_2 &= 4 \\ -y_2 + y_3 + x_3 &= 5 \\ -y_2 + y_4 + x_4 &= 6 \\ +12y_1 + 25y_2 - 20y_3 - 15y_4 + P &= 0 \end{aligned}$$

The simplex tableau for this problem is:

$$\begin{array}{c} \\ x_1 \\ x_2 \\ x_3 \\ x_4 \\ P \end{array} \begin{array}{c} \begin{array}{ccccccccc} y_1 & y_2 & y_3 & y_4 & x_1 & x_2 & x_3 & x_4 & P \end{array} \\ \left[\begin{array}{ccccccccc|c} -1 & 0 & 1 & 0 & 1 & 0 & 0 & 0 & 0 & 5 \\ -1 & 0 & 0 & 1 & 0 & 1 & 0 & 0 & 0 & 4 \\ 0 & -1 & \textcircled{1} & 0 & 0 & 0 & 1 & 0 & 0 & 5 \\ 0 & -1 & 0 & 1 & 0 & 0 & 0 & 1 & 0 & 6 \\ \hline 12 & 25 & -20 & -15 & 0 & 0 & 0 & 0 & 1 & 0 \end{array}\right] \end{array} \begin{array}{l} \frac{5}{1} = 5 \\ \\ \frac{5}{1} = 5 \\ \\ \\ \end{array}$$

$(-1)R_3 + R_1 \to R_1$ and $20R_3 + R_5 \to R_5$

$$\sim \left[\begin{array}{ccccccccc|c} -1 & 1 & 0 & 0 & 1 & 0 & -1 & 0 & 0 & 0 \\ -1 & 0 & 0 & \textcircled{1} & 0 & 1 & 0 & 0 & 0 & 4 \\ 0 & -1 & 1 & 0 & 0 & 0 & 1 & 0 & 0 & 5 \\ 0 & -1 & 0 & 1 & 0 & 0 & 0 & 1 & 0 & 6 \\ \hline 12 & 5 & 0 & -15 & 0 & 0 & 20 & 0 & 1 & 100 \end{array}\right] \begin{array}{l} \\ \frac{4}{1} = 4 \\ \\ \frac{6}{1} = 6 \\ \\ \end{array}$$

$(-1)R_2 + R_4 \to R_4$ and $15R_2 + R_5 \to R_5$

$$\sim \left[\begin{array}{ccccccccc|c} -1 & 1 & 0 & 0 & 1 & 0 & -1 & 0 & 0 & 0 \\ -1 & 0 & 0 & 1 & 0 & 1 & 0 & 0 & 0 & 4 \\ 0 & -1 & 1 & 0 & 0 & 0 & 1 & 0 & 0 & 5 \\ \textcircled{1} & -1 & 0 & 0 & 0 & -1 & 0 & 1 & 0 & 2 \\ \hline -3 & 5 & 0 & 0 & 0 & 15 & 20 & 0 & 1 & 160 \end{array}\right]$$

$3R_4 + R_5 \to R_5$, $R_4 + R_1 \to R_1$, and $R_4 + R_2 \to R_2$

$$\begin{array}{c} \\ x_1 \\ y_4 \\ \sim y_3 \\ y_1 \\ P \end{array} \begin{array}{c} \begin{array}{ccccccccc} y_1 & y_2 & y_3 & y_4 & x_1 & x_2 & x_3 & x_4 & P \end{array} \\ \left[\begin{array}{ccccccccc|c} 0 & 0 & 0 & 0 & 1 & -1 & -1 & 1 & 0 & 2 \\ 0 & -1 & 0 & 1 & 0 & 0 & 0 & 1 & 0 & 6 \\ 0 & -1 & 1 & 0 & 0 & 0 & 1 & 0 & 0 & 5 \\ 1 & -1 & 0 & 0 & 0 & -1 & 0 & 1 & 0 & 2 \\ \hline 0 & 2 & 0 & 0 & 0 & 12 & 20 & 3 & 1 & 166 \end{array}\right] \end{array}$$

Optimal solution: min $C = 166$ at $x_1 = 0$, $x_2 = 12$, $x_3 = 20$, $x_4 = 3$.

45. (A) Let x_1 = the number of hours the Cedarburg plant is operated,
x_2 = the number of hours the Grafton plant is operated,
and x_3 = the number of hours the West Bend plant is operated.

The mathematical model for this problem is:

Minimize $C = 70x_1 + 75x_2 + 90x_3$

Subject to:
$$\begin{aligned} 20x_1 + 10x_2 + 20x_3 &\geq 300 \\ 10x_1 + 20x_2 + 20x_3 &\geq 200 \\ x_1, x_2, x_3 &\geq 0 \end{aligned}$$

Divide each of the problem constraint inequalities by 10 to simplify the calculations. The matrices corresponding to the given problem and the dual problem are:

$$A = \left[\begin{array}{ccc|c} 2 & 1 & 2 & 30 \\ 1 & 2 & 2 & 20 \\ \hline 70 & 75 & 90 & 1 \end{array}\right] \text{ and } A^T = \left[\begin{array}{cc|c} 2 & 1 & 70 \\ 1 & 2 & 75 \\ 2 & 2 & 90 \\ \hline 30 & 20 & 1 \end{array}\right] \text{ respectively.}$$

Thus, the dual problem is:

Maximize $P = 30y_1 + 20y_2$

Subject to:
$$\begin{aligned} 2y_1 + y_2 &\leq 70 \\ y_1 + 2y_2 &\leq 75 \\ 2y_1 + 2y_2 &\leq 90 \\ y_1, y_2 &\geq 0 \end{aligned}$$

We introduce slack variables x_1, x_2, and x_3 to obtain the initial system:

$$\begin{aligned} 2y_1 + y_2 + x_1 \qquad\qquad &= 70 \\ y_1 + 2y_2 \quad + x_2 \qquad &= 75 \\ 2y_1 + 2y_2 \qquad + x_3 \quad &= 90 \\ -30y_1 - 20y_2 \qquad\qquad + P &= 0 \end{aligned}$$

The simplex tableau for this problem is:

$$\begin{array}{c} \\ x_1 \\ x_2 \\ x_3 \\ P \end{array}\begin{array}{c} \begin{array}{cccccc} y_1 & y_2 & x_1 & x_2 & x_3 & P \end{array} \\ \left[\begin{array}{cccccc|c} \textcircled{2} & 1 & 1 & 0 & 0 & 0 & 70 \\ 1 & 2 & 0 & 1 & 0 & 0 & 75 \\ 2 & 2 & 0 & 0 & 1 & 0 & 90 \\ \hline -30 & -20 & 0 & 0 & 0 & 1 & 0 \end{array}\right] \end{array} \begin{array}{l} \frac{70}{2} = 35 \\ \frac{75}{1} = 75 \\ \frac{90}{2} = 45 \end{array} \sim \left[\begin{array}{cccccc|c} \textcircled{1} & \frac{1}{2} & \frac{1}{2} & 0 & 0 & 0 & 35 \\ 1 & 2 & 0 & 1 & 0 & 0 & 75 \\ 2 & 2 & 0 & 0 & 1 & 0 & 90 \\ \hline -30 & -20 & 0 & 0 & 0 & 1 & 0 \end{array}\right]$$

$\frac{1}{2}R_1 \to R_1$

$(-1)R_1 + R_2 \to R_2$, $(-2)R_1 + R_3 \to R_3$, and $30R_1 + R_4 \to R_4$

$$\sim \left[\begin{array}{cccccc|c} 1 & \frac{1}{2} & \frac{1}{2} & 0 & 0 & 0 & 35 \\ 0 & \frac{3}{2} & -\frac{1}{2} & 1 & 0 & 0 & 40 \\ 0 & \textcircled{1} & -1 & 0 & 1 & 0 & 20 \\ \hline 0 & -5 & 15 & 0 & 0 & 1 & 1050 \end{array}\right] \begin{array}{l} \frac{35}{1/2} = 70 \\ \frac{40}{3/2} = \frac{80}{3} \approx 26.67 \\ \frac{20}{1} = 20 \end{array}$$

$\left(-\frac{1}{2}\right)R_3 + R_1 \to R_1$, $\left(-\frac{3}{2}\right)R_3 + R_2 \to R_2$, and $5R_3 + R_4 \to R_4$

$$\sim \begin{array}{c} \\ y_1 \\ x_2 \\ y_2 \\ P \end{array}\begin{array}{c} \begin{array}{cccccc} y_1 & y_2 & x_1 & x_2 & x_3 & P \end{array} \\ \left[\begin{array}{cccccc|c} 1 & 0 & 1 & 0 & -\frac{1}{2} & 0 & 25 \\ 0 & 0 & 1 & 1 & -\frac{3}{2} & 0 & 10 \\ 0 & 1 & -1 & 0 & 1 & 0 & 20 \\ \hline 0 & 0 & 10 & 0 & 5 & 1 & 1150 \end{array}\right] \end{array}$$

The minimal production cost is \$1150 when the Cedarburg plant is operated 10 hours per day, the West Bend plant is operated 5 hours per day, and the Grafton plant is not used.

(B) If the demand for deluxe ice cream increases to 300 gallons per day and all other data remains the same, then the matrices for this problem and the dual problem are:

$$A = \left[\begin{array}{ccc|c} 2 & 1 & 2 & 30 \\ 1 & 2 & 2 & 30 \\ \hline 70 & 75 & 90 & 1 \end{array}\right] \text{ and } A^T = \left[\begin{array}{cc|c} 2 & 1 & 70 \\ 1 & 2 & 75 \\ 2 & 2 & 90 \\ \hline 30 & 30 & 1 \end{array}\right] \text{ respectively.}$$

Thus, the dual problem is:

Maximize $P = 30y_1 + 30y_2$

Subject to:
$$\begin{aligned} 2y_1 + y_2 &\le 70 \\ y_1 + 2y_2 &\le 75 \\ 2y_1 + 2y_2 &\le 90 \\ y_1, y_2 &\ge 0 \end{aligned}$$

We introduce slack variables x_1, x_2, and x_3 to obtain the initial system:

$$
\begin{aligned}
2y_1 + y_2 + x_1 \qquad\qquad &= 70 \\
y_1 + 2y_2 \qquad + x_2 \qquad &= 75 \\
2y_1 + 2y_2 \qquad\qquad + x_3 &= 90 \\
-30y_1 - 30y_2 \qquad\qquad + P &= 0
\end{aligned}
$$

The simplex tableau for this problem is:

$$
\begin{array}{c} \\ x_1 \\ x_2 \\ x_3 \\ P \end{array}
\begin{array}{c}
\begin{array}{cccccc} y_1 & y_2 & x_1 & x_2 & x_3 & P \end{array} \\
\left[\begin{array}{cccccc|c}
\textcircled{2} & 1 & 1 & 0 & 0 & 0 & 70 \\
1 & 2 & 0 & 1 & 0 & 0 & 75 \\
2 & 2 & 0 & 0 & 1 & 0 & 90 \\
\hdashline
-30 & -30 & 0 & 0 & 0 & 1 & 0
\end{array}\right]
\end{array}
\begin{array}{l} \\ \frac{70}{2} = 35 \\ \frac{75}{1} = 75 \\ \frac{90}{2} = 45 \\ \\ \end{array}
$$

$\frac{1}{2}R_1 \to R_1$

Note: Either column 1 or column 2 can be used as the pivot column. We chose column 1.

$$
\sim \left[\begin{array}{cccccc|c}
\textcircled{1} & \frac{1}{2} & \frac{1}{2} & 0 & 0 & 0 & 35 \\
1 & 2 & 0 & 1 & 0 & 0 & 75 \\
2 & 2 & 0 & 0 & 1 & 0 & 90 \\
\hdashline
-30 & -30 & 0 & 0 & 0 & 1 & 0
\end{array}\right]
\sim \left[\begin{array}{cccccc|c}
1 & \frac{1}{2} & \frac{1}{2} & 0 & 0 & 0 & 35 \\
0 & \frac{3}{2} & -\frac{1}{2} & 1 & 0 & 0 & 40 \\
0 & \textcircled{1} & -1 & 0 & 1 & 0 & 20 \\
\hdashline
0 & -15 & 15 & 0 & 0 & 1 & 1050
\end{array}\right]
\begin{array}{l} \frac{35}{1/2} = 70 \\ \frac{40}{3/2} = \frac{80}{3} \approx 26.67 \\ \frac{20}{1} = 20 \\ \\ \end{array}
$$

$(-1)R_1 + R_2 \to R_2$, $\qquad \left(-\frac{1}{2}\right)R_3 + R_1 \to R_1$,

$(-2)R_1 + R_3 \to R_3$, $\qquad \left(-\frac{3}{2}\right)R_3 + R_2 \to R_2$,

$30R_1 + R_4 \to R_4$ $\qquad 15R_3 + R_4 \to R_4$

$$
\sim \begin{array}{c} \\ y_1 \\ x_2 \\ y_2 \\ P \end{array}
\begin{array}{c}
\begin{array}{cccccc} y_1 & y_2 & x_1 & x_2 & x_3 & P \end{array} \\
\left[\begin{array}{cccccc|c}
1 & 0 & 1 & 0 & -\frac{1}{2} & 0 & 25 \\
0 & 0 & 1 & 1 & -\frac{3}{2} & 0 & 10 \\
0 & 1 & -1 & 0 & 1 & 0 & 20 \\
\hdashline
0 & 0 & 0 & 0 & 15 & 1 & 1350
\end{array}\right]
\end{array}
$$

The minimal production cost is \$1350 when the West Bend plant is operated 15 hours per day, and the Cedarburg and Grafton plants are not used.

(C) If the demand for deluxe ice cream increases to 400 gallons per day and all other data remains the same, then the matrices for this problem and the dual are:

$$
A = \left[\begin{array}{ccc|c}
2 & 1 & 2 & 30 \\
1 & 2 & 2 & 40 \\
\hline
70 & 75 & 90 & 1
\end{array}\right]
\text{ and } \quad
A^T = \left[\begin{array}{cc|c}
2 & 1 & 70 \\
1 & 2 & 75 \\
2 & 2 & 90 \\
\hline
30 & 40 & 1
\end{array}\right]
\text{ respectively.}
$$

Thus, the dual problem is:

Maximize $P = 30y_1 + 40y_2$

Subject to: $2y_1 + y_2 \le 70$

$y_1 + 2y_2 \le 75$

$2y_1 + 2y_2 \le 90$

$y_1, y_2 \ge 0$

We introduce slack variables x_1, x_2, and x_3 to obtain the initial system:

$$\begin{aligned} 2y_1 + y_2 + x_1 \quad &= 70 \\ y_1 + 2y_2 \quad + x_2 \quad &= 75 \\ 2y_1 + 2y_2 \quad + x_3 \quad &= 90 \\ -30y_1 - 40y_2 \quad + P &= 0 \end{aligned}$$

The simplex tableau for this problem is:

$$\begin{array}{c} \\ x_1 \\ x_2 \\ x_3 \\ P \end{array}\begin{array}{c} \begin{matrix} y_1 & y_2 & x_1 & x_2 & x_3 & P \end{matrix} \\ \left[\begin{array}{cccccc|c} 2 & 1 & 1 & 0 & 0 & 0 & 70 \\ 1 & \textcircled{2} & 0 & 1 & 0 & 0 & 75 \\ 2 & 2 & 0 & 0 & 1 & 0 & 90 \\ \hdashline -30 & -40 & 0 & 0 & 0 & 1 & 0 \end{array}\right] \end{array} \begin{array}{l} \frac{70}{1} = 70 \\ \frac{75}{2} = 37.5 \\ \frac{90}{2} = 45 \end{array}$$

$\frac{1}{2}R_2 \to R_2$

$$\sim \left[\begin{array}{cccccc|c} 2 & 1 & 1 & 0 & 0 & 0 & 70 \\ \frac{1}{2} & \textcircled{1} & 0 & \frac{1}{2} & 0 & 0 & \frac{75}{2} \\ 2 & 2 & 0 & 0 & 1 & 0 & 90 \\ \hdashline -30 & -40 & 0 & 0 & 0 & 1 & 0 \end{array}\right] \sim \left[\begin{array}{cccccc|c} \frac{3}{2} & 0 & 1 & -\frac{1}{2} & 0 & 0 & \frac{65}{2} \\ \frac{1}{2} & 1 & 0 & \frac{1}{2} & 0 & 0 & \frac{75}{2} \\ \textcircled{1} & 0 & 0 & -1 & 1 & 0 & 15 \\ \hdashline -10 & 0 & 0 & 20 & 0 & 1 & 1500 \end{array}\right] \begin{array}{l} \frac{65/2}{3/2} \approx 21.67 \\ \frac{75/2}{1/2} = 75 \\ \frac{15}{1} = 15 \end{array}$$

$(-1)R_2 + R_1 \to R_1,$

$(-2)R_2 + R_3 \to R_3,$

$40R_2 + R_4 \to R_4$

$\left(-\frac{3}{2}\right)R_3 + R_1 \to R_1,$

$\left(-\frac{1}{2}\right)R_3 + R_2 \to R_2,$

$10R_3 + R_4 \to R_4$

$$\sim \begin{array}{c} \\ x_1 \\ y_2 \\ y_1 \\ P \end{array}\begin{array}{c} \begin{matrix} y_1 & y_2 & x_1 & x_2 & x_3 & P \end{matrix} \\ \left[\begin{array}{cccccc|c} 0 & 0 & 1 & 1 & -\frac{3}{2} & 0 & 10 \\ 0 & 1 & 0 & 1 & -\frac{1}{2} & 0 & 30 \\ 1 & 0 & 0 & -1 & 1 & 0 & 15 \\ \hdashline 0 & 0 & 0 & 10 & 10 & 1 & 1650 \end{array}\right] \end{array}$$

The minimal production cost is \$1650 when the Grafton plant and West Bend plant are each operated 10 hours per day, and the Cedarburg plant is not used.

47. Let x_1 = the number of single-sided drives from Associated Electronics,
x_2 = the number of double-sided drives from Associated Electronics,
x_3 = the number of single-sided drives from Digital Drives,
and x_4 = the number of double-sided drives from Digital Drives.

The mathematical model is: Minimize $C = 250x_1 + 350x_2 + 290x_3 + 320x_4$

$$\begin{aligned} \text{Subject to: } x_1 + x_2 &\le 1000 \\ x_3 + x_4 &\le 2000 \\ x_1 + x_3 &\ge 1200 \\ x_2 + x_4 &\ge 1600 \\ x_1, x_2, x_3, x_4 &\ge 0 \end{aligned}$$

We multiply the first two problem constraints by -1 to obtain inequalities of the $\ge$ type. The model becomes:

$$\begin{aligned} &\text{Minimize } C = 250x_1 + 350x_2 + 290x_3 + 320x_4 \\ &\text{Subject to: } -x_1 - x_2 \ge -1000 \\ &\qquad -x_3 - x_4 \ge -2000 \\ &\qquad x_1 + x_3 \ge 1200 \\ &\qquad x_2 + x_4 \ge 1600 \\ &\qquad x_1, x_2, x_3, x_4 \ge 0 \end{aligned}$$

The matrices for this problem and the dual problem are:

$$A = \left[\begin{array}{cccc|c} -1 & -1 & 0 & 0 & -1000 \\ 0 & 0 & -1 & -1 & -2000 \\ 1 & 0 & 1 & 0 & 1200 \\ 0 & 1 & 0 & 1 & 1600 \\ \hline 250 & 350 & 290 & 320 & 1 \end{array}\right] \text{ and } A^T = \left[\begin{array}{cccc|c} -1 & 0 & 1 & 0 & 250 \\ -1 & 0 & 0 & 1 & 350 \\ 0 & -1 & 1 & 0 & 290 \\ 0 & -1 & 0 & 1 & 320 \\ \hline -1000 & -2000 & 1200 & 1600 & 1 \end{array}\right]$$

Thus, the dual problem is:

$$\begin{aligned} &\text{Maximize } P = -1000y_1 - 2000y_2 + 1200y_3 + 1600y_4 \\ &\text{Subject to: } -y_1 + y_3 \le 250 \\ &\qquad -y_1 + y_4 \le 350 \\ &\qquad -y_2 + y_3 \le 290 \\ &\qquad -y_2 + y_4 \le 320 \\ &\qquad y_1, y_2, y_3, y_4 \ge 0 \end{aligned}$$

We introduce slack variables x_1, x_2, x_3, and x_4 to obtain the initial system:

$$\begin{aligned} -y_1 + y_3 + x_1 &= 250 \\ -y_1 + y_4 + x_2 &= 350 \\ -y_2 + y_3 + x_3 &= 290 \\ -y_2 + y_4 + x_4 &= 320 \\ 1000y_1 + 2000y_2 - 1200y_3 - 1600y_4 + P &= 0 \end{aligned}$$

The simplex tableau for this problem is:

$$\begin{array}{c} \\ x_1 \\ x_2 \\ x_3 \\ x_4 \\ P \end{array} \begin{array}{c} \begin{array}{ccccccccc} y_1 & y_2 & y_3 & y_4 & x_1 & x_2 & x_3 & x_4 & P \end{array} \\ \left[\begin{array}{ccccccccc|c} -1 & 0 & 1 & 0 & 1 & 0 & 0 & 0 & 0 & 250 \\ -1 & 0 & 0 & 1 & 0 & 1 & 0 & 0 & 0 & 350 \\ 0 & -1 & 1 & 0 & 0 & 0 & 1 & 0 & 0 & 290 \\ 0 & -1 & 0 & \textcircled{1} & 0 & 0 & 0 & 1 & 0 & 320 \\ \hline 1000 & 2000 & -1200 & -1600 & 0 & 0 & 0 & 0 & 1 & 0 \end{array}\right] \end{array} \begin{array}{l} \\ \frac{350}{1} = 350 \\ \\ \frac{320}{1} = 320 \\ \\ \end{array}$$

$(-1)R_4 + R_2 \to R_2$ and $1600R_4 + R_5 \to R_5$

$$\sim \left[\begin{array}{ccccccccc|c} -1 & 0 & \textcircled{1} & 0 & 1 & 0 & 0 & 0 & 0 & 250 \\ -1 & 1 & 0 & 0 & 0 & 1 & 0 & -1 & 0 & 30 \\ 0 & -1 & 1 & 0 & 0 & 0 & 1 & 0 & 0 & 290 \\ 0 & -1 & 0 & 1 & 0 & 0 & 0 & 1 & 0 & 320 \\ \hdashline 1000 & 400 & -1200 & 0 & 0 & 0 & 0 & 1600 & 1 & 512{,}000 \end{array}\right] \begin{array}{l} \frac{250}{1} = 250 \\ \\ \frac{290}{1} = 290 \\ \\ \\ \end{array}$$

$(-1)R_1 + R_3 \rightarrow R_3$ and $1200R_1 + R_5 \rightarrow R_5$

$$\sim \left[\begin{array}{ccccccccc|c} -1 & 0 & 1 & 0 & 1 & 0 & 0 & 0 & 0 & 250 \\ -1 & 1 & 0 & 0 & 0 & 1 & 0 & -1 & 0 & 30 \\ \textcircled{1} & -1 & 0 & 0 & -1 & 0 & 1 & 0 & 0 & 40 \\ 0 & -1 & 0 & 1 & 0 & 0 & 0 & 1 & 0 & 320 \\ \hdashline -200 & 400 & 0 & 0 & 1200 & 0 & 0 & 1600 & 1 & 812{,}000 \end{array}\right]$$

$R_3 + R_1 \rightarrow R_1$, $R_3 + R_2 \rightarrow R_2$, and $200R_3 + R_5 \rightarrow R_5$

$$\sim \begin{array}{c} \\ y_3 \\ x_2 \\ y_1 \\ y_4 \\ P \end{array} \begin{array}{c} \begin{array}{ccccccccc} y_1 & y_2 & y_3 & y_4 & x_1 & x_2 & x_3 & x_4 & P \end{array} \\ \left[\begin{array}{ccccccccc|c} 0 & -1 & 1 & 0 & 0 & 0 & 1 & 0 & 0 & 290 \\ 0 & 0 & 0 & 0 & -1 & 1 & 1 & -1 & 0 & 70 \\ 1 & -1 & 0 & 0 & -1 & 0 & 1 & 0 & 0 & 40 \\ 0 & -1 & 0 & 1 & 0 & 0 & 0 & 1 & 0 & 320 \\ \hdashline 0 & 200 & 0 & 0 & 1000 & 0 & 200 & 1600 & 1 & 820{,}000 \end{array}\right] \end{array}$$

The minimal purchase cost is $820,000 when 1000 single-sided and no double-sided drives are ordered from Associated Electronics, and 200 single-sided and 1600 double-sided drives are ordered from Digital Drives.

49. Let x_1 = the number of ounces of food L,
x_2 = the number of ounces of food M,
x_3 = the number of ounces of food N.

Mathematical model: Minimize $C = 20x_1 + 24x_2 + 18x_3$

Subject to:
$$\begin{aligned} 20x_1 + 10x_2 + 10x_3 &\geq 300 \\ 10x_1 + 10x_2 + 10x_3 &\geq 200 \\ 10x_1 + 15x_2 + 10x_3 &\geq 240 \\ x_1, x_2, x_3 &\geq 0 \end{aligned}$$

Divide the first two problem constraints by 10 and the third by 5. This will simplify the calculations.

$$A = \left[\begin{array}{ccc|c} 2 & 1 & 1 & 30 \\ 1 & 1 & 1 & 20 \\ 2 & 3 & 2 & 48 \\ \hline 20 & 24 & 18 & 1 \end{array}\right] \quad \text{and} \quad A^T = \left[\begin{array}{ccc|c} 2 & 1 & 2 & 20 \\ 1 & 1 & 3 & 24 \\ 1 & 1 & 2 & 18 \\ \hline 30 & 20 & 48 & 1 \end{array}\right]$$

The dual problem is: Maximize $P = 30y_1 + 20y_2 + 48y_3$

Subject to:
$$\begin{aligned} 2y_1 + y_2 + 2y_3 &\le 20 \\ y_1 + y_2 + 3y_3 &\le 24 \\ y_1 + y_2 + 2y_3 &\le 18 \\ y_1, y_2, y_3 &\ge 0 \end{aligned}$$

We introduce slack variables x_1, x_2, and x_3 to obtain the initial system:

$$\begin{aligned} 2y_1 + y_2 + 2y_3 + x_1 &= 20 \\ y_1 + y_2 + 3y_3 + x_2 &= 24 \\ y_1 + y_2 + 2y_3 + x_3 &= 18 \\ -30y_1 - 20y_2 - 24y_3 + P &= 0 \end{aligned}$$

The simplex tableau for this problem is:

$$\begin{array}{c} \\ x_1 \\ x_2 \\ x_3 \\ P \end{array} \begin{array}{c} \begin{array}{ccccccc|c} y_1 & y_2 & y_3 & x_1 & x_2 & x_3 & P & \end{array} \\ \left[\begin{array}{ccccccc|c} 2 & 1 & 2 & 1 & 0 & 0 & 0 & 20 \\ 1 & 1 & \textcircled{3} & 0 & 1 & 0 & 0 & 24 \\ 1 & 1 & 2 & 0 & 0 & 1 & 0 & 18 \\ \hdashline -30 & -20 & -48 & 0 & 0 & 0 & 1 & 0 \end{array}\right] \end{array} \begin{array}{l} \frac{20}{2} = 10 \\ \frac{24}{3} = 8 \\ \frac{18}{2} = 9 \end{array}$$

$\frac{1}{3}R_2 \to R_2$

$$\sim \left[\begin{array}{ccccccc|c} 2 & 1 & 2 & 1 & 0 & 0 & 0 & 20 \\ \frac{1}{3} & \frac{1}{3} & \textcircled{1} & 0 & \frac{1}{3} & 0 & 0 & 8 \\ 1 & 1 & 2 & 0 & 0 & 1 & 0 & 18 \\ \hdashline -30 & -20 & -48 & 0 & 0 & 0 & 1 & 0 \end{array}\right]$$

$(-2)R_2 + R_1 \to R_1$, $(-2)R_2 + R_3 \to R_3$, and $48R_2 + R_4 \to R_4$

$$\sim \left[\begin{array}{ccccccc|c} \textcircled{\frac{4}{3}} & \frac{1}{3} & 0 & 1 & -\frac{2}{3} & 0 & 0 & 4 \\ \frac{1}{3} & \frac{1}{3} & 1 & 0 & \frac{1}{3} & 0 & 0 & 8 \\ \frac{1}{3} & \frac{1}{3} & 0 & 0 & -\frac{2}{3} & 1 & 0 & 2 \\ \hdashline -14 & -4 & 0 & 0 & 16 & 0 & 1 & 384 \end{array}\right] \begin{array}{l} \frac{4}{4/3} = 3 \\ \frac{8}{1/3} = 24 \\ \frac{2}{1/3} = 6 \end{array}$$

$\frac{3}{4}R_1 \to R_1$

$$\sim\left[\begin{array}{ccccccc|c} \textcircled{1} & \frac{1}{4} & 0 & \frac{3}{4} & -\frac{1}{2} & 0 & 0 & 3 \\ \frac{1}{3} & \frac{1}{3} & 1 & 0 & \frac{1}{3} & 0 & 0 & 8 \\ \frac{1}{3} & \frac{1}{3} & 0 & 0 & -\frac{2}{3} & 1 & 0 & 2 \\ \hdashline -14 & -4 & 0 & 0 & 16 & 0 & 1 & 384 \end{array}\right]$$

$\left(-\frac{1}{3}\right)R_1 + R_2 \to R_2$, $\left(-\frac{1}{3}\right)R_1 + R_3 \to R_3$, and $14R_1 + R_4 \to R_4$

$$\sim\left[\begin{array}{ccccccc|c} 1 & \frac{1}{4} & 0 & \frac{3}{4} & -\frac{1}{2} & 0 & 0 & 3 \\ 0 & \frac{1}{4} & 1 & -\frac{1}{4} & \frac{1}{2} & 0 & 0 & 7 \\ 0 & \textcircled{\frac{1}{4}} & 0 & -\frac{1}{4} & -\frac{1}{2} & 1 & 0 & 1 \\ \hdashline 0 & -\frac{1}{2} & 0 & \frac{21}{2} & 9 & 0 & 1 & 426 \end{array}\right] \quad \begin{array}{l} \frac{3}{1/4} = 12 \\ \frac{7}{1/4} = 28 \\ \frac{1}{1/4} = 4 \end{array}$$

$4R_3 \to R_3$

$$\sim\left[\begin{array}{ccccccc|c} 1 & \frac{1}{4} & 0 & \frac{3}{4} & -\frac{1}{2} & 0 & 0 & 3 \\ 0 & \frac{1}{4} & 1 & -\frac{1}{4} & \frac{1}{2} & 0 & 0 & 7 \\ 0 & \textcircled{1} & 0 & -1 & -2 & 4 & 0 & 4 \\ \hdashline 0 & -\frac{1}{2} & 0 & \frac{21}{2} & 9 & 0 & 1 & 426 \end{array}\right] \sim \begin{array}{c} \\ y_1 \\ y_3 \\ y_2 \\ P \end{array} \left[\begin{array}{ccccccc|c} y_1 & y_2 & y_3 & x_1 & x_2 & x_3 & P & \\ 1 & 0 & 0 & 1 & 0 & -1 & 0 & 2 \\ 0 & 0 & 1 & 0 & 1 & -1 & 0 & 6 \\ 0 & 1 & 0 & -1 & -2 & 4 & 0 & 4 \\ \hdashline 0 & 0 & 0 & 10 & 8 & 2 & 1 & 428 \end{array}\right]$$

$\left(-\frac{1}{4}\right)R_3 + R_1 \to R_1$, $\left(-\frac{1}{4}\right)R_3 + R_2 \to R_2$, and $\frac{1}{2}R_3 + R_4 \to R_4$

The minimal cholesterol intake is 428 units when 10 ounces of food *L*, 8 ounces of food *M*, and 2 ounces of food *N* are used.

51. Let x_1 = the number of students bused from North Division to Central,
x_2 = the number of students bused from North Division to Washington,
x_3 = the number of students bused from South Division to Central,
and x_4 = the number of students bused from South Division to Washington.
The mathematical model for this problem is:

Minimize $C = 5x_1 + 2x_2 + 3x_3 + 4x_4$

Subject to:
$$\begin{aligned} x_1 + x_2 &\ge 300 \\ x_3 + x_4 &\ge 500 \\ x_1 + x_3 &\le 400 \\ x_2 + x_4 &\le 500 \\ x_1, x_2, x_3, x_4 &\ge 0 \end{aligned}$$

We multiply the last two problem constraints by -1 so that all the constraints are of the ≥ type. The model becomes:

Minimize $C = 5x_1 + 2x_2 + 3x_3 + 4x_4$

Subject to:
$$\begin{aligned} x_1 + x_2 &\ge 300 \\ x_3 + x_4 &\ge 500 \\ -x_1 - x_3 &\ge -400 \\ -x_2 - x_4 &\ge -500 \\ x_1, x_2, x_3, x_4 &\ge 0 \end{aligned}$$

The matrices for this problem and the dual problem are:

$$A = \left[\begin{array}{rrrr|r} 1 & 1 & 0 & 0 & 300 \\ 0 & 0 & 1 & 1 & 500 \\ -1 & 0 & -1 & 0 & -400 \\ 0 & -1 & 0 & -1 & -500 \\ \hline 5 & 2 & 3 & 4 & 1 \end{array}\right] \quad \text{and} \quad A^T = \left[\begin{array}{rrrr|r} 1 & 0 & -1 & 0 & 5 \\ 1 & 0 & 0 & -1 & 2 \\ 0 & 1 & -1 & 0 & 3 \\ 0 & 1 & 0 & -1 & 4 \\ \hline 300 & 500 & -400 & -500 & 1 \end{array}\right]$$

The dual problem is: Maximize $P = 300y_1 + 500y_2 - 400y_3 - 500y_4$

$$\begin{array}{lrrrrl} \text{Subject to:} & y_1 & & - y_3 & & \le 5 \\ & y_1 & & & - y_4 & \le 2 \\ & & y_2 & - y_3 & & \le 3 \\ & & y_2 & & - y_4 & \le 4 \\ & & & & y_1, y_2, y_3, y_4 & \ge 0 \end{array}$$

We introduce slack variables x_1, x_2, x_3, and x_4 to obtain the initial system:

$$\begin{array}{rrrrrrrrrl} y_1 & & - y_3 & & + x_1 & & & & & = 5 \\ y_1 & & & - y_4 & & + x_2 & & & & = 2 \\ & y_2 & - y_3 & & & & + x_3 & & & = 3 \\ & y_2 & & - y_4 & & & & + x_4 & & = 4 \\ -300y_1 & - 500y_2 & + 400y_3 & + 500y_4 & & & & & + P & = 0 \end{array}$$

The simplex tableau for this problem is:

$$\begin{array}{c} \\ x_1 \\ x_2 \\ x_3 \\ x_4 \\ P \end{array} \begin{array}{c} \begin{array}{ccccccccc} y_1 & y_2 & y_3 & y_4 & x_1 & x_2 & x_3 & x_4 & P \end{array} \\ \left[\begin{array}{rrrrrrrrr|r} 1 & 0 & -1 & 0 & 1 & 0 & 0 & 0 & 0 & 5 \\ 1 & 0 & 0 & -1 & 0 & 1 & 0 & 0 & 0 & 2 \\ 0 & \textcircled{1} & -1 & 0 & 0 & 0 & 1 & 0 & 0 & 3 \\ 0 & 1 & 0 & -1 & 0 & 0 & 0 & 1 & 0 & 4 \\ \hdashline -300 & -500 & 400 & 500 & 0 & 0 & 0 & 0 & 1 & 0 \end{array}\right] \end{array} \begin{array}{l} \\ \\ \\ \frac{3}{1} = 3 \\ \frac{4}{1} = 4 \\ \\ \end{array}$$

$(-1)R_3 + R_4 \to R_4$ and $500R_4 + R_5 \to R_5$

$$\sim \left[\begin{array}{rrrrrrrrr|r} 1 & 0 & -1 & 0 & 1 & 0 & 0 & 0 & 0 & 5 \\ \textcircled{1} & 0 & 0 & -1 & 0 & 1 & 0 & 0 & 0 & 2 \\ 0 & 1 & -1 & 0 & 0 & 0 & 1 & 0 & 0 & 3 \\ 0 & 0 & 1 & -1 & 0 & 0 & -1 & 1 & 0 & 1 \\ \hdashline -300 & 0 & -100 & 500 & 0 & 0 & 500 & 0 & 1 & 1500 \end{array}\right] \begin{array}{l} \frac{5}{1} = 5 \\ \frac{2}{1} = 2 \\ \\ \\ \\ \end{array}$$

$(-1)R_2 + R_1 \to R_1$ and $300R_2 + R_5 \to R_5$

$$\sim \left[\begin{array}{ccccccccc|c} 0 & 0 & -1 & 1 & 1 & -1 & 0 & 0 & 0 & 3 \\ 1 & 0 & 0 & -1 & 0 & 1 & 0 & 0 & 0 & 2 \\ 0 & 1 & -1 & 0 & 0 & 0 & 1 & 0 & 0 & 3 \\ 0 & 0 & \textcircled{1} & -1 & 0 & 0 & -1 & 1 & 0 & 1 \\ \hline 0 & 0 & -100 & 200 & 0 & 300 & 500 & 0 & 1 & 2100 \end{array}\right]$$

$R_4 + R_1 \rightarrow R_1$, $R_4 + R_3 \rightarrow R_3$, and $100R_4 + R_5 \rightarrow R_5$

$$\begin{array}{c} \\ x_1 \\ y_1 \\ \sim y_2 \\ y_3 \\ P \end{array} \begin{array}{c} \begin{array}{cccccccccc} y_1 & y_2 & y_3 & y_4 & x_1 & x_2 & x_3 & x_4 & P & \end{array} \\ \left[\begin{array}{ccccccccc|c} 0 & 0 & 0 & 0 & 1 & -1 & -1 & 1 & 0 & 4 \\ 1 & 0 & 0 & -1 & 0 & 1 & 0 & 0 & 0 & 2 \\ 0 & 1 & 0 & -1 & 0 & 0 & 0 & 1 & 0 & 4 \\ 0 & 0 & 1 & -1 & 0 & 0 & -1 & 1 & 0 & 1 \\ \hline 0 & 0 & 0 & 100 & 0 & 300 & 400 & 100 & 1 & 2200 \end{array}\right] \end{array}$$

The minimal cost is $2200 when 300 students are bused from North Division to Washington, 400 students are bused from South Division to Central, and 100 students are bused from South Division to Washington. No students are bused from North Division to Central.

EXERCISE 5-6

Things to remember:

Given a linear programming problem with an objective function to be maximized and problem constraints that are a combination of ≥ and ≤ inequalities as well as equations. The solution method is called the BIG M method.

1. THE BIG M METHOD—INTRODUCING SLACK, SURPLUS AND ARTIFICIAL VARIABLES TO FORM THE MODIFIED PROBLEM

   STEP 1. If any problem constraints have negative constants on the right-hand side, multiply both sides by -1 to obtain a constraint with a nonnegative constant. [If the constraint is an inequality, this will reverse the direction of the inequality.]

   STEP 2. Introduce a SLACK VARIABLE in each ≤ constraint.

   STEP 3. Introduce a SURPLUS VARIABLE and an ARTIFICIAL VARIABLE in each ≥ constraint.

   STEP 4. Introduce an artificial variable in each = constraint.

   STEP 5. For each artificial variable a_i, add $-Ma_i$ to the objective function. Use the same constant M for all artificial variables.

2. THE BIG M METHOD—SOLVING THE PROBLEM

STEP 1. Form the preliminary simplex tableau for the modified problem.

STEP 2. Use row operations to eliminate the M's in the bottom row of the preliminary simplex tableau in the columns corresponding to the artificial variables. The resulting tableau is the initial simplex tableau.

STEP 3. Solve the modified problem by applying the simplex method to the initial simplex tableau found in Step 2.

STEP 4. Relate the optimal solution of the modified problem to the original problem.

(a) If the modified problem has no optimal solution, then the original problem has no optimal solution.

(b) If all artificial variables are zero in the solution to the modified problem, then delete the artificial variables to find an optimal solution to the original problem.

(c) If any artificial variables are nonzero in the optimal solution to the modified problem, then the original problem has no optimal solution.

---

1. (A) We introduce a slack variable s_1 to convert the first inequality ($\le$) into an equation, and we use a surplus variable s_2 and an artificial variable a_1 to convert the second inequality ($\ge$) into an equation. The modified problem is: Maximize $P = 5x_1 + 2x_2 - Ma_1$

$$\begin{aligned} \text{Subject to: } x_1 + 2x_2 + s_1 \qquad\qquad &= 12 \\ x_1 + x_2 \qquad - s_2 + a_1 &= 4 \\ x_1, x_2, s_1, s_2, a_1 &\ge 0 \end{aligned}$$

(B) The preliminary simplex tableau for the modified problem is:

$$\begin{array}{c} \begin{matrix} x_1 & x_2 & s_1 & s_2 & a_1 & P \end{matrix} \\ \left[\begin{array}{cccccc|c} 1 & 2 & 1 & 0 & 0 & 0 & 12 \\ 1 & 1 & 0 & -1 & 1 & 0 & 4 \\ \hline -5 & -2 & 0 & 0 & M & 1 & 0 \end{array}\right] \end{array} \sim \left[\begin{array}{cccccc|c} 1 & 2 & 1 & 0 & 0 & 0 & 12 \\ 1 & 1 & 0 & -1 & 1 & 0 & 4 \\ \hline -M-5 & -M-2 & 0 & M & 0 & 1 & -4M \end{array}\right]$$

$(-M)R_2 + R_3 \to R_3$

Thus, the initial simplex tableau is:

$$\begin{array}{c} \begin{matrix} x_1 & x_2 & s_1 & s_2 & a_1 & P \end{matrix} \\ \left[\begin{array}{cccccc|c} 1 & 2 & 1 & 0 & 0 & 0 & 12 \\ 1 & 1 & 0 & -1 & 1 & 0 & 4 \\ \hline -M-5 & -M-2 & 0 & M & 0 & 1 & -4M \end{array}\right] \end{array}$$

(C) We use the simplex method to solve the modified problem.

$$\begin{array}{c} \\ s_1 \\ a_1 \\ P \end{array}\begin{array}{c} \begin{array}{cccccc} x_1 & x_2 & s_1 & s_2 & a_1 & \end{array} \\ \left[\begin{array}{cccccc|c} 1 & 2 & 1 & 0 & 0 & 0 & 12 \\ \textcircled{1} & 1 & 0 & -1 & 1 & 0 & 4 \\ \hline -M-5 & -M-2 & 0 & M & 0 & 1 & -4M \end{array}\right] \end{array} \begin{array}{l} \frac{12}{1} = 12 \\ \frac{4}{1} = 4 \end{array}$$

$(-1)R_2 + R_1 \rightarrow R_1$ and $(M + 5)R_2 + R_3 \rightarrow R_3$

$$\sim \left[\begin{array}{cccccc|c} 0 & 1 & 1 & \textcircled{1} & -1 & 0 & 8 \\ 1 & 1 & 0 & -1 & 1 & 0 & 4 \\ \hline 0 & 3 & 0 & -5 & M+5 & 1 & 20 \end{array}\right] \sim \begin{array}{c} \\ s_2 \\ x_1 \\ P \end{array}\begin{array}{c} \begin{array}{cccccc} x_1 & x_2 & s_1 & s_2 & a_1 & P \end{array} \\ \left[\begin{array}{cccccc|c} 0 & 1 & 1 & 1 & -1 & 0 & 8 \\ 1 & 2 & 1 & 0 & 0 & 0 & 12 \\ \hline 0 & 8 & 5 & 0 & M & 1 & 60 \end{array}\right] \end{array}$$

$R_1 + R_2 \rightarrow R_2$ and $5R_1 + R_3 \rightarrow R_3$

Thus, the optimal solution of the modified problem is: max $P = 60$ at $x_1 = 12$, $x_2 = 0$, $s_1 = 0$, $s_2 = 8$, $a_1 = 0$.

(D) The optimal solution of the original problem is: max $P = 60$ at $x_1 = 12$, $x_2 = 0$.

3. (A) We introduce the slack variable s_1 and the artificial variable a_1 to obtain the modified problem: Maximize $P = 3x_1 + 5x_2 - Ma_1$

$$\begin{aligned} \text{Subject to: } 2x_1 + x_2 + s_1 \quad &= 8 \\ x_1 + x_2 \quad + a_1 &= 6 \\ x_1, x_2, s_1, a_1 &\geq 0 \end{aligned}$$

(B) The preliminary simplex tableau for the modified problem is:

$$\begin{array}{c} \begin{array}{ccccc} x_1 & x_2 & s_1 & a_1 & P \end{array} \\ \left[\begin{array}{ccccc|c} 2 & 1 & 1 & 0 & 0 & 8 \\ 1 & 1 & 0 & 1 & 0 & 6 \\ \hline -3 & -5 & 0 & M & 1 & 0 \end{array}\right] \end{array} \sim \left[\begin{array}{ccccc|c} 2 & 1 & 1 & 0 & 0 & 8 \\ 1 & 1 & 0 & 1 & 0 & 6 \\ \hline -M-3 & -M-5 & 0 & 0 & 1 & -6M \end{array}\right]$$

$(-M)R_2 + R_3 \rightarrow R_3$

Thus, the initial simplex tableau is:

$$\begin{array}{c} \\ s_1 \\ a_1 \\ P \end{array}\begin{array}{c} \begin{array}{ccccc} x_1 & x_2 & s_1 & a_1 & P \end{array} \\ \left[\begin{array}{ccccc|c} 2 & 1 & 1 & 0 & 0 & 8 \\ 1 & 1 & 0 & 1 & 0 & 6 \\ \hline -M-3 & -M-5 & 0 & 0 & 1 & -6M \end{array}\right] \end{array}$$

(C) We use the simplex method to solve the modified problem.

$$\begin{array}{c} \\ s_1 \\ a_1 \\ P \end{array}\begin{array}{c}\begin{array}{cccccc} x_1 & x_2 & s_1 & a_1 & & \end{array}\\ \left[\begin{array}{ccccc|c} 2 & 1 & 1 & 0 & 0 & 8 \\ 1 & \textcircled{1} & 0 & 1 & 0 & 6 \\ \hdashline -M-3 & -M-5 & 0 & 0 & 1 & -6M \end{array}\right]\end{array} \begin{array}{l} \frac{8}{1} = 8 \\ \frac{6}{1} = 6 \end{array} \sim \begin{array}{c} \\ s_1 \\ x_2 \\ P \end{array}\begin{array}{c}\begin{array}{cccccc} x_1 & x_2 & s_1 & a_1 & P & \end{array}\\ \left[\begin{array}{ccccc|c} 1 & 0 & 1 & -1 & 0 & 2 \\ 1 & 1 & 0 & 1 & 0 & 6 \\ \hdashline 2 & 0 & 0 & M+5 & 1 & 30 \end{array}\right]\end{array}$$

$(-1)R_2 + R_1 \to R_1$ and $(M+5)R_2 + R_3 \to R_3$

Thus, the optimal solution of the modified problem is max $P = 30$ at $x_1 = 0$, $x_2 = 6$, $s_1 = 2$, $a_1 = 0$.

(D) The optimal solution of the original problem is: max $P = 30$ at $x_1 = 0$, $x_2 = 6$.

5. (A) We introduce slack, surplus, and artificial variables to obtain the modified problem: Maximize $P = 4x_1 + 3x_2 - Ma_1$

$$\begin{aligned} \text{Subject to: } -x_1 + 2x_2 + s_1 &= 2 \\ x_1 + x_2 - s_2 + a_1 &= 4 \\ x_1, x_2, s_1, s_2, a_1 &\ge 0 \end{aligned}$$

(B) The preliminary simplex tableau for the modified problem is:

$$\begin{array}{c}\begin{array}{ccccccc} x_1 & x_2 & s_1 & s_2 & a_1 & P & \end{array}\\ \left[\begin{array}{cccccc|c} -1 & 2 & 1 & 0 & 0 & 0 & 2 \\ 1 & 1 & 0 & -1 & 1 & 0 & 4 \\ \hdashline -4 & -3 & 0 & 0 & M & 1 & 0 \end{array}\right]\end{array} \sim \left[\begin{array}{cccccc|c} -1 & 2 & 1 & 0 & 0 & 0 & 2 \\ 1 & 1 & 0 & -1 & 1 & 0 & 4 \\ \hdashline -M-4 & -M-3 & 0 & M & 0 & 1 & -4M \end{array}\right]$$

$(-M)R_2 + R_3 \to R_3$

Thus, the initial simplex tableau is:

$$\begin{array}{c} \\ s_1 \\ a_1 \\ P \end{array}\begin{array}{c}\begin{array}{ccccccc} x_1 & x_2 & s_1 & s_2 & a_1 & P & \end{array}\\ \left[\begin{array}{cccccc|c} -1 & 2 & 1 & 0 & 0 & 0 & 2 \\ 1 & 1 & 0 & -1 & 1 & 0 & 4 \\ \hdashline -M-4 & -M-3 & 0 & M & 0 & 1 & -4M \end{array}\right]\end{array}$$

(C) We use the simplex method to solve the modified problem:

$$\begin{array}{c}\begin{array}{ccccccc} x_1 & x_2 & s_1 & s_2 & a_1 & P & \end{array}\\ \left[\begin{array}{cccccc|c} -1 & 2 & 1 & 0 & 0 & 0 & 2 \\ \textcircled{1} & 1 & 0 & -1 & 1 & 0 & 4 \\ \hdashline -M-4 & -M-3 & 0 & M & 0 & 1 & -4M \end{array}\right]\end{array} \sim \begin{array}{c}\begin{array}{ccccccc} x_1 & x_2 & s_1 & s_2 & a_1 & P & \end{array}\\ \left[\begin{array}{cccccc|c} 0 & 3 & 1 & -1 & 1 & 0 & 6 \\ 1 & 1 & 0 & -1 & 1 & 0 & 4 \\ \hdashline 0 & 1 & 0 & -4 & M+4 & 1 & 16 \end{array}\right]\end{array}$$

$R_2 + R_1 \to R_1$, $(M+4)R_2 + R_3 \to R_3$

No optimal solution exists because the elements in the pivot column (the s_2 column) above the dashed line are negative.

7. (A) We introduce slack, surplus, and artificial variables to obtain the modified problem: Maximize $P = 5x_1 + 10x_2 - Ma_1$

$$\begin{aligned} \text{Subject to: } x_1 + x_2 + s_1 \qquad\qquad &= 3 \\ 2x_1 + 3x_2 \qquad - s_2 + a_1 &= 12 \\ x_1, x_2, s_1, s_2, a_1 &\geq 0 \end{aligned}$$

(B) The preliminary simplex tableau for the modified problem is:

$$\begin{array}{c} \\ s_1 \\ a_1 \\ P \end{array} \begin{array}{c} \begin{matrix} x_1 & x_2 & s_1 & s_2 & a_1 & P \end{matrix} \\ \left[\begin{array}{cccccc|c} 1 & 1 & 1 & 0 & 0 & 0 & 3 \\ 2 & 3 & 0 & -1 & 1 & 0 & 12 \\ \hline -5 & -10 & 0 & 0 & M & 1 & 0 \end{array}\right] \end{array} \sim \left[\begin{array}{cccccc|c} 1 & 1 & 1 & 0 & 0 & 0 & 3 \\ 2 & 3 & 0 & -1 & 1 & 0 & 12 \\ \hline -2M - 5 & -3M - 10 & 0 & M & 0 & 1 & -12M \end{array}\right]$$

$(-M)R_2 + R_3 \rightarrow R_3$

Thus, the initial simplex tableau is:

$$\begin{array}{c} \\ s_1 \\ a_1 \\ P \end{array} \begin{array}{c} \begin{matrix} x_1 & x_2 & s_1 & s_2 & a_1 & P \end{matrix} \\ \left[\begin{array}{cccccc|c} 1 & 1 & 1 & 0 & 0 & 0 & 3 \\ 2 & 3 & 0 & -1 & 1 & 0 & 12 \\ \hline -2M - 5 & -3M - 10 & 0 & M & 0 & 1 & -12M \end{array}\right] \end{array}$$

(C) Applying the simplex method to the initial tableau, we have:

$$\left[\begin{array}{cccccc|c} 1 & \textcircled{1} & 1 & 0 & 0 & 0 & 3 \\ 2 & 3 & 0 & -1 & 1 & 0 & 12 \\ \hline -2M - 5 & -3M - 10 & 0 & M & 0 & 1 & -12M \end{array}\right]$$

$(-3)R_1 + R_2 \rightarrow R_2$, $(3M + 10)R_1 + R_3 \rightarrow R_3$

$$\sim \begin{array}{c} \\ x_2 \\ a_1 \\ P \end{array} \begin{array}{c} \begin{matrix} x_1 & x_2 & s_1 & s_2 & a_1 & P \end{matrix} \\ \left[\begin{array}{cccccc|c} 1 & 1 & 1 & 0 & 0 & 0 & 3 \\ -1 & 0 & -3 & -1 & 1 & 0 & 3 \\ \hline M + 5 & 0 & 3M + 10 & M & 0 & 1 & -3M + 30 \end{array}\right] \end{array}$$

The optimal solution of the modified problem is: max $P = -3M + 30$ at $x_1 = 0$, $x_2 = 3$, $s_1 = 0$, $s_2 = 0$, and $a_1 = 3$.

(D) The original problem does not have an optimal solution, since the artificial variable a_1 in the solution of the modified problem has a nonzero value.

9. To minimize $P = 2x_1 - x_2$, we maximize $T = -P = -2x_1 + x_2$. Introducing slack, surplus, and artificial variables, we obtain the modified problem:

Maximize $T = -2x_1 + x_2 - Ma_1$

Subject to:
$$\begin{aligned} x_1 + x_2 + s_1 \qquad &= 8 \\ 5x_1 + 3x_2 - s_2 + a_1 &= 30 \\ x_1, x_2, s_1, s_2, a_1 &\geq 0 \end{aligned}$$

The preliminary simplex tableau for this problem is:

$$\begin{array}{cccccc|c} x_1 & x_2 & s_1 & s_2 & a_1 & T & \\ 1 & 1 & 1 & 0 & 0 & 0 & 8 \\ 5 & 3 & 0 & -1 & 1 & 0 & 30 \\ \hline 2 & -1 & 0 & 0 & M & 1 & 0 \end{array}$$

$(-M)R_2 + R_3 \to R_3$

$$\sim \begin{array}{c} \\ s_1 \\ a_1 \\ T \end{array}\begin{array}{cccccc|c} x_1 & x_2 & s_1 & s_2 & a_1 & T & \\ 1 & 1 & 1 & 0 & 0 & 0 & 8 \\ \textcircled{5} & 3 & 0 & -1 & 1 & 0 & 30 \\ \hline -5M + 2 & -3M - 1 & 0 & M & 0 & 1 & -30M \end{array} \quad \begin{array}{l} \frac{8}{1} = 8 \\ \frac{30}{5} = 6 \end{array}$$

(This is the initial simplex tableau.) $\quad \frac{1}{5}R_2 \to R_2$

$$\sim \left[\begin{array}{cccccc|c} 1 & 1 & 1 & 0 & 0 & 0 & 8 \\ \textcircled{1} & \frac{3}{5} & 0 & -\frac{1}{5} & \frac{1}{5} & 0 & 6 \\ \hline -5M + 2 & -3M - 1 & 0 & M & 0 & 1 & -30M \end{array}\right]$$

$(-1)R_2 + R_1 \to R_1$, $(5M - 2)R_2 + R_3 \to R_3$

$$\sim \left[\begin{array}{cccccc|c} 0 & \textcircled{\frac{2}{5}} & 1 & \frac{1}{5} & -\frac{1}{5} & 0 & 2 \\ 1 & \frac{3}{5} & 0 & -\frac{1}{5} & \frac{1}{5} & 0 & 6 \\ \hline 0 & -\frac{11}{5} & 0 & \frac{2}{5} & M - \frac{2}{5} & 1 & -12 \end{array}\right] \quad \begin{array}{l} \frac{2}{2/5} = 5 \\ \frac{6}{3/5} = 10 \end{array}$$

$\frac{5}{2}R_1 \to R_1$

$$\sim \left[\begin{array}{cccccc|c} 0 & \textcircled{1} & \frac{5}{2} & \frac{1}{2} & -\frac{1}{2} & 0 & 5 \\ 1 & \frac{3}{5} & 0 & -\frac{1}{5} & \frac{1}{5} & 0 & 6 \\ \hline 0 & -\frac{11}{5} & 0 & \frac{2}{5} & M - \frac{2}{5} & 1 & -12 \end{array}\right] \sim \begin{array}{c} \\ x_2 \\ x_1 \\ T \end{array}\begin{array}{cccccc|c} x_1 & x_2 & s_1 & s_2 & a_1 & T & \\ 0 & 1 & \frac{5}{2} & \frac{1}{2} & -\frac{1}{2} & 0 & 5 \\ 1 & 0 & -\frac{3}{2} & -\frac{1}{2} & \frac{1}{2} & 0 & 3 \\ \hline 0 & 0 & \frac{11}{2} & \frac{3}{2} & M - \frac{3}{2} & 1 & -1 \end{array}$$

$\left(-\frac{3}{5}\right)R_1 + R_2 \to R_2$, $\left(\frac{11}{5}\right)R_1 + R_3 \to R_3$

Thus, the optimal solution is: max $T = -1$ at $x_1 = 3$, $x_2 = 5$, and min $P = -\text{max } T = 1$.

The modified problem for maximizing $P = 2x_1 - x_2$ subject to the given constraints is: Maximize $P = 2x_1 - x_2 - Ma_1$

$$\text{Subject to: } \begin{aligned} x_1 + x_2 + s_1 \qquad\qquad &= 8 \\ 5x_1 + 3x_2 - s_2 + a_1 &= 30 \\ x_1, x_2, s_1, s_2, a_1 &\geq 0 \end{aligned}$$

The preliminary simplex tableau for the modified problem is:

$$\begin{matrix} x_1 & x_2 & s_1 & s_2 & a_1 & P \end{matrix}$$
$$\left[\begin{array}{cccccc|c} 1 & 1 & 1 & 0 & 0 & 0 & 8 \\ 5 & 3 & 0 & -1 & 1 & 0 & 30 \\ \hdashline -2 & 1 & 0 & 0 & M & 1 & 0 \end{array}\right]$$

$(-M)R_2 + R_3 \rightarrow R_3$

$$\sim \begin{array}{c} \\ s_1 \\ a_1 \\ P \end{array} \left[\begin{array}{cccccc|c} x_1 & x_2 & s_1 & s_2 & a_1 & P & \\ 1 & 1 & 1 & 0 & 0 & 0 & 8 \\ \textcircled{5} & 3 & 0 & -1 & 1 & 0 & 30 \\ \hdashline -5M - 2 & -3M + 1 & 0 & M & 0 & 1 & -30M \end{array}\right] \begin{array}{l} \\ \frac{8}{1} = 8 \\ \frac{30}{5} = 6 \\ \\ \end{array}$$

(This is the initial simplex tableau.) $\frac{1}{5}R_2 \rightarrow R_2$

$$\sim \left[\begin{array}{cccccc|c} 1 & 1 & 1 & 0 & 0 & 0 & 8 \\ \textcircled{1} & \frac{3}{5} & 0 & -\frac{1}{5} & \frac{1}{5} & 0 & 6 \\ \hdashline -5M - 2 & -3M + 1 & 0 & M & 0 & 1 & -30M \end{array}\right] \sim \left[\begin{array}{cccccc|c} 0 & \frac{2}{5} & 1 & \textcircled{\frac{1}{5}} & -\frac{1}{5} & 0 & 2 \\ 1 & \frac{3}{5} & 0 & -\frac{1}{5} & \frac{1}{5} & 0 & 6 \\ \hdashline 0 & \frac{11}{5} & 0 & -\frac{2}{5} & M + \frac{2}{5} & 1 & 12 \end{array}\right]$$

$(-1)R_2 + R_1 \rightarrow R_1$, $(5M + 2)R_2 + R_3 \rightarrow R_3$ $\qquad$ $5R_1 \rightarrow R_1$

$$\sim \left[\begin{array}{cccccc|c} 0 & 2 & 5 & 1 & -1 & 0 & 10 \\ 1 & \frac{3}{5} & 0 & -\frac{1}{5} & \frac{1}{5} & 0 & 6 \\ \hdashline 0 & \frac{11}{5} & 0 & -\frac{2}{5} & M + \frac{2}{5} & 1 & 12 \end{array}\right] \sim \begin{array}{c} \\ s_2 \\ x_2 \\ P \end{array} \left[\begin{array}{cccccc|c} x_1 & x_2 & s_1 & s_2 & a_1 & P & \\ 0 & 2 & 5 & 1 & -1 & 0 & 10 \\ 1 & 1 & 1 & 0 & 0 & 0 & 8 \\ \hdashline 0 & 3 & 2 & 0 & M & 1 & 16 \end{array}\right]$$

$\frac{1}{5}R_1 + R_2 \rightarrow R_2$ and $\frac{2}{5}R_1 + R_3 \rightarrow R_3$

Thus, the optimal solution is: max $P = 16$ at $x_1 = 8$, $x_2 = 0$.

11. We introduce slack, surplus, and artificial variables to obtain the modified problem: Maximize $P = 2x_1 + 5x_2 - Ma_1$

$$\text{Subject to: } \begin{aligned} x_1 + 2x_2 + s_1 \qquad\qquad\qquad &= 18 \\ 2x_1 + x_2 \qquad + s_2 \qquad\qquad &= 21 \\ x_1 + x_2 \qquad\qquad - s_3 + a_1 &= 10 \\ x_1, x_2, s_1, s_2, s_3, a_1 &\geq 0 \end{aligned}$$

The preliminary simplex tableau for this problem is:

$$\begin{array}{c} \\ \\ \end{array}\begin{array}{ccccccc|c} x_1 & x_2 & s_1 & s_2 & s_3 & a_1 & P & \\ \end{array}$$

$$\left[\begin{array}{ccccccc|c} 1 & 2 & 1 & 0 & 0 & 0 & 0 & 18 \\ 2 & 1 & 0 & 1 & 0 & 0 & 0 & 21 \\ 1 & 1 & 0 & 0 & -1 & 1 & 0 & 10 \\ \hdashline -2 & -5 & 0 & 0 & 0 & M & 1 & 0 \end{array}\right]$$

$(-M)R_3 + R_4 \to R_4$

$$\sim \begin{array}{c} \\ s_1 \\ s_2 \\ a_1 \\ P \end{array} \left[\begin{array}{ccccccc|c} x_1 & x_2 & s_1 & s_2 & s_3 & a_1 & P & \\ 1 & \textcircled{2} & 1 & 0 & 0 & 0 & 0 & 18 \\ 2 & 1 & 0 & 1 & 0 & 0 & 0 & 21 \\ 1 & 1 & 0 & 0 & -1 & 1 & 0 & 10 \\ \hdashline -M - 2 & -M - 5 & 0 & 0 & M & 0 & 1 & -10M \end{array}\right] \begin{array}{l} \\ \frac{18}{2} = 9 \\ \frac{21}{1} = 21 \\ \frac{10}{1} = 10 \\ \\ \end{array}$$

(This is the initial simplex tableau.)

$\frac{1}{2}R_1 \to R_1$

$$\sim \left[\begin{array}{ccccccc|c} \frac{1}{2} & \textcircled{1} & \frac{1}{2} & 0 & 0 & 0 & 0 & 9 \\ 2 & 1 & 0 & 1 & 0 & 0 & 0 & 21 \\ 1 & 1 & 0 & 0 & -1 & 1 & 0 & 10 \\ \hdashline -M - 2 & -M - 5 & 0 & 0 & M & 0 & 1 & -10M \end{array}\right]$$

$(-1)R_1 + R_2 \to R_2$, $(-1)R_1 + R_3 \to R_3$, and $(M+5)R_1 + R_4 \to R_4$

$$\sim \left[\begin{array}{ccccccc|c} \frac{1}{2} & 1 & \frac{1}{2} & 0 & 0 & 0 & 0 & 9 \\ \frac{3}{2} & 0 & -\frac{1}{2} & 1 & 0 & 0 & 0 & 12 \\ \textcircled{\frac{1}{2}} & 0 & -\frac{1}{2} & 0 & -1 & 1 & 0 & 1 \\ \hdashline -\frac{1}{2}M + \frac{1}{2} & 0 & \frac{1}{2}M + \frac{5}{2} & 0 & M & 0 & 1 & -M + 45 \end{array}\right] \begin{array}{l} \frac{9}{1/2} = 18 \\ \frac{12}{3/2} = 8 \\ \frac{1}{1/2} = 2 \\ \\ \end{array}$$

$2R_3 \to R_3$

$$\sim \left[\begin{array}{ccccccc|c} \frac{1}{2} & 1 & \frac{1}{2} & 0 & 0 & 0 & 0 & 9 \\ \frac{3}{2} & 0 & -\frac{1}{2} & 1 & 0 & 0 & 0 & 12 \\ \textcircled{1} & 0 & -1 & 0 & -2 & 2 & 0 & 2 \\ \hdashline -\frac{1}{2}M + \frac{1}{2} & 0 & \frac{1}{2}M + \frac{5}{2} & 0 & M & 0 & 1 & -M + 45 \end{array}\right]$$

$\left(-\frac{1}{2}\right)R_3 + R_1 \to R_1$, $\left(-\frac{3}{2}\right)R_3 + R_2 \to R_2$, and $\left(\frac{1}{2}M - \frac{1}{2}\right)R_3 + R_4 \to R_4$

$$\begin{array}{c} \\ x_2 \\ s_2 \\ x_1 \\ P \end{array} \sim \left[\begin{array}{ccccccc|c} x_1 & x_2 & s_1 & s_2 & s_3 & a_1 & P & \\ 0 & 1 & 1 & 0 & 1 & -1 & 0 & 8 \\ 0 & 0 & 1 & 1 & 3 & -3 & 0 & 9 \\ 1 & 0 & -1 & 0 & -2 & 2 & 0 & 2 \\ \hline 0 & 0 & 3 & 0 & 1 & M-1 & 1 & 44 \end{array}\right]$$

Optimal solution: max $P = 44$ at $x_1 = 2$, $x_2 = 8$.

13. We introduce surplus and artificial variables to obtain the modified problem: Maximize $P = 10x_1 + 12x_2 + 20x_3 - Ma_1 - Ma_2$

$$\begin{aligned} \text{Subject to: } 3x_1 + x_2 + 2x_3 - s_1 + a_1 \quad &= 12 \\ x_1 - x_2 + 2x_3 \qquad\quad + a_2 &= 6 \\ x_1, x_2, x_3, s_1, a_1, a_2 &\geq 0 \end{aligned}$$

The preliminary simplex tableau for the modified problem is:

$$\left[\begin{array}{ccccccc|c} x_1 & x_2 & x_3 & s_1 & a_1 & a_2 & P & \\ 3 & 1 & 2 & -1 & 1 & 0 & 0 & 12 \\ 1 & -1 & 2 & 0 & 0 & 1 & 0 & 6 \\ \hline -10 & -12 & -20 & 0 & M & M & 1 & 0 \end{array}\right]$$

$(-M)R_1 + R_3 \to R_3$

$$\sim \left[\begin{array}{ccccccc|c} 3 & 1 & 2 & -1 & 1 & 0 & 0 & 12 \\ 1 & -1 & 2 & 0 & 0 & 1 & 0 & 6 \\ \hline -3M-10 & -M-12 & -2M-20 & M & 0 & M & 1 & -12M \end{array}\right]$$

$(-M)R_2 + R_3 \to R_3$

$$\begin{array}{c} \\ a_1 \\ a_2 \\ P \end{array} \sim \left[\begin{array}{ccccccc|c} x_1 & x_2 & x_3 & s_1 & a_1 & a_2 & P & \\ 3 & 1 & 2 & -1 & 1 & 0 & 0 & 12 \\ 1 & -1 & \textcircled{2} & 0 & 0 & 1 & 0 & 6 \\ \hline -4M-10 & -12 & -4M-20 & M & 0 & 0 & 1 & -18M \end{array}\right] \begin{array}{l} \\ \frac{12}{2} = 6 \\ \frac{6}{2} = 3 \\ \\ \end{array}$$

$\frac{1}{2}R_2 \to R_2$

$$\sim \left[\begin{array}{ccccccc|c} 3 & 1 & 2 & -1 & 1 & 0 & 0 & 12 \\ \frac{1}{2} & -\frac{1}{2} & \textcircled{1} & 0 & 0 & \frac{1}{2} & 0 & 3 \\ \hline -4M-10 & -12 & -4M-20 & M & 0 & 0 & 1 & -18M \end{array}\right]$$

$(-2)R_2 + R_1 \to R_1$ and $(4M+20)R_2 + R_3 \to R_3$

$$\sim \left[\begin{array}{ccccccc|c} 2 & \textcircled{2} & 0 & -1 & 1 & -1 & 0 & 6 \\ \frac{1}{2} & -\frac{1}{2} & 1 & 0 & 0 & \frac{1}{2} & 0 & 3 \\ \hline -2M & -2M-22 & 0 & M & 0 & 2M+10 & 1 & -6M+60 \end{array}\right]$$

$\frac{1}{2}R_1 \to R_1$

$$\sim \left[\begin{array}{ccccccc|c} 1 & \textcircled{1} & 0 & -\frac{1}{2} & \frac{1}{2} & -\frac{1}{2} & 0 & 3 \\ \frac{1}{2} & -\frac{1}{2} & 1 & 0 & 0 & \frac{1}{2} & 0 & 3 \\ \hdashline -2M & -2M-22 & 0 & M & 0 & 2M+10 & 1 & -6M+60 \end{array}\right]$$

$\frac{1}{2}R_1 + R_2 \to R_2$, $(2M+22)R_1 + R_3 \to R_3$

$$\begin{array}{c} \\ x_2 \\ \sim x_3 \\ P \end{array} \begin{array}{c} \begin{array}{ccccccc} x_1 & x_2 & x_3 & s_1 & a_1 & a_2 & P \end{array} \\ \left[\begin{array}{ccccccc|c} 1 & 1 & 0 & -\frac{1}{2} & \frac{1}{2} & -\frac{1}{2} & 0 & 3 \\ 1 & 0 & 1 & -\frac{1}{4} & \frac{1}{4} & \frac{1}{4} & 0 & \frac{9}{2} \\ \hdashline 22 & 0 & 0 & -11 & M+11 & M-1 & 1 & 126 \end{array}\right] \end{array}$$

No optimal solution exists because there are no positive numbers in the pivot column.

15. We will maximize $P = -C = 5x_1 + 12x_2 - 16x_3$ subject to the given constraints. Introduce slack, surplus, and artificial variables to obtain the modified problem:

Maximize $P = 5x_1 + 12x_2 - 16x_3 - Ma_1 - Ma_2$

Subject to:
$$\begin{aligned} x_1 + 2x_2 + x_3 + s_1 \qquad\qquad &= 10 \\ 2x_1 + 3x_2 + x_3 \qquad - s_2 + a_1 \qquad &= 6 \\ 2x_1 + x_2 - x_3 \qquad\qquad\qquad + a_2 &= 1 \\ x_1, x_2, x_3, s_1, s_2, a_1, a_2 &\geq 0 \end{aligned}$$

The preliminary simplex tableau for the modified problem is:

$$\begin{array}{c} \begin{array}{cccccccc} x_1 & x_2 & x_3 & s_1 & s_2 & a_1 & a_2 & P \end{array} \\ \left[\begin{array}{cccccccc|c} 1 & 2 & 1 & 1 & 0 & 0 & 0 & 0 & 10 \\ 2 & 3 & 1 & 0 & -1 & 1 & 0 & 0 & 6 \\ 2 & 1 & -1 & 0 & 0 & 0 & 1 & 0 & 1 \\ \hdashline -5 & -12 & 16 & 0 & 0 & M & M & 1 & 0 \end{array}\right] \end{array}$$

$(-M)R_2 + R_4 \to R_4$

$$\sim \left[\begin{array}{cccccccc|c} 1 & 2 & 1 & 1 & 0 & 0 & 0 & 0 & 10 \\ 2 & 3 & 1 & 0 & -1 & 1 & 0 & 0 & 6 \\ 2 & 1 & -1 & 0 & 0 & 0 & 1 & 0 & 1 \\ \hdashline -2M-5 & -3M-12 & -M+16 & 0 & M & 0 & M & 1 & -6M \end{array}\right]$$

$(-M)R_3 + R_4 \to R_4$

$$\sim \begin{array}{c} \\ s_1 \\ a_1 \\ a_2 \\ P \end{array}\left[\begin{array}{cccccccc|c} x_1 & x_2 & x_3 & s_1 & s_2 & a_1 & a_2 & P & \\ 1 & 2 & 1 & 1 & 0 & 0 & 0 & 0 & 10 \\ 2 & 3 & 1 & 0 & -1 & 1 & 0 & 0 & 6 \\ 2 & \textcircled{1} & -1 & 0 & 0 & 0 & 1 & 0 & 1 \\ \hdashline -4M-5 & -4M-12 & 16 & 0 & M & 0 & 0 & 1 & -7M \end{array}\right] \begin{array}{l} \\ \frac{10}{2} = 5 \\ \frac{6}{3} = 2 \\ \frac{1}{1} = 1 \\ \\ \end{array}$$

$(-2)R_3 + R_1 \to R_1$, $(-3)R_3 + R_2 \to R_2$, and $(4M+12)R_3 + R_4 \to R_4$

$$\sim \left[\begin{array}{cccccccc|c} -3 & 0 & 3 & 1 & 0 & 0 & -2 & 0 & 8 \\ -4 & 0 & \textcircled{4} & 0 & -1 & 1 & -3 & 0 & 3 \\ 2 & 1 & -1 & 0 & 0 & 0 & 1 & 0 & 1 \\ \hdashline 4M+19 & 0 & -4M+4 & 0 & M & 0 & 4M+12 & 1 & -3M+12 \end{array}\right] \begin{array}{l} \frac{8}{3} \approx 2.67 \\ \frac{3}{4} = .75 \\ \\ \\ \end{array}$$

$\frac{1}{4}R_2 \to R_2$

$$\sim \left[\begin{array}{cccccccc|c} -3 & 0 & 3 & 1 & 0 & 0 & -2 & 0 & 8 \\ -1 & 0 & \textcircled{1} & 0 & -\frac{1}{4} & \frac{1}{4} & -\frac{3}{4} & 0 & \frac{3}{4} \\ 2 & 1 & -1 & 0 & 0 & 0 & 1 & 0 & 1 \\ \hdashline 4M+19 & 0 & -4M+4 & 0 & M & 0 & 4M+12 & 1 & -3M+12 \end{array}\right]$$

$(-3)R_2 + R_1 \to R_1$, $R_2 + R_3 \to R_3$, and $(4M-4)R_2 + R_4 \to R_4$

$$\sim \begin{array}{c} \\ s_1 \\ x_3 \\ x_2 \\ P \end{array}\left[\begin{array}{cccccccc|c} x_1 & x_2 & x_3 & s_1 & s_2 & a_1 & a_2 & P & \\ 0 & 0 & 0 & 1 & \frac{3}{4} & -\frac{3}{4} & \frac{1}{4} & 0 & \frac{23}{4} \\ -1 & 0 & 1 & 0 & -\frac{1}{4} & \frac{1}{4} & -\frac{3}{4} & 0 & \frac{3}{4} \\ 1 & 1 & 0 & 0 & -\frac{1}{4} & \frac{1}{4} & \frac{1}{4} & 0 & \frac{7}{4} \\ \hdashline 23 & 0 & 0 & 0 & 1 & M-1 & M+15 & 1 & 9 \end{array}\right]$$

Optimal solution: min $C = -9$ at $x_1 = 0$, $x_2 = \frac{7}{4}$, and $x_3 = \frac{3}{4}$.

17. We introduce a slack and an artificial variable to obtain the modified problem: Maximize $P = 3x_1 + 5x_2 + 6x_3 - Ma_1$

$$\begin{aligned} \text{Subject to: } 2x_1 + x_2 + 2x_3 + s_1 \quad &= 8 \\ 2x_1 + x_2 - 2x_3 \quad + a_1 &= 0 \\ x_1, x_2, x_3, s_1, a_1 &\geq 0 \end{aligned}$$

The preliminary simplex tableau for the modified problem is:

$$\begin{array}{c} \\ s_1 \\ a_1 \\ P \end{array}\left[\begin{array}{cccccc|c} x_1 & x_2 & x_3 & s_1 & a_1 & P & \\ 2 & 1 & 2 & 1 & 0 & 0 & 8 \\ 2 & 1 & -2 & 0 & 1 & 0 & 0 \\ \hdashline -3 & -5 & -6 & 0 & M & 1 & 0 \end{array}\right] \sim \left[\begin{array}{cccccc|c} 2 & 1 & 2 & 1 & 0 & 0 & 8 \\ \textcircled{2} & 1 & -2 & 0 & 1 & 0 & 0 \\ \hdashline -3-2M & -5-M & -6+2M & 0 & 0 & 1 & 0 \end{array}\right]$$

$(-M)R_2 + R_3 \to R_3$ $\qquad\qquad$ $\frac{1}{2}R_2 \to R_2$

$$\sim\left[\begin{array}{cccccc|c} 2 & 1 & 2 & 1 & 0 & 0 & 8 \\ \textcircled{1} & \frac{1}{2} & -1 & 0 & \frac{1}{2} & 0 & 0 \\ \hdashline -3-2M & -5-M & -6+2M & 0 & 0 & 1 & 0 \end{array}\right] \sim \left[\begin{array}{cccccc|c} 0 & 0 & \textcircled{4} & 1 & -1 & 0 & 8 \\ 1 & \frac{1}{2} & -1 & 0 & \frac{1}{2} & 0 & 0 \\ \hdashline 0 & -\frac{7}{2} & -9 & 0 & \frac{3}{2}+M & 1 & 0 \end{array}\right]$$

$(-2)R_2 + R_1 \to R_1$, $(3 + 2M)R_2 + R_3 \to R_3$ $\qquad$ $\frac{1}{4}R_1 \to R_1$

$$\sim\left[\begin{array}{cccccc|c} 0 & 0 & \textcircled{1} & \frac{1}{4} & -\frac{1}{4} & 0 & 2 \\ 1 & \frac{1}{2} & -1 & 0 & \frac{1}{2} & 0 & 0 \\ \hdashline 0 & -\frac{7}{2} & -9 & 0 & \frac{3}{2}+M & 1 & 0 \end{array}\right] \sim \left[\begin{array}{cccccc|c} 0 & 0 & 1 & \frac{1}{4} & -\frac{1}{4} & 0 & 2 \\ 1 & \textcircled{$\frac{1}{2}$} & 0 & \frac{1}{4} & \frac{1}{4} & 0 & 2 \\ \hdashline 0 & -\frac{7}{2} & 0 & \frac{9}{4} & -\frac{3}{4}+M & 1 & 18 \end{array}\right]$$

$R_1 + R_2 \to R_2$, $9R_1 + R_3 \to R_3$ $\qquad$ $2R_2 \to R_2$

$$\sim\left[\begin{array}{cccccc|c} 0 & 0 & 1 & \frac{1}{4} & -\frac{1}{4} & 0 & 2 \\ 2 & \textcircled{1} & 0 & \frac{1}{2} & \frac{1}{2} & 0 & 4 \\ \hdashline 0 & -\frac{7}{2} & 0 & \frac{9}{4} & -\frac{3}{4}+M & 1 & 18 \end{array}\right] \sim \begin{array}{c} \\ x_3 \\ x_2 \\ P \end{array}\!\!\begin{array}{c} \begin{array}{cccccc} x_1 & x_2 & x_3 & s_1 & a_1 & P \end{array} \\ \left[\begin{array}{cccccc|c} 0 & 0 & 1 & \frac{1}{4} & -\frac{1}{4} & 0 & 2 \\ 2 & 1 & 0 & \frac{1}{2} & \frac{1}{2} & 0 & 4 \\ \hdashline 7 & 0 & 0 & 4 & 1+M & 1 & 32 \end{array}\right] \end{array}$$

$\frac{7}{2}R_2 + R_3 \to R_3$

Optimal solution: max $P = 32$ at $x_1 = 0$, $x_2 = 4$, $x_3 = 2$.

19. We introduce slack, surplus, and artificial variables to obtain the modified problem: Maximize $P = 2x_1 + 3x_2 + 4x_3 - Ma_1$

$$\begin{aligned} \text{Subject to: } \quad x_1 + 2x_2 + x_3 + s_1 &= 25 \\ 2x_1 + x_2 + 2x_3 + s_2 &= 60 \\ x_1 + 2x_2 - x_3 - s_3 + a_1 &= 10 \\ x_1, x_2, x_3, s_1, s_2, s_3, a_1 &\geq 0 \end{aligned}$$

The preliminary simplex tableau for the modified problem is:

$$\begin{array}{c} \begin{array}{cccccccc} x_1 & x_2 & x_3 & s_1 & s_2 & s_3 & a_1 & P \end{array} \\ \left[\begin{array}{cccccccc|c} 1 & 2 & 1 & 1 & 0 & 0 & 0 & 0 & 25 \\ 2 & 1 & 2 & 0 & 1 & 0 & 0 & 0 & 60 \\ 1 & 2 & -1 & 0 & 0 & -1 & 1 & 0 & 10 \\ \hdashline -2 & -3 & -4 & 0 & 0 & 0 & M & 1 & 0 \end{array}\right] \end{array}$$

$(-M)R_3 + R_4 \to R_4$

$$\sim \begin{array}{c} \\ s_1 \\ s_2 \\ a_1 \\ P \end{array}\!\!\begin{array}{c} \begin{array}{cccccccc} x_1 & x_2 & x_3 & s_1 & s_2 & s_3 & a_1 & P \end{array} \\ \left[\begin{array}{cccccccc|c} 1 & 2 & 1 & 1 & 0 & 0 & 0 & 0 & 25 \\ 2 & 1 & 2 & 0 & 1 & 0 & 0 & 0 & 60 \\ 1 & \textcircled{2} & -1 & 0 & 0 & -1 & 1 & 0 & 10 \\ \hdashline -2-M & -3-2M & -4+M & 0 & 0 & M & 0 & 1 & -10M \end{array}\right] \end{array}$$

$\frac{1}{2}R_3 \to R_3$

$$\sim\left[\begin{array}{cccccccc|c}
1 & 2 & 1 & 1 & 0 & 0 & 0 & 0 & 25 \\
2 & 1 & 2 & 0 & 1 & 0 & 0 & 0 & 60 \\
\frac{1}{2} & \textcircled{1} & -\frac{1}{2} & 0 & 0 & -\frac{1}{2} & \frac{1}{2} & 0 & 5 \\
\hline
-2-M & -3-2M & -4+M & 0 & 0 & M & 0 & 1 & -10M
\end{array}\right]$$

$(-2)R_3 + R_1 \to R_1, \quad (-1)R_3 + R_2 \to R_2, \quad (3 + 2M)R_3 + R_4 \to R_4$

$$\sim\left[\begin{array}{cccccccc|c}
0 & 0 & \textcircled{2} & 1 & 0 & 1 & -1 & 0 & 15 \\
\frac{3}{2} & 0 & \frac{5}{2} & 0 & 1 & \frac{1}{2} & -\frac{1}{2} & 0 & 55 \\
\frac{1}{2} & 1 & -\frac{1}{2} & 0 & 0 & -\frac{1}{2} & \frac{1}{2} & 0 & 5 \\
\hline
-\frac{1}{2} & 0 & -\frac{11}{2} & 0 & 0 & -\frac{3}{2} & \frac{3}{2}+M & 1 & 15
\end{array}\right]$$

$\frac{1}{2}R_1 \to R_1$

$$\sim\left[\begin{array}{cccccccc|c}
0 & 0 & \textcircled{1} & \frac{1}{2} & 0 & \frac{1}{2} & -\frac{1}{2} & 0 & \frac{15}{2} \\
\frac{3}{2} & 0 & \frac{5}{2} & 0 & 1 & \frac{1}{2} & -\frac{1}{2} & 0 & 55 \\
\frac{1}{2} & 1 & -\frac{1}{2} & 0 & 0 & -\frac{1}{2} & \frac{1}{2} & 0 & 5 \\
\hline
-\frac{1}{2} & 0 & -\frac{11}{2} & 0 & 0 & -\frac{3}{2} & \frac{3}{2}+M & 1 & 15
\end{array}\right]$$

$\left(-\frac{5}{2}\right)R_1 + R_2 \to R_2, \quad \frac{1}{2}R_1 + R_3 \to R_3, \quad \frac{11}{2}R_1 + R_4 \to R_4$

$$\sim\left[\begin{array}{cccccccc|c}
0 & 0 & 1 & \frac{1}{2} & 0 & \frac{1}{2} & -\frac{1}{2} & 0 & \frac{15}{2} \\
\frac{3}{2} & 0 & 0 & -\frac{5}{4} & 1 & -\frac{3}{4} & \frac{3}{4} & 0 & \frac{145}{4} \\
\textcircled{\tfrac{1}{2}} & 1 & 0 & \frac{1}{4} & 0 & -\frac{1}{4} & \frac{1}{4} & 0 & \frac{35}{4} \\
\hline
-\frac{1}{2} & 0 & 0 & \frac{11}{4} & 0 & \frac{5}{4} & -\frac{5}{4}+M & 1 & \frac{225}{4}
\end{array}\right]$$

$2R_3 \to R_3$

$$\sim\left[\begin{array}{cccccccc|c}
0 & 0 & 1 & \frac{1}{2} & 0 & \frac{1}{2} & -\frac{1}{2} & 0 & \frac{15}{2} \\
\frac{3}{2} & 0 & 0 & -\frac{5}{4} & 1 & -\frac{3}{4} & \frac{3}{4} & 0 & \frac{145}{4} \\
\textcircled{1} & 2 & 0 & \frac{1}{2} & 0 & -\frac{1}{2} & \frac{1}{2} & 0 & \frac{35}{2} \\
\hline
-\frac{1}{2} & 0 & 0 & \frac{11}{4} & 0 & \frac{5}{4} & -\frac{5}{4}+M & 1 & \frac{225}{4}
\end{array}\right]$$

$\left(-\frac{3}{2}\right)R_3 + R_2 \to R_2, \quad \frac{1}{2}R_3 + R_4 \to R_4$

$$
\sim \begin{array}{c} \\ x_3 \\ s_2 \\ x_1 \\ P \end{array}
\begin{array}{c}
\begin{array}{cccccccc|c} x_1 & x_2 & x_3 & s_1 & s_2 & s_3 & a_1 & P & \end{array} \\
\left[\begin{array}{cccccccc|c}
0 & 0 & 1 & \frac{1}{2} & 0 & \frac{1}{2} & -\frac{1}{2} & 0 & \frac{15}{2} \\
0 & -3 & 0 & -2 & 1 & 0 & 0 & 0 & 10 \\
1 & 2 & 0 & \frac{1}{2} & 0 & -\frac{1}{2} & \frac{1}{2} & 0 & \frac{35}{2} \\
\hdashline
0 & 1 & 0 & 3 & 0 & 1 & -1 + M & 1 & 65
\end{array}\right]
\end{array}
$$

Optimal solution: max $P = 65$ at $x_1 = \frac{35}{2}$, $x_2 = 0$, $x_3 = \frac{15}{2}$.

21. We introduce slack, surplus, and artificial variables to obtain the modified problem:

Maximize $P = x_1 + 2x_2 + 5x_3 - Ma_1 - Ma_2$

Subject to:

$$
\begin{aligned}
x_1 + 3x_2 + 2x_3 + s_1 \qquad\qquad\qquad\qquad &= 60 \\
2x_1 + 5x_2 + 2x_3 \qquad - s_2 \qquad + a_1 \qquad &= 50 \\
x_1 - 2x_2 + x_3 \qquad\qquad - s_3 \qquad + a_2 &= 40 \\
x_1, x_2, x_3, s_1, s_2, s_3, a_1, a_2 &\ge 0
\end{aligned}
$$

The preliminary simplex tableau for the modified problem is:

$$
\begin{array}{c}
\begin{array}{ccccccccc|c} x_1 & x_2 & x_3 & s_1 & s_2 & a_1 & s_3 & a_2 & P & \end{array} \\
\left[\begin{array}{ccccccccc|c}
1 & 3 & 2 & 1 & 0 & 0 & 0 & 0 & 0 & 60 \\
2 & 5 & 2 & 0 & -1 & 1 & 0 & 0 & 0 & 50 \\
1 & -2 & 1 & 0 & 0 & 0 & -1 & 1 & 0 & 40 \\
\hdashline
-1 & -2 & -5 & 0 & 0 & M & 0 & M & 1 & 0
\end{array}\right]
\end{array}
$$

$(-M)R_2 + R_4 \to R_4$

$$
\sim \left[\begin{array}{ccccccccc|c}
1 & 3 & 2 & 1 & 0 & 0 & 0 & 0 & 0 & 60 \\
2 & 5 & 2 & 0 & -1 & 1 & 0 & 0 & 0 & 50 \\
1 & -2 & 1 & 0 & 0 & 0 & -1 & 1 & 0 & 40 \\
\hdashline
-2M - 1 & -5M - 2 & -2M - 5 & 0 & M & 0 & 0 & M & 1 & -50M
\end{array}\right]
$$

$(-M)R_3 + R_4 \to R_4$

$$
\sim \left[\begin{array}{ccccccccc|c}
1 & 3 & 2 & 1 & 0 & 0 & 0 & 0 & 0 & 60 \\
2 & 5 & \textcircled{2} & 0 & -1 & 1 & 0 & 0 & 0 & 50 \\
1 & -2 & 1 & 0 & 0 & 0 & -1 & 1 & 0 & 40 \\
\hdashline
-3M - 1 & -3M - 2 & -3M - 5 & 0 & M & 0 & M & 0 & 1 & -90M
\end{array}\right]
$$

$\frac{1}{2}R_2 \to R_2$

$$
\sim \left[\begin{array}{ccccccccc|c}
1 & 3 & 2 & 1 & 0 & 0 & 0 & 0 & 0 & 60 \\
1 & \frac{5}{2} & \textcircled{1} & 0 & -\frac{1}{2} & \frac{1}{2} & 0 & 0 & 0 & 25 \\
1 & -2 & 1 & 0 & 0 & 0 & -1 & 1 & 0 & 40 \\
\hdashline
-3M - 1 & -3M - 2 & -3M - 5 & 0 & M & 0 & M & 0 & 1 & -90M
\end{array}\right]
$$

$(-2)R_2 + R_1 \to R_1$, $(-1)R_2 + R_3 \to R_3$, $(3M + 5)R_2 + R_4 \to R_4$

$$\sim\left[\begin{array}{ccccccccc|c} -1 & -2 & 0 & 1 & \textcircled{1} & -1 & 0 & 0 & 0 & 10 \\ 1 & \frac{5}{2} & 1 & 0 & -\frac{1}{2} & \frac{1}{2} & 0 & 0 & 0 & 25 \\ 0 & -\frac{9}{2} & 0 & 0 & \frac{1}{2} & -\frac{1}{2} & -1 & 1 & 0 & 15 \\ \hdashline 4 & \frac{9}{2}M + \frac{21}{2} & 0 & 0 & -\frac{1}{2}M - \frac{5}{2} & \frac{3}{2}M + \frac{5}{2} & M & 0 & 1 & -15M + 125 \end{array}\right]$$

$$\frac{1}{2}R_1 + R_2 \to R_2, \quad \left(-\frac{1}{2}\right)R_1 + R_3 \to R_3, \quad \left(\frac{1}{2}M + \frac{5}{2}\right)R_1 + R_4 \to R_4$$

$$\sim\left[\begin{array}{ccccccccc|c} -1 & -2 & 0 & 1 & 1 & -1 & 0 & 0 & 0 & 10 \\ \frac{1}{2} & \frac{3}{2} & 1 & \frac{1}{2} & 0 & 0 & 0 & 0 & 0 & 30 \\ \textcircled{\frac{1}{2}} & -\frac{7}{2} & 0 & -\frac{1}{2} & 0 & 0 & -1 & 1 & 0 & 10 \\ \hdashline -\frac{1}{2}M + \frac{3}{2} & \frac{7}{2}M + \frac{11}{2} & 0 & \frac{1}{2}M + \frac{5}{2} & 0 & M & M & 0 & 1 & -10M + 150 \end{array}\right]$$

$$2R_3 + R_1 \to R_1, \quad (-1)R_3 + R_2 \to R_2, \quad (M-3)R_3 + R_4 \to R_4, \quad 2R_3 \to R_3$$

$$\sim\begin{array}{c} \\ s_2 \\ x_3 \\ x_1 \\ \\ \end{array}\begin{array}{c}\begin{array}{ccccccccc|c} x_1 & x_2 & x_3 & s_1 & s_2 & a_1 & s_3 & a_2 & P & \end{array} \\ \left[\begin{array}{ccccccccc|c} 0 & -9 & 0 & 0 & 1 & -1 & -2 & 2 & 0 & 30 \\ 0 & 5 & 1 & 1 & 0 & 0 & 1 & -1 & 0 & 20 \\ 1 & -7 & 0 & -1 & 0 & 0 & -2 & 2 & 0 & 20 \\ \hdashline 0 & 16 & 0 & 4 & 0 & M & 3 & M - 3 & 1 & 120 \end{array}\right]\end{array}$$

Optimal solution: max $P = 120$ at $x_1 = 20$, $x_2 = 0$, $x_3 = 20$.

23. (A) Refer to Problem 5.
The graph of the feasible region is shown at the right. Since it is unbounded, $P = 4x_1 + 3x_2$ does not have a maximum value by Theorem 2(B) in Section 5.2.

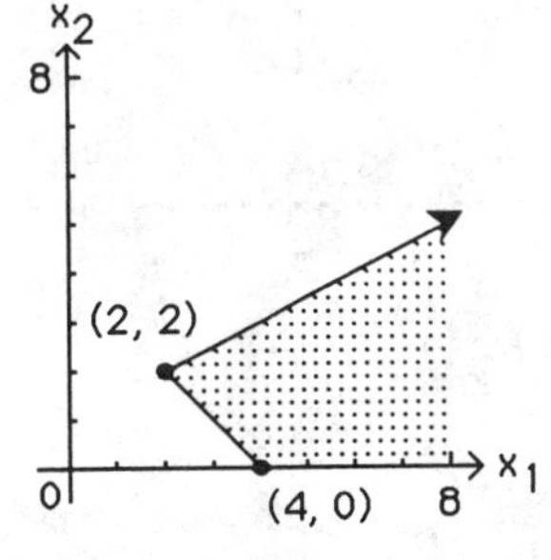

(B) Refer to Problem 7.
The graph of the feasible region is empty. Therefore, $P = 5x_1 + 10x_2$ does not have a maximum value, by Theorem 2(C) in Section 5.2.

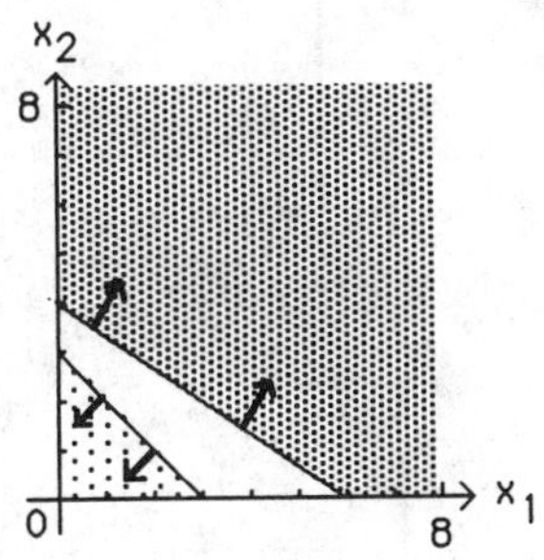

25. We will maximize $P = -C = -10x_1 + 40x_2 + 5x_3$

$$\begin{aligned} \text{Subject to: } x_1 + 3x_2 \qquad + s_1 \qquad &= 6 \\ 4x_2 + x_3 \qquad + s_2 &= 3 \\ x_1, x_2, x_3, s_1, s_2 &\geq 0 \end{aligned}$$

where s_1, s_2 are slack variables. The simplex tableau for this problem is:

$$\begin{array}{c} \\ s_1 \\ s_2 \\ P \end{array}\begin{array}{c} \begin{array}{cccccc} x_1 & x_2 & x_3 & s_1 & s_2 & P \end{array} \\ \left[\begin{array}{cccccc|c} 1 & 3 & 0 & 1 & 0 & 0 & 6 \\ 0 & \textcircled{4} & 1 & 0 & 1 & 0 & 3 \\ \hdashline 10 & -40 & -5 & 0 & 0 & 1 & 0 \end{array}\right] \end{array} \begin{array}{l} \frac{6}{3} = 2 \\ \frac{3}{4} = .75 \end{array} \sim \left[\begin{array}{cccccc|c} 1 & 3 & 0 & 1 & 0 & 0 & 6 \\ 0 & \textcircled{1} & \frac{1}{4} & 0 & \frac{1}{4} & 0 & \frac{3}{4} \\ \hdashline 10 & -40 & -5 & 0 & 0 & 1 & 0 \end{array}\right]$$

$\frac{1}{4}R_2 \to R_2$ $\qquad$ $(-3)R_2 + R_1 \to R_1$ and $40R_2 + R_3 \to R_3$

$$\sim \begin{array}{c} \\ x_1 \\ x_2 \\ P \end{array}\begin{array}{c} \begin{array}{cccccc} x_1 & x_2 & x_3 & s_1 & s_2 & P \end{array} \\ \left[\begin{array}{cccccc|c} 1 & 0 & -\frac{3}{4} & 1 & -\frac{3}{4} & 0 & \frac{15}{4} \\ 0 & 1 & \frac{1}{4} & 0 & \frac{1}{4} & 0 & \frac{3}{4} \\ \hdashline 10 & 0 & 5 & 0 & 10 & 1 & 30 \end{array}\right] \end{array}$$

Optimal solution: min $C = -30$ at $x_1 = 0$, $x_2 = \frac{3}{4}$, $x_3 = 0$.

27. Introduce slack, surplus, and artificial variables to obtain the modified problem: Maximize $P = -5x_1 + 10x_2 + 15x_3 - Ma_1$

Subject to:
$$\begin{aligned} 2x_1 + 3x_2 + x_3 + s_1 &= 24 \\ x_1 - 2x_2 - 2x_3 - s_2 + a_1 &= 1 \\ x_1, x_2, x_3, s_1, s_2, a_1 &\geq 0 \end{aligned}$$

The preliminary simplex tableau for the modified problem is:

$$\begin{array}{c} \begin{array}{ccccccc} x_1 & x_2 & x_3 & s_1 & s_2 & a_1 & P \end{array} \\ \left[\begin{array}{ccccccc|c} 2 & 3 & 1 & 1 & 0 & 0 & 0 & 24 \\ 1 & -2 & -2 & 0 & -1 & 1 & 0 & 1 \\ \hdashline 5 & -10 & -15 & 0 & 0 & M & 1 & 0 \end{array}\right] \end{array}$$

$(-M)R_2 + R_3 \to R_3$

$$\sim \begin{array}{c} \\ s_1 \\ a_1 \\ P \end{array}\begin{array}{c} \begin{array}{ccccccc} x_1 & x_2 & x_3 & s_1 & s_2 & a_1 & P \end{array} \\ \left[\begin{array}{ccccccc|c} 2 & 3 & 1 & 1 & 0 & 0 & 0 & 24 \\ \textcircled{1} & -2 & -2 & 0 & -1 & 1 & 0 & 1 \\ \hdashline -M + 5 & 2M - 10 & 2M - 15 & 0 & M & 0 & 1 & -M \end{array}\right] \end{array} \begin{array}{l} \frac{24}{2} = 12 \\ \frac{1}{1} = 1 \end{array}$$

$(-2)R_2 + R_1 \to R_1$ and $(M - 5)R_2 + R_3 \to R_3$

$$\sim \left[\begin{array}{ccccccc|c} 0 & 7 & \textcircled{5} & 1 & 2 & -2 & 0 & 22 \\ 1 & -2 & -2 & 0 & -1 & 1 & 0 & 1 \\ \hdashline 0 & 0 & -5 & 0 & 5 & M - 5 & 1 & -5 \end{array}\right] \sim \left[\begin{array}{ccccccc|c} 0 & \frac{7}{5} & \textcircled{1} & \frac{1}{5} & \frac{2}{5} & -\frac{2}{5} & 0 & \frac{22}{5} \\ 1 & -2 & -2 & 0 & -1 & 1 & 0 & 1 \\ \hdashline 0 & 0 & -5 & 0 & 5 & M - 5 & 1 & -5 \end{array}\right]$$

$\frac{1}{5}R_1 \to R_1$ $\qquad$ $2R_1 + R_2 \to R_2$ and $5R_1 + R_3 \to R_3$

$$\begin{array}{c} \\ x_3 \\ \sim x_1 \\ P \end{array}\begin{array}{c} \begin{array}{cccccccc} x_1 & x_2 & x_3 & s_1 & s_2 & a_1 & P & \end{array} \\ \left[\begin{array}{ccccccc|c} 0 & \frac{7}{5} & 1 & \frac{1}{5} & \frac{2}{5} & -\frac{2}{5} & 0 & \frac{22}{5} \\ 1 & \frac{4}{5} & 0 & \frac{2}{5} & -\frac{1}{5} & \frac{1}{5} & 0 & \frac{49}{5} \\ \hdashline 0 & 7 & 0 & 1 & 7 & M-7 & 1 & 17 \end{array}\right] \end{array}$$

Optimal solution: max $P = 17$ at $x_1 = \frac{49}{5}$, $x_2 = 0$, $x_3 = \frac{22}{5}$.

29. The matrices corresponding to the given problem and the dual problem are:

$$A = \left[\begin{array}{ccc|c} 1 & 3 & 0 & 6 \\ 0 & 4 & 1 & 3 \\ \hline 10 & 40 & 5 & 1 \end{array}\right] \quad \text{and} \quad A^T = \left[\begin{array}{cc|c} 1 & 0 & 10 \\ 3 & 4 & 40 \\ 0 & 1 & 5 \\ \hline 6 & 3 & 1 \end{array}\right]$$

Thus, the dual problem is: Maximize $P = 6y_1 + 3y_2$

Subject to:
$$\begin{aligned} y_1 \qquad &\le 10 \\ 3y_1 + 4y_2 &\le 40 \\ y_2 &\le 5 \\ y_1, y_2 &\ge 0 \end{aligned}$$

We introduce the slack variables x_1, x_2, and x_3 to obtain the initial system:

$$\begin{aligned} y_1 \qquad\qquad + x_1 \qquad\qquad\qquad &= 10 \\ 3y_1 + 4y_2 \qquad + x_2 \qquad\qquad &= 40 \\ y_2 \qquad\qquad + x_3 \qquad &= 5 \\ -6y_1 - 3y_2 \qquad\qquad\qquad + P &= 0 \end{aligned}$$

The simplex tableau for this problem is:

$$\begin{array}{c} \\ x_1 \\ x_2 \\ x_3 \\ P \end{array}\begin{array}{c} \begin{array}{cccccc} y_1 & y_2 & x_1 & x_2 & x_3 & P \end{array} \\ \left[\begin{array}{cccccc|c} \textcircled{1} & 0 & 1 & 0 & 0 & 0 & 10 \\ 3 & 4 & 0 & 1 & 0 & 0 & 40 \\ 0 & 1 & 0 & 0 & 1 & 0 & 5 \\ \hdashline -6 & -3 & 0 & 0 & 0 & 1 & 0 \end{array}\right] \end{array} \begin{array}{l} \frac{10}{1} = 10 \\ \frac{40}{3} \approx 13.33 \\ \\ \\ \end{array}$$

$(-3)R_1 + R_2 \to R_2$ and $6R_1 + R_4 \to R_4$

$$\sim \left[\begin{array}{cccccc|c} 1 & 0 & 1 & 0 & 0 & 0 & 10 \\ 0 & \textcircled{4} & -3 & 1 & 0 & 0 & 10 \\ 0 & 1 & 0 & 0 & 1 & 0 & 5 \\ \hdashline 0 & -3 & 6 & 0 & 0 & 1 & 60 \end{array}\right] \begin{array}{l} \\ \frac{10}{4} = 2.5 \\ \frac{5}{1} = 5 \\ \\ \end{array} \sim \left[\begin{array}{cccccc|c} 1 & 0 & 1 & 0 & 0 & 0 & 10 \\ 0 & \textcircled{1} & -\frac{3}{4} & \frac{1}{4} & 0 & 0 & \frac{5}{2} \\ 0 & 1 & 0 & 0 & 1 & 0 & 5 \\ \hdashline 0 & -3 & 6 & 0 & 0 & 1 & 60 \end{array}\right]$$

$\frac{1}{4}R_2 \to R_2$

$(-1)R_2 + R_3 \to R_3$ and $3R_2 + R_4 \to R_4$

$$\sim\begin{array}{c} \\ y_1 \\ y_2 \\ x_3 \\ P \end{array}\begin{array}{c} \begin{array}{cccccc} y_1 & y_2 & x_1 & x_2 & x_3 & P \end{array} \\ \left[\begin{array}{cccccc|c} 1 & 0 & 1 & 0 & 0 & 0 & 10 \\ 0 & 1 & -\frac{3}{4} & \frac{1}{4} & 0 & 0 & \frac{5}{2} \\ 0 & 0 & \frac{3}{4} & -\frac{1}{4} & 1 & 0 & \frac{5}{2} \\ \hline 0 & 0 & \frac{15}{4} & \frac{3}{4} & 0 & 1 & \frac{135}{2} \end{array}\right] \end{array}$$

Optimal solution: min $C = \frac{135}{2}$, $x_1 = \frac{15}{4}$, $x_2 = \frac{3}{4}$, $x_3 = 0$.

31. We introduce the slack variables s_1 and s_2 to obtain the initial system:

$$\begin{aligned} x_1 + 3x_2 + x_3 + s_1 \qquad &= 40 \\ 2x_1 + x_2 + 3x_3 \qquad + s_2 \qquad &= 60 \\ -12x_1 - 9x_2 - 5x_3 \qquad\qquad + P &= 0 \end{aligned}$$

The simplex tableau for this problem is:

$$\begin{array}{c} \\ s_1 \\ s_2 \\ \\ \end{array}\begin{array}{c} \begin{array}{cccccc} x_1 & x_2 & x_3 & s_1 & s_2 & P \end{array} \\ \left[\begin{array}{cccccc|c} 1 & 3 & 1 & 1 & 0 & 0 & 40 \\ \textcircled{2} & 1 & 3 & 0 & 1 & 0 & 60 \\ \hline -12 & -9 & -5 & 0 & 0 & 1 & 0 \end{array}\right] \end{array} \begin{array}{l} \frac{40}{1} = 40 \\ \frac{60}{2} = 30 \end{array} \sim \left[\begin{array}{cccccc|c} 1 & 3 & 1 & 1 & 0 & 0 & 40 \\ \textcircled{1} & \frac{1}{2} & \frac{3}{2} & 0 & \frac{1}{2} & 0 & 30 \\ \hline -12 & -9 & -5 & 0 & 0 & 1 & 0 \end{array}\right]$$

$\frac{1}{2}R_2 \to R_2$ $\qquad$ $(-1)R_2 + R_1 \to R_1$ and $12R_2 + R_3 \to R_3$

$$\sim\left[\begin{array}{cccccc|c} 0 & \textcircled{\frac{5}{2}} & -\frac{1}{2} & 1 & -\frac{1}{2} & 0 & 10 \\ 1 & \frac{1}{2} & \frac{3}{2} & 0 & \frac{1}{2} & 0 & 30 \\ \hline 0 & -3 & 13 & 0 & 6 & 1 & 360 \end{array}\right] \begin{array}{l} \frac{10}{5/2} = 4 \\ \frac{30}{1/2} = 60 \end{array} \sim \left[\begin{array}{cccccc|c} 0 & \textcircled{1} & -\frac{1}{5} & \frac{2}{5} & -\frac{1}{5} & 0 & 4 \\ 1 & \frac{1}{2} & \frac{3}{2} & 0 & \frac{1}{2} & 0 & 30 \\ \hline 0 & -3 & 13 & 0 & 6 & 1 & 360 \end{array}\right]$$

$\frac{2}{5}R_1 \to R_1$ $\qquad$ $\left(-\frac{1}{2}\right)R_1 + R_2 \to R_2$ and $3R_1 + R_3 \to R_3$

$$\sim\begin{array}{c} \\ x_2 \\ x_1 \\ P \end{array}\begin{array}{c} \begin{array}{cccccc} x_1 & x_2 & x_3 & s_1 & s_2 & P \end{array} \\ \left[\begin{array}{cccccc|c} 0 & 1 & -\frac{1}{5} & \frac{2}{5} & -\frac{1}{5} & 0 & 4 \\ 1 & 0 & \frac{8}{5} & -\frac{1}{5} & \frac{3}{5} & 0 & 28 \\ \hline 0 & 0 & \frac{62}{5} & \frac{6}{5} & \frac{27}{5} & 1 & 372 \end{array}\right] \end{array}$$

Optimal solution: max $P = 372$ at $x_1 = 28$, $x_2 = 4$, $x_3 = 0$.

33. Let x_1 = the number of 16K modules
and x_2 = the number of 64K modules.
The mathematical model is: Maximize $P = 18x_1 + 30x_2$

$$\begin{aligned} \text{Subject to: } 10x_1 + 15x_2 &\le 2200 \\ 2x_1 + 4x_2 &\le 500 \\ x_1 &\ge 50 \\ x_1, x_2 &\ge 0 \end{aligned}$$

We introduce slack, surplus, and artificial variables to obtain the modified problem: Maximize $P = 18x_1 + 30x_2 - Ma_1$

$$\begin{array}{rl} \text{Subject to: } & 10x_1 + 15x_2 + s_1 = 2200 \\ & 2x_1 + 4x_2 + s_2 = 500 \\ & x_1 - s_3 + a_1 = 50 \\ & x_1, x_2, s_1, s_2, s_3, a_1 \geq 0 \end{array}$$

The preliminary simplex tableau for the modified problem is:

$$\begin{array}{c} \begin{array}{ccccccc} x_1 & x_2 & s_1 & s_2 & s_3 & a_1 & P \end{array} \\ \left[\begin{array}{ccccccc|c} 10 & 15 & 1 & 0 & 0 & 0 & 0 & 2200 \\ 2 & 4 & 0 & 1 & 0 & 0 & 0 & 500 \\ 1 & 0 & 0 & 0 & -1 & 1 & 0 & 50 \\ \hline -18 & -30 & 0 & 0 & 0 & M & 1 & 0 \end{array}\right] \end{array}$$

$(-M)R_3 + R_4 \rightarrow R_4$

$$\sim \begin{array}{c} \\ s_1 \\ s_2 \\ a_1 \\ P \end{array} \begin{array}{c} \begin{array}{ccccccc} x_1 & x_2 & s_1 & s_2 & s_3 & a_1 & P \end{array} \\ \left[\begin{array}{ccccccc|c} 10 & 15 & 1 & 0 & 0 & 0 & 0 & 2200 \\ 2 & 4 & 0 & 1 & 0 & 0 & 0 & 500 \\ \textcircled{1} & 0 & 0 & 0 & -1 & 1 & 0 & 50 \\ \hline -M - 18 & -30 & 0 & 0 & M & 0 & 1 & -50M \end{array}\right] \end{array} \begin{array}{l} \frac{2200}{10} = 220 \\ \frac{500}{2} = 250 \\ \frac{50}{1} = 50 \\ \end{array}$$

$(-10)R_3 + R_1 \rightarrow R_1$, $(-2)R_3 + R_2 \rightarrow R_2$, and $(M + 18)R_3 + R_4 \rightarrow R_4$

$$\sim \left[\begin{array}{ccccccc|c} 0 & 15 & 1 & 0 & 10 & -10 & 0 & 1700 \\ 0 & \textcircled{4} & 0 & 1 & 2 & -2 & 0 & 400 \\ 1 & 0 & 0 & 0 & -1 & 1 & 0 & 50 \\ \hline 0 & -30 & 0 & 0 & -18 & M + 18 & 1 & 900 \end{array}\right] \begin{array}{l} \frac{1700}{15} = 113.33 \\ \frac{400}{4} = 100 \\ \\ \\ \end{array}$$

$\frac{1}{4}R_2 \rightarrow R_2$

$$\sim \left[\begin{array}{ccccccc|c} 0 & 15 & 1 & 0 & 10 & -10 & 0 & 1700 \\ 0 & \textcircled{1} & 0 & \frac{1}{4} & \frac{1}{2} & -\frac{1}{2} & 0 & 100 \\ 1 & 0 & 0 & 0 & -1 & 1 & 0 & 50 \\ \hline 0 & -30 & 0 & 0 & -18 & M + 18 & 1 & 900 \end{array}\right]$$

$(-15)R_2 + R_1 \rightarrow R_1$, $30R_2 + R_4 \rightarrow R_4$

$$\sim \left[\begin{array}{ccccccc|c} 0 & 0 & 1 & -\frac{15}{4} & \textcircled{\frac{5}{2}} & -\frac{5}{2} & 0 & 200 \\ 0 & 1 & 0 & \frac{1}{4} & \frac{1}{2} & -\frac{1}{2} & 0 & 100 \\ 1 & 0 & 0 & 0 & -1 & 1 & 0 & 50 \\ \hline 0 & 0 & 0 & \frac{15}{2} & -3 & M + 3 & 1 & 3900 \end{array}\right] \begin{array}{l} \frac{200}{5/2} = 80 \\ \frac{100}{1/2} = 200 \\ \\ \\ \end{array}$$

$\frac{2}{5}R_1 \rightarrow R_1$

$$\sim \left[\begin{array}{ccccccc|c} 0 & 0 & \frac{2}{5} & -\frac{3}{2} & \textcircled{1} & -1 & 0 & 80 \\ 0 & 1 & 0 & \frac{1}{4} & \frac{1}{2} & -\frac{1}{2} & 0 & 100 \\ 1 & 0 & 0 & 0 & -1 & 1 & 0 & 50 \\ \hdashline 0 & 0 & 0 & \frac{15}{2} & -3 & M+3 & 1 & 3900 \end{array}\right]$$

$\left(-\frac{1}{2}\right)R_1 + R_2 \to R_2,\ R_1 + R_3 \to R_3,\ 3R_1 + R_4 \to R_4$

$$\begin{array}{c} \\ s_3 \\ x_2 \\ x_1 \\ \\ \end{array} \sim \begin{array}{c} \begin{array}{ccccccc} x_1 & x_2 & s_1 & s_2 & s_3 & a_1 & P \end{array} \\ \left[\begin{array}{ccccccc|c} 0 & 0 & \frac{2}{5} & -\frac{3}{2} & 1 & -1 & 0 & 80 \\ 0 & 1 & -\frac{1}{5} & 1 & 0 & 0 & 0 & 60 \\ 1 & 0 & \frac{2}{5} & -\frac{3}{2} & 0 & 0 & 0 & 130 \\ \hdashline 0 & 0 & \frac{6}{5} & 3 & 0 & M & 1 & 4140 \end{array}\right] \end{array}$$

The maximum profit is \$4140 when 130 16K modules and 60 64K modules are manufactured each day.

35. Let x_1 = the number of ads placed in the *Sentinel*,
x_2 = the number of ads placed in the *Journal*,
and x_3 = the number of ads placed in the *Tribune*.

The mathematical model is: Minimize $C = 200x_1 + 200x_2 + 100x_3$

Subject to:
$$\begin{aligned} x_1 + x_2 + x_3 &\le 10 \\ 2000x_1 + 500x_2 + 1500x_3 &\ge 16{,}000 \\ x_1, x_2, x_3 &\ge 0 \end{aligned}$$

Divide the second constraint inequality by 100 to simplify the calculations, and introduce slack, surplus, and artificial variables to obtain the equivalent form:

Maximize $P = -C = -200x_1 - 200x_2 - 100x_3 - Ma_1$

Subject to:
$$\begin{aligned} x_1 + x_2 + x_3 + s_1 &= 10 \\ 20x_1 + 5x_2 + 15x_3 - s_2 + a_1 &= 160 \\ x_1, x_2, x_3, s_1, s_2, a_1 &\ge 0 \end{aligned}$$

The simplex tableau for the modified problem is:

$$\begin{array}{c} \begin{array}{ccccccc} x_1 & x_2 & x_3 & s_1 & s_2 & a_1 & P \end{array} \\ \left[\begin{array}{ccccccc|c} 1 & 1 & 1 & 1 & 0 & 0 & 0 & 10 \\ 20 & 5 & 15 & 0 & -1 & 1 & 0 & 160 \\ \hdashline 200 & 200 & 100 & 0 & 0 & M & 1 & 0 \end{array}\right] \end{array}$$

$(-M)R_2 + R_3 \to R_3$

$$
\begin{array}{c}
\\ s_1 \\ \sim a_1 \\ P
\end{array}
\begin{array}{ccccccc|c}
x_1 & x_2 & x_3 & s_1 & s_2 & a_1 & P & \\
1 & 1 & 1 & 1 & 0 & 0 & 0 & 10 \\
\textcircled{20} & 5 & 15 & 0 & -1 & 1 & 0 & 160 \\
\hline
-20M + 200 & -5M + 200 & -15M + 100 & 0 & M & 0 & 1 & -160M
\end{array}
\quad
\begin{array}{l}
\frac{10}{1} = 10 \\
\frac{160}{20} = 8
\end{array}
$$

$\frac{1}{20}R_2 \rightarrow R_2$

$$
\sim \left[\begin{array}{ccccccc|c}
1 & 1 & 1 & 1 & 0 & 0 & 0 & 10 \\
\textcircled{1} & \frac{1}{4} & \frac{3}{4} & 0 & -\frac{1}{20} & \frac{1}{20} & 0 & 8 \\
\hline
-20M + 200 & -5M + 200 & -15M + 100 & 0 & M & 0 & 1 & -160M
\end{array}\right]
$$

$(-1)R_2 + R_1 \rightarrow R_1$ and $(20M - 200)R_2 + R_3 \rightarrow R_3$

$$
\sim \left[\begin{array}{ccccccc|c}
0 & \frac{3}{4} & \textcircled{$\frac{1}{4}$} & 1 & \frac{1}{20} & -\frac{1}{20} & 0 & 2 \\
1 & \frac{1}{4} & \frac{3}{4} & 0 & -\frac{1}{20} & \frac{1}{20} & 0 & 8 \\
\hline
0 & 150 & -50 & 0 & 10 & M - 10 & 1 & -1600
\end{array}\right]
\begin{array}{l}
\frac{2}{1/4} = 8 \\
\frac{8}{3/4} = \frac{32}{3} \approx 10.67
\end{array}
$$

$4R_1 \rightarrow R_1$

$$
\sim \left[\begin{array}{ccccccc|c}
0 & 3 & \textcircled{1} & 4 & \frac{1}{5} & -\frac{1}{5} & 0 & 8 \\
1 & \frac{1}{4} & \frac{3}{4} & 0 & -\frac{1}{20} & \frac{1}{20} & 0 & 8 \\
\hline
0 & 150 & -50 & 0 & 10 & M - 10 & 1 & -1600
\end{array}\right]
$$

$\left(-\frac{3}{4}\right)R_1 + R_2 \rightarrow R_2$ and $50R_1 + R_3 \rightarrow R_3$

$$
\begin{array}{c}
\\ x_3 \\ \sim x_1 \\ P
\end{array}
\begin{array}{ccccccc|c}
x_1 & x_2 & x_3 & s_1 & s_2 & a_1 & P & \\
0 & 3 & 1 & 4 & \frac{1}{5} & -\frac{1}{5} & 0 & 8 \\
1 & -2 & 0 & -3 & -\frac{1}{5} & \frac{1}{5} & 0 & 2 \\
\hline
0 & 300 & 0 & 200 & 20 & M - 20 & 1 & -1200
\end{array}
$$

The minimal cost is \$1200 when two ads are placed in the *Sentinel*, no ads are placed in the *Journal*, and eight ads are placed in the *Tribune*.

37. Let x_1 = the number of bottles of brand A,
x_2 = the number of bottles of brand B,
and x_3 = the number of bottles of brand C.

The mathematical model is: Minimize $C = 0.6x_1 + 0.4x_2 + 0.9x_3$

Subject to:
$$10x_1 + 10x_2 + 20x_3 \ge 100$$
$$2x_1 + 3x_2 + 4x_3 \le 24$$
$$x_1, x_2, x_3 \ge 0$$

Divide the first inequality by 10, and introduce slack, surplus, and artificial variables to obtain the equivalent form:

Maximize $P = -10C = -6x_1 - 4x_2 - 9x_3 - Ma_1$

Subject to:
$$x_1 + x_2 + 2x_3 - s_1 + a_1 = 10$$
$$2x_1 + 3x_2 + 4x_3 + s_2 = 24$$
$$x_1, x_2, x_3, s_1, s_2, a_1 \ge 0$$

The simplex tableau for the modified problem is:

$$\begin{array}{c} \\ \\ \\ \end{array}\begin{array}{ccccccc|c} x_1 & x_2 & x_3 & s_1 & a_1 & s_2 & P & \\ 1 & 1 & 2 & -1 & 1 & 0 & 0 & 10 \\ 2 & 3 & 4 & 0 & 0 & 1 & 0 & 24 \\ \hline 6 & 4 & 9 & 0 & M & 0 & 1 & 0 \end{array}$$

$(-M)R_1 + R_3 \to R_3$

$$\sim \begin{array}{c} \\ a_1 \\ s_2 \\ \\ \end{array}\begin{array}{ccccccc|c} x_1 & x_2 & x_3 & s_1 & a_1 & s_2 & P & \\ 1 & 1 & \textcircled{2} & -1 & 1 & 0 & 0 & 10 \\ 2 & 3 & 4 & 0 & 0 & 1 & 0 & 24 \\ \hline -M+6 & -M+4 & -2M+9 & M & 0 & 0 & 1 & -10M \end{array} \quad \begin{array}{l} \frac{10}{2} = 5 \\ \frac{24}{4} = 6 \end{array}$$

$\frac{1}{2}R_1 \to R_1$

$$\sim \left[\begin{array}{ccccccc|c} \frac{1}{2} & \frac{1}{2} & \textcircled{1} & -\frac{1}{2} & \frac{1}{2} & 0 & 0 & 5 \\ 2 & 3 & 4 & 0 & 0 & 1 & 0 & 24 \\ \hline -M+6 & -M+4 & -2M+9 & M & 0 & 0 & 1 & -10M \end{array}\right]$$

$(-4)R_1 + R_2 \to R_2$ and $(2M - 9)R_1 + R_3 \to R_3$

$$\sim \left[\begin{array}{ccccccc|c} \frac{1}{2} & \frac{1}{2} & 1 & -\frac{1}{2} & \frac{1}{2} & 0 & 0 & 5 \\ 0 & \textcircled{1} & 0 & 2 & -2 & 1 & 0 & 4 \\ \hline \frac{3}{2} & -\frac{1}{2} & 0 & \frac{9}{2} & M - \frac{9}{2} & 0 & 1 & -45 \end{array}\right] \quad \begin{array}{l} \frac{5}{1/2} = 10 \\ \frac{4}{1} = 4 \end{array}$$

$\left(-\frac{1}{2}\right)R_2 + R_1 \to R_1$ and $\frac{1}{2}R_2 + R_3 \to R_3$

$$\begin{array}{c} \\ x_3 \\ \sim x_2 \\ P \end{array}\begin{array}{c} \begin{array}{ccccccc} x_1 & x_2 & x_3 & s_1 & a_1 & s_2 & P \end{array} \\ \left[\begin{array}{ccccccc|c} \frac{1}{2} & 0 & 1 & -\frac{3}{2} & \frac{3}{2} & -\frac{1}{2} & 0 & 3 \\ 0 & 1 & 0 & 2 & -2 & 1 & 0 & 4 \\ \hline \frac{3}{2} & 0 & 0 & \frac{11}{2} & M - \frac{11}{2} & \frac{1}{2} & 1 & -43 \end{array}\right] \end{array}$$

The minimal cost is \$4.30 when 0 bottles of brand A, 4 bottles of brand B and 3 bottles of brand C are consumed.

39. Let x_1 = the number of cubic yards of mix A,
x_2 = the number of cubic yards of mix B,
and x_3 = the number of cubic yards of mix C.

The mathematical model is: Maximize $P = 12x_1 + 16x_2 + 8x_3$

Subject to:
$$16x_1 + 8x_2 + 16x_3 \geq 800$$
$$12x_1 + 8x_2 + 16x_3 \leq 700$$
$$x_1, x_2, x_3 \geq 0$$

We simplify the inequalities, and introduce slack, surplus, and artificial variables to obtain the modified problem:

Maximize $P = 12x_1 + 16x_2 + 8x_3 - Ma_1$

Subject to:
$$4x_1 + 2x_2 + 4x_3 - s_1 + a_1 = 200$$
$$3x_1 + 2x_2 + 4x_3 + s_2 = 175$$
$$x_1, x_2, x_3, s_1, s_2, a_1 \geq 0$$

The simplex tableau for the modified problem is:

$$\begin{array}{c} \begin{array}{ccccccc} x_1 & x_2 & x_3 & s_1 & a_1 & s_2 & P \end{array} \\ \left[\begin{array}{ccccccc|c} 4 & 2 & 4 & -1 & 1 & 0 & 0 & 200 \\ 3 & 2 & 4 & 0 & 0 & 1 & 0 & 175 \\ \hline -12 & -16 & -8 & 0 & M & 0 & 1 & 0 \end{array}\right] \end{array}$$

$(-M)R_1 + R_3 \to R_3$

$$\begin{array}{c} \\ a_1 \\ \sim s_2 \\ \\ \end{array}\begin{array}{c} \begin{array}{ccccccc} x_1 & x_2 & x_3 & s_1 & a_1 & s_2 & P \end{array} \\ \left[\begin{array}{ccccccc|c} \textcircled{4} & 2 & 4 & -1 & 1 & 0 & 0 & 200 \\ 3 & 2 & 4 & 0 & 0 & 1 & 0 & 175 \\ \hline -4M - 12 & -2M - 16 & -4M - 8 & M & 0 & 0 & 1 & -200M \end{array}\right] \end{array} \begin{array}{l} \frac{200}{4} = 50 \\ \frac{175}{3} \approx 58.33 \end{array}$$

$\frac{1}{4}R_1 \to R_1$

$$\sim \left[\begin{array}{ccccccc|c} \textcircled{1} & \frac{1}{2} & 1 & -\frac{1}{4} & \frac{1}{4} & 0 & 0 & 50 \\ 3 & 2 & 4 & 0 & 0 & 1 & 0 & 175 \\ \hline -4M - 12 & -2M - 16 & -4M - 8 & M & 0 & 0 & 1 & -200M \end{array}\right]$$

$(-3)R_1 + R_2 \to R_2$ and $(4M + 12)R_1 + R_3 \to R_3$

$$\sim\left[\begin{array}{ccccccc|c} 1 & \frac{1}{2} & 1 & -\frac{1}{4} & \frac{1}{4} & 0 & 0 & 50 \\ 0 & \textcircled{\frac{1}{2}} & 1 & \frac{3}{4} & -\frac{3}{4} & 1 & 0 & 25 \\ \hdashline 0 & -10 & 4 & -3 & M+3 & 0 & 1 & 600 \end{array}\right] \begin{array}{l} \frac{50}{1/2} = 100 \\ \frac{25}{1/2} = 50 \\ \end{array}$$

$2R_2 \rightarrow R_2$

$$\sim\left[\begin{array}{ccccccc|c} 1 & \frac{1}{2} & 1 & -\frac{1}{4} & \frac{1}{4} & 0 & 0 & 50 \\ 0 & \textcircled{1} & 2 & \frac{3}{2} & -\frac{3}{2} & 2 & 0 & 50 \\ \hdashline 0 & -10 & 4 & -3 & M+3 & 0 & 1 & 600 \end{array}\right]$$

$\left(-\frac{1}{2}\right)R_2 + R_1 \rightarrow R_1$ and $10R_2 + R_3 \rightarrow R_3$

$$\begin{array}{c} \\ x_1 \\ \sim x_2 \\ \\ \end{array} \begin{array}{c} \begin{array}{ccccccc} x_1 & x_2 & x_3 & s_1 & a_1 & s_2 & P \end{array} \\ \left[\begin{array}{ccccccc|c} 1 & 0 & 0 & -1 & 1 & -1 & 0 & 25 \\ 0 & 1 & 2 & \frac{3}{2} & -\frac{3}{2} & 2 & 0 & 50 \\ \hdashline 0 & 0 & 24 & 12 & M-12 & 20 & 1 & 1100 \end{array}\right] \end{array}$$

The maximum amount of nitrogen is 1100 pounds when 25 cubic yards of mix A, 50 cubic yards of mix B, and 0 cubic yards of mix C are used.

41. Let x_1 = the number of car frames produced in Milwaukee,
x_2 = the number of truck frames produced in Milwaukee,
x_3 = the number of car frames produced in Racine,
and x_4 = the number of truck frames produced in Racine.

The mathematical model for this problem is:

Maximize $P = 50x_1 + 70x_2 + 50x_3 + 70x_4$

Subject to:
$$\begin{aligned} x_1 + x_3 &\le 250 \\ x_2 + x_4 &\le 350 \\ x_1 + x_2 &\le 300 \\ x_3 + x_4 &\le 200 \\ 150x_1 + 200x_2 &\le 50{,}000 \\ 135x_3 + 180x_4 &\le 35{,}000 \\ x_1, x_2, x_3, x_4 &\ge 0 \end{aligned}$$

43. Let x_1 = the number of barrels of A used in regular gasoline,
x_2 = the number of barrels of A used in premium gasoline,
x_3 = the number of barrels of B used in regular gasoline,
x_4 = the number of barrels of B used in premium gasoline,
x_5 = the number of barrels of C used in regular gasoline,
and x_6 = the number of barrels of C used in premium gasoline.

Cost $C = 28(x_1 + x_2) + 30(x_3 + x_4) + 34(x_5 + x_6)$
Revenue $R = 38(x_1 + x_3 + x_5) + 46(x_2 + x_4 + x_6)$
Profit $P = R - C = 10x_1 + 18x_2 + 8x_3 + 16x_4 + 4x_5 + 12x_6$

Thus, the mathematical model for this problem is:

Maximize $P = 10x_1 + 18x_2 + 8x_3 + 16x_4 + 4x_5 + 12x_6$

Subject to:

$$
\begin{aligned}
x_1 + x_2 &\le 40{,}000 \\
x_3 + x_4 &\le 25{,}000 \\
x_5 + x_6 &\le 15{,}000 \\
x_1 + x_3 + x_5 &\ge 30{,}000 \\
x_2 + x_4 + x_6 &\ge 25{,}000 \\
-5x_1 + 5x_3 + 15x_5 &\ge 0 \\
-15x_2 - 5x_4 + 5x_6 &\ge 0 \\
x_1, x_2, x_3, x_4, x_5, x_6 &\ge 0
\end{aligned}
$$

45. Let x_1 = percentage of funds invested in high-tech funds
x_2 = percentage of funds invested in global funds
x_3 = percentage of funds invested in corporate bonds
x_4 = percentage of funds invested in municipal bonds
x_5 = percentage of funds invested in CD's

Risk levels:
High-tech funds: $2.7x_1$
Global funds: $1.8x_2$
Corporate Bonds: $1.2x_3$
Municipal Bonds: $0.5x_4$
CD's: $0x_5$

Total risk level: $2.7x_1 + 1.8x_2 + 1.2x_3 + 0.5x_4$
Return: $0.11x_1 + 0.1x_2 + 0.09x_3 + 0.08x_4 + 0.05x_5$

The mathematical model for this problem is:
Maximize $P = 0.11x_1 + 0.1x_2 + 0.09x_3 + 0.08x_4 + 0.05x_5$

Subject to:

$$
\begin{aligned}
x_1 + x_2 + x_3 + x_4 + x_5 &= 1 \\
2.7x_1 + 1.8x_2 + 1.2x_3 + 0.5x_4 &\le 1.8 \\
x_5 &\ge 0.2 \\
x_1, x_2, x_3, x_4, x_5 &\ge 0
\end{aligned}
$$

47. Let x_1 = the number of ounces of food L,
x_2 = the number of ounces of food M,
and x_3 = the number of ounces of food N.

The mathematical model for this problem is:

Minimize $C = 0.4x_1 + 0.6x_2 + 0.8x_3$

Subject to:

$$
\begin{aligned}
30x_1 + 10x_2 + 30x_3 &\ge 400 \\
10x_1 + 10x_2 + 10x_3 &\ge 200 \\
10x_1 + 30x_2 + 20x_3 &\ge 300 \\
8x_1 + 4x_2 + 6x_3 &\le 150 \\
60x_1 + 40x_2 + 50x_3 &\le 900 \\
x_1, x_2, x_3 &\ge 0
\end{aligned}
$$

49. Let x_1 = the number of students from A enrolled in school I,
x_2 = the number of students from A enrolled in school II,
x_3 = the number of students from B enrolled in school I,
x_4 = the number of students from B enrolled in school II,
x_5 = the number of students from C enrolled in school I,
and x_6 = the number of students from C enrolled in school II.

The mathematical model for this problem is:

Minimize $C = 4x_1 + 8x_2 + 6x_3 + 4x_4 + 3x_5 + 9x_6$

Subject to:
$$
\begin{aligned}
x_1 + x_2 &= 500\\
x_3 + x_4 &= 1200\\
x_5 + x_6 &= 1800\\
x_1 + x_3 + x_5 &\geq 1400\\
x_2 + x_4 + x_6 &\geq 1400\\
x_1 + x_3 + x_5 &\leq 2000\\
x_2 + x_4 + x_6 &\leq 2000\\
x_1 &\leq 300\\
x_2 &\leq 300\\
x_3 &\leq 720\\
x_4 &\leq 720\\
x_5 &\leq 1080\\
x_6 &\leq 1080\\
x_1, x_2, x_3, x_4, x_5, x_6 &\geq 0
\end{aligned}
$$

CHAPTER 5 REVIEW

1. $2x_1 + x_2 \leq 8$
$3x_1 + 9x_2 \leq 27$
$x_1, x_2 \geq 0$

The graphs of the inequalities are shown at the right. The solution region is shaded; it is *bounded*.

The corner points are:
(0, 0), (0, 3), (3, 2), (4, 0)

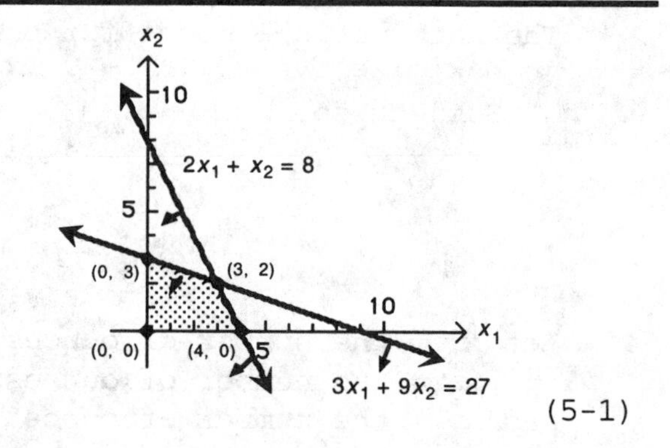

(5-1)

2. $3x_1 + x_2 \geq 9$
$2x_1 + 4x_2 \geq 16$
$x_1, x_2 \geq 0$

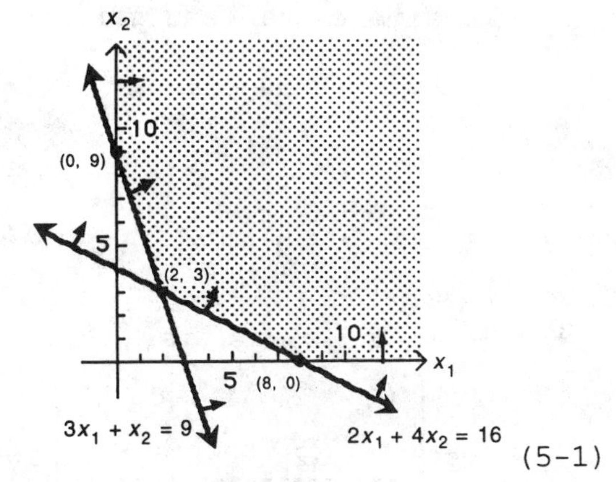

The graphs of the inequalities are shown at the right. The solution region is shaded; it is *unbounded.*

The corner points are:
(0, 9), (2, 3), (8, 0)

(5-1)

3. The feasible region is the solution set of the given inequalities, and is indicated by the shaded region in the graph at the right.

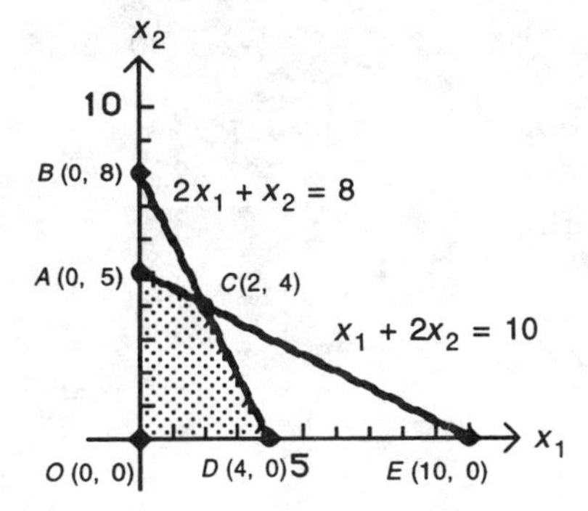

The corner points are (0, 0), (0, 5), (2, 4), and (4, 0).

The value of P at each corner point is:

Corner Point	$P = 6x_1 + 2x_2$
(0, 0)	$P = 6(0) + 2(0) = 0$
(0, 5)	$P = 6(0) + 2(5) = 10$
(2, 4)	$P = 6(2) + 2(4) = 20$
(4, 0)	$P = 6(4) + 2(0) = 24$

Thus, the maximum occurs at $x_1 = 4$, $x_2 = 0$, and the maximum value is $P = 24$.

(5-2)

4. We introduce the slack variables s_1 and s_2 to obtain the system of equations:

$$\begin{aligned} 2x_1 + x_2 + s_1 \quad &= 8 \\ x_1 + 2x_2 \quad + s_2 &= 10 \end{aligned}$$

(5-3)

5. There are 2 basic and 2 nonbasic variables. (5-3)

6. The basic solutions are given in the following table.

x_1	x_2	s_1	s_2	Intersection Point	Feasible?
0	0	8	10	O	Yes
0	8	0	-6	B	No
0	5	3	0	A	Yes
4	0	0	6	D	Yes
10	0	-12	0	E	No
2	4	0	0	C	Yes

(5-3)

7. The simplex tableau for Problem 3 is:

$$\begin{array}{c} \\ \\ \text{Exit } s_1 \\ s_2 \\ P \end{array} \begin{array}{c} \text{Enter} \\ \begin{array}{ccccc|c} x_1 & x_2 & s_1 & s_2 & P & \end{array} \\ \left[\begin{array}{ccccc|c} \textcircled{2} & 1 & 1 & 0 & 0 & 8 \\ 1 & 2 & 0 & 1 & 0 & 10 \\ \hdashline -6 & -2 & 0 & 0 & 1 & 0 \end{array}\right] \end{array} \begin{array}{l} \\ \\ \frac{8}{2} = 4 \\ \frac{10}{1} = 10 \\ \\ \end{array}$$

(5-4)

8.

$$\begin{array}{c} \\ s_1 \\ s_2 \\ P \end{array} \begin{array}{c} \begin{array}{ccccc} x_1 & x_2 & s_1 & s_2 & P \end{array} \\ \left[\begin{array}{ccccc|c} \textcircled{2} & 1 & 1 & 0 & 0 & 8 \\ 1 & 2 & 0 & 1 & 0 & 10 \\ \hdashline -6 & -2 & 0 & 0 & 1 & 0 \end{array}\right] \end{array} \sim \left[\begin{array}{ccccc|c} \textcircled{1} & \frac{1}{2} & \frac{1}{2} & 0 & 0 & 4 \\ 1 & 2 & 0 & 1 & 0 & 10 \\ \hdashline -6 & -2 & 0 & 0 & 1 & 0 \end{array}\right]$$

$\frac{1}{2}R_1 \to R_1$ $\qquad$ $(-1)R_1 + R_2 \to R_2$ and $6R_1 + R_3 \to R_3$

$$\sim \begin{array}{c} \\ x_1 \\ s_2 \\ P \end{array} \begin{array}{c} \begin{array}{ccccc} x_1 & x_2 & s_1 & s_2 & P \end{array} \\ \left[\begin{array}{ccccc|c} 1 & \frac{1}{2} & \frac{1}{2} & 0 & 0 & 4 \\ 0 & \frac{3}{2} & -\frac{1}{2} & 1 & 0 & 6 \\ \hdashline 0 & 1 & 3 & 0 & 1 & 24 \end{array}\right] \end{array}$$

Optimal solution: max $P = 24$ at $x_1 = 4$, $x_2 = 0$.

(5-4)

9.

$$\begin{array}{c} \\ \\ x_2 \\ s_2 \\ \text{Exit } s_3 \\ P \end{array} \begin{array}{c} \text{Enter} \\ \begin{array}{ccccccc} x_1 & x_2 & x_3 & s_1 & s_2 & s_3 & P \end{array} \\ \left[\begin{array}{ccccccc|c} 2 & 1 & 3 & -1 & 0 & 0 & 0 & 20 \\ 3 & 0 & 4 & 1 & 1 & 0 & 0 & 30 \\ \textcircled{2} & 0 & 5 & 2 & 0 & 1 & 0 & 10 \\ \hdashline -8 & 0 & -5 & 3 & 0 & 0 & 1 & 50 \end{array}\right] \end{array} \begin{array}{l} \\ \\ \frac{20}{2} = 10 \\ \frac{30}{3} = 10 \\ \frac{10}{2} = 5 \\ \\ \end{array}$$

The basic variables are x_2, s_2, s_3, and P, and the nonbasic variables are x_1, x_3, and s_1.

The first column is the pivot column and the third row is the pivot row. The pivot element is circled.

$$\sim\left[\begin{array}{ccccccc|c} 2 & 1 & 3 & -1 & 0 & 0 & 0 & 20 \\ 3 & 0 & 4 & 1 & 1 & 0 & 0 & 30 \\ \textcircled{1} & 0 & \frac{5}{2} & 1 & 0 & \frac{1}{2} & 0 & 5 \\ \hdashline -8 & 0 & -5 & 3 & 0 & 0 & 1 & 50 \end{array}\right]$$

$(-2)R_3 + R_1 \to R_1$, $(-3)R_3 + R_2 \to R_2$, $8R_3 + R_4 \to R_4$

$$\sim \begin{array}{c} \\ x_2 \\ s_2 \\ x_1 \\ P \end{array} \begin{array}{c} \begin{array}{ccccccc} x_1 & x_2 & x_3 & s_1 & s_2 & s_3 & P \end{array} \\ \left[\begin{array}{ccccccc|c} 0 & 1 & -2 & -3 & 0 & -1 & 0 & 10 \\ 0 & 0 & -\frac{7}{2} & -2 & 1 & -\frac{3}{2} & 0 & 15 \\ 1 & 0 & \frac{5}{2} & 1 & 0 & \frac{1}{2} & 0 & 5 \\ \hdashline 0 & 0 & 15 & 11 & 0 & 4 & 1 & 90 \end{array}\right] \end{array}$$

(5-4)

10. (A) The basic feasible solution is: $x_1 = 0$, $x_2 = 2$, $s_1 = 0$, $s_2 = 5$, $P = 12$. Additional pivoting is required because the last row contains a negative indicator.

(B) The basic feasible solution is: $x_1 = 0$, $x_2 = 0$, $s_1 = 0$, $s_2 = 7$, $P = 22$. There is no optimal solution because there are no positive elements above the dashed line in the pivot column, column 1.

(C) The basic feasible solution is: $x_1 = 6$, $x_2 = 0$, $s_1 = 15$, $s_2 = 0$, $P = 10$. This is the optimal solution. (5-4)

11. The feasible region is the solution set of the given inequalities and is indicated by the shaded region in the graph at the right.

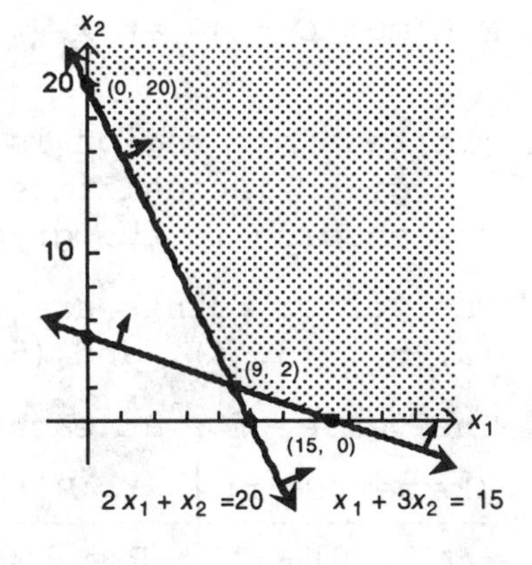

The corner points are (0, 20), (9, 2), and (15, 0).

The value of C at each corner point is:

Corner Point	$C = 5x_1 + 2x_2$
(0, 20)	$C = 5(0) + 2(20) = 40$
(9, 2)	$C = 5(9) + 2(2) = 49$
(15, 0)	$C = 5(15) + 2(0) = 75$

The minimum occurs at $x_1 = 0$, $x_2 = 20$, and the minimum value is $C = 40$.

(5-2)

12. The matrices corresponding to the given problem and the dual problem are:

$$A = \left[\begin{array}{cc|c} 1 & 3 & 15 \\ 2 & 1 & 20 \\ \hline 5 & 2 & 1 \end{array}\right] \quad \text{and} \quad A^T = \left[\begin{array}{cc|c} 1 & 2 & 5 \\ 3 & 1 & 2 \\ \hline 15 & 20 & 1 \end{array}\right]$$

Thus, the dual problem is: Maximize $P = 15y_1 + 20y_2$

Subject to:

$$y_1 + 2y_2 \le 5$$
$$3y_1 + y_2 \le 2$$
$$y_1, y_2 \ge 0$$

(5-5)

13. Introduce the slack variables x_1 and x_2 to obtain the initial system:

$$\begin{aligned} y_1 + 2y_2 + x_1 &= 5 \\ 3y_1 + y_2 + x_2 &= 2 \\ -15y_1 - 20y_2 + P &= 0 \end{aligned}$$

(5-5)

14. The first simplex tableau for the dual problem, Problem 12, is:

$$\begin{matrix} & y_1 & y_2 & x_1 & x_2 & P & \\ x_1 & 1 & 2 & 1 & 0 & 0 & 5 \\ x_2 & 3 & 1 & 0 & 1 & 0 & 2 \\ P & -15 & -20 & 0 & 0 & 1 & 0 \end{matrix}$$

(5-5)

15. Using the simplex method, we have:

$$\left[\begin{array}{rrrrr|r} 1 & 2 & 1 & 0 & 0 & 5 \\ 3 & \textcircled{1} & 0 & 1 & 0 & 2 \\ \hline -15 & -20 & 0 & 0 & 1 & 0 \end{array}\right] \begin{matrix} \frac{5}{2} = 2.5 \\ \frac{2}{1} = 2 \\ \end{matrix}$$

$(-2)R_2 + R_1 \to R_1$, $20R_2 + R_3 \to R_3$

$$\sim \begin{matrix} & y_1 & y_2 & x_1 & x_2 & P & \\ x_1 & -5 & 0 & 1 & -2 & 0 & 1 \\ y_2 & 3 & 1 & 0 & 1 & 0 & 2 \\ P & 45 & 0 & 0 & 20 & 1 & 40 \end{matrix}$$

Optimal solution: max $P = 40$ at $y_1 = 0$ and $y_2 = 2$. (5-4)

16. Minimum $C = 40$ at $x_1 = 0$ and $x_2 = 20$. (5-5)

17. The feasible region is the solution set of the given inequalities and is indicated by the shading in the graph at the right.

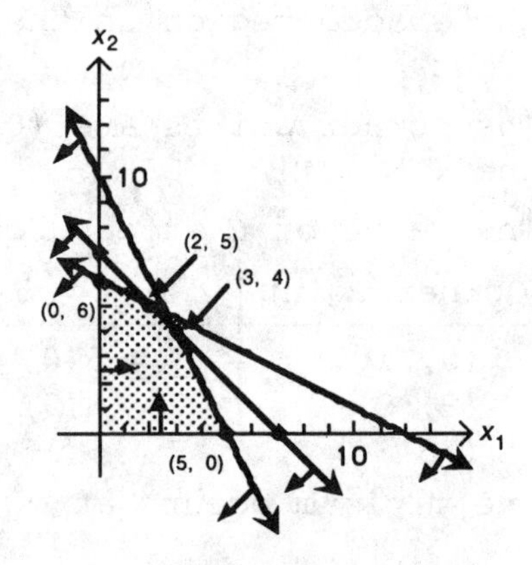

The corner points are (0, 0), (0, 6), (2, 5), (3, 4), and (5, 0).

The value of P at each corner point is:

Corner Point	$P = 3x_1 + 4x_2$
(0, 0)	$P = 3(0) + 4(0) = 0$
(0, 6)	$P = 3(0) + 4(6) = 24$
(2, 5)	$P = 3(2) + 4(5) = 26$
(3, 4)	$P = 3(3) + 4(4) = 25$
(5, 0)	$P = 3(5) + 4(0) = 15$

Thus, the maximum occurs at $x_1 = 2$, $x_2 = 5$, and the maximum value is $P = 26$.

(5-2)

18. We simplify the inequalities and introduce the slack variables s_1, s_2, and s_3 to obtain the equivalent form:

$$\begin{aligned} x_1 + 2x_2 + s_1 \quad\quad\quad &= 12 \\ x_1 + x_2 \quad + s_2 \quad\quad &= 7 \\ 2x_1 + x_2 \quad\quad + s_3 \quad &= 10 \\ -3x_1 - 4x_2 \quad\quad\quad + P &= 0 \end{aligned}$$

The simplex tableau for this problem is:

$$\begin{array}{c} \\ \\ \text{Exit } s_1 \\ s_2 \\ s_3 \\ P \end{array} \begin{array}{c} \quad\ \text{Enter} \\ \begin{matrix} x_1 & x_2 & s_1 & s_2 & s_3 & P \end{matrix} \\ \left[\begin{array}{cccccc|c} 1 & \textcircled{2} & 1 & 0 & 0 & 0 & 12 \\ 1 & 1 & 0 & 1 & 0 & 0 & 7 \\ 2 & 1 & 0 & 0 & 1 & 0 & 10 \\ \hline -3 & -4 & 0 & 0 & 0 & 1 & 0 \end{array}\right] \end{array} \begin{array}{l} \frac{12}{2} = 6 \\ \frac{7}{1} = 7 \\ \frac{10}{1} = 10 \end{array}$$

$\frac{1}{2}R_1 \to R_1$

$$\sim \left[\begin{array}{cccccc|c} \frac{1}{2} & \textcircled{1} & \frac{1}{2} & 0 & 0 & 0 & 6 \\ 1 & 1 & 0 & 1 & 0 & 0 & 7 \\ 2 & 1 & 0 & 0 & 1 & 0 & 10 \\ \hline -3 & -4 & 0 & 0 & 0 & 1 & 0 \end{array}\right] \sim \left[\begin{array}{cccccc|c} \frac{1}{2} & 1 & \frac{1}{2} & 0 & 0 & 0 & 6 \\ \textcircled{\frac{1}{2}} & 0 & -\frac{1}{2} & 1 & 0 & 0 & 1 \\ \frac{3}{2} & 0 & -\frac{1}{2} & 0 & 1 & 0 & 4 \\ \hline -1 & 0 & 2 & 0 & 0 & 1 & 24 \end{array}\right] \begin{array}{l} \frac{6}{1/2} = 12 \\ \frac{1}{1/2} = 2 \\ \frac{4}{3/2} \approx 2.67 \end{array}$$

$(-1)R_1 + R_2 \to R_2$, $(-1)R_1 + R_3 \to R_3$, and $4R_1 + R_4 \to R_4$; $2R_2 \to R_2$

$$\sim \left[\begin{array}{cccccc|c} \frac{1}{2} & 1 & \frac{1}{2} & 0 & 0 & 0 & 6 \\ \textcircled{1} & 0 & -1 & 2 & 0 & 0 & 2 \\ \frac{3}{2} & 0 & -\frac{1}{2} & 0 & 1 & 0 & 4 \\ \hline -1 & 0 & 2 & 0 & 0 & 1 & 24 \end{array}\right] \sim \begin{array}{c} \\ x_2 \\ x_1 \\ s_3 \\ P \end{array} \begin{array}{c} \begin{matrix} x_1 & x_2 & s_1 & s_2 & s_3 & P \end{matrix} \\ \left[\begin{array}{cccccc|c} 0 & 1 & 1 & -1 & 0 & 0 & 5 \\ 1 & 0 & -1 & 2 & 0 & 0 & 2 \\ 0 & 0 & 1 & -3 & 1 & 0 & 1 \\ \hline 0 & 0 & 1 & 2 & 0 & 1 & 26 \end{array}\right] \end{array}$$

$\left(-\frac{1}{2}\right)R_2 + R_1 \to R_1$, $\left(-\frac{3}{2}\right)R_2 + R_3 \to R_3$, and $R_2 + R_4 \to R_4$

Optimal solution: max $P = 26$ at $x_1 = 2$, $x_2 = 5$.

(5-4)

19. The feasible region is the solution set of the given inequalities and is indicated by the shaded region in the graph at the right.

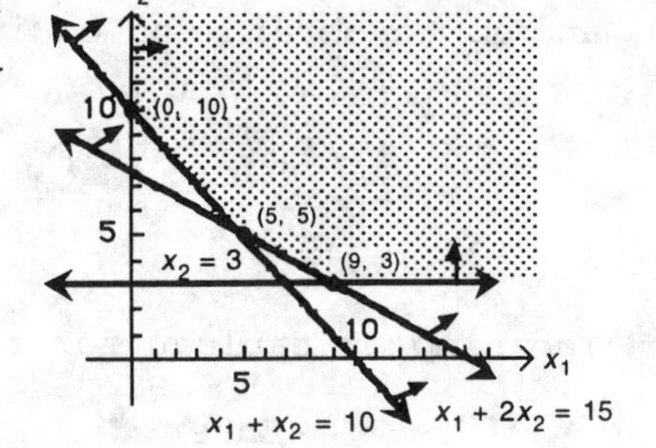

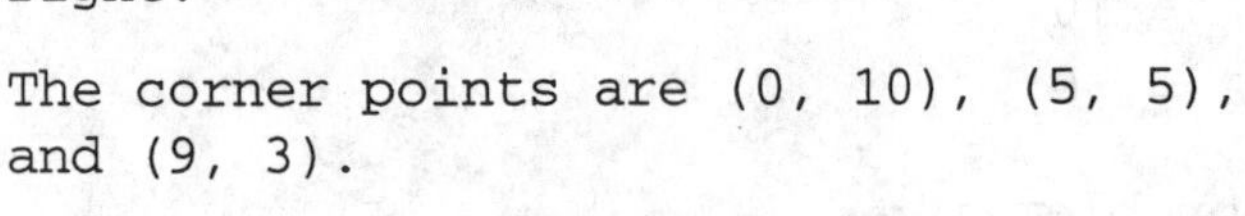

The corner points are (0, 10), (5, 5), and (9, 3).

The value of C at each corner point is:

Corner Point	$C = 3x_1 + 8x_2$
(0, 10)	$C = 3(0) + 8(10) = 80$
(5, 5)	$C = 3(5) + 8(5) = 55$
(9, 3)	$C = 3(9) + 8(3) = 51$

Thus, the minimum occurs at $x_1 = 9$, $x_2 = 3$, and the minimum value is $C = 51$.

(5-2)

20. The matrices corresponding to the given problem and the dual problem are:

$$A = \left[\begin{array}{cc|c} 1 & 1 & 10 \\ 1 & 2 & 15 \\ 0 & 1 & 3 \\ \hline 3 & 8 & 1 \end{array}\right] \quad \text{and} \quad A^T = \left[\begin{array}{ccc|c} 1 & 1 & 0 & 3 \\ 1 & 2 & 1 & 8 \\ \hline 10 & 15 & 3 & 1 \end{array}\right]$$

Thus, the dual problem is: Maximize $P = 10y_1 + 15y_2 + 3y_3$

Subject to: $y_1 + y_2 \le 3$

$y_1 + 2y_2 + y_3 \le 8$

$y_1, y_2, y_3 \ge 0$

(5-5)

21. Introduce the slack variables x_1 and x_2 to obtain the initial system:

$$\begin{aligned} y_1 + y_2 + x_1 &= 3 \\ y_1 + 2y_2 + y_3 + x_2 &= 8 \\ -10y_1 - 15y_2 - 3y_3 + P &= 0 \end{aligned}$$

The simplex tableau for this problem is:

$$\begin{array}{c} \\ x_1 \\ x_2 \\ P \end{array} \begin{array}{c} \begin{array}{cccccc} y_1 & y_2 & y_3 & x_1 & x_2 & P \end{array} \\ \left[\begin{array}{cccccc|c} 1 & \textcircled{1} & 0 & 1 & 0 & 0 & 3 \\ 1 & 2 & 1 & 0 & 1 & 0 & 8 \\ \hline -10 & -15 & -3 & 0 & 0 & 1 & 0 \end{array}\right] \end{array} \begin{array}{l} \frac{3}{1} = 3 \\ \frac{8}{2} = 4 \\ \end{array} \sim \left[\begin{array}{cccccc|c} 1 & 1 & 0 & 1 & 0 & 0 & 3 \\ -1 & 0 & \textcircled{1} & -2 & 1 & 0 & 2 \\ \hline 5 & 0 & -3 & 15 & 0 & 1 & 45 \end{array}\right]$$

$(-2)R_1 + R_2 \to R_2$ and $15R_1 + R_3 \to R_3$ $\qquad$ $3R_2 + R_3 \to R_3$

$$\begin{array}{c} \\ y_2 \\ \sim y_3 \\ \\ \end{array}\left[\begin{array}{cccccc|c} y_1 & y_2 & y_3 & x_1 & x_2 & P & \\ 1 & 1 & 0 & 1 & 0 & 0 & 3 \\ -1 & 0 & 1 & -2 & 1 & 0 & 2 \\ \hline 2 & 0 & 0 & 9 & 3 & 1 & 51 \end{array}\right]$$

Optimal solution: min $C = 51$ at $x_1 = 9$, $x_2 = 3$.

(5-5)

22. Introduce slack variables s_1 and s_2 to obtain the equivalent form:

$$\begin{aligned} x_1 - x_2 - 2x_3 + s_1 \quad\quad &= 3 \\ 2x_1 + 2x_2 - 5x_3 \quad + s_2 \quad &= 10 \\ -5x_1 - 3x_2 + 3x_3 \quad\quad + P &= 0 \end{aligned}$$

The simplex tableau for this problem is:

Enter

$$\begin{array}{c} \\ \text{Exit } s_1 \\ s_2 \\ P \end{array}\left[\begin{array}{cccccc|c} x_1 & x_2 & x_3 & s_1 & s_2 & P & \\ \textcircled{1} & -1 & -2 & 1 & 0 & 0 & 3 \\ 2 & 2 & -5 & 0 & 1 & 0 & 10 \\ \hline -5 & -3 & 3 & 0 & 0 & 1 & 0 \end{array}\right] \begin{array}{l} \\ \frac{3}{1} = 3 \\ \frac{10}{2} = 5 \\ \\ \end{array} \sim \left[\begin{array}{cccccc|c} 1 & -1 & -2 & 1 & 0 & 0 & 3 \\ 0 & \textcircled{4} & -1 & -2 & 1 & 0 & 4 \\ \hline 0 & -8 & -7 & 5 & 0 & 1 & 15 \end{array}\right]$$

$(-2)R_1 + R_2 \to R_2$ and $5R_1 + R_3 \to R_3$ $\qquad \frac{1}{4}R_2 \to R_2$

$$\sim \left[\begin{array}{cccccc|c} 1 & -1 & -2 & 1 & 0 & 0 & 3 \\ 0 & \textcircled{1} & -\frac{1}{4} & -\frac{1}{2} & \frac{1}{4} & 0 & 1 \\ \hline 0 & -8 & -7 & 5 & 0 & 1 & 15 \end{array}\right] \sim \left[\begin{array}{cccccc|c} x_1 & x_2 & x_3 & s_1 & s_2 & P & \\ 1 & 0 & -\frac{9}{4} & \frac{1}{2} & \frac{1}{4} & 0 & 4 \\ 0 & 1 & -\frac{1}{4} & -\frac{1}{2} & \frac{1}{4} & 0 & 1 \\ \hline 0 & 0 & -9 & 1 & 2 & 1 & 23 \end{array}\right]$$

$R_2 + R_1 \to R_1$ and $8R_2 + R_3 \to R_3$

No optimal solution exists; the elements in the pivot column (the x_3 column) above the dashed line are negative. (5-4)

23. Introduce slack variables s_1 and s_2 to obtain the equivalent form:

$$\begin{aligned} x_1 - x_2 - 2x_3 + s_1 \quad\quad &= 3 \\ x_1 + x_2 \quad\quad + s_2 \quad &= 5 \\ -5x_1 - 3x_2 + 3x_3 \quad\quad + P &= 0 \end{aligned}$$

The simplex tableau for this problem is:

Enter

$$\begin{array}{c} \\ \text{Exit } s_1 \\ s_2 \\ P \end{array}\left[\begin{array}{cccccc|c} x_1 & x_2 & x_3 & s_1 & s_2 & P & \\ \textcircled{1} & -1 & -2 & 1 & 0 & 0 & 3 \\ 1 & 1 & 0 & 0 & 1 & 0 & 5 \\ \hline -5 & -3 & 3 & 0 & 0 & 1 & 0 \end{array}\right] \begin{array}{l} \\ \frac{3}{1} = 3 \\ \frac{5}{1} = 5 \\ \\ \end{array} \sim \left[\begin{array}{cccccc|c} 1 & -1 & -2 & 1 & 0 & 0 & 3 \\ 0 & \textcircled{2} & 2 & -1 & 1 & 0 & 2 \\ \hline 0 & -8 & -7 & 5 & 0 & 1 & 15 \end{array}\right]$$

$(-1)R_1 + R_2 \to R_2$ and $5R_1 + R_3 \to R_3$ $\qquad \frac{1}{2}R_2 \to R_2$

$$\sim \left[\begin{array}{cccccc|c} 1 & -1 & -2 & 1 & 0 & 0 & 3 \\ 0 & \textcircled{1} & 1 & -\frac{1}{2} & \frac{1}{2} & 0 & 1 \\ \hline 0 & -8 & -7 & 5 & 0 & 1 & 15 \end{array}\right] \sim \begin{array}{c} \\ x_1 \\ x_2 \\ P \end{array} \begin{array}{c} \begin{array}{cccccc} x_1 & x_2 & x_3 & s_1 & s_2 & P \end{array} \\ \left[\begin{array}{cccccc|c} 1 & 0 & -1 & \frac{1}{2} & \frac{1}{2} & 0 & 4 \\ 0 & 1 & 1 & -\frac{1}{2} & \frac{1}{2} & 0 & 1 \\ \hline 0 & 0 & 1 & 1 & 4 & 1 & 23 \end{array}\right] \end{array}$$

$R_2 + R_1 \rightarrow R_1$ and $8R_2 + R_3 \rightarrow R_3$ Optimal solution: max $P = 23$ at $x_1 = 4$, $x_2 = 1$, $x_3 = 0$. (5-4)

24. (A) We introduce a surplus variable s_1 and an artificial variable a_1 to convert the first inequality ($\geq$) into an equation; we introduce a slack variable s_2 to convert the second inequality ($\leq$) into an equation.

The modified problem is: Maximize $P = x_1 + 3x_2 - Ma_1$

$$\begin{aligned} \text{Subject to: } x_1 + x_2 - s_1 + a_1 &= 6 \\ x_1 + 2x_2 + s_2 &= 8 \\ x_1, x_2, s_1, s_2, a_1 &\geq 0 \end{aligned}$$

(B) The preliminary simplex tableau is:

$$\begin{array}{c} \begin{array}{cccccc} x_1 & x_2 & s_1 & a_1 & s_2 & P \end{array} \\ \left[\begin{array}{cccccc|c} 1 & 1 & -1 & 1 & 0 & 0 & 6 \\ 1 & 2 & 0 & 0 & 1 & 0 & 8 \\ \hline -1 & -3 & 0 & M & 0 & 1 & 0 \end{array}\right] \end{array}$$

Now

$$\begin{array}{c} \\ a_1 \\ s_2 \\ P \end{array} \begin{array}{c} \begin{array}{cccccc} x_1 & x_2 & s_1 & a_1 & s_2 & P \end{array} \\ \left[\begin{array}{cccccc|c} 1 & 1 & -1 & 1 & 0 & 0 & 6 \\ 1 & 2 & 0 & 0 & 1 & 0 & 8 \\ \hline -1 & -3 & 0 & M & 0 & 1 & 0 \end{array}\right] \end{array} \sim \left[\begin{array}{cccccc|c} 1 & 1 & -1 & 1 & 0 & 0 & 6 \\ 1 & 2 & 0 & 0 & 1 & 0 & 8 \\ \hline -M-1 & -M-3 & M & 0 & 0 & 1 & -6M \end{array}\right]$$

$(-M)R_1 + R_3 \rightarrow R_3$

Thus, the initial simplex tableau is:

$$\sim \begin{array}{c} \\ a_1 \\ s_1 \\ P \end{array} \begin{array}{c} \begin{array}{cccccc} x_1 & x_2 & s_1 & a_1 & s_2 & P \end{array} \\ \left[\begin{array}{cccccc|c} 1 & 1 & -1 & 1 & 0 & 0 & 6 \\ 1 & 2 & 0 & 0 & 1 & 0 & 8 \\ \hline -M-1 & -M-3 & M & 0 & 0 & 1 & -6M \end{array}\right] \end{array}$$

(C)

$$\sim \begin{array}{c} \\ a_1 \\ s_2 \\ P \end{array} \begin{array}{c} \begin{array}{cccccc} x_1 & x_2 & s_1 & a_1 & s_2 & P \end{array} \\ \left[\begin{array}{cccccc|c} 1 & 1 & -1 & 1 & 0 & 0 & 6 \\ 1 & \textcircled{2} & 0 & 0 & 1 & 0 & 8 \\ \hline -M-1 & -M-3 & M & 0 & 0 & 1 & -6M \end{array}\right] \end{array} \begin{array}{l} \\ \frac{6}{1} = 6 \\ \frac{8}{2} = 4 \\ \\ \end{array}$$

$\frac{1}{2}R_2 \rightarrow R_2$

$$\sim\left[\begin{array}{cccccc|c} 1 & 1 & -1 & 1 & 0 & 0 & 6 \\ \frac{1}{2} & \textcircled{1} & 0 & 0 & \frac{1}{2} & 0 & 4 \\ \hdashline -M-1 & -M-3 & M & 0 & 0 & 1 & -6M \end{array}\right]$$

$(-1)R_2 + R_1 \rightarrow R_1$, $(M + 3)R_2 + R_3 \rightarrow R_3$

$$\sim\left[\begin{array}{cccccc|c} \textcircled{\tfrac{1}{2}} & 0 & -1 & 1 & -\frac{1}{2} & 0 & 2 \\ \frac{1}{2} & 1 & 0 & 0 & \frac{1}{2} & 0 & 4 \\ \hdashline -\frac{1}{2}M + \frac{1}{2} & 0 & M & 0 & \frac{1}{2}M + \frac{3}{2} & 1 & -2M + 12 \end{array}\right] \begin{array}{l} \frac{2}{1/2} = 4 \\ \frac{4}{1/2} = 8 \\ \\ \end{array}$$

$2R_1 \rightarrow R_1$

$$\sim\left[\begin{array}{cccccc|c} \textcircled{1} & 0 & -2 & 2 & -1 & 0 & 4 \\ \frac{1}{2} & 1 & 0 & 0 & \frac{1}{2} & 0 & 4 \\ \hdashline -\frac{1}{2}M + \frac{1}{2} & 0 & M & 0 & \frac{1}{2}M + \frac{3}{2} & 1 & -2M + 12 \end{array}\right]$$

$\left(-\frac{1}{2}\right)R_1 + R_2 \rightarrow R_2$, $\left(\frac{1}{2}M - \frac{1}{2}\right)R_1 + R_3 \rightarrow R_3$

$$\begin{array}{c} \\ x_1 \\ \sim x_2 \\ P \end{array} \begin{array}{c} \begin{array}{cccccc} x_1 & x_2 & s_1 & a_1 & s_2 & P \end{array} \\ \left[\begin{array}{cccccc|c} 1 & 0 & -2 & 2 & -1 & 0 & 4 \\ 0 & 1 & 1 & -1 & 1 & 0 & 2 \\ \hdashline 0 & 0 & 1 & M-1 & 2 & 1 & 10 \end{array}\right] \end{array}$$

The optimal solution to the modified problem is: Maximum $P = 10$ at $x_1 = 4$, $x_2 = 2$, $s_1 = 0$, $a_1 = 0$, $s_2 = 0$.

(D) Since $a_1 = 0$, the solution to the original problem is: Maximum $P = 10$ at $x_1 = 4$, $x_2 = 2$. (5-4)

25. (A) We introduce a surplus variable s_1 and an artificial variable a_1 to convert the first inequality ($\geq$) into an equation; we introduce a slack variable s_2 to convert the second inequality ($\leq$) into an equation.

The modified problem is: Maximize $P = x_1 + x_2 - Ma_1$

$$\begin{aligned} \text{Subject to: } x_1 + x_2 - s_1 + a_1 &= 5 \\ x_1 + 2x_2 + s_2 &= 4 \\ x_1, x_2, s_1, s_2, a_1 &\geq 0 \end{aligned}$$

(B) The preliminary simplex tableau is:

$$\begin{array}{c} \\ \\ \\ \end{array}\begin{matrix} x_1 & x_2 & s_1 & a_1 & s_2 & P \end{matrix}$$
$$\left[\begin{array}{cccccc|c} 1 & 1 & -1 & 1 & 0 & 0 & 5 \\ 1 & 2 & 0 & 0 & 1 & 0 & 4 \\ \hline -1 & -1 & 0 & M & 0 & 1 & 0 \end{array}\right]$$

Now

$$\left[\begin{array}{cccccc|c} 1 & 1 & -1 & 1 & 0 & 0 & 5 \\ 1 & 2 & 0 & 0 & 1 & 0 & 4 \\ \hline -1 & -1 & 0 & M & 0 & 1 & 0 \end{array}\right] \sim \begin{array}{c} a_1 \\ s_2 \\ P \end{array}\left[\begin{array}{cccccc|c} x_1 & x_2 & s_1 & a_1 & s_2 & P & \\ 1 & 1 & -1 & 1 & 0 & 0 & 5 \\ 1 & 2 & 0 & 0 & 1 & 0 & 4 \\ \hline -M-1 & -M-1 & M & 0 & 0 & 1 & -5M \end{array}\right]$$

$(-M)R_1 + R_3 \to R_3$ Initial simplex tableau

(C)
$$\left[\begin{array}{cccccc|c} 1 & 1 & -1 & 1 & 0 & 0 & 5 \\ \textcircled{1} & 2 & 0 & 0 & 1 & 0 & 4 \\ \hline -M-1 & -M-1 & M & 0 & 0 & 1 & -5M \end{array}\right]$$

$(-1)R_2 + R_1 \to R_1$, $(M+1)R_2 + R_3 \to R_3$

$$\sim \begin{array}{c} \\ a_1 \\ x_1 \\ P \end{array}\left[\begin{array}{cccccc|c} x_1 & x_2 & s_1 & a_1 & s_2 & P & \\ 0 & -1 & -1 & 1 & -1 & 0 & 1 \\ 1 & 2 & 0 & 0 & 1 & 0 & 4 \\ \hline 0 & M+1 & M & 0 & M+1 & 1 & -M+4 \end{array}\right]$$

The optimal solution to the modified problem is: $x_1 = 4$, $x_2 = 0$, $s_1 = 0$, $a_1 = 1$, $s_2 = 0$, $P = -M + 4$. (5-6)

(D) Since $a_1 \neq 0$, the original problem does not have a solution.

26. Multiply the second inequality by -1 to obtain a positive number on the right-hand side. This yields the problem:

Maximize $P = 2x_1 + 3x_2 + x_3$

Subject to:
$$\begin{aligned} x_1 - 3x_2 + x_3 &\le 7 \\ x_1 + x_2 - 2x_3 &\ge 2 \quad \text{(Note: Direction of inequality is reversed.)} \\ 3x_1 + 2x_2 - x_3 &= 4 \\ x_1, x_2, x_3 &\ge 0 \end{aligned}$$

Now, introduce a slack variable s_1 to convert the first inequality ($\le$) into an equation; introduce a surplus variable s_2 and an artificial variable a_1 to convert the second inequality (≥ 0) into an equation; introduce an artificial variable a_2 into the equation.

The modified problem is:

Maximize $P = 2x_1 + 3x_2 + x_3 - Ma_1 - Ma_2$

Subject to:
$$\begin{aligned} x_1 - 3x_2 + x_3 + s_1 &= 7 \\ x_1 + x_2 - 2x_3 - s_2 + a_1 &= 2 \\ 3x_1 + 2x_2 - x_3 + a_2 &= 4 \\ x_1, x_2, x_3, s_1, s_2, a_1, a_2 &\ge 0 \end{aligned}$$
(5-6)

27. The geometric method solves maximization and minimization problems involving two decision variables. If the feasible region is bounded, there are no restrictions on the coefficients or constants. If the feasible region is unbounded and the coefficients of the objective function are positive, then a minimization problem has a solution, but a maximization does not. (5-2)

28. The basic simplex method with slack variables solves standard maximization problems involving ≤ constraints with nonnegative constants on the right side. (5-4)

29. The dual method solves minimization problems with positive coefficients in the objective function. (5-5)

30. The big M method solves any linear programming problem. (5-6)

31. Introduce slack variables s_1, s_2, s_3, and s_4 to obtain:
Maximize $P = 2x_1 + 3x_2$
Subject to:

$$\begin{aligned} x_1 + 2x_2 + s_1 \qquad\qquad\quad &= 22 \\ 3x_1 + x_2 \qquad + s_2 \qquad\quad &= 26 \\ x_1 \qquad\qquad\qquad + s_3 \quad &= 8 \\ x_2 \qquad\qquad\qquad + s_4 &= 10 \\ x_1, x_2, s_1, s_2, s_3, s_4 &\geq 0 \end{aligned}$$

This system can be written in the initial form:

$$\begin{aligned} x_1 + 2x_2 + s_1 \qquad\qquad\qquad &= 22 \\ 3x_1 + x_2 \qquad + s_2 \qquad\qquad &= 26 \\ x_1 \qquad\qquad\qquad + s_3 \qquad &= 8 \\ x_2 \qquad\qquad\qquad + s_4 \quad &= 10 \\ -2x_1 - 3x_2 \qquad\qquad\qquad + P &= 0 \end{aligned}$$

The simplex tableau for this problem is:

$$\begin{array}{c} \\ s_1 \\ s_2 \\ s_3 \\ s_4 \\ P \end{array}\begin{array}{c} \begin{array}{ccccccc} x_1 & x_2 & s_1 & s_2 & s_3 & s_4 & P \end{array} \\ \left[\begin{array}{ccccccc|c} 1 & 2 & 1 & 0 & 0 & 0 & 0 & 22 \\ 3 & 1 & 0 & 1 & 0 & 0 & 0 & 26 \\ 1 & 0 & 0 & 0 & 1 & 0 & 0 & 8 \\ 0 & \textcircled{1} & 0 & 0 & 0 & 1 & 0 & 10 \\ \hline -2 & -3 & 0 & 0 & 0 & 0 & 1 & 0 \end{array}\right] \end{array} \begin{array}{l} \\ \frac{22}{2} = 11 \\ \frac{26}{1} = 26 \\ \\ \frac{10}{1} = 10 \\ \\ \end{array}$$

$(-2)R_4 + R_1 \rightarrow R_1$, $(-1)R_4 + R_2 \rightarrow R_2$,
$3R_4 + R_5 \rightarrow R_5$

Basic Solution						Corner Point
x_1	x_2	s_1	s_2	s_3	s_4	(0, 0)
0	0	22	26	8	10	

$$\begin{array}{c} \\ s_1 \\ s_2 \\ s_3 \\ x_2 \\ \\ \end{array}\begin{array}{c} \begin{array}{ccccccc} x_1 & x_2 & s_1 & s_2 & s_3 & s_4 & P \end{array} \\ \left[\begin{array}{ccccccc|c} \textcircled{1} & 0 & 1 & 0 & 0 & -2 & 0 & 2 \\ 3 & 0 & 0 & 1 & 0 & -1 & 0 & 16 \\ 1 & 0 & 0 & 0 & 1 & 0 & 0 & 8 \\ 0 & 1 & 0 & 0 & 0 & 1 & 0 & 10 \\ \hline -2 & 0 & 0 & 0 & 0 & 3 & 1 & 30 \end{array}\right] \end{array} \begin{array}{l} \frac{2}{1} = 2 \\ \frac{16}{3} = 5.33 \\ \frac{8}{1} = 8 \end{array}$$

$(-3)R_1 + R_2 \to R_2$, $(-1)R_1 + R_3 \to R_3$,
$2R_1 + R_5 \to R_5$

Basic Solution						Corner Point
x_1	x_2	s_1	s_2	s_3	s_4	(0, 10)
0	10	2	16	8	0	

$$\sim \begin{array}{c} \\ x_1 \\ s_2 \\ s_3 \\ x_2 \\ P \end{array}\begin{array}{c} \begin{array}{ccccccc} x_1 & x_2 & s_1 & s_2 & s_3 & s_4 & P \end{array} \\ \left[\begin{array}{ccccccc|c} 1 & 0 & 1 & 0 & 0 & -2 & 0 & 2 \\ 0 & 0 & -3 & 1 & 0 & \textcircled{5} & 0 & 10 \\ 0 & 0 & -1 & 0 & 1 & 2 & 0 & 6 \\ 0 & 1 & 0 & 0 & 0 & 1 & 0 & 10 \\ \hline 0 & 0 & 2 & 0 & 0 & -1 & 1 & 34 \end{array}\right] \end{array} \begin{array}{l} \frac{10}{5} = 2 \\ \frac{6}{2} = 3 \\ \frac{10}{1} = 10 \end{array}$$

$\frac{1}{5}R_2 \to R_2$

Basic Solution						Corner Point
x_1	x_2	s_1	s_2	s_3	s_4	(2, 10)
2	10	0	10	6	0	

$$\sim \left[\begin{array}{ccccccc|c} 1 & 0 & 1 & 0 & 0 & -2 & 0 & 2 \\ 0 & 0 & -\frac{3}{5} & \frac{1}{5} & 0 & \textcircled{1} & 0 & 2 \\ 0 & 0 & -1 & 0 & 1 & 2 & 0 & 6 \\ 0 & 1 & 0 & 0 & 0 & 1 & 0 & 10 \\ \hline 0 & 0 & 2 & 0 & 0 & -1 & 1 & 34 \end{array}\right]$$

$2R_2 + R_1 \to R_1$, $(-2)R_2 + R_3 \to R_3$, $(-1)R_2 + R_4 \to R_4$, $R_2 + R_5 \to R_5$

$$
\begin{array}{c}
 \\ x_1 \\ s_4 \\ \sim\ s_3 \\ x_2 \\ P
\end{array}
\begin{array}{c}
\begin{array}{ccccccc} x_1 & x_2 & s_1 & s_2 & s_3 & s_4 & P \end{array} \\
\left[\begin{array}{ccccccc|c}
1 & 0 & -\frac{1}{5} & \frac{2}{5} & 0 & 0 & 0 & 6 \\
0 & 0 & -\frac{3}{5} & \frac{1}{5} & 0 & 1 & 0 & 2 \\
0 & 0 & \frac{1}{5} & -\frac{2}{5} & 1 & 0 & 0 & 2 \\
0 & 1 & \frac{3}{5} & -\frac{1}{5} & 0 & 0 & 0 & 8 \\
\hdashline
0 & 0 & \frac{7}{5} & \frac{1}{5} & 0 & 0 & 1 & 36
\end{array}\right]
\end{array}
$$

Basic Solution						Corner Point
x_1	x_2	s_1	s_2	s_3	s_4	(6, 8)
6	8	0	0	2	2	

Optimal solution:
max $P = 36$ at $x_1 = 6$, $x_2 = 8$.

The graph of the feasible region and the path to the optimal solution is shown at the right.

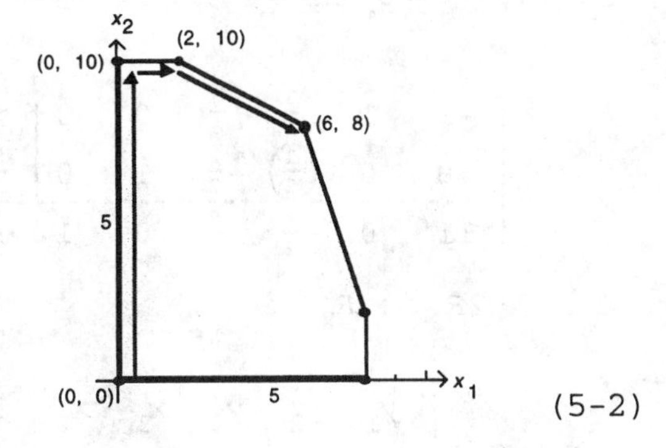

(5-2)

32. Multiply the first constraint inequality by -1 to transform it into a $\geq$ inequality. Now the problem now is: Minimize $C = 3x_1 + 2x_2$

$$
\begin{aligned}
\text{Subject to: } -2x_1 - x_2 &\geq -20 \\
2x_1 + x_2 &\geq 9 \\
x_1 + x_2 &\geq 6 \\
x_1, x_2 &\geq 0
\end{aligned}
$$

The matrices corresponding to this problem and its dual are, respectively:

$$
A = \left[\begin{array}{cc|c}
-2 & -1 & -20 \\
2 & 1 & 9 \\
1 & 1 & 6 \\
\hline
3 & 2 & 1
\end{array}\right]
\quad \text{and} \quad
A^T = \left[\begin{array}{ccc|c}
-2 & 2 & 1 & 3 \\
-1 & 1 & 1 & 2 \\
\hline
-20 & 9 & 6 & 1
\end{array}\right]
$$

Thus, the dual problem is: Maximize $P = -20y_1 + 9y_2 + 6y_3$

$$
\begin{aligned}
\text{Subject to: } -2y_1 + 2y_2 + y_3 &\leq 3 \\
-y_1 + y_2 + y_3 &\leq 2 \\
y_1, y_2, y_3 &\geq 0
\end{aligned}
$$

We introduce slack variables x_1 and x_2 to obtain the initial system for the dual problem:

$$
\begin{aligned}
-2y_1 + 2y_2 + y_3 + x_1 \qquad &= 3 \\
-y_1 + y_2 + y_3 \qquad + x_2 &= 2 \\
20y_1 - 9y_2 - 6y_3 \qquad + P &= 0
\end{aligned}
$$

The simplex tableau for this problem is:

$$\begin{array}{c} \\ x_1 \\ x_2 \\ P \end{array}\begin{array}{c} \begin{array}{cccccc} y_1 & y_2 & y_3 & x_1 & x_2 & P \end{array} \\ \left[\begin{array}{cccccc|c} -2 & \textcircled{2} & 1 & 1 & 0 & 0 & 3 \\ -1 & 1 & 1 & 0 & 1 & 0 & 2 \\ \hdashline 20 & -9 & -6 & 0 & 0 & 1 & 0 \end{array}\right] \end{array} \begin{array}{l} \frac{3}{2} = 1.5 \\ \frac{2}{1} = 2 \\ \end{array} \sim \left[\begin{array}{cccccc|c} -1 & \textcircled{1} & \frac{1}{2} & \frac{1}{2} & 0 & 0 & \frac{3}{2} \\ -1 & 1 & 1 & 0 & 1 & 0 & 2 \\ \hdashline 20 & -9 & -6 & 0 & 0 & 1 & 0 \end{array}\right]$$

$\frac{1}{2}R_1 \to R_1$ $\qquad$ $(-1)R_1 + R_2 \to R_2,\ 9R_1 + R_3 \to R_3$

$$\sim \left[\begin{array}{cccccc|c} -1 & 1 & \frac{1}{2} & \frac{1}{2} & 0 & 0 & \frac{3}{2} \\ 0 & 0 & \textcircled{\tfrac{1}{2}} & -\frac{1}{2} & 1 & 0 & \frac{1}{2} \\ \hdashline 11 & 0 & -\frac{3}{2} & \frac{9}{2} & 0 & 1 & \frac{27}{2} \end{array}\right] \begin{array}{l} \frac{3/2}{1/2} = 3 \\ \frac{1/2}{1/2} = 1 \\ \end{array} \sim \left[\begin{array}{cccccc|c} -1 & 1 & \frac{1}{2} & \frac{1}{2} & 0 & 0 & \frac{3}{2} \\ 0 & 0 & \textcircled{1} & -1 & 2 & 0 & 1 \\ \hdashline 11 & 0 & -\frac{3}{2} & \frac{9}{2} & 0 & 1 & \frac{27}{2} \end{array}\right]$$

$2R_2 \to R_2$ $\qquad$ $\left(-\frac{1}{2}\right)R_2 + R_1 \to R_1,\ \frac{3}{2}R_2 + R_3 \to R_3$

$$\sim \begin{array}{c} \\ y_2 \\ y_3 \\ P \end{array}\begin{array}{c} \begin{array}{cccccc} y_1 & y_2 & y_3 & x_1 & x_2 & P \end{array} \\ \left[\begin{array}{cccccc|c} -1 & 1 & 0 & 1 & -1 & 0 & 1 \\ 0 & 0 & 1 & -1 & 2 & 0 & 1 \\ \hdashline 11 & 0 & 0 & 3 & 3 & 1 & 15 \end{array}\right] \end{array}$$

Optimal solution: min $C = 15$ at $x_1 = 3$ and $x_2 = 3$. (5-5)

33. First convert the problem to a maximization problem by seeking the maximum of $P = -C = -3x_1 - 2x_2$. Next, introduce a slack variable s_1 into the first inequality to convert it into an equation; introduce surplus variables s_2 and s_3 and artificial variables a_1 and a_2 into the second and third inequalities to convert them into equations.

The modified problem is: Maximize $P = -3x_1 - 2x_2 - Ma_1 - Ma_2$

$$\begin{aligned} \text{Subject to: } 2x_1 + x_2 + s_1 &= 20 \\ 2x_1 + x_2 - s_2 + a_1 &= 9 \\ x_1 + x_2 - s_3 + a_2 &= 6 \\ x_1, x_2, s_1, s_2, s_3, a_1, a_2 &\ge 0 \end{aligned}$$

The preliminary simplex tableau is:

$$\begin{array}{c} \begin{array}{cccccccc} x_1 & x_2 & s_1 & s_2 & a_1 & s_3 & a_2 & P \end{array} \\ \left[\begin{array}{cccccccc|c} 2 & 1 & 1 & 0 & 0 & 0 & 0 & 0 & 20 \\ 2 & 1 & 0 & -1 & 1 & 0 & 0 & 0 & 9 \\ 1 & 1 & 0 & 0 & 0 & -1 & 1 & 0 & 6 \\ \hdashline 3 & 2 & 0 & 0 & M & 0 & M & 1 & 0 \end{array}\right] \end{array}$$

$(-M)R_2 + R_4 \to R_4$

$$\sim\left[\begin{array}{cccccccc|c} 2 & 1 & 1 & 0 & 0 & 0 & 0 & 0 & 20 \\ 2 & 1 & 0 & -1 & 1 & 0 & 0 & 0 & 9 \\ 1 & 1 & 0 & 0 & 0 & -1 & 1 & 0 & 6 \\ \hdashline -2M+3 & -M+2 & 0 & M & 0 & 0 & M & 1 & -9M \end{array}\right]$$

$(-M)R_3 + R_4 \rightarrow R_4$

$$\begin{array}{c} \\ s_1 \\ a_1 \\ a_2 \\ P \end{array} \sim \begin{array}{cccccccc|c} x_1 & x_2 & s_1 & s_2 & a_1 & s_3 & a_2 & P & \\ 2 & 1 & 1 & 0 & 0 & 0 & 0 & 0 & 20 \\ \textcircled{2} & 1 & 0 & -1 & 1 & 0 & 0 & 0 & 9 \\ 1 & 1 & 0 & 0 & 0 & -1 & 1 & 0 & 6 \\ \hdashline -3M+3 & -2M+2 & 0 & M & 0 & M & 0 & 1 & -15M \end{array} \quad \begin{array}{l} \\ \frac{20}{2} = 10 \\ \frac{9}{2} = 4.5 \\ \frac{6}{1} = 6 \\ \\ \end{array}$$

$\frac{1}{2}R_2 \rightarrow R_2$

$$\sim\left[\begin{array}{cccccccc|c} 2 & 1 & 1 & 0 & 0 & 0 & 0 & 0 & 20 \\ \textcircled{1} & \frac{1}{2} & 0 & -\frac{1}{2} & \frac{1}{2} & 0 & 0 & 0 & \frac{9}{2} \\ 1 & 1 & 0 & 0 & 0 & -1 & 1 & 0 & 6 \\ \hdashline -3M+3 & -2M+2 & 0 & M & 0 & M & 0 & 1 & -15M \end{array}\right]$$

$(-2)R_2 + R_1 \rightarrow R_1$, $(-1)R_2 + R_3 \rightarrow R_3$, $(3M - 3)R_2 + R_4 \rightarrow R_4$

$$\sim\left[\begin{array}{cccccccc|c} 0 & 0 & 1 & 1 & -1 & 0 & 0 & 0 & 11 \\ 1 & \frac{1}{2} & 0 & -\frac{1}{2} & \frac{1}{2} & 0 & 0 & 0 & \frac{9}{2} \\ 0 & \textcircled{\tfrac{1}{2}} & 0 & \frac{1}{2} & -\frac{1}{2} & -1 & 1 & 0 & \frac{3}{2} \\ \hdashline 0 & -\frac{1}{2}M+\frac{1}{2} & 0 & -\frac{1}{2}M+\frac{3}{2} & \frac{3}{2}M-\frac{3}{2} & M & 0 & 1 & -\frac{3}{2}M-\frac{27}{2} \end{array}\right] \begin{array}{l} \\ \frac{9}{1/2} = 18 \\ \frac{3/2}{1/2} = 3 \\ \\ \end{array}$$

$2R_3 \rightarrow R_3$

$$\sim\left[\begin{array}{cccccccc|c} 0 & 0 & 1 & 1 & -1 & 0 & 0 & 0 & 11 \\ 1 & \frac{1}{2} & 0 & -\frac{1}{2} & \frac{1}{2} & 0 & 0 & 0 & \frac{9}{2} \\ 0 & \textcircled{1} & 0 & 1 & -1 & -2 & 2 & 0 & 3 \\ \hdashline 0 & -\frac{1}{2}M+\frac{1}{2} & 0 & -\frac{1}{2}M+\frac{3}{2} & \frac{3}{2}M-\frac{3}{2} & M & 0 & 1 & -\frac{3}{2}M-\frac{27}{2} \end{array}\right]$$

$\left(-\frac{1}{2}\right)R_3 + R_2 \rightarrow R_2$, $\left(\frac{1}{2}M - \frac{1}{2}\right)R_3 + R_4 \rightarrow R_4$

$$\begin{array}{c} \\ s_1 \\ x_1 \\ x_2 \\ P \end{array} \sim \begin{array}{cccccccc|c} x_1 & x_2 & s_1 & s_2 & a_1 & s_3 & a_2 & P & \\ 0 & 0 & 1 & 1 & -1 & 0 & 0 & 0 & 11 \\ 1 & 0 & 0 & -1 & 1 & 1 & -1 & 0 & 3 \\ 0 & 1 & 0 & 1 & -1 & -2 & 2 & 0 & 3 \\ \hdashline 0 & 0 & 0 & 1 & M-1 & 1 & M-1 & 1 & -15 \end{array}$$

Optimal solution: Max $P = -15$ at $x_1 = 3$, $x_2 = 3$. Thus, the optimal solution of the original problem is: Min $C = 15$ at $x_1 = 3$, $x_2 = 3$. (5-6)

34. Multiply the first two constraint inequalities by -1 to transform them into ≥ inequalitites. The problem now is:

Minimize $C = 15x_1 + 12x_2 + 15x_3 + 18x_4$

Subject to:
$$\begin{aligned} -x_1 - x_2 \qquad\qquad &\geq -240 \\ - x_3 - x_4 &\geq -500 \\ x_1 \qquad + x_3 \qquad &\geq 400 \\ x_2 \qquad + x_4 &\geq 300 \\ x_1, x_2, x_3, x_4 &\geq 0 \end{aligned}$$

The matrices corresponding to this problem and its dual are, respectively:

$$A = \left[\begin{array}{cccc|c} -1 & -1 & 0 & 0 & -240 \\ 0 & 0 & -1 & -1 & -500 \\ 1 & 0 & 1 & 0 & 400 \\ 0 & 1 & 0 & 1 & 300 \\ \hline 15 & 12 & 15 & 18 & 1 \end{array}\right] \text{ and } A^T = \left[\begin{array}{cccc|c} -1 & 0 & 1 & 0 & 15 \\ -1 & 0 & 0 & 1 & 12 \\ 0 & -1 & 1 & 0 & 15 \\ 0 & -1 & 0 & 1 & 18 \\ \hline -240 & -500 & 400 & 300 & 1 \end{array}\right]$$

Thus, the dual problem is: Maximize $P = -240y_1 - 500y_2 + 400y_3 + 300y_4$

Subject to:
$$\begin{aligned} -y_1 \qquad + y_3 \qquad &\leq 15 \\ -y_1 \qquad\qquad + y_4 &\leq 12 \\ - y_2 + y_3 \qquad &\leq 15 \\ - y_2 \qquad + y_4 &\leq 18 \\ y_1, y_2, y_3, y_4 &\geq 0 \end{aligned}$$

We introduce the slack variables x_1, x_2, x_3, and x_4 to obtain the initial system for the dual problem:

$$\begin{aligned} -y_1 \qquad + y_3 \qquad + x_1 \qquad\qquad\qquad &= 15 \\ -y_1 \qquad\qquad + y_4 \qquad + x_2 \qquad\qquad &= 12 \\ - y_2 + y_3 \qquad\qquad\qquad + x_3 \qquad &= 15 \\ - y_2 \qquad + y_4 \qquad\qquad\qquad + x_4 &= 18 \\ 240y_1 + 500y_2 - 400y_3 - 300y_4 \qquad\qquad\qquad\qquad + P &= 0 \end{aligned}$$

The simplex tableau for this problem is:

	y_1	y_2	y_3	y_4	x_1	x_2	x_3	x_4	P		
x_1	-1	0	(1)	0	1	0	0	0	0	15	$\frac{15}{1} = 15$
x_2	-1	0	0	1	0	1	0	0	0	12	
x_3	0	-1	1	0	0	0	1	0	0	15	$\frac{15}{1} = 15$
x_4	0	-1	0	1	0	0	0	1	0	18	
P	240	500	-400	-300	0	0	0	0	1	0	

[Note: Either element can be chosen as the pivot; we choose the element in the first row.]

$(-1)R_1 + R_3 \rightarrow R_3$, $400R_1 + R_5 \rightarrow R_5$

$$\sim \left[\begin{array}{rrrrrrrrr|r} -1 & 0 & 1 & 0 & 1 & 0 & 0 & 0 & 0 & 15 \\ -1 & 0 & 0 & \textcircled{1} & 0 & 1 & 0 & 0 & 0 & 12 \\ 1 & -1 & 0 & 0 & -1 & 0 & 1 & 0 & 0 & 0 \\ 0 & -1 & 0 & 1 & 0 & 0 & 0 & 1 & 0 & 18 \\ \hdashline -160 & 500 & 0 & -300 & 400 & 0 & 0 & 0 & 1 & 6000 \end{array}\right]$$

$(-1)R_2 + R_4 \to R_4$, $300R_2 + R_5 \to R_5$

$$\sim \left[\begin{array}{rrrrrrrrr|r} -1 & 0 & 1 & 0 & 1 & 0 & 0 & 0 & 0 & 15 \\ -1 & 0 & 0 & 1 & 0 & 1 & 0 & 0 & 0 & 12 \\ \textcircled{1} & -1 & 0 & 0 & -1 & 0 & 1 & 0 & 0 & 0 \\ 1 & -1 & 0 & 0 & 0 & -1 & 0 & 1 & 0 & 6 \\ \hdashline -460 & 500 & 0 & 0 & 400 & 300 & 0 & 0 & 1 & 9600 \end{array}\right]$$

$R_3 + R_1 \to R_1$, $R_3 + R_2 \to R_2$, $(-1)R_3 + R_4 \to R_4$, $460R_3 + R_5 \to R_5$

$$\sim \left[\begin{array}{rrrrrrrrr|r} 0 & -1 & 1 & 0 & 0 & 0 & 1 & 0 & 0 & 15 \\ 0 & -1 & 0 & 1 & -1 & 1 & 1 & 0 & 0 & 12 \\ 1 & -1 & 0 & 0 & -1 & 0 & 1 & 0 & 0 & 0 \\ 0 & 0 & 0 & 0 & \textcircled{1} & -1 & -1 & 1 & 0 & 6 \\ \hdashline 0 & 40 & 0 & 0 & -60 & 300 & 460 & 0 & 1 & 9600 \end{array}\right]$$

$60R_4 + R_5 \to R_5$

$$\sim \begin{array}{c} \\ y_3 \\ y_4 \\ y_1 \\ x_4 \\ P \end{array} \begin{array}{c} \begin{array}{ccccccccc} y_1 & y_2 & y_3 & y_4 & x_1 & x_2 & x_3 & x_4 & P \end{array} \\ \left[\begin{array}{rrrrrrrrr|r} 0 & -1 & 1 & 0 & 0 & 0 & 1 & 0 & 0 & 15 \\ 0 & -1 & 0 & 1 & -1 & 1 & 1 & 0 & 0 & 12 \\ 1 & -1 & 0 & 0 & -1 & 0 & 1 & 0 & 0 & 0 \\ 0 & 0 & 0 & 0 & 1 & -1 & -1 & 1 & 0 & 6 \\ \hdashline 0 & 40 & 0 & 0 & 0 & 240 & 400 & 60 & 1 & 9960 \end{array}\right] \end{array}$$

Optimal solution: min $C = 9960$ at $x_1 = 0$, $x_2 = 240$, $x_3 = 400$, $x_4 = 60$.

35. (A) Let x_1 = the number of regular sails
and x_2 = the number of competition sails.
The mathematical model for this problem is:

Maximize $P = 100x_1 + 200x_2$

Subject to: $2x_1 + 3x_2 \le 150$

$4x_1 + 10x_2 \le 380$

$x_1, x_2 \ge 0$

The feasible region is indicated by the shading in the graph below.

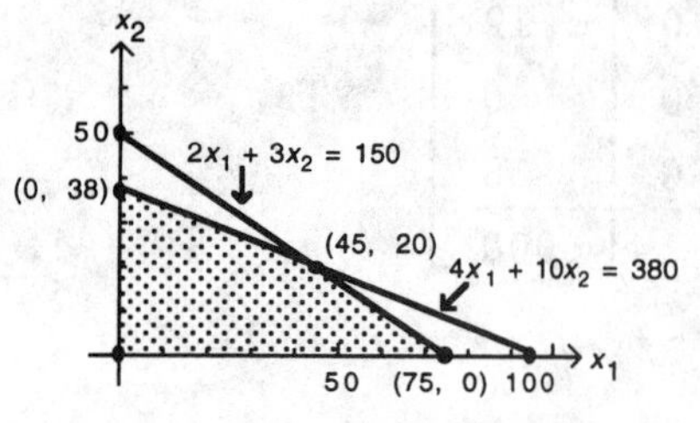

The corner points are (0, 0), (0, 38), (45, 20), (75, 0).

The value P at each corner point is:

Corner point	$P = 100x_1 + 200x_2$
(0, 0)	$P = 100(0) + 200(0) = 0$
(0, 38)	$P = 100(0) + 200(38) = 7{,}600$
(45, 20)	$P = 100(45) + 200(20) = 8{,}500$
(75, 0)	$P = 100(75) + 200(0) = 7{,}500$

Optimal solution: max P = \$8,500 when 45 regular and 20 competition sails are produced.

(B) The mathematical model for this problem is:

$$\begin{aligned} \text{Maximize } P &= 100x_1 + 260x_2 \\ \text{Subject to: } 2x_1 + 3x_2 &\le 150 \\ 4x_1 + 10x_2 &\le 380 \\ x_1, x_2 &\ge 0 \end{aligned}$$

The feasible region and the corner points are the same as in part (A). The value of P at each corner point is:

Corner point	$P = 100x_1 + 260x_2$
(0, 0)	$P = 100(0) + 260(0) = 0$
(0, 38)	$P = 100(0) + 260(38) = 9{,}880$
(45, 20)	$P = 100(45) + 260(20) = 9{,}700$
(75, 0)	$P = 100(75) + 260(0) = 7{,}500$

The maximum profit increases to \$9,880 when 38 competition and 0 regular sails are produced.

(C) The mathematical model for this problem is:

$$\begin{aligned} \text{Maximize } P &= 100x_1 + 140x_2 \\ \text{Subject to: } 2x_1 + 3x_2 &\le 150 \\ 4x_1 + 10x_2 &\le 380 \\ x_1, x_2 &\ge 0 \end{aligned}$$

The feasible region and the corner points are the same as in parts (A) and (B). The value of P at each corner point is:

Corner point	$P = 100x_1 + 140x_2$
(0, 0)	$P = 100(0) + 140(0) = 0$
(0, 38)	$P = 100(0) + 140(38) = 5{,}320$
(45, 20)	$P = 100(45) + 140(20) = 7{,}300$
(75, 0)	$P = 100(75) + 140(0) = 7{,}500$

The maximum profit decreases to \$7,500 when 0 competition and 75 regular sails are produced. (5-2)

36. (A) Let x_1 = amount invested in oil stock
x_2 = amount invested in steel stock
x_3 = amount invested in government bonds

The mathematical model for this problem is:

$$\begin{aligned} &\text{Maximize } P = 0.12x_1 + 0.09x_2 + 0.05x_3 \\ &\text{Subject to: } x_1 + x_2 + x_3 \le 150{,}000 \\ &\qquad x_1 \le 50{,}000 \\ &\qquad x_1 + x_2 - x_3 \le 25{,}000 \\ &\qquad x_1, x_2, x_3 \ge 0 \end{aligned}$$

Introduce slack variables s_1, s_2, s_3 to obtain the initial system:

$$\begin{aligned} x_1 + x_2 + x_3 + s_1 &= 150{,}000 \\ x_1 + s_2 &= 50{,}000 \\ x_1 + x_2 - x_3 + s_3 &= 25{,}000 \\ -0.12x_1 - 0.09x_2 - 0.05x_3 + P &= 0 \\ x_1, x_2, x_3, s_1, s_2, s_3 &\ge 0 \end{aligned}$$

The simplex tableau for this problem is:

$$\begin{array}{c} \\ s_1 \\ s_2 \\ s_3 \\ P \end{array} \begin{array}{c} \begin{array}{ccccccc} x_1 & x_2 & x_3 & s_1 & s_2 & s_3 & P \end{array} \\ \left[\begin{array}{ccccccc|c} 1 & 1 & 1 & 1 & 0 & 0 & 0 & 150{,}000 \\ 1 & 0 & 0 & 0 & 1 & 0 & 0 & 50{,}000 \\ \textcircled{1} & 1 & -1 & 0 & 0 & 1 & 0 & 25{,}000 \\ \hline -0.12 & -0.09 & -0.05 & 0 & 0 & 0 & 1 & 0 \end{array}\right] \end{array}$$

$(-1)R_3 + R_1 \to R_1$, $(-1)R_3 + R_2 \to R_1$, $0.12R_3 + R_4 \to R_4$

$$\sim \left[\begin{array}{ccccccc|c} 0 & 0 & 2 & 1 & 0 & -1 & 0 & 125{,}000 \\ 0 & -1 & \textcircled{1} & 0 & 1 & -1 & 0 & 25{,}000 \\ 1 & 1 & -1 & 0 & 0 & 1 & 0 & 25{,}000 \\ \hline 0 & 0.03 & -0.17 & 0 & 0 & 0.12 & 1 & 3{,}000 \end{array}\right] \begin{array}{l} \frac{125{,}000}{2} = 62{,}500 \\ \frac{25{,}000}{1} = 25{,}000 \\ \\ \\ \end{array}$$

$(-2)R_2 + R_1 \to R_1$, $R_2 + R_3 \to R_3$, $(0.17)R_2 + R_4 \to R_4$

$$\sim \left[\begin{array}{ccccccc|c} 0 & \textcircled{2} & 0 & 1 & -2 & 1 & 0 & 75{,}000 \\ 0 & -1 & 1 & 0 & 1 & -1 & 0 & 25{,}000 \\ 1 & 0 & 0 & 0 & 1 & 0 & 0 & 50{,}000 \\ \hline 0 & -0.14 & 0 & 0 & 0.17 & -0.05 & 1 & 7{,}250 \end{array}\right]$$

$\frac{1}{2}R_1 \to R_1$

$$\sim \left[\begin{array}{ccccccc|c} 0 & \textcircled{1} & 0 & \frac{1}{2} & -1 & \frac{1}{2} & 0 & 37{,}500 \\ 0 & -1 & 1 & 0 & 1 & -1 & 0 & 25{,}000 \\ 1 & 0 & 0 & 0 & 1 & 0 & 0 & 50{,}000 \\ \hline 0 & -0.14 & 0 & 0 & 0.17 & -0.05 & 1 & 7{,}250 \end{array}\right]$$

$R_1 + R_2 \to R_2$, $0.14R_1 + R_4 \to R_4$

$$\sim \begin{array}{c} \\ x_2 \\ x_3 \\ x_1 \\ P \end{array} \left[\begin{array}{ccccccc|c} x_1 & x_2 & x_3 & s_1 & s_2 & s_3 & P & \\ 0 & 1 & 0 & \frac{1}{2} & -1 & \frac{1}{2} & 0 & 37{,}500 \\ 0 & 0 & 1 & \frac{1}{2} & 0 & -\frac{1}{2} & 0 & 62{,}500 \\ 1 & 0 & 0 & 0 & 1 & 0 & 0 & 50{,}000 \\ \hline 0 & 0 & 0 & 0.07 & 0.03 & 0.02 & 1 & 12{,}500 \end{array}\right]$$

The maximum return is \$12,500 when \$50,000 is invested in oil stock, \$37,500 in steel stock, and \$62,500 in government bonds.

(B) The mathematical model for this problem is:

Maximize $P = 0.09x_1 + 0.12x_2 + 0.05x_3$

Subject to:
$$\begin{aligned} x_1 + x_2 + x_3 &\le 150{,}000 \\ x_1 &\le 50{,}000 \\ x_1 + x_2 - x_3 &\le 25{,}000 \\ x_1, x_2, x_3 &\ge 0 \end{aligned}$$

Introduce slack variables s_1, s_2, s_3 to obtain the initial system:

$$\begin{aligned} x_1 + x_2 + x_3 + s_1 \qquad\qquad &= 150{,}000 \\ x_1 \qquad + s_2 \qquad &= 50{,}000 \\ x_1 + x_2 - x_3 \qquad + s_3 &= 25{,}000 \\ -0.09x_1 - 0.12x_2 - 0.05x_3 \qquad + P &= 0 \end{aligned}$$

The simplex tableau for this problem is:

$$\begin{array}{c} \\ s_1 \\ s_2 \\ s_3 \\ P \end{array} \left[\begin{array}{ccccccc|c} x_1 & x_2 & x_3 & s_1 & s_2 & s_3 & P & \\ 1 & 1 & 1 & 1 & 0 & 0 & 0 & 150{,}000 \\ 1 & 0 & 0 & 0 & 1 & 0 & 0 & 50{,}000 \\ 1 & \textcircled{1} & -1 & 0 & 0 & 1 & 0 & 25{,}000 \\ \hline -0.09 & -0.12 & -0.05 & 0 & 0 & 0 & 1 & 0 \end{array}\right]$$

$(-1)R_3 + R_1 \to R_1$, $0.12R_3 + R_4 \to R_4$

$$\sim \left[\begin{array}{ccccccc|c} 0 & 0 & \textcircled{2} & 1 & 0 & -1 & 0 & 125{,}000 \\ 1 & 0 & 0 & 0 & 1 & 0 & 0 & 50{,}000 \\ 1 & 1 & -1 & 0 & 0 & 1 & 0 & 25{,}000 \\ \hline 0.03 & 0 & -0.17 & 0 & 0 & 0.12 & 1 & 3{,}000 \end{array}\right]$$

$\frac{1}{2}R_1 \to R_1$

$$\sim \left[\begin{array}{ccccccc|c} 0 & 0 & \textcircled{1} & \frac{1}{2} & 0 & -\frac{1}{2} & 0 & 62{,}500 \\ 1 & 0 & 0 & 0 & 1 & 0 & 0 & 50{,}000 \\ 1 & 1 & -1 & 0 & 0 & 1 & 0 & 25{,}000 \\ \hline 0.03 & 0 & -0.17 & 0 & 0 & 0.12 & 1 & 3{,}000 \end{array}\right]$$

$R_1 + R_3 \to R_3$, $0.17R_1 + R_4 \to R_4$

$$\sim \begin{array}{c} x_3 \\ s_2 \\ x_2 \\ P \end{array}\left[\begin{array}{ccccccc|c} x_1 & x_2 & x_3 & s_1 & s_2 & s_3 & P & \\ 0 & 0 & 1 & \frac{1}{2} & 0 & -\frac{1}{2} & 0 & 62,500 \\ 1 & 0 & 0 & 0 & 1 & 0 & 0 & 50,000 \\ 1 & 1 & 0 & \frac{1}{2} & 0 & \frac{1}{2} & 0 & 87,500 \\ \hline 0.03 & 0 & 0 & 0.085 & 0 & 0.35 & 1 & 13,625 \end{array}\right]$$

The maximum return is \$13,625 when \$0 is invested in oil stock, \$87,500 is invested in steel stock, and \$62,500 in invested in government bonds. (5-4)

37. Let x_1 = the number of motors from A to X,
x_2 = the number of motors from A to Y,
x_3 = the number of motors from B to X,
x_4 = the number of motors from B to Y.

The mathematical model for this problem is:

Minimize $C = 5x_1 + 8x_2 + 9x_3 + 7x_4$

Subject to:
$$\begin{aligned} x_1 + x_2 &\le 1,500 \\ x_3 + x_4 &\le 1,000 \\ x_1 + x_3 &\ge 900 \\ x_2 + x_4 &\ge 1,200 \end{aligned}$$

Multiply the first two inequalities by -1 to obtain $\ge$ inequalities. The model then becomes:

Minimize $C = 5x_1 + 8x_2 + 9x_3 + 7x_4$

Subject to:
$$\begin{aligned} -x_1 - x_2 &\ge -1,500 \\ -x_3 - x_4 &\ge -1,000 \\ x_1 + x_3 &\ge 900 \\ x_2 + x_4 &\ge 1,200 \\ x_1, x_2, x_3, x_4 &\ge 0 \end{aligned}$$

The matrices for this problem and the dual problem are:

$$A = \left[\begin{array}{cccc|c} -1 & -1 & 0 & 0 & -1,500 \\ 0 & 0 & -1 & -1 & -1,000 \\ 1 & 0 & 1 & 0 & 900 \\ 0 & 1 & 0 & 1 & 1,200 \\ \hline 5 & 8 & 9 & 7 & 1 \end{array}\right] \text{ and } A^T = \left[\begin{array}{cccc|c} -1 & 0 & 1 & 0 & 5 \\ -1 & 0 & 0 & 1 & 8 \\ 0 & -1 & 1 & 0 & 9 \\ 0 & -1 & 0 & 1 & 7 \\ \hline -1,500 & -1,000 & 900 & 1,200 & 1 \end{array}\right]$$

The dual problem is:

Maximize $P = -1,500y_1 - 1,000y_2 + 900y_3 + 1,200y_4$

Subject to:
$$\begin{aligned} -y_1 + y_3 &\le 5 \\ -y_1 + y_4 &\le 8 \\ -y_2 + y_3 &\le 9 \\ -y_2 + y_4 &\le 7 \\ y_1, y_2, y_3, y_4 &\ge 0 \end{aligned}$$

Introduce slack variables x_1, x_2, x_3, x_4 to obtain the initial system:

$$\begin{aligned}
-y_1 \quad\quad + y_3 \quad\quad + x_1 \quad\quad\quad\quad\quad\quad &= 5 \\
-y_1 \quad\quad\quad\quad + y_4 \quad\quad + x_2 \quad\quad\quad\quad &= 8 \\
-y_2 + y_3 \quad\quad\quad\quad\quad\quad + x_3 \quad\quad &= 9 \\
-y_2 \quad\quad + y_4 \quad\quad\quad\quad\quad\quad + x_4 &= 7 \\
1{,}500y_1 + 1{,}000y_2 - 900y_3 - 1{,}200y_4 \quad + P &= 0
\end{aligned}$$

The simplex tableau for this problem is:

$$\begin{array}{c}
\\ x_1 \\ x_2 \\ x_3 \\ x_4 \\ P
\end{array}
\begin{array}{c}
\begin{array}{ccccccccc|c} y_1 & y_2 & y_3 & y_4 & x_1 & x_2 & x_3 & x_4 & P & \end{array} \\
\left[\begin{array}{ccccccccc|c}
-1 & 0 & 1 & 0 & 1 & 0 & 0 & 0 & 0 & 5 \\
-1 & 0 & 0 & 1 & 0 & 1 & 0 & 0 & 0 & 8 \\
0 & -1 & 1 & 0 & 0 & 0 & 1 & 0 & 0 & 9 \\
0 & -1 & 0 & \textcircled{1} & 0 & 0 & 0 & 1 & 0 & 7 \\
\hdashline
1{,}500 & 1{,}000 & -900 & -1{,}200 & 0 & 0 & 0 & 0 & 1 & 0
\end{array}\right]
\end{array}$$

$(-1)R_4 + R_2 \to R_2$, $1{,}200R_4 + R_5 \to R_5$

$$\sim \left[\begin{array}{ccccccccc|c}
-1 & 0 & \textcircled{1} & 0 & 1 & 0 & 0 & 0 & 0 & 5 \\
-1 & 1 & 0 & 0 & 0 & 1 & 0 & -1 & 0 & 1 \\
0 & -1 & 1 & 0 & 0 & 0 & 1 & 0 & 0 & 9 \\
0 & -1 & 0 & 1 & 0 & 0 & 0 & 1 & 0 & 7 \\
\hdashline
1{,}500 & -200 & -900 & 0 & 0 & 0 & 0 & 1{,}200 & 1 & 8{,}400
\end{array}\right]
\begin{array}{l} \frac{5}{1} = 5 \\ \\ \frac{9}{1} = 9 \\ \\ \\ \end{array}$$

$(-1)R_1 + R_3 \to R_3$, $900R_1 + R_5 \to R_5$

$$\sim \left[\begin{array}{ccccccccc|c}
-1 & 0 & 1 & 0 & 1 & 0 & 0 & 0 & 0 & 5 \\
-1 & \textcircled{1} & 0 & 0 & 0 & 1 & 0 & -1 & 0 & 1 \\
1 & -1 & 0 & 0 & -1 & 0 & 1 & 0 & 0 & 4 \\
0 & -1 & 0 & 1 & 0 & 0 & 0 & 1 & 0 & 7 \\
\hdashline
600 & -200 & 0 & 0 & 900 & 0 & 0 & 1{,}200 & 1 & 12{,}900
\end{array}\right]$$

$R_2 + R_3 \to R_3$, $R_2 + R_4 \to R_4$, $200R_2 + R_5 \to R_5$

$$\sim
\begin{array}{c}
\\ y_3 \\ y_2 \\ x_3 \\ y_4 \\ P
\end{array}
\begin{array}{c}
\begin{array}{ccccccccc|c} y_1 & y_2 & y_3 & y_4 & x_1 & x_2 & x_3 & x_4 & P & \end{array} \\
\left[\begin{array}{ccccccccc|c}
-1 & 0 & 1 & 0 & 1 & 0 & 0 & 0 & 0 & 5 \\
-1 & 1 & 0 & 0 & 0 & 1 & 0 & -1 & 0 & 1 \\
0 & 0 & 0 & 0 & -1 & 1 & 1 & -1 & 0 & 5 \\
-1 & 0 & 0 & 1 & 0 & 1 & 0 & 0 & 0 & 8 \\
\hdashline
400 & 0 & 0 & 0 & 900 & 200 & 0 & 1{,}000 & 1 & 13{,}100
\end{array}\right]
\end{array}$$

Optimal soltuion: min C = \$13,100 when 900 motors are shipped from factory *A* to plant *X*, 200 motors are shipped from factory *A* to plant *Y*, 0 motors are shipped from factory *B* to plant *X*, and 1,000 motors are shipped from factory *B* to plant *Y*. (5-5)

38. Let x_1 = number of pounds of long grain rice in Brand A
Let x_2 = number of pounds of long grain rice in Brand B
Let x_3 = number of pounds of wild rice in Brand A
Let x_4 = number of pounds of wild rice in Brand B

The mathematical model for this problem is:

Maximize $P = 0.8x_1 + 0.5x_2 - 1.9x_3 - 2.2x_4$

Subject to:
$$x_3 \ge 0.1(x_1 + x_3)$$
$$x_4 \ge 0.05(x_2 + x_4)$$
$$x_1 + x_2 \le 8{,}000$$
$$x_3 + x_4 \le 500$$
$$x_1,\ x_2,\ x_3,\ x_4 \ge 0$$

which is the same as:

Maximize $P = 0.8x_1 + 0.5x_2 - 1.9x_3 - 2.2x_4$

Subject to:
$$\begin{aligned} 0.1x_1 \qquad\qquad - 0.9x_3 \qquad\qquad &\le 0 \\ 0.05x_2 \qquad\qquad - 0.95x_4 &\le 0 \\ x_1 + x_2 \qquad\qquad\qquad &\le 8{,}000 \\ x_3 + x_4 &\le 500 \\ x_1,\ x_2,\ x_3,\ x_4 &\ge 0 \end{aligned}$$

Introduce slack variables s_1, s_2, s_3, and s_4 to obtain the initial system:

$$\begin{array}{c} \\ y_3 \\ y_2 \\ x_3 \\ y_4 \\ P \end{array} \begin{array}{c} \begin{array}{cccccccccc} x_1 & x_2 & x_3 & x_4 & s_1 & s_2 & s_3 & s_4 & P & \end{array} \\ \left[\begin{array}{ccccccccc|c} \textcircled{0.1} & 0 & -0.9 & 0 & 1 & 0 & 0 & 0 & 0 & 0 \\ 0 & 0.05 & 0 & -0.95 & 0 & 1 & 0 & 0 & 0 & 0 \\ 1 & 1 & 0 & 0 & 0 & 0 & 1 & 0 & 0 & 8{,}000 \\ 0 & 0 & 1 & 1 & 0 & 0 & 0 & 1 & 0 & 500 \\ \hdashline -0.8 & -0.5 & 1.9 & 2.2 & 0 & 0 & 0 & 0 & 1 & 0 \end{array}\right] \end{array} \begin{array}{l} \frac{0}{0.1} = 1 \\ \\ \frac{8000}{1} = 8{,}000 \\ \\ \\ \end{array}$$

$10R_1 \to R_1$

$$\sim \left[\begin{array}{ccccccccc|c} \textcircled{1} & 0 & -9 & 0 & 10 & 0 & 0 & 0 & 0 & 0 \\ 0 & 0.05 & 0 & -0.95 & 0 & 1 & 0 & 0 & 0 & 0 \\ 1 & 1 & 0 & 0 & 0 & 0 & 1 & 0 & 0 & 8{,}000 \\ 0 & 0 & 1 & 1 & 0 & 0 & 0 & 1 & 0 & 500 \\ \hdashline -0.8 & -0.5 & 1.9 & 2.2 & 0 & 0 & 0 & 0 & 1 & 0 \end{array}\right]$$

$(-1)R_1 + R_3 \to R_3,\ 0.8R_1 + R_5 \to R_5$

$$\sim \left[\begin{array}{ccccccccc|c} 1 & 0 & -9 & 0 & 10 & 0 & 0 & 0 & 0 & 0 \\ 0 & 0.05 & 0 & -0.95 & 0 & 1 & 0 & 0 & 0 & 0 \\ 0 & 1 & 9 & 0 & -10 & 0 & 1 & 0 & 0 & 8{,}000 \\ 0 & 0 & \textcircled{1} & 1 & 0 & 0 & 0 & 1 & 0 & 500 \\ \hdashline 0 & -0.5 & -5.3 & 2.2 & 8 & 0 & 0 & 0 & 1 & 0 \end{array}\right] \begin{array}{l} \\ \\ \frac{8000}{9} \approx 888.9 \\ \frac{500}{1} = 500 \\ \\ \end{array}$$

$9R_4 + R_1 \to R_1,\ (-9)R_4 + R_3 \to R_3,\ 5.3R_4 + R_5 \to R_5$

$$\sim\left[\begin{array}{ccccccccc|c} 1 & 0 & 0 & 9 & 10 & 0 & 0 & 9 & 0 & 4{,}500 \\ 0 & \textcircled{0.05} & 0 & -0.95 & 0 & 1 & 0 & 0 & 0 & 0 \\ 0 & 1 & 0 & -9 & -10 & 0 & 1 & -9 & 0 & 3{,}500 \\ 0 & 0 & 1 & 1 & 0 & 0 & 0 & 1 & 0 & 500 \\ \hdashline 0 & -0.5 & 0 & 7.5 & 8 & 0 & 0 & 5.3 & 1 & 2{,}650 \end{array}\right] \quad \begin{array}{l} \frac{0}{0.05} = 0 \\ \frac{3{,}500}{1} = 3{,}500 \end{array}$$

$20R_2 \to R_2$

$$\sim\left[\begin{array}{ccccccccc|c} 1 & 0 & 0 & 9 & 10 & 0 & 0 & 9 & 0 & 4{,}500 \\ 0 & \textcircled{1} & 0 & -19 & 0 & 20 & 0 & 0 & 0 & 0 \\ 0 & 1 & 0 & -9 & -10 & 0 & 1 & -9 & 0 & 3{,}500 \\ 0 & 0 & 1 & 1 & 0 & 0 & 0 & 1 & 0 & 500 \\ \hdashline 0 & -0.5 & 0 & 7.5 & 8 & 0 & 0 & 5.3 & 1 & 2{,}650 \end{array}\right]$$

$(-1)R_2 + R_3 \to R_3,\ (0.5)R_2 + R_5 \to R_5$

$$\sim\left[\begin{array}{ccccccccc|c} 1 & 0 & 0 & 9 & 10 & 0 & 0 & 9 & 0 & 4{,}500 \\ 0 & 1 & 0 & -19 & 0 & 20 & 0 & 0 & 0 & 0 \\ 0 & 0 & 0 & \textcircled{10} & -10 & -20 & 1 & -9 & 0 & 3{,}500 \\ 0 & 0 & 1 & 1 & 0 & 0 & 0 & 1 & 0 & 500 \\ \hdashline 0 & 0 & 0 & -2 & 8 & 10 & 0 & 5.3 & 1 & 2{,}650 \end{array}\right] \quad \begin{array}{l} \frac{4{,}500}{9} = 500 \\ \\ \frac{3{,}500}{10} = 350 \\ \frac{500}{1} = 500 \end{array}$$

$\frac{1}{10}R_3 \to R_3$

$$\sim\left[\begin{array}{ccccccccc|c} 1 & 0 & 0 & 9 & 10 & 0 & 0 & 9 & 0 & 4{,}500 \\ 0 & 1 & 0 & -19 & 0 & 20 & 0 & 0 & 0 & 0 \\ 0 & 0 & 0 & \textcircled{1} & -1 & -2 & \frac{1}{10} & -\frac{9}{10} & 0 & 350 \\ 0 & 0 & 1 & 1 & 0 & 0 & 0 & 1 & 0 & 500 \\ \hdashline 0 & 0 & 0 & -2 & 8 & 10 & 0 & 5.3 & 1 & 2{,}650 \end{array}\right]$$

$(-9)R_3 + R_1 \to R_1,\ 19R_3 + R_2 \to R_2,\ (-1)R_3 + R_4 \to R_4,\ 2R_3 + R_5 \to R_5$

$$\sim \begin{array}{c} \\ x_1 \\ x_2 \\ x_4 \\ x_3 \\ P \end{array} \begin{array}{c} \begin{array}{ccccccccc} x_1 & x_2 & x_3 & x_4 & s_1 & s_2 & s_{31} & s_4 & P \end{array} \\ \left[\begin{array}{ccccccccc|c} 1 & 0 & 0 & 0 & 19 & 18 & \frac{9}{10} & 0 & \frac{171}{10} & 1{,}350 \\ 0 & 1 & 0 & 0 & -19 & -18 & -\frac{19}{10} & 0 & \frac{171}{10} & 6{,}650 \\ 0 & 0 & 0 & 1 & -1 & -2 & -\frac{1}{10} & 0 & \frac{9}{10} & 350 \\ 0 & 0 & 1 & 0 & 1 & 2 & \frac{1}{10} & 0 & \frac{19}{10} & 150 \\ \hdashline 0 & 0 & 0 & 0 & 6 & 6 & \frac{1}{5} & 1 & \frac{7}{2} & 3{,}350 \end{array}\right] \end{array}$$

The maximum profit is $3,350 when 1,350 pounds of long grain rice and 150 pounds of wild rice are used to produce 1,500 pounds of brand *A*, and 6,650 pounds of long grain rice and 350 pounds of wild rice are used to produce 7,000 pounds of brand *B*. (5-5)

39. Let x_1 = number of grams of mix A
x_2 = number of grams of mix B

The constraints are:

vitamins: $2x_1 + 5x_2 \geq 850$
$2x_1 + 4x_2 \geq 800$
$4x_1 + 5x_2 \geq 1,150$
$x_1, x_2 \geq 0$

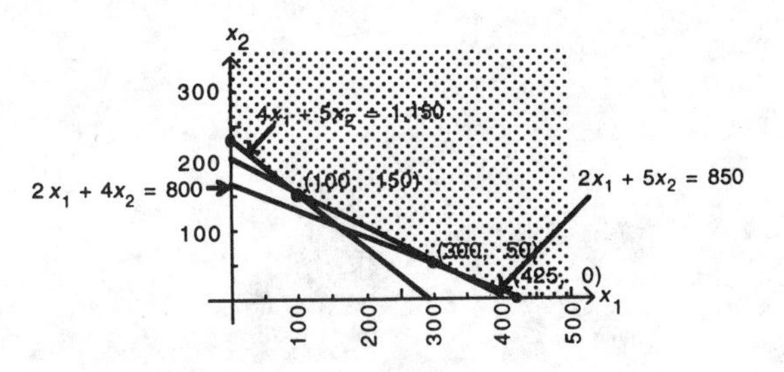

The feasible region is indicated by the shading in the graph at the right. The corner points are: (0, 230), (100, 150), (300, 50), (425, 0).

(A) The mathematical model for this problem is:
minimize $C = 0.04x_1 + 0.09x_2$ subject to the contraints given above.

The value of C at each corner point is:

Corner Point	$C = 0.04x_1 + 0.09x_2$
(0, 230)	$C = 0.04(0) + 0.09(230) = 20.70$
(100, 150)	$C = 0.04(100) + 0.09(150) = 17.50$
(300, 50)	$C = 0.04(300) + 0.09(50) = 16.50$
(425, 0)	$C = 0.04(425) + 0.09(0) = 17.00$

The minimum cost is \$16.50 when 300 grams of mix A and 50 grams of mix B are used.

(B) The mathematical model for this problem is:
minimize $C = 0.04x_1 + 0.06x_2$ subject to the constraints given above.

The value of C at each corner point is:

Corner Point	$C = 0.04x_1 + 0.06x_2$
(0, 230)	$C = 0.04(0) + 0.06(230) = 13.80$
(100, 150)	$C = 0.04(100) + 0.06(150) = 13.00$
(300, 50)	$C = 0.04(300) + 0.06(50) = 15.00$
(425, 0)	$C = 0.04(425) + 0.06(0) = 17.00$

The minimum cost decreases to \$13.00 when 100 grams of mix A and 150 grams of mix B are used.

(C) The mathematical model for this problem is:
minimize $C = 0.04x_1 + 0.12x_2$ subject to the constraints given above.

The value of C at each corner point is:

Corner Point	$C = 0.04x_1 + 0.12x_2$
(0, 230)	$C = 0.04(0) + 0.12(230) = 27.60$
(100, 150)	$C = 0.04(100) + 0.12(150) = 22.00$
(300, 50)	$C = 0.04(300) + 0.12(50) = 18.00$
(425, 0)	$C = 0.04(425) + 0.12(0) = 17.00$

The minimum cost increases to \$17.00 when 425 grams of mix A and 0 grams of mix B are used. (5-2)

6 PROBABILITY

EXERCISE 6-1

Things to remember:

1. Let A be a set with finitely many elements. Then $n(A)$ denotes the number of elements in A.

2. ADDITION PRINCIPLE (for counting)

   For any two sets A and B,

   $$n(A \cup B) = n(A) + n(B) - n(A \cap B)$$

   If A and B are disjoint, i.e., if $A \cap B = \varnothing$, then

   $$n(A \cup B) = n(A) + n(B)$$

3. MULTIPLICATION PRINCIPLE (for counting)

   (a) If two operations O_1 and O_2 are performed in order, with N_1 possible outcomes for the first operation and N_2 possible outcomes for the second operation, then there are

   $$N_1 \cdot N_2$$

   possible combined outcomes of the first operation followed by the second.

   (b) In general, if n operations $O_1, O_2, \ldots, O_n$ are performed in order with possible number of outcomes $N_1, N_2, \ldots, N_n$, respectively, then there are

   $$N_1 \cdot N_2 \cdot \ldots \cdot N_n$$

   possible combined outcomes of the operations performed in the given order.

---

1. $n(A) = 75 + 40 = 115$

3. $n(A \cup B) = 75 + 40 + 95 = 210$

$n[(A \cup B)'] = 90$

Thus, $n(U) = n(A \cup B) + n[(A \cup B)'] = 210 + 90 = 300$

[Note: $U = (A \cup B) \cup (A \cup B)'$ and $(A \cup B) \cap (A \cup B)' = \varnothing$]

5. $B' = A \cap B' \cup (A \cup B)'$ and $(A \cap B') \cap (A \cup B)' = \varnothing$.

Thus, $n(B') = n(A \cap B') + n[(A \cup B)']$

$= 75 + 90 = 165$

7. $n(A \cup B) = n(A) + n(B) - n(A \cap B)$
$= 115 + 135 - 40 = 210$

Note, also, that $A \cup B = (A \cap B') \cup (A \cap B) \cup (A' \cap B)$, and
$(A \cap B') \cap (A \cap B) = \varnothing$, $(A \cap B') \cap (A' \cap B) = \varnothing$,
$(A \cap B) \cap (A' \cap B) = \varnothing$.

So, $n(A \cup B) = n(A \cap B') + n(A \cap B) + n(A' \cap B)$
$= 75 + 40 + 95 = 210$

9. $n(A' \cap B) = 95$

11. $(A \cap B) \cup (A \cap B)' = U$ and $(A \cap B) \cap (A \cap B)' = \varnothing$

Thus, $n(U) = n(A \cap B) + n[(A \cap B)']$
or $n[(A \cap B)'] = n(U) - n(A \cap B) = 300 - 40 = 260$

13. (A) Tree Diagram

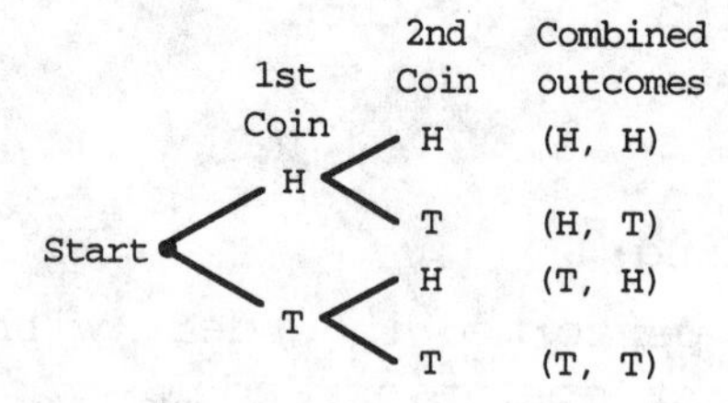

Thus, there are 4 ways.

(B) Multiplication Principle
O_1: 1st coin
N_1: 2 ways
O_2: 2nd coin
N_2: 2 ways

Thus, there are
$N_1 \cdot N_2 = 2 \cdot 2 = 4$ ways.

15. (A) Tree Diagram

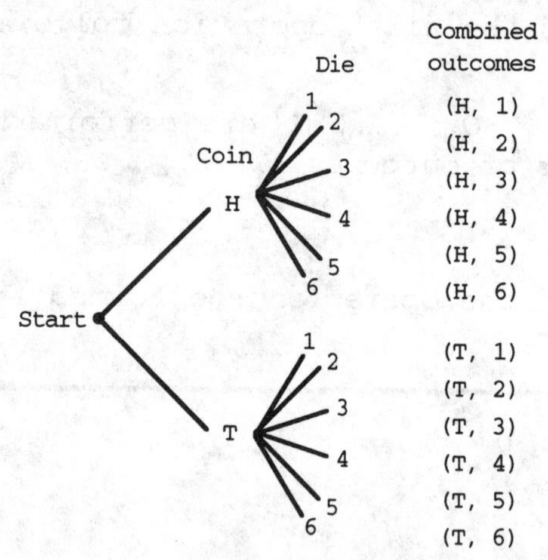

Thus, there are 12 combined outcomes.

(B) Multiplication Principle
O_1: Coin
N_1: 2 outcomes
O_2: Die
N_2: 6 outcomes

Thus, there are
$N_1 \cdot N_2 = 2 \cdot 6 = 12$ combined outcomes

17. $A = (A \cap B') \cup (A \cap B)$ and $B = (A' \cap B) \cup (A \cap B)$

Thus, $n(A) = n(A \cap B') + n(A \cap B)$, so
$n(A \cap B') = n(A) - n(A \cap B) = 80 - 20 = 60$
$n(A \cap B) = 20$.
$n(B) = n(A' \cap B) + n(A \cap B)$ so
$n(A' \cap B) = n(B) - n(A \cap B) = 50 - 20 = 30$

Also, $A \cup B = (A' \cap B) \cup (A \cap B) \cup (A \cap B')$ and

$U = (A \cup B) \cup (A' \cap B')$ where $(A \cup B) \cap (A' \cap B') = \varnothing$

Thus, $n(U) = n(A \cup B) + n(A' \cap B')$ so

$n(A' \cap B') = n(U) - n(A \cup B)$

$= 200 - (60 + 20 + 30) = 200 - 110 = 90$

19. $n(A \cup B) = n(A) + n(B) - n(A \cap B)$

So $n(A \cap B) = n(A) + n(B) - n(A \cup B)$

$= 25 + 55 - 60 = 20$

$n(A \cap B') = n(A) - n(A \cap B) = 25 - 20 = 5$

$n(A' \cap B) = n(B) - n(A \cap B) = 55 - 20 = 35$

and $n(A' \cap B') = n(U) - n(A \cup B)$

$= 100 - 60 = 40$

21. $n(A \cap B) = 30$

$n(A \cap B') = n(A) - n(A \cap B) = 70 - 30 = 40$

$n(A' \cap B) = n(B) - n(A \cap B) = 90 - 30 = 60$

$n(A' \cap B') = n(U) - [n(A \cap B') + n(A \cap B) + n(A' \cap B)]$

$= 200 - [40 + 30 + 60] = 200 - 130 = 70$

Therefore,

	A	A'	Totals
B	30	60	90
B'	40	70	110
Totals	70	130	200

23. $n(A \cap B) = n(A) + n(B) - n(A \cup B) = 45 + 55 - 80 = 20$

$n(A \cap B') = n(A) - n(A \cap B) = 45 - 20 = 25$

$n(A' \cap B) = n(B) - n(A \cap B) = 55 - 20 = 35$

$n(A' \cap B') = n(U) - n(A \cup B) = 100 - 80 = 20$

Therefore,

	A	A'	Totals
B	20	35	55
B'	25	20	45
Totals	45	55	100

25. (A) True. Suppose $A = \varnothing$ is the empty set. Then $A \cap B = \varnothing \cap B = \varnothing$.
Similarly if $B = \varnothing$ is the empty set.
Thus, if either A or B is the empty set, then $A \cap B = \varnothing$, and A and B are disjoint.

(B) False. Let $A = \{1, 2, 3\}$ and $B = \{4, 5, 6\}$. Then $A \cap B = \varnothing$ and neither A nor B is the empty set.

27. Using the Multiplication Principle:

O_1: Choose the color
N_1: 5 ways

O_2: Choose the transmission
N_2: 3 ways

O_3: Choose the interior
N_3: 4 ways

O_4: Choose the engine
N_4: 2 ways

Thus, there are

$N_1 \cdot N_2 \cdot N_3 \cdot N_4 = 5 \cdot 3 \cdot 4 \cdot 2 = 120$ different variations of this model car.

29. (A) Number of four-letter code words, no letter repeated.

O_1: Selecting the first letter
N_1: 6 ways

O_2: Selecting the second letter
N_2: 5 ways

O_3: Selecting the third letter
N_3: 4 ways

O_4: Selecting the fourth letter
N_4: 3 ways

Thus, there are

$N_1 \cdot N_2 \cdot N_3 \cdot N_4 = 6 \cdot 5 \cdot 4 \cdot 3 = 360$

possible code words. Note that this is the number of permutations of 6 objects taken 4 at a time:

$$P_{6,4} = \frac{6!}{(6 - 4)!} = \frac{6 \cdot 5 \cdot 4 \cdot 3 \cdot 2!}{2!} = 360$$

(B) Number of four-letter code words, allowing repetition.

O_1: Selecting the first letter
N_1: 6 ways

O_2: Selecting the second letter
N_2: 6 ways

O_3: Selecting the third letter
N_3: 6 ways

O_4: Selecting the fourth letter
N_4: 6 ways

Thus, there are

$N_1 \cdot N_2 \cdot N_3 \cdot N_4 = 6 \cdot 6 \cdot 6 \cdot 6 = 6^4 = 1296$

possible code words.

(C) Number of four-letter code words, adjacent letters different.

O_1: Selecting the first letter
N_1: 6 ways

O_2: Selecting the second letter
N_2: 5 ways

O_3: Selecting the third letter
N_3: 5 ways

O_4: Selecting the fourth letter
N_4: 5 ways

Thus, there are

$N_1 \cdot N_2 \cdot N_3 \cdot N_4 = 6 \cdot 5 \cdot 5 \cdot 5 = 6 \cdot 5^3 = 750$

possible code words.

31. (A) Number of five-digit combinations, no digit repeated.

O_1: Selecting the first digit
N_1: 10 ways

O_2: Selecting the second digit
N_2: 9 ways

O_3: Selecting the third digit
N_3: 8 ways

O_4: Selecting the fourth digit
N_4: 7 ways

O_5: Selecting the fifth digit
N_5: 6 ways

Thus, there are

$N_1 \cdot N_2 \cdot N_3 \cdot N_4 \cdot N_5 = 10 \cdot 9 \cdot 8 \cdot 7 \cdot 6 = 30{,}240$

possible combinations

(B) Number of five-digit combinations, allowing repetition.

O_1: Selecting the first digit
N_1: 10 ways

O_2: Selecting the second digit
N_2: 10 ways

O_3: Selecting the third digit
N_3: 10 ways

O_4: Selecting the fourth digit
N_4: 10 ways

O_5: Selecting the fifth digit
N_5: 10 ways

Thus, there are

$N_1 \cdot N_2 \cdot N_3 \cdot N_4 \cdot N_5 = 10 \cdot 10 \cdot 10 \cdot 10 \cdot 10 = 10^5 = 100{,}000$

possible combinations

(C) Number of five digit combinations, if successive digits must be different.

O_1: Selecting the first digit
N_1: 10 ways

O_2: Selecting the second digit
N_2: 9 ways

O_3: Selecting the third digit
N_3: 9 ways

O_4: Selecting the fourth digit
N_4: 9 ways

O_5: Selecting the fifth digit
N_5: 9 ways

Thus, there are

$N_1 \cdot N_2 \cdot N_3 \cdot N_4 \cdot N_5 = 10 \cdot 9 \cdot 9 \cdot 9 \cdot 9 = 10 \cdot 9^4 = 65{,}610$ possible combinations

33. (A) Letters and/or digits may be repeated.

O_1: Selecting the first letter
N_1: 26 ways

O_2: Selecting the second letter
N_2: 26 ways

O_3: Selecting the third letter
N_3: 26 ways

O_4: Selecting the first digit
N_4: 10 ways

O_5: Selecting the second digit
N_5: 10 ways

O_6: Selecting the third digit
N_6: 10 ways

Thus, there are

$N_1 \cdot N_2 \cdot N_3 \cdot N_4 \cdot N_5 \cdot N_6 = 26 \cdot 26 \cdot 26 \cdot 10 \cdot 10 \cdot 10 = 17{,}576{,}000$

different license plates.

(B) No repeated letters and no repeated digits are allowed.

O_1: Select the three letters, no letter repeated
N_1: $26 \cdot 25 \cdot 24 = 15{,}600$ ways

O_2: Select the three numbers, no number repeated
N_2: $10 \cdot 9 \cdot 8 = 720$ ways

Thus, there are

$N_1 \cdot N_2 = 15{,}600 \cdot 720 = 11{,}232{,}000$

different license plates with no letter or digit repeated.

35.

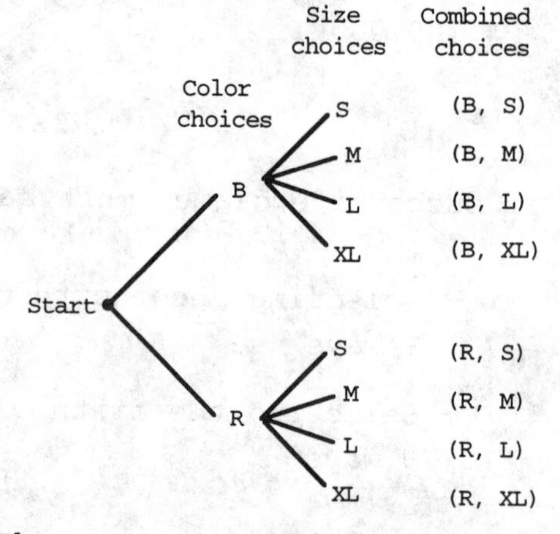

There are 8 combined choices; the color can be chosen in 2 ways and the size can be chosen in 4 ways. The number of possible combined choices is $2 \cdot 4 = 8$ just as in Example 3.

37. Let T = the people who play tennis, and
G = the people who play golf.

Then $n(T) = 32$, $n(G) = 37$, $n(T \cap G) = 8$ and $n(U) = 75$.

Thus, $n(T \cup G) = n(T) + n(G) - n(T \cap G) = 32 + 37 - 8 = 61$

The set of people who play neither tennis nor golf is represented by $T' \cap G'$. Since $U = (T \cup G) \cup (T' \cap G')$ and $(T \cup G) \cap (T' \cap G') = \emptyset$, it follows that $n(T' \cap G') = n(U) - n(T \cup G) = 75 - 61 = 14$.

There are 14 people who play neither tennis nor golf.

39. Let F = the people who speak French, and
G = the people who speak German.

Then $n(F) = 42$, $n(G) = 55$, $n(F' \cap G') = 17$ and $n(U) = 100$. Since $U = (F \cup G) \cup (F' \cap G')$ and $(F \cup G) \cap (F' \cap G') = \emptyset$, it follows that

$$n(F \cup G) = n(U) - n(F' \cap G') = 100 - 17 = 83$$

Now $n(F \cup G) = n(F) + n(G) - n(F \cap G)$, so

$$n(F \cap G) = n(F) + n(G) - n(F \cup G)$$
$$= 42 + 55 - 83 = 14$$

There are 14 people who speak both French and German.

41. (A)

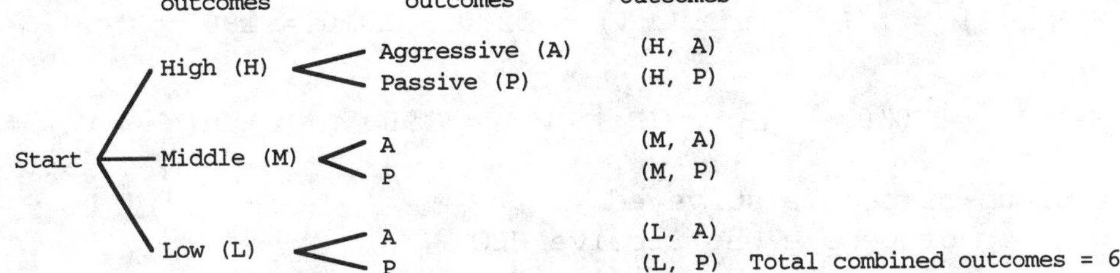

(B) Operation 1: Test scores can be classified into three groups, high, middle, or low:
$N_1 = 3$

Operation 2: Interviews can be classified into two groups, aggressive or passive:
$N_2 = 2$

The total possible combined classifications is:
$N_1 \cdot N_2 = 3 \cdot 2 = 6$

43. O_1: Travel from home to airport and back
N_1: 2 ways

O_2: Fly to first city
N_2: 3 ways

O_3: Fly to second city
N_3: 2 ways

O_4: Fly to third city
N_4: 1 way

Thus, there are
$N_1 \cdot N_2 \cdot N_3 \cdot N_4 = 2 \cdot 3 \cdot 2 \cdot 1 = 12$
different travel plans.

45. Let U = the group of people surveyed
M = people who own a microwave oven, and
V = people who own a VCR.

Then $n(U) = 1200$, $n(M) = 850$, $n(V) = 740$ and $n(M \cap V) = 580$. Now draw a Venn diagram.

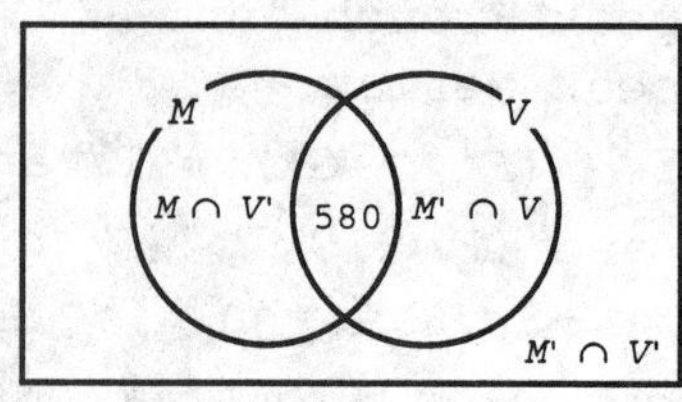

From this diagram, we see that

$$n(M \cap V') = n(M) - n(M \cap V) = 850 - 580 = 270$$
$$n(M' \cap V) = n(V) - n(M \cap V) = 740 - 580 = 160$$
$$n(M \cup V) = n(M \cap V') + n(M \cap V) + n(M' \cap V)$$
$$= 580 + 270 + 160 = 1010$$

and $n(M' \cap V') = n(U) - n(M \cup V) = 1200 - 1010 = 190$

Thus,
(A) $n(M \cup V) = 1010$ (B) $n(M' \cap V') = 190$ (C) $n(M \cap V') = 270$

47. Let U = group of people surveyed
H = group of people who receive HBO
S = group of people who receive Showtime.

Then, $n(U) = 8{,}000$, $n(H) = 2{,}450$, $n(S) = 1{,}940$ and $n(H' \cap S') = 5{,}180$

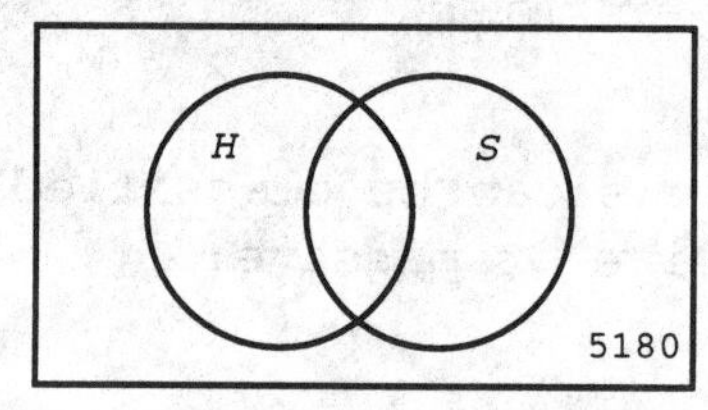

Now, $n(H \cup S) = n(U) - n(H' \cap S') = 8{,}000 - 5{,}180 = 2{,}820$

Since $n(H \cup S) = n(H) + n(S) - n(H \cap S)$, we have

$$n(H \cap S) = n(H) + n(S) - n(H \cup S) = 2{,}450 + 1{,}940 - 2{,}820 = 1{,}570$$

Thus, 1,570 subscribers receive both channels.

49. From the table:

(A) The number of males aged 20-24 *and* below minimum wage is: 102 (the element in the (2, 2) position in the body of the table.)

(B) The number of females aged 20 or older *and* at minimum wage is: 186 + 503 = 689 (the sum of the elements in the (3, 2) and (3, 3) positions.)

(C) The number of workers who are *either* aged 16-19 *or* are males at minimum wage is:

$$343 + 118 + 367 + 251 + 154 + 237 = 1{,}470$$

(D) The number of workers below minimum wage is: 379 + 993 = 1,372.

51. (A)

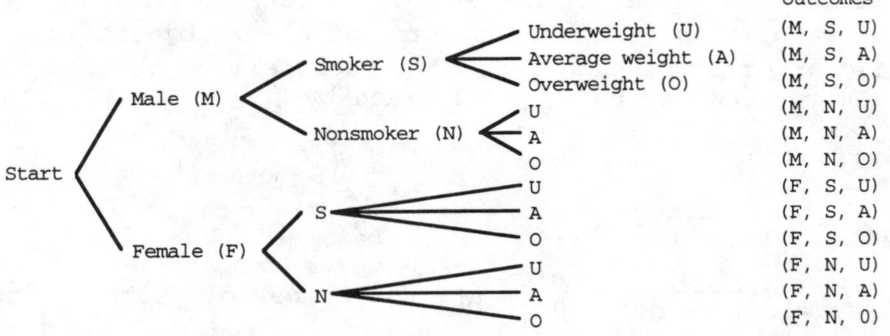

(B) Operation 1: Two classifications, male and female; $N_1 = 2$.

Operation 2: Two classifications, smoker and nonsmoker; $N_2 = 2$.

Operation 3: Three classifications, underweight, average weight, and overweight; $N_3 = 3$.

Thus the total possible combined classifications $= N_1 \cdot N_2 \cdot N_3 = 2 \cdot 2 \cdot 3 = 12$

53. F = number of individuals who contributed to the first campaign
S = number of individuals who contributed to the second campaign.

Then $n(F) = 1,475$, $n(S) = 2,350$ and $n(F \cap S) = 920$.

Now, $n(F \cup S) = n(F) + n(S) - n(F \cap S)$
$= 1,475 + 2,350 - 920$
$= 2,905$

Thus, 2,905 individuals contributed to either the first campaign or the second campaign.

EXERCISE 6-2

Things to remember:

1. FACTORIAL

   For n a natural number,

   $n! = n(n - 1)(n - 2) \cdot \ldots \cdot 3 \cdot 2 \cdot 1$
   $0! = 1$
   $n! = n(n - 1)!$

   [NOTE: Many calculators have an $\boxed{n!}$ key or its equivalent.]

2. PERMUTATIONS

   A PERMUTATION of a set of distinct objects is an arrangement of the objects in a specific order, without repetitions. The number of permutations of n distinct objects without repetition, denoted by $P_{n,n}$, is:

   $P_{n,n} = n(n - 1) \cdot \ldots \cdot 3 \cdot 2 \cdot 1 = n!$ (n factors)

3. PERMUTATIONS OF n OBJECTS TAKEN r AT A TIME

A permutation of a set of n distinct objects taken r at a time without repetition is an arrangement of the r objects in a specific order. The number of permutations of n objects taken r at a time, denoted by $P_{n,r}$, is given by:

$$P_{n,r} = n(n - 1)(n - 2)\cdot \ldots \cdot(n - r + 1) \quad (r \text{ factors})$$

or $$P_{n,r} = \frac{n!}{(n - r)!} \qquad 0 \le r \le n$$

[Note: $P_{n,n} = \frac{n!}{(n - n)!} = \frac{n!}{0!} = n!$, the number of permutations of n objects taken n at a time. Remember, by definition, $0! = 1$.]

4. COMBINATIONS OF n OBJECTS TAKEN r AT A TIME

A combination of a set of n distinct objects taken r at a time without repetition is an r-element subset of the set of n objects. (The arrangement of the elements in the subset does not matter.) The number of combinations of n objects taken r at a time, denoted by $C_{n,r}$ or by $\binom{n}{r}$ is given by:

$$C_{n,r} = \binom{n}{r} = \frac{P_{n,r}}{r!} = \frac{n!}{r!(n - r)!} \qquad 0 \le r \le n$$

5. NOTE: In a permutation, the ORDER of the objects counts. In a combination, order does not count.

---

1. $4! = 4\cdot3\cdot2\cdot1 = 24$

3. $\frac{9!}{8!} = \frac{9\cdot8!}{8!} = 9$

5. $\frac{11!}{8!} = \frac{11\cdot10\cdot9\cdot8!}{8!} = 990$

7. $\frac{5!}{2!3!} = \frac{5\cdot4\cdot3!}{2\cdot1\cdot3!} = 10$

9. $\frac{7!}{4!(7 - 4)!} = \frac{7!}{4!3!} = \frac{7\cdot6\cdot5\cdot4!}{4!\cdot3\cdot2\cdot1} = 35$

11. $\frac{7!}{7!(7 - 7)!} = \frac{7!}{7!0!} = \frac{1}{1} = 1$

13. $P_{5,3} = \frac{5!}{(5 - 3)!} = \frac{5!}{2!} = \frac{5\cdot4\cdot3\cdot2!}{2!} = 60$

15. $P_{52,4} = \frac{52!}{(52 - 4)!} = \frac{52!}{48!} = \frac{52\cdot51\cdot50\cdot49\cdot48!}{48!} = 6,497,400$

17. $C_{5,3} = \frac{5!}{3!(5 - 3)!} = \frac{5!}{3!2!} = \frac{5\cdot4\cdot3!}{3!\cdot2\cdot1} = 10$

19. $C_{52,4} = \frac{52!}{4!(52 - 4)!} = \frac{52!}{4!48!} = \frac{52\cdot51\cdot50\cdot49\cdot48!}{4\cdot3\cdot2\cdot1\cdot48!} = 270,725$

21. (A) Permutation: the order of selection--finance, academic affairs, student affairs--matters.

(B) Combination: the order of selection does not matter.

23. The number of different finishes (win, place, show) for the ten horses is the number of permutations of 10 objects 3 at a time. This is:

$$P_{10,3} = \frac{10!}{(10 - 3)!} = \frac{10!}{7!} = \frac{10\cdot 9\cdot 8\cdot 7!}{7!} = 720$$

25. (A) The number of ways that a three-person subcommittee can be selected from a seven-member committee is the number of combinations (since order *is not* important in selecting a subcommittee) of 7 objects 3 at a time. This is:

$$C_{7,3} = \frac{7!}{3!(7 - 3)!} = \frac{7!}{3!4!} = \frac{7\cdot 6\cdot 5\cdot 4!}{3\cdot 2\cdot 1\cdot 4!} = 35$$

(B) The number of ways a president, vice-president, and secretary can be chosen from a committee of 7 people is the number of permutations (since order *is* important in choosing 3 people for the positions) of 7 objects 3 at a time. This is:

$$P_{7,3} = \frac{7!}{(7 - 3)!} = \frac{7!}{4!} = \frac{7\cdot 6\cdot 5\cdot 4!}{4!} = 7\cdot 6\cdot 5 = 210$$

27. Calculate $x!$, 3^x, x^3 for $x = 1, 2, 3, \ldots$
For $x \geq 7$, $x! > 3^x$; for $x \geq 1$, $3^x > x^3$ (except for $x = 3$). In general, $x!$ grows much faster than 3^x and 3^x grows much faster than x^3.

29. This is a "combinations" problem. We want the number of ways to select 5 objects from 13 objects with order not counting. This is:

$$C_{13,5} = \frac{13!}{5!(13 - 5)!} = \frac{13!}{5!8!} = \frac{13\cdot 12\cdot 11\cdot 10\cdot 9\cdot 8!}{5\cdot 4\cdot 3\cdot 2\cdot 1\cdot 8!} = 1287$$

31. The five spades can be selected in $C_{13,5}$ ways and the two hearts can be selected in $C_{13,2}$ ways. Applying the Multiplication Principle, we have:

$$\text{Total number of hands} = C_{13,5}\cdot C_{13,2} = \frac{13!}{5!(13 - 5)!}\cdot\frac{13!}{2!(13 - 2)!}$$
$$= \frac{13!}{5!8!}\cdot\frac{13!}{2!11!} = 100{,}386$$

33. The three appetizers can be selected in $C_{8,3}$ ways. The four main courses can be selected in $C_{10,4}$ ways. The two desserts can be selected in $C_{7,2}$ ways. Now, applying the Multiplication Principle, the total number of ways in which the above can be selected is given by:

$$C_{8,3}\cdot C_{10,4}\cdot C_{7,2} = \frac{8!}{3!(8 - 3)!}\cdot\frac{10!}{4!(10 - 4)!}\cdot\frac{7!}{2!(7 - 2)!} = 246{,}960$$

35. $C_{n,\,n-r} = \dfrac{n!}{(n - r)!(n - [n - r])!} = \dfrac{n!}{(n - r)!r!} = \dfrac{n!}{r!(n - r)!} = C_{n,r}$
The sequence of numbers is the same read top to bottom or bottom to top.

37. (A) A chord joins two distinct points. Thus, the total number of chords is given by:

$$C_{8,2} = \frac{8!}{2!(8-2)!} = \frac{8!}{2!6!} = \frac{8\cdot7\cdot6!}{2\cdot1\cdot6!} = 28$$

(B) Each triangle requires three distinct points. Thus, there are

$$C_{8,3} = \frac{8!}{3!(8-3)!} = \frac{8!}{3!5!} = \frac{8\cdot7\cdot6\cdot5!}{3\cdot2\cdot1\cdot5!} = 56 \text{ triangles.}$$

(C) Each quadrilateral requires four distinct points. Thus, there are

$$C_{8,4} = \frac{8!}{4!(8-4)!} = \frac{8!}{4!4!} = \frac{8\cdot7\cdot6\cdot5\cdot4!}{4\cdot3\cdot2\cdot1\cdot4!} = 70 \text{ quadrilaterals.}$$

39. (A) Two people.

O_1: First person selects a chair O_2: Second person selects a chair
N_1: 5 ways N_2: 4 ways

Thus, there are

$N_1 \cdot N_2 = 5\cdot4 = 20$

ways to seat two people in a row of 5 chairs. Note that this is $P_{5,2}$.

(B) Three people. There will be $P_{5,3}$ ways to seat 3 people in a row of 5 chairs:

$$P_{5,3} = \frac{5!}{(5-3)!} = \frac{5!}{2!} = \frac{5\cdot4\cdot3\cdot2!}{2!} = 60$$

(C) Four people. The number of ways to seat 4 people in a row of 5 chairs is given by:

$$P_{5,4} = \frac{5!}{(5-4)!} = \frac{5!}{1!} = 5\cdot4\cdot3\cdot2 = 120$$

(D) Five people. The number of ways to seat 5 people in a row of 5 chairs is given by:

$$P_{5,5} = \frac{5!}{(5-1)!} = \frac{5!}{0!} = 5! = 120$$

41. (A) The distinct positions are taken into consideration. The number of starting teams is given by:

$$P_{8,5} = \frac{8!}{(8-5)!} = \frac{8!}{3!} = \frac{8\cdot7\cdot6\cdot5\cdot4\cdot3!}{3!} = 6720$$

(B) The distinct positions are not taken into consideration. The number of starting teams is given by:

$$C_{8,5} = \frac{8!}{5!(8-5)!} = \frac{8!}{5!3!} = \frac{8\cdot7\cdot6\cdot5!}{5!\cdot3\cdot2\cdot1} = 56$$

(C) Either Mike or Ken, but not both, must start; distinct positions are not taken into consideration.

O_1: Select either Mike or Ken
N_1: 2 ways

O_2: Select 4 players from the remaining 6

N_2: $C_{6,4}$

Thus, the number of starting teams is given by:

$$N_1 \cdot N_2 = 2 \cdot C_{6,4} = 2 \cdot \frac{6!}{4!(6-4)!} = 2 \cdot \frac{6 \cdot 5 \cdot 4!}{4! \cdot 2 \cdot 1} = 30$$

43. For many calculators, $k = 69$, but your calculator may be different. Note that $k!$ may also be calculated as $P_{k,k}$. On a TI-85, the largest integer k for which $k!$ can be calculated using $P_{k,k}$ is 449.

45. The largest value will be $C_{24,12} = \frac{24!}{12!12!} = 2{,}704{,}156$.

47. (A) Three printers are to be selected for the display. The *order* of selection does not count. Thus, the number of ways to select the 3 printers from 24 is:

$$C_{24,3} = \frac{24!}{3!(24-3)!} = \frac{24 \cdot 23 \cdot 22(21!)}{3 \cdot 2 \cdot 1(21!)} = 2{,}024$$

(B) Nineteen of the 24 printers are not defective. Thus, the number of ways to select 3 non-defective printers is:

$$C_{19,3} = \frac{19!}{3!(19-3)!} = \frac{19 \cdot 18 \cdot 17(16!)}{3 \cdot 2 \cdot 1(16!)} = 969$$

49. (A) There are $8 + 12 + 10 = 30$ stores in all. The jewelry store chain will select 10 of these stores to close. Since order does not count here, the total number of ways to select the 10 stores to close is:

$$C_{30,10} = \frac{30!}{10!(30-10)!} = \frac{30 \cdot 29 \cdot 28 \cdot 27 \cdot 26 \cdot 25 \cdot 24 \cdot 23 \cdot 22 \cdot 21(20!)}{10 \cdot 9 \cdot 8 \cdot 7 \cdot 6 \cdot 5 \cdot 4 \cdot 3 \cdot 2 \cdot 1(20!)}$$
$$= 30{,}045{,}015$$

(B) The number of ways to close 2 stores in Georgia is: $C_{8,2}$

The number of ways to close 5 stores in Florida is: $C_{12,5}$

The number of ways to close 3 stores in Alabama is: $C_{10,3}$

By the multiplication principle, the total number of ways to select the 10 stores for closing is:

$$C_{8,2} \cdot C_{12,5} \cdot C_{10,3} = \frac{8!}{2!(8-2)!} \cdot \frac{12!}{5!(12-5)!} \cdot \frac{10!}{3!(10-3)!}$$
$$= \frac{8 \cdot 7 \cdot 6!}{2 \cdot 1 \cdot 6!} \cdot \frac{12 \cdot 11 \cdot 10 \cdot 9 \cdot 8(7!)}{5 \cdot 4 \cdot 3 \cdot 2 \cdot 1(7!)} \cdot \frac{10 \cdot 9 \cdot 8(7!)}{3 \cdot 2 \cdot 1(7!)}$$
$$= 28 \cdot 792 \cdot 120 = 2{,}661{,}120$$

51. (A) Three females can be selected in $C_{6,3}$ ways. Two males can be selected in $C_{5,2}$ ways. Applying the Multiplication Principle, we have:

$$\text{Total number of ways} = C_{6,3} \cdot C_{5,2} = \frac{6!}{3!(6-3)!} \cdot \frac{5!}{2!(5-2)!} = 200$$

(B) Four females and one male can be selected in $C_{6,4} \cdot C_{5,1}$ ways. Thus,

$$C_{6,4} \cdot C_{5,1} = \frac{6!}{4!(6-4)!} \cdot \frac{5!}{1!(5-1)!} = 75$$

(C) Number of ways in which 5 females can be selected is:

$$C_{6,5} = \frac{6!}{5!(6 - 5)!} = 6$$

(D) Number of ways in which 5 people can be selected is:

$$C_{6+5,5} = C_{11,5} = \frac{11!}{5!(11 - 5)!} = 462$$

(E) At least four females includes four females and five females. Four females and one male can be selected in 75 ways [see part (B)]. Five females can be selected in 6 ways [see part (C)]. Thus,

Total number of ways = $C_{6,4} \cdot C_{5,1} + C_{6,5} = 75 + 6 = 81$

53. (A) Select 3 samples from 8 blood types, no two samples having the same type. This is a permutation problem. The number of different examinations is:

$$P_{8,3} = \frac{8!}{(8 - 3)!} = \frac{8!}{5!} = \frac{8 \cdot 7 \cdot 6 \cdot 5!}{5!} = 336$$

(B) Select 3 samples from 8 blood types, repetition is allowed.

O_1: Select the first sample
N_1: 8 ways

O_2: Select the second sample
N_2: 8 ways

O_3: Select the third sample
N_3: 8 ways

Thus, the number of different examinations in this case is:

$N_1 \cdot N_2 \cdot N_3 = 8 \cdot 8 \cdot 8 = 83 = 512$

55. This is a permutations problem. The number of buttons is given by:

$$P_{4,2} = \frac{4!}{(4 - 2)!} = \frac{4!}{2!} = \frac{4 \cdot 3 \cdot 2!}{2!} = 12$$

EXERCISE 6-3

Things to remember:

1. SAMPLE SPACE

A set S is a SAMPLE SPACE for an experiment if:

(a) Each element of S is an outcome of the experiment.

(b) Each outcome of the experiment corresponds to one and only one element of S.

Each element in the sample space is called a SIMPLE OUTCOME or SIMPLE EVENT.

2. EVENT

Given a sample space S. An EVENT E is any subset of S (including the empty set $\varnothing$ and the sample space S). An event with only one element is called a SIMPLE EVENT; an event with more than one element is a COMPOUND EVENT. An event E *occurs* if the result of performing the experiment is one of the simple events in E.

3. There is no one correct sample space for a given experiment. When specifying a sample space for an experiment, include as much detail as necessary to answer all questions of interest regarding the outcomes of the experiment. When in doubt, choose a sample space with more elements rather than fewer.

4. PROBABILITIES FOR SIMPLE EVENTS

Given a sample space

$$S = \{e_1, e_2, \ldots, e_n\}.$$

To each simple event e_i assign a real number denoted by $P(e_i)$, called the PROBABILITY OF THE EVENT e_i. These numbers can be assigned in an arbitrary manner provided the following two conditions are satisfied:

(a) $0 \le P(e_i) \le 1$

(The probability of a simple event is a number between 0 and 1, inclusive.)

(b) $P(e_1) + P(e_2) + \ldots + P(e_n) = 1$

(The sum of the probabilities of all simple events in the sample space is 1.)

Any probability assignment that meets these two conditions is called an ACCEPTABLE PROBABILITY ASSIGNMENT.

5. PROBABILITY OF AN EVENT E

Given an acceptable probability assignment for the simple events in a sample space S, the probability of an arbitrary event E, denoted $P(E)$, is defined as follows:

(a) $P(E) = 0$ if E is the empty set.

(b) If E is a simple event, then $P(E)$ has already been assigned.

(c) If E is a compound event, then $P(E)$ is the sum of the probabilities of all the simple events in E.

(d) If $E = S$, then $P(E) = P(S) = 1$ [this follows from 4(b)].

6. STEPS FOR FINDING THE PROBABILITY OF AN EVENT E

(a) Set up an appropriate sample space S for the experiment.

(b) Assign acceptable probabilities to the simple events in S.

(c) To obtain the probability of an arbitrary event E, add the probabilities of the simple events in E.

7. EMPIRICAL PROBABILITY

If an experiment is conducted n times and event E occurs with FREQUENCY $f(E)$, then the ratio $f(E)/n$ is called the RELATIVE FREQUENCY of the occurrence of event E in n trials. The EMPIRICAL PROBABILITY of E, denoted by $P(E)$, is given by the number (if it exists) that the relative frequency $f(E)/n$ approaches as n gets larger and larger. For any particular n, the relative frequency $f(E)/n$ is also called the APPROXIMATE EMPIRICAL PROBABILITY of event E:

$$P(E) \approx \frac{\text{Frequency of occurrence of } E}{\text{Total number of trials}} = \frac{f(E)}{n}$$

(The larger n is, the better the approximation.)

8. PROBABILITIES UNDER AN EQUALLY LIKELY ASSUMPTION

If, in a sample space

$$S = \{e_1, e_2, \ldots, e_n\},$$

each simple event is as likely to occur as any other, then $P(e_i) = \frac{1}{n}$, for $i = 1, 2, \ldots, n$, i.e. assign the same probability, $1/n$, to each simple event. The probability of an arbitrary event E in this case is:

$$P(E) = \frac{\text{Number of elements in } E}{\text{Number of elements in } S} = \frac{n(E)}{n(S)}$$

---

1. $P(E) = 1$ means that the occurrence of E is certain.

3. Let B = boy and G = girl. Then

$S = \{(B, B), (B, G), (G, B), (G, G)\}$

where (B, B) means both children are boys, (B, G) means the first child is a boy, the second is a girl, and so on. The event E corresponding to having two children of opposite sex is $E = \{(B, G), (G, B)\}$. Since the simple events are equally likely,

$$P(E) = \frac{n(E)}{n(S)} = \frac{2}{4} = \frac{1}{2}.$$

5. (A) Reject; $P(G) = -0.35$—no probability can be negative.

(B) Reject; $P(J) + P(G) + P(P) + P(S) = 0.32 + 0.28 + 0.24 + 0.30 = 1.14 \neq 1$

(C) Acceptable; each probability is between 0 and 1 (inclusive), and the sum of the probabilities is 1.

7. $P(\{J, P\}) = P(J) + P(P) = 0.26 + 0.30 = 0.56.$

9. $S = \{(B, B, B), (B, B, G), (B, G, B), (B, G, G), (G, B, B), (G, B, G), (G, G, B), (G, G, G)\}$

$E = \{(B, B, G)\}$

Since the events are equally likely and $n(S) = 8$, $P(E) = \frac{1}{8}$.

11. The number of three-digit sequences with no digit repeated is $P_{10,3}$. Since the possible opening combinations are equally likely, the probability of guessing the right combination is:

$$\frac{1}{P_{10,3}} = \frac{1}{10\cdot 9\cdot 8} = \frac{1}{720} \approx 0.0014$$

13. Let S = the set of five-card hands. Then $n(S) = C_{52,5}$.
Let A = "five black cards." Then $n(A) = C_{26,5}$.
Since individual hands are equally likely to occur:

$$P(A) = \frac{n(A)}{n(S)} = \frac{C_{26,5}}{C_{52,5}} = \frac{\frac{26!}{5!21!}}{\frac{52!}{5!47!}} = \frac{26\cdot 25\cdot 24\cdot 23\cdot 22}{52\cdot 51\cdot 50\cdot 49\cdot 48} \approx 0.025$$

15. S = set of five-card hands; $n(S) = C_{52,5}$.
F = "five face cards"; $n(F) = C_{12,5}$.
Since individual hands are equally likely to occur:

$$P(F) = \frac{n(F)}{n(S)} = \frac{C_{12,5}}{C_{52,5}} = \frac{\frac{12!}{5!7!}}{\frac{52!}{5!47!}} = \frac{12\cdot 11\cdot 10\cdot 9\cdot 8}{52\cdot 51\cdot 50\cdot 49\cdot 48} \approx 0.000305$$

17. S = {all the days in a year} (assume 365 days; exclude leap year); Equivalently, number the days of the year, beginning with January 1. Then $S = \{1, 2, 3, \ldots, 365\}$.

Assume that each day is as likely as any other day for a person to be born. Then the probability of each simple event is: $\frac{1}{365}$.

19. $n(S) = P_{5,5} = 5! = 120$
Let A = all notes inserted into the correct envelopes. Then $n(A) = 1$ and

$$P(A) = \frac{n(A)}{n(S)} = \frac{1}{120} \approx 0.00833$$

21. Using the sample space shown in Figure 3, we have
$n(S) = 36$, $n(A) = 1$,
where Event A = "Sum being 2":

$$P(A) = \frac{n(A)}{n(S)} = \frac{1}{36}$$

23. Let E = "Sum being 6." Then $n(E) = 5$. Thus, $P(E) = \frac{n(E)}{n(S)} = \frac{5}{36}$.

25. Let E = "Sum being less than 5." Then $n(E) = 6$. Thus,

$$P(E) = \frac{n(E)}{n(S)} = \frac{6}{36} = \frac{1}{6}.$$

27. Let E = "Sum not 7 or 11." Then $n(E) = 28$ and $P(E) = \frac{n(E)}{n(S)} = \frac{28}{36} = \frac{7}{9}$.

29. E = "Sum being 1" is not possible. Thus, $P(E) = 0$.

31. Let E = "Sum is divisible by 3" = "Sum is 3, 6, 9, or 12." Then $n(E) = 12$ and $P(E) = \frac{n(E)}{n(S)} = \frac{12}{36} = \frac{1}{3}$.

33. Let E = "Sum is 7 or 11." Then $n(E) = 8$. Thus, $P(E) = \frac{n(E)}{n(S)} = \frac{8}{36} = \frac{2}{9}$.

35. Let E = "Sum is divisible by 2 or 3" = "Sum is 2, 3, 4, 6, 8, 9, 10, 12." Then $n(E) = 24$, and $P(E) = \frac{n(E)}{n(S)} = \frac{24}{36} = \frac{2}{3}$.

For Problems 37–41, the sample space S is given by:

$$S = \{(H, H, H), (H, H, T), (H, T, H), (H, T, T)\}$$

The outcomes are equally likely and $n(S) = 4$.

37. Let E = "1 head." Then $n(E) = 1$ and $P(E) = \frac{n(E)}{n(S)} = \frac{1}{4}$.

39. Let E = "3 heads." Then $n(E) = 1$ and $P(E) = \frac{n(E)}{n(S)} = \frac{1}{4}$.

41. Let E = "More than 1 head." Then $n(E) = 3$ and $P(E) = \frac{n(E)}{n(S)} = \frac{3}{4}$.

43. (A) Yes. If we flip a fair coin 20 times, a representation of the sample space is {0, 1, 2, …, 19, 20} where each element denotes the number of heads. Each of these outcomes is possible; 10 is the outcome with the highest probability.

(B) Yes. The probability of getting 37 heads in 40 tosses of a fair coin is very small. Based on the evidence, it would appear that $P(H) = \frac{37}{40}$ and $P(T) = \frac{3}{40}$.

For Problems 45–51, the sample space S is given by:

$$S = \begin{Bmatrix} (1, 1), & (1, 2), & (1, 3) \\ (2, 1), & (2, 2), & (2, 3) \\ (3, 1), & (3, 2), & (3, 3) \end{Bmatrix}$$

The outcomes are equally likely and $n(S) = 9$.

45. Let E = "Sum is 2." Then $n(E) = 1$ and $P(E) = \frac{n(E)}{n(S)} = \frac{1}{9}$.

47. Let E = "Sum is 4." Then $n(E) = 3$ and $P(E) = \frac{n(E)}{n(S)} = \frac{3}{9} = \frac{1}{3}$.

49. Let E = "Sum is 6." Then $n(E) = 1$ and $P(E) = \frac{n(E)}{n(S)} = \frac{1}{9}$.

51. Let E = "Sum is odd" = "Sum is 3 or 5." Then $n(E) = 4$ and $P(E) = \frac{n(E)}{n(S)} = \frac{4}{9}$.

For Problems 53–59, the sample space S is the set of all 5-card hands. Then $n(S) = C_{52,5}$. The outcomes are equally likely.

53. Let E = "5 cards, jacks through aces." Then $n(E) = C_{16,5}$. Thus,

$$P(E) = \frac{C_{16,5}}{C_{52,5}} = \frac{\frac{16!}{5!11!}}{\frac{52!}{5!47!}} = \frac{16\cdot15\cdot14\cdot13\cdot12}{52\cdot51\cdot50\cdot49\cdot48} \approx 0.00168.$$

55. Let E = "4 aces." Then $n(E) = 48$ (the remaining card can be any one of the 48 cards which are not aces). Thus,

$$P(E) = \frac{48}{C_{52,5}} = \frac{48}{\frac{52!}{5!47!}} = \frac{48\cdot5!}{52\cdot51\cdot50\cdot49\cdot48} = \frac{5\cdot4\cdot3\cdot2}{52\cdot51\cdot50\cdot49} \approx 0.0000185$$

57. Let E = "Straight flush, ace high." Then $n(E) = 4$ (one such hand in each suit). Thus,

$$P(E) = \frac{4}{C_{52,5}} = \frac{4\cdot5!}{52\cdot51\cdot50\cdot49\cdot48} = \frac{480}{52\cdot51\cdot50\cdot49\cdot48} \approx 0.0000015$$

59. Let E = "2 aces and 3 queens." The number of ways to get 2 aces is $C_{4,2}$ and the number of ways to get 3 queens is $C_{4,3}$.

Thus, $n(E) = C_{4,2}\cdot C_{4,3} = \frac{4!}{2!2!}\cdot\frac{4!}{3!1!} = \frac{4\cdot3}{2}\cdot\frac{4}{1} = 24$

and

$$P(E) = \frac{n(E)}{n(S)} = \frac{24}{C_{52,5}} = \frac{24\cdot5!}{52\cdot51\cdot50\cdot59\cdot48} \approx 0.000009.$$

61. (A) From the plot, $P(6) = \frac{7}{50} = 0.14$.

(B) If the outcomes are equally likely, $P(6) = \frac{1}{6} = 0.167$.

(C) The answer here depends on the results of your simulation.

63. (A) Represent the outcomes H and T by the numbers 1 and 2, respectively, and select 500 random integers form the set {1, 2}.

(B) The answer depends on the results of your simulation.

(C) If the outcomes are equally likely, then $P(H) = P(T) = \frac{1}{2}$.

65. (A) The sample space S is the set of all possible permutations of the 12 brands taken 4 at a time, and $n(S) = P_{12,4}$. Thus, the probability of selecting 4 brands and identifying them correctly, with no answer repeated, is:

$$P(E) = \frac{1}{P_{12,4}} = \frac{1}{\frac{12!}{(12-4)!}} = \frac{1}{12\cdot11\cdot10\cdot9} \approx 0.000084$$

(B) Allowing repetition, $n(S) = 12^4$ and the probability of identifying them correctly is:

$$P(F) = \frac{1}{12^4} \approx 0.000048$$

67. (A) Total number of applicants = 6 + 5 = 11.

$$n(S) = C_{11,5} = \frac{11!}{5!(11-5)!} = 462$$

The number of ways that three females and two males can be selected is:

$$C_{6,3} \cdot C_{5,2} = \frac{6!}{3!(6-3)!} \cdot \frac{5!}{2!(5-2)!} = 20 \cdot 10 = 200$$

Thus, $P(A) = \dfrac{C_{6,3} \cdot C_{5,2}}{C_{11,5}} = \dfrac{200}{462} = 0.433$

(B) $P(\text{4 females and 1 male}) = \dfrac{C_{6,4} \cdot C_{5,1}}{C_{11,5}} = 0.162$

(C) $P(\text{5 females}) = \dfrac{C_{6,5}}{C_{11,5}} = 0.013$

(D)
$$\begin{aligned} P(\text{at least four females}) &= P(\text{4 females and 1 male}) + P(\text{5 females}) \\ &= \frac{C_{6,4} \cdot C_{5,1}}{C_{11,5}} + \frac{C_{6,5}}{C_{11,5}} \\ &= 0.162 + 0.013 \text{ [refer to parts (B) and (C)]} \\ &= 0.175 \end{aligned}$$

69. (A) The sample space S consists of the number of permutations of the 8 blood types chosen 3 at a time. Thus, $n(S) = P_{8,3}$ and the probability of guessing the three types in a sample correctly is:

$$P(E) = \frac{1}{P_{8,3}} = \frac{1}{\dfrac{8!}{(8-3)!}} = \frac{1}{8 \cdot 7 \cdot 6} \approx 0.0030$$

(B) Allowing repetition, $n(S) = 8^3$ and the probabilty of guessing the three types in a sample correctly is:

$$P(E) = \frac{1}{8^3} \approx 0.0020$$

71. (A) The total number of ways of selecting a president and a vice-president from the 11 members of the council is:

$P_{11,2}$, i.e., $n(S) = P_{11,2}$.

The total number of ways of selecting the president and the vice-president from the 6 democrats is $P_{6,2}$. Thus, if E is the event "The president and vice-president are both Democrats," then

$$P(E) = \frac{P_{6,2}}{P_{11,2}} = \frac{\dfrac{6!}{(6-2)!}}{\dfrac{11!}{(11-2)!}} = \frac{6 \cdot 5}{11 \cdot 10} = \frac{30}{110} \approx 0.273.$$

(B) The total number of ways of selecting a committee of 3 from the 11 members of the council is:

$C_{11,3}$, i.e., $n(S) = C_{11,3} = \dfrac{11!}{3!(11-3)!} = \dfrac{11\cdot10\cdot9\cdot8!}{3\cdot2\cdot1\cdot8!} = 165$

If we let F be the event "The majority are Republicans," which is the same as having either 2 Republicans and 1 Democrat or all 3 Republicans, then

$$n(F) = C_{5,2}\cdot C_{6,1} + C_{5,3} = \frac{5!}{2!(5-2)!}\cdot\frac{6!}{1!(6-1)!} + \frac{5!}{3!(5-3)!}$$
$$= 10\cdot6 + 10 = 70.$$

Thus,

$$P(F) = \frac{n(F)}{n(S)} = \frac{70}{165} \approx 0.424.$$

EXERCISE 6-4

Things to remember:

1. UNION AND INTERSECTION OF EVENTS

If A and B are two events in a sample space S, then the UNION of A and B, denoted by $A \cup B$, and the INTERSECTION of A and B, denoted by $A \cap B$, are defined as follows:

$A \cup B = \{e \in S \mid e \in A \text{ OR } e \in B\}$ $\quad$ $A \cap B = \{e \in S \mid e \in A \text{ AND } e \in B\}$

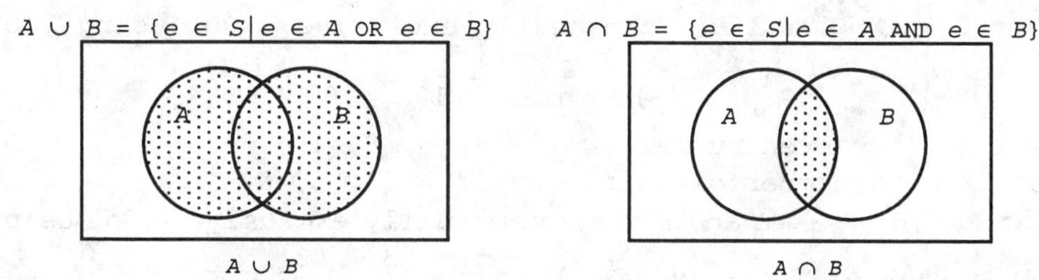

$A \cup B$

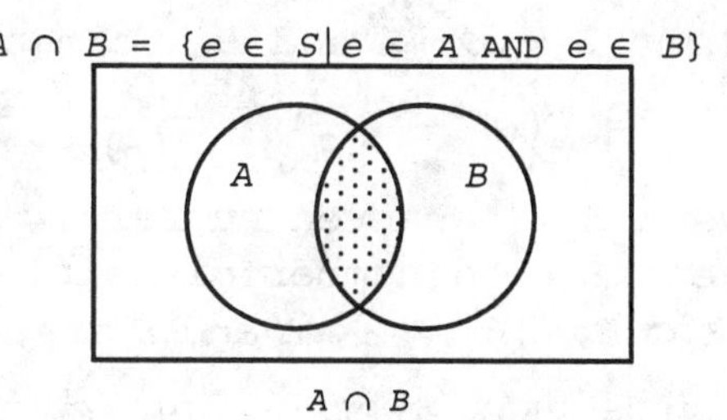

$A \cap B$

Furthermore, we define:

The **event A or B** to be $A \cup B$.

The **event A and B** to be $A \cap B$.

2. PROBABILITY OF A UNION OF TWO EVENTS

For any events A and B,

(a) $P(A \cup B) = P(A) + P(B) - P(A \cap B)$.

If A and B are MUTUALLY EXCLUSIVE ($A \cap B = \varnothing$), then

(b) $P(A \cup B) = P(A) + P(B)$.

3. PROBABILITY OF COMPLEMENTS

For any event E, $E \cup E' = S$ and $E \cap E' = \varnothing$. Thus,

$P(E) = 1 - P(E')$

$P(E') = 1 - P(E)$

4. PROBABILITY TO ODDS

If $P(E)$ is the probability of the event E, then:

(a) Odds for $E = \dfrac{P(E)}{1 - P(E)} = \dfrac{P(E)}{P(E')}$ $[P(E) \neq 1]$

(b) Odds against $E = \dfrac{P(E')}{P(E)}$ $[P(E) \neq 0]$

[NOTE: When possible, odds are expressed as ratios of whole numbers.]

5. ODDS TO PROBABILTY

If the odds for an event E are $\frac{a}{b}$, then the probability of E is:

$$P(E) = \frac{a}{a + b}$$

---

1. Let E be the event "failing within 90 days."
Then E' = "not failing within 90 days."

$P(E') = 1 - P(E)$ (using 2)
$= 1 - .003$
$= .997$

3. Let Event A = "a number less than 3" = {1, 2}.
Let Event B = "a number greater than 7" = {8, 9, 10}.

Since $A \cap B = \varnothing$, A and B are mutually exclusive. So, using 1(b),

$P(A \cup B) = P(A) + P(B) = \dfrac{n(A)}{n(S)} + \dfrac{n(B)}{n(S)} = \dfrac{2}{10} + \dfrac{3}{10} = \dfrac{1}{2}$

5. Let Event A = "an even number" = {2, 4, 6, 8, 10}.
Let Event B = "a number divisible by 3" = {3, 6, 9}.
Since $A \cap B = \{6\} \neq \varnothing$, A and B are not mutually exclusive. So, using 1(a),

$P(A \cup B) = P(A) + P(B) - P(A \cap B) = \dfrac{5}{10} + \dfrac{3}{10} - \dfrac{1}{10} = \dfrac{7}{10}$

7. $P(A) = \dfrac{35 + 5}{35 + 5 + 20 + 40}$
$= \dfrac{40}{100} = .4$

9. $P(B) = \dfrac{5 + 20}{35 + 5 + 20 + 40}$
$= \dfrac{25}{100} = .25$

11. $P(A \cap B) = \dfrac{5}{35 + 5 + 20 + 40}$
$= \dfrac{5}{100} = .05$

13. $P(A' \cap B) = \dfrac{20}{35 + 5 + 20 + 40}$
$= \dfrac{20}{100} = .2$

15. $P(A \cup B) = \dfrac{35 + 5 + 20}{35 + 5 + 20 + 40}$
$= \dfrac{60}{100} = .6$

17. $P(A' \cup B) = \dfrac{20 + 40 + 5}{35 + 5 + 20 + 40}$
$= \dfrac{65}{100} = .65$

19. $P(\text{sum of 5 or 6}) = P(\text{sum of 5}) + P(\text{sum of 6})$ [using 1(b)]
$= \dfrac{4}{36} + \dfrac{5}{36} = \dfrac{9}{36} = \dfrac{1}{4}$ or .25

21. $P(1 \text{ on first die or } 1 \text{ on second die})$ [using 1(a)]

$= P(1 \text{ on first die}) + P(1 \text{ on second die}) - P(1 \text{ on both dice})$

$= \frac{6}{36} + \frac{6}{36} - \frac{1}{36} = \frac{11}{36}$

23. Use 3 to find the odds for Event E.

(A) $P(E) = \frac{3}{8}$, $P(E') = 1 - P(E) = \frac{5}{8}$

Odds for $E = \frac{P(E)}{P(E')}$

$= \frac{3/8}{5/8} = \frac{3}{5}$ (3 to 5)

Odds against $E = \frac{P(E')}{P(E)}$

$= \frac{5/8}{3/8} = \frac{5}{3}$ (5 to 3)

(B) $P(E) = \frac{1}{4}$, $P(E') = 1 - P(E) = \frac{3}{4}$

Odds for $E = \frac{P(E)}{P(E')}$

$= \frac{1/4}{3/4} = \frac{1}{3}$ (1 to 3)

Odds against $E = \frac{P(E')}{P(E)}$

$= \frac{3/4}{1/4} = \frac{3}{1}$ (3 to 1)

(C) $P(E) = .4$, $P(E') = 1 - P(E) = .6$

Odds for $E = \frac{P(E)}{P(E')}$

$= \frac{.4}{.6} = \frac{2}{3}$ (2 to 3)

Odds against $E = \frac{P(E')}{P(E)}$

$= \frac{.6}{.4} = \frac{3}{2}$ (3 to 2)

(D) $P(E) = .55$, $P(E') = 1 - P(E) - .45$

Odds for $E = \frac{P(E)}{P(E')}$

$= \frac{.55}{.45} = \frac{11}{9}$ (11 to 9)

Odds against $E = \frac{P(E')}{P(E)}$

$= \frac{.45}{.55} = \frac{9}{11}$ (9 to 11)

25. Use 4 to find the probabilty of event E.

(A) Odds for $E = \frac{3}{8}$

$P(E) = \frac{3}{3 + 8} = \frac{3}{11}$

(B) Odds for $E = \frac{11}{7}$

$P(E) = \frac{11}{11 + 7} = \frac{11}{18}$

(C) Odds for $E = \frac{4}{1}$

$P(E) = \frac{4}{4 + 1} = \frac{4}{5} = .8$

(D) Odds for $E = \frac{49}{51}$

$P(E) = \frac{49}{49 + 51} = \frac{49}{100} = .49$

27. (A) False: A coin flipped 40 times and comes up heads 37 times:

empirical probability: $P(H) = \frac{37}{40}$

theoretical probability: $P(H) = \frac{1}{2} = \frac{20}{40}$

(For example, see Problem 43, Exercise 6-3.)

(B) True: In general

$$P(E \cup F) = P(E) + P(F) - P(E \cap F)$$

Therefore,

$$P(E) + P(F) = P(E \cup F) + P(E \cap F)$$

whether or not E and F are mutually exclusive.

29. Odds for $E = \frac{P(E)}{P(E')} = \frac{1/2}{1/2} = 1.$

The odds in favor of getting a head in a single toss of a coin are 1 to 1.

31. The sample space for this problem is:

S = {HHH, HHT, THH, HTH, TTH, HTT, THT, TTT}

Let Event E = "getting at least 1 head."

Let Event E' = "getting no heads."

Thus, $\frac{P(E)}{P(E')} = \frac{7/8}{1/8} = \frac{7}{1}$

The odds in favor of getting at least 1 head are 7 to 1.

33. Let Event E = "getting a number greater than 4."

Let Event E' = "not getting a number greater than 4."

Thus, $\frac{P(E')}{P(E)} = \frac{4/6}{2/6} = \frac{2}{1}$

The odds against getting a number greater than 4 in a single roll of a die are 2 to 1.

35. Let Event E = "getting 3 or an even number" = {2, 3, 4, 6}.

Let Event E' = "not getting 3 or an even number" = {1, 5}.

Thus, $\frac{P(E')}{P(E)} = \frac{2/6}{4/6} = \frac{1}{2}$

The odds against getting 3 or an even number are 1 to 2.

37. Let E = "rolling a five." Then $P(E) = \frac{n(E)}{n(S)} = \frac{4}{36} = \frac{1}{9}$ and $P(E') = \frac{8}{9}$.

(A) Odds for $E = \frac{1/9}{8/9} = \frac{1}{8}$ (1 to 8)

(B) The house should pay $8 for the game to be fair (see Example 6).

39. (A) Let E = "sum is less than 4 or greater than 9." Then

$$P(E) = \frac{10 + 30 + 120 + 80 + 70}{1000} = \frac{310}{1000} = \frac{31}{100} = .31 \text{ and } P(E') = \frac{69}{100}.$$

Thus,

$$\text{Odds for } E = \frac{31/100}{69/100} = \frac{31}{69}$$

(B) Let F = "sum is even or divisible by 5." Then

$$P(F) = \frac{10 + 50 + 110 + 170 + 120 + 70 + 70}{1000} = \frac{600}{1000} = \frac{6}{10} = .6$$

and $P(F') = \frac{4}{10}$. Thus,

$$\text{Odds for } F = \frac{6/10}{4/10} = \frac{6}{4} = \frac{3}{2}$$

41. Let A = "drawing a face card" (Jack, Queen, King) and B = "drawing a club."

Then $P(A \cup B) = P(A) + P(B) - P(A \cap B) = \frac{12}{52} + \frac{13}{52} - \frac{3}{52} = \frac{22}{52} = \frac{11}{26}$

$P[(A \cup B)'] = \frac{15}{26}$

Odds for $A \cup B = \frac{11/26}{15/26} = \frac{11}{15}$

43. Let A = "drawing a black card" and B = "drawing an ace."

$P(A \cup B) = P(A) + P(B) - P(A \cap B) = \frac{26}{52} + \frac{4}{52} - \frac{2}{52} = \frac{28}{52} = \frac{7}{13}$

$P[(A \cup B)'] = \frac{6}{13}$

Odds for $A \cup B = \frac{7/13}{6/13} = \frac{7}{6}$

45. The sample space S is the set of all 5-card hands and $n(S) = C_{52,5}$

Let E = "getting at least one diamond."
Then E' = "no diamonds" and $n(E) = C_{39,5}$.

Thus, $P(E') = \frac{C_{39,5}}{C_{52,5}}$, and

$$P(E) = 1 - \frac{C_{39,5}}{C_{52,5}} = 1 - \frac{\frac{39!}{5!34!}}{\frac{52!}{5!47!}} = 1 - \frac{39\cdot38\cdot37\cdot36\cdot35}{52\cdot51\cdot50\cdot49\cdot48} \approx 1 - .22 = .78.$$

47. The number of numbers less than or equal to 1000 which are divisible by 6 is the largest integer in $\frac{1000}{6}$ or 166.

The number of numbers less than or equal to 1000 which are divisible by 8 is the largest integer in $\frac{1000}{8}$ or 125.

The number of numbers less than or equal to 1000 which are divisible by both 6 and 8 is the same as the number of numbers which are divisible by 24. This is the largest integer in $\frac{1000}{24}$ or 41.

Thus, if A is the event "selecting a number which is divisible by either 6 or 8," then

$n(A) = 166 + 125 - 41 = 250$ and $P(A) = \frac{250}{1000} = .25$.

49. In general, for three events A, B, and C,

$$P(A \cup B \cup C) = P(A) + P(B) + P(C) - P(A \cap B) - P(A \cap C) - P(B \cap C) + P(A \cap B \cap C)$$

Therefore,

(*) $P(A \cup B \cup C) = P(A) + P(B) + P(C) - P(A \cap B)$

will hold if A and C, and B and C are mutually exclusive. Note that, if either $A \cap C = \emptyset$, or $B \cap C = \emptyset$, then $A \cap B \cap C = \emptyset$.

Equation (*) will also hold if A, B, and C are mutually exclusive, in which case

$$P(A \cup B \cup C) = P(A) + P(B) + P(C)$$

51. From Example 5,

$$P(E) = 1 - \frac{365!}{365^n(365 - n)!} = 1 - \frac{1}{365^n} \cdot \frac{365!}{(365 - n)!}$$
$$= 1 - \frac{1}{(365)^n} \cdot P_{365,n}$$
$$= 1 - \frac{P_{365,n}}{(365)^n}$$

For calculators with a $P_{n,r}$ key, this form involves fewer calculator steps. Also, 365! produces an overflow error on many calculators, while $P_{365,n}$ does not produce an overflow error for many values of n.

53. S = set of all lists of n birth months, $n \le 12$. Then $n(S) = 12 \cdot 12 \cdot \ldots \cdot 12$ (n times) $= 12^n$.

Let E = "at least two people have the same birth month."
Then E' = "no two people have the same birth month."

$$n(E') = 12 \cdot 11 \cdot 10 \cdot \ldots \cdot [12 - (n - 1)]$$
$$= \frac{12 \cdot 11 \cdot 10 \cdot \ldots \cdot [12 - (n - 1)](12 - n)[12 - (n + 1)] \cdot \ldots \cdot 3 \cdot 2 \cdot 1}{(12 - n)[12 - (n + 1)] \cdot \ldots \cdot 3 \cdot 2 \cdot 1}$$
$$= \frac{12!}{(12 - n)!}$$

Thus, $P(E') = \dfrac{\frac{12!}{(12 - n)!}}{12^n} = \dfrac{12!}{12^n(12 - n)!}$ and $P(E) = 1 - \dfrac{12!}{12^n(12 - n)!}$.

55. Odds for $E = \dfrac{P(E)}{P(E')} = \dfrac{P(E)}{1 - P(E)} = \dfrac{a}{b}$. Therefore,

$bP(E) = a[1 - P(E)] = a - aP(E)$.

Thus, $aP(E) + bP(E) = a$

$$(a + b)P(E) = a$$
$$P(E) = \frac{a}{a + b}$$

57. (A) From the plot, 7 and 8 each came up 10 times.

Therefore, $P(7 \text{ or } 8) = \frac{10}{50} + \frac{10}{50} = \frac{20}{50} = 0.4$.

(B) Theoretical probability: $P(7) = \frac{1}{6}$, $P(8) = \frac{5}{36}$

so $P(7 \text{ or } 8) = \frac{1}{6} + \frac{5}{36} = \frac{6}{36} + \frac{5}{36} = \frac{11}{36} \approx 0.306$.

(C) The answer depends on the results of your simulation.

59. Venn diagram:

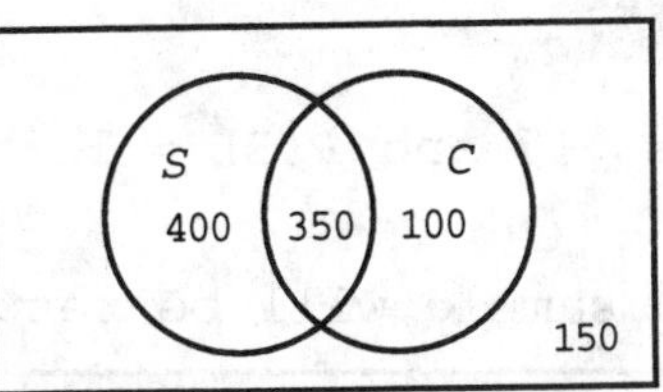

Let S be the event that the student owns a stereo and C be the event that the student owns a car.

The table corresponding to the given data is as follows:

	C	C'	Total
S	350	400	750
S'	100	150	250
Total	450	550	1000

The corresponding probabilities are:

	C	C'	Total
S	.35	.40	.75
S'	.10	.15	.25
Total	.45	.55	1.00

From the above table:

(A) $P(C \text{ or } S) = P(C \cup S) = P(C) + P(S) - P(C \cap S)$
$= .45 + .75 - .35$
$= .85$

(B) $P(C' \cap S') = .15$

61. (A) Using the table, we have:

$P(M_1 \text{ or } A) = P(M_1 \cup A) = P(M_1) + P(A) - P(M_1 \cap A)$
$= .2 + .3 - .05$
$= .45$

(B) $P[(M_2 \cap A') \cup (M_3 \cap A')] = P(M_2 \cap A') + P(M_3 \cap A')$
$= .2 + .35$ (from the table)
$= .55$

63. Let K = "defective keyboard"
and D = "defective disk drive."

Then $K \cup D$ = "either a defective keyboard or a defective disk drive"
and $(K \cup D)'$ = "neither the keyboard nor the disk drive is defective"
$= K' \cap D'$

$P(K \cup D) = P(K) + P(D) - P(K \cap D) = .06 + .05 - .01 = .1$

Thus, $P(K' \cap D') = 1 - .1 = .9$.

65. The sample space S is the set of all possible 10-element samples from the 60 watches, and $n(S) = C_{60,10}$. Let E be the event that a sample contains at least one defective watch. Then E' is the event that a sample contains no defective watches. Now, $n(E') = C_{51,10}$.

Thus, $P(E') = \frac{C_{51,10}}{C_{60,10}} = \frac{\frac{51!}{10!41!}}{\frac{60!}{10!50!}} \approx .17$ and $P(E) \approx 1 - .17 = .83$.

Therefore, the probabilty that a sample will be returned is .83.

67. The given information is displayed in the Venn diagram:

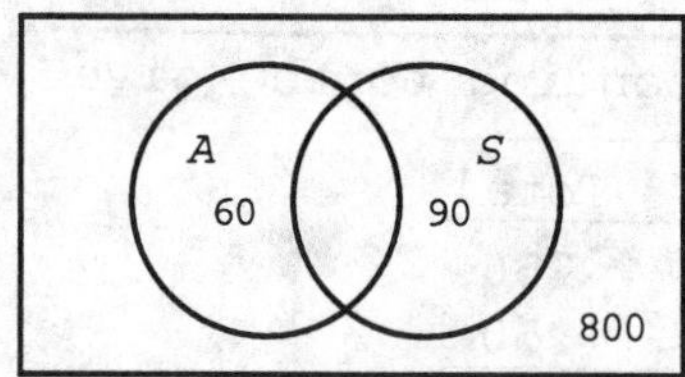

A = suffers from loss of appetite
S = suffers from loss of sleep

Thus, we can conclude that $n(A \cap S) = 1000 - (60 + 90 + 800) = 50$.

$P(A \cap S) = \frac{50}{1000} = .05$

69. (A) "Unaffiliated or no preference" = $U \cup N$.

$P(U \cup N) = P(U) + P(N) - P(U \cap N)$

$= \frac{150}{1000} + \frac{85}{1000} - \frac{15}{1000} = \frac{220}{1000} = \frac{11}{50} = .22$

Therefore, $P[(U \cup N)'] = 1 - \frac{11}{50} = \frac{39}{50}$ and

Odds for $U \cup N = \frac{11/50}{39/50} = \frac{11}{39}$

(B) "Affliated with a party and prefers candidate A" = $(D \cup R) \cap A$.

$P[(D \cup R) \cap A] = \frac{300}{1000} = \frac{3}{10} = .3$

The odds against this event are:

$\frac{1 - 3/10}{3/10} = \frac{7/10}{3/10} = \frac{7}{3}$

71. Let S = the set of all three-person groups from the total group.
Let E = the set of all three-person groups with at least one black.
Let E' = the set of all three-person groups with no blacks.
First, find $P(E')$, then use $P(E) = 1 - P(E')$ to find $P(E)$.

$P(E') = \frac{n(E')}{n(E)} = \frac{C_{15,3}}{C_{20,3}} \approx .4$

$P(E) = 1 - P(E') \approx .6$

EXERCISE 6-5

Things to remember:

1. CONDITIONAL PROBABILITY

For events A and B in a sample space S, the CONDITIONAL PROBABILITY of A given B, denoted $P(A \mid B)$, is defined by

$$P(A \mid B) = \frac{P(A \cap B)}{P(B)}, \quad P(B) \neq 0$$

2. PRODUCT RULE

For events A and B, $P(A) \neq 0$, $P(B) \neq 0$, in a sample space S,

$$P(A \cap B) = P(A) \cdot P(B \mid A) = P(B) \cdot P(A \mid B).$$

[Note: We can use either $P(A) \cdot P(B \mid A)$ or $P(B) \cdot P(A \mid B)$ to compute $P(A \cap B)$.]

3. PROBABILITY TREES

Given a sequence of probability experiments. To compute the probabilities of combined outcomes:

Step 1. Draw a tree diagram corresponding to all combined outcomes of the sequence of experiments.

Step 2. Assign a probability to each tree branch. (This is the probability of the occurrence of the event on the right end of the branch subject to the occurrence of all events on the path leading to the event on the right end of the branch. The probability of the occurrence of a combined outcome that corresponds to a path through the tree is the product of all branch probabilities on the path.*)

Step 3. Use the results in steps 1 and 2 to answer various questions related to the sequence of experiments as a whole.

If A and B are independent events with nonzero probabilities in a sample space S, then

(*) $P(A \mid B) = P(A)$ and $P(B \mid A) = P(B)$

If either equation in (*) holds, then A and B are independent.

4. INDEPENDENCE

Let A and B be any events in a sample space S. Then A and B are INDEPENDENT if and only if

$$P(A \cap B) = P(A) \cdot P(B).$$

Otherwise, A and B are DEPENDENT.

5. INDEPENDENT SET OF EVENTS

A set of events is said to be INDEPENDENT if for each finite subset $\{E_1, E_2, \ldots, E_k\}$

$$P(E_1 \cap E_2 \cap \ldots \cap E_k) = P(E_1)P(E_2) \ldots P(E_k)$$

---

1. $P(A) = .50$
See the given table.

3. $P(D) = .20$
See the given table.

5. $P(A \cap D) = .10$
See the given table for occurrences of both A and D.

7. $P(C \cap D)$ = probability of occurrences of both C and D = .06.

9. $P(A \mid D) = \dfrac{P(A \cap D)}{P(D)} = \dfrac{0.10}{0.20} = .50$

11. $P(C \mid D) = \dfrac{P(C \cap D)}{P(D)} = \dfrac{0.06}{0.20} = .30$

13. Events A and D are independent if $P(A \cap D) = P(A) \cdot P(D)$:
$P(A \cap D) = .10$
$P(A) \cdot P(D) = (.50)(.20) = .10$
Thus, A and D are independent.

15. $P(C \cap D) = .06$
$P(C) \cdot P(D) = (.20)(.20) = .04$
Since $P(C \cap D) \neq P(C) \cdot P(D)$, C and D are dependent.

17. (A) Let H_8 = "a head on the eighth toss." Since each toss is independent of the other tosses, $P(H_8) = \frac{1}{2}$.

(B) Let H_i = "a head on the *i*th toss." Since the tosses are independent,
$P(H_1 \cap H_2 \cap \cdots \cap H_8) = P(H_1)P(H_2) \cdots P(H_8) = \left(\frac{1}{2}\right)^8 = \frac{1}{2^8} = \frac{1}{256}$.
Similarly, if T_i = "a tail on the *i*th toss," then
$P(T_1 \cap T_2 \cap \cdots \cap T_8) = P(T_1)P(T_2) \cdots P(T_8) = \frac{1}{2^8} = \frac{1}{256}$. Finally, if
H = "all heads" and T = "all tails," then $H \cap T = \varnothing$ and
$P(H \cup T) = P(H) + P(T) = \frac{1}{256} + \frac{1}{256} = \frac{2}{256} = \frac{1}{128} \approx .00781$.

19. Given the table:

e_i	1	2	3	4	5
p_i	.3	.1	.2	.3	.1

E = "pointer lands on an even number" = {2, 4}.
F = "pointer lands on a number less than 4" = {1, 2, 3}.

(A) $P(F \mid E) = \dfrac{P(F \cap E)}{P(E)} = \dfrac{P(2)}{P(2) + P(4)} = \dfrac{.1}{.1 + .3} = \dfrac{.1}{.4} = \dfrac{1}{4}$

(B) $P(E \cap F) = P(2) = .1$
$P(E) = .4$,
$P(F) = P(1) + P(2) + P(3) = .3 + .1 + .2 = .6$,
and
$P(E)P(F) = (.4)(.6) = .24 \neq P(E \cap F)$.
Thus, E and F are dependent.

21. From the probability tree,

(A) $P(M \cap S) = (.3)(.6) = .18$

(B) $P(R) = P(N \cap R) + P(M \cap R) = (.7)(.2) + (.3)(.4)$
$= .14 + .12$
$= .26$

23. $E_1 = \{HH, HT\}$ and $P(E_1) = \frac{1}{2}$

$E_2 = \{TH, TT\}$ and $P(E_2) = \frac{1}{2}$

$E_4 = \{HH, TH\}$ and $P(E_4) = \frac{1}{2}$

(A) Since $E_1 \cap E_4 = \{HH\} \neq \varnothing$, E_1 and E_4 **are not** mutually exclusive.
Since $P(E_1 \cap E_4) = P(HH) = \frac{1}{4} = P(E_1) \cdot P(E_4)$, E_1 and E_4 are independent.

(B) Since $E_1 \cap E_2 = \varnothing$, E_1 and E_2 **are** mutually exclusive.
Since $P(E_1 \cap E_2) = 0$ and $P(E_1) \cdot P(E_2) = \frac{1}{4}$, $P(E_1 \cap E_2) \neq P(E_1) \cdot P(E_2)$.
Therefore, E_1 and E_2 are dependent.

25. Let E_i = "even number on the ith throw," $i = 1,2$,
and O_i = "odd number on the ith throw," $i = 1, 2$.
Then $P(E_i) = \frac{1}{2}$ and $P(O_i) = \frac{1}{2}$, $i = 1, 2$.
The probability tree for this experiment is shown at the right.

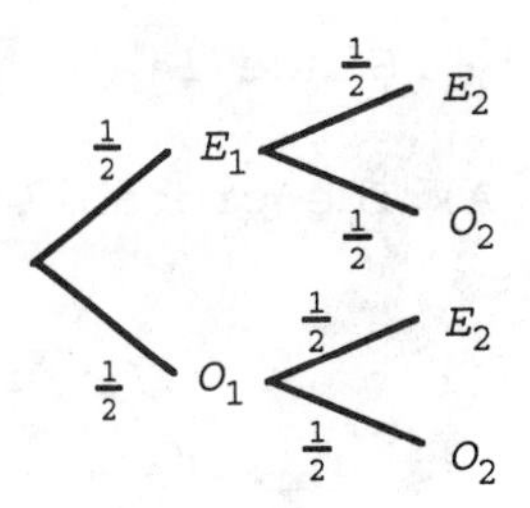

$$P(E_1 \cap E_2) = \left(\frac{1}{2}\right)\left(\frac{1}{2}\right) = \frac{1}{4}$$

$$P(E_1 \cup E_2) = P(E_1) + P(E_2) - P(E_1 \cap E_2) = \frac{1}{2} + \frac{1}{2} - \frac{1}{4} = \frac{3}{4}.$$

27. Let C = "first card is a club,"
and H = "second card is a heart."

(A) Without replacement, the probability tree is as shown at the right.
Thus, $P(C \cap H) = \left(\frac{1}{4}\right)\left(\frac{13}{51}\right) \approx .0637$.

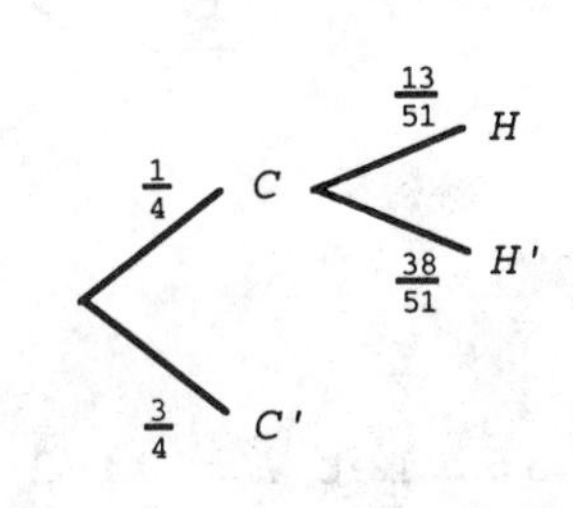

(B) With replacement, the draws are independent and
$P(C \cap H) = \left(\frac{1}{4}\right)\left(\frac{1}{4}\right) = \frac{1}{16} = .0625$.

29. G = "the card is black" = {spade or club} and $P(G) = \frac{1}{2}$.
H = "the card is divisible by 3" = {3, 6, or 9}. $P(H) = \frac{12}{52} = \frac{3}{13}$
$P(H \cap G)$ = {3, 6, or 9 of clubs or spades} $= \frac{6}{52} = \frac{3}{26}$

(A) $P(H \mid G) = \dfrac{P(H \cap G)}{P(G)} = \dfrac{3/26}{1/2} = \dfrac{6}{26} = \dfrac{3}{13}$

(B) $P(H \cap G) = \frac{3}{26} = P(H) \cdot P(G)$
Thus, H and G *are* independent.

31. (A) $S = \{BB, BG, GB, GG\}$

$A = \{BB, GG\}$ and $P(A) = \frac{2}{4} = \frac{1}{2}$

$B = \{BG, GB, GG\}$ and $P(B) = \frac{3}{4}$

$A \cap B = \{GG\}$.

$P(A \cap B) = \frac{1}{4}$ and $P(A) \cdot P(B) = \frac{1}{2} \cdot \frac{3}{4} = \frac{3}{8}$

Thus, $P(A \cap B) \neq P(A) \cdot P(B)$ and the events are dependent.

(B) $S = \{BBB, BBG, BGB, BGG, GBB, GBG, GGB, GGG\}$
$A = \{BBB, GGG\}$
$B = \{BGG, GBG, GGB, GGG\}$
$A \cap B = \{GGG\}$

$P(A) = \frac{2}{8} = \frac{1}{4}$, $P(B) = \frac{4}{8} = \frac{1}{2}$, and $P(A \cap B) = \frac{1}{8}$

Since $P(A \cap B) = \frac{1}{8} = P(A) \cdot P(B)$, A and B are independent.

33. (A) The probability tree with replacement is as follows:

Start → R_1 ($\frac{2}{7}$): → R_2 ($\frac{2}{7}$): $P(R_1 \cap R_2) = \frac{4}{49}$; → W_2 ($\frac{5}{7}$): $P(R_1 \cap W_2) = \frac{10}{49}$

Start → W_1 ($\frac{5}{7}$): → R_2 ($\frac{2}{7}$): $P(W_1 \cap R_2) = \frac{10}{49}$; → W_2 ($\frac{5}{7}$): $P(W_1 \cap W_2) = \frac{25}{49}$

(B) The probability tree without replacement is as follows:

Start → R_1 ($\frac{2}{7}$): → R_2 ($\frac{1}{6}$): $P(R_1 \cap R_2) = \frac{1}{21}$; → W_2 ($\frac{5}{6}$): $P(R_1 \cap W_2) = \frac{5}{21}$

Start → W_1 ($\frac{5}{7}$): → R_2 ($\frac{2}{6}$): $P(W_1 \cap R_2) = \frac{5}{21}$; → W_2 ($\frac{4}{6}$): $P(W_1 \cap W_2) = \frac{10}{21}$

35. Let E = At least one ball was red = $\{R_1 \cap R_2, R_1 \cap W_2, W_1 \cap R_2\}$.

(A) With replacement [see the probability tree in Problem 33(A)]:

$$P(E) = P(R_1 \cap R_2) + P(R_1 \cap W_2) + P(W_1 \cap R_2) = \frac{4}{49} + \frac{10}{49} + \frac{10}{49} = \frac{24}{49}$$

(B) Without replacement [see the probability tree in Problem 33(B)]:

$$P(E) = P(R_1 \cap R_2) + P(R_1 \cap W_2) + P(W_1 \cap R_2) = \frac{1}{21} + \frac{5}{21} + \frac{5}{21} = \frac{11}{21}$$

37. (A) False. Consider the experiment of flipping a coin and rolling a die. Let A = "The coin is a head" and B = "The die is a 6". Then A and B are independent, so

$$P(A \mid B) = P(A) = \frac{1}{2}, \quad P(B \mid A) = P(B) = \frac{1}{6}$$

(B) False. By definition, $P(B \mid A) = \dfrac{P(A \cap B)}{P(A)}$ which implies $P(A \cap B) = P(A) \cdot P(B \mid A)$. That is, this equation holds whether or not A and B are independent, so *any* pair of events, A and B, which are not independent provides a counter-example.

39. $n(S) = C_{9,2} = \dfrac{9!}{2!(9-2)!}$ (total number of balls $= 2 + 3 + 4 = 9$)

$$= \frac{9 \cdot 8 \cdot 7!}{2 \cdot 1 \cdot 7!} = 36$$

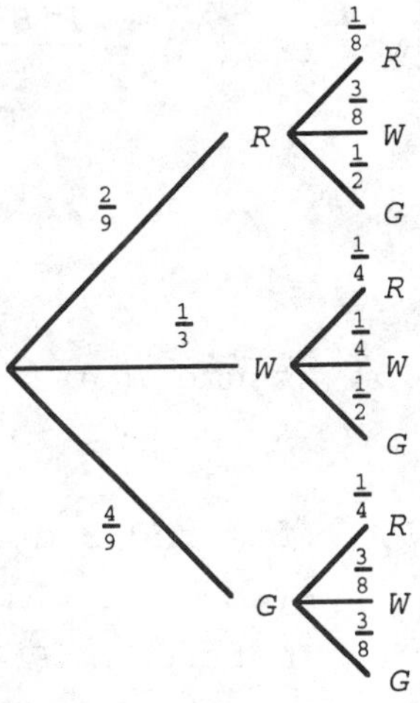

Let A = Both balls are the same color.

$n(A)$ = (No. of ways 2 red balls are selected) + (No. of ways 2 white balls are selected) + (No. of ways 2 green balls are selected)

$$= C_{2,2} + C_{3,2} + C_{4,2}$$

$$= \frac{2!}{2!(2-2)!} + \frac{3!}{2!(3-2)!} + \frac{4!}{2!(4-2)!}$$

$$= 1 + 3 + 6 = 10$$

$$P(A) = \frac{n(A)}{n(S)} = \frac{10}{36} = \frac{5}{18}$$

Alternatively, the probability tree for this experiment is shown at the right.

And $P(RR, WW, \text{ or } GG) = P(RR) + P(WW) + P(GG)$

$$= \left(\frac{2}{9}\right)\left(\frac{1}{8}\right) + \left(\frac{1}{3}\right)\left(\frac{1}{4}\right) + \left(\frac{4}{9}\right)\left(\frac{3}{8}\right) = \frac{2}{72} + \frac{1}{12} + \frac{12}{72} = \frac{20}{72} = \frac{5}{18}$$

41. The probability tree for this experiment is:

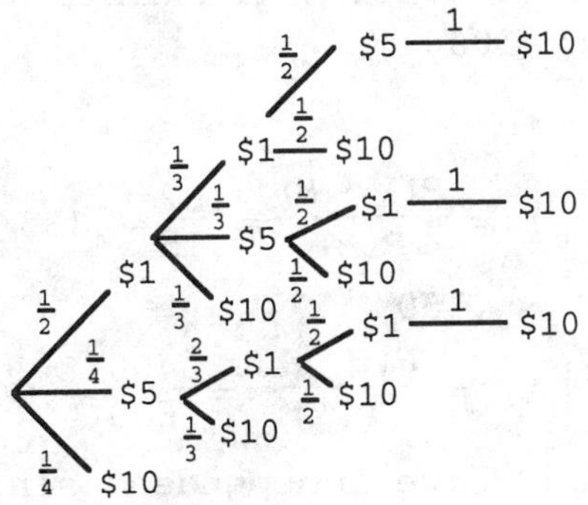

(A) $P(\$16) = \left(\frac{1}{4}\right)\left(\frac{2}{3}\right)\left(\frac{1}{2}\right) + \left(\frac{1}{2}\right)\left(\frac{1}{3}\right)\left(\frac{1}{2}\right)$

$$= \frac{1}{12} + \frac{1}{12} = \frac{1}{6} \approx .167$$

(B) $P(\$17) = \left(\frac{1}{2}\right)\left(\frac{1}{3}\right)\left(\frac{1}{2}\right) + \left(\frac{1}{2}\right)\left(\frac{1}{3}\right)\left(\frac{1}{2}\right) + \left(\frac{1}{4}\right)\left(\frac{2}{3}\right)\left(\frac{1}{2}\right)$

$$= \frac{1}{12} + \frac{1}{12} + \frac{1}{12} = \frac{1}{4} = .25$$

(C) Let A = "\$10 on second draw." Then

$$P(A) = \left(\frac{1}{4}\right)\left(\frac{1}{3}\right) + \left(\frac{1}{2}\right)\left(\frac{1}{3}\right) = \frac{1}{12} + \frac{1}{6} = \frac{1}{4} = .25$$

43. Assume that A and B are independent events with $P(A) \neq 0$, $P(B) \neq 0$. Then, by definition

(*) $P(A \cap B) = P(A) \cdot P(B)$.

Now,

$$P(A \mid B) = \frac{P(A \cap B)}{P(B)} \quad \text{(definition of conditional probability)}$$
$$= \frac{P(A) \cdot P(B)}{P(B)} \quad \text{(by *)}$$
$$= P(A)$$

Also,

$$P(B \mid A) = \frac{P(B \cap A)}{P(A)} = \frac{P(A \cap B)}{P(A)}$$
$$= \frac{P(A) \cdot P(B)}{P(A)}$$
$$= P(B)$$

45. Assume $P(A) \neq 0$. Then $P(A \mid A) = \frac{P(A \cap A)}{P(A)} = \frac{P(A)}{P(A)} = 1$.

47. If A and B are mutually exclusive, then $A \cap B = \varnothing$ and $P(A \cap B) = P(\varnothing) = 0$. Also, if $P(A) \neq 0$ and $P(B) \neq 0$, then $P(A) \cdot P(B) \neq 0$. Therefore, $P(A \cap B) = 0 \neq P(A) \cdot P(B)$, and events A and B are dependent.

49. (A)

To strike	Hourly H	Salary S	Salary + bonus B	Total
Yes (Y)	.400	.180	.020	.600
No (N)	.150	.120	.130	.400
Total	.550	.300	.150	1.000

[Note: The probability table above was derived from the table given in the problem by dividing each entry by 1000.]

Referring to the table in part (A):

(B) $P(Y \mid H) = \frac{P(Y \cap H)}{P(H)} = \frac{.400}{.55} \approx .727$

(C) $P(Y \mid B) = \frac{P(Y \cap B)}{P(B)} = \frac{.02}{.15} \approx .133$

(D) $P(S) = .300$

$P(S \mid Y) = \frac{P(S \cap Y)}{P(Y)} = \frac{.180}{.60} = .300$

(E) $P(H) = .550$

$P(H \mid Y) = \frac{P(H \cap Y)}{P(Y)} = \frac{.400}{.600} \approx .667$

(F) $P(B \cap N) = .130$

(G) S and Y are independent since $P(S \mid Y) = P(S) = .300$

(H) H and Y are dependent since $P(H \mid Y) \approx .667$ is not equal to $P(H) = .550$.

(I) $P(B \mid N) = \dfrac{P(B \cap N)}{P(N)}$
$= \dfrac{.130}{.400}$ (from table)
$= .325$

and $P(B) = .150$. Since $P(B \mid N) \neq P(B)$, B and N are dependent.

51. The probability tree for this experiment is:

- $\frac{1}{4}$ $20
- $\frac{1}{2}$ $5
  - $\frac{1}{3}$ $20
    - $\frac{1}{2}$ $5 —— $20
    - $\frac{1}{2}$ $20
  - $\frac{1}{3}$ $1
    - $20
  - $\frac{1}{3}$ $5
    - $\frac{1}{2}$ $20
    - $\frac{1}{2}$ $1 —— $20
- $\frac{1}{4}$ $1
  - $\frac{1}{3}$ $20
  - $\frac{2}{3}$ $5
    - $\frac{1}{2}$ $5 —— $20
    - $\frac{1}{2}$ $20

(A) $P(\$26{,}000) = \left(\frac{1}{2}\right)\left(\frac{1}{3}\right)\left(\frac{1}{2}\right) + \left(\frac{1}{4}\right)\left(\frac{2}{3}\right)\left(\frac{1}{2}\right)$
$= \frac{1}{12} + \frac{1}{12} = \frac{1}{6} \approx .167$

(B) $P(\$31{,}000) = \left(\frac{1}{2}\right)\left(\frac{1}{3}\right)\left(\frac{1}{2}\right) + \left(\frac{1}{2}\right)\left(\frac{1}{3}\right)\left(\frac{1}{2}\right) + \left(\frac{1}{4}\right)\left(\frac{2}{3}\right)\left(\frac{1}{2}\right)$
$= \frac{3}{12} = \frac{1}{4} = .25$

(C) Let A = "$20 on third draw." Then
$P(A) = \left(\frac{1}{4}\right)\left(\frac{2}{3}\right)\left(\frac{1}{2}\right) + \left(\frac{1}{2}\right)\left(\frac{1}{3}\right)\left(\frac{1}{2}\right) + \left(\frac{1}{2}\right)\left(\frac{1}{3}\right)\left(\frac{1}{2}\right)$
$= \frac{3}{12} = \frac{1}{4} = .25$

53. (A)

	C	C'	Totals
R	0.06	0.44	0.50
R'	0.02	0.48	0.50
Total	0.08	0.92	1.00

(B) $P(C) = 0.08$, $P(R) = 0.50$, $P(R \cap C) = 0.06$
Since $0.06 = P(R \cap C) \neq P(R) \cdot P(C) = 0.04$, R and C are **dependent.**

(C) $P(C \mid R) = \dfrac{P(C \cap R)}{P(R)} = \dfrac{0.06}{0.50} = 0.12$ and $P(C) = 0.08$
Since $P(C \mid R) > P(C)$, cancer is more likely to be developed if the red die is used. The FDA should ban the use of the red die.

(D) The new probability table is

	C	C'	Totals
R	0.02	0.48	0.50
R'	0.06	0.44	0.50
Total	0.08	0.92	1.00

Now $P(C \mid R) = \frac{P(C \cap R)}{P(R)} = \frac{0.02}{0.5} = 0.04$ and $P(C) = 0.08$

Since $P(C \mid R) < P(C)$ it appears that the red die reduces the development of cancer. Therefore, the use of the die should not be banned.

55. (A)

	Below 90 A	90–120 B	Above 120 C	Total
Female (F)	.130	.286	.104	.520
Male (F')	.120	.264	.096	.480
Total	.250	.550	.200	1.000

[Note: The probability table above was derived from the table given in the problem by dividing each entry by 1000.]

Referring to the table in part (A):

(B) $P(A \mid F) = \frac{P(A \cap F)}{P(F)} = \frac{.130}{.520} \approx .250$

$P(A \mid F') = \frac{P(A \cap F')}{P(F')} = \frac{.120}{.480} = .250$

(C) $P(C \mid F) = \frac{P(C \cap F)}{P(F)} = \frac{.104}{.520} \approx .200$

$P(C \mid F') = \frac{P(C \cap F')}{P(F')} = \frac{.096}{.480} = .200$

(D) $P(A) = .25$

$P(A \mid F) = \frac{P(A \cap F)}{P(F)} = \frac{.130}{.520} = .250$

(E) $P(B) = .55$

$P(B \mid F') = \frac{P(B \cap F')}{P(F')} = \frac{.264}{.480} = .550$

(F) $P(F \cap C) = .104$

(G) No, the results in parts (B), (C), (D), and (E) imply that A, B, and C are independent of F and F'.

EXERCISE 6-6

Things to remember:

1. BAYES' FORMULA

   Let $U_1, U_2, \ldots, U_n$ be n mutually exclusive events whose union is the sample space S. Let E be an arbitrary event in S such that $P(E) \neq 0$. Then

$$P(U_1 \mid E) = \frac{P(U_1 \cap E)}{P(E)}$$

$$= \frac{P(U_1 \cap E)}{P(U_1 \cap E) + P(U_2 \cap E) + \cdots + P(U_n \cap E)}$$

$$= \frac{P(E \mid U_1)P(U_1)}{P(E \mid U_1)P(U_1) + \cdots + P(E \mid U_n)P(U_n)}$$

   Similar results hold for $U_2, U_3, \ldots, U_n$.

2. BAYES' FORMULA AND PROBABILITY TREES

$$P(U_1 \mid E) = \frac{\text{Product of branch probabilities leading to } E \text{ through } U_1}{\text{Sum of all branch probabilities leading to } E}$$

Similar results hold for $U_2, U_3, \ldots, U_n$.

---

1. $P(M \cap A) = P(M) \cdot P(A \mid M) = (.6)(.8) = .48$

3. $P(A) = P(M \cap A) + P(N \cap A) = P(M)P(A \mid M) + P(N)P(A \mid N)$
$= (.6)(.8) + (.4)(.3) = .60$

5. $P(M \mid A) = \frac{P(M \cap A)}{P(M \cap A) + P(N \cap A)} = \frac{.48}{.60}$ (see Problems 1 and 3)
$= .80$

7. Referring to the Venn diagram:

$$P(U_1 \mid R) = \frac{P(U_1 \cap R)}{P(R)} = \frac{\frac{25}{100}}{\frac{60}{100}} = \frac{25}{60} = \frac{5}{12} \approx .417$$

Using Bayes' formula:

$$P(U_1 \mid R) = \frac{P(U_1 \cap R)}{P(U_1 \cap R) + P(U_2 \cap R)} = \frac{P(U_1)P(R \mid U_1)}{P(U_1)P(R \mid U_1) + P(U_2)P(R \mid U_2)}$$

$$= \frac{\left(\frac{40}{100}\right)\left(\frac{25}{40}\right)}{\left(\frac{40}{100}\right)\left(\frac{25}{40}\right) + \left(\frac{60}{100}\right)\left(\frac{35}{60}\right)} = \frac{.25}{.25 + .35} = \frac{.25}{.60} = \frac{5}{12} \approx .417$$

9. $P(U_1 \mid R') = \frac{P(U_1 \cap R')}{P(R')} = \frac{\frac{15}{100}}{1 - P(R)}$ (from the Venn diagram)

$$= \frac{\frac{15}{100}}{1 - \frac{60}{100}} = \frac{\frac{15}{100}}{\frac{40}{100}} = \frac{3}{8} = .375$$

Using Bayes' formula:

$$P(U_1 \mid R') = \frac{P(U_1 \cap R')}{P(R')} = \frac{P(U_1)P(R' \mid U_1)}{P(U_1 \cap R') + P(U_2 \cap R')}$$

$$= \frac{P(U_1)P(R' \mid U_1)}{P(U_1)P(R' \mid U_1) + P(U_2)P(R' \mid U_2)} = \frac{\left(\frac{40}{100}\right)\left(\frac{15}{40}\right)}{\left(\frac{40}{100}\right)\left(\frac{15}{40}\right) + \left(\frac{60}{100}\right)\left(\frac{25}{60}\right)}$$

$$= \frac{.15}{.15 + .25} = \frac{15}{40} = \frac{3}{8} = .375$$

11. $P(U \mid C) = \dfrac{P(U \cap C)}{P(C)} = \dfrac{P(U \cap C)}{P(U \cap C) + P(V \cap C) + P(W \cap C)}$

$= \dfrac{(.2)(.4)}{(.2)(.4) + (.5)(.2) + (.3)(.6)}$ [Note: Recall $P(A \cap B) = P(A) \cdot P(B \mid A)$.]

$= \dfrac{.08}{.36} \approx .222$

13. $P(W \mid C) = \dfrac{P(W \cap C)}{P(C)} = \dfrac{P(W \cap C)}{P(W \cap C) + P(V \cap C) + P(U \cap C)}$

$= \dfrac{(.3)(.6)}{(.3)(.6) + (.5)(.2) + (.2)(.4)}$ (see Problem 11)

$= \dfrac{.18}{.36} = .5$

15. $P(V \mid C) = \dfrac{P(V \cap C)}{P(C)} = \dfrac{P(V \cap C)}{P(V \cap C) + P(W \cap C) + P(U \cap C)}$

$= \dfrac{(.5)(.2)}{(.5)(.2) + (.3)(.6) + (.2)(.4)} = \dfrac{.1}{.36} = .278$

17. From the Venn diagram,

$P(U_1 \mid R) = \dfrac{5}{5 + 15 + 20} = \dfrac{5}{40} = \dfrac{1}{8} = .125$

or $= \dfrac{P(U_1 \cap R)}{P(R)} = \dfrac{\frac{5}{100}}{\frac{40}{100}} = .125$

Using Bayes' formula:

$P(U_1 \mid R) = \dfrac{P(U_1 \cap R)}{P(U_1 \cap R) + P(U_2 \cap R) + P(U_3 \cap R)} = \dfrac{\frac{5}{100}}{\frac{5}{100} + \frac{15}{100} + \frac{20}{100}}$

$= \dfrac{.05}{.05 + .15 + .2} = \dfrac{.05}{.40} = .125$

19. From the Venn diagram,

$P(U_3 \mid R) = \dfrac{20}{5 + 15 + 20} = \dfrac{20}{40} = .5$

Using Bayes' formula:

$P(U_3 \mid R) = \dfrac{P(U_3 \cap R)}{P(U_1 \cap R) + P(U_2 \cap R) + P(U_3 \cap R)} = \dfrac{\frac{20}{100}}{\frac{5}{100} + \frac{15}{100} + \frac{20}{100}}$

$= \dfrac{.2}{.05 + .15 + .2} = \dfrac{.2}{.4} = .5$

21. From the Venn diagram,

$$P(U_2 \mid R) = \frac{15}{5 + 15 + 20} = \frac{15}{40} = .375$$

Using Bayes' formula:

$$P(U_2 \mid R) = \frac{P(U_2 \cap R)}{P(U_1 \cap R) + P(U_2 \cap R) + P(U_3 \cap R)} = \frac{\frac{15}{100}}{\frac{5}{100} + \frac{15}{100} + \frac{20}{100}}$$

$$= \frac{.15}{.05 + .15 + .2} = \frac{.15}{.40}$$

$$= \frac{3}{8} = .375$$

23. From the given tree diagram, we have:

$$P(A) = \frac{1}{4} \qquad P(A') = \frac{3}{4}$$

$$P(B \mid A) = \frac{1}{5} \qquad P(B \mid A') = \frac{3}{5}$$

$$P(B' \mid A) = \frac{4}{5} \qquad P(B' \mid A') = \frac{2}{5}$$

We want to find the following:

$$P(B) = P(B \cap A) + P(B \cap A') = P(A)P(B \mid A) + P(A')P(B \mid A')$$

$$= \left(\frac{1}{4}\right)\left(\frac{1}{5}\right) + \left(\frac{3}{4}\right)\left(\frac{3}{5}\right) = \frac{1}{20} + \frac{9}{20} = \frac{10}{20} = \frac{1}{2}$$

$$P(B') = 1 - P(B) = 1 - \frac{1}{2} = \frac{1}{2}$$

$$P(A \mid B) = \frac{P(A \cap B)}{P(B)} = \frac{P(A)P(B \mid A)}{P(B)} = \frac{\left(\frac{1}{4}\right)\left(\frac{1}{5}\right)}{\frac{1}{2}} = \frac{\frac{1}{20}}{\frac{1}{2}} = \frac{1}{10}$$

Thus, $P(A' \mid B) = 1 - P(A \mid B) = 1 - \frac{1}{10} = \frac{9}{10}$.

$$P(A \mid B') = \frac{P(A \cap B')}{P(B')} = \frac{P(A)P(B' \mid A)}{P(B')} = \frac{\left(\frac{1}{4}\right)\left(\frac{4}{5}\right)}{\frac{1}{2}} = \frac{\frac{4}{20}}{\frac{1}{2}} = \frac{2}{5}$$

Thus, $P(A' \mid B') = 1 - P(A \mid B') = 1 - \frac{2}{5} = \frac{3}{5}$.

Therefore, the tree diagram for this problem is as shown at the right.

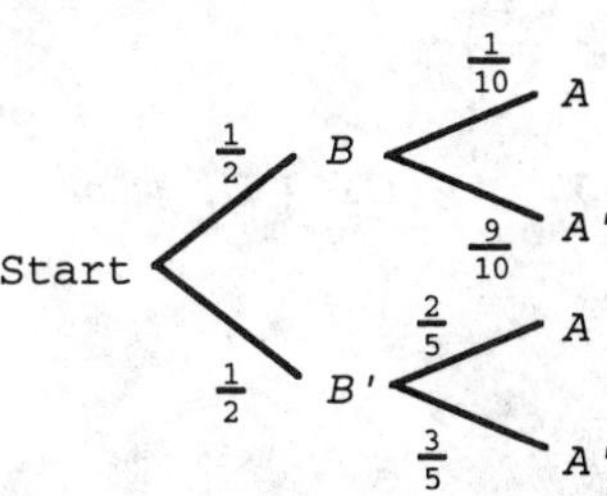

The following tree diagram is to be used for Problems 25 and 27.

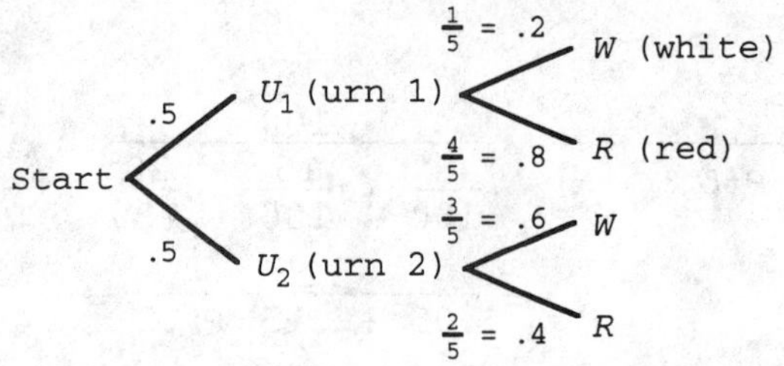

Start → .5 → U_1 (urn 1) → $\frac{1}{5} = .2$ → W (white); $\frac{4}{5} = .8$ → R (red)

Start → .5 → U_2 (urn 2) → $\frac{3}{5} = .6$ → W; $\frac{2}{5} = .4$ → R

25. $$P(U_1 \mid W) = \frac{P(U_1 \cap W)}{P(W)} = \frac{P(U_1 \cap W)}{P(U_1 \cap W) + P(U_2 \cap W)}$$
$$= \frac{P(U_1)P(W \mid U_1)}{P(U_1)P(W \mid U_1) + P(U_2)P(W \mid U_2)} = \frac{(.5)(.2)}{(.5)(.2) + (.5)(.6)}$$
$$= \frac{.1}{.4} = .25$$

27. $$P(U_2 \mid R) = \frac{P(U_2 \cap R)}{P(R)} = \frac{P(U_2 \cap R)}{P(U_2 \cap R) + P(U_1 \cap R)}$$
$$= \frac{P(U_2)P(R \mid U_2)}{P(U_2)P(R \mid U_2) + P(U_1)P(R \mid U_1)} = \frac{(.5)(.4)}{(.5)(.4) + (.5)(.8)}$$
$$= \frac{.4}{1.2} = \frac{1}{3} \approx .333$$

29. $$P(W_1 \mid W_2) = \frac{P(W_1 \cap W_2)}{P(W_2)} = \frac{P(W_1)P(W_2 \mid W_1)}{P(R_1 \cap W_2) + P(W_1 \cap W_2)}$$
$$= \frac{P(W_1)P(W_2 \mid W_1)}{P(R_1)P(W_2 \mid R_1) + P(W_1)P(W_2 \mid W_1)} = \frac{\left(\frac{5}{9}\right)\left(\frac{4}{8}\right)}{\left(\frac{4}{9}\right)\left(\frac{5}{8}\right) + \left(\frac{5}{9}\right)\left(\frac{4}{8}\right)} = \frac{\frac{20}{72}}{\frac{20}{72} + \frac{20}{72}}$$
$$= \frac{20}{40} = \frac{1}{2} \text{ or } .5$$

31. $$P(U_{R_1} \mid U_{R_2}) = \frac{P(U_{R_1} \cap U_{R_2})}{P(U_{R_2})} = \frac{P(U_{R_1})P(U_{R_2} \mid U_{R_1})}{P(U_{W_1} \cap U_{R_2}) + P(U_{R_1} \cap U_{R_2})}$$
$$= \frac{P(U_{R_1})P(U_{R_2} \mid U_{R_1})}{P(U_{W_1})P(U_{R_2} \mid U_{W_1}) + P(U_{R_1})P(U_{R_2} \mid U_{R_1})}$$
$$= \frac{\left(\frac{7}{10}\right)\left(\frac{5}{10}\right)}{\left(\frac{3}{10}\right)\left(\frac{4}{10}\right) + \left(\frac{7}{10}\right)\left(\frac{5}{10}\right)} = \frac{.35}{.12 + .35}$$
$$= \frac{.35}{.47} = \frac{35}{47} \approx .745$$

The tree diagram follows:

$$\text{Start} \begin{cases} \frac{7}{10}\ U_{R_1} \begin{cases} \frac{5}{10}\ U_{R_2} \\ \frac{5}{10}\ U_{W_2} \end{cases} \\ \frac{3}{10}\ U_{W_1} \begin{cases} \frac{4}{10}\ U_{R_2} \\ \frac{6}{10}\ U_{W_2} \end{cases} \end{cases}$$

where U_{R_1} is red from urn one,
U_{R_2} is red from urn two,
U_{W_1} is white from urn one,
and U_{W_2} is white from urn two.

33. Suppose $c = e$. then

$$P(M) = ac + be = ac + bc = c(a + b) = c \quad (a + b = 1)$$

and

$$P(M \mid U) = \frac{P(M \cap U)}{P(U)} = \frac{ac}{a} = c$$

Therefore, M and U are independent.
Alternatively, note that

$$P(M) = c, \; P(U) = a \text{ and}$$
$$P(M \cap U) = ac = P(M) \cdot P(U),$$

which implies that M and U are independent.

35. Draw a tree diagram to verify the probabilities given below.

(A) With replacement--True.

$$P(B_2 \mid B_1) = \frac{m}{m + n}$$

$$P(B_1 \mid B_2) = \frac{P(B_1 \cap B_2)}{P(B_1 \cap B_2) + P(W_1 \cap B_2)}$$

$$= \frac{\frac{m^2}{(m + n)^2}}{\frac{m^2}{(m + n)^2} + \frac{mn}{(m + n)^2}} = \frac{m^2}{m^2 + mn} = \frac{m}{m + n} = P(B_2 \mid B_1)$$

(B) Without replacement--True.

$$P(B_2 \mid B_1) = \frac{m - 1}{m + n - 1}$$

$$P(B_1 \mid B_2) = \frac{P(B_1 \cap B_2)}{P(B_1 \cap B_2) + P(W_1 \cap B_2)}$$

$$= \frac{\frac{m(m - 1)}{(m + n)(m + n - 1)}}{\frac{m(m - 1)}{(m + n)(m + n - 1)} + \frac{nm}{(m + n)(m + n - 1)}}$$

$$= \frac{m(m - 1)}{m(m - 1) + mn} = \frac{m - 1}{m + n - 1} = P(B_2 \mid B_1)$$

37.

Start
- $\frac{13}{52}$ H_1 (heart)
  - $\frac{12}{51}$ H_2 (heart)
  - $\frac{39}{51}$ $\overline{H}_2$ (not a heart)
- $\frac{39}{52}$ $\overline{H}_1$ (not a heart)
  - $\frac{13}{51}$ H_2
  - $\frac{38}{51}$ $\overline{H}_2$

$$P(H_1 \mid H_2) = \frac{P(H_1 \cap H_2)}{P(H_2)} = \frac{P(H_1 \cap H_2)}{P(H_1 \cap H_2) + P(\overline{H}_1 \cap H_2)}$$

$$= \frac{P(H_1)P(H_2 \mid H_1)}{P(H_1)P(H_2 \mid H_1) + P(\overline{H}_1)P(H_2 \mid \overline{H}_1)} = \frac{\frac{13}{52} \cdot \frac{12}{51}}{\frac{13}{52} \cdot \frac{12}{51} + \frac{39}{52} \cdot \frac{13}{51}}$$

$$= \frac{13(12)}{13(12) + 39(13)} = \frac{12}{51} \approx .235$$

39. Consider the following Venn diagram:

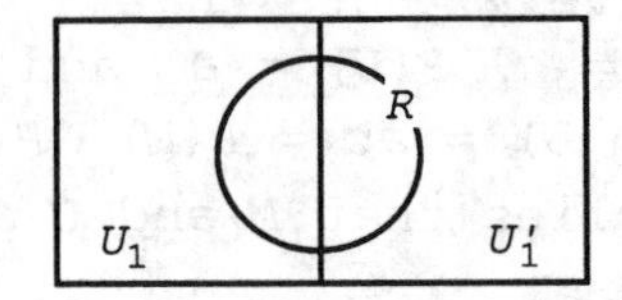

$$P(U_1 \mid R) = \frac{P(U_1 \cap R)}{P(U_1 \cap R) + P(U_1' \cap R)} \text{ and}$$

$$P(U_1' \mid R) = \frac{P(U_1' \cap R)}{P(U_1 \cap R) + P(U_1' \cap R)}$$

Adding these two equations, we obtain:

$$P(U_1 \mid R) + P(U_1' \mid R) = \frac{P(U_1 \cap R)}{P(U_1 \cap R) + P(U_1' \cap R)} + \frac{P(U_1' \cap R)}{P(U_1 \cap R) + P(U_1' \cap R)}$$

$$= \frac{P(U_1 \cap R) + P(U_1' \cap R)}{P(U_1 \cap R) + P(U_1' \cap R)} = 1$$

41. Consider the following tree diagram:

Start
- .7 S (satisfactory)
  - .9 P (pass)
  - .1 NP (not pass)
- .3 NS (not satisfactory)
  - .2 P
  - .8 NP

$$P(S \mid P) = \frac{P(S \cap P)}{P(P)} = \frac{P(S \cap P)}{P(S \cap P) + P(NS \cap P)} = \frac{P(S)P(P \mid S)}{P(S)P(P \mid S) + P(NS)P(P \mid NS)}$$

$$= \frac{(.7)(.9)}{(.7)(.9) + (.3)(.2)} = \frac{.63}{.69} \approx .913$$

$$P(S \mid NP) = \frac{P(S \cap NP)}{P(NP)} = \frac{P(S \cap NP)}{P(S \cap NP) + P(NS \cap NP)} = \frac{(.7)(.1)}{(.7)(.1) + (.3)(.8)}$$

$$= \frac{.07}{.31} \approx .226$$

43. Consider the following tree diagram:

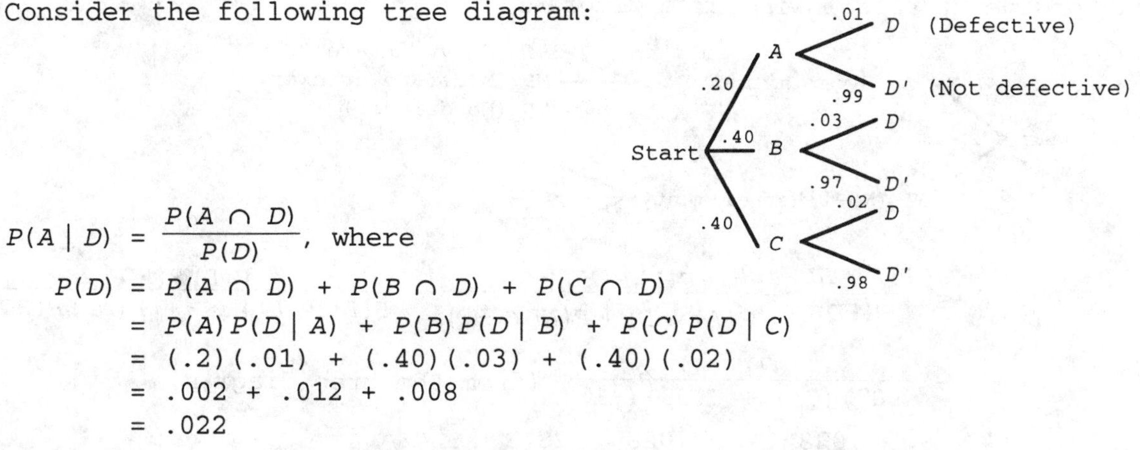

$$P(A \mid D) = \frac{P(A \cap D)}{P(D)}, \text{ where}$$

$$\begin{aligned} P(D) &= P(A \cap D) + P(B \cap D) + P(C \cap D) \\ &= P(A)P(D \mid A) + P(B)P(D \mid B) + P(C)P(D \mid C) \\ &= (.2)(.01) + (.40)(.03) + (.40)(.02) \\ &= .002 + .012 + .008 \\ &= .022 \end{aligned}$$

Thus, $P(A \mid D) = \dfrac{P(A \cap D)}{P(D)} = \dfrac{P(A)P(D \mid A)}{P(D)} = \dfrac{(.20)(.01)}{.022} = \dfrac{.002}{.022} = \dfrac{2}{22}$ or .091

Similarly,

$$P(B \mid D) = \frac{P(B \cap D)}{P(D)} = \frac{P(B)P(D \mid B)}{P(D)} = \frac{(.40)(.03)}{.022} = \frac{.012}{.022} = \frac{6}{11} \text{ or } .545,$$

and $P(C \mid D) = \dfrac{P(C \cap D)}{P(D)} = \dfrac{P(C)P(D \mid C)}{P(D)} = \dfrac{(.40)(.02)}{.022} = \dfrac{.008}{.022} = \dfrac{4}{11}$ or .364.

45. Consider the following tree diagram:

Start
.02 C (Cancer)
.98 CT (Cancer by new test)
.02 NCT (No cancer by new test)
.98 NC (No cancer)
.01 CT
.99 NCT

$$P(C \mid CT) = \frac{P(C \cap CT)}{P(CT)} = \frac{P(C)P(CT \mid C)}{P(C \cap CT) + P(NC \cap CT)} = \frac{P(C)P(CT \mid C)}{P(C)P(CT \mid C) + P(NC)P(CT \mid NC)}$$

$$= \frac{(.02)(.98)}{(.02)(.98) + (.98)(.01)} = \frac{.0196}{.0196 + .0098} = \frac{.0196}{.0294} = .6667$$

$$P(C \mid NCT) = \frac{P(C \cap NCT)}{P(NCT)} = \frac{P(C)P(NCT \mid C)}{P(C)P(NCT \mid C) + P(NC)P(NCT \mid NC)}$$

$$= \frac{(.02)(.02)}{(.02)(.02) + (.98)(.99)} \approx .000412$$

47. Consider the following tree diagram.

Start
- .07 → L (Liver ailment)
  - .4 → HD (Heavy drinker)
  - .5 → MD (Moderate drinker)
  - .1 → ND (Nondrinker)
- .93 → NL (No liver ailment)
  - .1 → HD
  - .7 → MD
  - .2 → ND

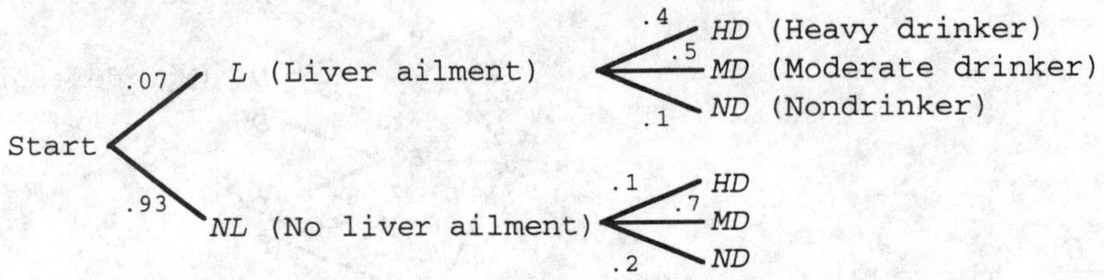

$$P(L \mid HD) = \frac{P(L \cap HD)}{P(HD)} = \frac{P(L)P(HD \mid L)}{P(L \cap HD) + P(NL \cap HD)} = \frac{P(L)P(HD \mid L)}{P(L)P(HD \mid L) + P(NL)P(HD \mid NL)}$$

$$= \frac{(.07)(.4)}{(.07)(.4) + (.93)(.1)} \quad \text{(from the tree diagram)}$$

$$= \frac{.028}{.028 + .093} = \frac{.028}{.121} = \frac{28}{121} \approx .231$$

$$P(L \mid ND) = \frac{P(L \cap ND)}{P(ND)} = \frac{P(L)P(ND \mid L)}{P(L \cap ND) + P(NL \cap ND)} = \frac{P(L)P(ND \mid L)}{P(L)P(ND \mid L) + P(NL)P(ND \mid NL)}$$

$$= \frac{(.07)(.1)}{(.07)(.1) + (.93)(.2)} \quad \text{(from the tree diagram)}$$

$$= \frac{.007}{.007 + .186} = \frac{.007}{.194} = \frac{7}{194} \approx .036$$

49. Consider the following tree diagram.

Start
- .5 → L (Lying)
  - .8 → LT (Test indicates lie)
  - .2 → NLT (Test indicates no lie)
- .5 → $\overline{L}$ (Not lying)
  - .05 → LT
  - .95 → NLT

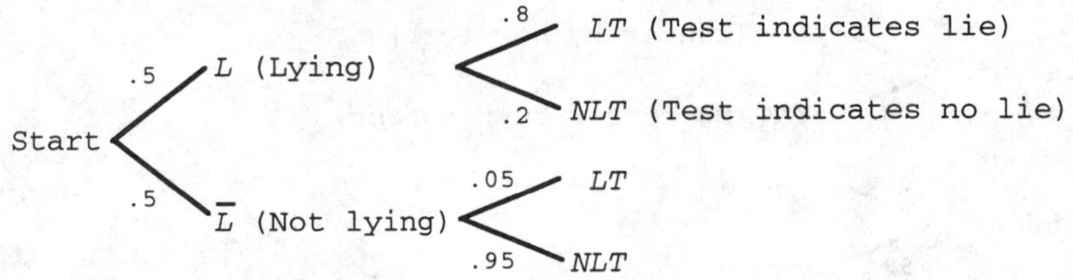

$$P(L \mid LT) = \frac{P(L \cap LT)}{P(LT)} = \frac{P(L \cap LT)}{P(L \cap LT) + P(\overline{L} \cap LT)} = \frac{(.5)(.8)}{(.5)(.8) + (.5)(.05)}$$

$$= \frac{.4}{.425} \approx .941$$

If the test indicates that the subject was lying, then he was lying with a probability of 0.941.

$$P(\overline{L} \mid LT) = \frac{P(\overline{L} \cap LT)}{P(LT)} = \frac{(.5)(.05)}{(.5)(.8) + (.5)(.05)}$$

$$= \frac{.05}{.85} \approx .0588$$

If the test indicates that the subject was lying, there is still a probability of 0.0588 that he was not lying.

EXERCISE 6-7

Things to remember:

1. RANDOM VARIABLE

   A random variable is a function that assigns a numerical value to each simple event in a sample space S.

2. PROBABILITY DISTRIBUTION OF A RANDOM VARIABLE X

A probability function $P(X = x) = p(x)$ is a PROBABILITY DISTRIBUTION OF THE RANDOM VARIABLE X if

(a) $0 \le p(x) \le 1$, $x \in \{x_1, x_2, \ldots, x_n\}$,

(b) $p(x_1) + p(x_2) + \cdots + p(x_n) = 1$,

where $\{x_1, x_2, \ldots, x_n\}$ are values of X.

3. EXPECTED VALUE OF A RANDOM VARIABLE X

Given the probability distribution for the random variable X:

$$\left.\begin{matrix} x_i: x_1, x_2, \ldots, x_m \\ p_i: p_1, p_2, \ldots, p_m \end{matrix}\right\} p_i = p(x_i)$$

The expected value of X, denoted by $E(X)$, is given by the formula:

$$E(X) = x_1p_1 + x_2p_2 + \cdots + x_mp_m$$

4. Steps for computing the expected value of a random variable X.

(a) Form the probability distribution for the random variable X.

(b) Multiply each image value of X, x_i, by its corresponding probability of occurrence, p_i, then add the results.

---

1. Expected value of X:

$E(X) = -3(.3) + 0(.5) + 4(.2) = -0.1$

3. Assign the number 0 to the event of observing zero heads, the number 1 to the event of observing one head, and the number 2 to the event of observing two heads. The probability distribution for X, then, is:

x_i	0	1	2
p_i	$\frac{1}{4}$	$\frac{1}{2}$	$\frac{1}{4}$

[Note: One head can occur two ways out of a total of four different ways (HT, TH).]

Hence, $E(X) = 0 \cdot \frac{1}{4} + 1 \cdot \frac{1}{2} + 2 \cdot \frac{1}{4} = 1$.

5. Assign a payoff of \$1 to the event of observing a head and -\$1 to the event of observing a tail. Thus, the probability distribution for X is:

x_i	1	-1
p_i	$\frac{1}{2}$	$\frac{1}{2}$

Hence, $E(X) = 1 \cdot \frac{1}{2} + (-1) \cdot \frac{1}{2} = 0$. The game is fair.

7. The table shows a payoff or probability distribution for the game.

Net gain	x_i	-3	-2	-1	0	1	2
	p_i	$\frac{1}{6}$	$\frac{1}{6}$	$\frac{1}{6}$	$\frac{1}{6}$	$\frac{1}{6}$	$\frac{1}{6}$

[Note: A payoff valued at -$3 is assigned to the event of observing a "1" on the die, resulting in a net gain of -$3, and so on.]

Hence, $E(X) = -3 \cdot \frac{1}{6} - 2 \cdot \frac{1}{6} - 1 \cdot \frac{1}{6} + 0 \cdot \frac{1}{6} + 1 \cdot \frac{1}{6} + 2 \cdot \frac{1}{6} = -\frac{1}{2}$ or -$0.50.

The game is not fair.

9. The probability distribution is:

Number of Heads	Gain, x_i	Probability, p_i
0	2	$\frac{1}{4}$
1	-3	$\frac{1}{2}$
2	2	$\frac{1}{4}$

The expected value is:

$E(X) = 2 \cdot \frac{1}{4} + (-3) \cdot \frac{1}{2} + 2 \cdot \frac{1}{4} = 1 - \frac{3}{2} = -\frac{1}{2}$ or -$0.50.

11. In 4 rolls of a die, the total number of possible outcomes is $6 \cdot 6 \cdot 6 \cdot 6 = 6^4$. Thus, $n(S) = 6^4 = 1296$. The total number of outcomes that contain *no* 6's is $5 \cdot 5 \cdot 5 \cdot 5 = 5^4$. Thus, if E is the event "At least one 6," then $n(E) = 6^4 - 5^4 = 671$ and

$P(E) = \frac{n(E)}{n(S)} = \frac{671}{1296} \approx 0.5178.$

First, we compute the expected value to you.

The payoff table is:

x_i	-$1	$1
P_i	0.5178	0.4822

The expected value to you is:

$E(X) = (-1)(0.5178) + 1(0.4822) = -0.0356$ or -$0.036

The expected value to her is:

$E(X) = 1(0.5178) + (-1)(0.4822) = 0.0356$ or $0.036

13. Let x = amount you should lose if a 6 turns up.
The payoff table is:

	1	2	3	4	5	6
x_i	$5	$5	$10	$10	$10	$$x$
p_i	$\frac{1}{6}$	$\frac{1}{6}$	$\frac{1}{6}$	$\frac{1}{6}$	$\frac{1}{6}$	$\frac{1}{6}$

Now $E(X) = 5\left(\frac{1}{6}\right) + 5\left(\frac{1}{6}\right) + 10\left(\frac{1}{6}\right) + 10\left(\frac{1}{6}\right) + 10\left(\frac{1}{6}\right) + x\left(\frac{1}{6}\right) = \frac{40}{6} + \frac{x}{6}$

The game is fair if and only if $E(X) = 0$:

solving $\frac{40}{6} + \frac{x}{6} = 0$

gives $x = -40$

Thus, you should **lose** $40 for the game to be fair.

15. $P(\text{sum} = 7) = \frac{6}{36} = \frac{1}{6}$

$P(\text{sum} = 11 \text{ or } 12) = P(\text{sum} = 11) + P(\text{sum} = 12) = \frac{2}{36} + \frac{1}{36} = \frac{3}{36} = \frac{1}{12}$

$P(\text{sum other than } 7, 11, \text{ or } 12) = 1 - P(\text{sum} = 7, 11, \text{ or } 12)$

$= 1 - \frac{9}{36} = \frac{27}{36} = \frac{3}{4}$

Let x_1 = sum is 7, x_2 = sum is 11 or 12, x_3 = sum is not 7, 11, or 12, and let t denote the amount you "win" if x_3 occurs. Then the payoff table is:

x_i	-$10	$11	t
p_i	$\frac{1}{6}$	$\frac{1}{12}$	$\frac{3}{4}$

The expected value is:

$E(X) = -10\left(\frac{1}{6}\right) + 11\left(\frac{1}{12}\right) + t\left(\frac{3}{4}\right) = \frac{-10}{6} + \frac{11}{12} + \frac{3t}{4}$

The game is fair if $E(X) = 0$, i.e., if

$\frac{-10}{6} + \frac{11}{12} + \frac{3}{4}t = 0$ or $\frac{3}{4}t = \frac{10}{6} - \frac{11}{12} = \frac{20}{12} - \frac{11}{12} = \frac{9}{12} = \frac{3}{4}$

Therefore, $t = \$1$.

17. Course A_1: $E(X) = (-200)(.1) + 100(.2) + 400(.4) + 100(.3)$

$= -20 + 20 + 160 + 30$

$= \$190$

Course A_2: $E(X) = (-100)(.1) + 200(.2) + 300(.4) + 200(.3)$

$= -10 + 40 + 120 + 60$

$= \$210$

A_2 will produce the largest expected value, and that value is $210.

19. The probability of winning $35 is $\frac{1}{38}$ and the probability of losing $1 is $\frac{37}{38}$. Thus, the payoff table is:

x_i	$35	-$1
p_i	$\frac{1}{38}$	$\frac{37}{38}$

The expected value of the game is:

$E(X) = 35\left(\frac{1}{38}\right) + (-1)\left(\frac{37}{38}\right) = \frac{35 - 37}{38} = \frac{-1}{19} \approx -0.0526$ or $E(X) = -5.26¢$ or $-5¢$

21. Let p = probability of winning. Then $1 - p$ is the probability of losing and the payoff table is:

	W	L
x_i	99,900	-100
p_i	p	$1 - p$

Since $E(X) = 100$, we have

$$\begin{aligned} 99{,}900(p) - 100(1 - p) &= 100 \\ 99{,}900p - 100 + 100p &= 100 \\ 100{,}000p &= 200 \\ p &= 0.002 \end{aligned}$$

The probability of winning is 0.002. Since the expected value is positive, you should play the game. In the long run, you will win $100 per game.

23. p_i $\quad$ x_i

$\frac{1}{5000}$ chance of winning $499

$\frac{3}{5000}$ chance of winning $99

$\frac{5}{5000}$ chance of winning $19

$\frac{20}{5000}$ chance of winning $4

$\frac{4971}{5000}$ chance of losing $1 [Note: 5000 - (1 + 3 + 5 + 20) = 4971.]

The payoff table is:

x_i	$499	$99	$19	$4	-$1
p_i	0.0002	0.0006	0.001	0.004	0.9942

Thus,

$$\begin{aligned} E(X) &= 499(0.0002) + 99(0.0006) + 19(0.001) + 4(0.004) - 1(0.9942) \\ &= -0.80 \end{aligned}$$

or

$E(X) = -\$0.80$ or $-80¢$

25. (A) Total number of simple events $= n(S) = C_{10,2} = \dfrac{10!}{2!(10-2)!}$

$$= \frac{10!}{2!8!} = \frac{10 \cdot 9}{2} = 45$$

$$P(\text{zero defective}) = P(0) = \frac{C_{7,2}}{45} = \frac{\frac{7!}{2!5!}}{45} = \frac{21}{45} = \frac{7}{15}$$

[Note: None defective means 2 selected fom 7 nondefective.]

$$P(\text{one defective}) = P(1) = \frac{C_{3,1} \cdot C_{7,1}}{45} = \frac{21}{45} = \frac{7}{15}$$

$$P(\text{two defective}) = P(2) = \frac{C_{3,2}}{45} = \frac{3}{45} = \frac{1}{15}$$

[Note: Two defectives selected from 3 defectives.]

The probability distribution is as follows:

x_i	0	1	2
P_i	$\frac{7}{15}$	$\frac{7}{15}$	$\frac{1}{15}$

(B) $E(X) = 0\left(\frac{7}{15}\right) + 1\left(\frac{7}{15}\right) + 2\left(\frac{1}{15}\right) = \frac{9}{15} = \frac{3}{5} = 0.6$

27. (A) The total number of simple events $= n(S) = C_{1000,5}$.

$$P(0 \text{ winning tickets}) = P(0) = \frac{C_{997,5}}{C_{1000,5}} = \frac{997\cdot996\cdot995\cdot994\cdot993}{1000\cdot999\cdot998\cdot997\cdot996} \approx 0.985$$

$$P(1 \text{ winning ticket}) = P(1) = \frac{C_{3,1}\cdot C_{997,4}}{C_{1000,5}} = \frac{3\cdot\frac{997!}{4!(993)!}}{\frac{1000!}{5!(995)!}} \approx 0.0149$$

$$P(2 \text{ winning tickets}) = P(2) = \frac{C_{3,2}\cdot C_{997,3}}{C_{1000,5}} = \frac{3\cdot\frac{997!}{3!(994)!}}{\frac{1000!}{5!(995)!}} \approx 0.0000599$$

$$P(3 \text{ winning tickets}) = P(3) = \frac{C_{3,3}\cdot C_{997,2}}{C_{1000,5}} = \frac{1\cdot\frac{997!}{2!(995)!}}{\frac{1000!}{5!(995)!}} \approx 0.00000006$$

The payoff table is as follows:

x_i	-\$5	\$195	\$395	\$595
P_i	0.985	0.0149	0.0000599	0.00000006

(B) The expected value to you is:

$$E(X) = (-5)(0.985) + 195(0.0149) + 395(0.0000599) + 595(0.00000006) \approx -\$2.00$$

29. (A) From the statistical plot, the number 13 came up 3 times in 200 games. If \$1 was bet on 13 in each of the 200 games, then the result is:

$$3(35) - 1(197) = 105 - 197 = -\$92$$

(B) Based on the simulation, the value per game is: $-\frac{92}{200} = -\$0.46$; you lose 46 cents per game. The (theoretical) expected value of the game is:

$$E(X) = \frac{1}{38}(35) + \frac{37}{38}(-1) = \frac{35}{38} - \frac{37}{38} = -\frac{1}{19} \approx \$0.0526;$$

You will lose 5 cents per game.

(C) The simulated gain or loss depends on the results of your simulation. From (B), the expected loss is: $-\frac{1}{19}(500) \approx -\$26.32$.

31. The payoff table is as follows:

Gain	x_i	\$4850	-\$150
	p_i	0.01	0.99

[Note: 5000 - 150 = 4850, the gain with a probability of 0.01 if stolen.]

Hence, $E(X) = 4850(0.01) - 150(0.99) = -\$100$

33. The payoff table for site A is as follows:

x_i	30 million	-3 million
p_i	0.2	0.8

Hence $E(X) = 30(0.2) - 3(0.8)$
$= 6 - 2.4$
$= \$3.6$ million

The payoff table for site B is as follows:

x_i	70 million	-4 million
p_i	0.1	0.9

Hence, $E(X) = 70(0.1) + (-4)(0.9)$
$= 7 - 3.6$
$= \$3.4$ million

The company should choose site A with $E(X) = \$3.6$ million.

35. Using 4,
$E(X) = 0(0.12) + 1(0.36) + 2(0.38) + 3(0.14) = 1.54$

37. Action A_1: $E(X) = 10(0.3) + 5(0.2) + 0(0.5) = \$4.00$
Action A_2: $E(X) = 15(0.3) + 3(0.1) + 0(0.6) = \$4.80$
Action A_2 is the better choice.

CHAPTER 6 REVIEW

1. (A) We construct the following tree diagram for the experiment:

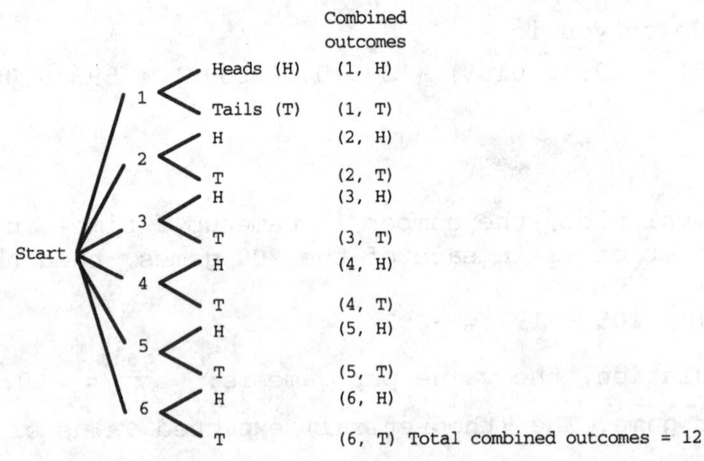

(B) Operation 1: Six possible outcomes, 1, 2, 3, 4, 5, or 6; $N_1 = 6$.
Operation 2: Two possible outcomes, heads (H) or tails (T); $N_2 = 2$.

Using the Multiplication Principle, the total combined outcomes $= N_1 \cdot N_2 = 6 \cdot 2 = 12$. (6-1)

2. (A) $n(A) = 30 + 35 = 65$
(B) $n(B) = 35 + 40 = 75$
(C) $n(A \cap B) = 35$
(D) $n(A \cup B) = 65 + 75 - 35 = 105$
or $n(A \cup B) = 30 + 35 + 40 = 105$
(E) $n(U) = 30 + 35 + 40 + 45 = 150$
(F) $n(A') = n(U) - n(A) = 150 - 65 = 85$
(G) $n([A \cap B]') = n(U) - n(A \cap B) = 150 - 35 = 115$
(H) $n([A \cup B]') = n(U) - n(A \cup B) = 150 - 105 = 45$
(6-1)

3. $C_{6,2} = \frac{6!}{2!(6-2)!} = \frac{6!}{2!4!} = \frac{6\cdot5\cdot4!}{2\cdot1\cdot4!} = 15$

$P_{6,2} = \frac{6!}{(6-2)!} = \frac{6!}{4!} = \frac{6\cdot5\cdot4!}{4!} = 30$ (6-2)

4. Operation 1: First person can choose the seat in 6 different ways; $N_1 = 6$.
Operation 2: Second person can choose the seat in 5 different ways; $N_2 = 5$.
Operation 3: Third person can choose the seat in 4 different ways; $N_3 = 4$.
Operation 4: Fourth person can choose the seat in 3 different ways; $N_4 = 3$.
Operation 5: Fifth person can choose the seat in 2 different ways; $N_5 = 2$.
Operation 6: Sixth person can choose the seat in 1 way; $N_6 = 1$.
Using the Multiplication Principle, the total number of different arrangements that can be made is $6\cdot5\cdot4\cdot3\cdot2\cdot1 = 720$. (6-1)

5. This is a permutations problem. The permutations of 6 objects taken 6 at a time is:
$P_{6,6} = \frac{6!}{(6-6)!} = 6! = 720$ (6-2)

6. First, we calculate the number of 5-card combinations that can be dealt from 52 cards:
$n(S) = C_{52,5} = \frac{52!}{5!\cdot47!} = 2{,}598{,}960$
We then calculate the number of 5-club combinations that can be obtained from 13 clubs:
$n(E) = C_{13,5} = \frac{13!}{5!\cdot8!} = 1287$
Thus,
$P(\text{5 clubs}) = P(E) = \frac{n(E)}{n(S)} = \frac{1287}{2{,}598{,}960} \approx 0.0005$ (6-3)

7. $n(S)$ is computed by using the permutation formula:
$n(S) = P_{15,2} = \frac{15!}{(15-2)!} = 15\cdot14 = 210$
Thus, the probability that Betty will be president and Bill will be treasurer is:
$\frac{n(E)}{n(S)} = \frac{1}{210} \approx 0.0048$ (6-3)

8. (A) The total number of ways of drawing 3 cards from 10 with order taken into account is given by:
$P_{10,3} = \frac{10!}{(10-3)!} = \frac{10\cdot9\cdot8\cdot7!}{7!} = 720$

Thus, the probability of drawing the code word "dig" is:

$$P(\text{"dig"}) = \frac{1}{720} \approx 0.0014$$

(B) The total number of ways of drawing 3 cards from 10 without regard to order is given by:

$$C_{10,3} = \frac{10!}{3!(10-3)!} = \frac{10 \cdot 9 \cdot 8 \cdot 7!}{3!7!} = 120$$

Thus, the probability of drawing the 3 cards "d," "i," and "g" (in some order) is:

$$P(\text{"d," "i," "g"}) = \frac{1}{120} \approx 0.0083$$ (6-3)

9. $P(\text{person having side effects}) = \frac{f(E)}{n} = \frac{50}{1000} = 0.05$ (6-3)

10. The payoff table is as follows:

x_i	-\$2	-\$1	\$0	\$1	\$2
p_i	$\frac{1}{5}$	$\frac{1}{5}$	$\frac{1}{5}$	$\frac{1}{5}$	$\frac{1}{5}$

Hence,

$$E(X) = (-2) \cdot \frac{1}{5} + (-1) \cdot \frac{1}{5} + 0 \cdot \frac{1}{5} + 1 \cdot \frac{1}{5} + 2 \cdot \frac{1}{5} = 0$$

The game is fair. (6-7)

11. $P(A) = .3,\ P(B) = .4,\ P(A \cap B) = .1$

(A) $P(A') = 1 - P(A) = 1 - .3 = .7$

(B) $P(A \cup B) = P(A) + P(B) - P(A \cap B) = .3 + .4 - .1 = .6$ (6-4)

12. Since the spinner cannot land on R and G simultaneously, $R \cap G = \varnothing$. Thus,

$P(R \cup G) = P(R) + P(G) = .3 + .5 = .8$

The odds for an event E are: $\frac{P(E)}{P(E')}$

Thus, the odds for landing on either R or G are: $\frac{P(R \cup G)}{P[(R \cup G)']} = \frac{.8}{.2} = \frac{8}{2}$ or the odds are 8 to 2. (6-4)

13. If the odds for an event E are a to b, then $P(E) = \frac{a}{a+b}$. Thus, the probability of rolling an 8 before rolling a 7 is: $\frac{5}{11} \approx .455$. (6-4)

14. $P(T) = .27$ (6-5)

15. $P(Z) = .20$ (6-5)

16. $P(T \cap Z) = .02$ (6-5)

17. $P(R \cap Z) = .03$ (6-5)

18. $P(R \mid Z) = \frac{P(R \cap Z)}{P(Z)} = \frac{.03}{.20} = .15$ (6-5)

19. $P(Z \mid R) = \dfrac{P(Z \cap R)}{P(R)} = \dfrac{.03}{.23} \approx .1304$ (6-5)

20. $P(T \mid Z) = \dfrac{P(T \cap Z)}{P(Z)} = \dfrac{.02}{.20} = .10$ (6-5)

21. No, because $P(T \cap Z) = .02 \neq P(T) \cdot P(Z) = (.27)(.20) = .054$. (6-5)

22. Yes, because $P(S \cap X) = .10 = P(S) \cdot P(X) = (.5)(.2)$. (6-5)

23. $P(A) = .4$ from the tree diagram. (6-5)

24. $P(B \mid A) = .2$ from the tree diagram. (6-5)

25. $P(B \mid A') = .3$ from the tree diagram. (6-5)

26. $P(A \cap B) = P(A)P(B \mid A) = (.4)(.2) = .08$ (6-5)

27. $P(A' \cap B) = P(A')P(B \mid A') = (.6)(.3) = .18$ (6-5)

28.
$$\begin{aligned} P(B) &= P(A \cap B) + P(A' \cap B) \\ &= P(A)P(B \mid A) + P(A')P(B \mid A') \\ &= (.4)(.2) + (.6)(.3) \\ &= .08 + .18 \\ &= .26 \end{aligned}$$
(6-5)

29.
$$\begin{aligned} P(A \mid B) &= \frac{P(A \cap B)}{P(B)} = \frac{P(A)P(B \mid A)}{P(A \cap B) + P(A' \cap B)} = \frac{P(A)P(B \mid A)}{P(A)P(B \mid A) + P(A')P(B \mid A')} \\ &= \frac{(.4)(.2)}{(.4)(.2) + (.6)(.3)} \quad \text{(from the tree diagram)} \\ &= \frac{.08}{.26} = \frac{8}{26} \text{ or } .307 \approx .31 \end{aligned}$$
(6-6)

30.
$$\begin{aligned} P(A \mid B') &= \frac{P(A \cap B')}{P(B')} = \frac{P(A)P(B' \mid A)}{1 - P(B)} = \frac{(.4)(.8)}{1 - .26} \quad [P(B) = .26, \text{ see Problem 28.}] \\ &= \frac{.32}{.74} = \frac{16}{37} \text{ or } .432 \end{aligned}$$
(6-6)

31. Let E = "born in June, July or August."

(A) Empirical Probability:
$$P(E) = \frac{f(E)}{n} = \frac{10}{32} = \frac{5}{16}$$

(B) Theoretical Probability:
$$P(E) = \frac{n(E)}{n(S)} = \frac{3}{12} = \frac{1}{4}$$

(C) As the sample size in part (A) increases, the approximate empirical probability of event E approaches the theoretical probability of event E. (6-3)

32. (A) True: If $A = \emptyset$ or $B = \emptyset$, then $A \cap B = \emptyset$ which implies that A and B are mutually exclusive.

(B) False: For a counter-example, choose mutually exclusive events A and B such that $P(A) \neq 0$ and $P(B) \neq 0$. If

$$P(A \cap B) = P(A) \cdot P(B)$$

Then we would have

$$0 = P(A \cap B) = P(A) \cdot P(B) \neq 0$$

(6-4)

33. (A) False. $P(A \cup B) = P(A) + P(B)$ implies that $P(A \cap B) = 0$; A and B independent implies $P(A) \cdot P(B) = P(A \cap B)$. Thus, for a counter-example, choose independent events A and B with $P(A) \neq 0$, $P(B) \neq 0$.

(B) False: For a counter-example, choose any event A such that $P(A) \neq 0$, $P(A') \neq 0$. Then $P(A \mid A') = 0 \neq P(A)$, $P(A' \mid A) = 0 \neq P(A')$. (6-5)

34. $S = \{HH, HT, TH, TT\}$.

The probabilities for 2 "heads," 1 "head," and 0 "heads" are, respectively, $\frac{1}{4}$, $\frac{1}{2}$, and $\frac{1}{4}$. Thus, the payoff table is:

x_i	\$5	-\$4	\$2
P_i	0.25	0.5	0.25

$E(X) = 0.25(5) + 0.5(-4) + 0.25(2) = -0.25$ or $-\$0.25$

The game is not fair. (6-7)

35. $S = \{(1,1), (2,2), (3,3), (1,2), (2,1), (1,3), (3,1), (2,3), (3,2)\}$

$n(S) = 3 \cdot 3 = 9$

(A) $P(A) = \frac{n(A)}{n(S)} = \frac{3}{9} = \frac{1}{3}$ $\quad [A = \{(1,1), (2,2), (3,3)\}]$

(B) $P(B) = \frac{n(B)}{n(S)} = \frac{2}{9}$ $\quad [B = \{(2,3), (3,2)\}]$ (6-5)

36. (A) $P(\text{jack or queen}) = P(\text{jack}) + P(\text{queen}) = \frac{4}{52} + \frac{4}{52} = \frac{8}{52} = \frac{2}{13}$

[Note: jack $\cap$ queen $= \emptyset$.]

The odds for drawing a jack or queen are 2 to 11.

(B) $P(\text{jack or spade}) = P(\text{jack}) + P(\text{spade}) - P(\text{jack and spade})$

$$= \frac{4}{52} + \frac{13}{52} - \frac{1}{52} = \frac{16}{52} = \frac{4}{13}$$

The odds for drawing a jack or a spade are 4 to 9.

(C) $P(\text{ace}) = \frac{4}{52} = \frac{1}{13}$. Thus,

$P(\text{card other than an ace}) = 1 - P(\text{ace}) = 1 - \frac{1}{13} = \frac{12}{13}$ (6-4)

37. (A) The probability of rolling a 5 is $\frac{4}{36} = \frac{1}{9}$.

Thus, the odds for rolling a five are 1 to 8.

(B) Let x = amount house should pay (and return the $1 bet). Then, for the game to be fair,

$$E(X) = x\left(\frac{1}{9}\right) + (-1)\left(\frac{8}{9}\right) = 0$$
$$\frac{x}{9} - \frac{8}{9} = 0$$
$$x = 8$$

Thus, the house should pay $8. (6-4)

38. Event E_1 = 2 heads; $f(E_1) = 210$.
Event E_2 = 1 head; $f(E_2) = 480$.
Event E_3 = 0 heads; $f(E_3) = 310$.
Total number of trials = 1000.

(A) The empirical probabilities for the events above are as follows:

$$P(E_1) = \frac{210}{1000} = 0.21$$
$$P(E_2) = \frac{480}{1000} = 0.48$$
$$P(E_3) = \frac{310}{1000} = 0.31$$

(B) Sample space S = {HH, HT, TH, TT}.

$$P(\text{2 heads}) = \frac{1}{4} = 0.25$$
$$P(\text{1 head}) = \frac{2}{4} = 0.5$$
$$P(\text{0 heads}) = \frac{1}{4} = 0.25$$

(C) Using part (B), the expected frequencies for each outcome are as follows:

$$\text{2 heads} = 1000 \cdot \frac{1}{4} = 250$$
$$\text{1 head} = 1000 \cdot \frac{2}{4} = 500$$
$$\text{0 heads} = 1000 \cdot \frac{1}{4} = 250$$

(6-3, 6-7)

39. Using the multiplication principle, the man has 5 children, $5 \cdot 3 = 15$ grandchildren, and $5 \cdot 3 \cdot 2 = 30$ greatgrandchildren, for a total of $5 + 15 + 30 = 50$ descendents. (6-1)

40. The individual tosses of a coin are independent events (the coin has no memory). Therefore, $P(H) = \frac{1}{2}$. (6-5)

41. (A) The sample space S is given by:

S = {(1,1), (1,2), (1,3), (1,4), (1,5), (1,6),
(2,1), (2,2), (2,3), (2,4), (2,5), (2,6),
(3,1), (3,2), (3,3), (3,4), (3,5), (3,6),
(4,1), (4,2), (4,3), (4,4), (4,5), (4,6),
(5,1), (5,2), (5,3), (5,4), (5,5), (5,6),
(6,1), (6,2), (6,3), (6,4), (6,5), (6,6)}

Sum 2, Sum 3, Sum 4, Sum 5

[Note: Event (2,3) means 2 on the the first die and 3 on the second die.]

The probability distribution corresponding to this sample space is:

Sum x_i	2	3	4	5	6	7	8	9	10	11	12
Probability p_i	$\frac{1}{36}$	$\frac{2}{36}$	$\frac{3}{36}$	$\frac{4}{36}$	$\frac{5}{36}$	$\frac{6}{36}$	$\frac{5}{36}$	$\frac{4}{36}$	$\frac{3}{36}$	$\frac{2}{36}$	$\frac{1}{36}$

(B) $E(X) = 2\left(\frac{1}{36}\right) + 3\left(\frac{2}{36}\right) + 4\left(\frac{3}{36}\right) + 5\left(\frac{4}{36}\right) + 6\left(\frac{5}{36}\right) + 7\left(\frac{6}{36}\right) + 8\left(\frac{5}{36}\right) + 9\left(\frac{4}{36}\right) + 10\left(\frac{3}{36}\right) + 11\left(\frac{2}{36}\right) + 12\left(\frac{1}{36}\right) = 7$ (6-7)

42. The event A that corresponds to the sum being divisible by 4 includes sums 4, 8, and 12. This set is:

A = {(1, 3), (2, 2), (3, 1), (2, 6), (3, 5), (4, 4), (5, 3), (6, 2), (6, 6)}

The event B that corresponds to the sum being divisible by 6 includes sums 6 and 12. This set is:

B = {(1, 5), (2, 4), (3, 3), (4, 2), (5, 1), (6, 6)}

$P(A) = \frac{n(A)}{n(S)} = \frac{9}{36} = \frac{1}{4}$

$P(B) = \frac{n(B)}{n(S)} = \frac{6}{36} = \frac{1}{6}$

$P(A \cap B) = \frac{1}{36}$ [Note: $A \cap B$ = {(6, 6)}]

$P(A \cup B) = \frac{14}{36}$ or $\frac{7}{18}$ [Note: $A \cup B$ = {(1, 3), (2, 2), (3, 1), (2, 6), (3, 5), (4, 4), (5, 3), (6, 2), (6, 6), (1, 5), (2, 4), (3, 3), (4, 2), (5, 1)}] (6-4)

43. The function P cannot be a probability function because:

(a) P cannot be negative. [Note: $P(e_2) = -0.2$.]

(b) P cannot have a value greater than 1. [Note: $P(e_4) = 2$.]

(c) The sum of the values of P must equal 1. [Note: $P(e_1) + P(e_2) + P(e_3) + P(e_4) = 0.1 + (-0.2) + 0.6 + 2 = 2.5 \neq 1$.] (6-3)

44. Since $n(A \cup B) = n(A) + n(B) - n(A \cap B)$, we have

$80 = 50 + 45 - n(A \cap B)$

and $n(A \cap B) = 15$

Now, $n(B') = n(U) - n(B) = 100 - 45 = 55$

$n(A') = n(U) - n(A) = 100 - 50 = 50$

$n(A \cap B') = 50 - 15 = 35$

$n(B \cap A') = 45 - 15 = 30$

$n(A' \cap B') = 55 - 35 = 20$

Thus,

	A	A'	Totals
B	15	30	45
B'	35	20	55
Totals	50	50	100

(6-4)

45. (A) $P(\text{odd number}) = P(1) + P(3) + P(5) = .2 + .3 + .1 = .6$

(B) Let E = "number less than 4,"
and F = "odd number."
Now, $E \cap F = \{1, 3\}$, $F = \{1, 3, 5\}$.

$$P(E \mid F) = \frac{P(E \cap F)}{P(F)} = \frac{.2 + .3}{.6} = \frac{5}{6}$$

(6-5)

46. Let E = "card is red" and F = "card is an ace." Then $F \cap E$ = "card is a red ace."

(A) $P(F \mid E) = \dfrac{P(F \cap E)}{P(E)} = \dfrac{2/52}{26/52} = \dfrac{1}{13}$

(B) $P(F \cap E) = \dfrac{1}{26}$, and $P(E) = \dfrac{1}{2}$, $P(F) = \dfrac{1}{13}$. Thus,

$P(F \cap E) = P(E) \cdot P(F)$, and E and F are independent.

(6-5)

47.

Operation	Number of ways of completing operation under condition: No letter repeated	Letters can be repeated	Adjacent letters not alike
O_1	8	8	8
O_2	7	8	7
O_3	6	8	7

Total outcomes, without repeating letters = $8 \cdot 7 \cdot 6 = 336$.
Total outcomes, with repeating letters = $8 \cdot 8 \cdot 8 = 512$.
Total outcomes, with adjacent letters not alike = $8 \cdot 7 \cdot 7 = 392$. (6-1)

48. (A) This is a permutations problem.

$$P_{6,3} = \frac{6!}{(6-3)!} = \frac{6\cdot5\cdot4\cdot3!}{3!} = 120$$

(B) This is a combinations problem.

$$C_{5,2} = \frac{5!}{2!(5-2)!} = \frac{5\cdot4\cdot3!}{2\cdot1\cdot3!} = 10 \qquad (6\text{-}2)$$

49. The largest value of $C_{25,r}$ is $C_{25,12} = C_{25,13} = 5{,}200{,}300$ (6-2)

50. (A) The tree diagram with replacement is:

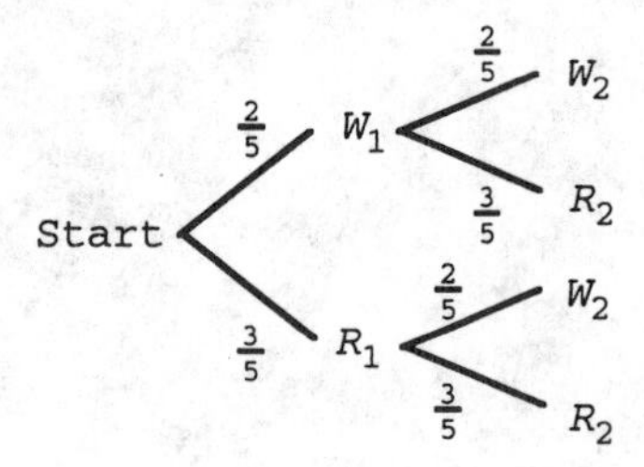

$$P(W_1 \cap R_2) = P(W_1)P(R_2 \mid W_1)$$
$$= \frac{2}{5}\cdot\frac{3}{5} = \frac{6}{25} \approx .24$$

(B) The tree diagram without replacement is:

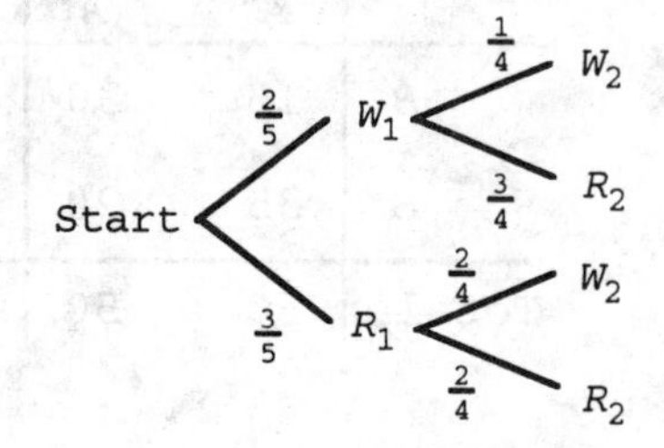

$$P(W_1 \cap R_2) = P(W_1)P(R_2 \mid W_1)$$
$$= \frac{2}{5}\cdot\frac{3}{4} = \frac{6}{20} = .3$$

(6-5)

51. Part (B) involves dependent events because

$$P(R_2 \mid W_1) = \frac{3}{4}$$
$$P(R_2) = P(W_1 \cap R_2) + P(R_1 \cap R_2) = \frac{6}{20} + \frac{6}{20} = \frac{12}{20} = \frac{3}{5}$$

and $P(R_2 \mid W_1) \neq P(R_2)$

The events in part (A) are independent. (6-5)

52. (A) Using the tree diagram in Problem 50(A), we have:

$$P(\text{zero red balls}) = P(W_1 \cap W_2) = P(W_1)P(W_2) = \frac{2}{5}\cdot\frac{2}{5} = \frac{4}{25} = .16$$
$$P(\text{one red ball}) = P(W_1 \cap R_2) + P(R_1 \cap W_2)$$
$$= P(W_1)P(R_2) + P(R_1)P(W_2)$$
$$= \frac{2}{5}\cdot\frac{3}{5} + \frac{3}{5}\cdot\frac{2}{5} = \frac{12}{25} = .48$$
$$P(\text{two red balls}) = P(R_1 \cap R_2) = P(R_1)P(R_2) = \frac{3}{5}\cdot\frac{3}{5} = \frac{9}{25} = .36$$

Thus, the probability distribution is:

Number of red balls x_i	Probability p_i
0	.16
1	.48
2	.36

The expected number of red balls is:

$E(X) = 0(.16) + 1(.48) + 2(.36) = .48 + .72 = 1.2$

(B) Using the tree diagram in Problem 50(B), we have:

$P(\text{zero red balls}) = P(W_1 \cap W_2) = P(W_1)P(W_2 \mid W_1) = \frac{2}{5} \cdot \frac{1}{4} = \frac{1}{10} = .1$

$$\begin{aligned} P(\text{one red ball}) &= P(W_1 \cap R_2) + P(R_1 \cap W_2) \\ &= P(W_1)P(R_2 \mid W_1) + P(R_1)P(W_2 \mid R_1) \\ &= \frac{2}{5} \cdot \frac{3}{4} + \frac{3}{5} \cdot \frac{2}{4} = \frac{12}{20} = \frac{3}{5} = .6 \end{aligned}$$

$P(\text{two red balls}) = P(R_1 \cap R_2) = P(R_1)P(R_2 \mid R_1) = \frac{3}{5} \cdot \frac{2}{4} = \frac{6}{20} = .3$

Thus, the probability distribution is:

Number of red balls x_i	Probability p_i
0	.1
1	.6
2	.3

The expected number of red balls is:

$E(X) = 0(.1) + 1(.6) + 2(.3) = 1.2$ (6-3)

53. The tree diagram for this problem is as follows:

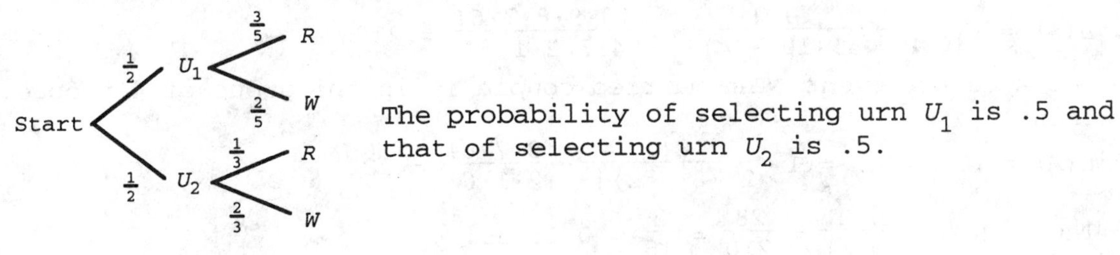

(A) $P(R \mid U_1) = \frac{3}{5}$ (B) $P(R \mid U_2) = \frac{1}{3}$

(C) $$\begin{aligned} P(R) &= P(R \cap U_1) + P(R \cap U_2) \\ &= P(U_1)P(R \mid U_1) + P(U_2)P(R \mid U_2) \\ &= \frac{1}{2} \cdot \frac{3}{5} + \frac{1}{2} \cdot \frac{1}{3} = \frac{28}{60} = \frac{7}{15} \approx .4667 \end{aligned}$$

(D) $$\begin{aligned} P(U_1 \mid R) &= \frac{P(U_1 \cap R)}{P(R)} = \frac{P(U_1)P(R \mid U_1)}{P(U_1)P(R \mid U_1) + P(U_2)P(R \mid U_2)} \\ &= \frac{\frac{1}{2} \cdot \frac{3}{5}}{\frac{1}{2} \cdot \frac{3}{5} + \frac{1}{2} \cdot \frac{1}{3}} = \frac{\frac{3}{10}}{\frac{7}{15}} = \frac{9}{14} \approx .6429 \end{aligned}$$

(E) $P(U_2 \mid W) = \frac{P(U_2 \cap W)}{P(W)} = \frac{P(U_2)P(W \mid U_2)}{P(U_2)P(W \mid U_2) + P(U_1)P(W \mid U_1)}$

$$= \frac{\frac{1}{2}\cdot\frac{2}{3}}{\frac{1}{2}\cdot\frac{2}{3} + \frac{1}{2}\cdot\frac{2}{5}} = \frac{\frac{2}{3}}{\frac{16}{15}} = \frac{5}{8} = .625$$

(F) $P(U_1 \cap R) = P(U_1)P(R \mid U_1) = \frac{1}{2}\cdot\frac{3}{5} = .3$

[Note: In parts (A)–(F), we derived the values of the probabilities from the tree diagram.] (6-5, 6-6)

54. No, because $P(R \mid U_1) \neq P(R)$. (See Problem 53.) (6-5)

55. $n(S) = C_{52,5}$

(A) Let A be the event "all diamonds." Then $n(A) = C_{13,5}$. Thus,

$$P(A) = \frac{n(A)}{n(S)} = \frac{C_{13,5}}{C_{52,5}}.$$

(B) Let B be the event "3 diamonds and 2 spades." Then $n(B) = C_{13,3}\cdot C_{13,2}$. Thus,

$$P(B) = \frac{n(B)}{n(S)} = \frac{C_{13,3}\cdot C_{13,2}}{C_{52,5}}.$$ (6-3)

56. $n(S) = C_{10,4} = \frac{10!}{4!(10-4)!} = \frac{10\cdot 9\cdot 8\cdot 7\cdot 6!}{4\cdot 3\cdot 2\cdot 1\cdot 6!} = 210$

Let A be the event "The married couple is in the group of 4 people." Then

$$n(A) = C_{2,2}\cdot C_{8,2} = 1\cdot\frac{8!}{2!(8-2)!} = \frac{8\cdot 7\cdot 6!}{2\cdot 1\cdot 6!} = 28.$$

Thus, $P(A) = \frac{n(A)}{n(S)} = \frac{28}{210} = \frac{2}{15} \approx 0.1333$. (6-3)

57. By the multiplication principle, there are

$$N_1 \cdot N_2 \cdot N_3$$

branches in the tree diagram. (6-1)

58. Events S and H are mutually exclusive. Hence, $P(S \cap H) = 0$, while $P(S) \neq 0$ and $P(H) \neq 0$. Therefore,

$$P(S \cap H) \neq P(S)\cdot P(H)$$

which implies that S and F are dependent. (6-5)

59. (A) From the plot, $P(2) = \frac{9}{20} = 0.18$.

(B) The event A = "the minimum of the two numbers is 2" contains the simple events (2, 2), (2, 3), (3, 2), (2, 4), (4, 2), (2, 5), (5, 2), (2, 6), (6, 2). Thus $n(A) = 9$ and $P(A) = \frac{9}{36} = \frac{1}{4} = 0.25$.

(C) The empirical probability depends on the results of your simulation. For the theoretical probability, let A = "minimum of the two numbers is 4". Then $A = \{(4, 4), (4, 5), (5, 4), (4, 6), (6, 4)\}$,
$n(A) = 5$ and $P(A) = \frac{5}{36} \approx 0.139$. (6-3)

60. The empirical probability depends on the results of your simulation. Since there are 2 black jacks in a standard 52-card deck, the theoretical probability of drawing a black jack is: $\frac{2}{52} = \frac{1}{26} \approx 0.038$. (6-3)

61. Let E_2 be the event "2 heads."

(A) From the table, $f(E_2) = 350$. Thus, the approximate empirical probability of obtaining 2 heads is:

$$P(E_2) \approx \frac{f(E_2)}{n} = \frac{350}{1000} = 0.350$$

(B) S = {HHH, HHT, HTH, HTT, THH, THT, TTH, TTT}
The theoretical probability of obtaining 2 heads is:

$$P(E_2) = \frac{n(E_2)}{n(S)} = \frac{3}{8} = 0.375$$

(C) The expected frequency of obtaining 2 heads in 1000 tosses of 3 fair coins is:
$f(E_2) = 1000(0.375) = 375$ (6-3, 6-7)

62. On one roll of the dice, the probability of getting a double six is $\frac{1}{36}$ and the probability of not getting a double six is $\frac{35}{36}$.

On two rolls of the dice there are $(36)^2$ possible outcomes. There are 71 ways to get at least one double six, namely a double six on the first roll and any one of the 35 other outcomes on the second roll, or a double six on the second roll and any one of the 35 other outcomes on the first roll, or a double six on both rolls. Thus, the probability of at least one double six on two rolls is $\frac{71}{(36)^2}$ and the probability of no double sixes is:

$$1 - \frac{71}{(36)^2} = \frac{(36)^2 - 2\cdot 36 + 1}{(36)^2} = \frac{(36 - 1)^2}{(36)^2} = \left(\frac{35}{36}\right)^2$$

Let E be the event "At least one double six." Then E' is the event "No double sixes." Continuing with the reasoning above, we conclude that, in 24 rolls of the die,

$$P(E') = \left(\frac{35}{36}\right)^{24} \approx 0.5086$$

Therefore, $P(E) = 1 - 0.5086 = 0.4914$.

The payoff table is:

x_i	1	-1
P_i	0.4914	0.5086

and $E(X) = 1(0.4914) + (-1)(0.5086)$
$= 0.4914 - 0.5086$
$= -0.0172$

Thus, your expectation is -\$0.0172.
Your friend's expectation is \$0.0172.
The game is not fair. (6-7)

63. (A) This is a permutations problem.

$$P_{10,3} = \frac{10!}{(10-3)!} = \frac{10!}{7!} = 10 \cdot 9 \cdot 8 = 720$$

(B) The number of ways in which women are selected for all three positions is given by:

$$P_{6,3} = \frac{6!}{(6-3)!} = \frac{6!}{3!} = 6 \cdot 5 \cdot 4 = 120$$

Thus, $P(\text{three women are selected}) = \frac{P_{6,3}}{P_{10,3}} = \frac{120}{720} = \frac{1}{6}$

(C) This is a combinations problem.

$$C_{10,3} = \frac{10!}{3!(10-3)!} = \frac{10 \cdot 9 \cdot 8 \cdot 7!}{3 \cdot 2 \cdot 1 \cdot 7!} = 120$$

(D) Let Event D = majority of team members will be women. Then
$n(D)$ = team has 3 women + team has 2 women
$= C_{6,3} + C_{6,2} \cdot C_{4,1}$
$= \frac{6!}{3!(6-3)!} + \frac{6!}{2!(6-2)!} \cdot \frac{4!}{1!(4-1)!} = 20 + 15 \cdot 4 = 80$

Thus,

$$P(D) = \frac{n(D)}{n(S)} = \frac{C_{6,3} + C_{6,2} \cdot C_{4,1}}{C_{10,3}} = \frac{80}{120} = \frac{2}{3}$$

(6-1, 6-2, 6-3)

64. Draw a Venn diagram with: A = Chess players, B = Checker players.

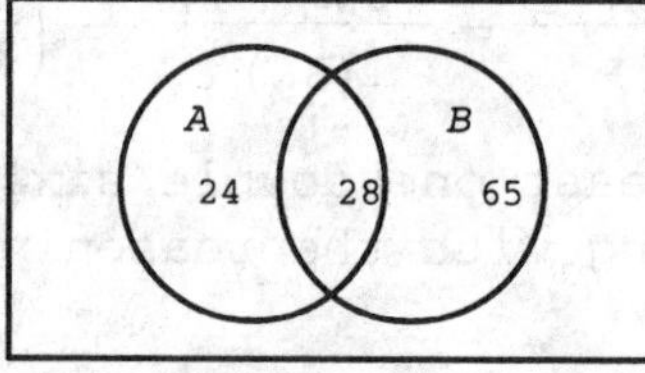

Now, $n(A \cap B) = 28$, $n(A \cap B') = n(A) - n(A \cap B) = 52 - 28 = 24$
$n(B \cap A') = n(B) - n(A \cap B) = 93 - 28 = 65$

Since there are 150 people in all,
$n(A' \cap B') = n(U) - [n(A \cap B') + n(A \cap B) + n(B \cap A')]$
$= 150 - (24 + 28 + 65) = 150 - 117 = 33$ (6-1)

65. The total number of ways that 3 people can be selected from a group of 10 is:

$$C_{10,3} = \frac{10!}{3!(10-3)!} = \frac{10\cdot 9\cdot 8\cdot 7!}{3\cdot 2\cdot 1\cdot 7!} = 120$$

The number of ways of selecting *no* women is:

$$C_{7,3} = \frac{7!}{3!(7-3)!} = \frac{7\cdot 6\cdot 5\cdot 4!}{3\cdot 2\cdot 1\cdot 4!} = 35$$

Thus, the number of samples of 3 people that contain at least one woman is $120 - 35 = 85$.

Therefore, if event A is "At least one woman is selected," then

$$P(A) = \frac{n(A)}{n(S)} = \frac{85}{120} = \frac{17}{24} \approx 0.708. \qquad (6\text{-}3)$$

66. $P(\text{second heart} \mid \text{first heart}) = P(H_2 \mid H_1) = \frac{12}{51} \approx .235$

[Note: One can see that $P(H_2 \mid H_1) = \frac{12}{51}$ directly.] (6-5)

67. $P(\text{first heart} \mid \text{second heart}) = P(H_1 \mid H_2)$

$$= \frac{P(H_1 \cap H_2)}{P(H_2)} = \frac{P(H_1)P(H_2 \mid H_1)}{P(H_2)}$$

$$= \frac{P(H_1)P(H_2 \mid H_1)}{P(H_1 \cap H_2) + P(H_1' \cap H_2)}$$

$$= \frac{P(H_1)P(H_2 \mid H_1)}{P(H_1)P(H_2 \mid H_1) + P(H_1')P(H_2 \mid H_1')}$$

$$= \frac{\frac{13}{52}\cdot\frac{12}{51}}{\frac{13}{52}\cdot\frac{12}{51} + \frac{39}{52}\cdot\frac{13}{51}} = \frac{12}{51} \approx .235 \qquad (6\text{-}5)$$

68. Since each die has 6 faces, there are $6\cdot 6 = 36$ possible pairs for the two up faces.

A sum of 2 corresponds to having (1, 1) as the up faces. This sum can be obtained in $3\cdot 3 = 9$ ways (3 faces on the first die, 3 faces on the second). Thus,

$$P(2) = \frac{9}{36} = \frac{1}{4}.$$

A sum of 3 corresponds to the two pairs (2, 1) and (1, 2). The number of such pairs is $2\cdot 3 + 3\cdot 2 = 12$. Thus,

$$P(3) = \frac{12}{36} = \frac{1}{3}.$$

A sum of 4 corresponds to the pairs (3, 1), (2, 2), (1, 3). There are $1\cdot 3 + 2\cdot 2 + 3\cdot 1 = 10$ such pairs. Thus,

$$P(4) = \frac{10}{36}.$$

A sum of 5 corresponds to the pairs (2, 3) and (3, 2). There are $2\cdot 1 + 1\cdot 2 = 4$ such pairs. Thus,

$$P(5) = \frac{4}{36} = \frac{1}{9}.$$

A sum of 6 corresponds to the pair (3, 3) and there is one such pair. Thus,

$P(6) = \frac{1}{36}$.

(A) The probability distribution for X is:

x_i	2	3	4	5	6
p_i	$\frac{9}{36}$	$\frac{12}{36}$	$\frac{10}{36}$	$\frac{4}{36}$	$\frac{1}{36}$

(B) The expected value is:

$$E(X) = 2\left(\frac{9}{36}\right) + 3\left(\frac{12}{36}\right) + 4\left(\frac{10}{36}\right) + 5\left(\frac{4}{36}\right) + 6\left(\frac{1}{36}\right) = \frac{120}{36} = \frac{10}{3} \quad (6\text{-}7)$$

69. The payoff table is:

x_i	-\$1.50	-\$0.50	\$0.50	\$1.50	\$2.50
p_i	$\frac{9}{36}$	$\frac{12}{36}$	$\frac{10}{36}$	$\frac{4}{36}$	$\frac{1}{36}$

and $E(X) = \frac{9}{36}(-1.50) + \frac{12}{36}(-0.50) + \frac{10}{36}(0.50) + \frac{4}{36}(1.50) + \frac{1}{36}(2.50)$

$= -0.375 - 0.167 + 0.139 + 0.167 + 0.069$

$= -0.167$ or $-\$0.167 \approx \$0.17$

The game is not fair. The game would be fair if you paid \$3.50 - \$0.17 = \$3.33 to play. (6-7)

70. Operation 1: Two possible outcomes, boy or girl, $N_1 = 2$.
Operation 2: Two possible outcomes, boy or girl, $N_2 = 2$.
Operation 3: Two possible outcomes, boy or girl, $N_3 = 2$.
Operation 4: Two possible outcomes, boy or girl, $N_4 = 2$.
Operation 5: Two possible outcomes, boy or girl, $N_5 = 2$.

Using the Multiplication Principle, the total combined outcomes is:
$N_1 \cdot N_2 \cdot N_3 \cdot N_4 \cdot N_5 = 2 \cdot 2 \cdot 2 \cdot 2 \cdot 2 = 32$.
If order pattern is not taken into account, there would be only 6 possible outcomes: families with 0, 1, 2, 3, 4, or 5 boys. (6-1)

71. The tree diagram for this experiment is:

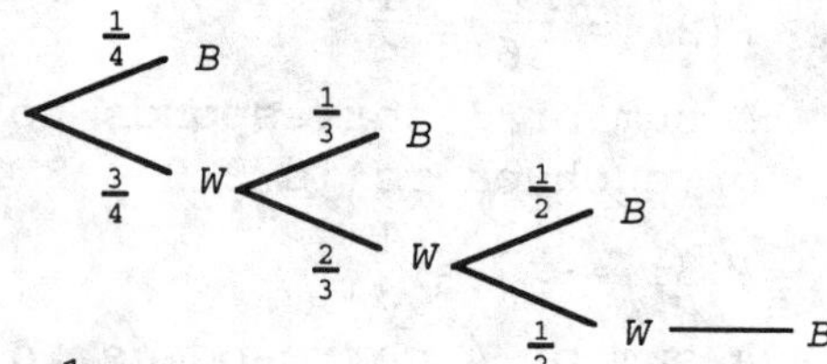

(A) $P(\text{black on the fourth draw}) = \frac{3}{4} \cdot \frac{2}{3} \cdot \frac{1}{2} = \frac{1}{4}$

The odds for black on the fourth draw are 1 to 3.

(B) Let x = amount house should pay (and return the \$1 bet). Then, for the game to be fair:

$$E(X) = x\left(\frac{1}{4}\right) + (-1)\left(\frac{3}{4}\right) = 0$$

$$\frac{x}{4} - \frac{3}{4} = 0$$

$$x = 3$$

Thus, the house should pay \$3. (6-4, 6-6)

72. $n(S) = 10\cdot10\cdot10\cdot10\cdot10 = 10^5$

Let event A = "at least two people identify the same book." Then A' = "each person identifies a different book," and

$$n(A') = 10\cdot9\cdot8\cdot7\cdot6 = \frac{10!}{5!}$$

Thus, $P(A') = \dfrac{\frac{10!}{5!}}{10^5} = \dfrac{10!}{5!10^5}$ and $P(A) = 1 - \dfrac{10!}{5!10^5} \approx 1 - .3 = .7.$ (6-4)

73. The number of routes starting from A and visiting each of the 5 stores exactly once is the number of permutations of 5 objects taken 5 at a time, i.e.,

$$P_{5,5} = \frac{5!}{(5-5)!} = 120.$$ (6-1)

74. Draw a Venn diagram with:

S = people who have invested in stocks, and
B = people who have invested in bonds.

Then $n(U) = 1000$, $n(S) = 340$, $n(B) = 480$ and $n(S \cap B) = 210$

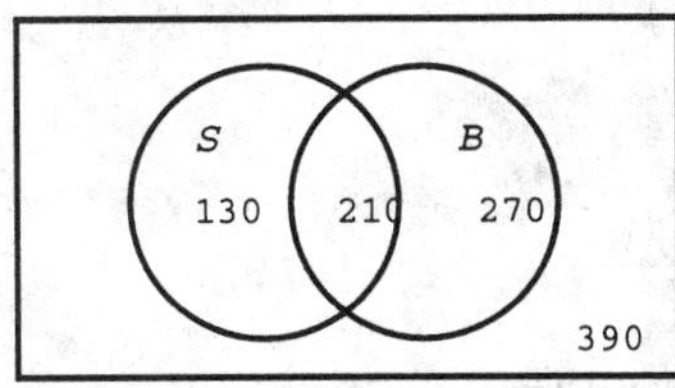

$n(S \cap B') = 340 - 210 = 130$ $n(B \cap S') = 480 - 210 = 270$

$n(S' \cap B') = 1000 - (130 + 210 + 270) = 1000 - 610 = 390$

(A) $n(S \cup B) = n(S) + n(B) - n(S \cap B) = 340 + 480 - 210 = 610.$

(B) $n(S' \cap B') = 390$

(C) $n(B \cap S') = 270$ (6-1)

75.

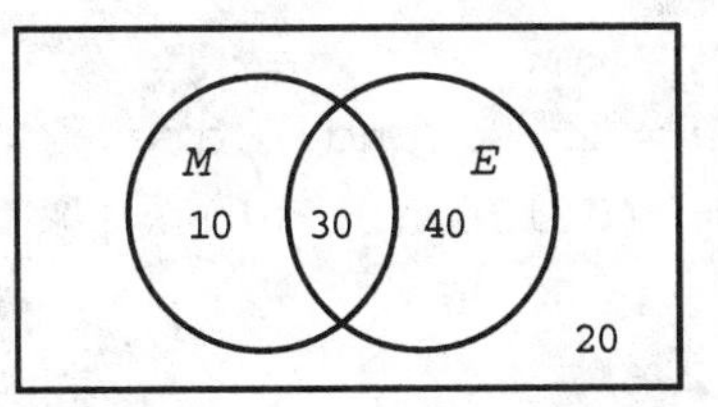

Event M = Reads the morning paper.
Event E = Reads the evening paper.

(A) $P(\text{reads a daily paper}) = P(M \text{ or } E) = P(M \cup E)$

$$= P(M) + P(E) - P(M \cap E)$$

$$= \frac{40}{100} + \frac{70}{100} - \frac{30}{100} = .8$$

(B) $P(\text{does not read a daily paper}) = \frac{20}{100}$ (from the Venn diagram)

$$= .2$$

or $= 1 - P(M \cup E)$ [i.e., $P((M \cup E)')$]

$$= 1 - .8 = .2$$

(C) $P(\text{reads exactly one daily paper}) = \frac{10 + 40}{100}$ (from the Venn diagram)

$$= .5$$

or $P((M \cap E') \text{ or } (M' \cap E))$

$$= P(M \cap E') + P(M' \cap E)$$

$$= \frac{10}{100} + \frac{40}{100} = .5$$

(6-1, 6-4)

76. Let A be the event that a person has seen the advertising and P be the event that the person purchased the product. Given:

$P(A) = .4$ and $P(P \mid A) = .85$

We want to find:

$P(A \cap P) = P(A)P(P \mid A) = (.4)(.85) = .34$ (6-5)

77. (A)

$$P(A) = \frac{290}{1000} = 0.290$$

$$P(B) = \frac{290}{1000} = 0.290$$

$$P(A \cap B) = \frac{100}{1000} = 0.100$$

$$P(A \mid B) = \frac{100}{290} = 0.345$$

$$P(B \mid A) = \frac{100}{290} = 0.345$$

(B) A and B are **not** independent because

$0.100 = P(A \cap B) \neq P(A) \cdot P(B) = (0.290)(0.290) = 0.084$

(C)

$$P(C) = \frac{880}{1000} = 0.880$$

$$P(D) = \frac{120}{1000} = 0.120$$

$$P(C \cap D) = 0$$

$$P(C \mid D) = P(D \mid C) = 0$$

(D) C and D are mutually exclusive since $C \cap D \neq \varnothing$. C and D are dependent since $0 = P(C \cap D) \neq P(C) \cdot P(D) = (0.120)(0.880) = 0.106$

(6-5)

78. The payoff table for plan A is:

x_i	10 million	-2 million
p_i	0.8	0.2

Hence, $E(X) = 10(0.8) - 2(0.2) = 8 - 0.4 = \$7.6$ million.

The payoff table for plan B is:

x_i	12 million	-2 million
p_i	0.7	0.3

Hence, $E(X) = 12(0.7) - 2(0.3) = 8.4 - 0.6 = \$7.8$ million.
Plan B should be chosen. (6-7)

79. The payoff table is:

Gain	x_i	\$270	-\$30
	p_i	0.08	0.92

[Note: 300 - 30 = 270 is the "gain" if the bicycle is stolen.]

Hence, $E(X) = 270(0.08) - 30(0.92) = 21.6 - 27.6 = -\$6$. (6-7)

80. $n(S) = C_{12,4} = \dfrac{12!}{4!(12-4)!} = \dfrac{12\cdot11\cdot10\cdot9\cdot8!}{4\cdot3\cdot2\cdot1\cdot8!} = 495$

The number of samples that contain *no* substandard parts is:

$C_{10,4} = \dfrac{10!}{4!(10-4)!} = \dfrac{10\cdot9\cdot8\cdot7\cdot6!}{4\cdot3\cdot2\cdot1\cdot6!} = 210$

Thus, the number of samples that have at least one defective part is 495 - 210 = 285. If E is the event "The shipment is returned," then

$P(E) = \dfrac{n(E)}{n(S)} = \dfrac{285}{495} \approx 0.576.$ (6-2, 6-4)

81. $n(S) = C_{12,3} = \dfrac{12!}{3!(12-3)!} = \dfrac{12\cdot11\cdot10\cdot9!}{3\cdot2\cdot1\cdot9!} = 220$

A sample will either have 0, 1, or 2 defective circuit boards.

$P(0) = \dfrac{C_{10,3}}{C_{12,3}} = \dfrac{\frac{10!}{3!(10-3)!}}{220} = \dfrac{\frac{10\cdot9\cdot8\cdot7!}{3\cdot2\cdot1\cdot7!}}{220} = \dfrac{120}{220} = \dfrac{12}{22}$

$P(1) = \dfrac{C_{2,1}\cdot C_{10,2}}{C_{12,3}} = \dfrac{2\cdot\frac{10!}{2!(10-2)!}}{220} = \dfrac{90}{220} = \dfrac{9}{22}$

$P(2) = \dfrac{C_{2,2}\cdot C_{10,1}}{220} = \dfrac{10}{220} = \dfrac{1}{22}$

(A) The probability distribution of X is:

x_i	0	1	2
p_i	$\frac{12}{22}$	$\frac{9}{22}$	$\frac{1}{22}$

(B) $E(X) = 0\left(\frac{12}{22}\right) + 1\left(\frac{9}{22}\right) + 2\left(\frac{1}{22}\right) = \frac{11}{22} = \frac{1}{2}$ (6-7)

82. Let Event NH = individual with normal heart,
Event MH = individual with minor heart problem,
Event SH = individual with severe heart problem,
and Event P = individual passes the cardiogram test.

Then, using the notation given above, we have:

$P(NH) = .82$
$P(MH) = .11$
$P(SH) = .07$
$P(P \mid NH) = .95$
$P(P \mid MH) = .30$
$P(P \mid SH) = .05$

We want to find $P(NH \mid P) = \dfrac{P(NH \cap P)}{P(P)} = \dfrac{P(NH)P(P \mid NH)}{P(NH \cap P) + P(MH \cap P) + P(SH \cap P)}$

$$= \frac{P(NH)P(P \mid NH)}{P(NH)P(P \mid NH) + P(MH)P(P \mid MH) + P(SH)P(P \mid SH)}$$

$$= \frac{(.82)(.95)}{(.82)(.95) + (.11)(.30) + (.07)(.05)} = .955$$

(6-6)

83. The tree diagram for this problem is as follows:

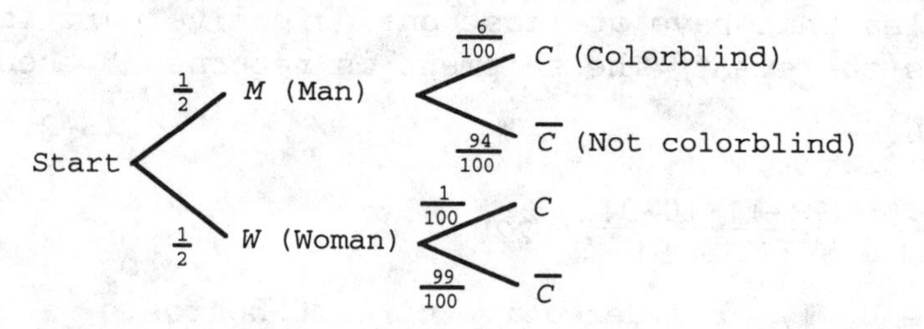

We now compute

$P(M \mid C) = \dfrac{P(M \cap C)}{P(C)} = \dfrac{P(M \cap C)}{P(M \cap C) + P(W \cap C)} = \dfrac{P(M)P(C \mid M)}{P(M)P(C \mid M) + P(W)P(C \mid W)}$

$$= \frac{\frac{1}{2} \cdot \frac{6}{100}}{\frac{1}{2} \cdot \frac{6}{100} + \frac{1}{2} \cdot \frac{1}{100}} = \frac{6}{7} \approx .857$$

(6-6)

84. According to the empirical probabilities, candidate A should have won the election. Since candidate B won the election one week later, either some of the students changed their minds during the week, or the 30 students in the math class were not representative of the entire student body, or both. (6-3)

7 DATA DESCRIPTION AND PROBABILITY DISTRIBUTIONS

EXERCISE 7-1

Things to remember:

1. The following techniques are used to graph qualitative data:
   (a) Vertical bar graphs.
   (b) Horizontal bar graphs.
   (c) Broken-line graphs.
   (d) Pie graphs.

1.

Gross National Product

Billions of dollars

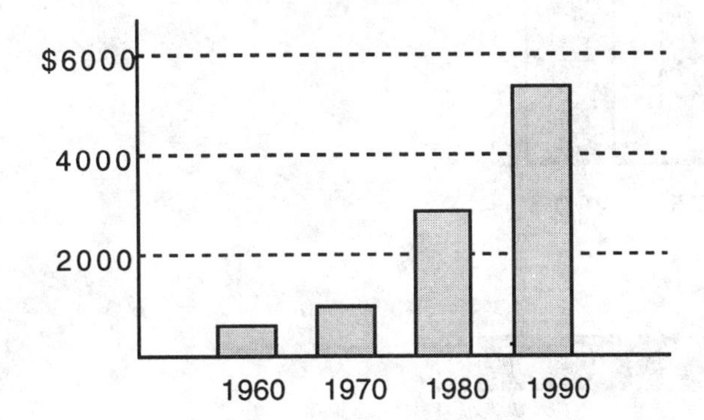

3. The mean family expenditure on housing in 1984 was $2200, and in 1994 it was $2500. The category of expenditure showing the greatest increase is Total taxes; approximately $1400. The category of expenditure with the greatest percentage increase is Recreation; approximately $\frac{5}{10} = \frac{1}{2}$ or 50%.

5.

Annual Railroad Carloadings in U. S.

(in millions)

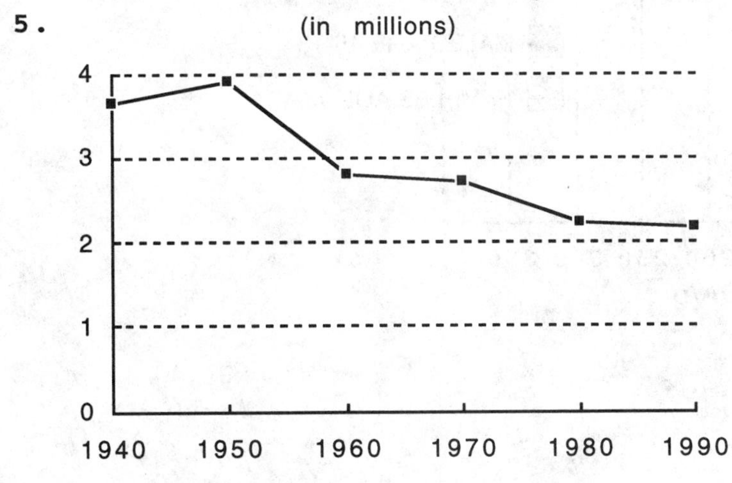

7.

Individual income 33%
Social insurance 30%
Sales and excise 14%
Property 10%
Corporate income 9%
All other 4%

9. Find the percentage of cars imported for each country:

			Percentage	Central Angle
Canada:	$\frac{1590}{3529} \approx 0.451$	or	45.1%	162.2°
Japan:	$\frac{1013}{3529} \approx 0.287$	or	28.7%	103.3°
Mexico:	$\frac{551}{3529} \approx 0.156$	or	15.6%	56.2°
Germany:	$\frac{234}{3529} \approx 0.066$	or	6.6%	23.9°
South Korea:	$\frac{141}{3529} \approx 0.40$	or	4%	14.4°

11.

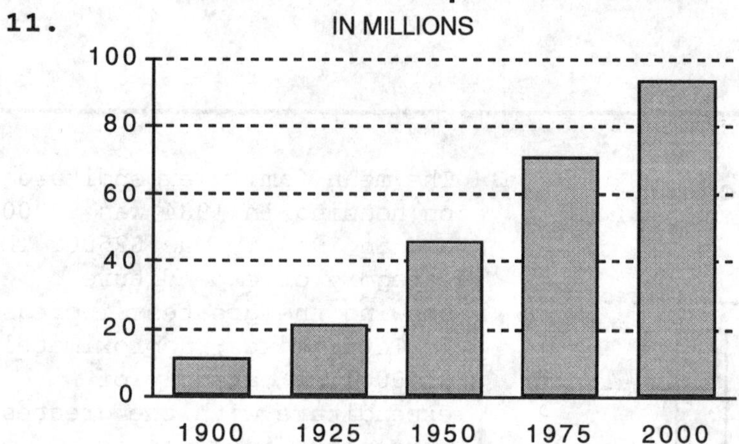

13.

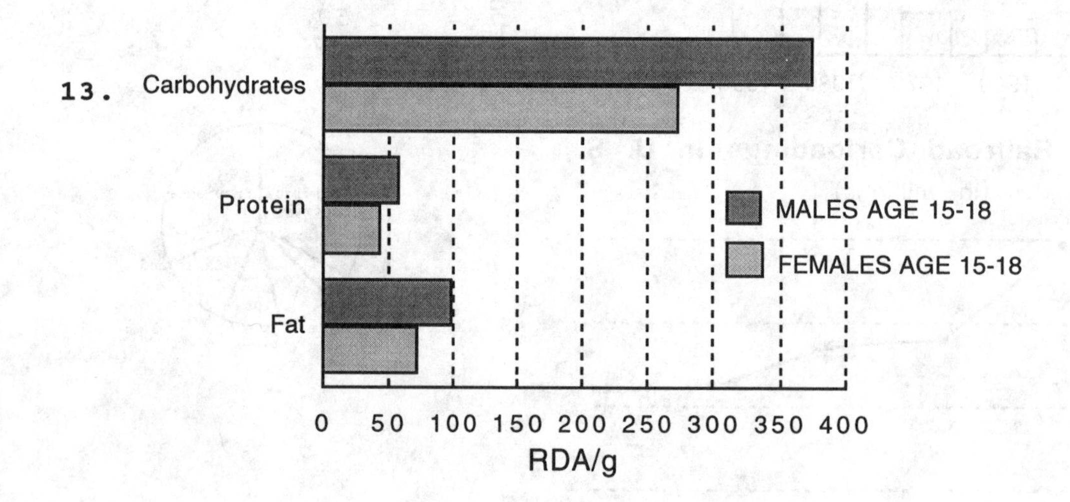

Fast Food Burgers--Nutritional Information

15. (A)

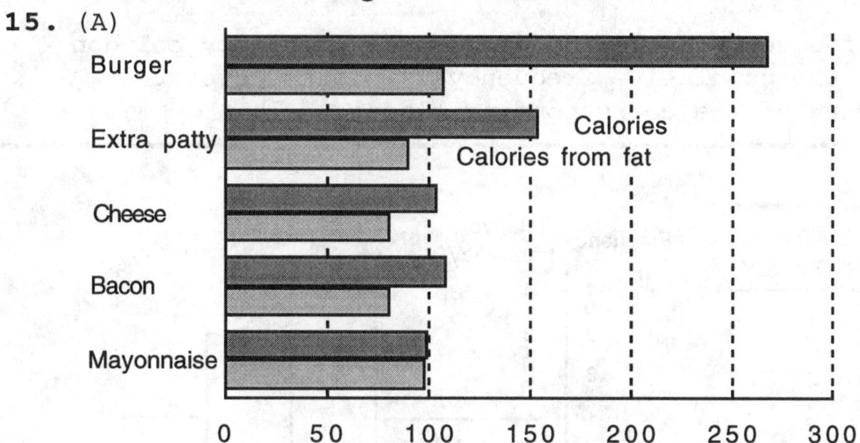

(B) The double bacon cheeseburger with mayo would supply 76.5% of the calories from fat allowed for the entire day.

17. **Prison Incarceration Rates (1992)**

NUMBER PER 100,000 POPULATION

United States
South Africa
Venezuela
Hungary
Canada

0 100 200 300 400 500

19. The median age in 1900 was approximately 23; in 1990 the median age was approximately 33.

The median age decreased in the 1950's and 1960's, but increased in the other decades.

EXERCISE 7-2

Things to remember:

1. The following techniques are used to depict quantitative data:

   (a) Frequency distribution: a table showing class intervals and their corresponding frequencies.

   (b) Histogram: graphic representation of a frequency distribution.

(c) Frequency polygon: a broken-line graph of a frequency distribution.

(d) Cumulative frequency table and cumulative frequency polygon (or ogive): the cumulative frequency is plotted over the upper boundary of the corresponding class.

---

1. (A)

CLASS INTERVAL	TALLY	FREQUENCY	RELATIVE FREQUENCY
0.5-2.5	I	1	.1
2.5-4.5	I	1	.1
4.5-6.5	IIII	4	.4
6.5-8.5	III	3	.3
8.5-10.5	I	1	.1

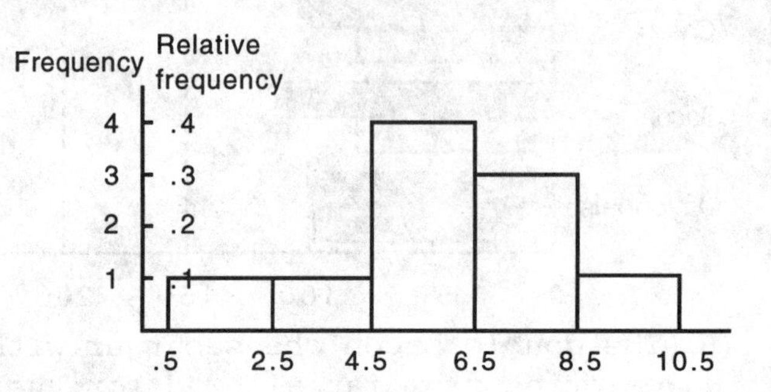

(B) The frequency table and histogram are the **same** as those in part (A).

(C) The data set in (B) is more spread out than the data set in (A).

3. (A) Let *Xmin* = 1.5, *Xmax* = 25.5, change *Xscl* from 1 to 2, and multiply *Ymax* and *Yscl* by 2; change *Xscl* from 1 to 4 and multiply *Ymax* and *Yscl* by 4.

(B) The shape become more symmetrical and more rectangular.

5. (A)

CLASS INTERVAL	TALLY	FREQUENCY	RELATIVE FREQUENCY
20.5-24.5	I	1	.05
24.5-28.5	III	3	.15
28.5-32.5	~~IIII~~ II	7	.35
32.5-36.5	~~IIII~~ I	6	.30
36.5-40.5	II	2	.10
40.5-44.5	I	1	.05

(B)

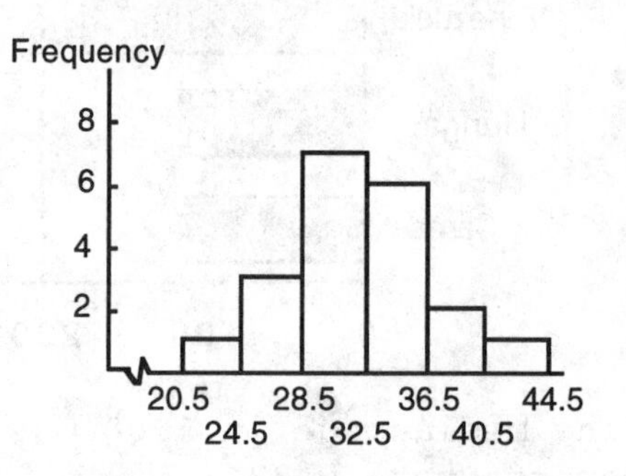

(C) Let A = salary above \$32,500;
Let B = salary below \$28,500.

From the data set itself

$$P(A) = \frac{n(A)}{n(S)} = \frac{9}{20} = 0.45;$$

$$P(B) = \frac{n(B)}{n(S)} = \frac{4}{20} = 0.20$$

From the frequency table

$P(A) = 0.30 + 0.10 + 0.05 = 0.45$

$P(B) = 0.05 + 0.15 = 0.20$

(D)

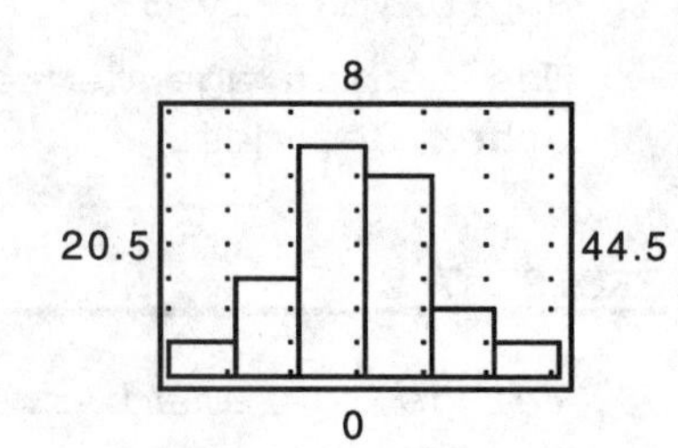

7. (A) The frequency and relative frequency table for the given data is shown below.

Class interval	Frequency	Relative frequency
-0.5- 4.5	5	.05
4.5- 9.5	54	.54
9.5-14.5	25	.25
14.5-19.5	13	.13
19.5-24.5	0	.00
24.5-29.5	1	.01
29.5-34.5	2	.02
	100	1.00

(B) The histogram below is a graphic representation of the tabulated data in part (A).

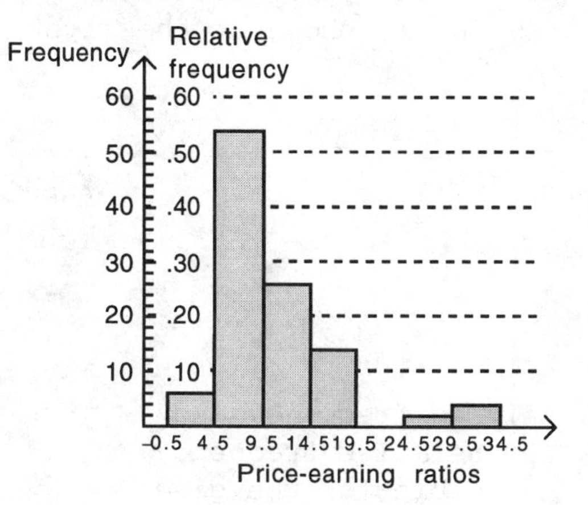

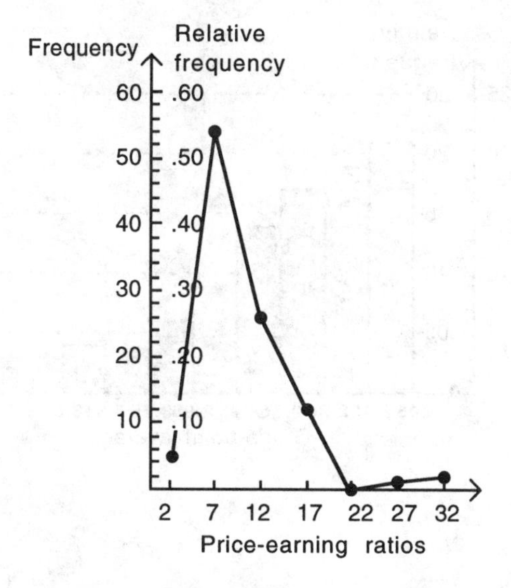

(C) A frequency polygon (broken-line graph) for the tabulated data in part (A) is shown at the left.

(D) An example of a cumulative and relative cumulative frequency table is shown below. From this table, we note that the probability of a price-earnings ratio drawn at random from the sample lying between 4.5 and 14.5 is 0.84 - 0.05 = 0.79.

Class interval	Frequency	Cumulative frequency	Relative cumulative frequency
-0.5- 4.5	5	5	.05
4.5- 9.5	54	59	.59
9.5-14.5	25	84	.84
14.5-19.5	13	97	.97
19.5-24.5	0	97	.97
24.5-29.5	1	98	.98
29.5-34.5	2	100	1.00

(E) A cumulative frequency polygon of the tabulated data in part (D) is shown at the right.

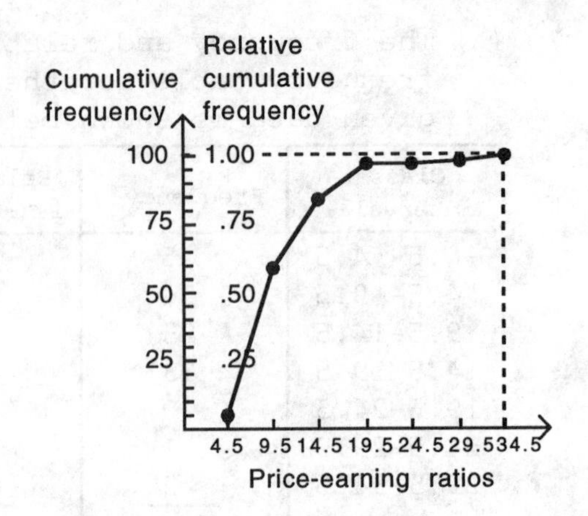

9. (A) The frequency and relative frequency table for the given data is shown below.

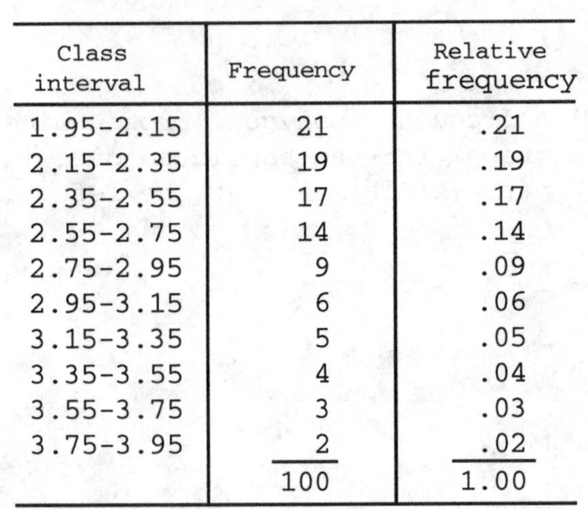

Class interval	Frequency	Relative frequency
1.95-2.15	21	.21
2.15-2.35	19	.19
2.35-2.55	17	.17
2.55-2.75	14	.14
2.75-2.95	9	.09
2.95-3.15	6	.06
3.15-3.35	5	.05
3.35-3.55	4	.04
3.55-3.75	3	.03
3.75-3.95	2	.02
	100	1.00

(B) A histogram of the tabulated data in part (A) is shown below.

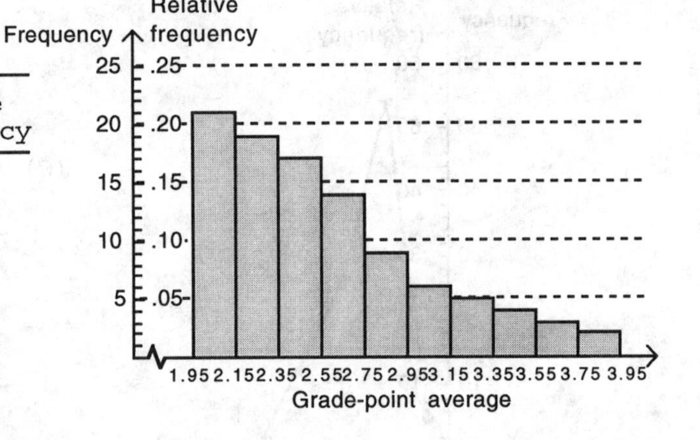

(C) A frequency polygon of the tabulated data in part (A) is shown at the right.

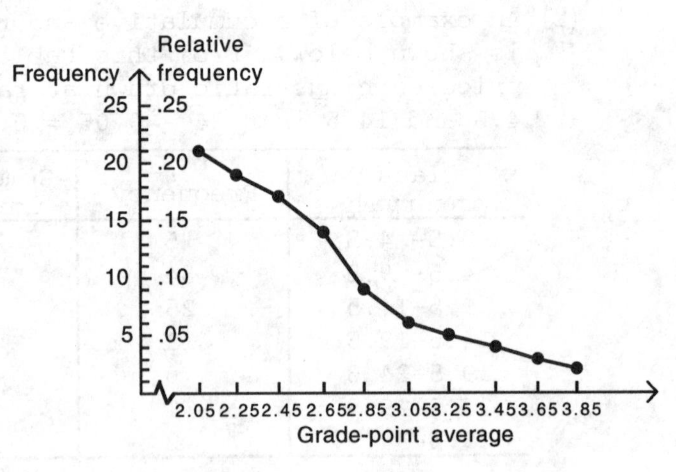

(D) A cumulative and relative cumulative frequency table for the given data is shown below. The probability of a GPA drawn at random from the sample being over 2.95 is 1 - 0.8 = 0.2.

Class interval	Frequency	Cumulative frequency	Relative cumulative frequency
1.95-2.15	21	21	.21
2.15-2.35	19	40	.40
2.35-2.55	17	57	.57
2.55-2.75	14	71	.71
2.75-2.95	9	80	.80
2.95-3.15	6	86	.86
3.15-3.35	5	91	.91
3.35-3.55	4	95	.95
3.55-3.75	3	98	.98
3.75-3.95	2	100	1.00

(E) A cumulative frequency polygon for the tabulated data in part (D) is shown at the right.

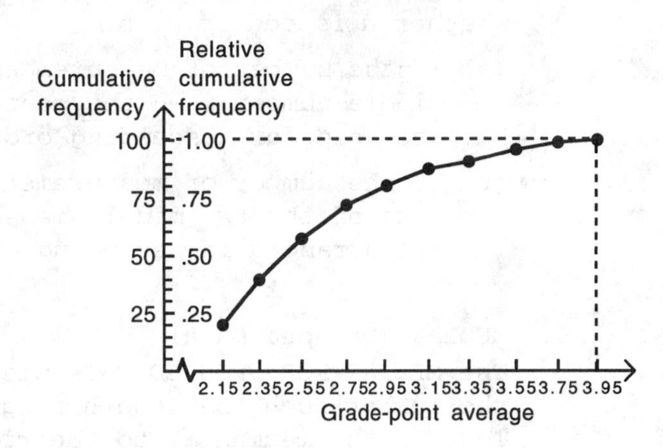

EXERCISE 7-3

Things to remember:

1. MEAN (Ungrouped Data)

   If $x_1, x_2, \ldots, x_n$ is a set of n measurements, then the MEAN of the set of measurements is given by:

$$[\text{mean}] = \frac{\sum_{i=1}^{n} x_i}{n} = \frac{x_1 + x_2 + \cdots + x_n}{n} \qquad (1)$$

   $\bar{x}$ = [mean] if the data set is a sample
   μ = [mean] if the data set is the population

2. MEAN (Grouped Data)

   A data set of n sample measurements is grouped into k classes in a frequency table. If x_i is the midpoint of the ith class interval and f_i is the ith class frequency, then the MEAN for the grouped data is given by:

$$[\text{mean}] = \frac{\sum_{i=1}^{k} x_i f_i}{n} = \frac{x_1 f_1 + x_2 f_2 + \cdots + x_k f_k}{n} \quad (2)$$

$$n = \sum_{i=1}^{k} f_i = (\text{Total number of measurements})$$

$\overline{x}$ = [mean] if data set is a sample
μ = [mean] if data set is the population

3. MEDIAN

The MEDIAN of a set of n measurements is a number such that half the n measurements fall below the median and half fall above. To find the median, we proceed as follows depending on whether n is odd or even.

(a) If the number of measurements n is odd, the median is the middle number when the measurements are arranged in ascending or descending order.

(b) If the number of measurements n is even, the median is the mean of the two middle measurements when the measurements are arranged in ascending or descending order.

4. MEDIAN (Grouped Data)
The MEDIAN FOR GROUPED DATA with no classes of frequency 0 is the number such that the histogram has the same area to the left of the median as to the right of the median.

5. MODE

The MODE is the most frequently occurring measurement in a data set. There may be a unique mode, several modes, or essentially no mode.

---

1. Arrange the given numbers in increasing order:
1, 2, 2, 3, [3, 3,] 3, 4, 4, 5
Using 1,

$$\text{Mean} = \overline{x} = \frac{1 + 2 + 2 + 3 + 3 + 3 + 3 + 4 + 4 + 5}{10} = \frac{30}{10} = 3$$

$$\text{Median} = \frac{3 + 3}{2} = 3$$

$$\text{Mode} = 3$$

3. The mean and median are not suitable for these data. The modal preference for flavor of ice cream is chocolate.

5. We construct a table indicating the class intervals, the class midpoints x_i, the frequencies f_i, and the products $x_i f_i$.

Class interval	Midpoint x_i	Frequency f_i	Product $x_i f_i$
0.5-2.5	1.5	2	3.0
2.5-4.5	3.5	5	17.5
4.5-6.5	5.5	7	38.5
6.5-8.5	7.5	1	7.5
		$n = \sum_{i=1}^{4} f_i = 15$	$\sum_{i=1}^{4} x_i f_i = 66.5$

Thus, $\bar{x} = \frac{\sum_{i=1}^{n} x_i f_i}{n} = \frac{66.5}{15} \approx 4.433 \approx 4.4$

7. The median; the measurement 81.2 is an extreme value which distorts the mean.

9. (A) We would expect the mean and the median to be close to 3.5.

(B) The answer depends on the results of your simulation.

11. (A) Let w, x, y, z be the four numbers, with $w \le x \le y \le z$.

Since the mode is 175, $w = x = 175$.
Since the median is 250,

$$\frac{175 + y}{2} = 250$$
$$175 + y = 500$$
$$y = 325$$

Since the mean is 300,

$$\frac{175 + 175 + 325 + z}{4} = 300$$
$$675 + z = 1200$$
$$z = 525$$

The set {175, 175, 325, 525} has mean 300, median 250, mode 175.

(B) Let the four numbers be u, v, w, x, where u, $v = m_3$. Choose w such that the mean of w and m_3 is m_2. Then choose x such that

$$\frac{u + v + w + x}{4} = m_1$$

13. Mean $= \frac{5.3 + 12.9 + 10.1 + 8.4 + 18.7 + 16.2 + 35.5 + 10.1}{8}$

$= \frac{117.2}{8} \approx 14.7$

Now arrange the data in order (we use increasing)

5.3, 8.4, 10.1, 10.1, 12.9, 16.2, 18.7, 35.5

Median $= \frac{10.1 + 12.9}{2} = 11.5$

Mode = 10.1

15.

Class interval	Midpoint x_i	Frequency f_i	Product $x_i f_i$
799.5- 899.5	849.5	3	2,548.5
899.5- 999.5	949.5	10	9,495
999.5-1099.5	1049.5	24	25,188
1099.5-1199.5	1149.5	12	13,794
1199.5-1299.5	1249.5	1	1,249.5
		$n = \sum_{i=1}^{5} f_i = 50$	$\sum_{i=1}^{5} x_i f_i = 52,275$

$$\text{Mean} = \overline{x} = \frac{\sum_{i=1}^{n} x_i f_i}{n} = \frac{52,275}{50} = 1045.5 \text{ hours.}$$

To find the median, we draw the histogram of the data

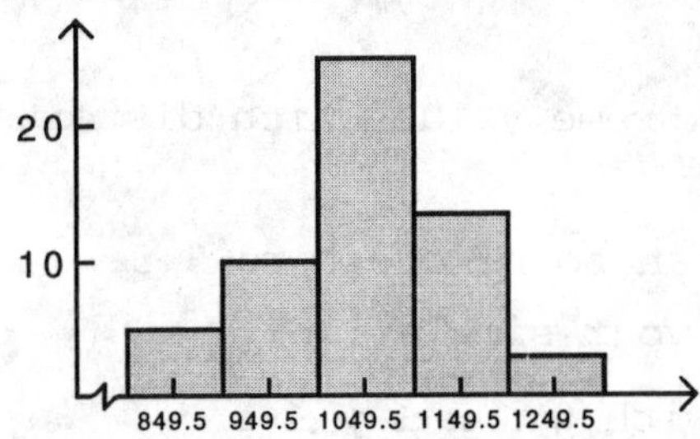

Total area = 300 + 1000 + 2400 + 1200 + 100 = 5000

Let M be the median. Then

$$100(3) + 100(10) + (M - 999.5)24 = 2500$$
$$(M - 999.5)24 = 1200$$
$$M - 999.5 = 50$$
$$M = 1049.5 \text{ hours}$$

17. Mean: $\sum_{i=1}^{6} x_i = 7,970,000$

$$\overline{x} = \frac{\sum_{i=1}^{6} x_i}{6} = \frac{7,970,000}{6} \approx 1,328,000$$

Median: arrange the numbers in order

1,100,000, 1,200,000, 1,200,000, 1,380,000, 1,490,000, 1,600,000

$$\text{Median} = \frac{1,200,000 + 1,380,000}{2} = 1,290,000$$

Mode: Mode = 1,200,000

19.

Class interval	Midpoint x_i	Frequency f_i	Product $x_i f_i$
41.5-43.5	42.5	3	127.5
43.5-45.5	44.5	7	311.5
45.5-47.5	46.5	13	604.5
47.5-49.5	48.5	17	824.5
49.5-51.5	50.5	19	959.5
51.5-53.5	52.5	17	892.5
53.5-55.5	54.5	15	817.5
55.5-57.5	56.5	7	395.5
57.5-59.5	58.5	2	117.0
		$n = \sum_{i=1}^{9} f_i = 100$	$\sum_{i=1}^{9} x_i f_i = 5050$

$$\text{Mean} = \overline{x} = \frac{\sum_{i=1}^{9} x_i f_i}{100} = \frac{5050}{100} = 50.5$$

The histogram for the data is:

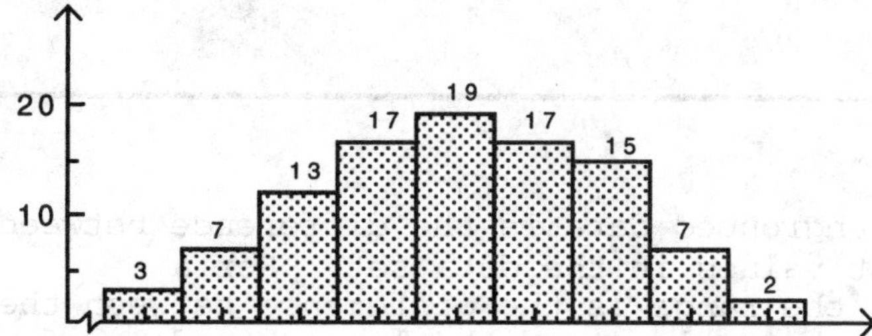

Total area = 6 + 14 + 26 + 34 + 38 + 34 + 30 + 14 + 4 = 200

Let M be the median. Then

$$6 + 14 + 26 + 34 + (M - 49.5)19 = 100$$

$$M - 49.5 = \frac{100 - 80}{19} \approx 1.05$$

$$M = 50.55$$

21. $\text{Mean} = \overline{x} = \frac{\sum_{i=1}^{7} x_i}{7} = \frac{188,000}{7} \approx 27,000$

Median: Arrange the numbers in order

11,000, 12,000, 20,000, 25,000, 37,000, 39,000, 44,000

Median = 25,000

Mode: no mode

23. The histogram for the data in Table 1 is:

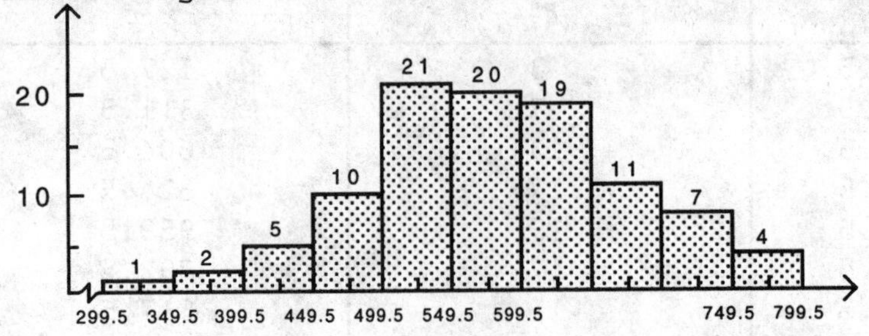

Total area = 50 + 100 + 250 + 500 + 1050 + 1000 + 950 + 550 + 350 + 200 = 5000

Let M be the median. Then

$$50 + 100 + 250 + 500 + 1050 + (M - 549.5)20 = 2500$$

$$M - 549.5 = \frac{2500 - 1950}{20} = \frac{550}{20} = 27.5$$

$$M = 577$$

EXERCISE 7-4

Things to remember:

1. The RANGE for a set of ungrouped data is the difference between the largest and smallest values in the data set. For a frequency distribution, the range is the difference between the upper boundary of the highest class and the lower boundary of the lowest class.

2. VARIANCE (Ungrouped Data)*

   The SAMPLE VARIANCE s^2 of a set of n sample measurements x_1, x_2, ..., x_n with mean $\overline{x}$ is given by

$$s^2 = \frac{\sum_{i=1}^{n} (x_i - \overline{x})^2}{n - 1} \tag{3}$$

   If x_1, x_2, ..., x_n is the whole population with mean μ, then the POPULATION VARIANCE σ^2 is given by

$$\sigma^2 = \frac{\sum_{i=1}^{n} (x_i - \mu)^2}{n}$$

   *In this section, our interest is restricted to sample variance.

3. STANDARD DEVIATION (Ungrouped Data)*

The SAMPLE STANDARD DEVIATION s of a set of n sample measurements $x_1, x_2, \ldots, x_n$ with mean $\bar{x}$ is given by

$$s = \sqrt{\frac{\sum_{i=1}^{n} (x_i - \bar{x})^2}{n - 1}} \tag{4}$$

If $x_1, x_2, \ldots, x_n$ is the whole popluation with mean μ, then the POPULATION STANDARD DEVIATION σ is given by

$$\sigma = \sqrt{\frac{\sum_{i=1}^{n} (x_i - \mu)^2}{n}}$$

*In this section, our interest is restricted to sample standard deviation.

4. STANDARD DEVIATION (Grouped Data)*

Suppose a data set of n sample measurements is grouped into k classes in a frequency table where x_i is the midpoint and f_i the frequency of the ith class interval. Then the SAMPLE STANDARD DEVIATION s for the grouped data is given by

$$s = \sqrt{\frac{\sum_{i=1}^{k} (x_i - \bar{x})^2 f_i}{n - 1}} \tag{5}$$

where $n = \sum_{i=1}^{k} f_i$ = Total number of measurements

If $x_1, x_2, \ldots, x_n$ is the whole population with mean μ, then the POPULATION STANDARD DEVIATION σ is given by

$$\sigma = \sqrt{\frac{\sum_{i=1}^{n} (x_i - \mu)^2 f_i}{n}}$$

*In this section, our interest is restricted to sample standard deviation.

---

1. $\bar{x} = 3$, $n = 10$ (see Problem 1, Exercise 7-3)

$$s = \sqrt{\frac{\sum_{i=1}^{n} (x_i - \bar{x})^2}{n - 1}}$$

$$= \sqrt{\frac{(1-3)^2 + (2-3)^2 + (2-3)^2 + (3-3)^2 + (3-3)^2 + (3-3)^2 + (3-3)^2 + (4-3)^2 + (4-3)^2 + (5-3)^2}{10 - 1}}$$

$$= \sqrt{\frac{4 + 1 + 1 + 0 + 0 + 0 + 0 + 1 + 1 + 4}{9}} = \sqrt{\frac{12}{9}} \approx 1.15$$

3. (A) $\overline{x} = \frac{\sum_{i=1}^{10} x_i}{10} = \frac{33}{10} = 3.3$

$$s = \sqrt{\frac{\sum_{i=1}^{n} (x_i - 3.3)^2}{10 - 1}}$$

$$= \sqrt{\frac{(4-3.3)^2 + (2-3.3)^2 + (3-3.3)^2 + (5-3.3)^2 + (3-3.3)^2 + (1-3.3)^2 + (6-3.3)^2 + (4-3.3)^2 + (2-3.3)^2 + (3-3.3)^2}{9}}$$

$$= \sqrt{\frac{20.10}{9}} = 1.49$$

The measurements that are in the interval (1.81, 4.79) are within one standard deviation of the mean; 70% of the measurements, {4, 2, 3, 3, 4, 2, 3}, are in this interval.

The measurements that are in the interval (0.32, 6.28) are within 2 standard deviations of the mean; 100% of the measurements are in this interval. It follows immediately that 100% of the measurements are within 3 standard deviations of the mean.

(B) Yes. If the histogram is bell-shaped, then approximately 68% of the data are within one standard deviation of the mean, approximately 95% are within 2 standard deviations of the mean, and almost all the data is within 3 standard deviations of the mean. The data set given here satisfies these criteria.

(C)

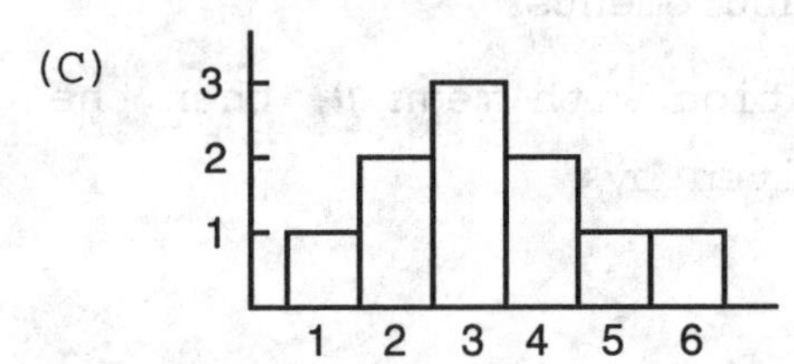

5.

Interval	Midpoint x_i	Frequency f_i	$x_i f_i$	$(x_i - \overline{x})^2$	$(x_i - \overline{x})^2 f_i$
0.5- 3.5	2	2	4	19.36	38.72
3.5- 6.5	5	5	25	1.96	9.80
6.5- 9.5	8	7	56	2.56	17.92
9.5-12.5	11	1	11	21.16	21.16
		$n = 15$	$\sum_{i=1}^{4} x_i f_i = 96$		$\sum_{i=1}^{4} (x_i - \overline{x})^2 f_i = 87.60$

$\overline{x} = \frac{96}{15} = 6.4$ $\qquad$ $s = \sqrt{\frac{87.60}{15 - 1}} \approx 2.5$

7. (A) False: If the sample variance is less than 1, then the sample standard deviation is greater than the sample variance.

(B) True: The expression $\dfrac{\sum_{i=1}^{n}(x_i - \mu)^2}{n}$ is always nonnegative.

9. (A) The first data set. The sums are not equally likely; a sum of 7 is much more likely than a sum of 2 or 12. In the second data set, the numbers 2 through 12 are equally likely and so the numbers will have greater dispersion.

(B) The answer depends on the results of your simulation.

11.

x_i	$(x_i - \bar{x})^2$
2.35	4.00
1.42	8.58
8.05	13.69
6.71	5.57
3.11	1.54
2.56	3.20
0.72	13.18
4.17	.03
5.33	.96
7.74	11.49
3.88	.22
6.21	3.46
$\sum_{i=1}^{12} x_i = 52.25$	$\sum_{i=1}^{12}(x_i - \bar{x})^2 = 65.92$

$$\bar{x} = \frac{\sum_{i=1}^{12} x_i}{12} = \frac{52.25}{12} \approx 4.35.$$ Thus, the mean $\bar{x}$ = \$4.35.

$$s = \sqrt{\frac{\sum_{i=1}^{12}(x_i - \bar{x})^2}{12 - 1}} = \sqrt{\frac{65.92}{11}} \approx \sqrt{5.99} \approx 2.45.$$

Thus, the standard deviation s = \$2.45.

13.

Interval	Midpt.	Freq.	$x_i f_i$	$(x_i - \bar{x})^2$	$(x_i - \bar{x})^2 f_i$
6.95- 7.45	7.20	2	14.40	2.25	4.50
7.45- 7.95	7.70	10	77.00	1.00	10.00
7.95- 8.45	8.20	23	188.60	.25	5.75
8.45- 8.95	8.70	30	261.00	.00	.00
8.95- 9.45	9.20	21	193.20	.25	5.25
9.45- 9.95	9.70	13	126.10	1.00	13.00
9.95-10.45	10.20	1	10.20	2.25	2.25
		$n = 100$	$\sum_{i=1}^{7} x_i f_i = 870.50$		$\sum_{i=1}^{7}(x_i - \bar{x})^2 f_i = 40.75$

$$\bar{x} = \frac{\sum_{i=1}^{7} x_i f_i}{100} = \frac{870.5}{100} = 8.705$$

Thus, the mean $\bar{x} = 8.7$ hours.

$$s = \sqrt{\frac{\sum_{i=1}^{7} (x_i - \bar{x})^2 f_i}{n - 1}} = \sqrt{\frac{40.75}{99}} \approx \sqrt{.4116} \approx .6$$

Thus, the standard deviation $s = .6$ hours.

15.

x_i	$(x_i - \bar{x})^2$
4.9	.04
5.1	.00
3.9	1.44
4.2	.81
6.4	1.69
3.4	2.89
5.8	.49
6.1	1.00
5.0	.01
5.6	.25
5.8	.49
4.6	.25
$\sum_{i=1}^{12} x_i = 60.8$	$\sum_{i=1}^{12} (x_i - \bar{x})^2 = 9.36$

$$\bar{x} = \frac{\sum_{i=1}^{12} x_i}{12} = \frac{60.8}{12} \approx 5.1$$

Thus, the mean $\bar{x} = 5.1$ minutes.

$$s = \sqrt{\frac{\sum_{i=1}^{12} (x_i - \bar{x})^2}{12 - 1}} = \sqrt{\frac{9.36}{11}}$$

$$\approx \sqrt{0.8509} \approx .92$$

Thus, the standard deviation $s = .9$ minutes.

17.

x_i	$(x_i - \bar{x})^2$
9	4
11	0
11	0
15	16
10	1
12	1
12	1
13	4
8	9
7	16
13	4
12	1
$\sum_{i=1}^{12} x_i = 133$	$\sum_{i=1}^{12} (x_i - \bar{x})^2 = 57$

$$\bar{x} = \frac{\sum_{i=1}^{12} x_i}{12} = \frac{133}{12} \approx 11.08$$

Thus, the mean $\bar{x} = 11$.

$$s = \sqrt{\frac{\sum_{i=1}^{12} (x_i - \bar{x})^2}{12 - 1}} = \sqrt{\frac{57}{11}}$$

$$\approx \sqrt{5.18} \approx 2.28$$

Thus, the standard deviation $s = 2$.

EXERCISE 7-5

Things to remember:

1. BERNOULLI TRIALS

   A sequence of experiments is called a SEQUENCE OF BERNOULLI TRIALS, or a BINOMIAL EXPERIMENT if:

   (a) Only two oucomes are possible on each trial.

   (b) The probability of success p for each trial is a constant (the probability of failure is $q = 1 - p$).

   (c) All trials are independent.

2. PROBABILITY OF x SUCCESSES IN n BERNOULLI TRIALS

The probability of exactly x successes in n independent repeated Bernoulli trials, with the probability of success of each trial p (and of failure q), is

$$P(x \text{ successes}) = C_{n,x}p^x q^{n-x}$$

where $C_{n,x}$ is the number of combinations of n objects taken x at a time.

3. BINOMIAL FORMULA

For n a natural number,

$$(a + b)^n = C_{n,0}a^n + C_{n,1}a^{n-1}b + C_{n,2}a^{n-2}b^2 + \cdots + C_{n,n}b^n$$

4. BINOMIAL DISTRIBUTION

$$P(X_n = x) = P(x \text{ successes in } n \text{ trials})$$
$$= C_{n,x}p^x q^{n-x} \qquad x \in \{0, 1, 2, \ldots, n\}$$

where p is the probability of success and q is the probability of failure on each trial.

Informally, $P(X_n = x)$ is written $P(x)$.

5. MEAN AND STANDARD DEVIATION (RANDOM VARIABLE IN A BINOMIAL DISTRIBUTION)

Mean: $\mu = np$

Standard deviation: $\sigma = \sqrt{npq}$

---

1. $p = \frac{1}{2}$

$q = 1 - \frac{1}{2} = \frac{1}{2}$

$C_{3,2}\left(\frac{1}{2}\right)^2\left(\frac{1}{2}\right)^{3-2} = \frac{3!}{2!1!} \cdot \frac{1}{8}$

$= \frac{3}{8} = .375$

3. $p = \frac{1}{2}$

$q = 1 - \frac{1}{2} = \frac{1}{2}$

$C_{3,0}\left(\frac{1}{2}\right)^0\left(\frac{1}{2}\right)^{3-0} = \frac{3!}{0!3!} \cdot \frac{1}{8}$

$= \frac{1}{8} = .125$

5. $p = .4$

$q = 1 - .4 = .6$

$C_{5,3}(.4)^3(.6)^{5-3} = \frac{5!}{3!2!}(.4)^3(.6)^2$

$= 10(.064)(.36) = .2304$

7. p = probability of getting heads $= \frac{1}{2}$

q = probability of getting tails $= \frac{1}{2}$

$x = 2$, $n = 3$

$P(2) = C_{3,2}\left(\frac{1}{2}\right)^2\left(\frac{1}{2}\right)^{3-2} = \frac{3!}{2!1!} \cdot \frac{1}{8} = \frac{3}{8} = .375$

9. $p = \frac{1}{2},\ q = \frac{1}{2},\ x = 0,\ n = 3$

$$P(0) = C_{3,0}\left(\frac{1}{2}\right)^0\left(\frac{1}{2}\right)^{3-0} = \frac{3!}{0!3!}\cdot\frac{1}{8} = \frac{1}{8} = .125$$

11. $P(\text{at least 2 heads}) = P(x \ge 2) = P(2) + P(3)$

$$= C_{3,2}\left(\frac{1}{2}\right)^2\left(\frac{1}{2}\right)^{3-2} + C_{3,3}\left(\frac{1}{2}\right)^3\left(\frac{1}{2}\right)^{3-3}$$

$$= \frac{3!}{2!1!}\cdot\frac{1}{8} + \frac{3!}{3!0!}\cdot\frac{1}{8} = \frac{3}{8} + \frac{1}{8} = .5$$

13. $P(x) = C_{2,x}(.3)^x(.7)^{2-x}$

x	0	1	2
$P(x)$	.49	.42	.09

The histogram for this distribution is shown at the right.

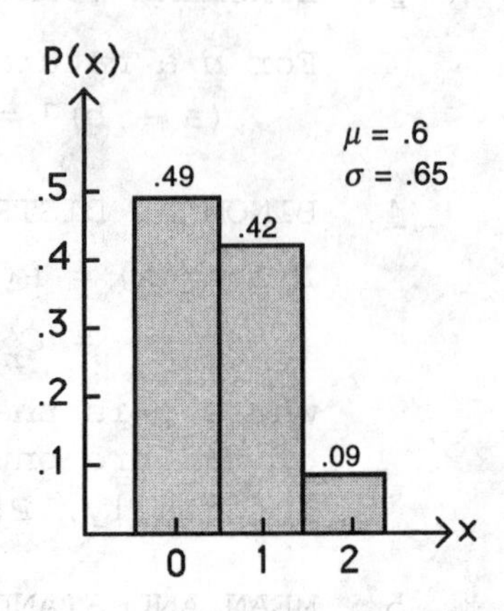

Mean $= np = 2(.3) = .6$ (using 5)

Standard deviation $= \sigma = \sqrt{npq}$ (using 5)

$$= \sqrt{2(.3)(.7)}$$

$$\approx .65$$

15. $P(x) = C_{4,x}(.5)^x(.5)^{4-x}$

x	$P(x)$
0	.06
1	.25
2	.38
3	.25
4	.06

The histogram for this distribution is shown at the right.

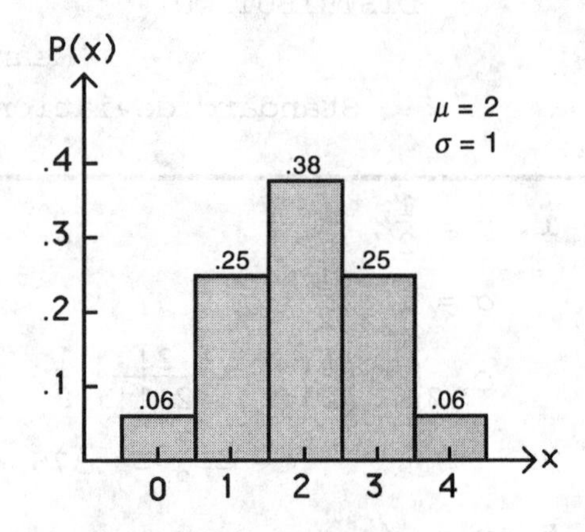

$\mu = np = 4 \times .5 = 2$

$\sigma = \sqrt{npq} = \sqrt{4 \times .5 \times .5} = 1$

17. Let p = probability of getting a "2" in one trial $= \frac{1}{6}$,

and q = probability of not getting a "2" in one trial $= \frac{5}{6}$.

$n = 4,\ x = 3$

$$P(3) = C_{4,3}\left(\frac{1}{6}\right)^3\left(\frac{5}{6}\right)^{4-3} = \frac{4!}{3!1!}\left(\frac{1}{6}\right)^3\left(\frac{5}{6}\right) \approx .0154$$

19. Let p = probability of getting a "1" = $\frac{1}{6}$,

and q = probability of not getting a "1" = $\frac{5}{6}$.

$n = 4$, $x = 0$

$$P(0) = C_{4,0}\left(\frac{1}{6}\right)^0\left(\frac{5}{6}\right)^{4-0} = \frac{4!}{0!4!}\left(\frac{5}{6}\right)^4 \approx .482$$

21. Let p = probability of getting a "6" = $\frac{1}{6}$,

and q = probability of not getting a "6" = $\frac{5}{6}$.

It is actually easier to compute the probability of the complement event, $P(x < 1)$:

$$P(x \geq 1) = 1 - P(x < 1) = 1 - P(0)$$
$$= 1 - C_{4,0}\left(\frac{1}{6}\right)^0\left(\frac{5}{6}\right)^4 = 1 - .4822 \approx .518$$

23. $p = .35$, $q = 1 - .35 = .65$, $n = 4$

(A) The probability of getting exactly two hits is given by:

$$P(x = 2) = C_{4,2}(.35)^2(.65)^2 \approx .311$$

(B) The probability of getting at least two hits is given by:

$$P(x \geq 2) = P(2) + P(3) + P(4)$$
$$= C_{4,2}(.35)^2(.65)^2 + C_{4,3}(.35)^3(.65) + C_{4,4}(.35)^4$$
$$= .3105 + .1115 + .0150 \approx .437$$

25. $p = \frac{1}{5}$, $q = \frac{4}{5}$

Let R denote the event "all answers are wrong" and let S denote the event "at least half of the answers are correct." Then

$$P(R) = C_{10,0}\left(\frac{1}{5}\right)^0\left(\frac{4}{5}\right)^{10} = \left(\frac{4}{5}\right)^{10} = 0.107$$

$$P(S) = \sum_{k=5}^{10} C_{10,k}\left(\frac{1}{5}\right)^k\left(\frac{4}{5}\right)^{10-k}$$

$$= \frac{C_{10,5}(4)^5 + C_{10,6}(4)^4 + C_{10,7}(4)^3 + C_{10,8}(4)^2 + C_{10,9}(4) + C_{10,10}}{5^{10}}$$

$$= \frac{252(1024) + 210(256) + 120(64) + 45(16) + 9(4) + 1}{5^{10}}$$

$$\approx 0.033$$

It is more likely that all guesses are wrong.

27. $P(x) = C_{6,x}(.4)^x(.6)^{6-x}$

x	$P(x)$
0	.05
1	.19
2	.31
3	.28
4	.14
5	.04
6	.004

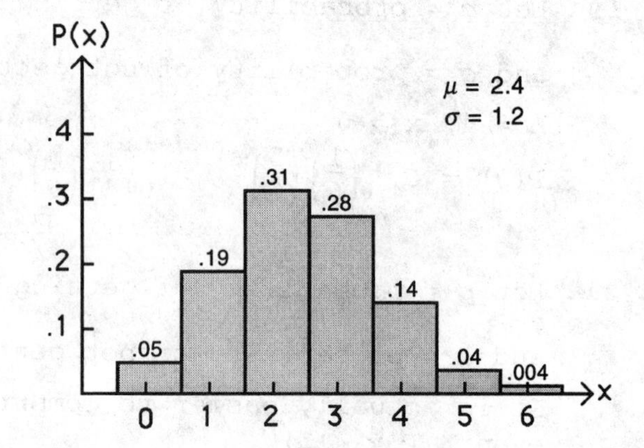

The histogram for this distribution is shown at the right.

$\mu = np = 6 \times .4 = 2.4$

$\sigma = \sqrt{npq} = \sqrt{6 \times .4 \times .6} = 1.2$

29. $P(x) = C_{8,x}(.3)^x(.7)^{8-x}$

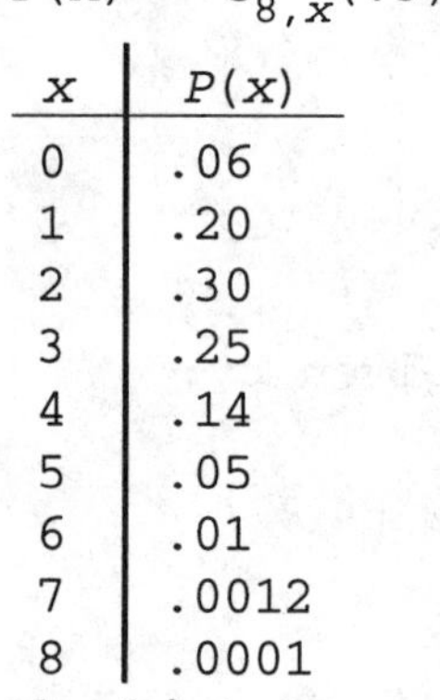

x	$P(x)$
0	.06
1	.20
2	.30
3	.25
4	.14
5	.05
6	.01
7	.0012
8	.0001

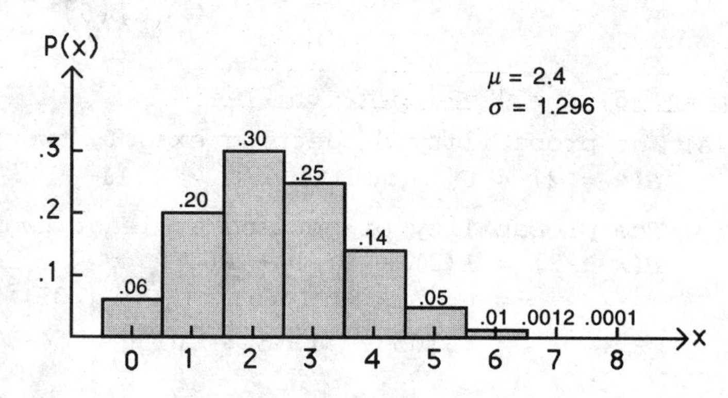

The histogram for this distribution is shown above.

$\mu = np = 8 \times .3 = 2.4$

$\sigma = \sqrt{npq} = \sqrt{8 \times .3 \times .7} \approx 1.296$

31. Given $p = 0.85$, $q = 0.15$, $n = 20$
Using a graphing utility, calculate the binomial probability distribution table.

(A) Mean: $\mu = np = 20(0.85) = 17$
Standard deviation: $\sigma = \sqrt{npq} = \sqrt{20(0.85)(0.15)} \approx 1.597$

(B) To be within one standard deviation of the mean, we must have
$17 - 1.597 < x < 17 + 1.597$ or $15.403 < x < 18.597$
Thus, $x = 16$, 17 or 18.

$P(x = 16) = C_{20,16}(0.85)^{16}(0.15)^4 = 4845(0.07425)(0.00051) \approx 0.18212$

$P(x = 17) = C_{20,17}(0.85)^{17}(0.15)^3 = 1140(0.06311)(0.00338) \approx 0.24283$

$P(x = 18) = C_{20,18}(0.85)^{18}(0.15)^2 = 190(0.05365)(0.0225) \approx 0.22934$

Therefore, $P(16 \le x \le 18) \approx 0.18212 + 0.24283 + 0.22934 = 0.65429$
≈ 0.654

33. Let p = probability of getting heads = $\frac{3}{4}$,

and q = probability of not getting heads = $\frac{1}{4}$.

$n = 5$, $x = 5$

The probability of getting all heads $P(5) = C_{5,5}\left(\frac{3}{4}\right)^5\left(\frac{1}{4}\right)^0 = .2373$.

The probability of getting all tails is the same as the probability of getting no heads. Thus,

$P(0) = C_{5,0}\left(\frac{3}{4}\right)^0\left(\frac{1}{4}\right)^5 = .00098$

Therefore,

$$P(\text{all heads or all tails}) = P(5) + P(0) = .2373 + .00098 = .23828 \approx .238$$

35. The theoretical probability distribution is obtained by using $P(x) = C_{3,x}(.5)^x(.5)^{3-x}$

Frequency of heads in 100 tosses of 3 coins

Number of heads x	$P(x)$	Theoretical frequency $100P(x)$	Actual frequency
0	.125	12.5	(List your
1	.375	37.5	experimental
2	.375	37.5	results here.)
3	.125	12.5	

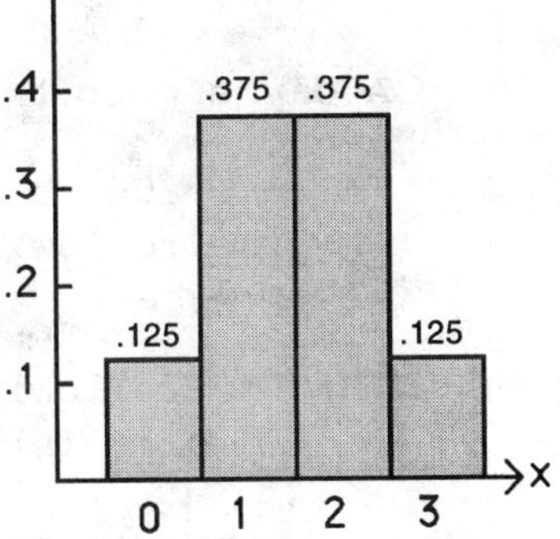

The histogram for the theoretical distribution is shown above.

37. To be symmetric about $x = \frac{n}{2}$, we must have $\mu = np = \frac{n}{2}$ which implies $p = \frac{1}{2}$. Alternatively, for the distribution to be symmetric we must have

$$C_{n,k}p^k q^{n-k} = C_{n,n-k}p^{n-k}q^k$$

for $k = 0, 1, 2, \ldots, n$. This implies

$$p^k q^{n-k} = p^{n-k}q^k$$

which implies

$$p = q = \frac{1}{2}$$

39. (A) $p = 0.5$, $q = 0.5$, $n = 10$

Mean: $\mu = np = 0.5(10) = 5$

Standard deviation: $\sigma = \sqrt{npq} = \sqrt{10(0.5)(0.5)} \approx 1.581$

(B) The answer depends on the results of your simulation.

41. (A) Let p = probability of completing the program = .7, and q = probability of not completing the program = .3.

$n = 7$, $x = 5$

$P(5) = C_{7,5}(.7)^5(.3)^2 = 21(.1681)(.09) = .318$

(B) $P(x \geq 5) = P(5) + P(6) + P(7)$

$= .318 + C_{7,6}(.7)^6(.3) + C_{7,7}(.7)^7(.3)^0$

$= .318 + 7(.1176)(.3) + 1(.0824)(1)$

$= .3180 + .2471 + .0824 \approx .647$

43. Let p = probability that an item is defective = .06, and q = probability that an item is not defective = .94.

$n = 10$

$P(x > 2) = 1 - P[x \leq 2] = 1 - [P(2) + P(1) + P(0)]$

$= 1 - [C_{10,2}(.06)^2(.94)^8 + C_{10,1}(.06)^1(.94)^9 + C_{10,0}(.06)^0(.94)^{10}]$

$= 1 - [.0988 + .3438 + .5386] = 1 - .9812 \approx .0188$

A day's output will be inspected with a probability of .0188.

45. (A) $p = .05$, $q = .95$, $n = 6$

The following function defines the distribution:

$P(x) = C_{6,x}(.05)^x(.95)^{6-x}$

(B) The following table is obtained by using the distribution function in part (A).

x	$P(x)$
0	.735
1	.232
2	.031
3	.002
4	.0001
5	.000
6	.000

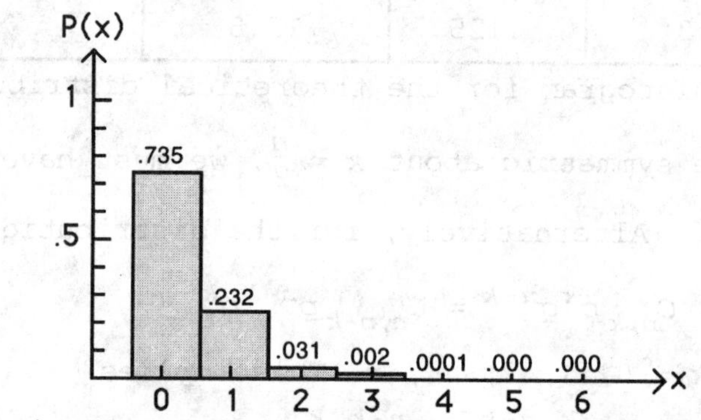

(C) The histogram for the distribution in part (B) is shown above.

(D) $\mu = np = 6 \times .05 = .30$

$\sigma = \sqrt{npq} = \sqrt{6 \times (.05) \times (.95)} = .53$

47. Let p = probability of detecting TB = .8, and q = probability of not detecting TB = .2.

$n = 4$

The probability that at least one of the specialists will detect TB is:

$P(x \geq 1) = 1 - P(x < 1) = 1 - P(0)$

$= 1 - C_{4,0}(.8)^0(.2)^4$

$= 1 - .0016 = .9984 \approx .998$

49. Let p = probability of having a child with brown eyes = .75,
and q = probability of not having a child with brown eyes (i.e., with blue eyes) = .25.

$n = 5$

(A) $x = 0$ (all blue-eyed children, i.e., no brown-eyed children)
$P(0) = C_{5,0}(.75)^0(.25)^5 = .00098 \approx .001$

(B) $x = 3$
$P(3) = C_{5,3}(.75)^3(.25)^2 \approx .264$

(C) $x \geq 3$

$$\begin{aligned} P(x \geq 3) &= P(3) + P(4) + P(5) \\ &\approx .264 + C_{5,4}(.75)^4(.25)^1 + C_{5,5}(.75)^5(.25)^0 \\ &\approx .2640 + .3955 + .2373 \\ &= .8968 \approx .897 \end{aligned}$$

51. (A) $p = .6$, $q = .4$, $n = 6$

The following function defines the distribution:
$P(x) = C_{6,x}(.6)^x(.4)^{6-x}$

(B) The following table is obtained by using the distribution function in part (A).

x	$P(x)$
0	.004
1	.037
2	.138
3	.276
4	.311
5	.187
6	.047

(C) The histogram for the distribution in part (B) is shown at the right.

(D) $\mu = np = 6(.6) = 3.6$

$\sigma = \sqrt{npq} = \sqrt{6 \times .4 \times .6} = 1.2$

53. Let p = probability of getting the right answer to a question = $\frac{1}{5}$,

and q = probability of not getting the right answer to a question = $\frac{4}{5}$.

$n = 10$, $x \geq 7$

$$\begin{aligned} P(x \geq 7) &= P(7) + P(8) + P(9) + P(10) \\ &= C_{10,7}\left(\frac{1}{5}\right)^7\left(\frac{4}{5}\right)^3 + C_{10,8}\left(\frac{1}{5}\right)^8\left(\frac{4}{5}\right)^2 + C_{10,9}\left(\frac{1}{5}\right)^9\left(\frac{4}{5}\right) + C_{10,10}\left(\frac{1}{5}\right)^{10} \\ &\approx .000864 \end{aligned}$$

55. (A) p = probability of answer being correct by guessing = $\frac{1}{5}$ = .2,

$q = .8$, $n = 5$

The following function defines the distribution:
$P(x) = C_{5,x}(.2)^x(.8)^{5-x}$

(B) The following table is obtained by using the distribution function in part (A):

x	$P(x)$
0	.328
1	.410
2	.205
3	.051
4	.006
5	.000

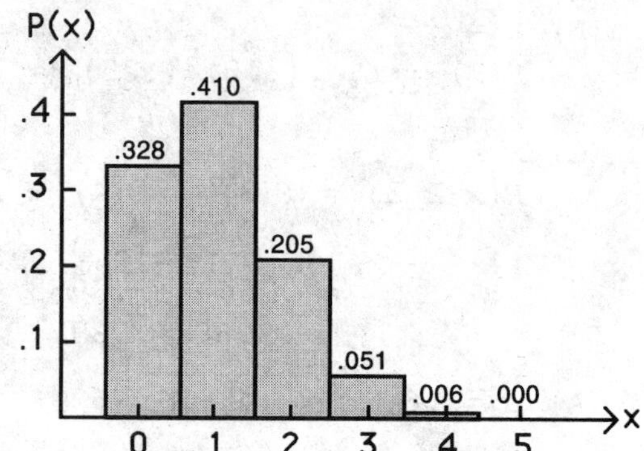

(C) The histogram for part (B) is shown at the right.

(D) $\mu = np = 5 \times .2 = 1.0$

$\sigma = \sqrt{npq} = \sqrt{5 \times .2 \times .8} \approx .894$

57. Let p = probability of a divorce within 20 years = .60, and q = probability of no divorce within 20 years = .40.

$n = 6$

(A) $P(x = 0) = C_{6,0}(.60)^0(.40)^6 \approx .0041$

(B) $P(x = 6) = C_{6,6}(.60)^6(.40)^0 \approx .0467$

(C) $P(x = 2) = C_{6,2}(.60)^2(.40)^4 \approx .138$

(D) $P(x \geq 2) = 1 - P(x < 2) = 1 - [P(0) + P(1)]$
$= 1 - [.0041 + C_{6,1}(.60)^1(.40)^5]$
$\approx 1 - [.0041 + .0369]$
$= .959$

EXERCISE 7-6

Things to remember:

1. NORMAL CURVE PROPERTIES

   1. Normal curves are bell-shaped and are symmetrical with respect to a vertical line.
   2. The mean is at the point where the axis of symmetry intersects the horizontal axis.
   3. The shape of a normal curve is completely determined by its mean and standard deviation—a small standard deviation indicates a tight clustering about the mean and thus a tall, narrow curve; a large standard deviation indicates a large deviation from the mean and thus a broad, flat curve.
   4. Irrespective of the shape, the area between the curve and the x axis is always 1.
   5. Irrespective of the shape, 68.26% of the area will lie within an interval of 1 standard deviation on either side of the mean, 95.44% within 2 standard deviations on either side, and 99.74% within 3 standard deviations on either side (see the figure).

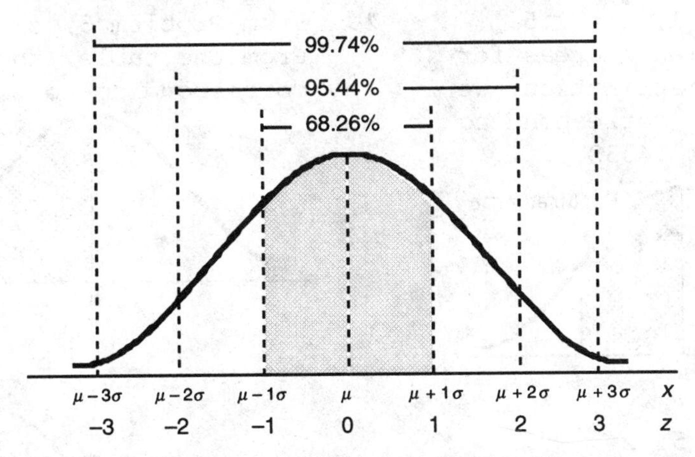

2. If μ and σ are the mean and standard deviation of a normal curve and x is a measurement, then the number of standard deviations that x is from the mean is given by:

$$z = \frac{x - \mu}{\sigma}$$

3. PROPERTIES OF A NORMAL PROBABILITY DISTRIBUTION

   (a) $P(a \le x \le b)$ = the area under the normal curve from a to b.

   (b) $P(-\infty < x < \infty) = 1$ = the total area under the normal curve.

   (c) $P(x = c) = 0$.

4. RULE OF THUMB TEST

Use a normal distribution to approximate a binomial distribution only if the interval $[\mu - 3\sigma, \mu + 3\sigma]$ lies entirely in the interval from 0 to n.

---

1. $x = 65$, $\mu = 50$, $\sigma = 10$

$z = \dfrac{65 - 50}{10}$ (using 2)

$= 1.5$

$x = 65$ is 1.5 standard deviations away from μ.

3. $x = 83$, $\mu = 50$, $\sigma = 10$

$x = \dfrac{83 - 50}{10}$ (using 2)

$= 3.3$

$x = 83$ is 3.3 standard deviations away from μ.

5. $x = 45$, $\mu = 50$, $\sigma = 10$

$z = \dfrac{45 - 50}{10} = -.5$

$x = 45$ is .5 standard deviations away from μ.

7. $x = 42$, $\mu = 50$, $\sigma = 10$

$z = \dfrac{42 - 50}{10} = -.8$

$x = 42$ is .8 standard deviations away from μ.

9. From Problem 1, $z = 1.5$. From the table of areas for the normal distribution, we have the area corresponding to $z = 1.5$ is .4332.

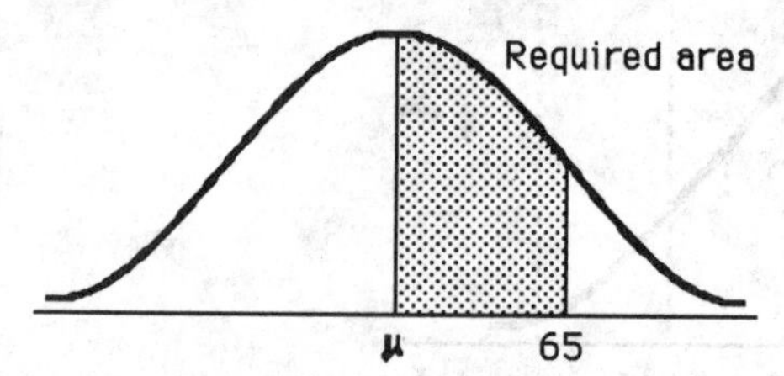

11. From Problem 3, $z = 3.3$. From the table, the area corresponding to $z = 3.3$ is .4995.

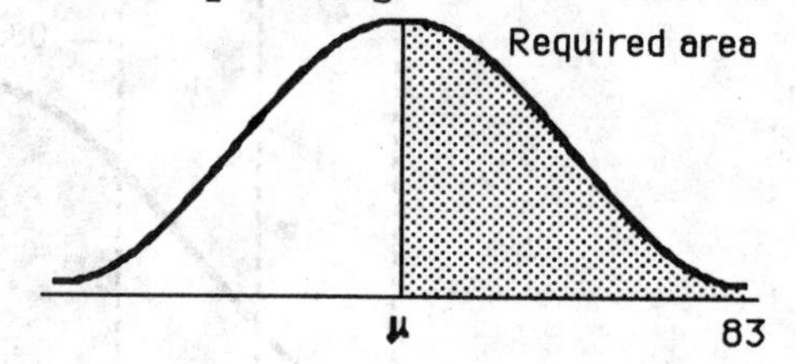

13. From Problem 5, $z = -.5$. From the table, the area corresponding to $z = .5$ is .1915.

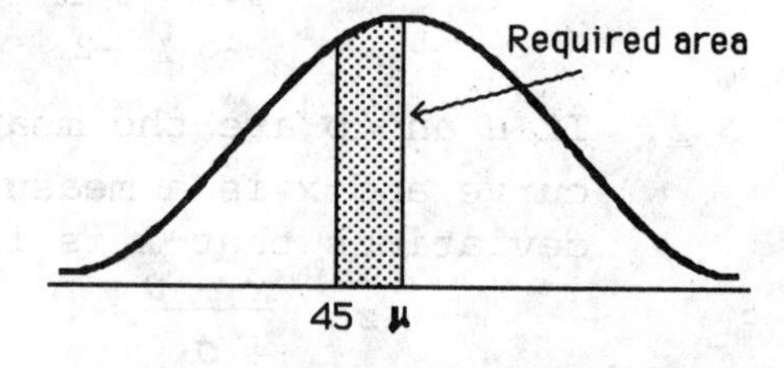

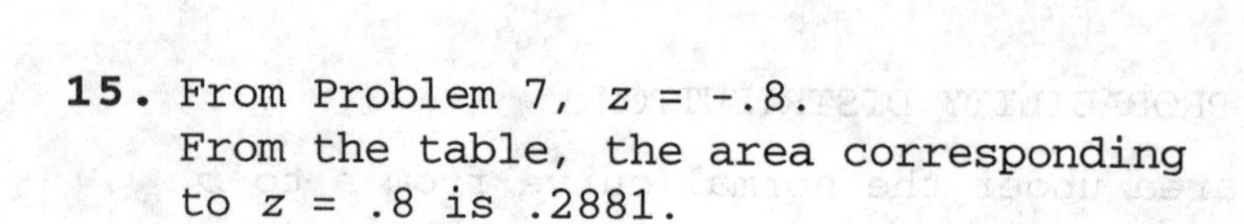

15. From Problem 7, $z = -.8$. From the table, the area corresponding to $z = .8$ is .2881.

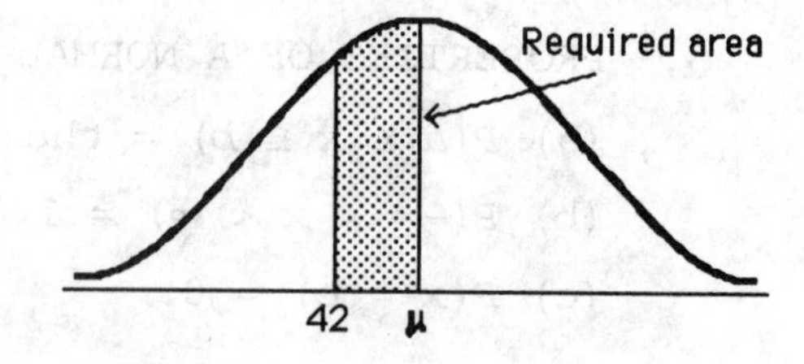

17. $\mu = 70$, $\sigma = 8$

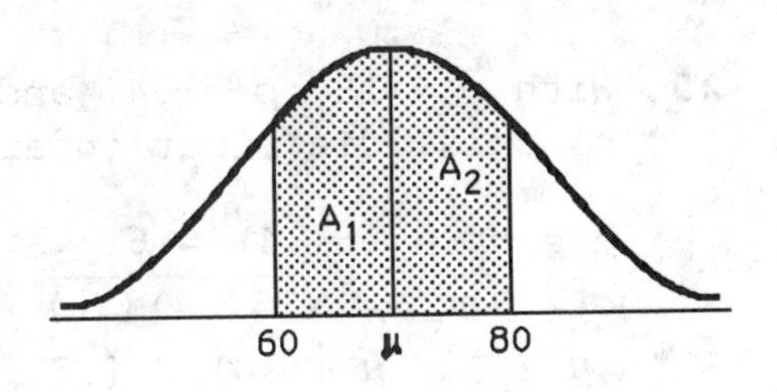

$$z \text{ (for } x = 60) = \frac{60 - 70}{8} = -1.25$$

$$z \text{ (for } x = 80) = \frac{80 - 70}{8} = 1.25$$

Area $A_1 = .3944$. Area $A_2 = .3944$.
Total area $= A = A_1 + A_2 = .7888$.

19. $\mu = 70$, $\sigma = 8$

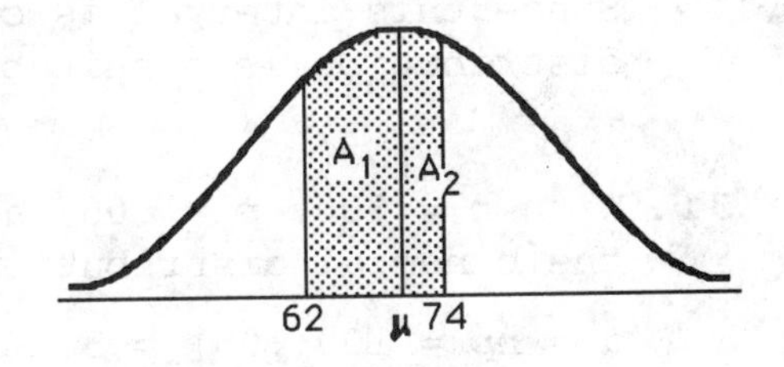

$$z \text{ (for } x = 62) = \frac{62 - 70}{8} = -1.00$$

$$z \text{ (for } x = 74) = \frac{74 - 70}{8} = .5$$

Area $A_1 = .3413$. Area $A_2 = .1915$.
Total area $= A = A_1 + A_2 = .5328$.

21. $\mu = 70$, $\sigma = 8$

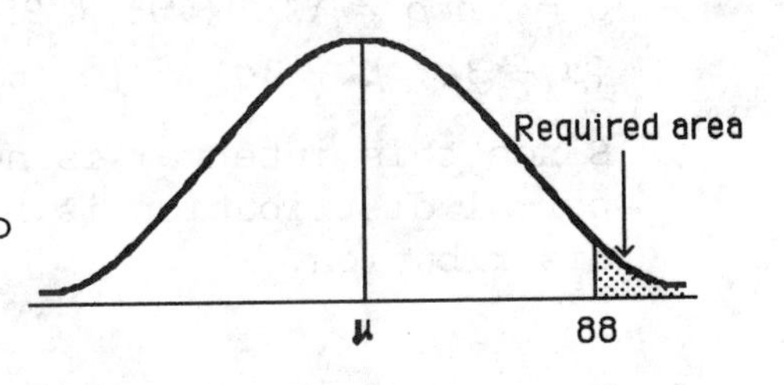

$$z \text{ (for } x = 88) = \frac{88 - 70}{8} = 2.25$$

Required area $= .5 -$ (area corresponding to $z = 2.25$)
$= .5 - .4878$
$= .0122$

23. $\mu = 70$, $\sigma = 8$

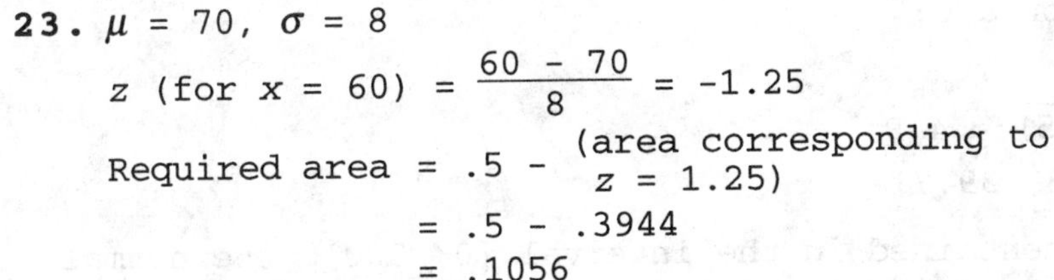

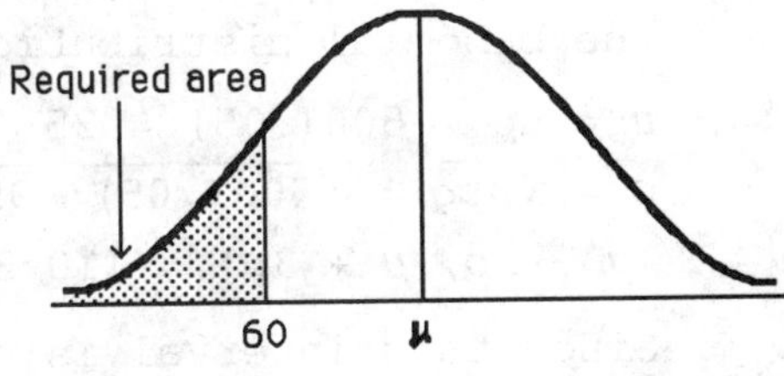

$$z \text{ (for } x = 60) = \frac{60 - 70}{8} = -1.25$$

Required area $= .5 -$ (area corresponding to $z = 1.25$)
$= .5 - .3944$
$= .1056$

25. (A) False: The shape of a normal distribution is determined by the mean, μ, and the standard deviation, σ. For example, a small standard deviation indicates a clustering around the mean and thus a tall, narrow curve; a large standard deviation indicates a large dispersion from the mean and thus a broad, flat curve.

(B) True: The area between the normal curve and the x axis is **1** for *all* normal distributions.

27. With $n = 15$, $p = .7$, and $q = .3$, the mean and standard deviation of the binomial distribution are:

$\mu = np = 10.5$

$\sigma = \sqrt{npq} = \sqrt{(15)(.7)(.3)} \approx 1.8$

$[\mu - 3\sigma, \mu + 3\sigma] = [5.1, 15.9]$

Since this interval is not contained in the interval [0, 15], the normal distribution should *not* be used to approximate the binomial distribution.

29. With $n = 15$, $p = .4$, and $q = .6$, the mean and standard deviation of the binomial distribution are:

$\mu = np = 15(.4) = 6$

$\sigma = \sqrt{npq} = \sqrt{15(.4)(.6)} \approx 1.9$

$[\mu - 3\sigma, \mu + 3\sigma] = [.3, 11.7]$

Since this interval is contained in the interval [0, 15], the normal distribution *is* a suitable approximation for the binomial distribution.

31. With $n = 100$, $p = .05$, and $q = .95$, the mean and standard deviation of the binomial distribution are:

$\mu = np = 100(.05) = 5$

$\sigma = \sqrt{npq} = \sqrt{100(.05)(.95)} \approx 2.2$

$[\mu - 3\sigma, \mu + 3\sigma] = [-1.6, 11.6]$

Since this interval is not contained in the interval [0, 100], the normal distribution is *not* a suitable approximation for the binomial distribution.

33. With $n = 500$, $p = .05$, and $q = .95$, the mean and standard deviation of the binomial distribution are:

$\mu = np = 500(.05) = 25$

$\sigma = \sqrt{npq} = \sqrt{500(.05)(.95)} \approx 4.9$

$[\mu - 3\sigma, \mu + 3\sigma] = [10.3, 39.7]$

Since this interval is contained in the interval [0, 500], the normal distribution *is* a suitable approximation for the binomial distribution.

35. We have $p = 0.1$, $q = 0.9$. If n is the number of trials, then the mean $\mu = np = 0.1n$ and the standard deviation $\sigma = \sqrt{npq} = \sqrt{0.09n} = 0.3\sqrt{n}$.

By the Rule of Thumb test, we must have

$$0.1n - 3(0.3\sqrt{n}) \geq 0$$

and

$$0.1n + 3(0.3\sqrt{n}) \leq n$$

The second inequality is automatically satisfied. The first inequality is:

$$0.1n - 0.9\sqrt{n} \geq 0$$
$$\sqrt{n} \geq 9$$
$$n \geq 81$$

In Problems 37–41, $\mu = 500(.4) = 200$, *and* $\sigma = \sqrt{npq} = \sqrt{500(.4)(.6)} \approx 10.95$. *The intervals are adjusted as in Examples 17 and 18.*

37. z (for $x = 184.5$) $= \dfrac{184.5 - 200}{10.95} \approx -1.42$

z (for $x = 220.5$) $= \dfrac{220.5 - 200}{10.95} \approx 1.87$

Thus, the probability that the number of successes will be between 185 and 220
= area A_1 + area A_2
= (area corresponding to $z = 1.42$) + (area corresponding to $z = 1.87$)
= .4222 + .4693
= .8915
≈ .89

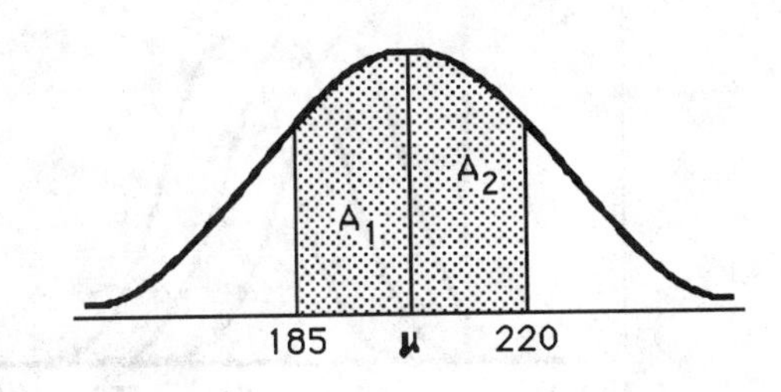

39. z (for $x = 209.5$) $= \dfrac{209.5 - 200}{10.95} \approx .87$

z (for $x = 220.5$) $= \dfrac{220.5 - 200}{10.95} \approx 1.87$

Thus, the probability that the number of successes will be between 210 and 220
= area A
= (area corresponding to $z = 1.87$) − (area corresponding to $z = .87$)
= .4693 − .3078
= .1615
≈ .16

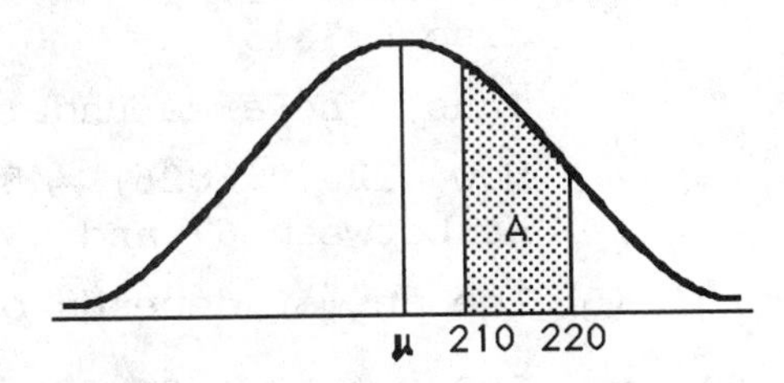

41. z (for $x = 224.5$) $= \dfrac{224.5 - 200}{10.95} \approx 2.24$

The probability that the number of successes will be 225 or more
= area A
= .5 − (area corresponding to $z = 2.24$)
= .5 − .4875
= .0125
≈ .01

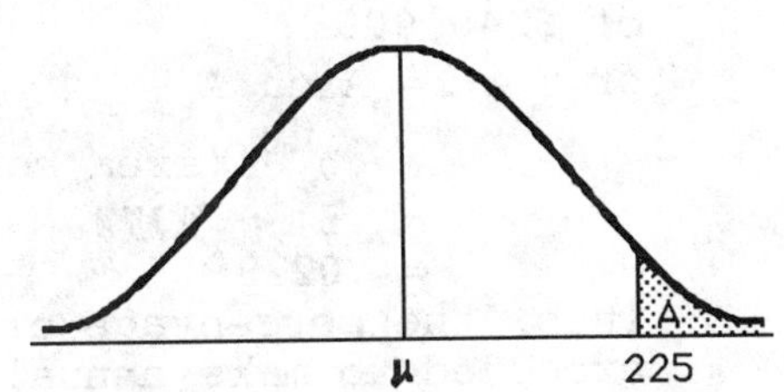

43. z (for $x = 175.5$) $= \dfrac{175.5 - 200}{10.95} \approx -2.24$

The probability that the number of successes will be 175 or less

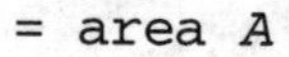

= area A
= .5 − (area corresponding to $z = 2.24$)
= .5 − .4875
= .0125
≈ .01

45.

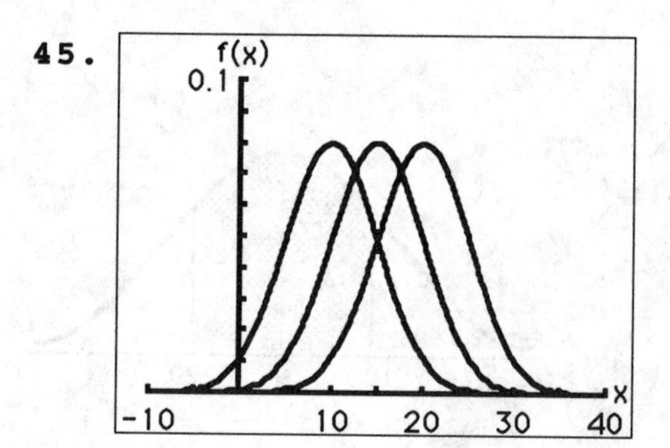

47.

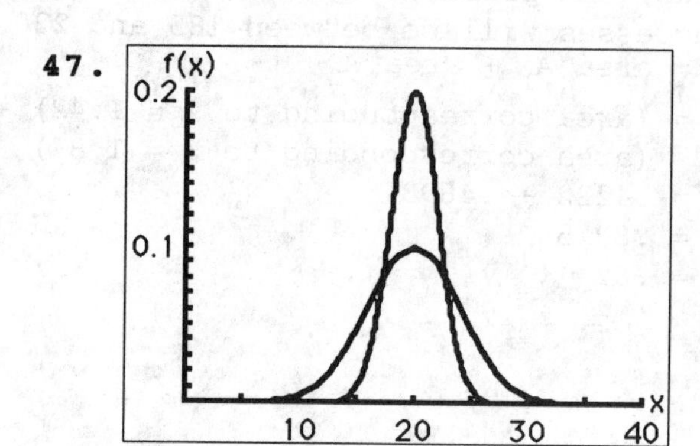

49. 120 scores are selected from a normal distribution with mean $\mu = 75$, and standard deviation $\sigma = 8$.

(A) The area under the normal curve between 75 and 83:

$$z = \frac{x - \mu}{\sigma} = \frac{83 - 75}{8} = 1; \text{ Area} = 0.3413 \quad \text{(Table 1)}$$

By symmetry, the area under the normal curve between 67 and 75 is also 0.3413.

Thus, the area under the normal curve between 67 and 83 is 0.6826.

Now, 120(0.6826) = 81.912; approximately 82 scores are expected to be between 67 and 83.

(B) The answer depends on the results of your simulation.

51. $\mu = 200{,}000$, $\sigma = 20{,}000$, $x \geq 240{,}000$

z (for $x = 240{,}000$) $= \dfrac{240{,}000 - 200{,}000}{20{,}000} = 2.0$

Fraction of the salespeople who would be expected to make annual sales of \$240,000
or more = Area A_1
= .5 − (area between μ and 240,000)
= .5 − .4772
= .0228

Thus, the percentage of salespeople expected to make annual sales of \$240,000 or more is 2.28%.

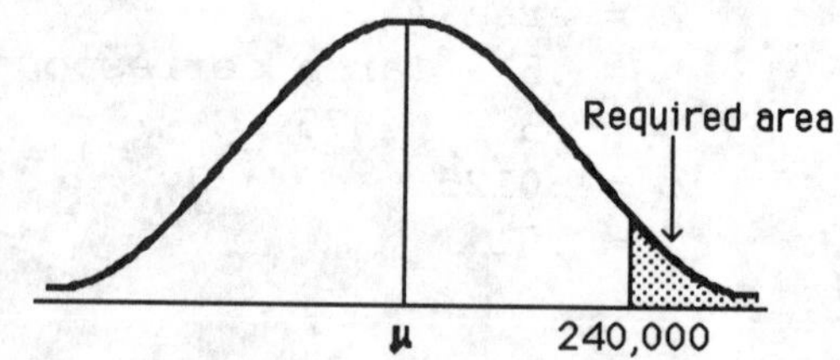

53. $x = 105$, $x = 95$, $\mu = 100$, $\sigma = 2$

$$z \text{ (for } x = 105) = \frac{105 - 100}{2} = 2.5$$

$$z \text{ (for } x = 95) = \frac{95 - 100}{2} = -2.5$$

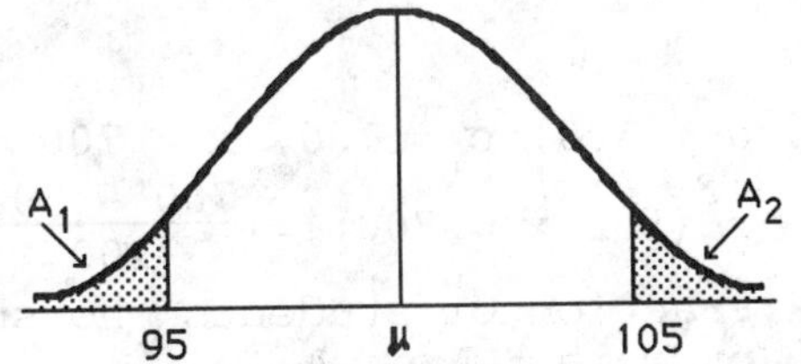

Fraction of parts to be rejected $= \text{Area } A_1 + A_2$
$= 1 - 2(\text{area corresponding to } z = 2.5)$
$= 1 - 2(.4938)$
$= .0124$

Thus, the percentage of parts to be rejected is 1.24%.

55. With $n = 40$, $p = .6$, and $q = .4$, the mean and standard deviation of the binomial distribution are:

$\mu = np = 40(.6) = 24$

$\sigma = \sqrt{npq} = \sqrt{40(.6)(.4)} \approx 3.10$

$$z \text{ (for } x = 15.5) = \frac{15.5 - 24}{3.1} = -2.74$$

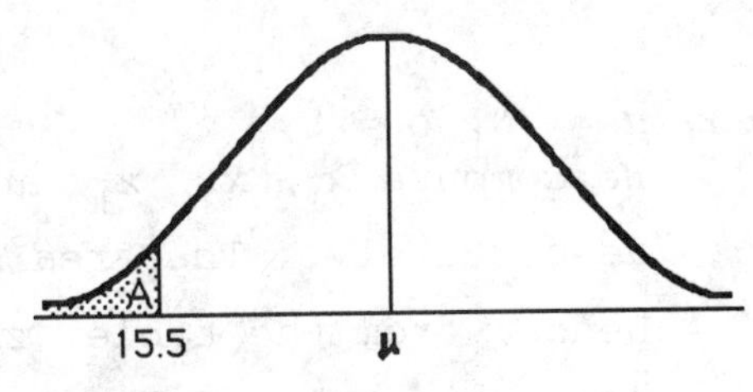

The probability that 15 or fewer households use the product
$= \text{area } A$
$= .5 - (\text{area corresponding to } z = 2.74)$
$= .5 - .4969$
$= .0031$

Either a rare event has occurred, e.g., the sample was not random, or the company's claim is false.

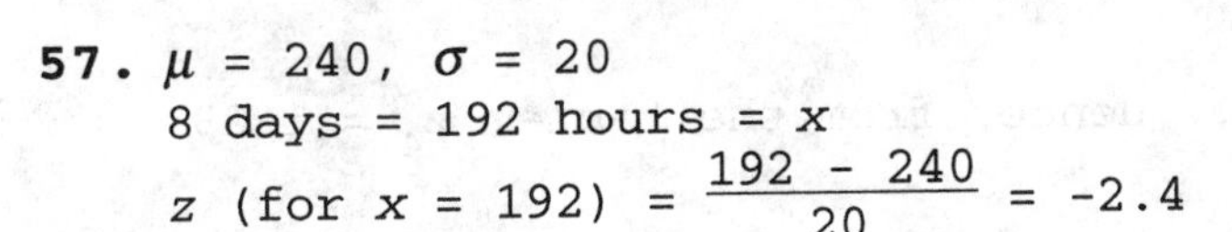

57. $\mu = 240$, $\sigma = 20$

8 days = 192 hours = x

$$z \text{ (for } x = 192) = \frac{192 - 240}{20} = -2.4$$

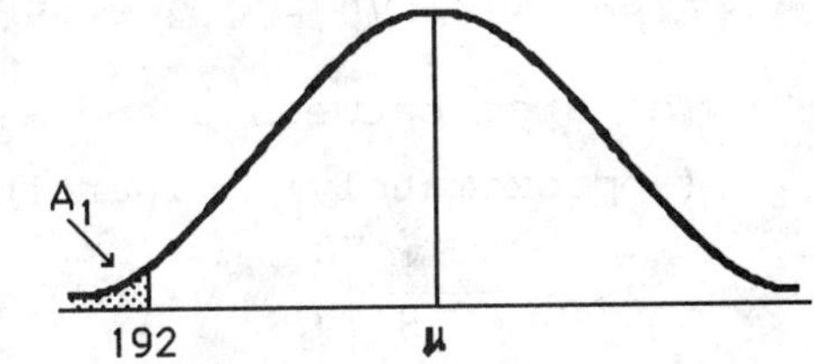

Fraction of people having this incision who would heal in 192 hours or less $= \text{Area } A_1$
$= .5 - (\text{area corresponding to } z = 2.4)$
$= .5 - .4918$
$= .0082$

Thus, the percentage of people who would heal in 8 days or less is .82%.

59. $p = .25$, $q = .75$, $n = 1000$

$\mu = np = (1000)(.25) = 250$

$\sigma = \sqrt{npq} = \sqrt{1000 \times .25 \times .75}$
$\approx 13.693 \approx 13.69$

$x = 220.5$ or less

$$z \text{ (for } x = 220.5) = \frac{220.5 - 250}{13.69} = -2.15$$

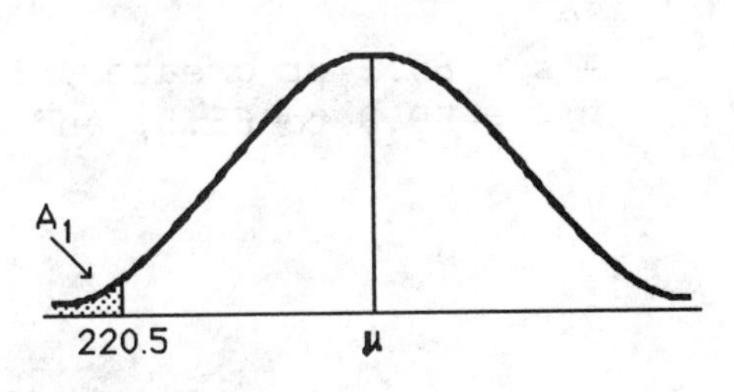

Probability that 220 or fewer families will have two girls $= \text{Area } A_1$
$= .5 - (\text{area corresponding to } z = 2.15)$
$= .5 - .4842$
$= .0158$

61. $\mu = 500$, $\sigma = 100$, $x = 700$ or more

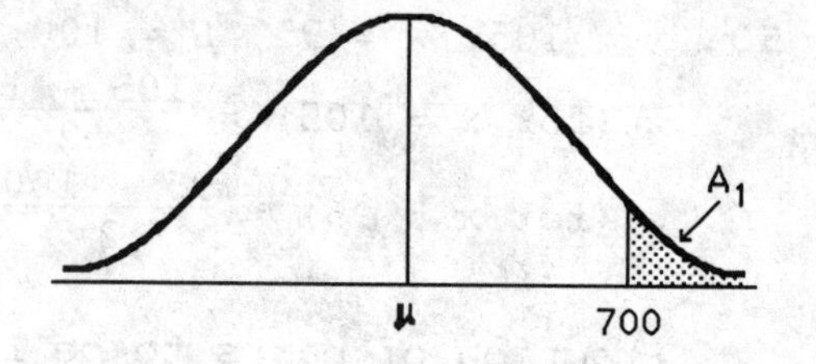

z (for $x = 700$) $= \dfrac{700 - 500}{100} = 2$

Fraction of students who should score 700 or more = Area A_1

$= .5 -$ (area corresponding to $z = 2$)
$= .5 - .4722$
$= .0228$

Thus, 2.28% should score 700 or more.

63. $\mu = 70$, $\sigma = 8$

We compute x_1, x_2, x_3, and x_4 corresponding to z_1, z_2, z_3, and z_4, respectively. The area between μ and x_3 is .2.

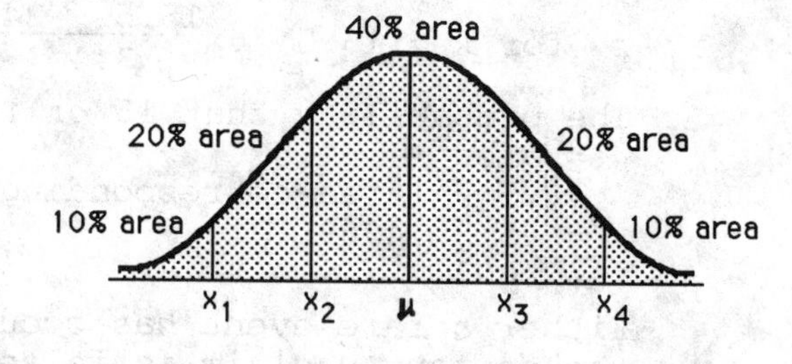

Hence, from the table, $z_3 = .52$ (approximately). Thus, we have:

$$.52 = \frac{x_3 - 70}{8}$$

$x_3 - 70 = 4.16$ $\left[\underline{\text{Note}}\text{: } z = \dfrac{x - \mu}{\sigma}.\right]$ ≈ 4.2

and $x_3 = 74.2$.

Also, $x_2 = 70 - 4.2 = 65.8$

The area between μ and x_4 is .4. Hence, from the table, $z_4 = 1.28$ (approximately). Therefore:

$$1.28 = \frac{x_4 - 70}{8}$$

$x_4 - 70 = 10.24 \approx 10.2$ and $x_4 = 70 + 10.2 = 80.2$.

Also, $x_1 = 70 - 10.2 = 59.8$.

Thus, we have $x_1 = 59.8$, $x_2 = 65.8$, $x_3 = 74.2$, $x_4 = 80.2$.

So,
A's = 80.2 or greater, *B*'s = 74.2 to 80.2, *C*'s = 65.8 to 74.2,
D's = 59.8 to 65.8, and *F*'s = 59.8 or lower.

1.

United Nations Peacekeeping

Number of Active Field Operations

25	
20	
15	
10	
5	
0	

1988 1989 1990 1991 1992 1993 1994

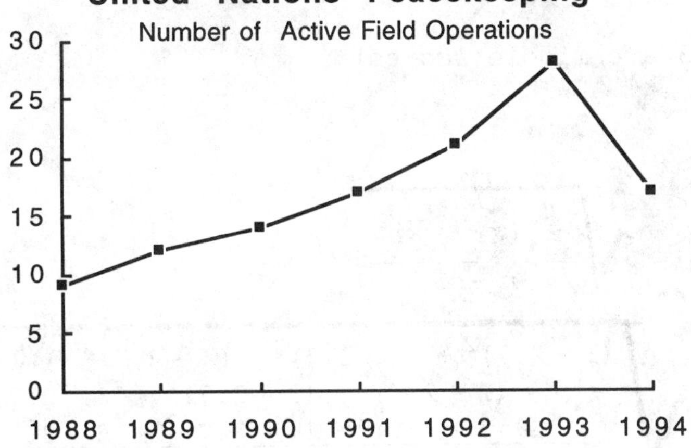

(7-1)

2.

Living Arrangements of the Elderly, 65 years and over

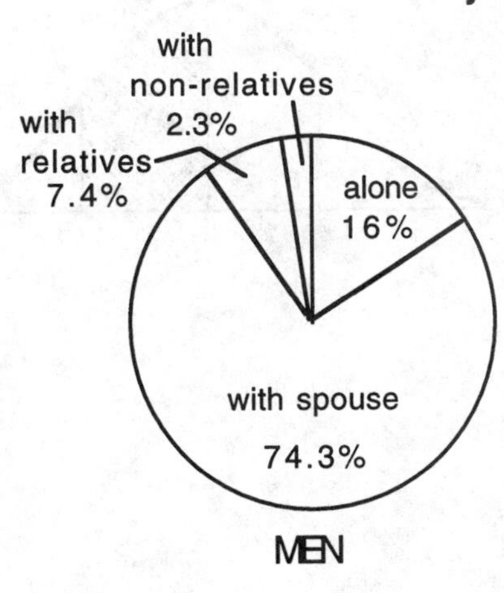

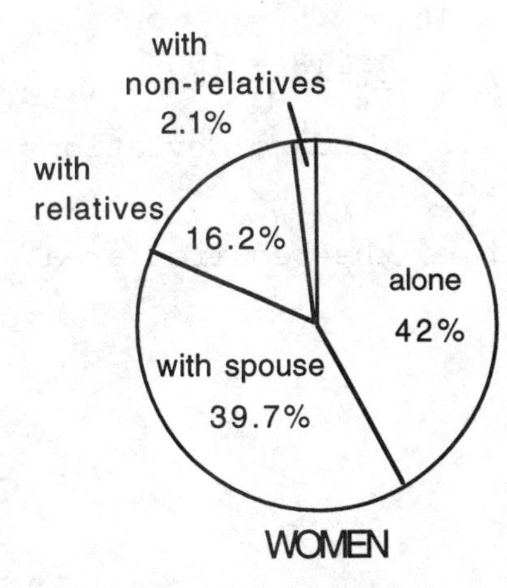

(7-1)

3. (A) $P(x) = C_{3,x}(.4)^x(.6)^{3-x}$

x	$P(x)$
0	.216
1	.432
2	.288
3	.064

The histogram for this distribution is shown at the right.

(B) $\mu = np = 3(.4) = 1.2$

$\sigma = \sqrt{npq} = \sqrt{3 \times .4 \times .6} \approx .85$

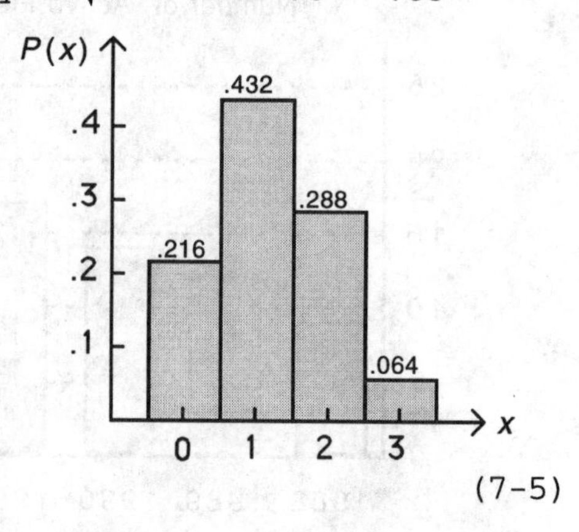

(7-5)

4. (A) Mean $= \bar{x} = \dfrac{1 + 1 + 2 + 2 + 2 + 3 + 3 + 4 + 4 + 5}{10} = \dfrac{27}{10} = 2.7$

(B) Median $= \dfrac{2 + 3}{2}$ (2 and 3 are middle scores)

$= 2.5$

(C) Mode $= 2$

(D) Standard deviation $= s = \sqrt{\dfrac{\sum_{i=1}^{n}(x_i - \bar{x})^2}{n - 1}}$

$$= \sqrt{\frac{(1 - 2.7)^2 + (1 - 2.7)^2 + (2 - 2.7)^2 + (2 - 2.7)^2 + (2 - 2.7)^2 + (3 - 2.7)^2 + (3 - 2.7)^2 + (4 - 2.7)^2 + (4 - 2.7)^2 + (5 - 2.7)^2}{10 - 1}}$$

≈ 1.34

(7-3, 7-4)

5. (A) $\mu = 100$, $\sigma = 10$

z (for $x = 118$) $= \dfrac{118 - 100}{10} = 1.8$

$x = 118$ is 1.8 standard deviations from the mean.

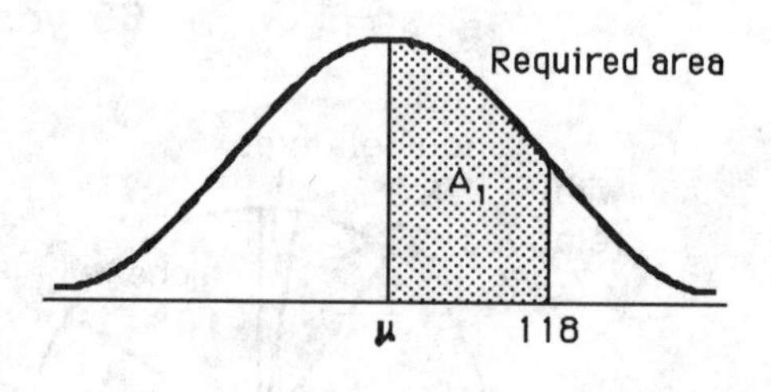

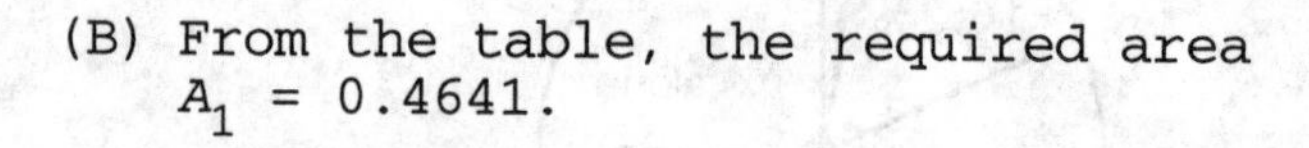

(B) From the table, the required area $A_1 = 0.4641$.

(7-6)

6. (A) The frequency and relative frequency table for the given data is shown below.

Interval	Frequency	Relative frequency
9.5-11.5	1	.04
11.5-13.5	5	.20
13.5-15.5	12	.48
15.5-17.5	6	.24
17.5-19.5	1	.04
	25	1.00

(B) The histogram below shows both frequency and relative frequency scales on the y axis.

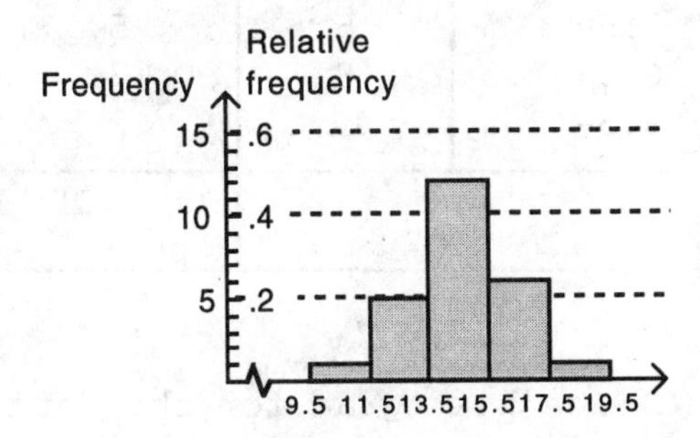

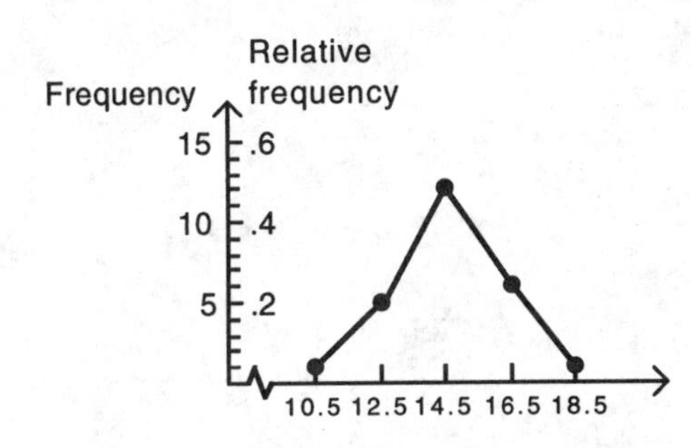

(C) The polygon graph to the left also shows both frequency and relative frequency scales on the y axis.

(D) A cumulative and relative cumulative table for the data is given below.

Interval	Frequency	Cumulative frequency	Relative cumulative frequency
9.5-11.5	1	1	.04
11.5-13.5	5	6	.24
13.5-15.5	12	18	.72
15.5-17.5	6	24	.96
17.5-19.5	1	25	1.00

(E) The polygon graph at the right shows both the cumulative frequency and relative cumulative frequency scales on the y axis.

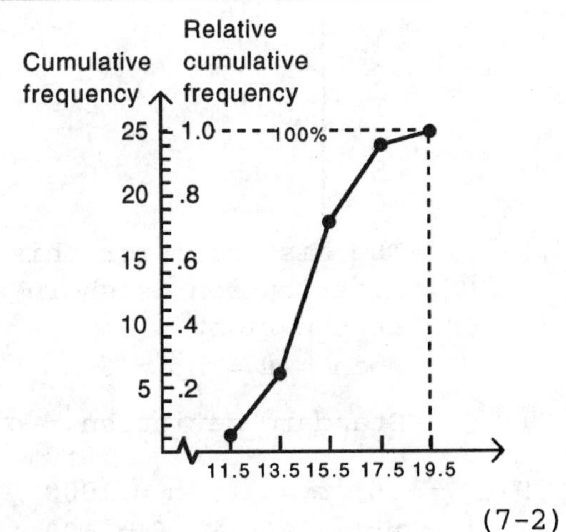

(7-2)

7.

Interval	Midpoint x_i	Frequency f_i	$x_i f_i$	$(x_i - \overline{x})^2$	$(x_i - \overline{x})^2 f_i$
0.5- 3.5	2	1	2	25	25
3.5- 6.5	5	5	25	4	20
6.5- 9.5	8	7	56	1	7
9.5-12.5	11	2	22	16	32
		$n = 15$	$\sum_{i=1}^{4} x_i f_i = 105$		$\sum_{i=1}^{4} (x_i - \overline{x})^2 f_i = 84$

(A) $\overline{x} = \frac{105}{15} = 7$ (B) $s = \sqrt{\frac{84}{14}} \approx 2.45$

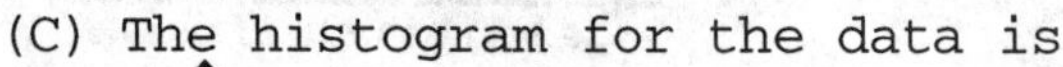

(C) The histogram for the data is

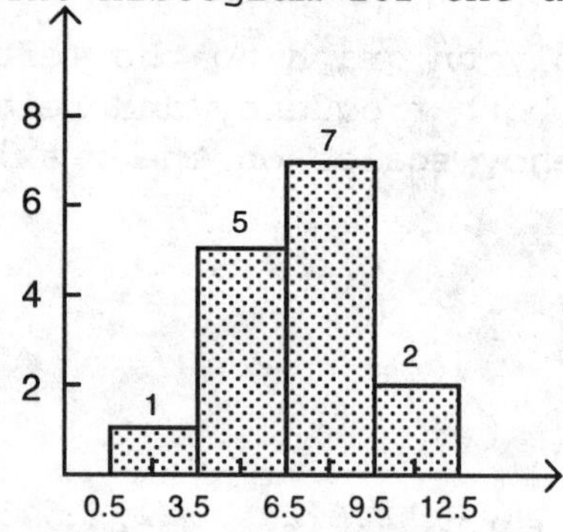

Total area = 3(1) + 3(5) + 3(7) + 3(2) = 45.

Let M be the median. Then

$$3(1) + 3(5) + (M - 6.5)7 = 22.5$$
$$(M - 6.5)7 = 4.5$$
$$M - 6.5 = 0.643$$
$$M = 7.143$$

(7-3, 7-4)

8. (A) $P(x) = C_{6,x}(.5)^x(.5)^{6-x}$

x	$P(x)$
0	.016
1	.094
2	.234
3	.313
4	.234
5	.094
6	.016

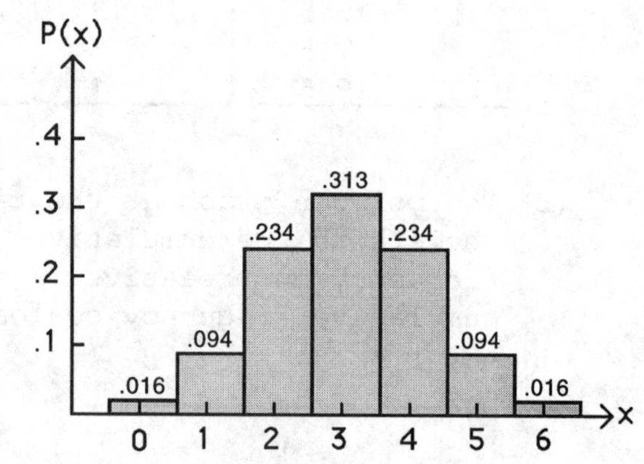

The histogram for this distribution is shown at the right.

(B) Mean = $\mu = np = 6 \times .5 = 3$

Standard deviation = $\sigma = \sqrt{npq} = \sqrt{6 \times .5 \times .5} \approx 1.225$ (7-5)

9. $p = .6,\ q = .4,\ n = 1000$

$\mu = np = 1000 \times .6 = 600$

$s = \sqrt{npq} = \sqrt{1000 \times .6 \times .4} = \sqrt{240} \approx 15.49$ (7-5)

10. (A) True: $\bar{x} = \frac{x_1 + x_2 + \dots + x_n}{n}$;

$$\frac{(x_1 + 5) + (x_2 + 5) + \dots + (x_n + 5)}{n} = \frac{x_1 + x_2 + \dots + x_n + 5n}{n}$$
$$= \frac{x_1 + x_2 + \dots + x_n}{n} + 5$$
$$= \bar{x} + 5$$

(B) False: Suppose that the data set $x_1, x_2, \dots, x_n$ has mean $\bar{x}$ and standard deviation 5. Then the data set $x_1 + 5, x_2 + 5, \dots, x_n + 5$ has mean $\bar{x} + 5$ [part (A)], but the standard deviation s' will remain the same, $s' = s$:

$$s' = \sqrt{\frac{\sum_{i=1}^{n} (\bar{x}_i + 5 - [\bar{x} + 5])^2}{n}} = \sqrt{\frac{\sum_{i=1}^{n} (x_i - \bar{x})^2}{n}} = s$$ (7-3, 7-4)

11. (A) False: If X is a binomial random variable with mean μ, then $P(X \geq \mu) = 0.5$ only if $p = q = 0.5$; any binomial distribution with $p \neq 0.5$ will serve as a counter-example.

(B) True: The normal distribution is symmetric about the mean.

(C) True: The area is 1 in each case. (7-5, 7-6)

12. The mean μ and the standard deviation σ for the binomial distribution are:

$\mu = np = 1000 \times .6 = 600$

$\sigma = \sqrt{npq} = \sqrt{1000 \times .6 \times .4} \approx 15.49 \approx 15.5$

Now, we approximate the binomial distribution with a normal distribution.

z (for $x = 550$) $= \frac{549.5 - 600}{15.5} = -3.26$

z (for $x = 650$) $= \frac{650.5 - 600}{15.5} = 3.26$

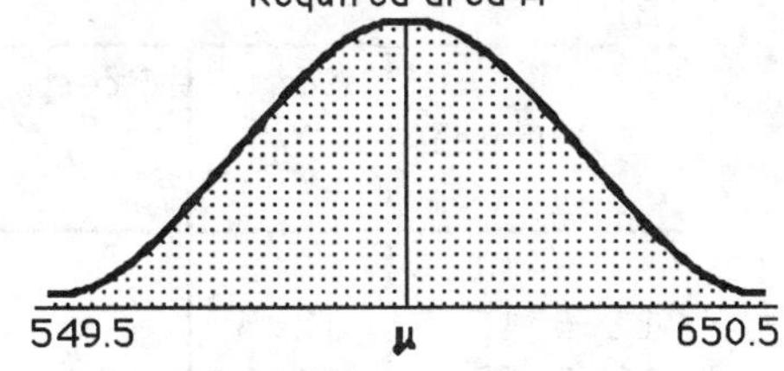

The probability of obtaining successes between 550 and 650 = Area A

= 2(area corresponding to $z = 3.26$)
= 2(.4994)
= .9988 ≈ .999 (7-6)

13. (A) $\mu = 50$, $\sigma = 6$

z (for $x = 41$) $= \frac{41 - 50}{6} = -1.5$

z (for $x = 62$) $= \frac{62 - 50}{6} = 2.0$

Required area $= A_1 + A_2$
= (area corresponding to $z = 1.5$) + (area corresponding to $z = 2$)
= .4332 + .4772 = .9104

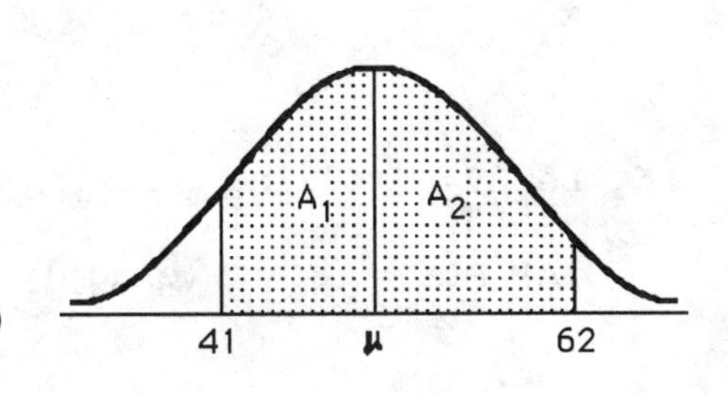

(B) z (for $x = 59$) $= \dfrac{59 - 50}{6} = 1.5$

Required area $= .5 -$ (area corresponding to $z = 1.5$)
$= .5 - .4332 = .0668$

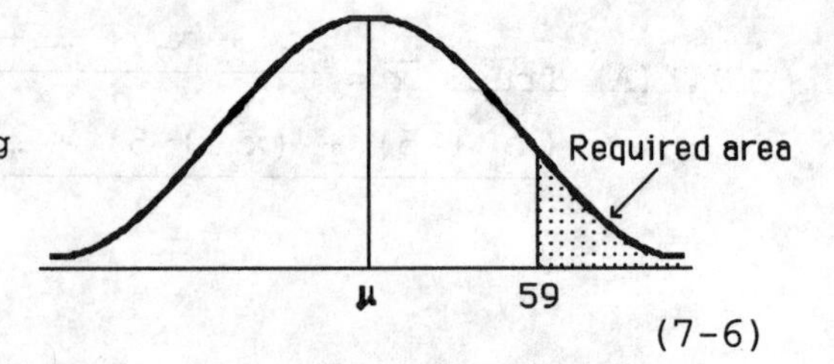

(7-6)

14. (A) We would expect the first data set to have smaller standard deviation because the values range between 2 and 12, while in the second data set, the values range between 1 and 36.

(B) The answer depends on the results of your simulation. (7-4)

15. (A)

x_i	Frequency f_i	$x_i f_i$	$(x_i - \bar{x})^2$ $(\bar{x} = 14.6)$	$(x_i - \bar{x})^2 f_i$
11	1	11	12.96	12.96
12	2	24	6.76	13.52
13	3	39	2.56	7.68
14	7	98	.36	2.52
15	5	75	.16	.80
16	3	48	1.96	5.88
17	3	51	5.76	17.28
19	1	19	19.36	19.36
	$n = 25$	$\sum_{i=1}^{8} x_i f_i = 365$		$\sum_{i=1}^{8} (x_i - \bar{x})^2 f_i = 80$

Thus, $\bar{x} = \dfrac{365}{25} = 14.6$ $\qquad s = \sqrt{\dfrac{80}{24}} \approx 1.83$

(B)

Interval	Midpt. x_i	Freq. f_i	$x_i f_i$	$(x_i - \bar{x})^2$ $(\bar{x} = 14.6)$	$(x_i - \bar{x})^2 f_i$
9.5-11.5	10.5	1	10.5	16.81	16.81
11.5-13.5	12.5	5	62.5	4.41	22.05
13.5-15.5	14.5	12	174.0	.01	.12
15.5-17.5	16.5	6	99.0	3.61	21.66
17.5-19.5	18.5	1	18.5	15.21	15.21
		$n = 25$	$\sum_{i=1}^{5} x_i f_i = 364.5$		$\sum_{i=1}^{5} (x_i - \bar{x})^2 f_i = 75.85$

$\bar{x} = \dfrac{364.5}{25} \approx 14.6$ $\qquad s = \sqrt{\dfrac{75.85}{24}} \approx 1.78$ (7-3, 7-4)

16. Let E = "rolling a six." Then $P(E) = \dfrac{1}{6}$, $P(E') = \dfrac{5}{6}$. Thus, $p = \dfrac{1}{6}$, $q = \dfrac{5}{6}$, $n = 5$.

(A) $P(X = 3) = P(\text{exactly three 6's}) = C_{5,3}\left(\dfrac{1}{6}\right)^3\left(\dfrac{5}{6}\right)^2 = \dfrac{5!}{3!2!}\left(\dfrac{25}{6^5}\right) \approx .0322$

(B) $P(\text{at least three 6's}) = P(x \geq 3) = P(3) + P(4) + P(5)$

$$= C_{5,3}\left(\frac{1}{6}\right)^3\left(\frac{5}{6}\right)^2 + C_{5,4}\left(\frac{1}{6}\right)^4\left(\frac{5}{6}\right) + C_{5,5}\left(\frac{1}{6}\right)^6$$

$$= \frac{5!}{3!2!}\left(\frac{25}{6^5}\right) + \frac{5!}{4!1!}\left(\frac{5}{6^5}\right) + \frac{5!}{5!0!}\left(\frac{1}{6^5}\right) \approx .0355 \quad (7\text{-}5)$$

17. The probability of getting a 7 is: $p = \frac{6}{36} = \frac{1}{6}$

Thus, the probability of not getting a 7 is: $q = \frac{5}{6}$

Now, $P(\text{at least one 7}) = P(x \geq 1) = 1 - P(x < 1)$

$$= 1 - P(0)$$

$$= 1 - C_{3,0}\left(\frac{1}{6}\right)^0\left(\frac{5}{6}\right)^3$$

$$= 1 - \frac{125}{6^3}$$

$$\approx 1 - .5787 = .4213 \quad (7\text{-}5)$$

18. (A) The set of scores {10, 10, 20, 20, 90, 90, 90, 90, 90, 90} has the specified property:

$$\overline{x} = \frac{2(10) + 2(20) + 6(90)}{10} = 60$$

Median = mode = 90

The scores {5, 10, 10, 15, 85, 85, 85, 85, 85, 85} also have this property:

$\overline{x} = 55$, Median = mode = 85.

(B) Let $x_1, x_2, x_3, \ldots, x_{10}$ be the 10 exam scores arranged in increasing order. Then

$$\overline{x} = \frac{x_1 + x_2 + x_3 + \cdots + x_{10}}{10} \quad \text{and} \quad M = \frac{x_5 + x_6}{2}$$

If $\overline{x} + 50 = M$, then

$$x_1 + x_2 + x_3 + x_4 + x_5 + x_6 + x_7 + x_8 + x_9 + x_{10} + 500 = 5x_5 + 5x_6$$

and

$$x_1 + x_2 + x_3 + x_4 + x_7 + x_8 + x_9 + x_{10} + 500 = 4x_5 + 4x_6$$

Now, $x_6 \leq x_7 \leq x_8 \leq x_9 \leq x_{10}$, so $4x_6 \leq x_7 + x_8 + x_9 + x_{10}$

Therefore, $x_1 + x_2 + x_3 + x_4 + 500 \leq 4x_5 \leq 400$

This implies $x_1 + x_2 + x_3 + x_4 \leq -100$

which is impossible, since $x_1, x_2, x_3, x_4 \geq 0$

Thus, the median cannot be 50 points greater than the mean.

The mode could be 50 points greater than the mean.

Consider the exam scores:

0, 1, 2, 3, 4, 5, 6, 7, 8, 100, 100

$\overline{x} = 23.6$, mode = 100 (7-3)

19. (A) This is a binomial experiment with $p = 0.39$, $q = 0.61$, and $n = 20$.

$$P(8) = C_{20,8}(0.39)^8(0.61)^{12} \approx 0.179$$

(B) $\mu = 7.8$, $\sigma = 2.18$; $\mu - 3\sigma = 1.26$, $\mu + 3\sigma = 14.34$

The interval [1.26, 14.34] lies in the interval [0, 20]. Therefore, the normal distribution can be used to approximate the binomial distribution.

(C) The normal distribution is continuous, not discrete. Therefore, the correct analogue of (A) is

$$P(7.5 \le x \le 8.5) \approx 0.18$$

using the table in Appendix C. (7-6)

20. Given $p = 0.9$, $q = 0.1$, $n = 12$

(A) Mean: $\mu = np = 12(0.9) = 10.8$

Standard deviation: $\sigma = \sqrt{npq} = \sqrt{12(0.9)(0.1)} \approx 1.039$.

(B) Probability of 12 wins:

$$P(12) = C_{12,12}(0.9)^{12}(0.1)^0 = (0.9)^{12} \approx 0.282$$

(C) The answer depends on the results of your simulation. (7-5)

21. Arrange the given numbers in increasing order:
4, 5, 5, 5, 8, 12, 13, 15, 16, 17

(A) Mean $\bar{x} = \dfrac{4 + 5 + 5 + 5 + 8 + 12 + 13 + 15 + 16 + 17}{10} = 10$

(B) Median $= \dfrac{8 + 12}{2} = 10$

(C) Mode = 5

(D) Standard deviation:

$$s = \sqrt{\frac{(4-10)^2 + 3(5-10)^2 + (8-10)^2 + (12-10)^2 + (13-10)^2 + (15-10)^2 + (16-10)^2 + (17-10)^2}{10-1}}$$

$$= \sqrt{\frac{36 + 75 + 4 + 4 + 9 + 25 + 36 + 49}{9}} = \sqrt{\frac{238}{9}} \approx 5.14 \quad (7\text{-}3,\ 7\text{-}4)$$

22. The mean and median are not suitable for these data. The modal preference is soft drink. (7-3)

23. (A) The frequency and relative frequency table is as follows:

Interval	Frequency f_i	Relative frequency
29.5-31.5	3	.086
31.5-33.5	7	.2
33.5-35.5	14	.4
35.5-37.5	7	.2
37.5-39.5	4	.114
	$\sum_{i=1}^{5} f_i = 35$	Sum = 1.00

(B)

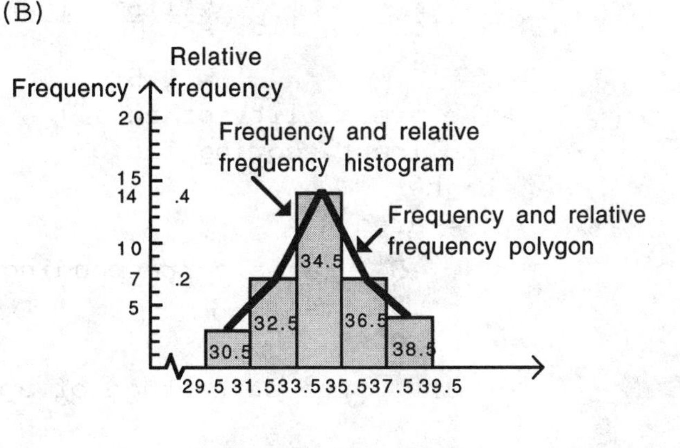

(C)

Interval	Midpt. x_i	Freq. f_i	$x_i f_i$	$(x_i - \bar{x})^2$ ($\bar{x}$ = 34.61)	$(x_i - \bar{x})^2 f_i$
29.5-31.5	30.5	3	91.5	16.89	50.67
31.5-33.5	32.5	7	227.5	4.45	31.15
33.5-35.5	34.5	14	483.0	.01	.14
35.5-37.5	36.5	7	255.5	3.57	24.99
37.5-39.5	38.5	4	154.0	15.13	60.53
		$n = 35$	$\sum_{i=1}^{5} x_i f_i = 1211.5$		$\sum_{i=1}^{5} (x_i - \bar{x})^2 f_i = 167.48$

$\bar{x} = \frac{1211.5}{35} = 34.61$ $\quad s = \sqrt{\frac{167.48}{34}} = 2.22$ (7-2, 7-3, 7-4)

24. $\mu = 100$, $\sigma = 10$

(A) $z \text{ (for } x = 92) = \frac{92 - 100}{10} = -.8$

$z \text{ (for } x = 108) = \frac{108 - 100}{10} = .8$

The probability of an applicant scoring between 92 and 108

= area A

= 2·area A_1

= 2(area corresponding to z = .8)

= 2(.2881) = .5762

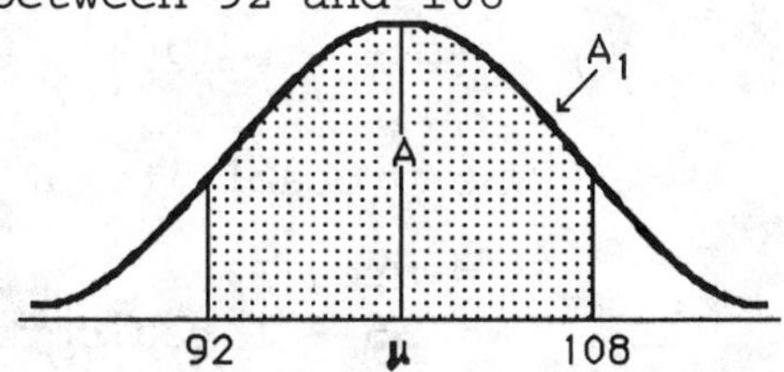

Thus, the percentage of applicants scoring between 92 and 108 is 57.62%.

(B) z (for x = 115) = $\frac{115 - 100}{10}$ = 1.5

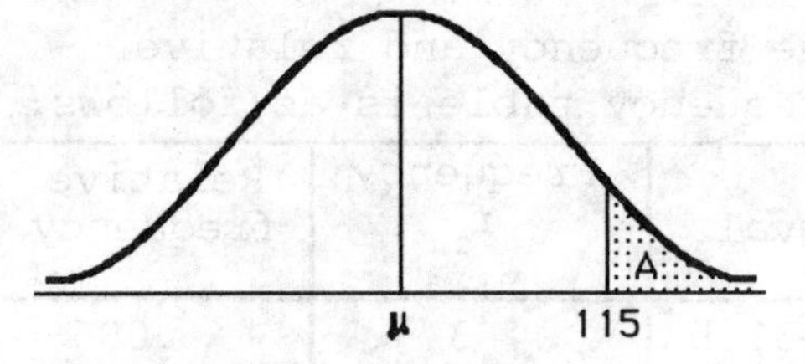

The probability of an applicant scoring 115 or higher
= area A
= .5 - (area corresponding to z = 1.5)
= .5 - .4332
= .0668

Thus, the percentage of applicants scoring 115 or higher is 6.68%. (7-6)

25. Based on the publisher's claim, the probability that a person selected at random reads the newspaper is: p = .7.

Thus, the probability that a randomly selected person does not read the newspaper is: q = .3.

(A) With n = 200, the mean μ = 200(.7) = 140, and standard deviation $\sigma = \sqrt{200(.7)(.3)} = \sqrt{42} \approx 6.48$.

(B) $[\mu - 3\sigma, \mu + 3\sigma]$ = [120.76, 159.24] and the interval lies entirely in the interval [0, 200]. Thus, the normal distribution *does* provide an adequate approximation to the binomial distribution.

(C) z (for x = 129.5) = $\frac{129.5 - 140}{6.48} = \frac{-10.5}{6.48}$ = -1.62

z (for x = 155.5) = $\frac{155.5 - 140}{6.48} = \frac{15.5}{6.48}$ = 2.39

The probability of finding between 130 and 155 readers in the sample
= area A_1 + area A_2
= (area corresponding to z = 1.62) + (area corresponding to z = 2.39)
= .4474 + .4916
= .9390

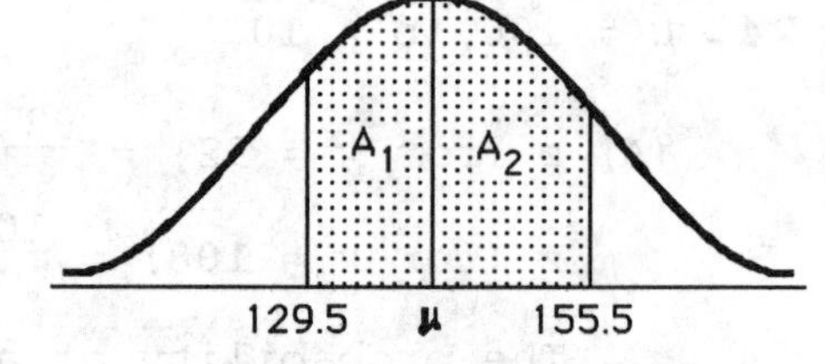

(D) z (for x = 125.5) = $\frac{125.5 - 140}{6.48} = \frac{-14.5}{6.48}$ = -2.24

The probability of finding 125 or fewer readers in the sample
= area A
= .5 - (area corresponding to z = 2.24)
= .5 - .4875
= .0125

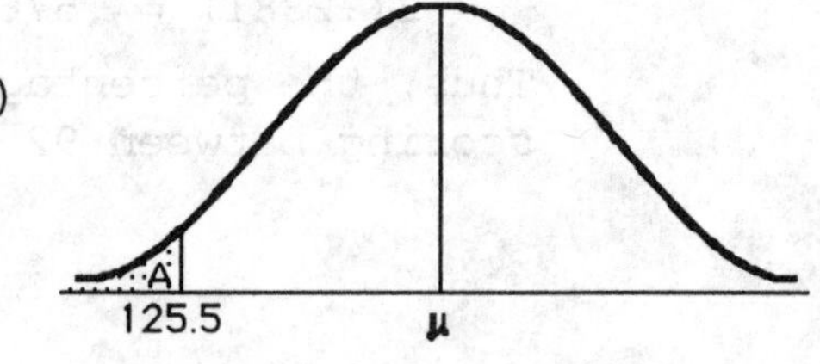

(E)

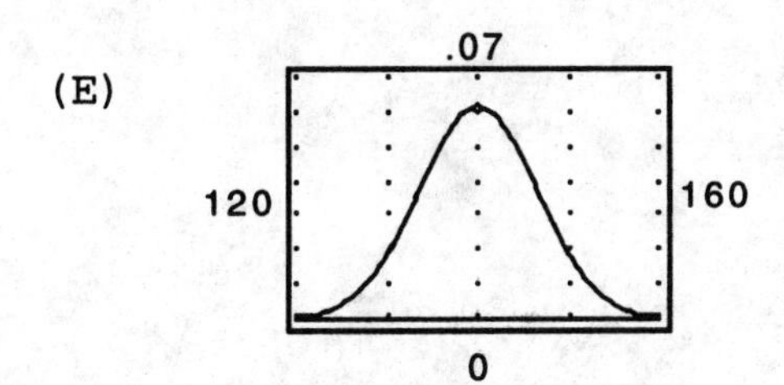

(7-5, 7-6)

26. $p = .9$, $q = .1$, and $n = 3$

$$\begin{aligned} P(x \geq 1) &= 1 - P(x < 1) \\ &= 1 - P(0) \\ &= 1 - C_{3,0}\left(\frac{9}{10}\right)^0\left(\frac{1}{10}\right)^3 \\ &= 1 - \frac{1}{1000} \\ &= .999 \end{aligned}$$

(7-5)

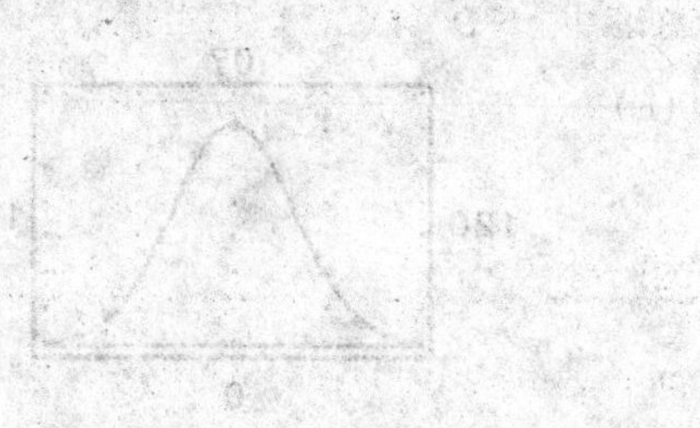

8 GAMES AND DECISIONS

EXERCISE 8-1

Things to remember:

1. STRICTLY DETERMINED MATRIX GAMES

   A matrix game is STRICTLY DETERMINED if a payoff value is simultaneously a row minimum and a column maximum. If such a value exists, it is called a SADDLE VALUE. In a strictly determined game, the OPTIMAL STRATEGIES are:

   R should choose any row containing a saddle value
   C should choose any column containing a saddle value

   A saddle value is called the VALUE of a strictly determined game. The game is FAIR if its value is zero. [Note: In a strictly determined game (assuming both players play their optimal strategy), knowledge of an opponent's move provides no advantage, since the payoff will always be a saddle value.]

2. LOCATING SADDLE VALUES:

   (a) Circle the minimum value in each row (it may occur in more than one place).
   (b) Place squares around the maximum value in each column (it may occur in more than one place).
   (c) Any entry with both a circle and a square around it is a saddle value.

3. If a game matrix has two or more saddle values, then they are equal.

4. A matrix game is said to be NONSTRICTLY DETERMINED if it does not have saddle values.

---

1. $\begin{bmatrix} \boxed{3} & \boxed{\textcircled{2}} \\ 2 & \textcircled{-1} \end{bmatrix}$ (using 4)

The game is strictly determined.

(A) 2 (upper right or 1,2 position)
(B) *R* plays row 1 and *C* plays column 2.
(C) The value of the game is 2.

3. 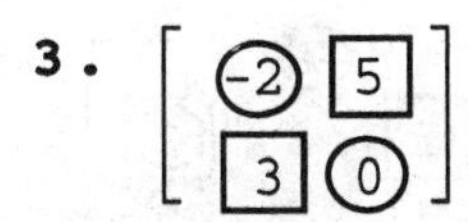

The game is not strictly determined.

5. $\begin{bmatrix} \textcircled{-3} & 0 \\ \boxed{4} & \boxed{\textcircled{1}} \end{bmatrix}$

The game is strictly determined.

(A) 1 (2,2 position)

(B) R plays row 2 and C plays column 2.

(C) The value of the game is 1.

7. $\begin{bmatrix} 1 & 1 \\ 1 & 1 \end{bmatrix}$

(A) Since all entries are equal, each entry is a saddle value.

(B) R plays either Row 1 or Row 2; C plays either Column 1 or Column 2.

(C) Value of the game: 1.

9. 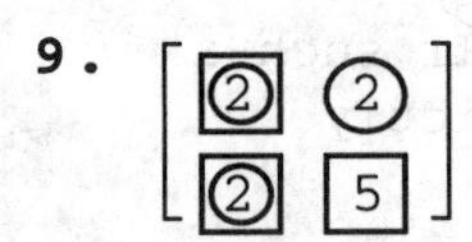

The game is strictly determined.

(A) Both 2's in column 1.

(B) R plays either row 1 or row 2 and C plays column 1.

(C) The value of the game is 2.

11. $\begin{bmatrix} \boxed{2} & -1 & \textcircled{-5} \\ 1 & \boxed{\textcircled{0}} & 3 \\ -3 & \textcircled{-7} & \boxed{8} \end{bmatrix}$

The game is strictly determined.

(A) 0 (in 2,2 position).

(B) R plays row 2 and C plays column 2.

(C) The value of the game is 0.

13. $\begin{bmatrix} 3 & \textcircled{-2} \\ \textcircled{1} & \boxed{5} \\ \textcircled{-4} & 0 \\ \boxed{5} & \textcircled{-3} \end{bmatrix}$

The game is not strictly determined.

15. $\begin{bmatrix} 3 & -1 & \boxed{4} & \textcircled{-7} \\ 1 & \boxed{\textcircled{0}} & 2 & 3 \\ \boxed{5} & -2 & \textcircled{-3} & 0 \\ 3 & \boxed{\textcircled{0}} & 1 & \boxed{5} \end{bmatrix}$

The game is strictly determined.

(A) Both 0's in column 2 (2,2 and 4,2 positions)

(B) R plays either row 2 or row 4 and C plays column 2.

(C) The value of the game is 0.

17. $\begin{bmatrix} 0 & 4 & -8 & -3 \\ 2 & 5 & 3 & 2 \\ 1 & -3 & -2 & -9 \\ 2 & 4 & 7 & 2 \end{bmatrix}$

(A) $\begin{bmatrix} 0 & 4 & \textcircled{-8} & -3 \\ \boxed{\textcircled{2}} & \boxed{5} & 3 & \boxed{\textcircled{2}} \\ 1 & -3 & -2 & \textcircled{-9} \\ \boxed{\textcircled{2}} & 4 & \boxed{7} & \boxed{\textcircled{2}} \end{bmatrix}$

Saddle value: 2

(B) Optimal strategy for R: play Row 2 or Row 4.
Optimal strategy for C: play Column 1 or Column 4.

(C) The value of the game: 2

19. Since the value of the game is positive, Player R has the advantage.

21. $\begin{bmatrix} -3 & m \\ \boxed{0} & 1 \end{bmatrix}$ No; 0 is a saddle value irrespective of the value of m.

23.

		Station C	
		\$1.00	\$1.05
Station R	\$1.05	50% (circled, boxed)	70% (boxed)
	\$1.10	40% (circled)	50%

The saddle value is 50%, as shown in the upper left hand corner (1,1 position).

Optimum strategies: R plays row 1 (\$1.05) and C plays column 1 (\$1.00).

25.

Store C

$$\text{Store } R \begin{array}{c} \text{T.C.} \\ \text{I.V.} \\ \text{S.L.T.} \end{array} \overset{\begin{array}{ccc} \text{T.C.} & \text{I.V.} & \text{S.L.T.} \end{array}}{\begin{bmatrix} 50\% & 20\% + \frac{50\%}{2} & 20\% + \frac{30\%}{2} \\ 30\% + \frac{50\%}{2} & 50\% & 30\% + \frac{20\%}{2} \\ 50\% + \frac{30\%}{2} & 50\% + \frac{20\%}{2} & 50\% \end{bmatrix}} = \begin{bmatrix} 50\% & 45\% & \textcircled{35\%} \\ 55\% & 50\% & \textcircled{40\%} \\ \boxed{65\%} & \boxed{60\%} & \boxed{\textcircled{50\%}} \end{bmatrix}$$

The optimal strategy for both stores is to locate in South Lake Tahoe and split the basin business equally.

EXERCISE 8-2

Things to remember:

1. Given the game matrix $M = \begin{bmatrix} a & b \\ c & d \end{bmatrix}$.

   R's strategy is denoted by a probability row matrix

   $$P = [p_1 \quad p_2],\ p_1 \geq 0,\ p_2 \geq 0,\ p_1 + p_2 = 1.$$

   C's strategy is denoted by a probability column matrix

   $$Q = \begin{bmatrix} q_1 \\ q_2 \end{bmatrix},\ q_1 \geq 0,\ q_2 \geq 0,\ q_1 + q_2 = 1.$$

2. EXPECTED VALUE OF A MATRIX GAME FOR R

   For the game matrix $M = \begin{bmatrix} a & b \\ c & d \end{bmatrix}$ and strategies $P = [p_1 \quad p_2]$, $Q = \begin{bmatrix} q_1 \\ q_2 \end{bmatrix}$ for R and C, respectively, the EXPECTED VALUE of the game for R is given by

   $$E(P,\ Q) = PMQ = ap_1q_1 + bp_1q_2 + cp_2q_1 + dp_2q_2.$$

3. FUNDAMENTAL THEOREM OF GAME THEORY: For every $m \times n$ game matrix M, there exists strategies P^* and Q^* (not necessarily unique) for R and C, respectively, and a unique number v such that:

$P^*MQ \geq v$ for every strategy Q of C;

$PMQ^* \leq v$ for every strategy P of R.

The number v is called the VALUE of the game. The triplet (v, P^*, Q^*) is called a SOLUTION OF THE GAME. The expected value of the game is

$E(P^*, Q^*) = P^*MQ^* = v.$

4. SOLUTION TO A 2 × 2 NONSTRICTLY DETERMINED MATRIX GAME

For the nonstrictly determined game matrix $M = \begin{bmatrix} a & b \\ c & d \end{bmatrix}$, the optimal strategies P^* and Q^* and the value v are given by

$$P^* = \begin{bmatrix} \frac{d - c}{D} & \frac{a - b}{D} \end{bmatrix}, \quad Q^* = \begin{bmatrix} \frac{d - b}{D} \\ \frac{a - c}{D} \end{bmatrix}, \text{ and } v = \frac{ad - bc}{D},$$

where $D = (a + d) - (b + c)$. [Note: Under the assumption that M is not strictly determined, it can be shown that $D = (a + d) - (b + c)$ will never be 0.]

5. RECESSIVE ROWS: A row in a game matrix is said to be recessive and may be deleted if there exists a (dominant) row with corresponding elements greater than or equal to those in the given row.

RECESSIVE COLUMNS: A column in a game matrix is said to be recessive and may be deleted if there exists a (dominant) column with corresponding elements less than or equal to those in the given column.

---

1. $\begin{bmatrix} \textcircled{-1} & \boxed{2} \\ \boxed{2} & \textcircled{-4} \end{bmatrix}$ The game is not strictly determined.

Using 4: The optimal strategy for R is

$$P^* = \begin{bmatrix} \frac{d - c}{D} & \frac{a - b}{D} \end{bmatrix} = \begin{bmatrix} \frac{-4 - 2}{-9} & \frac{-1 - 2}{-9} \end{bmatrix} = \begin{bmatrix} \frac{2}{3} & \frac{1}{3} \end{bmatrix}$$

[Note: $D = (a + d) - (b + c) = (-1 - 4) - (2 + 2) = -9$.]

The optimal strategy for C is

$$Q^* = \begin{bmatrix} \frac{d - b}{D} \\ \frac{a - c}{D} \end{bmatrix} = \begin{bmatrix} \frac{-4 - 2}{-9} \\ \frac{-1 - 2}{-9} \end{bmatrix} = \begin{bmatrix} \frac{2}{3} \\ \frac{1}{3} \end{bmatrix}$$

The value of the game, v, is: $\frac{ad - bc}{D} = \frac{(-1)(-4) - (2)(2)}{-9} = 0.$

3. $\begin{bmatrix} \textcircled{-1} & \boxed{2} \\ \boxed{1} & \textcircled{0} \end{bmatrix}$ The game is not strictly determined.

Using 4:

$P^* = \begin{bmatrix} \frac{0-1}{-4} & \frac{-1-2}{-4} \end{bmatrix} = \begin{bmatrix} \frac{1}{4} & \frac{3}{4} \end{bmatrix}$ [Note: $D = (-1 + 0) - (2 + 1) = -4$.]

$Q^* = \begin{bmatrix} \frac{0-2}{-4} \\ \frac{-1-1}{-4} \end{bmatrix} = \begin{bmatrix} \frac{1}{2} \\ \frac{1}{2} \end{bmatrix}$

The value of the game, v, is: $\dfrac{(-1)(0) - (2)(1)}{-4} = \dfrac{1}{2}$.

5. $\begin{bmatrix} \boxed{4} & \textcircled{-6} \\ \textcircled{-2} & \boxed{3} \end{bmatrix}$ The game is not strictly determined.

Using 4:

$P^* = \begin{bmatrix} \frac{3+2}{15} & \frac{4+6}{15} \end{bmatrix} = \begin{bmatrix} \frac{1}{3} & \frac{2}{3} \end{bmatrix}$ $Q^* = \begin{bmatrix} \frac{3+6}{15} \\ \frac{4+2}{15} \end{bmatrix} = \begin{bmatrix} \frac{3}{5} \\ \frac{2}{5} \end{bmatrix}$

The value of the game, v, is: $\dfrac{(4)(3) - (-6)(-2)}{15} = 0$.

7. $\begin{bmatrix} \boxed{5} & \textcircled{-1} \\ 4 & \boxed{\textcircled{1}} \end{bmatrix}$ The game is strictly determined.

$P^* = [0 \quad 1]$ $Q^* = \begin{bmatrix} 0 \\ 1 \end{bmatrix}$

The value of the game, v = saddle value = 1.

9. Using 5, we can eliminate the recessive columns $\begin{bmatrix} 0 \\ 1 \end{bmatrix}$ and $\begin{bmatrix} 2 \\ -1 \end{bmatrix}$. Thus, we obtain the 2×2 matrix:

$\begin{bmatrix} \boxed{2} & \textcircled{-1} \\ \textcircled{-2} & \boxed{1} \end{bmatrix}$ The game is not strictly determined.

Now, using 4:

$P^* = \begin{bmatrix} \frac{1+2}{6} & \frac{2+1}{6} \end{bmatrix} = \begin{bmatrix} \frac{3}{6} & \frac{3}{6} \end{bmatrix} = \begin{bmatrix} \frac{1}{2} & \frac{1}{2} \end{bmatrix}$ [Note: $D = (2 + 1) - (-1 - 2) = 6$.]

$Q^* = \begin{bmatrix} \frac{1+1}{6} \\ \frac{2+2}{6} \end{bmatrix} = \begin{bmatrix} \frac{2}{6} \\ \frac{4}{6} \end{bmatrix} = \begin{bmatrix} \frac{1}{3} \\ \frac{2}{3} \end{bmatrix}$ or $\begin{bmatrix} 0 \\ \frac{1}{3} \\ \frac{2}{3} \\ 0 \end{bmatrix}$, $v = \dfrac{(2)(1) - (-1)(-2)}{6} = 0$

11. Using $\underline{5}$, we eliminate the recessive row [1 -4 -1] and the recessive column $\begin{bmatrix} -1 \\ 0 \\ 2 \end{bmatrix}$. Thus, we obtain the 2 × 2 matrix: $\begin{bmatrix} \boxed{2} & \textcircled{-3} \\ \textcircled{-1} & \boxed{2} \end{bmatrix}$.

The game is nonstrictly determined. Now, using $\underline{4}$:

$$P^* = \begin{bmatrix} \frac{2+1}{8} & \frac{2+3}{8} \end{bmatrix} = \begin{bmatrix} \frac{3}{8} & \frac{5}{8} \end{bmatrix} \text{ or } \begin{bmatrix} 0 & \frac{3}{8} & \frac{5}{8} \end{bmatrix}.$$ [Note: $D = (2 + 2) - (-3 - 1) = 8$.]

$$Q^* = \begin{bmatrix} \frac{2+3}{8} \\ \frac{2+1}{8} \end{bmatrix} = \begin{bmatrix} \frac{5}{8} \\ \frac{3}{8} \end{bmatrix} \text{ or } \begin{bmatrix} \frac{5}{8} \\ \frac{3}{8} \\ 0 \end{bmatrix}, \quad v = \frac{(2)(2) - (-3)(-1)}{8} = \frac{1}{8}$$

13. $\begin{bmatrix} 2 & \boxed{\textcircled{1}} & 2 \\ \boxed{3} & 0 & \textcircled{-5} \\ 1 & \textcircled{-2} & \boxed{7} \end{bmatrix}$ The game is strictly determined.

$P^* = [1 \quad 0 \quad 0]$ $Q^* = \begin{bmatrix} 0 \\ 1 \\ 0 \end{bmatrix}$ v = saddle value = 1

15. $M = \begin{bmatrix} a & b \\ c & d \end{bmatrix}$

(A) $M = \begin{bmatrix} a & a \\ c & d \end{bmatrix}$

True: Player R circles both entries in Row 1 and at least one of the entries in Row 2--a total of 3 entries. Since Player C "squares" at least two entries, at least one entry will have both a circle and a square.

(B) $M = \begin{bmatrix} a & b \\ a & d \end{bmatrix}$

True: Same argument as above with Player R and Player C interchanged.

(C) $M = \begin{bmatrix} a & b \\ c & a \end{bmatrix}$

False: Consider $M = \begin{bmatrix} \boxed{1} & \textcircled{-1} \\ \textcircled{-1} & \boxed{1} \end{bmatrix}$

M has no saddle value.

17. (A) $P = \begin{bmatrix} \frac{1}{4} & \frac{1}{4} & \frac{1}{4} & \frac{1}{4} \end{bmatrix}$, $Q = \begin{bmatrix} \frac{1}{4} \\ \frac{1}{4} \\ \frac{1}{4} \\ \frac{1}{4} \end{bmatrix}$

$$PMQ = \begin{bmatrix} \frac{1}{4} & \frac{1}{4} & \frac{1}{4} & \frac{1}{4} \end{bmatrix}\begin{bmatrix} 0 & 2 & 1 & 0 \\ 4 & 3 & 5 & 4 \\ 0 & 2 & 6 & 1 \\ 0 & 1 & 0 & 3 \end{bmatrix}\begin{bmatrix} \frac{1}{4} \\ \frac{1}{4} \\ \frac{1}{4} \\ \frac{1}{4} \end{bmatrix} = [1 \quad 2 \quad 3 \quad 2]\begin{bmatrix} \frac{1}{4} \\ \frac{1}{4} \\ \frac{1}{4} \\ \frac{1}{4} \end{bmatrix}$$

= 2 or \$2

Since you pay C \$3 to play, the expected value is \$2 - \$3 = -\$1.

(B) The largest of the row minimums is 3 in row 2. Thus, you should play row 2, i.e., $P = [0 \quad 1 \quad 0 \quad 0]$. Now

$$PMQ = [0 \quad 1 \quad 0 \quad 0]\begin{bmatrix} 0 & 2 & 1 & 0 \\ 4 & 3 & 5 & 4 \\ 0 & 2 & 6 & 1 \\ 0 & 1 & 0 & 3 \end{bmatrix}\begin{bmatrix} \frac{1}{4} \\ \frac{1}{4} \\ \frac{1}{4} \\ \frac{1}{4} \end{bmatrix} = [4 \quad 3 \quad 5 \quad 4]\begin{bmatrix} \frac{1}{4} \\ \frac{1}{4} \\ \frac{1}{4} \\ \frac{1}{4} \end{bmatrix}$$

= 4 or \$4

Since you pay C \$3 to play, the expected value is \$4 - \$3 = \$1.

(C) We eliminate the recessive rows [0 2 1 0] and [0 1 0 3], and the recessive columns

$\begin{bmatrix} 1 \\ 5 \\ 6 \\ 0 \end{bmatrix}$ and $\begin{bmatrix} 0 \\ 4 \\ 1 \\ 3 \end{bmatrix}$. The resulting matrix is: $\begin{bmatrix} \boxed{4} & \boxed{③} \\ ⓪ & 2 \end{bmatrix}$

and the game is strictly determined with a value of 3. You should play row 2 (in the original matrix) and C should play column 2 (in the original matrix).

Thus,

$P = [0 \quad 1 \quad 0 \quad 0]$ and $Q = \begin{bmatrix} 0 \\ 1 \\ 0 \\ 0 \end{bmatrix}$

$$PMQ = [0 \quad 1 \quad 0 \quad 0]\begin{bmatrix} 0 & 2 & 1 & 0 \\ 4 & 3 & 5 & 4 \\ 0 & 2 & 6 & 1 \\ 0 & 1 & 0 & 3 \end{bmatrix}\begin{bmatrix} 0 \\ 1 \\ 0 \\ 0 \end{bmatrix} = [4 \quad 3 \quad 5 \quad 4]\begin{bmatrix} 0 \\ 1 \\ 0 \\ 0 \end{bmatrix}$$

= 3 or \$3

Since you pay C \$3 to play, the expected value \$3 - \$3 = \$0.

19. $PMQ = [p_1 \quad p_2]\begin{bmatrix} a & b \\ c & d \end{bmatrix}\begin{bmatrix} q_1 \\ q_2 \end{bmatrix} = [p_1 \quad p_2]\begin{bmatrix} [a \quad b]\cdot\begin{bmatrix} q_1 \\ q_2 \end{bmatrix} \\ [c \quad d]\cdot\begin{bmatrix} q_1 \\ q_2 \end{bmatrix} \end{bmatrix}$

$$= [p_1 \quad p_2]\begin{bmatrix} aq_1 + bq_2 \\ cq_1 + dq_2 \end{bmatrix} = [p_1(aq_1 + bq_2) + p_2(cq_1 + dq_2)]$$

$$= [ap_1q_1 + bp_1q_2 + cp_2q_1 + dp_2q_2] = E(P, Q)$$

21. $P^*MQ = \left[\frac{d-c}{D} \quad \frac{a-b}{D}\right]\begin{bmatrix} a & b \\ c & d \end{bmatrix}\begin{bmatrix} q_1 \\ q_2 \end{bmatrix}$

In Problem 19, substitute $p_1 = \dfrac{d-c}{D}$ and $p_2 = \dfrac{a-b}{D}$. Thus,

$$P^*MQ = \frac{a(d-c)}{D}q_1 + \frac{b(d-c)}{D}q_2 + \frac{c(a-b)}{D}q_1 + \frac{d(a-b)}{D}q_2$$

$$= \frac{1}{D}[adq_1 - caq_1 + bdq_2 - bcq_2 + caq_1 - cbq_1 + daq_2 - dbq_2]$$

$$= \frac{1}{D}[(ad - cb)q_1 + (ad - bc)q_2]$$

$$= \frac{ad - cb}{D}[q_1 + q_2] \quad [\underline{\text{Note}}\text{: } q_1 + q_2 = 1.]$$

$$= \frac{ad - cb}{D}$$

Thus,

(A) $P^*MQ = \dfrac{ad - cb}{D}$

Similarly, $PMQ^* = [p_1 \quad p_2]\begin{bmatrix} a & b \\ c & d \end{bmatrix}\begin{bmatrix} \frac{d-b}{D} \\ \frac{a-c}{D} \end{bmatrix}$

$$= ap_1\left(\frac{d-b}{D}\right) + bp_1\left(\frac{a-c}{D}\right) + cp_2\left(\frac{d-b}{D}\right) + dp_2\left(\frac{a-c}{D}\right)$$

$$= \frac{1}{D}(adp_1 - abp_1 + bap_1 - bcp_1 + cdp_2 - cbp_2 + dap_2 - dcp_2)$$

$$= \frac{1}{D}(adp_1 - bcp_1 - cbp_2 + dap_2) = \frac{1}{D}[(ad - bc)p_1 + (da - cb)p_2]$$

$$= \frac{ad - bc}{D}(p_1 + p_2) = \frac{ad - bc}{D} \quad [\underline{\text{Note}}\text{: } p_1 + p_2 = 1.]$$

Thus,

(B) $PMQ^* = \dfrac{ad - bc}{D}$

From (A) and (B), we get $v = \dfrac{ad - bc}{D}$, which satisfies $P^*MQ \geq v$ and $PMQ^* \leq v$.

23. Solve the linear system

$$\frac{d - c}{D} = 0.9$$
$$\frac{a - b}{D} = 0.1$$
$$\frac{d - b}{D} = 0.3$$
$$\frac{a - c}{D} = 0.7$$

where $D = (a + d) - (b + c)$. In standard form, this system is:

$$\begin{aligned} 0.9a - 0.9b + 0.1c - 0.1d &= 0 \\ -0.9a + 0.9b - 0.1c + 0.1d &= 0 \\ 0.3a + 0.7b - 0.3c - 0.7d &= 0 \\ -0.3a - 0.7b + 0.3c + 0.7d &= 0 \end{aligned}$$

which reduces to

$$\begin{aligned} 0.9a - 0.9b + 0.1c - 0.1d &= 0 \\ 0.3a + 0.7b - 0.3c - 0.7d &= 0 \end{aligned}$$

Choose two of the unknowns arbitrarily and solve for the other two unknowns. For example, setting $d = 0$ and $c = 10$ gives

$$\begin{aligned} 0.9a - 0.9b &= -1 \\ 0.3a + 0.7b &= 3 \end{aligned}$$

whose solution is $a = \frac{20}{9}$, $b = \frac{10}{3}$

$$M = \begin{bmatrix} \frac{20}{9} & \frac{10}{3} \\ 10 & 0 \end{bmatrix}$$

25. Given:

$$\begin{array}{c|cccc} & TV & R & P & M \\ \hline TV & 0 & -1 & -1 & 0 \\ R & 1 & 2 & -1 & -1 \\ P & 0 & -1 & 0 & 1 \\ M & -1 & -1 & -1 & 0 \end{array}$$

Eliminate the recessive rows [0 -1 -1 0] and [-1 -1 -1 0] and the recessive columns $\begin{bmatrix} 0 \\ 1 \\ 0 \\ -1 \end{bmatrix}$ and $\begin{bmatrix} 0 \\ -1 \\ 1 \\ 0 \end{bmatrix}$ to obtain the 2×2 matrix $\begin{bmatrix} \boxed{2} & \textcircled{-1} \\ \textcircled{-1} & \boxed{0} \end{bmatrix}$.

The game is not strictly determined.

(A) The optimal strategies for R and C are

$P^* = \begin{bmatrix} \frac{0 + 1}{4} & \frac{2 + 1}{4} \end{bmatrix} = \begin{bmatrix} \frac{1}{4} & \frac{3}{4} \end{bmatrix}$ [Note: $D = 2 + 0 - (-1 - 1) = 4$.]

and $Q^* = \begin{bmatrix} \frac{0 + 1}{4} \\ \frac{2 + 1}{4} \end{bmatrix} = \begin{bmatrix} \frac{1}{4} \\ \frac{3}{4} \end{bmatrix}$, respectively.

In terms of the original problem:

$$P^* = \begin{bmatrix} 0 & \frac{1}{4} & \frac{3}{4} & 0 \end{bmatrix} \text{ and } Q^* = \begin{bmatrix} 0 \\ \frac{1}{4} \\ \frac{3}{4} \\ 0 \end{bmatrix}$$

The value of the game is: $v = \dfrac{2(0) - (-1)(-1)}{4} = -\dfrac{1}{4}$

(B) $P = [1 \quad 0 \quad 0 \quad 0]$ and $Q = \begin{bmatrix} 0 \\ \frac{1}{4} \\ \frac{3}{4} \\ 0 \end{bmatrix}$

$$PMQ = [1 \quad 0 \quad 0 \quad 0]\begin{bmatrix} 0 & -1 & -1 & 0 \\ 1 & 2 & -1 & -1 \\ 0 & -1 & 0 & 1 \\ -1 & -1 & -1 & 0 \end{bmatrix}\begin{bmatrix} 0 \\ \frac{1}{4} \\ \frac{3}{4} \\ 0 \end{bmatrix} = [0 \quad -1 \quad -1 \quad 0]\begin{bmatrix} 0 \\ \frac{1}{4} \\ \frac{3}{4} \\ 0 \end{bmatrix} = -1$$

The expected value is -1.

(C) $P = \begin{bmatrix} 0 & \frac{1}{4} & \frac{3}{4} & 0 \end{bmatrix}$ and $Q^* = \begin{bmatrix} 0 \\ 1 \\ 0 \\ 0 \end{bmatrix}$

$$PMQ = \begin{bmatrix} 0 & \frac{1}{4} & \frac{3}{4} & 0 \end{bmatrix}\begin{bmatrix} 0 & -1 & -1 & 0 \\ 1 & 2 & -1 & -1 \\ 0 & -1 & 0 & 1 \\ -1 & -1 & -1 & 0 \end{bmatrix}\begin{bmatrix} 0 \\ 1 \\ 0 \\ 0 \end{bmatrix} = \begin{bmatrix} 0 & \frac{1}{4} & \frac{3}{4} & 0 \end{bmatrix}\begin{bmatrix} -1 \\ 2 \\ -1 \\ -1 \end{bmatrix} = -\frac{1}{4}$$

The expected value is $-\frac{1}{4}$.

(D) $P = [0 \quad 0 \quad 1 \quad 0]$ and $Q = \begin{bmatrix} 0 \\ 0 \\ 1 \\ 0 \end{bmatrix}$

$$PMQ = [0 \quad 0 \quad 1 \quad 0]\begin{bmatrix} 0 & -1 & -1 & 0 \\ 1 & 2 & -1 & -1 \\ 0 & -1 & 0 & 1 \\ -1 & -1 & -1 & 0 \end{bmatrix}\begin{bmatrix} 0 \\ 0 \\ 1 \\ 0 \end{bmatrix} = [0 \quad -1 \quad 0 \quad 1]\begin{bmatrix} 0 \\ 0 \\ 1 \\ 0 \end{bmatrix} = 0$$

The expected value is 0.

27. Player R (You)

	Player C (Fate) Republican	Democrat
Solar energy	(\$1000)	[\$4000]
Oil	[\$5000]	(\$3000)

The game is not strictly determined.

Thus, using 4:

$$P^* = \begin{bmatrix} \frac{3000 - 5000}{-5000} & \frac{1000 - 4000}{-5000} \end{bmatrix} = \begin{bmatrix} \frac{2}{5} & \frac{3}{5} \end{bmatrix}$$

[Note: $D = (1000 + 3000) - (4000 + 5000) = -5000$.]

$$Q^* = \begin{bmatrix} \frac{3000 - 4000}{-5000} \\ \frac{1000 - 5000}{-5000} \end{bmatrix} = \begin{bmatrix} \frac{1}{5} \\ \frac{4}{5} \end{bmatrix}$$

$$v = \frac{(1000)(3000) - (4000)(5000)}{-5000} = \frac{-17{,}000}{-5} = \$3400$$

This means that you should invest $\frac{2}{5}(10{,}000) = \$4000$ in solar energy stocks and $\frac{3}{5}(10{,}000) = \$6000$ in oil stocks. You would then have an expected gain of \$3400 no matter how the election turns out.

EXERCISE 8-3

Things to remember:

2 × 2 MATRIX GAMES AND LINEAR PROGRAMMING—GEOMETRIC APPROACH

Given the nonstrictly determined matrix game with payoff matrix

$$M = \begin{bmatrix} a & b \\ c & d \end{bmatrix}.$$

To find $P^* = [p_1 \quad p_2]$, $Q^* = \begin{bmatrix} q_1 \\ q_2 \end{bmatrix}$, and v, proceed as follows:

STEP 1. If M is not a positive matrix, convert it into a positive matrix M_1 by adding a suitable positive constant k to each element. Let

$$M_1 = \begin{bmatrix} e & f \\ g & h \end{bmatrix}, \quad e = a + k,\ f = b + k,\ g = c + k,\ h = d + k.$$

This matrix game M_1 has the same optimal strategies P^* and Q^* as M, and if v_1 is the value of the game M_1, then $v = v_1 - k$ is the value of the original game M.

STEP 2. Set up the two corresponding linear programming problems:

(A) Minimize $y = x_1 + x_2$

Subject to: $ex_1 + gx_2 \geq 1$

$fx_1 + hx_2 \geq 1$

$x_1, x_2 \geq 0$

(B) Maximize $y = z_1 + z_2$

Subject to: $ez_1 + fz_2 \leq 1$

$gz_1 + hz_2 \leq 1$

$z_1, z_2 \geq 0$

STEP 3. Solve each linear programming problem geometrically. [Note: Since (B) is the dual of (A), both problems have the same optimal value, i.e., min y in (A) equals max y in (B).]

STEP 4. Using the solutions in Step 3:

$$v_1 = \frac{1}{y} = \frac{1}{x_1 + x_2} \quad \text{or} \quad v_1 = \frac{1}{y} = \frac{1}{z_1 + z_2}$$

$$P^* = [p_1 \quad p_2] = [v_1x_1 \quad v_1x_2]$$

$$Q^* = \begin{bmatrix} q_1 \\ q_2 \end{bmatrix} = \begin{bmatrix} v_1z_1 \\ v_1z_2 \end{bmatrix}$$

$$v = v_1 - k$$

STEP 5. The solution found in Step 4 can be checked by showing that

$P^*MQ^* = v$. (See Theorem 3 of Section 8.2.)

---

1. Convert $\begin{bmatrix} 2 & -3 \\ -1 & 2 \end{bmatrix}$ into a positive payoff matrix by adding 4 to each payoff.

$$M_1 = \begin{bmatrix} 6 & 1 \\ 3 & 6 \end{bmatrix}$$

Set up the two corresponding linear programming problems:

(A) Minimize $y = x_1 + x_2$

Subject to $6x_1 + 3x_2 \geq 1$

$x_1 + 6x_2 \geq 1$

$x_1 \geq 0,\ x_2 \geq 0$

(B) Maximize $y = z_1 + z_2$

Subject to $6z_1 + z_2 \leq 1$

$3z_1 + 6z_2 \leq 1$

$z_1 \geq 0,\ z_2 \geq 0$

Solve the linear programming problems geometrically, as shown in the following figures.

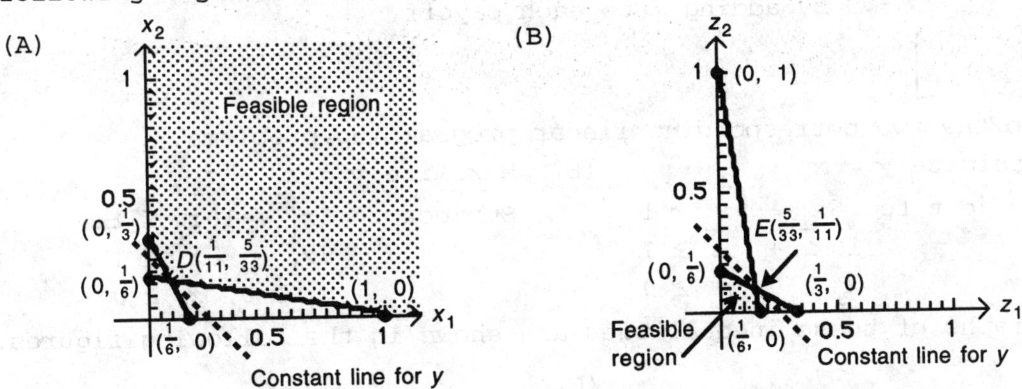

In figure (A), since the minimum $y = x_1 + x_2$ occurs at the point of intersection D, we solve

$$6x_1 + 3x_2 = 1$$
$$x_1 + 6x_2 = 1$$

to obtain the coordinates for D: $x_1 = \frac{3}{33} = \frac{1}{11}$

$$x_2 = \frac{5}{33}$$

In figure (B), since the maximum $y = z_1 + z_2$ occurs at the point of intersection E, we solve

$$6z_1 + z_2 = 1$$
$$3z_1 + 6z_2 = 1$$

to obtain the coordinates for E: $z_1 = \frac{5}{33}$

$$z_2 = \frac{3}{33} = \frac{1}{11}$$

Thus,

$$v_1 = \frac{1}{x_1 + x_2} = \frac{1}{z_1 + z_2} = \frac{1}{\frac{3}{33} + \frac{5}{33}} = \frac{33}{8}$$

$$p_1 = v_1x_1 = \frac{33}{8} \cdot \frac{3}{33} = \frac{3}{8} \qquad q_1 = v_1z_1 = \frac{33}{8} \cdot \frac{5}{33} = \frac{5}{8}$$

$$p_2 = v_1x_2 = \frac{33}{8} \cdot \frac{5}{33} = \frac{5}{8} \qquad q_2 = v_1z_2 = \frac{33}{8} \cdot \frac{3}{33} = \frac{3}{8}$$

The value v of the original matrix game is found by subtracting 4 from the value v_1 of M_1. Thus,

$$v = v_1 - 4 = \frac{33}{8} - 4 = \frac{1}{8}$$

The optimal strategies and the value of the game are given by:

$$P^* = \begin{bmatrix} \frac{3}{8} & \frac{5}{8} \end{bmatrix}, \qquad Q^* = \begin{bmatrix} \frac{5}{8} \\ \frac{3}{8} \end{bmatrix}, \qquad v = \frac{1}{8}$$

3. Convert $\begin{bmatrix} -1 & 3 \\ 2 & -6 \end{bmatrix}$ into a positive payoff matrix by adding 7 to each payoff.

$$M_1 = \begin{bmatrix} 6 & 10 \\ 9 & 1 \end{bmatrix}$$

Set up the two corresponding linear programming problems.

(A) Minimize $y = x_1 + x_2$
Subject to $6x_1 + 9x_2 \ge 1$
$10x_1 + x_2 \ge 1$
$x_1 \ge 0,\ x_2 \ge 0$

(B) Maximize $y = z_1 + z_2$
Subject to $6z_1 + 10z_2 \le 1$
$9z_1 + z_2 \le 1$
$z_1 \ge 0,\ z_2 \ge 0$

The graphs of these inequalities are shown in the following figures.

(A)

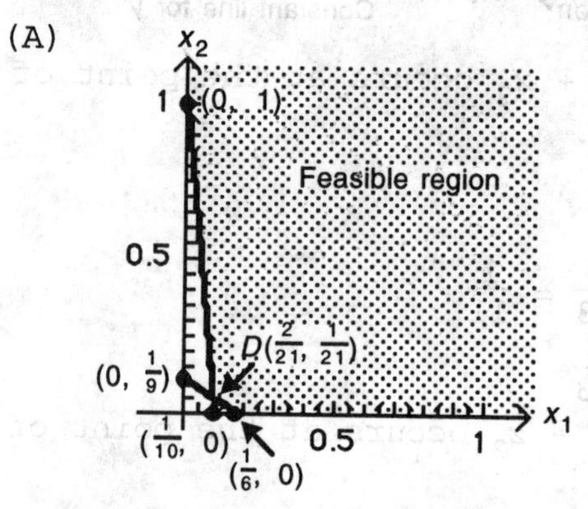

(B)

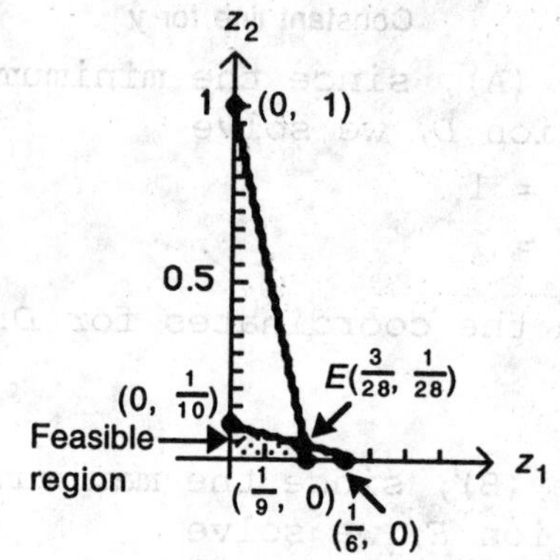

In figure (A), since the minimum $y = x_1 + x_2$ occurs at the point of intersection D, we solve

$6x_1 + 9x_2 = 1$
$10x_1 + x_2 = 1$

to obtain the coordinates for D: $x_1 = \frac{2}{21},\ x_2 = \frac{1}{21}$

Similarly, in figure (B), since the maximum $y = z_1 + z_2$ occurs at the point of intersection E, we solve

$6z_1 + 10z_2 = 1$
$9z_1 + z_2 = 1$

to obtain the coordinates for E: $z_1 = \frac{3}{28},\ z_2 = \frac{1}{28}$

Thus,

$$v_1 = \frac{1}{x_1 + x_2} = \frac{1}{z_1 + z_2} = \frac{1}{\frac{2}{21} + \frac{1}{21}} = 7$$

$p_1 = v_1x_1 = 7 \cdot \frac{2}{21} = \frac{2}{3}$ $\qquad q_1 = v_1z_1 = 7 \cdot \frac{3}{28} = \frac{3}{4}$

$p_2 = v_1x_2 = 7 \cdot \frac{1}{21} = \frac{1}{3}$ $\qquad q_2 = v_1z_2 = 7 \cdot \frac{1}{28} = \frac{1}{4}$

The value v of the original matrix is given by $v_1 - 7 = 7 - 7 = 0$. The optimal strategies and the value of the game are given by

$$P^* = \begin{bmatrix} \frac{2}{3} & \frac{1}{3} \end{bmatrix}, \quad Q^* = \begin{bmatrix} \frac{3}{4} \\ \frac{1}{4} \end{bmatrix}, \quad v = 0$$

5. Convert $\begin{bmatrix} -2 & -1 \\ 5 & 6 \end{bmatrix}$ into a positive payoff matrix by adding 3 to each payoff

$M_1 = \begin{bmatrix} 1 & 2 \\ 8 & 9 \end{bmatrix}$

Set up the two corresonding linear programming problems.

(A) Minimize $y = x_1 + x_2$
Subject to: $x_1 + 8x_2 \geq 1$
$2x_1 + 9x_2 \geq 1$
$x_1, x_2 \geq 0$

(B) Maximize $y = z_1 + z_2$
Subject to: $z_1 + 2z_2 \leq 1$
$8z_1 + 9z_2 \leq 1$
$z_1, z_2 \geq 0$

The graphs of these inequalities are shown in the following figures:

(A)

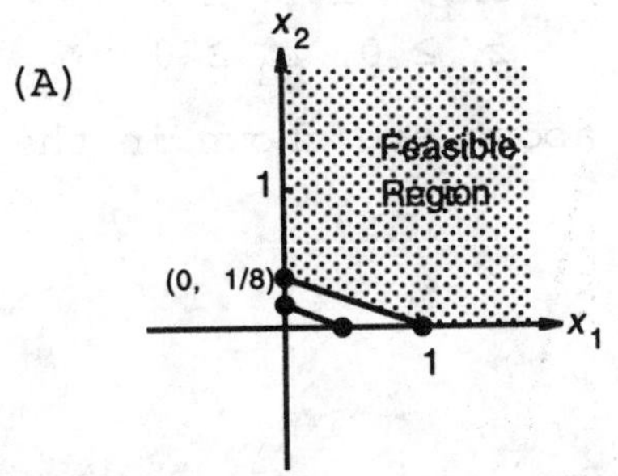

(B)

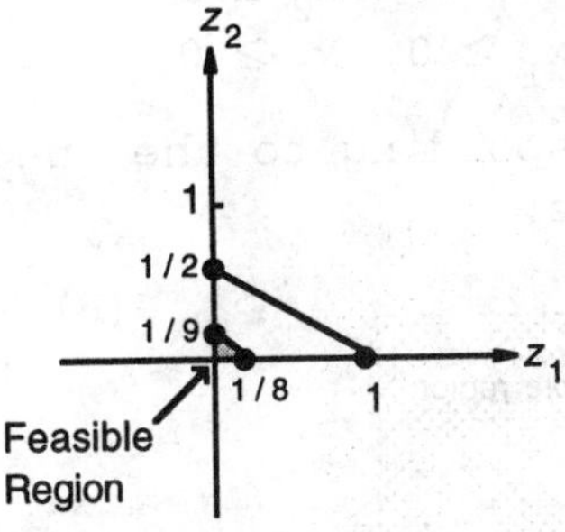

In Figure (A), the minimum $y = x_1 + x_2$ occurs at the point $\left(0, \frac{1}{8}\right)$.
So $x_1 = 0$, $x_2 = \frac{1}{8}$.

In Figure (B), the maximum $y = z_1 + z_2$ occurs at the point $\left(\frac{1}{8}, 0\right)$,
so $z_1 = \frac{1}{8}$, $z_2 = 0$.

Thus, $v_1 = \dfrac{1}{x_1 + x_2} = \dfrac{1}{0 + 1/8} = 8$

$p_1 = v_1x_1 = 8(0) = 0$ $\qquad q_1 = v_1z_1 = 8\left(\frac{1}{8}\right) = 1$

$p_2 = v_1x_2 = 8\left(\frac{1}{8}\right) = 1$ $\qquad q_2 = v_1z_2 = 8(0) = 0$

The value v of the original matrix is $v = v_1 - 3 = 8 - 3 = 5$. The optimal strategies and the value of the game are given by

$$P^* = [0, 1],\ Q^* = \begin{bmatrix} 1 \\ 0 \end{bmatrix},\ v = 5.$$

7. $M = \begin{bmatrix} -2 & -1 \\ 5 & 6 \end{bmatrix}$

Note that the matrix game is strictly determined:

$\begin{bmatrix} \textcircled{-2} & -1 \\ \boxed{\textcircled{5}} & \boxed{6} \end{bmatrix}$; 5 is a saddle value.

Thus, the method of Section 8-1 applies.

9. In the game matrix $\begin{bmatrix} 3 & -6 \\ 4 & -6 \\ -2 & 3 \\ -3 & 0 \end{bmatrix}$ row 1 and row 4 are recessive rows and should be eliminated. Thus, we have

$\begin{bmatrix} 4 & -6 \\ -2 & 3 \end{bmatrix}$. We then obtain the positive payoff matrix by adding 7 to each payoff:

$$M_1 = \begin{bmatrix} 11 & 1 \\ 5 & 10 \end{bmatrix}$$

The linear programming problems corresponding to M_1 are as follows:

(A) Minimize $y = x_1 + x_2$
Subject to $11x_1 + 5x_2 \geq 1$
$x_1 + 10x_2 \geq 1$
$x_1 \geq 0,\ x_2 \geq 0$

(B) Maximize $y = z_1 + z_2$
Subject to $11z_1 + z_2 \leq 1$
$5z_1 + 10z_2 \leq 1$
$z_1 \geq 0,\ z_2 \geq 0$

The graphs corresponding to the inequalities above are shown in the following figures.

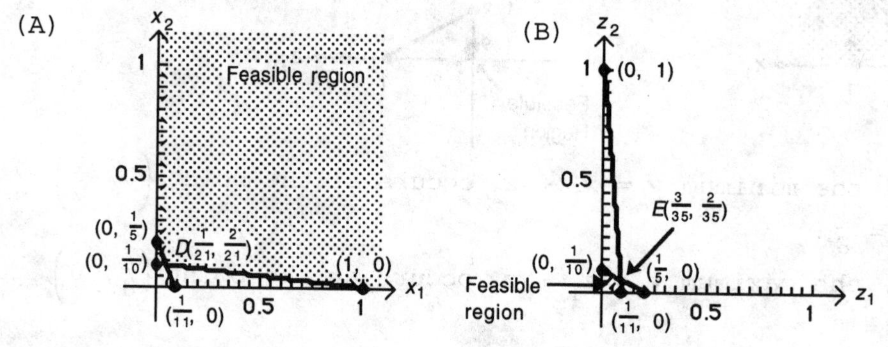

In figure (A), since the minimum $y = x_1 + x_2$ occurs at the point of intersection D, we solve

$$11x_1 + 5x_2 = 1$$
$$x_1 + 10x_2 = 1$$

to obtain the coordinates for D: $x_1 = \frac{1}{21}$

$$x_2 = \frac{2}{21}$$

Similarly, in figure (B), we solve

$$11z_1 + z_2 = 1$$
$$5z_1 + 10z_2 = 1$$

to obtain the coordinates for E: $z_1 = \frac{3}{35}$

$$z_2 = \frac{2}{35}$$

Thus,

$$v_1 = \frac{1}{x_1 + x_2} = \frac{1}{z_1 + z_2} = \frac{1}{\frac{1}{21} + \frac{2}{21}} = \frac{21}{3} = 7$$

$$p_1 = v_1x_1 = 7 \cdot \frac{1}{21} = \frac{1}{3} \qquad q_1 = v_1z_1 = 7 \cdot \frac{3}{35} = \frac{3}{5}$$

$$p_2 = v_1x_2 = 7 \cdot \frac{2}{21} = \frac{2}{3} \qquad q_2 = v_1z_2 = 7 \cdot \frac{2}{35} = \frac{2}{5}$$

The value v of the original matrix is given by $v = v_1 - 7 = 7 - 7 = 0$. Thus,

$$P^* = \begin{bmatrix} 0 & \frac{1}{3} & \frac{2}{3} & 0 \end{bmatrix}, \quad Q^* = \begin{bmatrix} \frac{3}{5} \\ \frac{2}{5} \end{bmatrix}, \quad v = 0$$

11. (A) Let $P = (p_1, p_2)$ and $Q = \begin{pmatrix} q_1 \\ q_2 \end{pmatrix}$, where $p_1 + p_2 = 1$, $q_1 + q_2 = 1$.

Then

$$P(kJ)Q = k(p_1, p_2)\begin{pmatrix} 1 & 1 \\ 1 & 1 \end{pmatrix}\begin{pmatrix} q_1 \\ q_2 \end{pmatrix} = k(p_1 + p_2, p_1 + p_2)\begin{pmatrix} q_1 \\ q_2 \end{pmatrix}$$

$$= k(1, 1))\begin{pmatrix} q_1 \\ q_2 \end{pmatrix}$$

$$= k(q_1 + q_2) = k$$

(B) Let $P = (p_1, p_2, \dots, p_m)$ and $Q = \begin{pmatrix} q_1 \\ q_2 \\ \vdots \\ q_n \end{pmatrix}$ where $\sum_{i=1}^{m} p_i = 1$, $\sum_{j=1}^{n} q_j = 1$

Then

$$P(kJ)Q = k(p_1, p_2, \dots, p_m)\underset{(m \times n)}{\begin{pmatrix} 1 & 1 & \dots & 1 \\ 1 & 1 & \dots & 1 \\ \vdots & \vdots & & \vdots \\ 1 & 1 & \dots & 1 \end{pmatrix}}\begin{pmatrix} q_1 \\ q_2 \\ \vdots \\ q_m \end{pmatrix}$$

$$= k\left(\sum_{i=1}^{m} p_i, \sum_{i=1}^{m} p_i, \dots, \sum_{i=1}^{m} p_i\right)\begin{pmatrix} q_1 \\ q_2 \\ \vdots \\ q_n \end{pmatrix}$$

$$= k(1, 1, \dots, 1)\begin{pmatrix} q_1 \\ q_2 \\ \vdots \\ q_n \end{pmatrix} = k\sum_{j=1}^{n} q_j = k$$

13. Eliminate the recessive rows [0 -1 -1 0] and [-1 -1 -1 0], and the recessive columns $\begin{bmatrix} 0 \\ 1 \\ 0 \\ -1 \end{bmatrix}$ and $\begin{bmatrix} 0 \\ -1 \\ 1 \\ 0 \end{bmatrix}$ to obtain the 2 × 2 matrix:

$$M = \begin{bmatrix} 2 & -1 \\ -1 & 0 \end{bmatrix}$$

Now, add 2 to each element to get the positive matrix:

$$M_1 = \begin{bmatrix} 4 & 1 \\ 1 & 2 \end{bmatrix}$$

The linear programming problems corresponding to M_1 are as follows:

(A) Minimize $y = x_1 + x_2$
Subject to $4x_1 + x_2 \geq 1$
$x_1 + 2x_2 \geq 1$
$x_1, x_2 \geq 0$

(B) Maximize $y = z_1 + z_2$
Subject to $4z_1 + z_2 \leq 1$
$z_1 + 2z_2 \leq 1$
$z_1, z_2 \geq 0$

The graphs corresponding to the inequalities above are shown in the following figures.

(A)

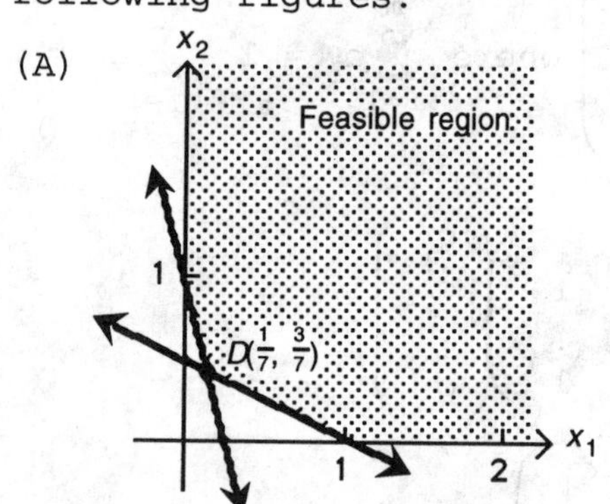

(B)

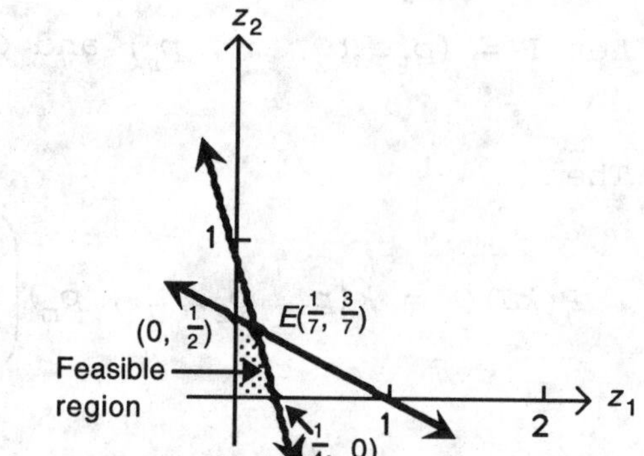

In figure (A), since the minimum $y = x_1 + x_2$ occurs at the point of intersection D, we solve

$$4x_1 + x_2 = 1$$
$$x_1 + 2x_2 = 1$$

to obtain $x_1 = \frac{1}{7}$ and $x_2 = \frac{3}{7}$

The system in (B) is the same as the system in (A) and so $z_1 = \frac{1}{7}$ and $z_2 = \frac{3}{7}$. Thus,

$$v = \frac{1}{x_1 + x_2} = \frac{1}{z_1 + z_2} = \frac{1}{\frac{1}{7} + \frac{3}{7}} = \frac{7}{4}.$$

$p_1 = v_1 x_1 = \frac{7}{4} \cdot \frac{1}{7} = \frac{1}{4}$ $q_1 = v_1 z_1 = \frac{7}{4} \cdot \frac{1}{7} = \frac{1}{4}$

$p_2 = v_1 x_2 = \frac{7}{4} \cdot \frac{3}{7} = \frac{3}{4}$ $q_2 = v_1 z_2 = \frac{7}{4} \cdot \frac{3}{7} = \frac{3}{4}$

The value v of the original matrix is given by $v = v_1 - 2 = \frac{7}{4} - 2 = -\frac{1}{4}$,

and $P^* = \begin{bmatrix} 0 & \frac{1}{4} & \frac{3}{4} & 0 \end{bmatrix}$, $Q^* = \begin{bmatrix} 0 \\ \frac{1}{4} \\ \frac{3}{4} \\ 0 \end{bmatrix}$.

15.

		Player C (Fate)	
		Republican	Democrat
Player R (You)	Solar energy	1000	4000
	Oil	5000	3000

This game is nonstrictly determined.
The linear programming problems corresponding to the matrix above are shown as follows:

(A) Minimize $y = x_1 + x_2$
Subject to $1000x_1 + 5000x_2 \geq 1$
$4000x_1 + 3000x_2 \geq 1$
$x_1, x_2 \geq 0$

(B) Maximize $y = z_1 + z_2$
Subject to $1000z_1 + 4000z_2 \leq 1$
$5000z_1 + 3000z_2 \leq 1$
$z_1, z_2 \geq 0$

The graphs corresponding to the inequalities are shown in the following figures.

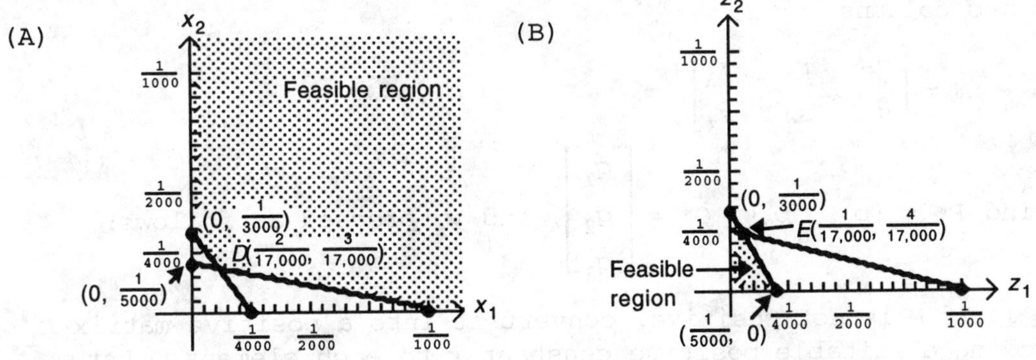

In figure (A), since the minimum $y = x_1 + x_2$ occurs at the point of intersection D, we solve

$1000x_1 + 5000x_2 = 1$
$4000x_1 + 3000x_2 = 1$

to obtain the coordinates of point D: $x_1 = \frac{2}{17,000}$; $x_2 = \frac{3}{17,000}$

Similarly, in figure (B), since the maximum $y = z_1 + z_2$ occurs at the point of intersection E, we solve

$1000z_1 + 4000z_2 = 1$
$5000z_1 + 3000z_2 = 1$

to obtain the coordinates of point E: $z_1 = \frac{1}{17,000}$

$z_2 = \frac{4}{17,000}$

Thus,

$$v = \frac{1}{x_1 + x_2} = \frac{1}{z_1 + z_2} = \frac{1}{\frac{2}{17,000} + \frac{3}{17,000}} = \frac{17,000}{5} = 3400$$

$$p_1 = vx_1 = 3400 \cdot \frac{2}{17,000} = \frac{2}{5} \qquad q_1 = vz_1 = 3400 \cdot \frac{1}{17,000} = \frac{1}{5}$$

$$p_2 = vx_2 = 3400 \cdot \frac{3}{17,000} = \frac{3}{5} \qquad q_2 = vz_2 = 3400 \cdot \frac{4}{17,000} = \frac{4}{5}$$

$$P^* = \begin{bmatrix} \frac{2}{5} & \frac{3}{5} \end{bmatrix}, \qquad Q^* = \begin{bmatrix} \frac{1}{5} \\ \frac{4}{5} \end{bmatrix}, \qquad v = 3400 \text{ or } v = \$3400$$

EXERCISE 8-4

Things to remember:

$m \times n$ MATRIX GAMES AND LINEAR PROGRAMMING—SIMPLEX METHOD AND THE DUAL

Given the nonstrictly determined matrix game M free of recessive rows and columns

$$M = \begin{bmatrix} r_1 & r_2 & r_3 \\ s_1 & s_2 & s_3 \end{bmatrix}.$$

To find $P^* = [p_1 \quad p_2]$, $Q^* = \begin{bmatrix} q_1 \\ q_2 \\ q_3 \end{bmatrix}$, and v, proceed as follows:

STEP 1. If M is not positive, convert it into a positive matrix M_1 by adding a suitable positive constant k to each element. Let

$$M_1 = \begin{bmatrix} a_1 & a_2 & a_3 \\ b_1 & b_2 & b_3 \end{bmatrix}$$

[Note: If v_1 is the value of the game M_1, then $v = v_1 - k$ is the value of the original game M.]

STEP 2. Set up the two linear programming problems:

(A) Minimize $y = x_1 + x_2$

Subject to: $a_1x_1 + b_1x_2 \geq 1$
$a_2x_1 + b_2x_2 \geq 1$
$a_3x_1 + b_3x_2 \geq 1$
$x_1, x_2 \geq 0$

(B) Maximize $y = z_1 + z_2 + z_3$

Subject to: $a_1z_1 + a_2z_2 + a_3z_3 \leq 1$
$b_1z_1 + b_2z_2 + b_3z_3 \leq 1$
$z_1, z_2, z_3 \geq 0$

[Note: (A) is the dual of (B).]

STEP 3. Solve the maximization problem (B) using the simplex method. Since (A) is the dual of (B), this will also produce the solution of (A).

STEP 4. Using the solutions in Step 3:

$$v_1 = \frac{1}{y} = \frac{1}{x_1 + x_2} \quad \text{or} \quad v_1 = \frac{1}{y} = \frac{1}{z_1 + z_2 + z_3}$$

$$P^* = [p_1 \quad p_2] = [v_1x_1 \quad v_1x_2]$$

$$Q^* = \begin{bmatrix} q_1 \\ q_2 \\ q_3 \end{bmatrix} = \begin{bmatrix} v_1z_1 \\ v_1z_2 \\ v_1z_3 \end{bmatrix}$$

$$v = v_1 - k$$

STEP 5. The solution found in Step 4 can be checked by showing that $P^*MQ^* = v$.

---

1. Convert $\begin{bmatrix} 1 & 4 & 0 \\ 0 & -1 & 2 \end{bmatrix}$ into a positive payoff matrix by adding 2 to each payoff:

$$M_1 = \begin{bmatrix} 3 & 6 & 2 \\ 2 & 1 & 4 \end{bmatrix}$$

The linear programming problems corresponding to M_1 are as follows:

(A) Minimize $y = x_1 + x_2$

Subject to $3x_1 + 2x_2 \geq 1$
$6x_1 + x_2 \geq 1$
$2x_1 + 4x_2 \geq 1$
$x_1 \geq 0, x_2 \geq 0$

(B) Maximize $y = z_1 + z_2 + z_3$

Subject to $3z_1 + 6z_2 + 2z_3 \leq 1$
$2z_1 + z_2 + 4z_3 \leq 1$
$z_1 \geq 0, z_2 \geq 0, z_3 \geq 0$

Solve (B) by using the simplex method. We introduce slack variables x_1 and x_2 to obtain:

$$\begin{aligned} 3z_1 + 6z_2 + 2z_3 + x_1 \qquad\qquad &= 1 \\ 2z_1 + z_2 + 4z_3 \qquad + x_2 \qquad &= 1 \\ -z_1 - z_2 - z_3 \qquad\qquad + y &= 0 \\ z_1, z_2, z_3 \geq 0,\ x_1, x_2 \geq 0 & \end{aligned}$$

The simplex tableau for this system is:

$$\begin{array}{c} \begin{array}{cccccc} z_1 & z_2 & z_3 & x_1 & x_2 & y \end{array} \\ \left[\begin{array}{cccccc|c} \textcircled{3} & 6 & 2 & 1 & 0 & 0 & 1 \\ 2 & 1 & 4 & 0 & 1 & 0 & 1 \\ \hdashline -1 & -1 & -1 & 0 & 0 & 1 & 0 \end{array}\right] \begin{array}{c} \frac{1}{3} \\ \frac{1}{2} \\ \\ \end{array} \\ \frac{1}{3}R_1 \to R_1 \end{array}$$

Choose the first column as the pivot column. Then, following the method outlined in Chapter 6, we get the pivot element in the 1,1 position, indicated by a circle.

$$\left[\begin{array}{cccccc|c} \textcircled{1} & 2 & \frac{2}{3} & \frac{1}{3} & 0 & 0 & \frac{1}{3} \\ 2 & 1 & 4 & 0 & 1 & 0 & 1 \\ \hdashline -1 & -1 & -1 & 0 & 0 & 1 & 0 \end{array}\right] \sim \left[\begin{array}{cccccc|c} 1 & 2 & \frac{2}{3} & \frac{1}{3} & 0 & 0 & \frac{1}{3} \\ 0 & -3 & \textcircled{\frac{8}{3}} & -\frac{2}{3} & 1 & 0 & \frac{1}{3} \\ \hdashline 0 & 1 & -\frac{1}{3} & \frac{1}{3} & 0 & 1 & \frac{1}{3} \end{array}\right] \begin{array}{c} \frac{1}{2} \\ \frac{1}{8} \\ \\ \end{array}$$

$R_2 + (-2)R_1 \to R_2$ and $R_3 + R_1 \to R_3$ $\qquad \frac{3}{8}R_2 \to R_2$

$$\sim \left[\begin{array}{cccccc|c} 1 & 2 & \frac{2}{3} & \frac{1}{3} & 0 & 0 & \frac{1}{3} \\ 0 & -\frac{9}{8} & \textcircled{1} & -\frac{1}{4} & \frac{3}{8} & 0 & \frac{1}{8} \\ \hdashline 0 & 1 & -\frac{1}{3} & \frac{1}{3} & 0 & 1 & \frac{1}{3} \end{array}\right] \sim \begin{array}{c} \begin{array}{cccccc} z_1 & z_2 & z_3 & x_1 & x_2 & y \end{array} \\ \left[\begin{array}{cccccc|c} 1 & \frac{11}{4} & 0 & \frac{1}{2} & -\frac{1}{4} & 0 & \frac{1}{4} \\ 0 & -\frac{9}{8} & 1 & -\frac{1}{4} & \frac{3}{8} & 0 & \frac{1}{8} \\ \hdashline 0 & \frac{5}{8} & 0 & \frac{1}{4} & \frac{1}{8} & 1 & \frac{9}{24} \end{array}\right] \end{array}$$

$$R_1 + \left(-\frac{2}{3}\right)R_2 \to R_1 \text{ and } R_3 + \frac{1}{3}R_2 \to R_3$$

The maximum, $y = z_1 + z_2 + z_3 = \frac{9}{24}$, occurs at $z_1 = \frac{1}{4}$, $z_2 = 0$, $z_3 = \frac{1}{8}$.

Thus, $v_1 = \dfrac{1}{z_1 + z_2 + z_3} = \dfrac{1}{y} = \dfrac{1}{\frac{9}{24}} = \dfrac{24}{9} = \dfrac{8}{3}$, and $q_1 = v_1 z_1 = \frac{8}{3} \cdot \frac{1}{4} = \frac{2}{3}$,

$q_2 = v_1 z_2 = \frac{8}{3} \cdot 0 = 0$, $q_3 = v_1 z_3 = \frac{8}{3} \cdot \frac{1}{8} = \frac{1}{3}$.

The solution to the minimization problem (A) can be read from the bottom row of the final simplex tableau for the dual problem above. Thus, from the row

$$\begin{array}{c} \begin{array}{cc} \qquad\qquad x_1 & x_2 \qquad\qquad \end{array} \\ \left[\begin{array}{cccccc|c} 0 & \frac{5}{8} & 0 & \frac{1}{4} & \frac{1}{8} & 1 & \frac{9}{24} \end{array}\right] \end{array}$$

we conclude that the solution to (A) is:

$$\text{Min } y = x_1 + x_2 = \frac{9}{24} = \frac{3}{8} \text{ at } x_1 = \frac{1}{4},\ x_2 = \frac{1}{8}.$$

Also,

$$p_1 = v_1x_1 = \frac{8}{3}\left(\frac{1}{4}\right) = \frac{2}{3}$$

$$p_2 = v_1x_2 = \frac{8}{3}\left(\frac{1}{8}\right) = \frac{1}{3}$$

Finally,

$$P^* = \begin{bmatrix} \frac{2}{3} & \frac{1}{3} \end{bmatrix}, \quad Q^* = \begin{bmatrix} \frac{2}{3} \\ 0 \\ \frac{1}{3} \end{bmatrix}, \quad \text{and} \quad v = \frac{8}{3} - 2 = \frac{2}{3}.$$

3. Convert $\begin{bmatrix} 0 & 1 & -2 \\ -1 & 0 & 3 \\ 2 & -3 & 0 \end{bmatrix}$ into a positive payoff matrix by adding 4 to each payoff:

$$M_1 = \begin{bmatrix} 4 & 5 & 2 \\ 3 & 4 & 7 \\ 6 & 1 & 4 \end{bmatrix}$$

The linear programming problems corresponding to M_1 are as follows:

(A) Minimize $y = x_1 + x_2 + x_3$

Subject to
$$\begin{aligned} 4x_1 + 3x_2 + 6x_3 &\geq 1 \\ 5x_1 + 4x_2 + x_3 &\geq 1 \\ 2x_1 + 7x_2 + 4x_3 &\geq 1 \\ x_1, x_2, x_3 &\geq 0 \end{aligned}$$

(B) Maximize $y = z_1 + z_2 + z_3$

Subject to
$$\begin{aligned} 4z_1 + 5z_2 + 2z_3 &\leq 1 \\ 3z_1 + 4z_2 + 7z_3 &\leq 1 \\ 6z_1 + z_2 + 4z_3 &\leq 1 \\ z_1, z_2, z_3 &\geq 0 \end{aligned}$$

We first solve (B) by using the simplex method. Introduce slack variables x_1, x_2, and x_3 to obtain:

$$\begin{aligned} 4z_1 + 5z_2 + 2z_3 + x_1 \qquad\qquad &= 1 \\ 3z_1 + 4z_2 + 7z_3 \qquad + x_2 \qquad &= 1 \\ 6z_1 + z_2 + 4z_3 \qquad\qquad + x_3 &= 1 \\ -z_1 - z_2 - z_3 \qquad\qquad\qquad + y &= 0 \\ z_1, z_2, z_3 \geq 0,\ x_1, x_2, x_3 &\geq 0 \end{aligned}$$

The simplex tableau for this system is as follows:

$$\begin{array}{c} \\ \\ \end{array}\begin{array}{ccccccc|c} z_1 & z_2 & z_3 & x_1 & x_2 & x_3 & y & \\ \hline 4 & 5 & 2 & 1 & 0 & 0 & 0 & 1 \\ 3 & 4 & \textcircled{7} & 0 & 1 & 0 & 0 & 1 \\ 6 & 1 & 4 & 0 & 0 & 1 & 0 & 1 \\ \hline -1 & -1 & -1 & 0 & 0 & 0 & 1 & 0 \end{array} \begin{array}{c} \frac{1}{2} \\ \frac{1}{7} \\ \frac{1}{4} \\ \\ \end{array} \sim \left[\begin{array}{ccccccc|c} 4 & 5 & 2 & 1 & 0 & 0 & 0 & 1 \\ \frac{3}{7} & \frac{4}{7} & \textcircled{1} & 0 & \frac{1}{7} & 0 & 0 & \frac{1}{7} \\ 6 & 1 & 4 & 0 & 0 & 1 & 0 & 1 \\ \hline -1 & -1 & -1 & 0 & 0 & 0 & 1 & 0 \end{array}\right]$$

$\frac{1}{7}R_2 \rightarrow R_2$ $\qquad$ $R_1 - 2R_2 \rightarrow R_1$, $R_3 - 4R_2 \rightarrow R_3$ and $R_4 + R_2 \rightarrow R_4$

$$
\sim \left[\begin{array}{ccccccc|c}
z_1 & z_2 & z_3 & x_1 & x_2 & x_3 & y & \\
\frac{22}{7} & \frac{27}{7} & 0 & 1 & -\frac{2}{7} & 0 & 0 & \frac{5}{7} \\
\frac{3}{7} & \frac{4}{7} & 1 & 0 & \frac{1}{7} & 0 & 0 & \frac{1}{7} \\
\textcircled{\frac{30}{7}} & -\frac{9}{7} & 0 & 0 & -\frac{4}{7} & 1 & 0 & \frac{3}{7} \\
\hdashline
-\frac{4}{7} & -\frac{3}{7} & 0 & 0 & \frac{1}{7} & 0 & 1 & \frac{1}{7}
\end{array}\right]
\begin{array}{l}
\\
\frac{5}{7} \div \frac{22}{7} = \frac{5}{22} \\
\frac{1}{7} \div \frac{3}{7} = \frac{1}{3} \\
\frac{3}{7} \div \frac{30}{7} = \frac{1}{10} \\
\\
\end{array}
$$

$\frac{7}{30}R_3 \to R_3$

$$
\sim \left[\begin{array}{ccccccc|c}
\frac{22}{7} & \frac{27}{7} & 0 & 1 & -\frac{2}{7} & 0 & 0 & \frac{5}{7} \\
\frac{3}{7} & \frac{4}{7} & 1 & 0 & \frac{1}{7} & 0 & 0 & \frac{1}{7} \\
\textcircled{1} & -\frac{3}{10} & 0 & 0 & -\frac{2}{15} & \frac{7}{30} & 0 & \frac{1}{10} \\
\hdashline
-\frac{4}{7} & -\frac{3}{7} & 0 & 0 & \frac{1}{7} & 0 & 1 & \frac{1}{7}
\end{array}\right]
$$

$R_1 + \left(-\frac{22}{7}\right)R_3 \to R_1$, $R_2 + \left(-\frac{3}{7}\right)R_3 \to R_2$, and $R_4 + \frac{4}{7}R_3 \to R_4$

$$
\sim \left[\begin{array}{ccccccc|c}
0 & \textcircled{\frac{24}{5}} & 0 & 1 & \frac{2}{15} & -\frac{11}{15} & 0 & \frac{2}{5} \\
0 & \frac{7}{10} & 1 & 0 & \frac{1}{5} & -\frac{1}{10} & 0 & \frac{1}{10} \\
1 & -\frac{3}{10} & 0 & 0 & -\frac{2}{15} & \frac{7}{30} & 0 & \frac{1}{10} \\
\hdashline
0 & -\frac{3}{5} & 0 & 0 & \frac{1}{15} & \frac{2}{15} & 1 & \frac{1}{5}
\end{array}\right]
\begin{array}{l}
\frac{2}{5} \div \frac{24}{5} = \frac{1}{12} \\
\frac{1}{10} \div \frac{7}{10} = \frac{1}{7} \\
\\
\\
\end{array}
$$

$\frac{5}{24}R_1 \to R_1$

$$
\sim \left[\begin{array}{ccccccc|c}
0 & \textcircled{1} & 0 & \frac{5}{24} & \frac{1}{36} & -\frac{11}{72} & 0 & \frac{1}{12} \\
0 & \frac{7}{10} & 1 & 0 & \frac{1}{5} & -\frac{1}{10} & 0 & \frac{1}{10} \\
1 & -\frac{3}{10} & 0 & 0 & -\frac{2}{15} & \frac{7}{30} & 0 & \frac{1}{10} \\
\hdashline
0 & -\frac{3}{5} & 0 & 0 & \frac{1}{15} & \frac{2}{15} & 1 & \frac{1}{5}
\end{array}\right]
\sim \left[\begin{array}{ccccccc|c}
z_1 & z_2 & z_3 & x_1 & x_2 & x_3 & y & \\
0 & 1 & 0 & \frac{5}{24} & \frac{1}{36} & -\frac{11}{72} & 0 & \frac{1}{12} \\
0 & 0 & 1 & -\frac{7}{48} & \frac{13}{72} & \frac{1}{144} & 0 & \frac{1}{24} \\
1 & 0 & 0 & \frac{1}{16} & -\frac{1}{8} & \frac{3}{16} & 0 & \frac{1}{8} \\
\hdashline
0 & 0 & 0 & \frac{1}{8} & \frac{1}{12} & \frac{1}{24} & 1 & \frac{1}{4}
\end{array}\right]
$$

$R_2 - \frac{7}{10}R_1 \to R_2$, $R_3 + \frac{3}{10}R_1 \to R_3$,
and $R_4 + \frac{3}{5}R_1 \to R_4$

We obtain: $z_1 = \frac{1}{8}$, $z_2 = \frac{1}{12}$, and $z_3 = \frac{1}{24}$. Thus, $v_1 = \dfrac{1}{\frac{1}{8} + \frac{1}{12} + \frac{1}{24}} = 4$,

and $q_1 = v_1 z_1 = 4 \cdot \frac{1}{8} = \frac{1}{2}$, $q_2 = v_1 z_2 = 4 \cdot \frac{1}{12} = \frac{1}{3}$, $q_3 = v_1 z_3 = 4 \cdot \frac{1}{24} = \frac{1}{6}$.

The solution to the minimization problem (A) can be read from the bottom row of the final simplex tableau for the dual problem above. Thus, from the row

$$
\left[\begin{array}{ccccccc|c}
 & & & x_1 & x_2 & x_3 & & \\
0 & 0 & 0 & \frac{1}{8} & \frac{1}{12} & \frac{1}{24} & 1 & \frac{1}{4}
\end{array}\right]
$$

we conclude that the solution to (A) is:

$$\text{Min } y = x_1 + x_2 + x_3 = \frac{1}{4} \text{ at } x_1 = \frac{1}{8},\ x_2 = \frac{1}{12},\ x_3 = \frac{1}{24}.$$

Also,

$$p_1 = v_1x_1 = 4\left(\frac{1}{8}\right) = \frac{1}{2}$$

$$p_2 = v_1x_2 = 4\left(\frac{1}{12}\right) = \frac{1}{3}$$

$$p_3 = v_1x_3 = 4\left(\frac{1}{24}\right) = \frac{1}{6}$$

Finally,

$$P^* = \begin{bmatrix} \frac{1}{2} & \frac{1}{3} & \frac{1}{6} \end{bmatrix}, \quad Q^* = \begin{bmatrix} \frac{1}{2} \\ \frac{1}{3} \\ \frac{1}{6} \end{bmatrix}, \quad \text{and} \quad v = 4 - 4 = 0.$$

5. $M = \begin{bmatrix} 4 & 2 & 2 & 1 \\ 0 & 1 & -1 & 3 \\ -2 & -1 & -3 & 2 \end{bmatrix}$

Eliminate the recessive row $[-2 \quad -1 \quad -3 \quad 2]$ and the recessive columns $\begin{bmatrix} 4 \\ 0 \\ -2 \end{bmatrix}$ and $\begin{bmatrix} 2 \\ 1 \\ -1 \end{bmatrix}$ to obtain the 2×2 matrix $\begin{bmatrix} 2 & 1 \\ -1 & 3 \end{bmatrix}$.

Convert this matrix into a positive payoff matrix by adding 2 to each payoff

$$M_1 = \begin{bmatrix} 4 & 3 \\ 1 & 5 \end{bmatrix}$$

The linear programming problems corresponding to M_1 are:

(A) Minimize $y = x_1 + x_2$

Subject to: $4x_1 + x_2 \ge 1$

$3x_1 + 5x_2 \ge 1$

$x_1, x_2 \ge 0$

(B) Maximize $y = z_1 + z_2$

Subject to: $4z_1 + 3z_2 \le 1$

$z_1 + 5z_2 \le 1$

$z_1, z_2 \ge 0$

Solve (*B*) by using the simplex method. We introduce slack variables x_1, x_2 to obtain:

$$\begin{aligned} 4z_1 + 3z_2 + x_1 \quad\quad &= 1 \\ z_1 + 5z_2 \quad\quad + x_2 &= 1 \\ z_1, z_2, x_1, x_2 &\ge 0 \end{aligned}$$

The simplex tableau for this system is:

$$\begin{array}{c} \begin{matrix} z_1 & z_2 & x_1 & x_2 & y \end{matrix} \\ \left[\begin{array}{ccccc|c} \textcircled{4} & 3 & 1 & 0 & 0 & 1 \\ 1 & 5 & 0 & 1 & 0 & 1 \\ \hline -1 & -1 & 0 & 0 & 1 & 0 \end{array}\right] \begin{matrix} \frac{1}{4} \\ 1 \\ \\ \end{matrix} \end{array} \quad \left(\frac{1}{4}\right)R_1 \to R_1$$

We choose the first column as the pivot column. Following the methods of Chapter 5, we get the pivot element in the 1,1 position (circled).

$$\left[\begin{array}{ccccc|c} 1 & \frac{3}{4} & \frac{1}{4} & 0 & 0 & \frac{1}{4} \\ 1 & 5 & 0 & 1 & 0 & 1 \\ \hline -1 & -1 & 0 & 0 & 1 & 0 \end{array}\right] \sim \left[\begin{array}{ccccc|c} 1 & \frac{3}{4} & \frac{1}{4} & 0 & 0 & \frac{1}{4} \\ 0 & \boxed{\frac{17}{4}} & -\frac{1}{4} & 1 & 0 & \frac{3}{4} \\ \hline 0 & -\frac{1}{4} & \frac{1}{4} & 0 & 1 & \frac{1}{4} \end{array}\right]$$

$(-1)R_1 + R_2 \to R_2$ $\quad\quad$ $\left(\frac{4}{17}\right)R_2 \to R_2$

$R_1 + R_3 \to R_3$

$$\sim \left[\begin{array}{ccccc|c} 1 & \frac{3}{4} & \frac{1}{4} & 0 & 0 & \frac{1}{4} \\ 0 & 1 & -\frac{1}{17} & \frac{4}{17} & 0 & \frac{3}{17} \\ \hline 0 & -\frac{1}{4} & \frac{1}{4} & 0 & 1 & \frac{1}{4} \end{array}\right]$$

$\left(-\frac{3}{4}\right)R_2 + R_1 \to R_1$, $\left(\frac{1}{4}\right)R_2 + R_3 \to R_3$

$$\begin{array}{cccccc} z_1 & z_2 & x_1 & x_2 & y & \\ \end{array}$$
$$\left[\begin{array}{ccccc|c} 1 & 0 & \frac{5}{17} & -\frac{3}{17} & 0 & \frac{2}{17} \\ 0 & 1 & -\frac{1}{17} & \frac{4}{17} & 0 & \frac{3}{17} \\ \hline 0 & 0 & \frac{4}{17} & \frac{1}{17} & 1 & \frac{5}{17} \end{array}\right]$$

The maximum $y = z_1 + z_2 = \frac{5}{17}$ occurs at $z_1 = \frac{2}{17}$, $z_2 = \frac{3}{17}$.

Thus, $v_1 = \frac{1}{z_1 + z_2} = \frac{1}{y} = \frac{17}{5}$ and $q_1 = v_1 z_1 = \frac{2}{5}$, $q_2 = v_1 z_2 = \frac{3}{5}$.

The solution to the minimization problem (A) can be read from the bottom row of the final simplex tableau for the dual problem (B). Thus, from the row

$$\begin{array}{ccccc|c} z_1 & z_2 & x_1 & x_2 & y & \\ {[0} & 0 & \frac{4}{17} & \frac{1}{17} & 1 & \frac{5}{17}] \end{array}$$

we find the solution to (A):

Minimum $y = x_1 + x_2 = \frac{5}{17}$ at $x_1 = \frac{4}{17}$, $x_2 = \frac{1}{17}$

Also, $p_1 = v_1 x_1 = \frac{4}{5}$, $p_2 = v_1 x_2 = \frac{1}{5}$

Finally,

$$P^* = \left[\frac{4}{5} \quad \frac{1}{5} \quad 0\right], \quad Q^* = \begin{bmatrix} 0 \\ 0 \\ \frac{2}{5} \\ \frac{3}{5} \end{bmatrix} \quad \text{and} \quad v = v_1 - 2 = \frac{17}{5} - 2 = \frac{7}{5}.$$

7. $M = \begin{bmatrix} -2 & -1 & 3 & -1 \\ 1 & 2 & 4 & 0 \\ -1 & 1 & -1 & 1 \\ 0 & 1 & -1 & 2 \end{bmatrix}$

Eliminate the recessive rows [-2 -1 3 -1], [-1 1 -1 1] and

recessive column $\begin{bmatrix} -1 \\ 2 \\ 1 \\ 1 \end{bmatrix}$ to obtain the 2 × 3 matrix $\begin{bmatrix} 1 & 4 & 0 \\ 0 & -1 & 2 \end{bmatrix}$.

This is the matrix of Problem 1. Thus,

$$P^* = [0 \quad \tfrac{2}{3} \quad 0 \quad \tfrac{1}{3}], \quad Q^* = \begin{bmatrix} \tfrac{2}{3} \\ 0 \\ 0 \\ \tfrac{1}{3} \end{bmatrix} \quad \text{and} \quad v = \frac{2}{3}.$$

9. (A) The matrix for this game is as follows:

	Paper	Stone	Scissors
Paper	0	1	-1
Stone	-1	0	1
Scissors	1	-1	0

(B) Convert this game matrix into a positive payoff matrix by adding 2 to each payoff:

$$M_1 = \begin{bmatrix} 2 & 3 & 1 \\ 1 & 2 & 3 \\ 3 & 1 & 2 \end{bmatrix}$$

The linear programming problems corresponding to M_1 are as follows:

(1) Minimize $y = x_1 + x_2 + x_3$

Subject to
$$\begin{aligned} 2x_1 + x_2 + 3x_3 &\ge 1 \\ 3x_1 + 2x_2 + x_3 &\ge 1 \\ x_1 + 3x_2 + 2x_3 &\ge 1 \\ x_1 \ge 0,\ x_2 \ge 0,\ x_3 &\ge 0 \end{aligned}$$

(2) Maximize $y = z_1 + z_2 + z_3$

Subject to
$$\begin{aligned} 2z_1 + 3z_2 + z_3 &\le 1 \\ z_1 + 2z_2 + 3z_3 &\le 1 \\ 3z_1 + z_2 + 2z_3 &\le 1 \\ z_1 \ge 0,\ z_2 \ge 0,\ z_3 &\ge 0 \end{aligned}$$

Solve (2) by using the simplex method. Introduce slack variables x_1, x_2, and x_3 to obtain:

$$\begin{aligned} 2z_1 + 3z_2 + z_3 + x_1 \qquad\qquad &= 1 \\ z_1 + 2z_2 + 3z_3 \qquad + x_2 \qquad &= 1 \\ 3z_1 + z_2 + 2z_3 \qquad\qquad + x_3 &= 1 \\ -z_1 - z_2 - z_3 \qquad\qquad\qquad + y &= 0 \\ z_1, z_2, z_3 \ge 0,\ x_1, x_2, x_3 &\ge 0 \end{aligned}$$

The simplex tableau for this system is given below:

$$\begin{array}{c} \begin{matrix} z_1 & z_2 & z_3 & x_1 & x_2 & x_3 & y \end{matrix} \\ \left[\begin{array}{ccccccc|c} 2 & 3 & 1 & 1 & 0 & 0 & 0 & 1 \\ 1 & 2 & \textcircled{3} & 0 & 1 & 0 & 0 & 1 \\ 3 & 1 & 2 & 0 & 0 & 1 & 0 & 1 \\ \hdashline -1 & -1 & -1 & 0 & 0 & 0 & 1 & 0 \end{array}\right] \begin{matrix} 1 \\ \frac{1}{3} \\ \frac{1}{2} \\ \\ \end{matrix} \end{array} \sim \left[\begin{array}{ccccccc|c} 2 & 3 & 1 & 1 & 0 & 0 & 0 & 1 \\ \frac{1}{3} & \frac{2}{3} & \textcircled{1} & 0 & \frac{1}{3} & 0 & 0 & \frac{1}{3} \\ 3 & 1 & 2 & 0 & 0 & 1 & 0 & 1 \\ \hdashline -1 & -1 & -1 & 0 & 0 & 0 & 1 & 0 \end{array}\right]$$

$\frac{1}{3}R_2 \to R_2$

$R_1 + (-1)R_2 \to R_1$, $R_3 + (-2)R_2 \to R_3$, and $R_4 + R_2 \to R_4$

$$\sim \left[\begin{array}{ccccccc|c} \frac{5}{3} & \frac{7}{3} & 0 & 1 & -\frac{1}{3} & 0 & 0 & \frac{2}{3} \\ \frac{1}{3} & \frac{2}{3} & 1 & 0 & \frac{1}{3} & 0 & 0 & \frac{1}{3} \\ \boxed{\frac{7}{3}} & -\frac{1}{3} & 0 & 0 & -\frac{2}{3} & 1 & 0 & \frac{1}{3} \\ \hline -\frac{2}{3} & -\frac{1}{3} & 0 & 0 & \frac{1}{3} & 0 & 1 & \frac{1}{3} \end{array}\right] \begin{array}{l} \frac{2}{3} \div \frac{5}{3} = \frac{2}{5} \\ \frac{1}{3} \div \frac{1}{3} = 1 \\ \frac{1}{3} \div \frac{7}{3} = \frac{1}{7} \\ \end{array}$$

$\frac{3}{7}R_3 \to R_3$

$$\begin{array}{c} \begin{array}{ccccccc} z_1 & z_2 & z_3 & x_1 & x_2 & x_3 & y \end{array} \\ \sim \left[\begin{array}{ccccccc|c} \frac{5}{3} & \frac{7}{3} & 0 & 1 & -\frac{1}{3} & 0 & 0 & \frac{2}{3} \\ \frac{1}{3} & \frac{2}{3} & 1 & 0 & \frac{1}{3} & 0 & 0 & \frac{1}{3} \\ \boxed{1} & -\frac{1}{7} & 0 & 0 & -\frac{2}{7} & \frac{3}{7} & 0 & \frac{1}{7} \\ \hline -\frac{2}{3} & -\frac{1}{3} & 0 & 0 & \frac{1}{3} & 0 & 1 & \frac{1}{3} \end{array}\right] \end{array}$$

$R_1 + \left(-\frac{5}{3}\right)R_3 \to R_1$, $R_2 + \left(-\frac{1}{3}\right)R_3 \to R_2$, and $R_4 + \frac{2}{3}R_3 \to R_4$

$$\sim \left[\begin{array}{ccccccc|c} 0 & \boxed{\frac{54}{21}} & 0 & 1 & \frac{3}{21} & -\frac{5}{7} & 0 & \frac{3}{7} \\ 0 & \frac{5}{7} & 1 & 0 & \frac{3}{7} & -\frac{1}{7} & 0 & \frac{2}{7} \\ 1 & -\frac{1}{7} & 0 & 0 & -\frac{2}{7} & \frac{3}{7} & 0 & \frac{1}{7} \\ \hline 0 & -\frac{3}{7} & 0 & 0 & \frac{1}{7} & \frac{2}{7} & 1 & \frac{3}{7} \end{array}\right] \begin{array}{l} \frac{3}{7} \div \frac{54}{21} = \frac{1}{6} \\ \frac{2}{7} \div \frac{5}{7} = \frac{2}{5} \\ \\ \end{array}$$

$\frac{21}{54}R_1 \to R_1$

$$\sim \left[\begin{array}{ccccccc|c} 0 & \boxed{1} & 0 & \frac{21}{54} & \frac{1}{18} & -\frac{5}{18} & 0 & \frac{1}{6} \\ 0 & \frac{5}{7} & 1 & 0 & \frac{3}{7} & -\frac{1}{7} & 0 & \frac{2}{7} \\ 1 & -\frac{1}{7} & 0 & 0 & -\frac{2}{7} & \frac{3}{7} & 0 & \frac{1}{7} \\ \hline 0 & -\frac{3}{7} & 0 & 0 & \frac{1}{7} & \frac{2}{7} & 1 & \frac{3}{7} \end{array}\right] \sim \begin{array}{c} \begin{array}{ccccccc} z_1 & z_2 & z_3 & x_1 & x_2 & x_3 & y \end{array} \\ \left[\begin{array}{ccccccc|c} 0 & 1 & 0 & \frac{21}{54} & \frac{1}{18} & -\frac{5}{18} & 0 & \frac{1}{6} \\ 0 & 0 & 1 & -\frac{5}{18} & \frac{7}{18} & \frac{7}{126} & 0 & \frac{1}{6} \\ 1 & 0 & 0 & \frac{1}{18} & -\frac{5}{18} & \frac{49}{126} & 0 & \frac{1}{6} \\ \hline 0 & 0 & 0 & \frac{1}{6} & \frac{1}{6} & \frac{1}{6} & 1 & \frac{1}{2} \end{array}\right] \end{array}$$

$R_2 + \left(-\frac{5}{7}\right)R_1 \to R_2$, $R_3 + \frac{1}{7}R_1 \to R_3$,

and $R_4 + \frac{3}{7}R_1 \to R_4$

Thus, $z_1 = \frac{1}{6}$, $z_2 = \frac{1}{6}$, $z_3 = \frac{1}{6}$, $v_1 = \frac{1}{z_1 + z_2 + z_3} = \frac{1}{\frac{1}{6} + \frac{1}{6} + \frac{1}{6}} = 2$, and

$q_1 = v_1 z_1 = 2 \cdot \frac{1}{6} = \frac{1}{3}$, $q_2 = v_1 z_2 = 2 \cdot \frac{1}{6} = \frac{1}{3}$, $q_3 = v_1 z_3 = 2 \cdot \frac{1}{6} = \frac{1}{3}$.

The solution to the minimization problem (A) can be read from the bottom row of the final simplex tableau for the dual problem shown on the previous page. Thus, from the row

$$\begin{array}{cccccccc} & & & x_1 & x_2 & x_3 & & \\ \left[0\right. & 0 & 0 & \frac{1}{6} & \frac{1}{6} & \frac{1}{6} & 1 & \left| \ \frac{1}{2}\right] \end{array}$$

we conclude that the solution to (A) is:

$$\text{Min } y = x_1 + x_2 + x_3 = \frac{1}{2} \text{ at } x_1 = \frac{1}{6},\ x_2 = \frac{1}{6},\ x_3 = \frac{1}{6}.$$

Also,

$$p_1 = v_1x_1 = 2\left(\frac{1}{6}\right) = \frac{1}{3}$$

$$p_2 = v_1x_2 = 2\left(\frac{1}{6}\right) = \frac{1}{3}$$

$$p_3 = v_1x_3 = 2\left(\frac{1}{6}\right) = \frac{1}{3}$$

Finally,

$$P^* = \begin{bmatrix} \frac{1}{3} & \frac{1}{3} & \frac{1}{3} \end{bmatrix}, \quad Q^* = \begin{bmatrix} \frac{1}{3} \\ \frac{1}{3} \\ \frac{1}{3} \end{bmatrix}, \quad \text{and} \quad v = 2 - 2 = 0.$$

11. (A)

$$\begin{array}{l} \\ \\ \text{Deluxe} \\ \text{Standard} \\ \text{Economy} \end{array} \begin{array}{c} \text{Economy} \\ \begin{array}{cc} \text{Up} & \text{Down} \end{array} \\ \begin{bmatrix} 2 & -1 \\ 1 & 1 \\ 0 & 4 \end{bmatrix} \end{array}$$

(B) Since the payoff matrix is not positive, we add 2 to each element to obtain the matrix:

$$M_1 = \begin{bmatrix} 4 & 1 \\ 3 & 3 \\ 2 & 6 \end{bmatrix}$$

The linear programming problems corresponding to M_1 are:

(A) Minimize $y = x_1 + x_2 + x_3$

Subject to
$$\begin{aligned} 4x_1 + 3x_2 + 2x_3 &\geq 1 \\ x_1 + 3x_2 + 6x_3 &\geq 1 \\ x_1, x_2, x_3 &\geq 0 \end{aligned}$$

(B) Maximize $y = z_1 + z_2$

Subject to
$$\begin{aligned} 4z_1 + z_2 &\leq 1 \\ 3z_1 + 3z_2 &\leq 1 \\ 2z_1 + 6z_2 &\leq 1 \\ z_1, z_2 &\geq 0 \end{aligned}$$

Solve (B) by using the simplex method. Introduce slack variables x_1, x_2, and x_3 to obtain:

$$\begin{aligned} 4z_1 + z_2 + x_1 \qquad\qquad\qquad &= 1 \\ 3z_1 + 3z_2 \qquad + x_2 \qquad\quad &= 1 \\ 2z_1 + 6z_2 \qquad\qquad + x_3 \quad &= 1 \\ -z_1 - z_2 \qquad\qquad\qquad + y &= 0 \end{aligned}$$

The simplex tableau for this system is:

$$\begin{array}{c} \\ \end{array}\begin{matrix} z_1 & z_2 & x_1 & x_2 & x_3 & y & \end{matrix}$$

$$\left[\begin{array}{cccccc|c} \textcircled{4} & 1 & 1 & 0 & 0 & 0 & 1 \\ 3 & 3 & 0 & 1 & 0 & 0 & 1 \\ 2 & 6 & 0 & 0 & 1 & 0 & 1 \\ \hdashline -1 & -1 & 0 & 0 & 0 & 1 & 0 \end{array}\right]\begin{matrix} \frac{1}{4} \\ \frac{1}{3} \\ \frac{1}{2} \\ \\ \end{matrix} \sim \left[\begin{array}{cccccc|c} \textcircled{1} & \frac{1}{4} & \frac{1}{4} & 0 & 0 & 0 & \frac{1}{4} \\ 3 & 3 & 0 & 1 & 0 & 0 & 1 \\ 2 & 6 & 0 & 0 & 1 & 0 & 1 \\ \hdashline -1 & -1 & 0 & 0 & 0 & 1 & 0 \end{array}\right]$$

$\frac{1}{4}R_1 \to R_1$ $\qquad$ $R_2 + (-3)R_1 \to R_2$, $R_3 + (-2)R_1 \to R_3$, $R_4 + R_1 \to R_4$

$$\sim \left[\begin{array}{cccccc|c} 1 & \frac{1}{4} & \frac{1}{4} & 0 & 0 & 0 & \frac{1}{4} \\ 0 & \frac{9}{4} & -\frac{3}{4} & 1 & 0 & 0 & \frac{1}{4} \\ 0 & \textcircled{\frac{11}{2}} & -\frac{1}{2} & 0 & 1 & 0 & \frac{1}{2} \\ \hdashline 0 & -\frac{3}{4} & \frac{1}{4} & 0 & 0 & 1 & \frac{1}{4} \end{array}\right]\begin{matrix} 1 \\ \frac{1}{9} \\ \frac{1}{11} \\ \\ \end{matrix} \sim \left[\begin{array}{cccccc|c} 1 & \frac{1}{4} & \frac{1}{4} & 0 & 0 & 0 & \frac{1}{4} \\ 0 & \frac{9}{4} & -\frac{3}{4} & 1 & 0 & 0 & \frac{1}{4} \\ 0 & \textcircled{1} & -\frac{1}{11} & 0 & \frac{2}{11} & 0 & \frac{1}{11} \\ \hdashline 0 & -\frac{3}{4} & \frac{1}{4} & 0 & 0 & 1 & \frac{1}{4} \end{array}\right]$$

$\frac{2}{11}R_3 \to R_3$ $\qquad$ $R_1 + \left(-\frac{1}{4}\right)R_3 \to R_1$, $R_2 + \left(-\frac{9}{4}\right)R_3 \to R_2$, $R_4 + \frac{3}{4}R_3 \to R_4$

$$\begin{matrix} z_1 & z_2 & x_1 & x_2 & x_3 & y \end{matrix}$$

$$\sim \left[\begin{array}{cccccc|c} 1 & 0 & \frac{12}{44} & 0 & -\frac{2}{44} & 0 & \frac{10}{44} \\ 0 & 0 & -\frac{24}{44} & 1 & -\frac{18}{44} & 0 & \frac{2}{44} \\ 0 & 1 & -\frac{1}{11} & 0 & \frac{2}{11} & 0 & \frac{1}{11} \\ \hdashline 0 & 0 & \frac{2}{11} & 0 & \frac{3}{22} & 1 & \frac{14}{44} \end{array}\right]$$

Thus, max $y = z_1 + z_2 = \frac{14}{44} = \frac{7}{22}$ occurs at $z_1 = \frac{10}{44} = \frac{5}{22}$, $z_2 = \frac{1}{11}$. Now $v_1 = \frac{1}{y} = \frac{1}{\frac{7}{22}} = \frac{22}{7}$ and $q_1 = v_1 z_1 = \frac{22}{7}\left(\frac{5}{22}\right) = \frac{5}{7}$, $q_2 = v_1 z_2 = \frac{22}{7}\left(\frac{2}{22}\right) = \frac{2}{7}$.

The solution to the minimization problem (A) can be read from the bottom row of the final simplex tableau for the dual problem above. Thus, from the row

$$\begin{matrix} & & x_1 & x_2 & x_3 & & \end{matrix}$$
$$\left[\begin{array}{ccccccc} 0 & 0 & \frac{2}{11} & 0 & \frac{3}{22} & 1 & \frac{7}{22} \end{array}\right]$$

we conclude that the solution to (A) is:

$$\text{Min } y = x_1 + x_2 + x_3 = \frac{7}{22} \text{ at } x_1 = \frac{2}{11},\ x_2 = 0,\ x_3 = \frac{3}{22}.$$

Also,

$$p_1 = v_1 x_1 = \frac{22}{7}\left(\frac{2}{11}\right) = \frac{4}{7}$$

$$p_2 = v_1 x_2 = \frac{22}{7}(0) = 0$$

$$p_3 = v_1 x_3 = \frac{22}{7}\left(\frac{3}{22}\right) = \frac{3}{7}$$

Finally, the optimal strategies are

$$P^* = \begin{bmatrix} \frac{4}{7} & 0 & \frac{3}{7} \end{bmatrix}, \quad Q^* = \begin{bmatrix} \frac{5}{7} \\ \frac{2}{7} \end{bmatrix};$$

and the value of the game is: $v = v_1 - 2 = \frac{22}{7} - 2 = \frac{8}{7}$ million dollars.

(C) Irrespective of what the economy does, the company's budget should be allocated as follows:

$\frac{4}{7}$ of budget on Deluxe, 0 on Standard, and $\frac{3}{7}$ on Economy.

(D) If the company orders only deluxe VCR's and fate plays the strategy "Down", then

$P = [1, 0, 0]$ and $Q = [0, 1]$

The expected value is

$$PMQ = \begin{bmatrix} 1 & 0 & 0 \end{bmatrix} \begin{bmatrix} 2 & -1 \\ 1 & 1 \\ 0 & 4 \end{bmatrix} \begin{bmatrix} 0 \\ 1 \end{bmatrix} = \begin{bmatrix} 2 & -1 \end{bmatrix} \begin{bmatrix} 0 \\ 1 \end{bmatrix} = -1$$

From part (B), the optimal strategy for the company is

$P^* = \begin{bmatrix} \frac{4}{7} & 0 & \frac{3}{7} \end{bmatrix}$. If the company uses this strategy and fate plays the strategy "Down", then the expected value of the game is

$$P^*MQ = \begin{bmatrix} \frac{4}{7} & 0 & \frac{3}{7} \end{bmatrix} \begin{bmatrix} 2 & -1 \\ 1 & 1 \\ 0 & 4 \end{bmatrix} \begin{bmatrix} 0 \\ 1 \end{bmatrix} = \begin{bmatrix} \frac{8}{7} & \frac{8}{7} \end{bmatrix} \begin{bmatrix} 0 \\ 1 \end{bmatrix} = \frac{8}{7}$$

CHAPTER 8 REVIEW

1. 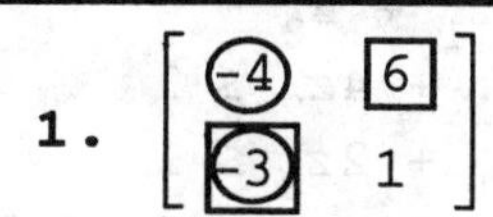

The game is strictly determined.

(A) Saddle value = -3 (2,1 position).

(B) R plays row 2 and C plays column 1.

(C) The value of the game is -3. (8-1)

2. $\begin{bmatrix} \textcircled{-5} & \boxed{2} \\ \boxed{3} & \textcircled{-1} \end{bmatrix}$

The game is not strictly determined. (8-1)

3. $\begin{bmatrix} -3 & -1 & \boxed{5} & \textcircled{-8} \\ 1 & \boxed{\textcircled{0}} & \textcircled{0} & \boxed{2} \\ 3 & \boxed{\textcircled{0}} & 1 & \textcircled{0} \\ \boxed{6} & -2 & \textcircled{-4} & \boxed{2} \end{bmatrix}$

The game is strictly determined.

(A) The two zeros in column 2 (2,2 and 3,2 positions).

(B) R plays either row 2 or row 3 and C plays column 2.

(C) The value of the game = 0. (8-1)

4. $\begin{bmatrix} 1 & \textcircled{-2} & \boxed{3} \\ \textcircled{-1} & \boxed{2} & 0 \\ \boxed{3} & 0 & \textcircled{-4} \end{bmatrix}$

The game is not strictly determined. (8-1)

5. The game matrix $\begin{bmatrix} -2 & 3 & 5 \\ -1 & -3 & 0 \\ 0 & -1 & 1 \end{bmatrix}$ has a recessive row 2 and a recessive column 3. Thus, the reduced game matrix is: $M_1 = \begin{bmatrix} -2 & 3 \\ 0 & -1 \end{bmatrix}$. (8-2)

6. $M = \begin{bmatrix} -2 & 1 \\ 0 & -1 \end{bmatrix}$

Optimal strategy for $R = P^* = \begin{bmatrix} \frac{d - c}{D} & \frac{a - b}{D} \end{bmatrix} = \begin{bmatrix} \frac{-1 - 0}{-4} & \frac{-2 - 1}{-4} \end{bmatrix}$

[<u>Note</u>: $D = (a + d) - (b + c)$ $= \begin{bmatrix} \frac{1}{4} & \frac{3}{4} \end{bmatrix}$
$= (-2 - 1) - (1 + 0) = -4.$]

Optimal strategy for $C = Q^* = \begin{bmatrix} \frac{d - b}{D} \\ \frac{a - c}{D} \end{bmatrix} = \begin{bmatrix} \frac{-1 - 1}{-4} \\ \frac{-2 - 0}{-4} \end{bmatrix} = \begin{bmatrix} \frac{1}{2} \\ \frac{1}{2} \end{bmatrix}$

Value of the game $= v = \dfrac{ad - bc}{D} = \dfrac{(-2)(-1) - (1)(0)}{-4} = -\dfrac{1}{2}$. (8-2)

7. Obtain the positive payoff matrix by adding 3 to each payoff:

$M_1 = \begin{bmatrix} 1 & 4 \\ 3 & 2 \end{bmatrix}$

The linear programming problems corresponding to M_1 are as follows:

(A) Minimize $y = x_1 + x_2$
Subject to $x_1 + 3x_2 \geq 1$
$4x_1 + 2x_2 \geq 1$
$x_1 \geq 0,\ x_2 \geq 0$

(B) Maximize $y = z_1 + z_2$
Subject to $z_1 + 4z_2 \leq 1$
$3z_1 + 2z_2 \leq 1$
$z_1 \geq 0,\ z_2 \geq 0$ (8-3)

8. The graphs corresponding to the linear programming problems given in the solution to Problem 7 are shown below.

(A)

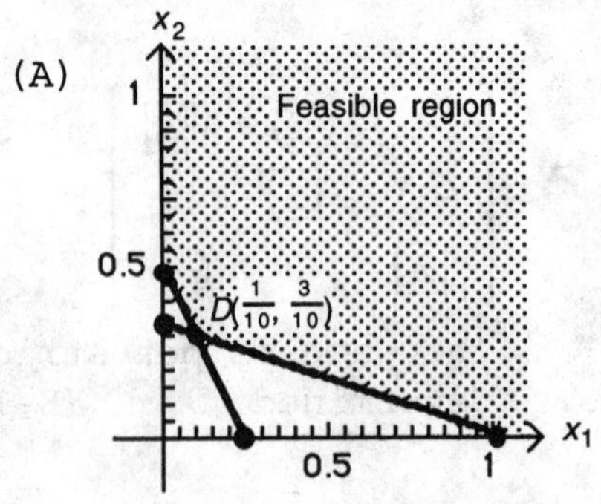

(B)

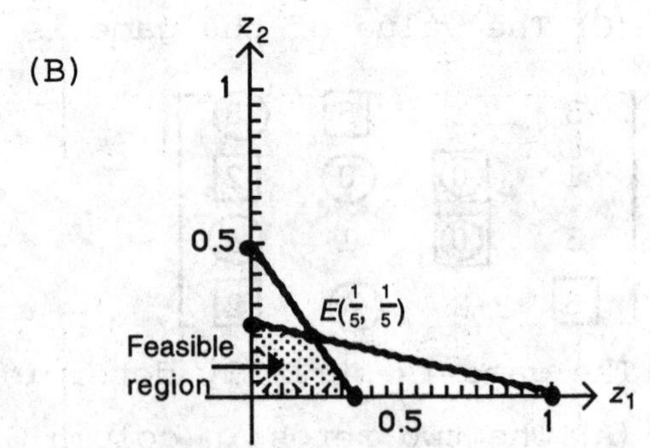

In (A), since the minimum $y = x_1 + x_2$ occurs at the point of intersection D, we solve

$x_1 + 3x_2 = 1$
$4x_1 + 2x_2 = 1$

to obtain the coordinates of D: $x_1 = \frac{1}{10}$ and $x_2 = \frac{3}{10}$.

Similarly, in (B), we solve

$$z_1 + 4z_2 = 1$$
$$3z_1 + 2z_2 = 1$$

to obtain the coordinates of E: $z_1 = \frac{1}{5}$ and $z_2 = \frac{1}{5}$.

Thus, $v_1 = \frac{1}{x_1 + x_2} = \frac{1}{\frac{1}{10} + \frac{3}{10}} = \frac{10}{4} = \frac{5}{2}$, and $v_1 = \frac{1}{z_1 + z_2} = \frac{1}{\frac{1}{5} + \frac{1}{5}} = \frac{5}{2}$.

$p_1 = v_1x_1 = \frac{5}{2} \cdot \frac{1}{10} = \frac{1}{4}$ $\qquad q_1 = v_1z_1 = \frac{5}{2} \cdot \frac{1}{5} = \frac{1}{2}$

$p_2 = v_1x_2 = \frac{5}{2} \cdot \frac{3}{10} = \frac{3}{4}$ $\qquad q_2 = v_1z_2 = \frac{5}{2} \cdot \frac{1}{5} = \frac{1}{2}$

Therefore, $P^* = \begin{bmatrix} \frac{1}{4} & \frac{3}{4} \end{bmatrix}$, $Q^* = \begin{bmatrix} \frac{1}{2} \\ \frac{1}{2} \end{bmatrix}$, and the value of the game is given by

$$v = v_1 - 3 = \frac{5}{2} - 3 = -\frac{1}{2}. \qquad (8\text{-}3)$$

9. First we solve Problem 7(B). Introduce slack variables x_1 and x_2 to obtain:

$$\begin{aligned} z_1 + 4z_2 + x_1 \quad &= 1 \\ 3z_1 + 2z_2 \quad + x_2 &= 1 \\ -z_1 - z_2 \quad + y &= 0 \end{aligned}$$
$$z_1 \ge 0,\ z_2 \ge 0,\ x_1,\ x_2 \ge 0$$

$$\begin{array}{c} \begin{matrix} z_1 & z_2 & x_1 & x_2 & y \end{matrix} \\ \left[\begin{array}{ccccc|c} 1 & 4 & 1 & 0 & 0 & 1 \\ \textcircled{3} & 2 & 0 & 1 & 0 & 1 \\ \hline -1 & -1 & 0 & 0 & 1 & 0 \end{array}\right] \begin{matrix} \frac{1}{1} \\ \frac{1}{3} \\ \end{matrix} \end{array} \sim \left[\begin{array}{ccccc|c} 1 & 4 & 1 & 0 & 0 & 1 \\ \textcircled{1} & \frac{2}{3} & 0 & \frac{1}{3} & 0 & \frac{1}{3} \\ \hline -1 & -1 & 0 & 0 & 1 & 0 \end{array}\right]$$

$\frac{1}{3}R_2 \to R_2$ $\qquad\qquad$ $R_1 + (-1)R_2 \to R_1$ and $R_3 + R_2 \to R_3$

$$\sim \left[\begin{array}{ccccc|c} 0 & \textcircled{\frac{10}{3}} & 1 & -\frac{1}{3} & 0 & \frac{2}{3} \\ 1 & \frac{2}{3} & 0 & \frac{1}{3} & 0 & \frac{1}{3} \\ \hline 0 & -\frac{1}{3} & 0 & \frac{1}{3} & 1 & \frac{1}{3} \end{array}\right] \begin{matrix} \frac{2}{3} \div \frac{10}{3} = \frac{2}{10} \\ \frac{1}{3} \div \frac{2}{3} = \frac{1}{2} \\ \end{matrix}$$

$\frac{3}{10}R_1 \to R_1$

$$\sim \left[\begin{array}{ccccc|c} 0 & \textcircled{1} & \frac{3}{10} & -\frac{1}{10} & 0 & \frac{1}{5} \\ 1 & \frac{2}{3} & 0 & \frac{1}{3} & 0 & \frac{1}{3} \\ \hline 0 & -\frac{1}{3} & 0 & \frac{1}{3} & 1 & \frac{1}{3} \end{array}\right] \sim \begin{array}{c} \begin{matrix} z_1 & z_2 & x_1 & x_2 & y \end{matrix} \\ \left[\begin{array}{ccccc|c} 0 & 1 & \frac{3}{10} & -\frac{1}{10} & 0 & \frac{1}{5} \\ 1 & 0 & -\frac{1}{5} & \frac{2}{5} & 0 & \frac{1}{5} \\ \hline 0 & 0 & \frac{1}{10} & \frac{3}{10} & 1 & \frac{2}{5} \end{array}\right] \end{array}$$

$R_2 + \left(-\frac{2}{3}\right)R_1 \to R_2$ and $R_3 + \frac{1}{3}R_1 \to R_3$

Thus, $z_1 = \frac{1}{5}$, $z_2 = \frac{1}{5}$, $y_{max} = \frac{2}{5}$, $v_1 = \frac{1}{z_1 + z_2} = \frac{1}{\frac{1}{5} + \frac{1}{5}} = \frac{5}{2}$, and

$q_1 = v_1 z_1 = \frac{5}{2} \cdot \frac{1}{5} = \frac{1}{2}$, $q_2 = v_1 z_2 = \frac{5}{2} \cdot \frac{1}{5} = \frac{1}{2}$.

The solution to the minimization problem (A) can be read from the bottom row of the final simplex tableau for the dual problem on the previous page. Thus, from the row

$$\begin{array}{ccccc|c} & & x_1 & x_2 & & \\ 0 & 0 & \frac{1}{10} & \frac{3}{10} & 1 & \frac{2}{5} \end{array}$$

we conclude that the solution is:

$$\text{Min } y = x_1 + x_2 = \frac{2}{5} \text{ at } x_1 = \frac{1}{10}, \ x_2 = \frac{3}{10}$$

Also,

$$p_1 = v_1 x_1 = \frac{5}{2}\left(\frac{1}{10}\right) = \frac{1}{4}$$

$$p_2 = v_1 x_2 = \frac{5}{2}\left(\frac{3}{10}\right) = \frac{3}{4}$$

Finally,

$$P^* = \begin{bmatrix} \frac{1}{4} & \frac{3}{4} \end{bmatrix}, \quad Q^* = \begin{bmatrix} \frac{1}{2} \\ \frac{1}{2} \end{bmatrix}, \quad \text{and} \quad v = \frac{5}{2} - 3 = -\frac{1}{2}.$$ (8-4)

10. (A) False: A strictly determined game is fair only if the value of the game (the saddle value) is 0; the saddle value of a strictly determined game is not necessarily 0. (See Problems 1, 5, Exercise 8-1).

(B) True: Theorem 1, Section 8-1 (8-1)

11. (A) True: Such a row is recessive and it is eliminated.

(B) False: A column is recessive if its entries are **greater** than or equal to the corresponding entries of another column. (8-2)

12. $\begin{bmatrix} \textcircled{-1} & \boxed{2} & \boxed{8} \\ \boxed{0} & \boxed{2} & \textcircled{-4} \\ \boxed{\textcircled{0}} & 1 & 3 \end{bmatrix}$ This game is strictly determined.

Also, $P^* = [0 \quad 0 \quad 1]$, $Q^* = \begin{bmatrix} 1 \\ 0 \\ 0 \end{bmatrix}$, and the value of the game, v, equals 0. (8-1)

13. $M = \begin{bmatrix} -1 & 5 & -3 & 7 \\ -4 & -3 & 2 & -2 \\ 3 & 0 & 2 & 1 \end{bmatrix}$

First, eliminate the recessive row $[-4 \quad -3 \quad 2 \quad -2]$, and the recessive columns

$\begin{bmatrix} -1 \\ -4 \\ 3 \end{bmatrix}$ and $\begin{bmatrix} 7 \\ -2 \\ 1 \end{bmatrix}$ to obtain the 2 × 2 matrix $\begin{bmatrix} \textcircled{5} & \boxed{-3} \\ \boxed{0} & \textcircled{2} \end{bmatrix}$.

The game is not strictly determined so the solution method of Section 8-2 can be applied. Let $D = (a + d) - (b + c) = (5 + 2) - (-3 + 0) = 10$. The optimal strategies are:

$$P^* = \left[\frac{d - c}{D}, \frac{a - b}{D}\right] = \left[\frac{2}{10}, \frac{8}{10}\right] = \left[\frac{1}{5}, \frac{4}{5}\right], \quad Q^* = \begin{bmatrix} \frac{d - b}{D} \\ \frac{a - c}{D} \end{bmatrix} = \begin{bmatrix} \frac{5}{10} \\ \frac{5}{10} \end{bmatrix} = \begin{bmatrix} \frac{1}{2} \\ \frac{1}{2} \end{bmatrix}$$

and $v = \dfrac{ad - bc}{D} = \dfrac{10}{10} = 1$.

Thus, the optimal strategies for the original matrix and the value v are:

$$P^* = \left[\tfrac{1}{5} \quad 0 \quad \tfrac{4}{5}\right], \quad Q^* = \begin{bmatrix} 0 \\ \frac{1}{2} \\ \frac{1}{2} \\ 0 \end{bmatrix}, \quad v = 1. \qquad (8\text{-}2)$$

14. $\begin{bmatrix} \boxed{0} & \boxed{3} & \textcircled{-1} \\ -1 & \textcircled{-2} & \boxed{1} \end{bmatrix}$ This game is nonstrictly determined.

We obtain a positive payoff matrix by adding 3 to each payoff:

$$M_1 = \begin{bmatrix} 3 & 6 & 2 \\ 2 & 1 & 4 \end{bmatrix}$$

The linear programming problems corresponding to M_1 are as follows:

(A) Minimize $y = x_1 + x_2$

Subject to
$$\begin{aligned} 3x_1 + 2x_2 &\ge 1 \\ 6x_1 + x_2 &\ge 1 \\ 2x_1 + 4x_2 &\ge 1 \\ x_1, x_2 &\ge 0 \end{aligned}$$

(B) Maximize $y = z_1 + z_2 + z_3$

Subject to
$$\begin{aligned} 3z_1 + 6z_2 + 2z_3 &\le 1 \\ 2z_1 + z_2 + 4z_3 &\le 1 \\ z_1, z_2, z_3 &\ge 0 \end{aligned}$$

Solve part (B) by using the simplex method. Introduce slack variables x_1 and x_2 to obtain:

$$\begin{aligned} 3z_1 + 6z_2 + 2z_3 + x_1 \qquad &= 1 \\ 2z_1 + z_2 + 4z_3 \qquad + x_2 &= 1 \\ -z_1 - z_2 - z_3 \qquad\qquad + y &= 0 \\ z_1, z_2, z_3 \ge 0, \; x_1, x_2 &\ge 0 \end{aligned}$$

$$\begin{array}{c} \begin{matrix} z_1 & z_2 & z_3 & x_1 & x_2 & y \end{matrix} \\ \left[\begin{array}{cccccc|c} \textcircled{3} & 6 & 2 & 1 & 0 & 0 & 1 \\ 2 & 1 & 4 & 0 & 1 & 0 & 1 \\ \hline -1 & -1 & -1 & 0 & 0 & 1 & 0 \end{array}\right] \begin{matrix} \frac{1}{3} \\ \frac{1}{2} \\ \\ \end{matrix} \end{array} \sim \left[\begin{array}{cccccc|c} \textcircled{1} & 2 & \frac{2}{3} & \frac{1}{3} & 0 & 0 & \frac{1}{3} \\ 2 & 1 & 4 & 0 & 1 & 0 & 1 \\ \hline -1 & -1 & -1 & 0 & 0 & 1 & 0 \end{array}\right]$$

$\frac{1}{3}R_1 \to R_1$ $\qquad$ $R_2 + (-2)R_1 \to R_2$ and $R_3 + R_1 \to R_3$

$$\sim \left[\begin{array}{cccccc|c} 1 & 2 & \frac{2}{3} & \frac{1}{3} & 0 & 0 & \frac{1}{3} \\ 0 & -3 & \textcircled{\frac{8}{3}} & -\frac{2}{3} & 1 & 0 & \frac{1}{3} \\ \hdashline 0 & 1 & -\frac{1}{3} & \frac{1}{3} & 0 & 1 & \frac{1}{3} \end{array}\right] \begin{array}{l} \frac{1}{3} \div \frac{2}{3} = \frac{1}{2} \\ \frac{1}{3} \div \frac{8}{3} = \frac{1}{8} \\ \\ \end{array}$$

$\frac{3}{8}R_2 \rightarrow R_2$

$$\sim \left[\begin{array}{cccccc|c} 1 & 2 & \frac{2}{3} & \frac{1}{3} & 0 & 0 & \frac{1}{3} \\ 0 & -\frac{9}{8} & \textcircled{1} & -\frac{1}{4} & \frac{3}{8} & 0 & \frac{1}{8} \\ \hdashline 0 & 1 & -\frac{1}{3} & \frac{1}{3} & 0 & 1 & \frac{1}{3} \end{array}\right] \sim \begin{array}{c} \begin{array}{cccccc} z_1 & z_2 & z_3 & x_1 & x_2 & y \end{array} \\ \left[\begin{array}{cccccc|c} 1 & \frac{11}{4} & 0 & \frac{1}{2} & -\frac{1}{4} & 0 & \frac{1}{4} \\ 0 & -\frac{9}{8} & 1 & -\frac{1}{4} & \frac{3}{8} & 0 & \frac{1}{8} \\ \hdashline 0 & \frac{5}{8} & 0 & \frac{1}{4} & \frac{1}{8} & 1 & \frac{3}{8} \end{array}\right] \end{array}$$

$R_1 + \left(-\frac{2}{3}\right)R_2 \rightarrow R_1$ and $R_3 + \frac{1}{3}R_2 \rightarrow R_3$

Thus, $z_1 = \frac{1}{4}$, $z_2 = 0$, $z_3 = \frac{1}{8}$, $y_{max} = \frac{3}{8}$, $v_1 = \dfrac{1}{\frac{1}{4} + 0 + \frac{1}{8}} = \frac{8}{3}$,

$q_1 = v_1z_1 = \frac{8}{3} \cdot \frac{1}{4} = \frac{2}{3}$, $q_2 = v_1z_2 = \frac{8}{3} \cdot 0 = 0$, $q_3 = v_1z_3 = \frac{8}{3} \cdot \frac{1}{8} = \frac{1}{3}$.

Therefore, $Q^* = \begin{bmatrix} \frac{2}{3} \\ 0 \\ \frac{1}{3} \end{bmatrix}$ and $v = v_1 - 3 = \frac{8}{3} - 3 = -\frac{1}{3}$.

The solution to the minimization problem (A) can be read from the bottom row of the final simplex tableau for the dual problem above. Thus, from the row

$$\begin{array}{c} \begin{array}{cccccc} & & & x_1 & x_2 & \end{array} \\ \left[\begin{array}{cccccc|c} 0 & \frac{5}{8} & 0 & \frac{1}{4} & \frac{1}{8} & 1 & \frac{3}{8} \end{array}\right] \end{array}$$

we conclude that the solution is:

Min $y = x_1 + x_2 = \frac{3}{8}$ at $x_1 = \frac{1}{4}$, $x_2 = \frac{1}{8}$

Now,

$$p_1 = v_1x_1 = \frac{8}{3}\left(\frac{1}{4}\right) = \frac{2}{3}$$

$$p_2 = v_1x_2 = \frac{8}{3}\left(\frac{1}{8}\right) = \frac{1}{3}$$

and

$$P^* = \begin{bmatrix} \frac{2}{3} & \frac{1}{3} \end{bmatrix}. \qquad (8\text{-}4)$$

15. $M = \begin{bmatrix} 2 & 6 & -4 & -7 \\ 4 & 7 & -3 & -5 \\ 3 & 3 & 9 & 8 \end{bmatrix}$

First eliminate the recessive row [2 6 -4 -7], and the recessive columns $\begin{bmatrix} 6 \\ 7 \\ 3 \end{bmatrix}$ and $\begin{bmatrix} -4 \\ -3 \\ 9 \end{bmatrix}$ to obtain the 2 × 2 matrix $\begin{bmatrix} \textcircled{4} & \boxed{-5} \\ \boxed{3} & \textcircled{8} \end{bmatrix}$.

The game is not strictly determined so the solution method of Section 8-2 can be applied. Let $D = (a + d) - (b + c) = (4 + 8) - (-5 + 3) = 14$. The optimal strategies are

$$P^* = \left[\frac{d - c}{D}, \frac{a - b}{D}\right] = \left[\frac{5}{14}, \frac{9}{14}\right]$$

$$Q^* = \begin{bmatrix} \frac{d - b}{D} \\ \frac{a - c}{D} \end{bmatrix} = \begin{bmatrix} \frac{13}{14} \\ \frac{1}{14} \end{bmatrix} \text{ and } v = \frac{ad - bc}{D} = \frac{47}{14}$$

Thus, the optimal strategies for the original matrix and the value v are:

$$P^* = \begin{bmatrix} 0 & \frac{5}{14} & \frac{9}{14} \end{bmatrix}, \quad Q^* = \begin{bmatrix} \frac{13}{14} \\ 0 \\ 0 \\ \frac{1}{14} \end{bmatrix}, \quad v = \frac{47}{14}$$ (8-2)

16. $\begin{bmatrix} -1 & 1 & \textcircled{-2} \\ \boxed{0} & \textcircled{-2} & \boxed{2} \\ \textcircled{-3} & \boxed{2} & -1 \end{bmatrix}$ This is a nonstrictly determined game.

We obtain the positive payoff matrix by adding 4 to each payoff:

$$M_1 = \begin{bmatrix} 3 & 5 & 2 \\ 4 & 2 & 6 \\ 1 & 6 & 3 \end{bmatrix}$$

The linear programming problems corresponding to M_1 are as follows:

(A) Minimize $y = x_1 + x_2 + x_3$

Subject to
$$3x_1 + 4x_2 + x_3 \ge 1$$
$$5x_1 + 2x_2 + 6x_3 \ge 1$$
$$2x_1 + 6x_2 + 3x_3 \ge 1$$
$$x_1, x_2, x_3 \ge 0$$

(B) Maximize $y = z_1 + z_2 + z_3$

Subject to
$$3z_1 + 5z_2 + 2z_3 \le 1$$
$$4z_1 + 2z_2 + 6z_3 \le 1$$
$$z_1 + 6z_2 + 3z_3 \le 1$$
$$z_1, z_2, z_3 \ge 0$$

We use the simplex method to solve part (B). Introduce slack variables x_1, x_2, and x_3 to obtain:

$$\begin{aligned} 3z_1 + 5z_2 + 2z_3 + x_1 \qquad\qquad &= 1 \\ 4z_1 + 2z_2 + 6z_3 \qquad + x_2 \qquad &= 1 \\ z_1 + 6z_2 + 3z_3 \qquad\qquad + x_3 &= 1 \\ -z_1 - z_2 - z_3 \qquad\qquad\qquad + y &= 0 \end{aligned}$$
$$z_1, z_2, z_3 \ge 0, \; x_1, x_2, x_3 \ge 0$$

$$\begin{array}{c} \begin{matrix} z_1 & z_2 & z_3 & x_1 & x_2 & x_3 & y \end{matrix} \\ \left[\begin{array}{ccccccc|c} 3 & 5 & 2 & 1 & 0 & 0 & 0 & 1 \\ \textcircled{4} & 2 & 6 & 0 & 1 & 0 & 0 & 1 \\ 1 & 6 & 3 & 0 & 0 & 1 & 0 & 1 \\ \hline -1 & -1 & -1 & 0 & 0 & 0 & 1 & 0 \end{array}\right] \begin{matrix} \frac{1}{3} \\ \frac{1}{4} \\ 1 \\ \end{matrix} \end{array} \sim \left[\begin{array}{ccccccc|c} 3 & 5 & 2 & 1 & 0 & 0 & 0 & 1 \\ \textcircled{1} & \frac{1}{2} & \frac{3}{2} & 0 & \frac{1}{4} & 0 & 0 & \frac{1}{4} \\ 1 & 6 & 3 & 0 & 0 & 1 & 0 & 1 \\ \hline -1 & -1 & -1 & 0 & 0 & 0 & 1 & 0 \end{array}\right]$$

$\frac{1}{4}R_2 \to R_2$

$R_1 + (-3)R_2 \to R_1$, $R_3 + (-1)R_2 \to R_3$, and $R_4 + R_2 \to R_4$

$$\sim \left[\begin{array}{ccccccc|c} 0 & \textcircled{\frac{7}{2}} & -\frac{5}{2} & 1 & -\frac{3}{4} & 0 & 0 & \frac{1}{4} \\ 1 & \frac{1}{2} & \frac{3}{2} & 0 & \frac{1}{4} & 0 & 0 & \frac{1}{4} \\ 0 & \frac{11}{2} & \frac{3}{2} & 0 & -\frac{1}{4} & 1 & 0 & \frac{3}{4} \\ \hdashline 0 & -\frac{1}{2} & \frac{1}{2} & 0 & \frac{1}{4} & 0 & 1 & \frac{1}{4} \end{array}\right] \quad \begin{array}{l} \frac{1}{4} \div \frac{7}{2} = \frac{1}{14} \\ \frac{1}{4} \div \frac{1}{2} = \frac{1}{2} \\ \frac{3}{4} \div \frac{11}{2} = \frac{3}{22} \end{array}$$

$\frac{2}{7}R_1 \to R_1$

$$\begin{array}{c} \begin{array}{ccccccc} z_1 & z_2 & z_3 & x_1 & x_2 & x_3 & y \end{array} \\ \sim \left[\begin{array}{ccccccc|c} 0 & \textcircled{1} & -\frac{5}{7} & \frac{2}{7} & -\frac{3}{14} & 0 & 0 & \frac{1}{14} \\ 1 & \frac{1}{2} & \frac{3}{2} & 0 & \frac{1}{4} & 0 & 0 & \frac{1}{4} \\ 0 & \frac{11}{2} & \frac{3}{2} & 0 & -\frac{1}{4} & 1 & 0 & \frac{3}{4} \\ \hdashline 0 & -\frac{1}{2} & \frac{1}{2} & 0 & \frac{1}{4} & 0 & 1 & \frac{1}{4} \end{array}\right] \end{array} \sim \begin{array}{c} \begin{array}{ccccccc} z_1 & z_2 & z_3 & x_1 & x_2 & x_3 & y \end{array} \\ \left[\begin{array}{ccccccc|c} 0 & 1 & -\frac{5}{7} & \frac{2}{7} & -\frac{3}{14} & 0 & 0 & \frac{1}{14} \\ 1 & 0 & \frac{13}{7} & -\frac{1}{7} & \frac{5}{14} & 0 & 0 & \frac{3}{14} \\ 0 & 0 & \frac{38}{7} & -\frac{11}{7} & \frac{13}{14} & 1 & 0 & \frac{5}{14} \\ \hdashline 0 & 0 & \frac{1}{7} & \frac{1}{7} & \frac{1}{7} & 0 & 1 & \frac{2}{7} \end{array}\right] \end{array}$$

$R_2 + \left(-\frac{1}{2}\right)R_1 \to R_2$, $R_3 + \left(-\frac{11}{2}\right)R_1 \to R_3$,
and $R_4 + \frac{1}{2}R_1 \to R_4$

Thus, $z_1 = \frac{3}{14}$, $z_2 = \frac{1}{14}$, $z_3 = 0$, and $y_{max} = \frac{2}{7}$.

$v_1 = \dfrac{1}{z_1 + z_2 + z_3} = \dfrac{1}{\frac{3}{14} + \frac{1}{14} + 0} = \dfrac{14}{4} = \dfrac{7}{2}$. Therefore,

$q_1 = v_1 z_1 = \frac{7}{2} \cdot \frac{3}{14} = \frac{3}{4}$, $q_2 = v_1 z_2 = \frac{7}{2} \cdot \frac{1}{14} = \frac{1}{4}$, $q_3 = v_1 z_3 = \frac{7}{2} \cdot 0 = 0$,

$Q^* = \begin{bmatrix} \frac{3}{4} \\ \frac{1}{4} \\ 0 \end{bmatrix}$, and $v = v_1 - 4 = \frac{7}{2} - 4 = -\frac{1}{2}$.

The solution to the minimization problem (A) can be read from the bottom row of the final simplex tableau for the dual problem above. Thus, from the row

$$\begin{array}{c} \begin{array}{ccc} x_1 & x_2 & x_3 \end{array} \\ \left[\begin{array}{ccccccc|c} 0 & 0 & \frac{1}{7} & \frac{1}{7} & \frac{1}{7} & 0 & 1 & \frac{2}{7} \end{array}\right] \end{array}$$

we conclude that the solution is:

Min $y = x_1 + x_2 + x_3 = \frac{2}{7}$ at $x_1 = \frac{1}{7}$, $x_2 = \frac{1}{7}$, $x_3 = 0$

Now,

$$p_1 = v_1 x_1 = \frac{7}{2}\left(\frac{1}{7}\right) = \frac{1}{2}$$

$$p_2 = v_1 x_2 = \frac{7}{2}\left(\frac{1}{7}\right) = \frac{1}{2}$$

$$p_3 = v_1 x_3 = \frac{7}{2}(0) = 0$$

and

$$P^* = \begin{bmatrix} \frac{1}{2} & \frac{1}{2} & 0 \end{bmatrix}. \qquad (8\text{-}4)$$

17. Yes. Let $M = \begin{bmatrix} a & b \\ c & d \end{bmatrix}$ be the payoff matrix for a strictly determined 2×2 matrix game, and suppose that a, the element in the (1, 1) position is a saddle value. Then $a \le b$ since a is the minimum in its row and $a \ge c$ since a is the maximum in its column.

If c, the element in the (2, 1) position is minimum in the second row, then $c \le d$ and the column $\begin{bmatrix} b \\ d \end{bmatrix}$ is recessive; $\begin{bmatrix} b \\ d \end{bmatrix} \ge \begin{bmatrix} a \\ c \end{bmatrix}$.

If d, the element in the (2, 2) position, is the minimum in the second row, then $b \ge a \ge c \ge d$ and $[c, d]$ is a recessive row; $[a \quad b] \ge [c \quad d]$.
(8-1, 8-2)

18. No. Assume that the matrix

$$\begin{bmatrix} 1 & 3 & 2 \\ 3 & 1 & 2 \\ 2 & 2 & 2 \end{bmatrix}$$

is the payoff matrix for a 3×3 matrix game. The minimum values in each row and the maximum values in each column are as indicated:

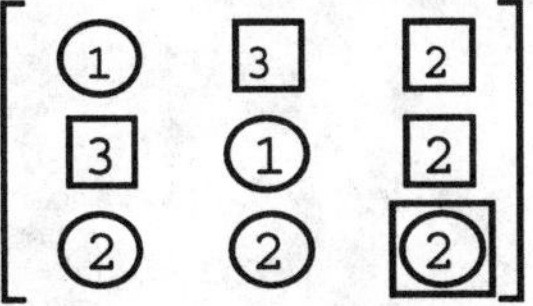

The 2 in the (3, 3) position is a saddle value so that the game is strictly determined. There are no recessive rows or columns. (8-1, 8-2)

19. (A) The payoff matrix M is:

		Cathy 1	Cathy 2
Ron	1	-2	3
	2	1	-2

(1) Use the method in Section 8-2.

$$\begin{bmatrix} \text{(-2)} & \boxed{3} \\ \boxed{1} & \text{(-2)} \end{bmatrix}$$ The game is not strictly determined.

Let $D = (a + d) - (b + c) = [-2 + (-2)] - (3 + 1) = -8$. The optimal strategies are:

$$P^* = [p_1 \quad p_2] = \begin{bmatrix} \frac{d-c}{D} & \frac{a-b}{D} \end{bmatrix} = \begin{bmatrix} \frac{-2-1}{-8} & \frac{-2-3}{-8} \end{bmatrix} = \begin{bmatrix} \frac{3}{8} & \frac{5}{8} \end{bmatrix}$$

$$Q^* = \begin{bmatrix} q_1 \\ q_2 \end{bmatrix} = \begin{bmatrix} \frac{d-b}{D} \\ \frac{a-c}{D} \end{bmatrix} = \begin{bmatrix} \frac{-2-3}{-8} \\ \frac{-2-1}{-8} \end{bmatrix} = \begin{bmatrix} \frac{5}{8} \\ \frac{3}{8} \end{bmatrix}$$

The expected value v of the game is given by:

$$v = \frac{ad - bc}{D} = \frac{4-3}{-8} = -\frac{1}{8}$$

(2) Use the method of Section 8-2: linear programming—geometric approach.

The payoff matrix is $M = \begin{bmatrix} -2 & 3 \\ 1 & -2 \end{bmatrix}$.

Add 3 to each entry to obtain

$$M_1 = \begin{bmatrix} 1 & 6 \\ 4 & 1 \end{bmatrix}$$

The corresponding linear programming problems are:

(i) Minimize $y = x_1 + x_2$

Subject to: $x_1 + 4x_2 \geq 1$

$6x_1 + x_2 \geq 1$

$x_1, x_2 \geq 0$

(ii) Maximize $y = z_1 + z_2$

Subject to: $z_1 + 6z_2 \leq 1$

$4z_1 + z_2 \leq 1$

$z_1, z_2 \geq 0$

Solving (i) geometrically, we have:

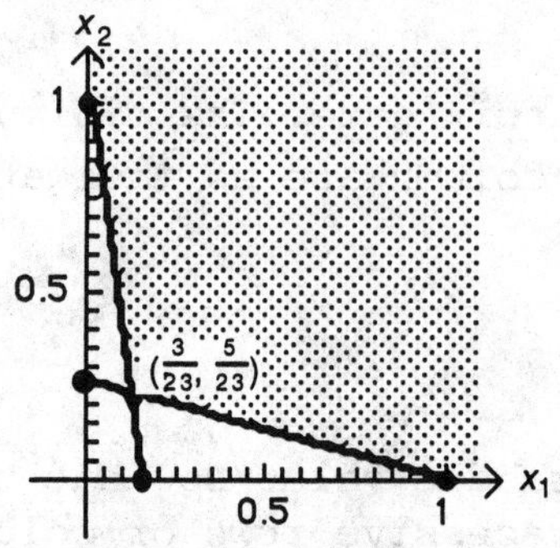

The corner points of the feasible region are $(0, 1)$, $\left(\frac{3}{23}, \frac{5}{23}\right)$, and $(1, 0)$. Evaluating the objective function $y = x_1 + x_2$ at the corner points, we have:

Corner Point	$y = x_1 + x_2$
$(0, 1)$	1
$\left(\frac{3}{23}, \frac{5}{23}\right)$	$\frac{8}{23}$
$(1, 0)$	1

Thus, the minimum value is $\frac{8}{23}$ and occurs at the corner point $\left(\frac{3}{23}, \frac{5}{23}\right)$.

Solving (ii) geometrically, we have:

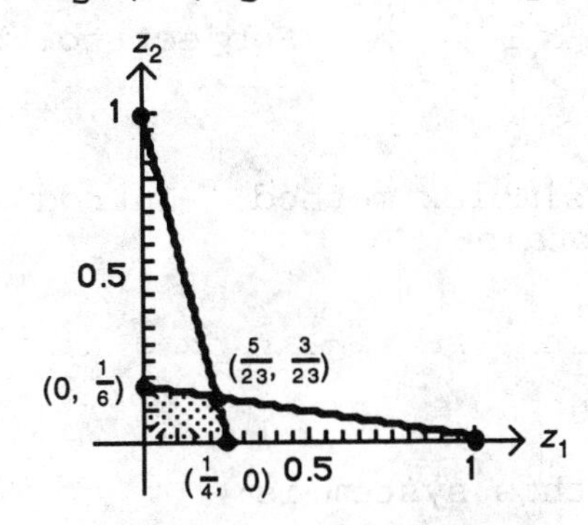

The corner points of the feasible region are $\left(0, \frac{1}{6}\right)$, $\left(\frac{5}{23}, \frac{3}{23}\right)$, and $\left(\frac{1}{4}, 0\right)$. Evaluating $y = z_1 + z_2$ at the corner points, we have:

Corner Point	$y = z_1 + z_2$
$\left(0, \frac{1}{6}\right)$	$\frac{1}{6}$
$\left(\frac{5}{23}, \frac{3}{23}\right)$	$\frac{8}{23}$
$\left(\frac{1}{4}, 0\right)$	$\frac{1}{4}$

Thus, the maximum value is $\frac{8}{23}$ and occurs at the corner point $\left(\frac{5}{23}, \frac{3}{23}\right)$.

Now, we have:

$$v_1 = \frac{1}{x_1 + x_2} = \frac{1}{z_1 + z_2} = \frac{1}{\frac{5}{23} + \frac{3}{23}} = \frac{23}{8}$$

$$p_1 = v_1 x_1 = \frac{23}{8} \cdot \frac{3}{23} = \frac{3}{8} \qquad q_1 = v_1 z_1 = \frac{23}{8} \cdot \frac{5}{23} = \frac{5}{8}$$

$$p_2 = v_1 x_2 = \frac{23}{8} \cdot \frac{5}{23} = \frac{5}{8} \qquad q_2 = v_1 z_2 = \frac{23}{8} \cdot \frac{3}{23} = \frac{3}{8}$$

The value v of the original matrix is:

$$v = v_1 - 3 = \frac{23}{8} - 3 = \frac{23}{8} - \frac{24}{8} = -\frac{1}{8}$$

Thus, the optimal strategies and the value of the game are:

$$P^* = \begin{bmatrix} \frac{3}{8} & \frac{5}{8} \end{bmatrix}, \qquad Q^* = \begin{bmatrix} \frac{5}{8} \\ \frac{3}{8} \end{bmatrix} \qquad \text{and} \qquad v = -\frac{1}{8}$$

(3) Use the method of Section 8-4: linear programming—the simplex method.

The linear programming problems are:

(i) Minimize $y = x_1 + x_2$

Subject to: $x_1 + 4x_2 \geq 1$
$6x_1 + x_2 \geq 1$
$x_1,\ x_2 \geq 0$

(ii) Maximize $y = z_1 + z_2$

Subject to: $z_1 + 6z_2 \leq 1$
$4z_1 + z_2 \leq 1$
$z_1,\ z_2 \geq 0$

We solve (ii) using the simplex method. Introduce slack variables x_1 and x_2 to obtain:

$$\begin{aligned} z_1 + 6z_2 + x_1 \qquad\quad &= 1 \\ 4z_1 + z_2 \qquad + x_2 \quad &= 1 \\ -z_1 - z_2 \qquad\qquad + y &= 0 \end{aligned}$$

The simplex tableau for this system is:

$$\begin{array}{c} \begin{matrix} z_1 & z_2 & x_1 & x_2 & y \end{matrix} \\ \left[\begin{array}{ccccc|c} 1 & 6 & 1 & 0 & 0 & 1 \\ \textcircled{4} & 1 & 0 & 1 & 0 & 1 \\ \hline -1 & -1 & 0 & 0 & 1 & 0 \end{array}\right] \end{array} \sim \left[\begin{array}{ccccc|c} 1 & 6 & 1 & 0 & 0 & 1 \\ \textcircled{1} & \frac{1}{4} & 0 & \frac{1}{4} & 0 & \frac{1}{4} \\ \hline -1 & -1 & 0 & 0 & 1 & 0 \end{array}\right]$$

$\frac{1}{4}R_2 \to R_2$ $\qquad$ $R_1 + (-1)R_2 \to R_1,\ R_3 + R_2 \to R_3$

$$\sim \left[\begin{array}{ccccc|c} 0 & \textcircled{\frac{23}{4}} & 1 & -\frac{1}{4} & 0 & \frac{3}{4} \\ 1 & \frac{1}{4} & 0 & \frac{1}{4} & 0 & \frac{1}{4} \\ \hline 0 & -\frac{3}{4} & 0 & \frac{1}{4} & 1 & \frac{1}{4} \end{array}\right] \sim \left[\begin{array}{ccccc|c} 0 & \textcircled{1} & \frac{4}{23} & -\frac{1}{23} & 0 & \frac{3}{23} \\ 1 & \frac{1}{4} & 0 & \frac{1}{4} & 0 & \frac{1}{4} \\ \hline 0 & -\frac{3}{4} & 0 & \frac{1}{4} & 1 & \frac{1}{4} \end{array}\right]$$

$\frac{4}{23}R_1 \to R_1$ $\qquad$ $R_2 + \left(-\frac{1}{4}\right)R_1 \to R_2,\ R_3 + \frac{3}{4}R_1 \to R_3$

$$\sim \begin{array}{c} \begin{matrix} z_1 & z_2 & x_1 & x_2 & y \end{matrix} \\ \left[\begin{array}{ccccc|c} 0 & 1 & \frac{4}{23} & -\frac{1}{23} & 0 & \frac{3}{23} \\ 1 & 0 & -\frac{1}{23} & \frac{24}{92} & 0 & \frac{5}{23} \\ \hline 0 & 0 & \frac{3}{23} & \frac{5}{23} & 1 & \frac{8}{23} \end{array}\right] \end{array}$$

The maximum $y = z_1 + z_2 = \frac{8}{23}$ occurs at $z_1 = \frac{5}{23}$, $z_2 = \frac{3}{23}$. Thus,

$$v_1 = \frac{1}{\frac{5}{23} + \frac{3}{23}} = \frac{23}{8}$$

and

$$q_1 = v_1 z_1 = \frac{23}{8} \cdot \frac{5}{23} = \frac{5}{8},\quad q_2 = v_1 z_2 = \frac{23}{8} \cdot \frac{3}{23} = \frac{3}{8}.$$

The solution to the minimization problem (i) can be read from the bottom row of the final tableau for the dual problem (ii).

Thus, the minimum $y = x_1 + x_2 = \frac{8}{23}$ occurs at $x_1 = \frac{3}{23}$, $x_2 = \frac{5}{23}$, and

$$p_1 = v_1 x_1 = \frac{23}{8} \cdot \frac{3}{23} = \frac{3}{8}, \quad p_2 = v_1 x_2 = \frac{23}{8} \cdot \frac{5}{23} = \frac{5}{8}.$$

Again,

$$P^* = \begin{bmatrix} \frac{3}{8} & \frac{5}{8} \end{bmatrix}, \quad Q^* = \begin{bmatrix} \frac{5}{8} \\ \frac{3}{8} \end{bmatrix}, \quad \text{and} \quad v = v_1 - 3 = \frac{23}{8} - 3 = -\frac{1}{8}.$$

(B) The optimal strategies for Ron and Cathy are:

Ron: $P^* = \begin{bmatrix} \frac{3}{8} & \frac{5}{8} \end{bmatrix}$; Cathy: $Q^* = \begin{bmatrix} \frac{5}{8} \\ \frac{3}{8} \end{bmatrix}$.

(C) The expected value of the game for Ron is: $-\$\frac{1}{8} = -12.5¢$.

The expected value of the game for Cathy is: $\$\frac{1}{8} = 12.5¢$.

(8-1, 8-2, 8-3, 8-4)

9 MARKOV CHAINS

EXERCISE 9-1

Things to remember:

1. MARKOV CHAINS

A MARKOV CHAIN, or PROCESS, is a sequence of experiments, trials, or observations such that the transition probability matrix from one state to the next is constant.

Given a Markov chain with n states, a kth STATE MATRIX is a matrix of the form

$$S_k = [s_{k1} \quad s_{k2} \quad \cdots \quad s_{kn}]$$

Each entry s_{ki} is the proportion of the population that are in state i after the kth trial, or, equivalently, the probability of a randomly selected element of the population being in state i after the kth trial. The sum of all the entries in the kth state matrix S_k must be 1.

A TRANSITION MATRIX is a constant square matrix P of order n such that the entry in the ith row and jth column indicates the probability of the system moving from the ith state to the jth state on the next observation or trial. The sum of the entries in each row must be 1.

2. COMPUTING STATE MATRICES FOR A MARKOV CHAIN

If S_0 is the initial state matrix and P is the transition matrix for a Markov chain, then the subsequent state matrices are given by:

$S_1 = S_0P$ First-state matrix

$S_2 = S_1P$ Second-state matrix

$S_3 = S_2P$ Third-state matrix

$\vdots$

$S_k = S_{k-1}P$ kth-state matrix

3. POWERS OF A TRANSITION MATRIX

If P is the transition matrix and S_0 is an initial state matrix for a Markov chain, then the kth state matrix is given by

$$S_k = S_0P^k$$

The entry in the ith row and jth column of P^k indicates the probability of the system moving from the ith state to the jth state in k observations or trials. The sum of the entries in each row of P^k is 1.

1. $S_1 = S_0P = [1 \quad 0]\begin{bmatrix} .8 & .2 \\ .4 & .6 \end{bmatrix} = \overset{\quad A \quad\ \ B}{[.8 \quad .2]}$

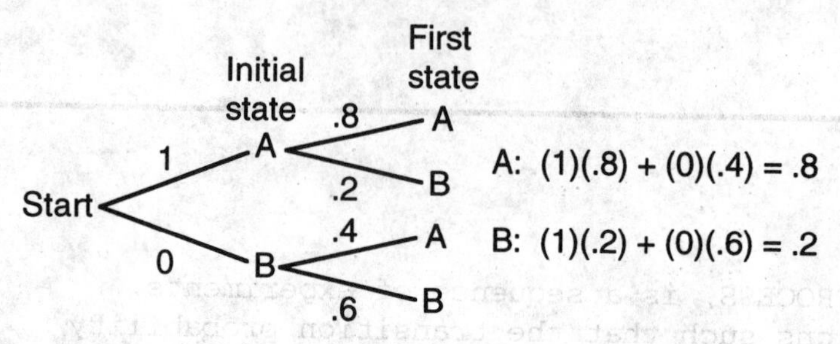

3. $S_1 = S_0P = [.5 \quad .5]\begin{bmatrix} .8 & .2 \\ .4 & .6 \end{bmatrix} = \overset{\quad A \quad\ \ B}{[.6 \quad .4]}$

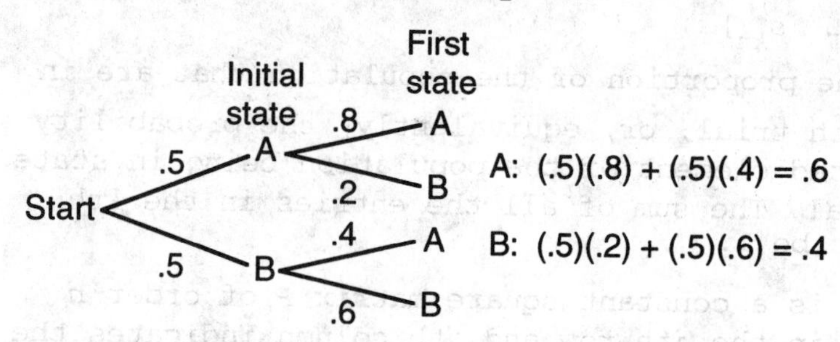

5. $S_2 = S_1P = [.8 \quad .2]\begin{bmatrix} .8 & .2 \\ .4 & .6 \end{bmatrix}$ (from Problem 1)

$= \overset{\quad A \quad\ \ B}{[.72 \quad .28]}$

The probability of being in state A after two trials is .72; the probability of being in state B after two trials is .28.

7. $S_2 = S_1P = [.6 \quad .4]\begin{bmatrix} .8 & .2 \\ .4 & .6 \end{bmatrix}$ (from Problem 3)

$= \overset{\quad A \quad\ \ B}{[.64 \quad .36]}$

The probability of being in state A after two trials is .64; the probability of being in state B after two trials is .36.

9. 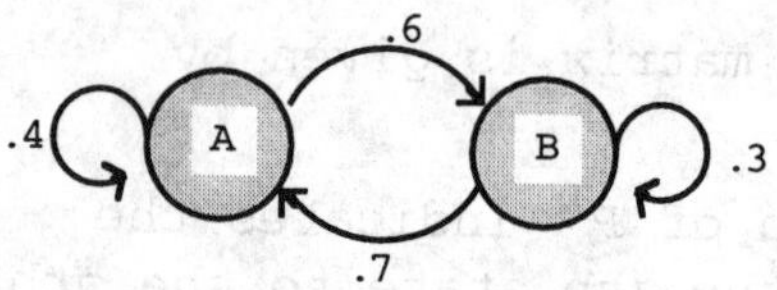 (Transition diagram: A → A .4, A → B .6, B → A .7, B → B .3)

$$\begin{array}{c} \\ A \\ B \end{array}\begin{array}{c} \begin{array}{cc} A & B \end{array} \\ \begin{bmatrix} .4 & .6 \\ .7 & .3 \end{bmatrix} \end{array}$$

11. No. Choose any x, $0 \le x \le 1$, then

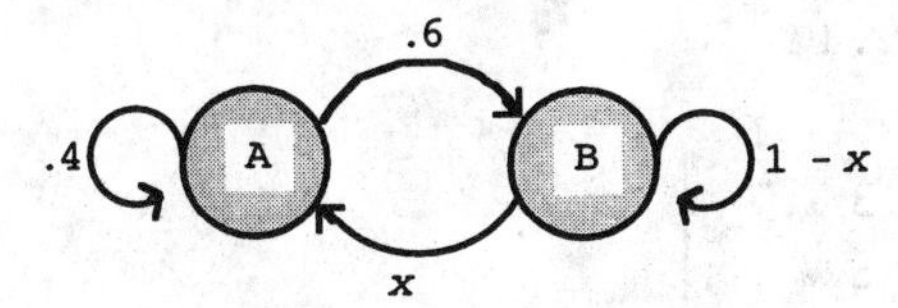

is an acceptable transition diagram, and

$$\begin{array}{c} \\ A \\ B \end{array}\begin{array}{c} \begin{array}{cc} A & B \end{array} \\ \begin{bmatrix} .4 & .6 \\ x & 1-x \end{bmatrix} \end{array}$$ is the corresponding transition matrix

13.

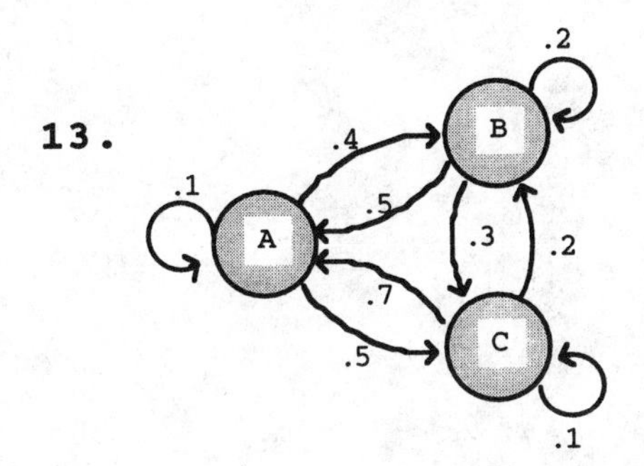

$$\begin{array}{c} \\ A \\ B \\ C \end{array}\begin{array}{c} \begin{array}{ccc} A & B & C \end{array} \\ \begin{bmatrix} .1 & .4 & .5 \\ .5 & .2 & .3 \\ .7 & .2 & .1 \end{bmatrix} \end{array}$$

15.

$0 + .5 + a = 1$ implies $a = .5$
$b + 0 + .4 = 1$ implies $b = .6$
$.2 + c + .1 = 1$ implies $c = .7$

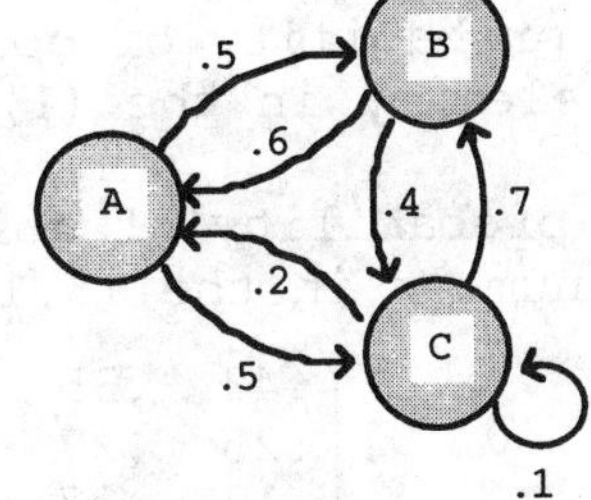

17.

$0 + a + .3 = 1$ implies $a = .7$
$0 + b + 0 = 1$ implies $b = 1$
$c + .8 + 0 = 1$ implies $c = .2$

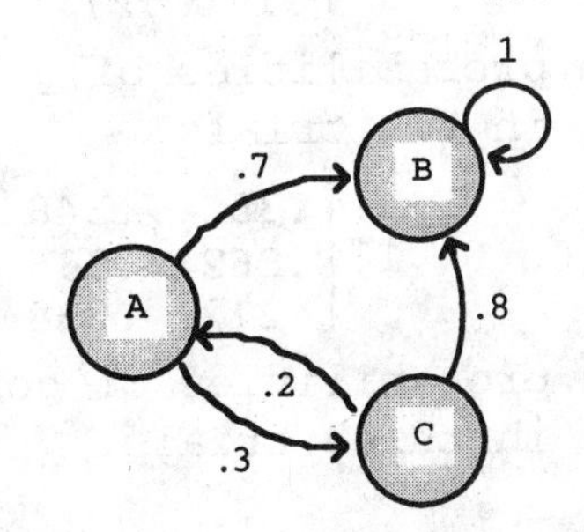

19. No. Choose any x, $0 \le x \le .4$ and let $a = x$, $b = .1$,
$c = 1 - (x + .4) = .6 - x$

$$\begin{array}{c} \\ A \\ B \\ C \end{array}\begin{array}{c} \begin{array}{ccc} A & B & C \end{array} \\ \begin{bmatrix} .2 & .1 & .7 \\ x & .4 & .6 - x \\ .5 & .1 & .4 \end{bmatrix} \end{array}$$ is a transition matrix.

21. The probability of staying in state A is .3.
The probability of staying in state B is .1.

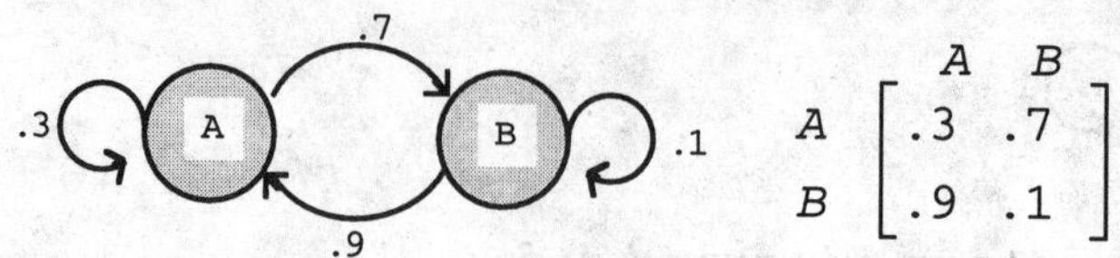

$$\begin{array}{c} \\ A \\ B \end{array}\begin{array}{c} A \quad B \\ \begin{bmatrix} .3 & .7 \\ .9 & .1 \end{bmatrix} \end{array}$$

23. The probability of staying in state A is .6.
The probability of staying in state B is .3.
Since the probability of staying in state C is 1, the probability of going from state C to state A is 0 and the probability of going from state C to state B is 0.

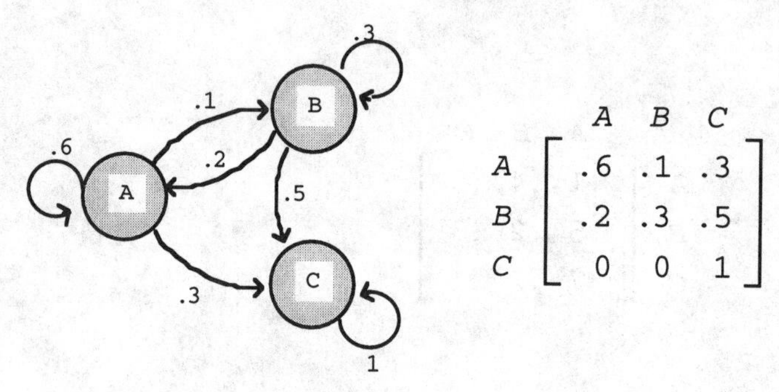

$$\begin{array}{c} \\ A \\ B \\ C \end{array}\begin{array}{c} A \quad B \quad C \\ \begin{bmatrix} .6 & .1 & .3 \\ .2 & .3 & .5 \\ 0 & 0 & 1 \end{bmatrix} \end{array}$$

25. Using P^2, the probability of going from state A to state B in two trials is the element in the (1,2) position—.35.

27. Using P^3, the probability of going from state C to state A in three trials is the number in the (3,1) position—.212.

29. $S_2 = S_0P^2 = [1 \quad 0 \quad 0]\begin{bmatrix} .43 & .35 & .22 \\ .25 & .37 & .38 \\ .17 & .27 & .56 \end{bmatrix} = \begin{array}{c} A \quad\ \ B \quad\ \ C \\ [.43 \quad .35 \quad .22] \end{array}$

These are the probabilities of going from state A to states A, B, and C, respectively, in two trials.

31. $S_3 = S_0P^3 = [0 \quad 0 \quad 1]\begin{bmatrix} .35 & .348 & .302 \\ .262 & .336 & .402 \\ .212 & .298 & .49 \end{bmatrix} = \begin{array}{c} A \quad\ \ B \quad\ \ C \\ [.212 \quad .298 \quad .49] \end{array}$

These are the probabilities of going from state C to states A, B, and C, respectively, in three trials.

33. $n = 9$

35. $P = \begin{bmatrix} .1 & .9 \\ .6 & .4 \end{bmatrix}$, $P^2 = \begin{bmatrix} .1 & .9 \\ .6 & .4 \end{bmatrix}\begin{bmatrix} .1 & .9 \\ .6 & .4 \end{bmatrix} = \begin{bmatrix} .55 & .45 \\ .3 & .7 \end{bmatrix}$;

$P^4 = P^2 \cdot P^2 = \begin{bmatrix} .55 & .45 \\ .3 & .7 \end{bmatrix}\begin{bmatrix} .55 & .45 \\ .3 & .7 \end{bmatrix} = \begin{array}{c} \\ A \\ B \end{array}\begin{array}{c} A \qquad B \\ \begin{bmatrix} .4375 & .5625 \\ .375 & .625 \end{bmatrix} \end{array}$

$S_4 = S_0P^4 = [.8 \quad .2]\begin{bmatrix} .4375 & .5625 \\ .375 & .625 \end{bmatrix} = \begin{array}{c} A \qquad B \\ [.425 \quad .575] \end{array}$

37. $P = \begin{bmatrix} 0 & .4 & .6 \\ 0 & 0 & 1 \\ 1 & 0 & 0 \end{bmatrix}$, $P^2 = \begin{bmatrix} 0 & .4 & .6 \\ 0 & 0 & 1 \\ 1 & 0 & 0 \end{bmatrix}\begin{bmatrix} 0 & .4 & .6 \\ 0 & 0 & 1 \\ 1 & 0 & 0 \end{bmatrix} = \begin{bmatrix} .6 & 0 & .4 \\ 1 & 0 & 0 \\ 0 & .4 & .6 \end{bmatrix}$;

$$P^4 = P^2 \cdot P^2 = \begin{bmatrix} .6 & 0 & .4 \\ 1 & 0 & 0 \\ 0 & .4 & .6 \end{bmatrix}\begin{bmatrix} .6 & 0 & .4 \\ 1 & 0 & 0 \\ 0 & .4 & .6 \end{bmatrix} = \begin{matrix} \\ A \\ B \\ C \end{matrix}\begin{matrix} A & B & C \\ \begin{bmatrix} .36 & .16 & .48 \\ .6 & 0 & .4 \\ .4 & .24 & .36 \end{bmatrix} \end{matrix}$$

$$S_4 = S_0P^4 = [.2 \quad .3 \quad .5]\begin{bmatrix} .36 & .16 & .48 \\ .6 & 0 & .4 \\ .4 & .24 & .36 \end{bmatrix} = \begin{matrix} A & B & C \\ [.452 & .152 & .396] \end{matrix}$$

39. $S_k = S_0P^k = [1 \quad 0]P^k$; The entries in S_k are the entries in the first row of P^k.

41. (A) $P^2 = \begin{bmatrix} .2 & .2 & .3 & .3 \\ 0 & 1 & 0 & 0 \\ .2 & .2 & .1 & .5 \\ 0 & 0 & 0 & 1 \end{bmatrix}\begin{bmatrix} .2 & .2 & .3 & .3 \\ 0 & 1 & 0 & 0 \\ .2 & .2 & .1 & .5 \\ 0 & 0 & 0 & 1 \end{bmatrix} = \begin{bmatrix} .1 & .3 & .09 & .51 \\ 0 & 1 & 0 & 0 \\ .06 & .26 & .07 & .61 \\ 0 & 0 & 0 & 1 \end{bmatrix}$

$$P^4 = \begin{bmatrix} .1 & .3 & .09 & .51 \\ 0 & 1 & 0 & 0 \\ .06 & .26 & .07 & .61 \\ 0 & 0 & 0 & 1 \end{bmatrix}\begin{bmatrix} .1 & .3 & .09 & .51 \\ 0 & 1 & 0 & 0 \\ .06 & .26 & .07 & .61 \\ 0 & 0 & 0 & 1 \end{bmatrix}$$

$$= \begin{matrix} \\ A \\ B \\ C \\ D \end{matrix}\begin{matrix} A & B & C & D \\ \begin{bmatrix} .0154 & .3534 & .0153 & .6159 \\ 0 & 1 & 0 & 0 \\ .0102 & .2962 & .0103 & .6833 \\ 0 & 0 & 0 & 1 \end{bmatrix} \end{matrix}$$

(B) The probability of going from state A to state D in 4 trials is the element in the (1,4) position: .6159.

(C) The element in the (3,2) position: .2962.

(D) The element in the (2,1) position: 0.

43. If $P = \begin{bmatrix} a & 1 - a \\ 1 - b & b \end{bmatrix}$ is a probability matrix

then $0 \le a \le 1$, $0 \le b \le 1$

$$P^2 = \begin{bmatrix} a & 1 - a \\ 1 - b & b \end{bmatrix}\begin{bmatrix} a & 1 - a \\ 1 - b & b \end{bmatrix}$$

$$= \begin{bmatrix} a^2 + (1 - a)(1 - b) & a(1 - a) + (1 - a)b \\ (1 - b)a + b(1 - b) & (1 - b)(1 - a) + b^2 \end{bmatrix}$$

Now, $a^2 + (1 - a)(1 - b) \ge 0$
and $a(1 - a) + (1 - a)b = (1 - a)(a + b) \ge 0$
since $0 \le a \le 1$ and $0 \le b \le 1$.

Also, $a^2 + (1 - a)(1 - b) + (1 - a)(a + b) = a^2 + (1 - a)[1 - b + a + b]$
$= a^2 + (1 - a)(1 + a)$
$= a^2 + 1 - a^2$
$= 1$

Therefore, the elements in the first row of P^2 are nonnegative and their sum is 1. The same arguments apply to the elements in the second row of P^2. Thus, P^2 is a probability matrix.

45. $P = \begin{bmatrix} .4 & .6 \\ .2 & .8 \end{bmatrix}$

(A) Let $S_0 = [0 \quad 10]$. Then

$S_2 = S_0P^2 = [.24 \quad .76]$

$S_4 = S_0P^4 = [.2496 \quad .7504]$

$S_8 = S_0P^8 = [.24999936 \quad .7500006]$

S_k is approaching $[.25 \quad .75]$

(B) Let $S_0 = [1 \quad 0]$. Then

$S_2 = S_0P^2 = [.28 \quad .72]$

$S_4 = S_0P^4 = [.2512 \quad .7488]$

$S_8 = S_0P^8 = [.25000192 \quad .7499980]$

S_k is approaching $[.25 \quad .75]$

(C) Let $S_0 = [.5 \quad .5]$. Then

$S_2 = S_0P^2 = [.26 \quad .74]$

$S_4 = S_0P^4 = [.2504 \quad .7496]$

$S_8 = S_0P^8 = [.25000064 \quad .74999936]$

S_k is approaching $[.25 \quad .75]$

(D) $[.25 \quad .75]\begin{bmatrix} .4 & .6 \\ .2 & .8 \end{bmatrix} = [.25 \quad .75]$

(E) The state matrices S_k appear to approach the same matrix $[.25 \quad .75]$, regardless of the values in the initial state matrix S_0.

47. $P^2 = \begin{bmatrix} .28 & .72 \\ .24 & .76 \end{bmatrix}$, $P^4 = \begin{bmatrix} .2512 & .7488 \\ .2496 & .7504 \end{bmatrix}$

$P^8 = \begin{bmatrix} .25000192 & .7499980 \\ .24999936 & .7500006 \end{bmatrix} \cdots$

The matrices P^k are approaching $Q = \begin{bmatrix} .25 & .75 \\ .25 & .75 \end{bmatrix}$; the rows of Q are the same as the matrix $S = [.25 \quad .75]$ in Problem 45.

49. Let R denote "rain" and R' "not rain".

(A)

(B) $\begin{matrix} & R & R' \\ R & .4 & .6 \\ R' & .06 & .94 \end{matrix}$

(C) Rain on Saturday: $P^2 = \begin{matrix} & R & R' \\ R & .196 & .804 \\ R' & .0804 & .9196 \end{matrix}$

The probability that it will rain on Saturday is .196.

$$\text{Rain on Sunday: } P^3 = \begin{matrix} & R & R' \\ R & .12664 & .87336 \\ R' & .087336 & .912664 \end{matrix}$$

The probability that it will rain on Sunday is .12664.

51. (A) .8 X .2 X' .8 .2

(B)
$$\begin{matrix} & X & X' \\ X & .8 & .2 \\ X' & .2 & .8 \end{matrix}$$

(C) $S = [.2 \quad .8]$ (columns X, X')

$$S_1 = SP = [.2 \quad .8]\begin{bmatrix} .8 & .2 \\ .2 & .8 \end{bmatrix} = [.32 \quad .68]$$

32% will be using brand X one week later.

$$S_2 = SP^2 = [.2 \quad .8]\begin{bmatrix} .68 & .32 \\ .32 & .68 \end{bmatrix} = [.392 \quad .608]$$

39.2% will be using brand X two weeks later.

53. (A)

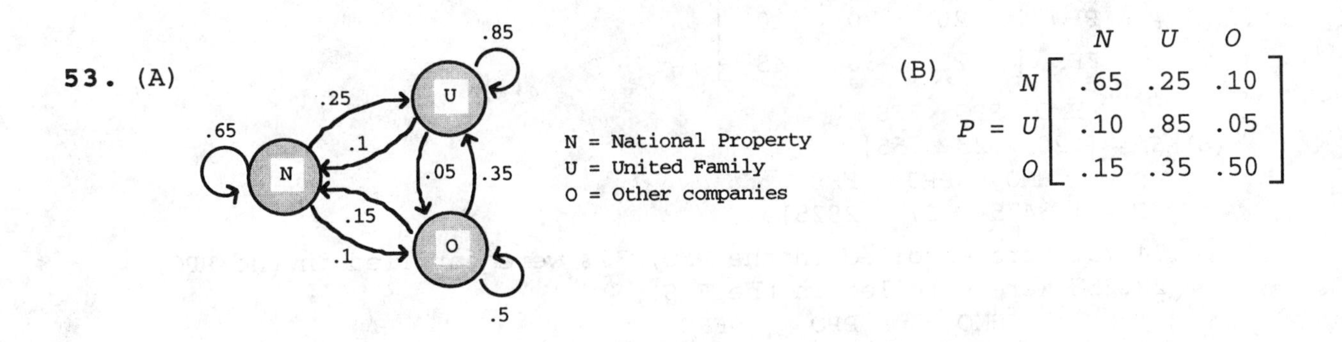

(B)
$$P = \begin{matrix} & N & U & O \\ N & .65 & .25 & .10 \\ U & .10 & .85 & .05 \\ O & .15 & .35 & .50 \end{matrix}$$

(C) $S = [.50 \quad .30 \quad .20]$ (columns N, U, O)

After one year: $SP = [.385 \quad .45 \quad .165]$ (columns N, U, O)
38.5% will be insured by National Property after one year.

After two years: $SP^2 = [.32 \quad .5365 \quad .1435]$ (columns N, U, O)
32% will be insured by National Property after two years.

(D) 45% of the homes will be insured by United Family after one year; 53.65% will be insured by United Family after two years.

55. (A)

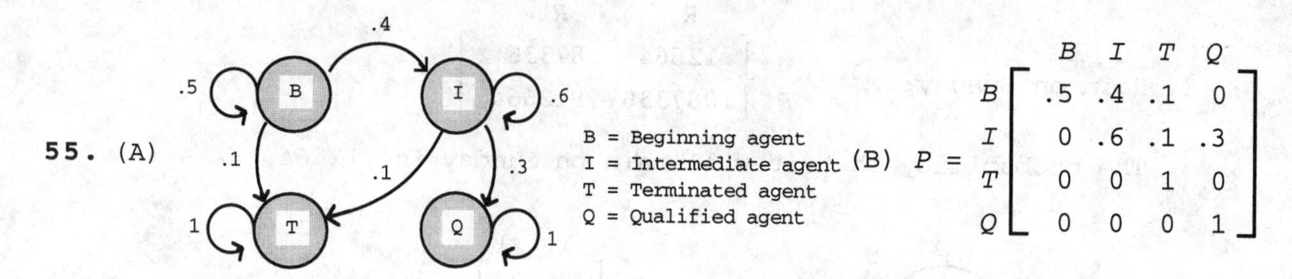

(B) $P = \begin{array}{c} \\ B \\ I \\ T \\ Q \end{array} \begin{array}{c} \begin{array}{cccc} B & I & T & Q \end{array} \\ \begin{bmatrix} .5 & .4 & .1 & 0 \\ 0 & .6 & .1 & .3 \\ 0 & 0 & 1 & 0 \\ 0 & 0 & 0 & 1 \end{bmatrix} \end{array}$

(C) $S = \overset{B}{[1} \quad \overset{I}{0} \quad \overset{T}{0} \quad \overset{Q}{0]}$

After one year: $SP^2 = [\overset{B}{.25} \quad \overset{I}{.44} \quad \overset{T}{.19} \quad \overset{Q}{.12}]$
The probability that a beginning agent will be promoted to qualified agent within one year (i.e., after 2 reviews) is: .12.

After two years: $SP^4 = [\overset{B}{.0625} \quad \overset{I}{.2684} \quad \overset{T}{.3079} \quad \overset{Q}{.3612}]$
The probability that a beginning agent will be promoted to qualified agent within two years (i.e., after 4 reviews) is: .3612.

57. (A) $P = \begin{array}{c} \\ \text{HMO} \\ \text{PPO} \\ \text{FFS} \end{array} \begin{array}{c} \begin{array}{ccc} \text{HMO} & \text{PPO} & \text{FFS} \end{array} \\ \begin{bmatrix} .80 & .15 & .05 \\ .20 & .70 & .10 \\ .25 & .30 & .45 \end{bmatrix} \end{array}$

(B) $S = [\overset{\text{HMO}}{.20} \quad \overset{\text{PPO}}{.25} \quad \overset{\text{FFS}}{.55}]$

$SP = [\overset{\text{HMO}}{.3475} \quad \overset{\text{PPO}}{.37} \quad \overset{\text{FFS}}{.2825}]$

34.75% were enrolled in the HMO; 37% were enrolled in the PPO; 28.25% were enrolled in the FFS.

(C) $SP^2 = [\overset{\text{HMO}}{.422625} \quad \overset{\text{PPO}}{.395875} \quad \overset{\text{FFS}}{.1815}]$

42.2625% will be enrolled in the HMO; 39.5875% will be enrolled in the PPO; 18.15% will be enrolled in the FFS.

59. (A) $P = \begin{array}{c} \\ H \\ R \end{array} \begin{array}{c} \begin{array}{cc} H & R \end{array} \\ \begin{bmatrix} .88 & .12 \\ .05 & .95 \end{bmatrix} \end{array}$ H = Homeowner, R = Renter

(B) $S = [\overset{H}{.451} \quad \overset{R}{.549}]$

1990: $SP = [.451 \quad .549]\begin{bmatrix} .88 & .12 \\ .05 & .95 \end{bmatrix} = [\overset{H}{.42433} \quad \overset{R}{.57567}]$

42.433% were homeowners in 1990.

2000: $SP^2 \approx [\overset{H}{.40219} \quad \overset{R}{.59781}]$

(C) 40.219% will be homeowners in 2000.

Things to remember:

1. STATIONARY MATRIX FOR A MARKOV CHAIN

   The state matrix $S = [s_1 \quad s_2 \quad \dots \quad s_n]$ is a STATIONARY MATRIX for a Markov chain with transition matrix P if

   $SP = S$

   where $s_i \geq 0$, $i = 1, \dots, n$ and $s_1 + s_2 + \dots + s_n = 1$.

2. REGULAR MARKOV CHAINS

   A transition matrix P is REGULAR if some power of P has only positive entries. A Markov chain is a REGULAR MARKOV CHAIN if its transition matrix is regular.

3. PROPERTIES OF REGULAR MARKOV CHAINS

   Let P be the transition matrix for a regular Markov chain.

   (A) There is a unique stationary matrix S which can be found by solving the equation
   $SP = S$

   (B) Given any initial state matrix S_0, the state matrices S_k approach the stationary matrix S.

   (C) The matrices P^k approach a limiting matrix $\bar{P}$ where each row of $\bar{P}$ is equal to the stationary matrix S.

---

1. $P = \begin{bmatrix} .1 & .9 \\ .7 & .3 \end{bmatrix}$ is regular; all entries are positive

3. $A = \begin{bmatrix} 1 & 0 \\ 0 & 1 \end{bmatrix}$ is not regular; $A^k = \begin{bmatrix} 1 & 0 \\ 0 & 1 \end{bmatrix}$ for all k.

5. $P = \begin{bmatrix} .2 & .8 \\ 1 & 0 \end{bmatrix}$, $P^2 = \begin{bmatrix} .84 & .16 \\ .2 & .8 \end{bmatrix}$

P is regular since P^2 has only positive entries.

7. $P = \begin{bmatrix} 1 & 0 \\ .6 & .4 \end{bmatrix}$, $P^2 = \begin{bmatrix} 1 & 0 \\ .84 & .16 \end{bmatrix}$, $P^3 = \begin{bmatrix} 1 & 0 \\ .936 & .064 \end{bmatrix}$, $P^4 = \begin{bmatrix} 1 & 0 \\ .9744 & .0256 \end{bmatrix}$, ...

It appears that P^k will have the form $\begin{bmatrix} 1 & 0 \\ a & b \end{bmatrix}$ for all positive integers k. Thus P is **not** regular.

9. $P = \begin{bmatrix} .3 & .4 & .3 \\ 0 & 0 & 1 \\ .4 & .2 & .4 \end{bmatrix}$, $P^2 = \begin{bmatrix} .21 & .18 & .61 \\ .4 & .2 & .4 \\ .28 & .24 & .48 \end{bmatrix}$

P is regular since P^2 has only positive entries.

11. $P = \begin{bmatrix} 0 & 1 & 0 \\ .4 & 0 & .6 \\ 0 & 1 & 0 \end{bmatrix}$, $P^2 = \begin{bmatrix} .4 & 0 & .6 \\ 0 & 1 & 0 \\ .4 & 0 & .6 \end{bmatrix}$, $P^3 = \begin{bmatrix} 0 & 1 & 0 \\ .4 & 0 & .6 \\ 0 & 1 & 0 \end{bmatrix}$, ...

In general, $P^{2k-1} = P$, $k = 1, 2, \ldots$
and $P^{2k} = P^2$, $k = 1, 2, \ldots$

Clearly, P is not regular.

13. $P = \begin{bmatrix} 0 & 0 & 1 \\ .8 & .1 & .1 \\ 0 & 1 & 0 \end{bmatrix}$, $P^2 = \begin{bmatrix} 0 & 1 & 0 \\ .08 & .11 & .81 \\ .8 & .1 & .1 \end{bmatrix}$, $P^3 = \begin{bmatrix} .8 & .1 & .1 \\ .088 & .821 & .091 \\ .08 & .11 & .81 \end{bmatrix}$

P is regular since P^3 has only positive entries.

15. Let $S = [s_1, s_2]$, and solve the system:

$$[s_1 \quad s_2]\begin{bmatrix} .1 & .9 \\ .6 & .4 \end{bmatrix} = [s_1 \quad s_2], \; s_1 + s_2 = 1$$

which is equivalent to

$$\begin{aligned} .1s_1 + .6s_2 &= s_1 \\ .9s_1 + .4s_2 &= s_2 \\ s_1 + s_2 &= 1 \end{aligned} \quad \text{or} \quad \begin{aligned} -.9s_1 + .6s_2 &= 0 \\ .9s_1 - .6s_2 &= 0 \\ s_1 + s_2 &= 1 \end{aligned}$$

The solution is: $s_1 = .4$, $s_2 = .6$

The stationary matrix $S = [.4 \quad .6]$; the limiting matrix $\overline{P} = \begin{bmatrix} .4 & .6 \\ .4 & .6 \end{bmatrix}$.

17. Let $S = [s_1, s_2]$, and solve the system:

$$[s_1 \quad s_2]\begin{bmatrix} .5 & .5 \\ .3 & .7 \end{bmatrix} = [s_1 \quad s_2], \; s_1 + s_2 = 1$$

which is equivalent to

$$\begin{aligned} .5s_1 + .3s_2 &= s_1 \\ .5s_1 + .7s_2 &= s_2 \\ s_1 + s_2 &= 1 \end{aligned} \quad \text{or} \quad \begin{aligned} -.5s_1 + .3s_2 &= 0 \\ .5s_1 - .3s_2 &= 0 \\ s_1 + s_2 &= 1 \end{aligned}$$

The solution is: $s_1 = \frac{3}{8} = .375$, $s_2 = \frac{5}{8} = .625$

The stationary matrix $S = [.375 \quad .625]$; the limiting matrix $\overline{P} = \begin{bmatrix} .375 & .625 \\ .375 & .625 \end{bmatrix}$.

19. Let $S = [s_1 \quad s_2 \quad s_3]$, and solve the system:

$$[s_1 \quad s_2 \quad s_3]\begin{bmatrix} .5 & .1 & .4 \\ .3 & .7 & 0 \\ 0 & .6 & .4 \end{bmatrix} = [s_1 \quad s_2 \quad s_3], \; s_1 + s_2 + s_3 = 1$$

which is equivalent to

$$\begin{aligned} .5s_1 + .3s_2 &= s_1 \\ .1s_1 + .7s_2 + .6s_3 &= s_2 \\ .4s_1 + .4s_3 &= s_3 \\ s_1 + s_2 + s_3 &= 1 \end{aligned} \quad \text{or} \quad \begin{aligned} -.5s_1 + .3s_2 &= 0 \\ .1s_1 - .3s_2 + .6s_3 &= 0 \\ .4s_1 - .6s_3 &= 0 \\ s_1 + s_2 + s_3 &= 1 \end{aligned}$$

From the first and third equations, we have $s_2 = \frac{5}{3}s_1$, and $s_3 = \frac{2}{3}s_1$.

Substituting these values into the fourth equation, we get:

$$s_1 + \frac{5}{3}s_1 + \frac{2}{3}s_1 = 1 \quad \text{or} \quad \frac{10}{3}s_1 = 1$$

Therefore, $s_1 = .3$, $s_2 = .5$, $s_3 = .2$.

The stationary matrix $S = [.3 \quad .5 \quad .2]$;

the limiting matrix $\bar{P} = \begin{bmatrix} .3 & .5 & .2 \\ .3 & .5 & .2 \\ .3 & .5 & .2 \end{bmatrix}$.

21. Let $S = [s_1 \quad s_2 \quad s_3]$, and solve the system:

$$[s_1 \quad s_2 \quad s_3]\begin{bmatrix} .8 & .2 & 0 \\ .5 & .1 & .4 \\ 0 & .6 & .4 \end{bmatrix} = [s_1 \quad s_2 \quad s_3], \; s_1 + s_2 + s_3 = 1$$

which is equivalent to

$$\begin{aligned} .8s_1 + .5s_2 &= s_1 \\ .2s_1 + .1s_2 + .6s_3 &= s_2 \\ .4s_2 + .4s_3 &= s_3 \\ s_1 + s_2 + s_3 &= 1 \end{aligned} \quad \text{or} \quad \begin{aligned} -.2s_1 + .5s_2 &= 0 \\ .2s_1 - .9s_2 + .6s_3 &= 0 \\ .4s_2 - .6s_3 &= 0 \\ s_1 + s_2 + s_3 &= 1 \end{aligned}$$

From the first and third equations, we have $s_1 = \frac{5}{2}s_2$, and $s_3 = \frac{2}{3}s_2$.

Substituting these values into the fourth equation, we get:

$$\frac{5}{2}s_2 + s_2 + \frac{2}{3}s_2 = 1 \quad \text{or} \quad \frac{25}{6}s_2 = 1$$

Therefore, $s_2 = \frac{6}{25} = .24$, $s_1 = .6$, $s_3 = .16$.

The stationary matrix $S = [.6 \quad .24 \quad .16]$;

the limiting matrix $\bar{P} = \begin{bmatrix} .6 & .24 & .16 \\ .6 & .24 & .16 \\ .6 & .24 & .16 \end{bmatrix}$.

23. (A) True: A 2×2 transition matrix with 2 entries equal to 0 will have one of the four forms

$$P_1 = \begin{pmatrix} 1 & 0 \\ 0 & 1 \end{pmatrix}, \; P_2 = \begin{pmatrix} 0 & 1 \\ 1 & 0 \end{pmatrix}, \; P_3 = \begin{pmatrix} 1 & 0 \\ 1 & 0 \end{pmatrix}, \; P_4 = \begin{pmatrix} 0 & 1 \\ 0 & 1 \end{pmatrix}.$$

Now, $P_1^k = P_1$, $P_3^k = P_3$, $P_4^k = P_4$ for all positive integers k;

$P_2^k = \begin{pmatrix} 1 & 0 \\ 0 & 1 \end{pmatrix}$ if k is even and $P_2^k = \begin{pmatrix} 0 & 1 \\ 1 & 0 \end{pmatrix}$ if k is odd.

Thus, none of the four transition matrices is regular.

(B) False: Let $P = \begin{bmatrix} 0 & .5 & .5 \\ .5 & 0 & .5 \\ .5 & .5 & 0 \end{bmatrix}$. Then $P^2 = \begin{bmatrix} .5 & .25 & .25 \\ .25 & .5 & .25 \\ .25 & .25 & .5 \end{bmatrix}$.

25. $P = \begin{bmatrix} .51 & .49 \\ .27 & .73 \end{bmatrix}$, $P^2 = \begin{bmatrix} .3924 & .6076 \\ .3348 & .6652 \end{bmatrix}$, $P^4 \approx \begin{bmatrix} .3574 & .6426 \\ .3541 & .6459 \end{bmatrix}$,

$P^8 \approx \begin{bmatrix} .3553 & .6447 \\ .3553 & .6447 \end{bmatrix}$, $P^{16} \approx \begin{bmatrix} .3553 & .6447 \\ .3553 & .6447 \end{bmatrix}$

Therefore, $S \approx [.3553 \quad .6447]$.

27. $P = \begin{bmatrix} .5 & .5 & 0 \\ 0 & .5 & .5 \\ .8 & .1 & .1 \end{bmatrix}$, $P^2 = \begin{bmatrix} .25 & .5 & .25 \\ .4 & .3 & .3 \\ .48 & .46 & .06 \end{bmatrix}$, $P^4 = \begin{bmatrix} .3825 & .39 & .2275 \\ .364 & .428 & .208 \\ .3328 & .4056 & .2616 \end{bmatrix}$,

$P^8 \approx \begin{bmatrix} .3640 & .4084 & .2277 \\ .3642 & .4095 & .2262 \\ .3620 & .4095 & .2285 \end{bmatrix}$, $P^{16} \approx \begin{bmatrix} .3636 & .4091 & .2273 \\ .3636 & .4091 & .2273 \\ .3636 & .4091 & .2273 \end{bmatrix}$

Therefore, $S \approx [.3636 \quad .4091 \quad .2273]$.

29. (A)

.6
.4 Red Blue .8
.2

(B) $P = \begin{matrix} \text{Red} \\ \text{Blue} \end{matrix} \overset{\begin{matrix} \text{Red} & \text{Blue} \end{matrix}}{\begin{bmatrix} .4 & .6 \\ .2 & .8 \end{bmatrix}}$

(C) Let $S = [s_1 \quad s_2]$ and solve the system:

$$[s_1 \quad s_2]\begin{bmatrix} .4 & .6 \\ .2 & .8 \end{bmatrix} = [s_1 \quad s_2],\ s_1 + s_2 = 1$$

which is equivalent to

$$\begin{aligned} .4s_1 + .2s_2 &= s_1 \\ .6s_1 + .8s_2 &= s_2 \\ s_1 + s_2 &= 1 \end{aligned} \qquad \text{or} \qquad \begin{aligned} -.6s_1 + .2s_2 &= 0 \\ .6s_1 - .2s_2 &= 0 \\ s_1 + s_2 &= 1 \end{aligned}$$

The solution is: $s_1 = .25$, $s_2 = .75$.
Thus, the stationary matrix $S = [.25 \quad .75]$. In the long run, the red urn will be selected 25% of the time and the blue urn 75% of the time.

31. (A) $S_1 = [.2 \quad .8]\begin{bmatrix} 0 & 1 \\ 1 & 0 \end{bmatrix} = [.8 \quad .2]$

$S_2 = S_1P = [.8 \quad .2]\begin{bmatrix} 0 & 1 \\ 1 & 0 \end{bmatrix} = [.2 \quad .8]$

$S_3 = S_2P = [.8 \quad .2]$

and so on.

The state matrices alternate between $[.2 \quad .8]$ and $[.8 \quad .2]$; they do not approach a "limiting" matrix.

(B) $S_1 = [.5 \quad .5]\begin{bmatrix} 0 & 1 \\ 1 & 0 \end{bmatrix} = [.5 \quad .5]$

$S_2 = S_1P = [.5 \quad .5]\begin{bmatrix} 0 & 1 \\ 1 & 0 \end{bmatrix} = [.5 \quad .5]$

and so on.

Thus, $S_1 = S_2 = S_3 = \ldots = [.5 \quad .5] = S_0$; S_0 is a stationary matrix.

(C) $P = \begin{bmatrix} 0 & 1 \\ 1 & 0 \end{bmatrix}$, $P^2 = \begin{bmatrix} 1 & 0 \\ 0 & 1 \end{bmatrix}$, $P^3 = \begin{bmatrix} 0 & 1 \\ 1 & 0 \end{bmatrix}$, ...

The powers of P alternate between P and the identity, I; they do not approach a limiting matrix.

(D) Parts (B) and (C) are not valid for this matrix. Since P is not regular, this does not contradict Theorem 1.

33. (A) $RP = [1 \quad 0 \quad 0]\begin{bmatrix} 1 & 0 & 0 \\ .2 & .2 & .6 \\ 0 & 0 & 1 \end{bmatrix} = [1 \quad 0 \quad 0]$

Therefore R is a stationary matrix for P.

$SP = [0 \quad 0 \quad 1]\begin{bmatrix} 1 & 0 & 0 \\ .2 & .2 & .6 \\ 0 & 0 & 1 \end{bmatrix} = [0 \quad 0 \quad 1]$

Therefore S is a stationary matrix for P.
The powers of P have the form

$$\begin{bmatrix} 1 & 0 & 0 \\ a & b & c \\ 0 & 0 & 1 \end{bmatrix}$$

Therefore P is not regular. As a result, P may have more than one stationary matrix.

(B) Following the hint, let

$T = a[1 \quad 0 \quad 0] + (1 - a)[0 \quad 0 \quad 1],\ 0 < a < 1.$
$= [a \quad 0 \quad 1 - a]$

Now, $TP = [a \quad 0 \quad 1 - a]\begin{bmatrix} 1 & 0 & 0 \\ .2 & .2 & .6 \\ 0 & 0 & 1 \end{bmatrix} = [a \quad 0 \quad 1 - a] = T$

Thus, $[a \quad 0 \quad 1 - a]$ is a stationary matrix for P for every a with $0 < a < 1$. Note that if $a = 1$, then $T = R$, and if $a = 0$, $T = S$. If we let $a = .5$, then $T = [.5 \quad 0 \quad .5]$ is a stationary matrix.

(C) P has infinitely many stationary matrices.

35. $\overline{P} = \begin{bmatrix} 1 & 0 & 0 \\ .25 & 0 & .75 \\ 0 & 0 & 1 \end{bmatrix}$

Each row of $\overline{P}$ is a stationary matrix for P. As we saw in Problem 33, part (B),

$T = [a \quad 0 \quad 1 - a]$

is a stationary matrix for P for each a where $0 \le a \le 1$; $a = 1$ gives the first row of $\overline{P}$, $a = .25$ gives the second row; $a = 0$ gives the third row.

37. (A) For P^2, $M_2 = .39$; for P^3, $M_3 = .3$; for P^4, $M_4 = .284$; for P^5, $M_5 = .277$.

Each entry of the second column of P^{k+1} is a product of the form

$ap_{12}^k + bp_{22}^k + cp_{32}^k$

where p_{12}^k, p_{22}^k, p_{32}^k are the entries in the second column of P^k;

$a, b, c \ge 0$ and $a + b + c = 1$. Thus,

$M_{k+1} = ap_{12}^k + bp_{22}^k + cp_{32}^k \le aM_k + bM_k + cM_k = (a + b + c)M_k = M_k$

Therefore, $M_k \ge M_{k+1}$ for all positive integers k.

39. The transition matrix is

$$P = \begin{matrix} & H & N \\ H & .89 & .11 \\ N & .29 & .71 \end{matrix} \quad \begin{matrix} H = \text{home trackage} \\ N = \text{national pool} \end{matrix}$$

Calculating powers of P, we have

$$P^2 = \begin{bmatrix} .824 & .176 \\ .464 & .536 \end{bmatrix}, \; P^4 \approx \begin{bmatrix} .7606 & .2394 \\ .6310 & .3690 \end{bmatrix}, \; P^8 \approx \begin{bmatrix} .7296 & .2704 \\ .7128 & .2872 \end{bmatrix},$$

$$P^{16} \approx \begin{bmatrix} .7251 & .2749 \\ .7248 & .2752 \end{bmatrix}$$

In the long run, 72.5% of the company's box cars will be on its home trackage.

41. (A) $S_1 = [.433 \quad .567]\begin{bmatrix} .93 & .07 \\ .2 & .8 \end{bmatrix} = [.51609 \quad .48391] \approx [.516 \quad .484]$

$$S_2 = S_0P^2 = [.433 \quad .567]\begin{bmatrix} .8789 & .1211 \\ .346 & .654 \end{bmatrix} = [.5767457 \quad .4232543] \approx [.577 \quad .423]$$

(B)

Year	Data %	Model %
1970	43.3	43.3
1980	51.5	51.6
1990	57.5	57.7

(C) $P^4 \approx \begin{bmatrix} .8143 & .1856 \\ .5303 & .4696 \end{bmatrix}, \; P^8 \approx \begin{bmatrix} .7616 & .2384 \\ .6810 & .3190 \end{bmatrix}, \; P^{16} \approx \begin{bmatrix} .7424 & .2576 \\ .7359 & .2641 \end{bmatrix},$

$$P^{32} \approx \begin{bmatrix} .7408 & .2592 \\ .7407 & .2593 \end{bmatrix}$$

In the long run, 74.1% of the female population will be in the labor force.

43. The transition matrix for this problem is:

$$\begin{matrix} & \text{GTT} & \text{NCJ} & \text{Dash} \\ \text{GTT} & .75 & .05 & .20 \\ \text{NCJ} & .15 & .75 & .1 \\ \text{Dash} & .05 & .10 & .85 \end{matrix}$$

To find the steady-state matrix, we solve the system

$$[s_1 \quad s_2 \quad s_3]\begin{bmatrix} .75 & .05 & .20 \\ .15 & .75 & .10 \\ .05 & .10 & .85 \end{bmatrix} = [s_1 \quad s_2 \quad s_3], \; s_1 + s_2 + s_3 = 1$$

which is equivalent to the system of equations

$$\begin{aligned} .75s_1 + .15s_2 + .05s_3 &= s_1 \\ .05s_1 + .75s_2 + .1s_3 &= s_2 \\ .20s_1 + .10s_2 + .85s_3 &= s_3 \\ s_1 + s_2 + s_3 &= 1 \end{aligned}$$

The solution of this system is $s_1 = .25$, $s_2 = .25$, $s_3 = .5$.

Thus, the expected market share of each company is:

GTT - 25%; NCJ - 25%; and Dash - 50%.

45. The transition matrix for this problem is:

	Poor	Satisfactory	Preferred
Poor	.60	.40	0
Satisfactory	.20	.60	.20
Preferred	0	.20	.80

To find the steady-state matrix, we solve the system

$$[s_1 \quad s_2 \quad s_3]\begin{bmatrix} .60 & .40 & 0 \\ .20 & .60 & .20 \\ 0 & .20 & .80 \end{bmatrix} = [s_1 \quad s_2 \quad s_3],\ s_1 + s_2 + s_3 = 1$$

which is equivalent to the system of equations:

$$\begin{aligned} .6s_1 + .2s_2 \qquad\qquad &= s_1 \\ .4s_1 + .6s_2 + .2s_3 &= s_2 \\ .2s_2 + .8s_3 &= s_3 \\ s_1 + \quad s_2 + \quad s_3 &= 1 \end{aligned}$$

The solution of this system is $s_1 = .20$, $s_2 = .40$, and $s_3 = .40$.

Thus, the expected percentage in each category is:

poor - 20%; satisfactory - 40%; and preferred - 40%.

47. The transition matrix is:

$$P = \begin{bmatrix} .4 & .1 & .3 & .2 \\ .3 & .2 & .2 & .3 \\ .1 & .2 & .2 & .5 \\ .3 & .3 & .1 & .3 \end{bmatrix}$$

$S_0P = [.3 \quad .3 \quad .4 \quad 0]P = [.25 \quad .17 \quad .23 \quad .35] = S_1$

$S_1P = [.25 \quad .17 \quad .23 \quad .35]P = [.28 \quad .21 \quad .19 \quad .32] = S_2$

$S_2P = [.28 \quad .21 \quad .19 \quad .32]P = [.29 \quad .20 \quad .20 \quad .31] = S_3$

$S_3P = [.29 \quad .20 \quad .20 \quad .31]P = [.29 \quad .20 \quad .20 \quad .31] = S_4$

Thus, $S = [.29 \quad .20 \quad .20 \quad .31]$ is the steady-state matrix. The expected market share for the two Acme soaps is .20 + .31 = .51 or 51%.

49. To find the stationary solution, we solve the system

$$[s_1 \quad s_2 \quad s_3]\begin{bmatrix} .5 & .5 & 0 \\ .25 & .5 & .25 \\ 0 & .5 & .5 \end{bmatrix} = [s_1 \quad s_2 \quad s_3],\ s_1 + s_2 + s_3 = 1,$$

which is equivalent to:

$$\begin{aligned} .5s_1 + .25s_2 \qquad\qquad &= s_1 \\ .5s_1 + .5s_2 + .5s_3 &= s_2 \\ .25s_2 + .5s_3 &= s_3 \\ s_1 + \quad s_2 + \quad s_3 &= 1 \end{aligned} \qquad \text{or} \qquad \begin{aligned} -.5s_1 + .25s_2 \qquad\qquad &= 0 \\ .5s_1 - .5s_2 + .5s_3 &= 0 \\ .25s_2 - .5s_3 &= 0 \\ s_1 + \quad s_2 + \quad s_3 &= 1 \end{aligned}$$

The solution of this system is $s_1 = .25$, $s_2 = .5$, $s_3 = .25$.

Thus, the stationary matrix is $S = [.25 \quad .5 \quad .25]$.

51. (A) Initial-state matrix = $\begin{matrix} \text{Rapid transit} & \text{Auto} \\ [.25 & .75] \end{matrix}$

(B) Second-state matrix = $[.25 \quad .75]\begin{bmatrix} .8 & .2 \\ .3 & .7 \end{bmatrix} = [.425 \quad .575]$

Thus, 42.5% will be using the new system after one month.

Third-state matrix = $[.425 \quad .575]\begin{bmatrix} .8 & .2 \\ .3 & .7 \end{bmatrix} = [.5125 \quad .4875]$

Thus, 51.25% will be using the new system after two months.

(C) To find the stationary solution, we solve the system

$[s_1 \quad s_2]\begin{bmatrix} .8 & .2 \\ .3 & .7 \end{bmatrix} = [s_1 \quad s_2]$, $s_1 + s_2 = 1$,

which is equivalent to: $\begin{aligned} .8s_1 + .3s_2 &= s_1 \\ .2s_1 + .7s_2 &= s_2 \\ s_1 + s_2 &= 1 \end{aligned}$ or $\begin{aligned} -.2s_1 + .3s_2 &= 0 \\ .2s_1 - .3s_2 &= 0 \\ s_1 + s_2 &= 1 \end{aligned}$

The solution of this system of linear equations is $s_1 = .6$ and $s_2 = .4$. Thus, the stationary solution is $S = [.6 \quad .4]$, which means that 60% of the commuters will use rapid transit and 40% will travel by automobile after the system has been in service for a long time.

53. (A) $s_1 = [.309 \quad .691]\begin{bmatrix} .61 & .39 \\ .21 & .79 \end{bmatrix} = [.3336 \quad .6664] \approx [.334 \quad .666]$

$s_2 = s_0P^2 = [.309 \quad .691]\begin{bmatrix} .454 & .546 \\ .294 & .706 \end{bmatrix} = [.34344 \quad .65656]$

$\approx [.343 \quad .657]$

(B)

Year	Data %	Model %
1970	30.9	30.9
1980	33.3	33.4
1990	34.4	34.3

(C) $P^4 \approx \begin{bmatrix} .367 & .633 \\ .341 & .659 \end{bmatrix}$, $P^8 \approx \begin{bmatrix} .350 & .650 \\ .350 & .650 \end{bmatrix}$, $P^{16} \approx \begin{bmatrix} .350 & .650 \\ .350 & .650 \end{bmatrix}$

In the long run, 35% of the population will live in the south region.

EXERCISE 9-3

Things to remember:

1. ABSORBING STATES AND TRANSIENT STATES

   A state in a Markov chain is an ABSORBING STATE if once the state is entered, it is impossible to leave. A nonabsorbing state is called a TRANSIENT STATE.

2. ABSORBING STATES AND TRANSITION MATRICES

A state in a Markov chain is ABSORBING if and only if the row of the transition matrix correspondig to the state has a 1 on the main diagonal and zeros elsewhere.

3. ABSORBING MARKOV CHAINS

A Markov chain is an ABSORBING CHAIN if

(A) There is at least one absorbing state.

(B) It is possible to go from each nonabsorbing state to at least one absorbing state in a finite number of steps.

4. STANDARD FORMS FOR ABSORBING MARKOV CHAINS

A transition matrix for an absorbing Markov chain is a STANDARD FORM if the rows and columns are labeled so that all the absorbing states precede all the nonabsorbing states. (There may be more than one standard form.) Any standard form can always be partitioned into four submatrices:

$$\begin{array}{c} \\ A \\ T \end{array}\begin{array}{c} \begin{array}{cc} A & T \end{array} \\ \left[\begin{array}{c|c} I & 0 \\ \hline R & Q \end{array}\right] \end{array} \quad \left[\begin{array}{l} A = \text{All absorbing states} \\ T = \text{All nonabsorbing states} \end{array}\right]$$

where I is an identity matrix and 0 is a zero matrix.

5. LIMITING MATRICES FOR ABSORBING MARKOV CHAINS

If a standard form P for an absorbing Markov chain is partitioned as

$$P = \left[\begin{array}{c|c} I & 0 \\ \hline R & Q \end{array}\right]$$

then P^k approaches a matrix $\bar{P}$ as k increases, where

$$\bar{P} = P = \left[\begin{array}{c|c} I & 0 \\ \hline FR & 0 \end{array}\right]$$

The matrix F is given by $F = (I - Q)^{-1}$ and is called the FUNDAMENTAL MATRIX for P.

The identity matrix used to form the fundamental matrix F must be the same size as the matrix Q.

6. PROPERTIES OF THE LIMITING MATRIX $\bar{P}$

If P is a standard form transition matrix for an absorbing Markov chain, F is the fundamental matrix, and $\bar{P}$ is the limiting matrix, then

(A) The entry in row i and column j of $\bar{P}$ is the long run probability of going from state i to state j. For the nonabsorbing states, these probabilities are also the entries in the matrix FR used to form $\bar{P}$.

(B) The sum of the entries in each row of the fundamental matrix F is the average number of trials it will take to go from each nonabsorbing state to some absorbing state.

[Note that the rows of both F and FR correspond to the nonabsorbing states in the order given in the standard form P.]

1. By $\underline{2}$, states B and C are absorbing states.

3. By $\underline{2}$, there are no absorbing states.

5. By $\underline{2}$, states A and D are absorbing states.

7. B is an absorbing state; the diagram represents an absorbing Markov chain since it is possible to go from states A and C to state B in a finite number of steps.

9. C is an absorbing state; the diagram does not represent an absorbing Markov chain since it is not possible to go from either states A or D to state C.

11. The transition diagram is represented by the matrix:

$$\begin{matrix} & \begin{matrix} A & B & C \end{matrix} \\ \begin{matrix} A \\ B \\ C \end{matrix} & \begin{bmatrix} .2 & .5 & .3 \\ 0 & 1 & 0 \\ .5 & .1 & .4 \end{bmatrix} \end{matrix}$$

A standard form for this matrix is:

$$\begin{matrix} & \begin{matrix} B & A & C \end{matrix} \\ \begin{matrix} B \\ A \\ C \end{matrix} & \begin{bmatrix} 1 & 0 & 0 \\ .5 & .2 & .3 \\ .1 & .5 & .4 \end{bmatrix} \end{matrix}$$

13. The transition diagram is represented by the matrix

$$\begin{matrix} & \begin{matrix} A & B & C & D \end{matrix} \\ \begin{matrix} A \\ B \\ C \\ D \end{matrix} & \begin{bmatrix} .3 & .4 & .2 & .1 \\ 0 & 1 & 0 & 0 \\ 0 & .4 & .3 & .3 \\ 0 & 0 & 0 & 1 \end{bmatrix} \end{matrix}$$

A standard form for this matrix

$$\begin{matrix} & \begin{matrix} B & D & A & C \end{matrix} \\ \begin{matrix} B \\ D \\ A \\ C \end{matrix} & \begin{bmatrix} 1 & 0 & 0 & 0 \\ 0 & 1 & 0 & 0 \\ .4 & .1 & .3 & .2 \\ .4 & .3 & 0 & .3 \end{bmatrix} \end{matrix}$$

15. A standard form for

$$P = \begin{matrix} & \begin{matrix} A & B & C \end{matrix} \\ \begin{matrix} A \\ B \\ C \end{matrix} & \begin{bmatrix} .2 & .3 & .5 \\ 1 & 0 & 0 \\ 0 & 0 & 1 \end{bmatrix} \end{matrix}$$

is:

$$\begin{matrix} & \begin{matrix} C & A & B \end{matrix} \\ \begin{matrix} C \\ A \\ B \end{matrix} & \begin{bmatrix} 1 & 0 & 0 \\ .5 & .2 & .3 \\ 0 & 1 & 0 \end{bmatrix} \end{matrix}$$

17. A standard form for

$$P = \begin{matrix} & \begin{matrix} A & B & C & D \end{matrix} \\ \begin{matrix} A \\ B \\ C \\ D \end{matrix} & \begin{bmatrix} .1 & .2 & .3 & .4 \\ 0 & 1 & 0 & 0 \\ .5 & .2 & .2 & .1 \\ 0 & 0 & 0 & 1 \end{bmatrix} \end{matrix}$$

is:

$$\begin{matrix} & \begin{matrix} B & D & A & C \end{matrix} \\ \begin{matrix} B \\ D \\ A \\ C \end{matrix} & \begin{bmatrix} 1 & 0 & 0 & 0 \\ 0 & 1 & 0 & 0 \\ .2 & .4 & .1 & .3 \\ .2 & .1 & .5 & .2 \end{bmatrix} \end{matrix}$$

19. For

$$P = \begin{matrix} & \begin{matrix} A & B & C \end{matrix} \\ \begin{matrix} A \\ B \\ C \end{matrix} & \begin{bmatrix} 1 & 0 & 0 \\ 0 & 1 & 0 \\ .1 & .4 & .5 \end{bmatrix} \end{matrix}$$

we have $R = [.1 \quad .4]$ and $Q = [.5]$.

The limiting matrix $\overline{P}$ has the form

$$\overline{P} = \left[\begin{array}{cc|c} 1 & 0 & 0 \\ 0 & 1 & 0 \\ \hline F & R & 0 \end{array}\right]$$

where $F = (I - Q)^{-1} = ([1] - [.5])^{-1} = [.5]^{-1} = [2]$ and $FR = [2][.1 \quad .4] = [.2 \quad .8]$.

Thus,

$$\bar{P} = \begin{matrix} & A & B & C \\ A & 1 & 0 & 0 \\ B & 0 & 1 & 0 \\ C & .2 & .8 & 0 \end{matrix}$$

Let $P(i \text{ to } j)$ denote the probability of going from state i to state j.
Then $P(C \text{ to } A) = .2$, $P(C \text{ to } B) = .8$
Since $F = [2]$, it will take an average of 2 trials to go from C to either A or B.

21. For

$$P = \begin{matrix} & A & B & C \\ A & 1 & 0 & 0 \\ B & .2 & .6 & .2 \\ C & .4 & .2 & .4 \end{matrix}$$

we have $R = \begin{bmatrix} .2 \\ .4 \end{bmatrix}$ and $Q = \begin{bmatrix} .6 & .2 \\ .2 & .4 \end{bmatrix}$

The limiting matrix $\bar{P}$ has the form

$$\bar{P} = \left[\begin{array}{cc|c} 1 & 0 & 0 \\ \hline F & R & 0 \end{array}\right] \text{ where } F = (I - Q)^{-1} = \left(\begin{bmatrix} 1 & 0 \\ 0 & 1 \end{bmatrix} - \begin{bmatrix} .6 & .2 \\ .2 & .4 \end{bmatrix}\right)^{-1}$$

$$= \begin{bmatrix} .4 & -.2 \\ -.2 & .6 \end{bmatrix}^{-1} = \begin{bmatrix} \frac{2}{5} & -\frac{1}{5} \\ -\frac{1}{5} & \frac{3}{5} \end{bmatrix}^{-1}$$

We use row operations to find the inverse:

$$\left[\begin{array}{cc|cc} \frac{2}{5} & -\frac{1}{5} & 1 & 0 \\ -\frac{1}{5} & \frac{3}{5} & 0 & 1 \end{array}\right] \sim \left[\begin{array}{cc|cc} 1 & -\frac{1}{2} & \frac{5}{2} & 0 \\ -\frac{1}{5} & \frac{3}{5} & 0 & 1 \end{array}\right] \sim \left[\begin{array}{cc|cc} 1 & -\frac{1}{2} & \frac{5}{2} & 0 \\ 0 & \frac{1}{2} & \frac{1}{2} & 1 \end{array}\right] \sim \left[\begin{array}{cc|cc} 1 & -\frac{1}{2} & \frac{5}{2} & 0 \\ 0 & 1 & 1 & 2 \end{array}\right]$$

$$\left(\frac{5}{2}\right)R_1 \to R_1 \qquad \left(\frac{1}{5}\right)R_1 + R_2 \to R_2 \qquad 2R_2 \to R_2 \qquad \left(\frac{1}{2}\right)R_2 + R_1 \to R_1$$

$$\sim \left[\begin{array}{cc|cc} 1 & 0 & 3 & 1 \\ 0 & 1 & 1 & 2 \end{array}\right]$$

Thus, $F = \begin{bmatrix} 3 & 1 \\ 1 & 2 \end{bmatrix}$ and $FR = \begin{bmatrix} 3 & 1 \\ 1 & 2 \end{bmatrix}\begin{bmatrix} .2 \\ .4 \end{bmatrix} = \begin{bmatrix} 1 \\ 1 \end{bmatrix}$

Now

$$\bar{P} = \begin{matrix} & A & B & C \\ A & 1 & 0 & 0 \\ B & 1 & 0 & 0 \\ C & 1 & 0 & 0 \end{matrix}$$

$P(B \text{ to } A) = 1$, $P(C \text{ to } A) = 1$

It will take an average of 4 trials to go from B to A; it will take an average of 3 trials to go from C to A.

23. For

$$P = \begin{matrix} & A & B & C & D \\ A & 1 & 0 & 0 & 0 \\ B & 0 & 1 & 0 & 0 \\ C & .1 & .2 & .6 & .1 \\ D & .2 & .2 & .3 & .3 \end{matrix}$$

we have $R = \begin{bmatrix} .1 & .2 \\ .2 & .2 \end{bmatrix}$ and $Q = \begin{bmatrix} .6 & .1 \\ .3 & .3 \end{bmatrix}$

The limiting matrix $\overline{P}$ has the form

$$\overline{P} = \left[\begin{array}{c|cc} I & 0 & 0 \\ & 0 & 0 \\ \hline FR & & 0 \end{array}\right]$$

where $F = (I - Q)^{-1} = \left(\begin{bmatrix} 1 & 0 \\ 0 & 1 \end{bmatrix} - \begin{bmatrix} .6 & .1 \\ .3 & .3 \end{bmatrix}\right)^{-1}$

$$= \begin{bmatrix} .4 & -.1 \\ -.3 & .7 \end{bmatrix}^{-1} = \begin{bmatrix} \frac{2}{5} & -\frac{1}{10} \\ -\frac{3}{10} & \frac{7}{10} \end{bmatrix}^{-1}$$

We use row operations to find the inverse:

$$\left[\begin{array}{cc|cc} \frac{2}{5} & -\frac{1}{10} & 1 & 0 \\ -\frac{3}{10} & \frac{7}{10} & 0 & 1 \end{array}\right] \sim \left[\begin{array}{cc|cc} 1 & -\frac{1}{4} & \frac{5}{2} & 0 \\ -\frac{3}{10} & \frac{7}{10} & 0 & 1 \end{array}\right] \sim \left[\begin{array}{cc|cc} 1 & -\frac{1}{4} & \frac{5}{2} & 0 \\ 0 & \frac{5}{8} & \frac{3}{4} & 1 \end{array}\right] \sim \left[\begin{array}{cc|cc} 1 & -\frac{1}{4} & \frac{5}{2} & 0 \\ 0 & 1 & \frac{6}{5} & \frac{8}{5} \end{array}\right]$$

$\left(\frac{5}{2}\right)R_1 \to R_1$ $\quad\left(\frac{3}{10}\right)R_1 + R_2 \to R_2$ $\quad\left(\frac{8}{5}\right)R_2 \to R_2$ $\quad\left(\frac{1}{4}\right)R_2 + R_1 \to R_1$

$$\sim \left[\begin{array}{cc|cc} 1 & 0 & \frac{14}{5} & \frac{2}{5} \\ 0 & 1 & \frac{6}{5} & \frac{8}{5} \end{array}\right]$$

Thus, $F = \begin{bmatrix} \frac{14}{5} & \frac{2}{5} \\ \frac{6}{5} & \frac{8}{5} \end{bmatrix} = \begin{bmatrix} 2.8 & .4 \\ 1.2 & 1.6 \end{bmatrix}$,

$$FR = \begin{bmatrix} 2.8 & .4 \\ 1.2 & 1.6 \end{bmatrix}\begin{bmatrix} .1 & .2 \\ .2 & .2 \end{bmatrix} = \begin{bmatrix} .36 & .64 \\ .44 & .56 \end{bmatrix}$$

and

$$\overline{P} = \begin{array}{c} \\ A \\ B \\ C \\ D \end{array}\begin{array}{c} \begin{array}{cccc} A & B & C & D \end{array} \\ \begin{bmatrix} 1 & 0 & 0 & 0 \\ 0 & 1 & 0 & 0 \\ .36 & .64 & 0 & 0 \\ .44 & .56 & 0 & 0 \end{bmatrix} \end{array}$$

$P(C \text{ to } A) = .36$, $P(C \text{ to } B) = .64$,
$P(D \text{ to } A) = .44$, $P(D \text{ to } B) = .56$

It will take an average of 3.2 trials to go from C to either A or B; it will take an average of 2.8 trials to go from D to either A or B.

25. (A) $S_0\overline{P} = [0 \quad 0 \quad 1]\begin{bmatrix} 1 & 0 & 0 \\ 0 & 1 & 0 \\ .2 & .8 & 0 \end{bmatrix} = [.2 \quad .8 \quad 0]$

(B) $S_0\overline{P} = [.2 \quad .5 \quad .3]\begin{bmatrix} 1 & 0 & 0 \\ 0 & 1 & 0 \\ .2 & .8 & 0 \end{bmatrix} = [.26 \quad .74 \quad 0]$

27. (A) $S_0\overline{P} = [0 \quad 0 \quad 1]\begin{bmatrix} 1 & 0 & 0 \\ 1 & 0 & 0 \\ 1 & 0 & 0 \end{bmatrix} = [1 \quad 0 \quad 0]$

(B) $S_0\overline{P} = [.2 \quad .5 \quad .3]\begin{bmatrix} 1 & 0 & 0 \\ 1 & 0 & 0 \\ 1 & 0 & 0 \end{bmatrix} = [1 \quad 0 \quad 0]$

29. (A) $S_0\overline{P} = [0 \quad 0 \quad 0 \quad 1]\begin{bmatrix} 1 & 0 & 0 & 0 \\ 0 & 1 & 0 & 0 \\ .36 & .64 & 0 & 0 \\ .44 & .56 & 0 & 0 \end{bmatrix} = [.44 \quad .56 \quad 0 \quad 0]$

(B) $S_0\overline{P} = [0 \quad 0 \quad 1 \quad 0]\begin{bmatrix} 1 & 0 & 0 & 0 \\ 0 & 1 & 0 & 0 \\ .36 & .64 & 0 & 0 \\ .44 & .56 & 0 & 0 \end{bmatrix} = [.36 \quad .64 \quad 0 \quad 0]$

(C) $S_0\overline{P} = [0 \quad 0 \quad .4 \quad .6]\begin{bmatrix} 1 & 0 & 0 & 0 \\ 0 & 1 & 0 & 0 \\ .36 & .64 & 0 & 0 \\ .44 & .56 & 0 & 0 \end{bmatrix} = [.408 \quad .592 \quad 0 \quad 0]$

(D) $S_0\overline{P} = [.1 \quad .2 \quad .3 \quad .4]\begin{bmatrix} 1 & 0 & 0 & 0 \\ 0 & 1 & 0 & 0 \\ .36 & .64 & 0 & 0 \\ .44 & .56 & 0 & 0 \end{bmatrix} = [.384 \quad .616 \quad 0 \quad 0]$

31. (A) True. If every state is absorbing, then the transition matrix is the identity matrix; there are no nonabsorbing states.

(B) False. For the transition matrix:

$$\begin{array}{c} \\ A \\ B \\ C \end{array}\begin{array}{c} \begin{array}{ccc} A & B & C \end{array} \\ \begin{bmatrix} 1 & 0 & 0 \\ 0 & 0 & 1 \\ 0 & 1 & 0 \end{bmatrix} \end{array}$$

A is an absorbing state; B and C are nonabsorbing states and it is impossible to go from either state B or state C to state A.

33. By Theorem 2, P has a limiting matrix:

$$P^4 \approx \begin{bmatrix} 1 & 0 & 0 & 0 \\ 0 & 1 & 0 & 0 \\ .6364 & .362 & 0 & 0 \\ .7364 & .262 & 0 & 0 \end{bmatrix}, \quad P^8 \approx \begin{bmatrix} 1 & 0 & 0 & 0 \\ 0 & 1 & 0 & 0 \\ .6375 & .3625 & 0 & 0 \\ .7375 & .2625 & 0 & 0 \end{bmatrix}$$

$$P^{16} \approx \begin{bmatrix} 1 & 0 & 0 & 0 \\ 0 & 1 & 0 & 0 \\ .6375 & .3625 & 0 & 0 \\ .7375 & .2625 & 0 & 0 \end{bmatrix}; \quad \overline{P} = \begin{bmatrix} 1 & 0 & 0 & 0 \\ 0 & 1 & 0 & 0 \\ .6375 & .3625 & 0 & 0 \\ .7375 & .2625 & 0 & 0 \end{bmatrix}$$

35. By Theorem 2, P has a limiting matrix:

$$P^4 \approx \begin{bmatrix} 1 & 0 & 0 & 0 & 0 \\ 0 & 1 & 0 & 0 & 0 \\ .0724 & .8368 & .0625 & .011 & .0173 \\ .174 & .7792 & 0 & .0279 & .0189 \\ .4312 & .5472 & 0 & .0126 & .009 \end{bmatrix},$$

$$P^{16} \approx \begin{bmatrix} 1 & 0 & 0 & 0 & 0 \\ 0 & 1 & 0 & 0 & 0 \\ .0875 & .9125 & 0 & 0 & 0 \\ .1875 & .8125 & 0 & 0 & 0 \\ .4375 & .5625 & 0 & 0 & 0 \end{bmatrix}, \quad P^{32} \approx \begin{bmatrix} 1 & 0 & 0 & 0 & 0 \\ 0 & 1 & 0 & 0 & 0 \\ .0875 & .9125 & 0 & 0 & 0 \\ .1875 & .8125 & 0 & 0 & 0 \\ .4375 & .5625 & 0 & 0 & 0 \end{bmatrix};$$

$$\overline{P} = \begin{bmatrix} 1 & 0 & 0 & 0 & 0 \\ 0 & 1 & 0 & 0 & 0 \\ .0875 & .9125 & 0 & 0 & 0 \\ .1875 & .8125 & 0 & 0 & 0 \\ .4375 & .5625 & 0 & 0 & 0 \end{bmatrix}$$

37. *Step 1*. Transition diagram:

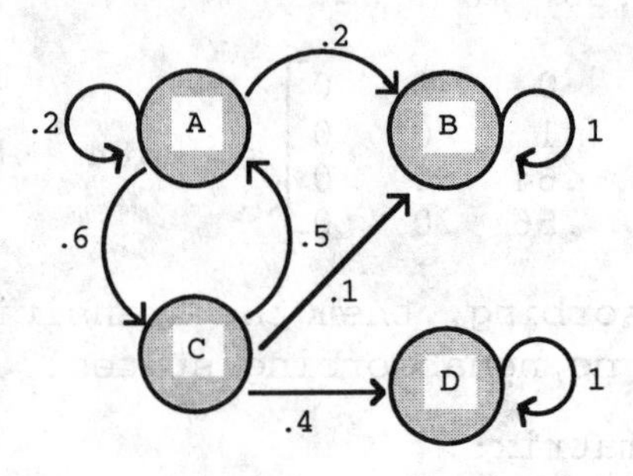

Standard form:

$$M = \begin{array}{c} \\ B \\ D \\ A \\ C \end{array} \begin{array}{c} \begin{array}{cccc} B & D & A & C \end{array} \\ \begin{bmatrix} 1 & 0 & 0 & 0 \\ 0 & 1 & 0 & 0 \\ .2 & 0 & .2 & .6 \\ .1 & .4 & .5 & 0 \end{bmatrix} \end{array}$$

Step 2. Limiting matrix:

For M, we have $R = \begin{bmatrix} .2 & 0 \\ .1 & .4 \end{bmatrix}$ and $Q = \begin{bmatrix} .2 & .6 \\ .5 & 0 \end{bmatrix}$

The limiting matrix $\overline{M}$ has the form:

$$\overline{M} = \left[\begin{array}{c|c} I & 0 \\ \hline FR & 0 \end{array}\right]$$

where $F = (I - Q)^{-1} = \left(\begin{bmatrix} 1 & 0 \\ 0 & 1 \end{bmatrix} - \begin{bmatrix} .2 & .6 \\ .5 & 0 \end{bmatrix}\right)^{-1} = \begin{bmatrix} .8 & -.6 \\ -.5 & 1 \end{bmatrix}^{-1}$

$$= \begin{bmatrix} \frac{4}{5} & -\frac{3}{5} \\ -\frac{1}{2} & 1 \end{bmatrix}^{-1}$$

We use row operations to find the inverse:

$$\left[\begin{array}{cc|cc} \frac{4}{5} & -\frac{3}{5} & 1 & 0 \\ -\frac{1}{2} & 1 & 0 & 1 \end{array}\right] \sim \left[\begin{array}{cc|cc} 1 & -\frac{3}{4} & \frac{5}{4} & 0 \\ -\frac{1}{2} & 1 & 0 & 1 \end{array}\right] \sim \left[\begin{array}{cc|cc} 1 & -\frac{3}{4} & \frac{5}{4} & 0 \\ 0 & \frac{5}{8} & \frac{5}{8} & 1 \end{array}\right]$$

$\left(\frac{5}{4}\right)R_1 \to R_1$ $\qquad \left(\frac{1}{2}\right)R_1 + R_2 \to R_2$ $\qquad \left(\frac{8}{5}\right)R_2 \to R_2$

$$\sim \left[\begin{array}{cc|cc} 1 & -\frac{3}{4} & \frac{5}{4} & 0 \\ 0 & 1 & 1 & \frac{8}{5} \end{array}\right] \sim \left[\begin{array}{cc|cc} 1 & 0 & 2 & \frac{6}{5} \\ 0 & 1 & 1 & \frac{8}{5} \end{array}\right]$$

$\left(\frac{3}{4}\right)R_2 + R_1 \to R_1$

Thus, $F = \begin{bmatrix} 2 & 1.2 \\ 1 & 1.6 \end{bmatrix}$ and $FR = \begin{bmatrix} 2 & 1.2 \\ 1 & 1.6 \end{bmatrix}\begin{bmatrix} .2 & 0 \\ .1 & .4 \end{bmatrix} = \begin{bmatrix} .52 & .48 \\ .36 & .64 \end{bmatrix}$

Therefore, $\overline{M} = \begin{matrix} & B & D & A & C \\ B & 1 & 0 & 0 & 0 \\ D & 0 & 1 & 0 & 0 \\ A & .52 & .48 & 0 & 0 \\ C & .36 & .64 & 0 & 0 \end{matrix}$

Step 3. Transition diagram for $\overline{M}$:

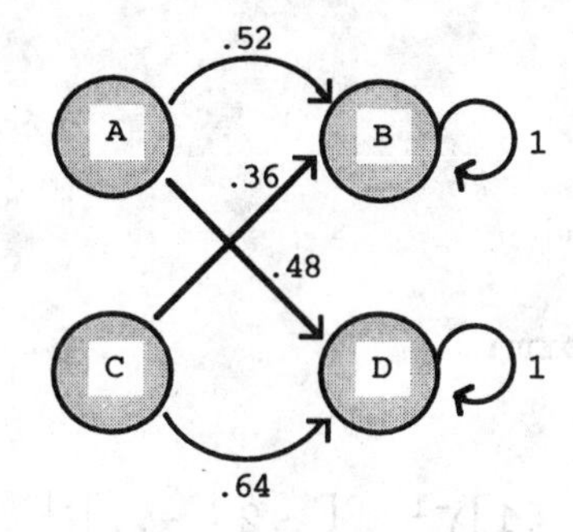

Limiting matrix for P:

$\overline{P} = \begin{matrix} & A & B & C & D \\ A & 0 & .52 & 0 & .48 \\ B & 0 & 1 & 0 & 0 \\ C & 0 & .36 & 0 & .64 \\ D & 0 & 0 & 0 & 1 \end{matrix}$

39. $P^4 \approx \begin{bmatrix} .1276 & .426 & .0768 & .3696 \\ 0 & 1 & 0 & 0 \\ .064 & .29 & .102 & .544 \\ 0 & 0 & 0 & 1 \end{bmatrix}$, $P^8 \approx \begin{bmatrix} .0212 & .5026 & .0176 & .4585 \\ 0 & 1 & 0 & 0 \\ .0147 & .3468 & .0153 & .6231 \\ 0 & 0 & 0 & 0 \end{bmatrix}$

$P^{32} \approx \begin{bmatrix} 0 & .52 & 0 & .48 \\ 0 & 1 & 0 & 0 \\ 0 & .36 & 0 & .64 \\ 0 & 0 & 0 & 1 \end{bmatrix}$

41. Let $S = [x \quad 1 - x \quad 0]$, $0 \le x \le 1$. Then

$$SP = [x \quad 1 - x \quad 0]\begin{bmatrix} 1 & 0 & 0 \\ 0 & 1 & 0 \\ .1 & .5 & .4 \end{bmatrix} = [x \quad 1 - x \quad 0]$$

Thus, S is a stationary matrix for P.

A stationary matrix for an absorbing Markov chain with two absorbing states and one nonabsorbing state will have one of the forms

$[x \quad 1 - x \quad 0]$, $[x \quad 0 \quad 1 - x]$, $[0 \quad x \quad 1 - x]$

43. (A) For P^2, $w_2 = .37$; for P^4, $w_4 = .297$; for P^8, $w_8 = .227$; for P^{16}, $w_{16} = .132$; for P^{32}, $w_{32} = .045$

(B) For large k, the entries of Q^k are close to 0.

45. A transition matrix for this problem is:

$$P = \begin{array}{c} \\ F \\ G \\ A \\ B \end{array}\begin{array}{c} \begin{array}{cccc} F & G & A & B \end{array} \\ \begin{bmatrix} 1 & 0 & 0 & 0 \\ .1 & .8 & .1 & 0 \\ .1 & .4 & .4 & .1 \\ 0 & 0 & 0 & 1 \end{bmatrix} \end{array}$$

A standard form for this matrix is:

$$M = \begin{array}{c} \\ F \\ B \\ G \\ A \end{array}\begin{array}{c} \begin{array}{cccc} F & B & G & A \end{array} \\ \begin{bmatrix} 1 & 0 & 0 & 0 \\ 0 & 1 & 0 & 0 \\ .1 & 0 & .8 & .1 \\ .1 & .1 & .4 & .4 \end{bmatrix} \end{array}$$

For this matrix, we have:

$$R = \begin{bmatrix} .1 & 0 \\ .1 & .1 \end{bmatrix} \text{ and } Q = \begin{bmatrix} .8 & .1 \\ .4 & .4 \end{bmatrix}$$

The limiting matrix for M has the form:

$$\overline{M} = \left[\begin{array}{c|c} I & 0 \\ \hline FR & 0 \end{array}\right]$$

where $F = (I - Q)^{-1} = \left(\begin{bmatrix} 1 & 0 \\ 0 & 1 \end{bmatrix} - \begin{bmatrix} .8 & .1 \\ .4 & .4 \end{bmatrix}\right)^{-1} = \begin{bmatrix} .2 & -.1 \\ -.4 & .6 \end{bmatrix}^{-1}$

$$= \begin{bmatrix} \frac{1}{5} & -\frac{1}{10} \\ -\frac{2}{5} & \frac{3}{5} \end{bmatrix}^{-1}$$

We use row operations to find the inverse:

$$\left[\begin{array}{cc|cc} \frac{1}{5} & -\frac{1}{10} & 1 & 0 \\ -\frac{2}{5} & \frac{3}{5} & 0 & 1 \end{array}\right] \sim \left[\begin{array}{cc|cc} 1 & -\frac{1}{2} & 5 & 0 \\ -\frac{2}{5} & \frac{3}{5} & 0 & 1 \end{array}\right] \sim \left[\begin{array}{cc|cc} 1 & -\frac{1}{2} & 5 & 0 \\ 0 & \frac{2}{5} & 2 & 1 \end{array}\right] \sim \left[\begin{array}{cc|cc} 1 & -\frac{1}{2} & 5 & 0 \\ 0 & 1 & 5 & \frac{5}{2} \end{array}\right]$$

$5R_1 \to R_1$ $\quad \left(\frac{2}{5}\right)R_1 + R_2 \to R_2$ $\quad \left(\frac{5}{2}\right)R_2 \to R_2$ $\quad \left(\frac{1}{2}\right)R_2 + R_1 \to R_1$

$$\sim \left[\begin{array}{cc|cc} 1 & 0 & \frac{15}{2} & \frac{5}{4} \\ 0 & 1 & 5 & \frac{5}{2} \end{array}\right]$$

Thus, $F = \begin{bmatrix} 7.5 & 1.25 \\ 5 & 2.5 \end{bmatrix}$ and $FR = \begin{bmatrix} .875 & .125 \\ .75 & .25 \end{bmatrix}$

Therefore,

$$\overline{M} = \begin{array}{c} \\ F \\ B \\ G \\ A \end{array}\begin{array}{c} \begin{array}{cccc} F & B & G & A \end{array} \\ \begin{bmatrix} 1 & 0 & 0 & 0 \\ 0 & 1 & 0 & 0 \\ .875 & .125 & 0 & 0 \\ .75 & .25 & 0 & 0 \end{bmatrix} \end{array}$$

(A) In the long run, 75% of the accounts in arrears will pay in full.

(B) In the long run, 12.5% of the accounts in good standing will become bad debts.

(C) The average number of months that an account in arrears will either be paid in full or classified as a bad debt is:

$5 + 2.5 = 7.5$ months

47. A transition matrix in standard form for this problem is:

$$P = \begin{matrix} & A & B & C & N \\ A & 1 & 0 & 0 & 0 \\ B & 0 & 1 & 0 & 0 \\ C & 0 & 0 & 1 & 0 \\ N & .6 & .3 & .11 & .8 \end{matrix}$$

For this matrix, we have $R = [.6 \quad .3 \quad .11]$ and $Q = [.8]$.

The limiting matrix for P has the form:

$$\overline{P} = \left[\begin{array}{c|c} I & 0 \\ \hline FR & 0 \end{array}\right]$$

where $F = (I - Q)^{-1} = ([1] - [.8])^{-1} = [.2]^{-1} = 5$

Now, $FR = [5][.6 \quad .3 \quad .11] = [.3 \quad .15 \quad .55]$

and

$$\overline{P} = \begin{matrix} & A & B & C & N \\ A & 1 & 0 & 0 & 0 \\ B & 0 & 1 & 0 & 0 \\ C & 0 & 0 & 1 & 0 \\ N & .3 & .15 & .55 & 0 \end{matrix}$$

(A) In the long run, the market share of each company is: Company A—30%; Company B—15%; and Company C—55%.

(B) On the average, it will take 5 years for a department to decide to use a calculator from one of these companies in their courses.

49. Let I denote ICU, C denote CCW, D denote "died", and R denote "released". A transition matrix in standard form for this problem is:

$$P = \begin{matrix} & D & R & I & C \\ D & 1 & 0 & 0 & 0 \\ R & 0 & 1 & 0 & 0 \\ I & .02 & 0 & .46 & .52 \\ C & .01 & .22 & .04 & .73 \end{matrix}$$

For this matrix, we have

$$R = \begin{bmatrix} .02 & 0 \\ .01 & .22 \end{bmatrix} \text{ and } Q = \begin{bmatrix} .46 & .52 \\ .04 & .73 \end{bmatrix}$$

The limiting matrix for P has the form:

$$\overline{P} = \left[\begin{array}{c|c} I & 0 \\ \hline FR & 0 \end{array}\right]$$

where $F = (I - Q)^{-1} = \left(\begin{bmatrix} 1 & 0 \\ 0 & 1 \end{bmatrix} - \begin{bmatrix} .46 & .52 \\ .04 & .73 \end{bmatrix}\right)^{-1}$

$$= \begin{bmatrix} .54 & -.52 \\ -.04 & .27 \end{bmatrix}^{-1} = \begin{bmatrix} 2.16 & 4.16 \\ .32 & 4.32 \end{bmatrix}$$

Now, $FR = \begin{bmatrix} 2.16 & 4.16 \\ .32 & 4.32 \end{bmatrix}\begin{bmatrix} .02 & 0 \\ .01 & .22 \end{bmatrix} = \begin{bmatrix} .0848 & .9152 \\ .0496 & .9504 \end{bmatrix}$

and

$$\overline{P} = \begin{array}{c} D \\ R \\ I \\ C \end{array}\begin{bmatrix} 1 & 0 & 0 & 0 \\ 0 & 1 & 0 & 0 \\ .0848 & .9152 & 0 & 0 \\ .0496 & .9504 & 0 & 0 \end{bmatrix}$$

with columns D, R, I, C.

(A) In the long run, 91.52% of the patients are released from the hospital.

(B) In the long run, 4.96% of the patients in the CCW die without being released from the hospital.

(C) The average number of days a patient in the ICU will stay in the hospital is:

$$2.16 + 4.16 = 6.32 \text{ days}$$

51. A transition matrix in standard form for this problem is:

$$P = \begin{array}{c} L \\ R \\ F \\ B \end{array}\begin{bmatrix} 1 & 0 & 0 & 0 \\ 0 & 1 & 0 & 0 \\ \frac{1}{4} & \frac{1}{4} & 0 & \frac{1}{2} \\ \frac{2}{5} & \frac{1}{5} & \frac{2}{5} & 0 \end{bmatrix}$$

with columns L, R, F, B.

For this matrix we have:

$$R = \begin{bmatrix} \frac{1}{4} & \frac{1}{4} \\ \frac{2}{5} & \frac{1}{5} \end{bmatrix} \text{ and } Q = \begin{bmatrix} 0 & \frac{1}{2} \\ \frac{2}{5} & 0 \end{bmatrix}$$

The limiting matrix for P has the form:

$$\overline{P} = \left[\begin{array}{c|c} I & 0 \\ FR & 0 \end{array}\right]$$

$$\text{where } F = (I - Q)^{-1} = \left(\begin{bmatrix} 1 & 0 \\ 0 & 1 \end{bmatrix} - \begin{bmatrix} 0 & \frac{1}{2} \\ \frac{2}{5} & 0 \end{bmatrix}\right)^{-1} = \begin{bmatrix} 1 & -\frac{1}{2} \\ -\frac{2}{5} & 1 \end{bmatrix}^{-1}$$

We use row operations to find the inverse:

$$\left[\begin{array}{cc|cc} 1 & -\frac{1}{2} & 1 & 0 \\ -\frac{2}{5} & 1 & 0 & 1 \end{array}\right] \sim \left[\begin{array}{cc|cc} 1 & -\frac{1}{2} & 1 & 0 \\ 0 & \frac{4}{5} & \frac{2}{5} & 1 \end{array}\right] \sim \left[\begin{array}{cc|cc} 1 & -\frac{1}{2} & 1 & 0 \\ 0 & 1 & \frac{1}{2} & \frac{5}{4} \end{array}\right] \sim \left[\begin{array}{cc|cc} 1 & 0 & \frac{5}{4} & \frac{5}{8} \\ 0 & 1 & \frac{1}{2} & \frac{5}{4} \end{array}\right]$$

$$\left(\frac{2}{5}\right)R_1 + R_2 \to R_2 \qquad \left(\frac{5}{4}\right)R_2 \to R_2 \qquad \left(\frac{1}{2}\right)R_2 + R_1 \to R_1$$

$$\text{Thus, } F = \begin{bmatrix} \frac{5}{4} & \frac{5}{8} \\ \frac{1}{2} & \frac{5}{4} \end{bmatrix} \text{ and } FR = \begin{bmatrix} \frac{5}{4} & \frac{5}{8} \\ \frac{1}{2} & \frac{5}{4} \end{bmatrix}\begin{bmatrix} \frac{1}{4} & \frac{1}{4} \\ \frac{2}{5} & \frac{1}{5} \end{bmatrix} = \begin{bmatrix} \frac{9}{16} & \frac{7}{16} \\ \frac{5}{8} & \frac{3}{8} \end{bmatrix}.$$

Now,

$$\overline{P} = \begin{array}{c} L \\ R \\ F \\ B \end{array}\begin{bmatrix} 1 & 0 & 0 & 0 \\ 0 & 1 & 0 & 0 \\ \frac{9}{16} & \frac{7}{16} & 0 & 0 \\ \frac{5}{8} & \frac{3}{8} & 0 & 0 \end{bmatrix}$$

with columns L, R, F, B.

(A) The long run probability that a rat placed in room B will end up in room R is $\frac{3}{8} = .375$.

(B) The average number of exits that a rat placed in room B will choose until it finds food is:

$$\frac{1}{2} + \frac{5}{4} = \frac{7}{4} = 1.75$$

CHAPTER 9 REVIEW

1. $S_1 = S_0P = [.3 \quad .7]\begin{bmatrix} .6 & .4 \\ .2 & .8 \end{bmatrix} = \overset{\quad A \qquad B}{[.32 \quad .68]}$

$S_2 = S_1P = [.32 \quad .68]\begin{bmatrix} .6 & .4 \\ .2 & .8 \end{bmatrix} = \overset{\quad A \qquad B}{[.328 \quad .672]}$

The probability of being in state A after one trial is .32; after two trials .328. The probability of being in state B after one trial is .68; after two trials .672. (9-1)

2. A is an absorbing state; the chain is absorbing since it is possible to go from state B to state A. (9-2, 9-3)

3. There are no absorbing states since there are no 1's on the main diagonal.

P is regular since $P^2 = \begin{bmatrix} .7 & .3 \\ .21 & .79 \end{bmatrix}$ has only positive entries. (9-2, 9-3)

4. $P = \begin{bmatrix} 0 & 1 \\ 1 & 0 \end{bmatrix}$ has no absorbing states. Since P^k, $k = 1, 2, 3, \ldots,$

alternates between $\begin{bmatrix} 0 & 1 \\ 1 & 0 \end{bmatrix}$ and $\begin{bmatrix} 1 & 0 \\ 0 & 1 \end{bmatrix}$, P is not regular. (9-2, 9-3)

5. States B and C are absorbing. The chain is absorbing; it is possible to go from the nonabsorbing state A to the absorbing state C in one step. (9-2, 9-3)

6. States A and B are absorbing. The chain is neither absorbing (it is not possible to go from States C and D to either state A or state B) nor regular (all powers of P will have the same form). (9-2, 9-3)

7. $P = \begin{array}{c} \\ A \\ B \\ C \end{array}\overset{\begin{array}{ccc} A & B & C \end{array}}{\begin{bmatrix} 0 & 1 & 0 \\ .1 & 0 & .9 \\ 0 & 1 & 0 \end{bmatrix}}$

There are no absorbing states.

P^k, $k = 1, 2, 3, \ldots,$ alternates between $\begin{bmatrix} 0 & 1 & 0 \\ .1 & 0 & .9 \\ 0 & 1 & 0 \end{bmatrix}$ and $\begin{bmatrix} .1 & 0 & .9 \\ 0 & 1 & 0 \\ .1 & 0 & .9 \end{bmatrix}$.

Thus, P is not regular. (9-1, 9-2, 9-3)

8. $P = \begin{array}{c} \\ A \\ B \\ C \end{array}\begin{array}{c} \begin{array}{ccc} A & B & C \end{array} \\ \begin{bmatrix} 0 & 1 & 0 \\ .1 & .2 & .7 \\ 0 & 0 & 1 \end{bmatrix} \end{array}$

C is an absorbing state. The chain is absorbing since it is possible to go from state A to state C (via B) and from state B to state C.

(9-1, 9-2, 9-3)

9. $P = \begin{array}{c} \\ A \\ B \\ C \end{array}\begin{array}{c} \begin{array}{ccc} A & B & C \end{array} \\ \begin{bmatrix} 0 & 0 & 1 \\ .1 & .2 & .7 \\ 0 & 1 & 0 \end{bmatrix} \end{array}$

There are no absorbing states since there are no 1's on the main diagonal.

P is regular since $P^3 = \begin{bmatrix} .1 & .2 & .7 \\ .074 & .388 & .538 \\ .02 & .74 & .24 \end{bmatrix}$ has only positive entries.

(9-1, 9-2, 9-3)

10.

$P = \begin{array}{c} \\ A \\ B \\ C \\ D \end{array}\begin{array}{c} \begin{array}{cccc} A & B & C & D \end{array} \\ \begin{bmatrix} .3 & .2 & 0 & .5 \\ 0 & 1 & 0 & 0 \\ 0 & 0 & .2 & .8 \\ 0 & 0 & .3 & .7 \end{bmatrix} \end{array}$

B is an absorbing state. The chain is not absorbing since it is not possible to go from state C to B, nor is it possible to go from state D to state B.

(9-1, 9-2, 9-3)

11.

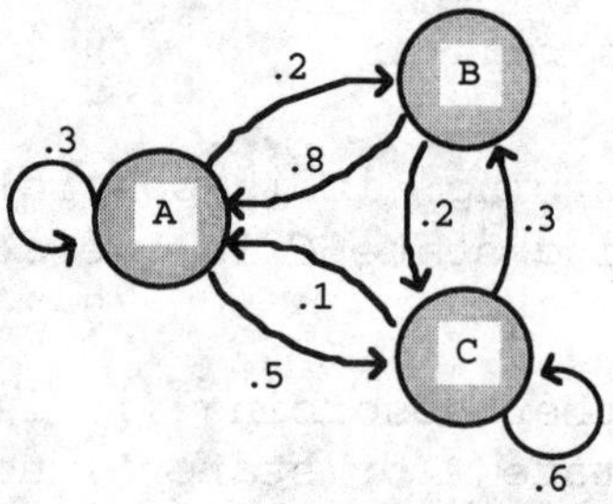

$P = \begin{array}{c} \\ A \\ B \\ C \end{array}\begin{array}{c} \begin{array}{ccc} A & B & C \end{array} \\ \begin{bmatrix} .3 & .2 & .5 \\ .8 & 0 & .2 \\ .1 & .3 & .6 \end{bmatrix} \end{array}$

(9-1)

12. $P = \begin{array}{c} \\ A \\ B \end{array}\begin{array}{c} \begin{array}{cc} A & B \end{array} \\ \begin{bmatrix} .4 & .6 \\ .9 & .1 \end{bmatrix} \end{array}$

(A) $P^2 = \begin{array}{c} \\ A \\ B \end{array}\begin{array}{c} \begin{array}{cc} A & B \end{array} \\ \begin{bmatrix} .7 & .3 \\ .45 & .55 \end{bmatrix} \end{array}$

The probability of going from state A to state B in two trials is .3.

(B) $P^3 = \begin{array}{c} \\ A \\ B \end{array}\begin{array}{c} \begin{array}{cc} A & B \end{array} \\ \begin{bmatrix} .55 & .45 \\ .675 & .325 \end{bmatrix} \end{array}$

The probability of going from state B to state A in three trials is .675.

(9-1)

13. Let $S = [s_1 \quad s_2]$ and solve the system:

$$[s_1 \quad s_2]\begin{bmatrix} .4 & .6 \\ .2 & .8 \end{bmatrix} = [s_1 \quad s_2], \quad s_1 + s_2 = 1$$

which is equivalent to

$$\begin{aligned} .4s_1 + .2s_2 &= s_1 \\ .6s_1 + .8s_2 &= s_2 \\ s_1 + s_2 &= 1 \end{aligned} \quad \text{or} \quad \begin{aligned} -.6s_1 + .2s_2 &= 0 \\ .6s_1 - .2s_2 &= 0 \\ s_1 + s_2 &= 1 \end{aligned}$$

The solution is: $s_1 = .25$, $s_2 = .75$

The stationary matrix $S = \begin{matrix} A & B \\ [.25 & .75] \end{matrix}$

The limiting matrix $\overline{P} = \begin{matrix} & A & B \\ A & .25 & .75 \\ B & .25 & .75 \end{matrix}$ (9-2)

14. Let $S = [s_1 \quad s_2 \quad s_3]$ and solve the system:

$$[s_1 \quad s_2 \quad s_3]\begin{bmatrix} .4 & .6 & 0 \\ .5 & .3 & .2 \\ 0 & .8 & .2 \end{bmatrix} = [s_1 \quad s_2 \quad s_3], \quad s_1 + s_2 + s_3 = 1$$

which is equivalent to

$$\begin{aligned} .4s_1 + .5s_2 &= s_1 \\ .6s_1 + .3s_2 + .8s_3 &= s_2 \\ .2s_2 + .2s_3 &= s_3 \\ s_1 + s_2 + s_3 &= 1 \end{aligned} \quad \text{or} \quad \begin{aligned} -.6s_1 + .5s_2 &= 0 \\ .6s_1 - .7s_2 + .8s_3 &= 0 \\ .2s_2 - .8s_3 &= 0 \\ s_1 + s_2 + s_3 &= 1 \end{aligned}$$

From the first and third equations, we have $s_1 = \frac{5}{6}s_2$ and $s_3 = \frac{1}{4}s_2$.
Substituting these values into the fourth equation, we get

$$\frac{5}{6}s_2 + s_2 + \frac{1}{4}s_2 = 1 \quad \text{or} \quad \frac{25}{12}s_2 = 1 \quad \text{and} \quad s_2 = .48$$

Therefore, $s_1 = .4$, $s_2 = .48$, $s_3 = .12$.

The stationary matrix $S = \begin{matrix} A & B & C \\ [.4 & .48 & .12] \end{matrix}$.

The limiting matrix $\overline{P} = \begin{matrix} & A & B & C \\ A & .4 & .48 & .12 \\ B & .4 & .48 & .12 \\ C & .4 & .48 & .12 \end{matrix}$ (9-2)

15. For $P = \begin{matrix} & A & B & C \\ A & 1 & 0 & 0 \\ B & 0 & 1 & 0 \\ C & .3 & .1 & .6 \end{matrix}$

we have $R = [.3 \quad .1]$ and $Q = [.6]$. The limiting matrix $\overline{P}$ has the form

$$\overline{P} = \left[\begin{array}{cc|c} 1 & 0 & 0 \\ 0 & 1 & 0 \\ \hline F & R & 0 \end{array}\right]$$

where $F = (I - Q)^{-1} = ([1] - [.6])^{-1} = [.4]^{-1} = \left[\frac{5}{2}\right]$

and $FR = \begin{bmatrix} 5 \\ 2 \end{bmatrix}[.3 \quad .1] = [.75 \quad .25]$.

Thus, $\bar{P} = \begin{matrix} & \begin{matrix} A & B & C \end{matrix} \\ \begin{matrix} A \\ B \\ C \end{matrix} & \begin{bmatrix} 1 & 0 & 0 \\ 0 & 1 & 0 \\ .75 & .25 & 0 \end{bmatrix} \end{matrix}$

$P(C \text{ to } A) = .75$, $P(C \text{ to } B) = .25$. Since $F = \begin{bmatrix} 5 \\ 2 \end{bmatrix}$ it will take an average of 2.5 trials to go from C to either A or B. (9-3)

16. For $P = \begin{matrix} & \begin{matrix} A & B & C & D \end{matrix} \\ \begin{matrix} A \\ B \\ C \\ D \end{matrix} & \begin{bmatrix} 1 & 0 & 0 & 0 \\ 0 & 1 & 0 & 0 \\ .1 & .5 & .2 & .2 \\ .1 & .1 & .4 & .4 \end{bmatrix} \end{matrix}$

we have $R = \begin{bmatrix} .1 & .5 \\ .1 & .1 \end{bmatrix}$ and $Q = \begin{bmatrix} .2 & .2 \\ .4 & .4 \end{bmatrix}$.

The limiting matrix $\bar{P}$ has the form

$$\bar{P} = \left[\begin{array}{c|c} I & 0 \\ \hline FR & 0 \end{array}\right]$$

where $F = (I - Q)^{-1} = \left(\begin{bmatrix} 1 & 0 \\ 0 & 1 \end{bmatrix} - \begin{bmatrix} .2 & .2 \\ .4 & .4 \end{bmatrix}\right)^{-1} = \begin{bmatrix} .8 & -.2 \\ -.4 & .6 \end{bmatrix}^{-1} = \begin{bmatrix} \frac{4}{5} & -\frac{1}{5} \\ -\frac{2}{5} & \frac{3}{5} \end{bmatrix}^{-1}$

We use row operations to find the inverse:

$$\left[\begin{array}{cc|cc} \frac{4}{5} & -\frac{1}{5} & 1 & 0 \\ -\frac{2}{5} & \frac{3}{5} & 0 & 1 \end{array}\right] \sim \left[\begin{array}{cc|cc} 1 & -\frac{1}{4} & \frac{5}{4} & 0 \\ -\frac{2}{5} & \frac{3}{5} & 0 & 1 \end{array}\right] \sim \left[\begin{array}{cc|cc} 1 & -\frac{1}{4} & \frac{5}{4} & 0 \\ 0 & \frac{1}{2} & \frac{1}{2} & 1 \end{array}\right] \sim \left[\begin{array}{cc|cc} 1 & -\frac{1}{4} & \frac{5}{4} & 0 \\ 0 & 1 & 1 & 2 \end{array}\right]$$

$$\left(\frac{5}{4}\right)R_1 \to R_1 \qquad \left(\frac{2}{5}\right)R_1 + R_2 \to R_2 \qquad 2R_2 \to R_2 \qquad \left(\frac{1}{4}\right)R_2 + R_1 \to R_1$$

$$\sim \left[\begin{array}{cc|cc} 1 & 0 & \frac{3}{2} & \frac{1}{2} \\ 0 & 1 & 1 & 2 \end{array}\right]$$

Thus, $F = \begin{bmatrix} \frac{3}{2} & \frac{1}{2} \\ 1 & 2 \end{bmatrix} = \begin{bmatrix} 1.5 & .5 \\ 1 & 2 \end{bmatrix}$, $FR = \begin{bmatrix} 1.5 & .5 \\ 1 & 2 \end{bmatrix}\begin{bmatrix} .1 & .5 \\ .1 & .1 \end{bmatrix} = \begin{bmatrix} .2 & .8 \\ .3 & .7 \end{bmatrix}$.

and $\bar{P} = \begin{matrix} & \begin{matrix} A & B & C & D \end{matrix} \\ \begin{matrix} A \\ B \\ C \\ D \end{matrix} & \begin{bmatrix} 1 & 0 & 0 & 0 \\ 0 & 1 & 0 & 0 \\ .2 & .8 & 0 & 0 \\ .3 & .7 & 0 & 0 \end{bmatrix} \end{matrix}$

$P(C \text{ to } A) = .2$, $P(C \text{ to } B) = .8$, $P(D \text{ to } A) = .3$, $P(D \text{ to } B) = .7$.
It takes an average of 2 trials to go from C to either A or B; it takes an average of three trials to go from D to A or B. (9-3)

17. $P = \begin{matrix} & \begin{matrix} A & B \end{matrix} \\ \begin{matrix} A \\ B \end{matrix} & \begin{bmatrix} .4 & .6 \\ .2 & .8 \end{bmatrix} \end{matrix}$, $P^4 \approx \begin{bmatrix} .2512 & .7488 \\ .2496 & .7504 \end{bmatrix}$, $P^8 \approx \begin{bmatrix} .2500 & .7499 \\ .2499 & .7500 \end{bmatrix}$; $\bar{P} = \begin{matrix} & \begin{matrix} A & B \end{matrix} \\ \begin{matrix} A \\ B \end{matrix} & \begin{bmatrix} .25 & .75 \\ .25 & .75 \end{bmatrix} \end{matrix}$

(9-3)

18. $$P = \begin{array}{c} \\ A \\ B \\ C \end{array}\!\!\begin{array}{c} \begin{array}{ccc} A & B & C \end{array} \\ \begin{bmatrix} .4 & .6 & 0 \\ .5 & .3 & .2 \\ 0 & .8 & .2 \end{bmatrix} \end{array}, \quad P^4 \approx \begin{bmatrix} .4066 & .4722 & .1212 \\ .3935 & .4895 & .117 \\ .404 & .468 & .128 \end{bmatrix},$$

$$P^8 \approx \begin{bmatrix} .4001 & .4799 & .1200 \\ .3999 & .4802 & .1199 \\ .4001 & .4798 & .1201 \end{bmatrix}; \quad \overline{P} = \begin{array}{c} \\ A \\ B \\ C \end{array}\!\!\begin{array}{c} \begin{array}{ccc} A & B & C \end{array} \\ \begin{bmatrix} .4 & .48 & .12 \\ .4 & .48 & .12 \\ .4 & .48 & .12 \end{bmatrix} \end{array} \qquad (9\text{-}3)$$

19. $$P = \begin{array}{c} \\ A \\ B \\ C \end{array}\!\!\begin{array}{c} \begin{array}{ccc} A & B & C \end{array} \\ \begin{bmatrix} 1 & 0 & 0 \\ 0 & 1 & 0 \\ .3 & .1 & .6 \end{bmatrix} \end{array}, \quad P^4 \approx \begin{bmatrix} 1 & 0 & 0 \\ 0 & 1 & 0 \\ .6528 & .2176 & .1296 \end{bmatrix},$$

$$P^8 \approx \begin{bmatrix} 1 & 0 & 0 \\ 0 & 1 & 0 \\ .7374 & .2458 & .01680 \end{bmatrix}, \quad P^{16} \approx \begin{bmatrix} 1 & 0 & 0 \\ 0 & 1 & 0 \\ .7498 & .2499 & 0 \end{bmatrix};$$

$$\overline{P} = \begin{array}{c} \\ A \\ B \\ C \end{array}\!\!\begin{array}{c} \begin{array}{ccc} A & B & C \end{array} \\ \begin{bmatrix} 1 & 0 & 0 \\ 0 & 1 & 0 \\ .75 & .25 & 0 \end{bmatrix} \end{array} \qquad (9\text{-}3)$$

20. $$P = \begin{array}{c} \\ A \\ B \\ C \\ D \end{array}\!\!\begin{array}{c} \begin{array}{cccc} A & B & C & D \end{array} \\ \begin{bmatrix} 1 & 0 & 0 & 0 \\ 0 & 1 & 0 & 0 \\ .1 & .5 & .2 & .2 \\ .1 & .1 & .4 & .4 \end{bmatrix} \end{array}, \quad P^4 \approx \begin{bmatrix} 1 & 0 & 0 & 0 \\ 0 & 1 & 0 & 0 \\ .1784 & .7352 & .0432 & .0432 \\ .2568 & .5704 & .0864 & .0864 \end{bmatrix},$$

$$P^8 \approx \begin{bmatrix} 1 & 0 & 0 & 0 \\ 0 & 1 & 0 & 0 \\ .1972 & .7916 & .0056 & .0056 \\ .2944 & .6832 & .0112 & .0112 \end{bmatrix}, \quad P^{16} \approx \begin{bmatrix} 1 & 0 & 0 & 0 \\ 0 & 1 & 0 & 0 \\ .2000 & .7999 & 0 & 0 \\ .2999 & .6997 & 0 & 0 \end{bmatrix},$$

$$\overline{P} = \begin{array}{c} \\ A \\ B \\ C \\ D \end{array}\!\!\begin{array}{c} \begin{array}{cccc} A & B & C & D \end{array} \\ \begin{bmatrix} 1 & 0 & 0 & 0 \\ 0 & 1 & 0 & 0 \\ .2 & .8 & 0 & 0 \\ .3 & .7 & 0 & 0 \end{bmatrix} \end{array} \qquad (9\text{-}3)$$

21. A standard form for the given matrix is:

$$P = \begin{array}{c} \\ B \\ D \\ A \\ C \end{array}\!\!\begin{array}{c} \begin{array}{cccc} B & D & A & C \end{array} \\ \begin{bmatrix} 1 & 0 & 0 & 0 \\ 0 & 1 & 0 & 0 \\ .1 & .1 & .6 & .2 \\ .2 & .2 & .3 & .3 \end{bmatrix} \end{array} \qquad (9\text{-}3)$$

22. We will determine the limiting matrix of:

$$P = \begin{array}{c} \\ A \\ B \\ C \end{array}\!\!\begin{array}{c} \begin{array}{ccc} A & B & C \end{array} \\ \begin{bmatrix} 0 & 1 & 0 \\ 0 & 0 & 1 \\ .2 & .6 & .2 \end{bmatrix} \end{array}$$

by solving

$$[s_1 \quad s_2 \quad s_3]\begin{bmatrix} 0 & 1 & 0 \\ 0 & 0 & 1 \\ .2 & .6 & .2 \end{bmatrix} = [s_1 \quad s_2 \quad s_3],\ s_1 + s_2 + s_3 = 1.$$

The corresponding system of equations is:

$$\begin{aligned} .2s_3 &= s_1 \\ s_1 + .6s_3 &= s_2 \\ s_2 + .2s_3 &= s_3 \\ s_1 + s_2 + s_3 &= 1 \end{aligned} \quad \text{or} \quad \begin{aligned} s_1 - .2s_3 &= 0 \\ s_1 - s_2 + .6s_3 &= 0 \\ s_2 - .8s_3 &= 0 \\ s_1 + s_2 + s_3 &= 0 \end{aligned}$$

From the first and third equations, we have $s_1 = .2s_3$ and $s_2 = .8s_3$. Substituting these values into the fourth equation gives

$.2s_3 + .8s_3 + s_3 = 1$ and $s_3 = .5$

It now follows that $s_1 = .1$ and $s_2 = .4$. Thus, $s = [.1 \quad .4 \quad .5]$ and

$$\bar{P} = \begin{matrix} & A & B & C \\ A & .1 & .4 & .5 \\ B & .1 & .4 & .5 \\ C & .1 & .4 & .5 \end{matrix}$$

(A) $[0 \quad 0 \quad 1]\begin{bmatrix} .1 & .4 & .5 \\ .1 & .4 & .5 \\ .1 & .4 & .5 \end{bmatrix} = \begin{matrix} A & B & C \\ [.1 & .4 & .5] \end{matrix}$

(B) $[.5 \quad .3 \quad .2]\begin{bmatrix} .1 & .4 & .5 \\ .1 & .4 & .5 \\ .1 & .4 & .5 \end{bmatrix} = \begin{matrix} A & B & C \\ [.1 & .4 & .5] \end{matrix}$

(9-3)

23. The transition matrix:

$$P = \begin{matrix} & A & B & C \\ A & 1 & 0 & 0 \\ B & 0 & 1 & 0 \\ C & .2 & .6 & .2 \end{matrix}$$

is the standard form for an absorbing Markov chain with two absorbing and one nonabsorbing states. For this matrix, we have:

$R = [.2 \quad .6]$ and $Q = [.2]$.

The limiting matrix has the form

$$\bar{P} = \left[\begin{array}{c|c} I & 0 \\ \hline FR & 0 \end{array}\right]$$

where $F = (I - Q)^{-1} = ([1] - [.2])^{-1} = [.8]^{-1} = [1.25]$

Thus, $FR = [1.25][.2 \quad .6] = [.25 \quad .75]$ and

$$\bar{P} = \begin{matrix} & A & B & C \\ A & 1 & 0 & 0 \\ B & 0 & 1 & 0 \\ C & .25 & .75 & 0 \end{matrix}$$

$$(A)\ [0 \quad 0 \quad 1]\begin{bmatrix} 1 & 0 & 0 \\ 0 & 1 & 0 \\ .25 & .75 & 0 \end{bmatrix} = \begin{matrix} A & B & C \\ [.25 & .75 & 0] \end{matrix}$$

$$(B)\ [.5 \quad .3 \quad .2]\begin{bmatrix} 1 & 0 & 0 \\ 0 & 1 & 0 \\ .25 & .75 & 0 \end{bmatrix} = \begin{matrix} A & B & C \\ [.55 & .45 & 0] \end{matrix}$$

(9-3)

24. No. If P is a transition matrix with 2 entries equal to 0, then P has one of the forms: $P_1 = \begin{pmatrix} 1 & 0 \\ 0 & 1 \end{pmatrix}$, $P_2 = \begin{pmatrix} 0 & 1 \\ 1 & 0 \end{pmatrix}$, $P_3 = \begin{pmatrix} 1 & 0 \\ 1 & 0 \end{pmatrix}$, $P_4 = \begin{pmatrix} 0 & 1 \\ 0 & 1 \end{pmatrix}$

$P_1{}^k = P_1$, $P_2{}^k = P_2$, $P_3{}^k = P_3$ for all k, and $P_2{}^k = \begin{pmatrix} 1 & 0 \\ 0 & 1 \end{pmatrix}$ if k is even and $P_2{}^k = P_2$ if k is odd. No power of P_i, $i = 1, 2, 3, 4$, has all positive entries. (9-2)

25. Yes; $P = \begin{bmatrix} 0 & .5 & .5 \\ .5 & 0 & .5 \\ .5 & .5 & 0 \end{bmatrix}$ is regular since $P^2 = \begin{bmatrix} .5 & .25 & .25 \\ .25 & .5 & .25 \\ .25 & .25 & .5 \end{bmatrix}$

$P = \begin{bmatrix} 0 & 0 & 1 \\ 0 & 0 & 1 \\ .2 & .3 & .5 \end{bmatrix}$ is regular since $P^2 = \begin{bmatrix} .2 & .3 & .5 \\ .2 & .3 & .5 \\ .1 & .15 & .75 \end{bmatrix}$ (9-2)

26. (A)

(B) $P = \begin{matrix} & R & B & G \\ R & .5 & .25 & .25 \\ B & .2 & .6 & .2 \\ G & .6 & .3 & .1 \end{matrix}$

(C) The chain is regular since it has only positive entries.

(D) Let $S = [s_1 \quad s_2 \quad s_3]$ and solve the system:

$$[s_1 \quad s_2 \quad s_3]\begin{bmatrix} .5 & .25 & .25 \\ .2 & .6 & .2 \\ .6 & .3 & .1 \end{bmatrix} = [s_1 \quad s_2 \quad s_3],\ s_1 + s_2 + s_3 = 1$$

which is equivalent to:

$$\begin{aligned} s_1 + s_2 + s_3 &= 1 \\ .5s_1 + .2s_2 + .6s_3 &= s_1 \\ .25s_1 + .6s_2 + .3s_3 &= s_2 \\ .25s_1 + .2s_2 + .1s_3 &= s_3 \end{aligned} \quad \text{or} \quad \begin{aligned} s_1 + s_2 + s_3 &= 1 \\ -.5s_1 + .2s_2 + .6s_3 &= 0 \\ .25s_1 - .4s_2 + .3s_3 &= 0 \\ .25s_1 + .2s_2 - .9s_3 &= 0 \end{aligned}$$

We use row operations to solve this system; but first multiply the second, third and fourth equations by 10 to simplify the calculations.

$$\left[\begin{array}{ccc|c} 1 & 1 & 1 & 1 \\ -5 & 2 & 6 & 0 \\ \frac{5}{2} & -4 & 3 & 0 \\ \frac{5}{2} & 2 & -9 & 0 \end{array}\right] \sim \left[\begin{array}{ccc|c} 1 & 1 & 1 & 1 \\ 0 & 7 & 11 & 5 \\ 0 & -\frac{13}{2} & \frac{1}{2} & -\frac{5}{2} \\ 0 & -\frac{1}{2} & -\frac{23}{2} & -\frac{5}{2} \end{array}\right] \sim \left[\begin{array}{ccc|c} 1 & 1 & 1 & 1 \\ 0 & 1 & 23 & 5 \\ 0 & -\frac{13}{2} & \frac{1}{2} & -\frac{5}{2} \\ 0 & 7 & 11 & 5 \end{array}\right]$$

$5R_1 + R_2 \to R_2$ $\quad -2R_4 \to R_4$ $\quad (-1)R_2 + R_1 \to R_1$

$\left(-\frac{5}{2}\right)R_1 + R_3 \to R_3$ $\quad R_2 \leftrightarrow R_4$ $\quad \left(\frac{13}{2}\right)R_2 + R_3 \to R_3$

$\left(-\frac{5}{2}\right)R_1 + R_4 \to R_4$ $\quad (-7)R_2 + R_4 \to R_4$

$$\sim \left[\begin{array}{ccc|c} 1 & 0 & -22 & -4 \\ 0 & 1 & 23 & 5 \\ 0 & 0 & 150 & 30 \\ 0 & 0 & -150 & -30 \end{array}\right] \sim \left[\begin{array}{ccc|c} 1 & 0 & -22 & -4 \\ 0 & 1 & 23 & 5 \\ 0 & 0 & 1 & \frac{1}{5} \\ 0 & 0 & -150 & -30 \end{array}\right] \sim \left[\begin{array}{ccc|c} 1 & 0 & 0 & \frac{2}{5} \\ 0 & 1 & 0 & \frac{2}{5} \\ 0 & 0 & 1 & \frac{1}{5} \\ 0 & 0 & 0 & 0 \end{array}\right]$$

$\frac{1}{150}R_3 \to R_3$ $\quad 22R_3 + R_1 \to R_1$

$(-23)R_3 + R_2 \to R_2$

$150R_3 + R_4 \to R_4$

The solution is $s_1 = 0.4$, $s_2 = 0.4$, $s_3 = 0.2$ and

$$\bar{P} = \begin{array}{c} \\ R \\ B \\ G \end{array}\begin{array}{c} \begin{array}{ccc} R & B & G \end{array} \\ \begin{bmatrix} .4 & .4 & .2 \\ .4 & .4 & .2 \\ .4 & .4 & .2 \end{bmatrix} \end{array}$$

In the long run, the red urn will be selected 40% of the time, the blue urn 40% of the time, and the green urn 20% of the time. (9-2)

27. (A)

(B) $P = \begin{array}{c} \\ R \\ B \\ G \end{array}\begin{array}{c} \begin{array}{ccc} R & B & G \end{array} \\ \begin{bmatrix} 1 & 0 & 0 \\ .2 & .6 & .2 \\ .6 & .3 & .1 \end{bmatrix} \end{array}$

(C) State R is an absorbing state. The chain is absorbing since it is possible to go from states B and G to state R in a finite number (namely 1) steps.

(D) For $P = \begin{bmatrix} 1 & 0 & 0 \\ .2 & .6 & .2 \\ .6 & .3 & .1 \end{bmatrix}$ we have $R = \begin{bmatrix} .2 \\ .6 \end{bmatrix}$ and $Q = \begin{bmatrix} .6 & .2 \\ .3 & .1 \end{bmatrix}$.

The limiting matrix $\bar{P}$ has the form:

$$\bar{P} = \left[\begin{array}{c|c} I & 0 \\ \hline FR & 0 \end{array}\right]$$

where $F = (I - Q)^{-1} = \left(\begin{bmatrix} 1 & 0 \\ 0 & 1 \end{bmatrix} - \begin{bmatrix} .6 & .2 \\ .3 & .1 \end{bmatrix}\right)^{-1} = \begin{bmatrix} .4 & -.2 \\ -.3 & .9 \end{bmatrix}^{-1}$

$$= \begin{bmatrix} \frac{2}{5} & -\frac{1}{5} \\ -\frac{3}{10} & \frac{9}{10} \end{bmatrix}^{-1}$$

We use row operations to find the inverse:

$$\left[\begin{array}{cc|cc} \frac{2}{5} & -\frac{1}{5} & 1 & 0 \\ -\frac{3}{10} & \frac{9}{10} & 0 & 1 \end{array}\right] \sim \left[\begin{array}{cc|cc} 1 & -\frac{1}{2} & \frac{5}{2} & 0 \\ -\frac{3}{10} & \frac{9}{10} & 0 & 1 \end{array}\right] \sim \left[\begin{array}{cc|cc} 1 & -\frac{1}{2} & \frac{5}{2} & 0 \\ 0 & \frac{3}{4} & \frac{3}{4} & 1 \end{array}\right] \sim \left[\begin{array}{cc|cc} 1 & -\frac{1}{2} & \frac{5}{2} & 0 \\ 0 & 1 & 1 & \frac{4}{3} \end{array}\right]$$

$\left(\frac{5}{2}\right)R_1 \to R_1$ $\quad \left(\frac{3}{10}\right)R_1 + R_2 \to R_2$ $\quad \frac{4}{3}R_2 \to R_2$ $\quad \left(\frac{1}{2}\right)R_2 + R_1 \to R_1$

$$\left[\begin{array}{cc|cc} 1 & 0 & 3 & \frac{2}{3} \\ 0 & 1 & 1 & \frac{4}{3} \end{array}\right]$$

Thus, $F = \begin{bmatrix} 3 & \frac{2}{3} \\ 1 & \frac{4}{3} \end{bmatrix}$ and $FR = \begin{bmatrix} 3 & \frac{2}{3} \\ 1 & \frac{4}{3} \end{bmatrix}\begin{bmatrix} \frac{1}{5} \\ \frac{3}{5} \end{bmatrix} = \begin{bmatrix} 1 \\ 1 \end{bmatrix}$.

Now, $\bar{P} = \begin{matrix} & R & B & G \\ R & 1 & 0 & 0 \\ B & 1 & 0 & 0 \\ G & 1 & 0 & 0 \end{matrix}$

Once the red urn is selected, the blue and green urns will never be selected again. It will take an average of 3.67 trials to reach the red urn from the blue urn and an average of 2.33 trials to reach the red urn from the green urn. (9-3)

28. $[x \quad y \quad z \quad 0]\begin{bmatrix} 1 & 0 & 0 & 0 \\ 0 & 1 & 0 & 0 \\ 0 & 0 & 1 & 0 \\ .1 & .3 & .4 & .2 \end{bmatrix} = [x \quad y \quad z \quad 0]$

Thus, $[x \quad y \quad z \quad 0]$ is a stationary matrix for P. If P is a transition matrix for an absorbing chain with three absorbing states and one nonabsorbing state, then P will have exactly three 1's on the main diagonal (and zeros elsewhere in the row containing the 1's) and one row with at least one nonzero entry off the main diagonal. One of the following matrices will be a stationary matrix for P:

$$[x \quad y \quad z \quad 0],\ [x \quad y \quad 0 \quad z],\ [x \quad 0 \quad y \quad z],\ [0 \quad x \quad y \quad z]$$

where $x + y + z = 1$. The position of the zero corresponds to the one row of P which has a nonzero entry off the main diagonal. (9-2, 9-3)

29. No such chain exists; if the chain has an absorbing state, then a corresponding transition matrix P will have a row containing a 1 on the main diagonal and zeros elsewhere. All powers of P will have the same row, and so the chain cannot be regular. (9-2, 9-3)

30. No such chain exists. The reasoning in Problem 29 applies here as well. (9-2, 9-3)

31. No such chain exists. By Theorem 1, Section 9-2 a regular Markov chain has a unique stationary matrix. (9-2)

32. $S = [1 \quad 0 \quad 0]$ and $S' = [0 \quad 1 \quad 0]$ are both stationary matrices for

$$P = \begin{array}{c} \\ A \\ B \\ C \end{array}\begin{array}{c} \begin{array}{ccc} A & B & C \end{array} \\ \begin{bmatrix} 1 & 0 & 0 \\ 0 & 1 & 0 \\ .6 & .3 & .1 \end{bmatrix} \end{array}$$

(9-3)

33. $P = \begin{array}{c} \\ A \\ B \end{array}\begin{array}{c} \begin{array}{cc} A & B \end{array} \\ \begin{bmatrix} 0 & 1 \\ 1 & 0 \end{bmatrix} \end{array}$ has no limiting matrix; $P^{2k} = \begin{bmatrix} 1 & 0 \\ 0 & 1 \end{bmatrix}$ and $P^{2k+1} = \begin{bmatrix} 0 & 1 \\ 1 & 0 \end{bmatrix}$ for all positive integers k. (9-2, 9-3)

34. No such chain exists. By Theorem 1, Section 9-2, a regular Markov chain has a unique limiting matrix. (9-2)

35. No such chain exists. By Theorem 2, Section 9-3, an absorbing Markov chain has a limiting matrix. (9-3)

36. $P = \begin{array}{c} \\ A \\ B \\ C \\ D \end{array}\begin{array}{c} \begin{array}{cccc} A & B & C & D \end{array} \\ \begin{bmatrix} .2 & .3 & .1 & .4 \\ 0 & 0 & 1 & 0 \\ 0 & .8 & 0 & .2 \\ 0 & 0 & 1 & 0 \end{bmatrix} \end{array}$ No limiting matrix

For example

$$P^{200} = P^{202} = P^{204} = \dots = \begin{bmatrix} 0 & .2 & .75 & .05 \\ 0 & .8 & 0 & .2 \\ 0 & 0 & 1 & 0 \\ 0 & .8 & 0 & .2 \end{bmatrix}$$

$$P^{201} = P^{203} = P^{205} = \dots = \begin{bmatrix} 0 & .6 & .25 & .15 \\ 0 & 0 & 1 & 0 \\ 0 & .8 & 0 & .2 \\ 0 & 0 & 1 & 0 \end{bmatrix}$$

(9-2, 9-3)

37. $P = \begin{array}{c} \\ A \\ B \\ C \\ D \end{array}\begin{array}{c} \begin{array}{cccc} A & B & C & D \end{array} \\ \begin{bmatrix} .1 & 0 & .3 & .6 \\ .2 & .4 & .1 & .3 \\ .3 & .5 & 0 & .2 \\ .9 & .1 & 0 & 0 \end{bmatrix} \end{array}$

Limiting matrix

$$\begin{array}{c} \\ A \\ B \\ C \\ D \end{array}\begin{array}{c} \begin{array}{cccc} A & B & C & D \end{array} \\ \begin{bmatrix} .392 & .163 & .134 & .311 \\ .392 & .163 & .134 & .311 \\ .392 & .163 & .134 & .311 \\ .392 & .163 & .134 & .311 \end{bmatrix} \end{array}$$

(9-2)

38. (A)

(B) $P = \begin{array}{c} \\ x \\ x' \end{array}\begin{array}{c} \begin{array}{cc} x & x' \end{array} \\ \begin{bmatrix} .7 & .3 \\ .5 & .5 \end{bmatrix} \end{array}$ (C) $S = [.2 \quad .8]$

(D) $S_1 = SP = [.2 \quad .8]\begin{bmatrix} .7 & .3 \\ .5 & .5 \end{bmatrix} = [.54 \quad .46]$

54% of the consumers will use brand x on the next purchase.

(E) To find the stationary matrix $S = [s_1 \quad s_2]$, we need to solve:

$$[s_1 \quad s_2]\begin{bmatrix} .7 & .3 \\ .5 & .5 \end{bmatrix} = [s_1 \quad s_2], \; s_1 + s_2 = 1$$

This yields the system of equations:

$$\begin{aligned} .7s_1 + .5s_2 &= s_1 \\ .3s_1 + .5s_2 &= s_2 \\ s_1 + s_2 &= 1 \end{aligned} \qquad \text{or} \qquad \begin{aligned} -.3s_1 + .5s_2 &= 0 \\ .3s_1 - .5s_2 &= 0 \\ s_1 + s_2 &= 1 \end{aligned}$$

The solution is $s_1 = .625$, $s_2 = .375$. Thus, $S = [.625 \quad .375]$.

(F) Brand X will have 62.5% of the market in the long run. (9-2)

39. A transition matrix in standard form for this problem is:

$$P = \begin{array}{c} \\ A \\ B \\ C \\ M \end{array} \begin{array}{c} \begin{array}{cccc} A & B & C & M \end{array} \\ \begin{bmatrix} 1 & 0 & 0 & 0 \\ 0 & 1 & 0 & 0 \\ 0 & 0 & 1 & 0 \\ .06 & .08 & .11 & .75 \end{bmatrix} \end{array}$$

For this matrix, $R = [.06 \quad .08 \quad .11]$ and $Q = [.75]$.
The limiting matrix for P has the form:

$$\overline{P} = \left[\begin{array}{c|c} I & 0 \\ \hline FR & 0 \end{array}\right]$$

where $F = (I - Q)^{-1} = ([1] - [.75])^{-1} = [.25]^{-1} = [4]$.
Thus, $FR = [4][.06 \quad .08 \quad .11] = [.24 \quad .32 \quad .44]$

$$\text{and} \quad \overline{P} = \begin{array}{c} \\ A \\ B \\ C \\ M \end{array} \begin{array}{c} \begin{array}{cccc} A & B & C & M \end{array} \\ \begin{bmatrix} 1 & 0 & 0 & 0 \\ 0 & 1 & 0 & 0 \\ 0 & 0 & 1 & 0 \\ .24 & .32 & .44 & 0 \end{bmatrix} \end{array}$$

(A) In the long run, brand A will have 24% of the market, brand B will have 32% and brand C will have 44%.

(B) A company will wait an average of 4 years before converting to one of the new milling machines. (9-3)

40. (A) $S_1 = [.106 \quad .894]\begin{bmatrix} .853 & .147 \\ .553 & .447 \end{bmatrix} = [.5848 \quad .4152]$

$S_2 = [.5848 \quad .4152]\begin{bmatrix} .853 & .147 \\ .553 & .447 \end{bmatrix} = [.72844 \quad .27156]$

Rounding to three decimal places, we have

$S_1 = [.585 \quad .415]$ and $S_2 = [.728 \quad .272]$

(B)

Year	Data %	Model %
1984	10.6	10.6
1988	58.0	58.5
1992	72.5	72.8

(C) To find the long term behavior, we will calculate the stationary matrix and the limiting matrix:

$$[s_1 \quad s_2]\begin{bmatrix} .853 & .147 \\ .553 & .447 \end{bmatrix} = [s_1 \quad s_2], \; s_1 + s_2 = 1$$

$$\begin{aligned} .853s_1 + .553s_2 &= s_1 & \qquad -.147s_1 + .553s_2 &= 0 \\ .147s_1 + .447s_2 &= s_2 \quad \text{or} & .147s_1 - .553s_2 &= 0 \\ s_1 + \quad s_2 &= 1 & s_1 + \quad s_2 &= 1 \end{aligned}$$

The solution is: $s_1 = .79$, $s_2 = .21$.

Thus, $S = [.79 \quad .21]$, and the limiting matrix is:

$$\bar{P} = \begin{matrix} & v & v' \\ v & .79 & .21 \\ v' & .79 & .21 \end{matrix}$$

79% of the households will own a VCR in the long run. (9-2)

41. A transition matrix in standard form for this problem is:

$$P = \begin{matrix} & F & L & T & A \\ F & 1 & 0 & 0 & 0 \\ L & 0 & 1 & 0 & 0 \\ T & 0 & .05 & .8 & .15 \\ A & .17 & .03 & 0 & .8 \end{matrix}$$

where F = "Fellow", A = "Associate", T = "Trainee", L = leaves

For this matrix, $R = \begin{bmatrix} 0 & .05 \\ .17 & .03 \end{bmatrix}$ and $Q = \begin{bmatrix} .8 & .15 \\ 0 & .8 \end{bmatrix}$.

The limiting matrix for P has the form

$$\bar{P} = \left[\begin{array}{c|c} I & 0 \\ \hline FR & 0 \end{array}\right]$$

where $F = (I - Q)^{-1} = \left(\begin{bmatrix} 1 & 0 \\ 0 & 1 \end{bmatrix} - \begin{bmatrix} .8 & .15 \\ 0 & .8 \end{bmatrix}\right)^{-1} = \begin{bmatrix} .2 & -.15 \\ 0 & .2 \end{bmatrix}^{-1}$

We use row operations to calculate the inverse:

$$\left[\begin{array}{cc|cc} .2 & -.15 & 1 & 0 \\ 0 & .2 & 0 & 1 \end{array}\right] \sim \left[\begin{array}{cc|cc} 1 & -.75 & 5 & 0 \\ 0 & 1 & 0 & 5 \end{array}\right] \sim \left[\begin{array}{cc|cc} 1 & 0 & 5 & 3.75 \\ 0 & 1 & 0 & 5 \end{array}\right]$$

$5R_1 \to R_1$ $\qquad (.75)R_2 + R_1 \to R_1$

$5R_2 \to R_2$

Thus, $F = \begin{bmatrix} 5 & 3.75 \\ 0 & 5 \end{bmatrix}$ and $FR = \begin{bmatrix} 5 & 3.75 \\ 0 & 5 \end{bmatrix}\begin{bmatrix} 0 & .05 \\ .17 & .03 \end{bmatrix} = \begin{bmatrix} .6375 & .3625 \\ .85 & .15 \end{bmatrix}$

The limiting matrix is:

$$\bar{P} = \begin{matrix} & F & L & T & A \\ F & 1 & 0 & 0 & 0 \\ L & 0 & 1 & 0 & 0 \\ T & .6375 & .3625 & 0 & 0 \\ A & .85 & .15 & 0 & 0 \end{matrix}$$

(A) In the long run, 63.75% of the trainees will become Fellows.

(B) In the long run, 15% of the Associates will leave the company.

(C) A trainee remains in the program an average of 5 + 3.75 = 8.75 years. (9-3)

42. We shall find the limiting matrix for:

$$P = \begin{array}{c} \\ R \\ P \\ W \end{array}\begin{array}{c} \begin{array}{ccc} R & P & W \end{array} \\ \begin{bmatrix} 1 & 0 & 0 \\ .5 & .5 & 0 \\ 0 & 1 & 0 \end{bmatrix} \end{array} \quad \text{We have } R = \begin{bmatrix} .5 \\ 0 \end{bmatrix} \text{ and } Q = \begin{bmatrix} .5 & 0 \\ 1 & 0 \end{bmatrix}.$$

The limiting matrix for P will have the form:

$$\overline{P} = \left[\begin{array}{c|c} I & 0 \\ \hline FR & 0 \end{array}\right]$$

where $F = (I - Q)^{-1} = \left(\begin{bmatrix} 1 & 0 \\ 0 & 1 \end{bmatrix} - \begin{bmatrix} .5 & 0 \\ 1 & 0 \end{bmatrix}\right)^{-1} = \begin{bmatrix} .5 & 0 \\ -1 & 1 \end{bmatrix}^{-1} = \begin{bmatrix} \frac{1}{2} & 0 \\ -1 & 1 \end{bmatrix}^{-1}$

We use row operations to find the inverse:

$$\left[\begin{array}{cc|cc} \frac{1}{2} & 0 & 1 & 0 \\ -1 & 1 & 0 & 1 \end{array}\right] \sim \left[\begin{array}{cc|cc} 1 & 0 & 2 & 0 \\ -1 & 1 & 0 & 1 \end{array}\right] \sim \left[\begin{array}{cc|cc} 1 & 0 & 2 & 0 \\ 0 & 1 & 2 & 1 \end{array}\right]$$

$2R_1 \to R_1$ $\qquad$ $R_1 + R_2 \to R_2$

Thus, $F = \begin{bmatrix} 2 & 0 \\ 2 & 1 \end{bmatrix}$ and $FR = \begin{bmatrix} 2 & 0 \\ 2 & 1 \end{bmatrix}\begin{bmatrix} .5 \\ 0 \end{bmatrix} = \begin{bmatrix} 1 \\ 1 \end{bmatrix}$

The limiting matrix $\overline{P}$ is:

$$\overline{P} = \begin{array}{c} \\ R \\ P \\ W \end{array}\begin{array}{c} \begin{array}{ccc} R & P & W \end{array} \\ \begin{bmatrix} 1 & 0 & 0 \\ 1 & 0 & 0 \\ 1 & 0 & 0 \end{bmatrix} \end{array}$$

From this matrix, we conclude that eventually all of the flowers will be red.
(9-3)

43. (A) $S_1 = [.028 \quad .972]\begin{bmatrix} .751 & .249 \\ .001 & .999 \end{bmatrix} = [.022 \quad .978]$

$S_2 = [.022 \quad .978]\begin{bmatrix} .751 & .249 \\ .001 & .999 \end{bmatrix} = [.0175 \quad .9825]$

(B)

Year	Data %	Model %
1980	2.8	2.8
1985	2.3	2.2
1990	1.9	1.75

(C) To find the long term behavior, we calculate the stationary matrix and the limiting matrix.

$$[s_1 \quad s_2]\begin{bmatrix} .751 & .249 \\ .001 & .999 \end{bmatrix} = [s_1 \quad s_2], \; s_1 + s_2 = 1$$

$$\begin{aligned} .751s_1 + .001s_2 &= s_1 \\ .249s_1 + .999s_2 &= s_2 \\ s_1 + s_2 &= 1 \end{aligned} \quad \text{or} \quad \begin{aligned} -.249s_1 + .001s_2 &= 0 \\ .249s_1 - .001s_2 &= 0 \\ s_1 + s_2 &= 1 \end{aligned}$$

The solution is: $s_1 = .004$, $s_2 = .996$.

The stationary matrix $S = [.004 \quad .996]$ and the limiting matrix:

$$P = \begin{bmatrix} .004 & .996 \\ .004 & .996 \end{bmatrix}$$

In the long run, only 0.4% of the population will live on a farm.
(9-2)

APPENDIX A BASIC ALGEBRA REVIEW

SELF TEST

1. (A) $7 \notin \{4, 6, 8\}$ True (B) $\{8\} \subset \{4, 6, 8\}$ True
(C) $\varnothing \in \{4, 6, 8\}$ False (D) $\varnothing \subset \{4, 6, 8\}$ True (A-1)

2. (A) Commutative $(\cdot)$: $x(y + z) = (y + z)x$
(B) Associative $(+)$: $2 + (x + y) = (2 + x) + y$
(C) Distributive: $(2 + 3)x = 2x + 3x$ (A-2)

3. $(3x - 4) + (x + 2) + (3x^2 + x - 8) + (x^3 + 8)$
$= 3x - 4 + x + 2 + 3x^2 + x - 8 + x^3 + 8 = x^3 + 3x^2 + 5x - 2$ (A-1)

4. $[(x + 2) + (x^3 + 8)] - [(3x - 4) + (3x^2 + x - 8)]$
$= x^3 + x + 10 - [3x^2 + 4x - 12]$
$= x^3 + x + 10 - 3x^2 - 4x + 12 = x^3 - 3x^2 - 3x + 22$ (A-1)

5. $(x^3 + 8)(3x^2 + x - 8) = x^3(3x^2 + x - 8) + 8(3x^2 + x - 8)$
$= 3x^5 + x^4 - 8x^3 + 24x^2 + 8x - 64$ (A-1)

6. $x^3 + 8$ has degree $\underline{\underline{3}}$ (A-1)

7. The coefficient of the second term in $3x^2 + x - 8$ is $\underline{\underline{1}}$ (A-1)

8. $5x^2 - 3x[4 - 3(x - 2)] = 5x^2 - 3x[4 - 3x + 6]$
$= 5x^2 - 3x(-3x + 10)$
$= 5x^2 + 9x^2 - 30x$
$= 14x^2 - 30x$ (A-1)

9. $(2x + y)(3x - 4y) = 6x^2 - 8xy + 3xy - 4y^2$
$= 6x^2 - 5xy - 4y^2$ (A-1)

10. $(2a - 3b)^2 = (2a)^2 - 2(2a)(3b) + (3b)^2$
$= 4a^2 - 12ab + 9b^2$ (A-1)

11. $(2x - y)(2x + y) - (2x - y)^2 = (2x)^2 - y^2 - (4x^2 - 4xy + y^2)$
$= 4x^2 - y^2 - 4x^2 + 4xy - y^2$
$= 4xy - 2y^2$ (A-1)

12. $(m^2 + 2mn - n^2)(m^2 - 2mn - n^2)$
$= m^2(m^2 - 2mn - n^2) + 2mn(m^2 - 2mn - n^2) - n^2(m^2 - 2mn - n^2)$
$= m^4 - 2m^3n - m^2n^2 + 2m^3n - 4m^2n^2 - 2mn^3 - m^2n^2 + 2mn^3 + n^4$
$= m^4 - 6m^2n^2 + n^4$ (A-1)

13. $(x - 2y)^3 = (x - 2y)(x - 2y)^2$
$= (x - 2y)(x^2 - 4xy + 4y^2)$
$= x(x^2 - 4xy + 4y^2) - 2y(x^2 - 4xy + 4y^2)$
$= x^3 - 4x^2y + 4xy^2 - 2x^2y + 8xy^2 - 8y^3$
$= x^3 - 6x^2y + 12xy^2 - 8y^3$ (A-1)

14. (A) $4{,}065{,}000{,}000{,}000 = 4.065 \times 10^{12}$
(B) $0.0073 = 7.3 \times 10^{-3}$ (A-4)

15. (A) $2.55 \times 10^8 = 255{,}000{,}000$ (B) $4.06 \times 10^{-4} = 0.000\,406$ (A-4)

16. (A) $M \cup N = \{2, 4, 5, 6\}$ (B) $M \cap N = \{5\}$
(C) $(M \cup N)' = \{8\}$ (D) $M \cap N' = \{2, 4\}$ (A-1)

17. (A) $n(A \cup B) = 28$ (B) $n(A \cap B) = 5$
(C) $n[(A \cup B)'] = 4$ (D) $n(A \cap B') = 10$ (A-1)

18. (A) True; if n is a natural number, then $n = \frac{n}{1}$
(B) False; a number with a repeating decimal expansion is a rational number. (A-2)

19. Integers that are not natural numbers: 0, -1, -2, ... (A-2)

20. $6(xy^3)^5 = 6x^5y^{15}$ (A-4) **21.** $\frac{9u^8v^6}{3u^4v^8} = \frac{3^2u^8v^6}{3u^4v^8} = \frac{3^{2-1}u^{8-4}}{v^{8-6}} = \frac{3u^4}{v^2}$ (A-4)

22. $(2 \times 10^5)(3 \times 10^{-3}) = 6 \times 10^{5-3} = 6 \times 10^2 = 600$ (A-4)

23. $(x^{-3}y^2)^{-2} = x^6y^{-4} = \frac{x^6}{y^4}$ (A-4) **24.** $u^{5/3}u^{2/3} = u^{5/3+2/3} = u^{7/3}$ (A-5)

25. $(9a^4b^{-2})^{1/2} = (3^2a^4b^{-2})^{1/2} = (3^2)^{1/2}(a^4)^{1/2}(b^{-2})^{1/2} = 3a^2b^{-1} = \frac{3a^2}{b}$ (A-5)

26. $\frac{5^0}{3^2} + \frac{3^{-2}}{2^{-2}} = \frac{1}{3^2} + \frac{\frac{1}{3^2}}{\frac{1}{2^2}} = \frac{1}{9} + \frac{1}{9} \cdot \frac{4}{1} = \frac{5}{9}$ (A-4)

27. $(x^{1/2} + y^{1/2})^2 = (x^{1/2})^2 + 2x^{1/2}y^{1/2} + (y^{1/2})^2 = x + 2x^{1/2}y^{1/2} + y$ (A-5)

28. $(3x^{1/2} - y^{1/2})(2x^{1/2} + 3y^{1/2}) = 6x + 9x^{1/2}y^{1/2} - 2x^{1/2}y^{1/2} - 3y$
$= 6x + 7x^{1/2}y^{1/2} - 3y$ (A-5)

29. $12x^2 + 5x - 3$
$a = 12,\ b = 5,\ c = -3$
Step 1. Use the ac-test
$ac = 12(-3) = -36$

pq
(1)(-36)
(-1)(36)
(2)(-18)
(-2)(18)
(3)(-12)
(-3)(12)
(4)(-9)
[(-4)(9)]
⋮

$-4 + 9 = 5 = b$

$$12x^2 + 5x - 3 = 12x^2 - 4x + 9x - 3$$
$$= (12x^2 - 4x) + (9x - 3)$$
$$= 4x(3x - 1) + 3(3x - 1)$$
$$= (3x - 1)(4x + 3) \qquad \text{(A-2)}$$

30. $8x^2 - 18xy + 9y^2$
$a = 8,\ b = -18,\ c = 9$
$ac = (8)(9) = 72$
Note that $(-6)(-12) = 72$ and $-12 - 6 = -18$. Thus
$$8x^2 - 18xy + 9y^2 = 8x^2 - 12xy - 6xy + 9y^2$$
$$= (8x^2 - 12xy) - (6xy - 9y^2)$$
$$= 4x(2x - 3y) - 3y(2x - 3y)$$
$$= (2x - 3y)(4x - 3y) \qquad \text{(A-2)}$$

31. $t^2 - 4t - 6$ This polynomial cannot be factored further. (A-2)

32. $6n^3 - 9n^2 - 15n = 3n(2n^2 - 3n - 5) = 3n(2n - 5)(n + 1)$ (A-2)

33. $(4x - y)^2 - 9x^2 = (4x - y - 3x)(4x - y + 3x) = (x - y)(7x - y)$ (A-2)

34. $2x^2 + 4xy - 5y^2$ cannot be factored further. (A-2)

35. $\dfrac{2}{5b} - \dfrac{4}{3a^3} - \dfrac{1}{6a^2b^2}$ LCD $= 30a^3b^2$
$$= \frac{6a^3b}{6a^3b} \cdot \frac{2}{5b} - \frac{10b^2}{10b^2} \cdot \frac{4}{3a^3} - \frac{5a}{5a} \cdot \frac{1}{6a^2b^2}$$
$$= \frac{12a^3b}{30a^3b^2} - \frac{40b^2}{30a^3b^2} - \frac{5a}{30a^3b^2} = \frac{12a^3b - 40b^2 - 5a}{30a^3b^2} \qquad \text{(A-3)}$$

36. $$\frac{3x}{3x^2 - 12x} + \frac{1}{6x} = \frac{\cancel{3x}}{\cancel{3x}(x - 4)} + \frac{1}{6x} = \frac{1}{x - 4} + \frac{1}{6x}$$
$$= \frac{6x}{6x(x - 4)} + \frac{x - 4}{6x(x - 4)} = \frac{6x + x - 4}{6x(x - 4)} = \frac{7x - 4}{6x(x - 4)} \qquad \text{(A-3)}$$

37. $$\frac{x}{x^2 - 16} - \frac{x + 4}{x^2 - 4x} = \frac{x}{(x - 4)(x + 4)} - \frac{x + 4}{x(x - 4)}$$
[LCD $= x(x - 4)(x + 4)$]
$$= \frac{x^2}{x(x - 4)(x + 4)} - \frac{(x + 4)^2}{x(x - 4)(x + 4)}$$
$$= \frac{x^2 - (x + 4)^2}{x(x - 4)(x + 4)}$$
$$= \frac{x^2 - x^2 - 8x - 16}{x(x - 4)(x + 4)} = \frac{-8(x + 2)}{x(x - 4)(x + 4)} \qquad \text{(A-3)}$$

38. $$\frac{y - 2}{y^2 - 4y + 4} \div \frac{y^2 + 2y}{y^2 + 4y + 4} = \frac{\cancel{y - 2}}{\underset{y - 2}{\cancel{(y - 2)^2}}} \cdot \frac{\overset{y + 2}{\cancel{(y + 2)^2}}}{y\cancel{(y + 2)}} = \frac{y + 2}{y(y - 2)}$$
(A-3)

39. $$\frac{\frac{1}{7 + h} - \frac{1}{7}}{h} = \frac{\frac{7 - (7 + h)}{7(7 + h)}}{h} = \frac{\frac{-h}{7(7 + h)}}{\frac{h}{1}} = \frac{-h}{7(7 + h)} \cdot \frac{1}{h} = \frac{-1}{7(7 + h)} \qquad \text{(A-3)}$$

40. $\dfrac{x^{-1} + y^{-1}}{x^{-2} - y^{-2}} = \dfrac{\dfrac{1}{x} + \dfrac{1}{y}}{\dfrac{1}{x^2} - \dfrac{1}{y^2}} = \dfrac{\dfrac{x + y}{xy}}{\dfrac{y^2 - x^2}{x^2y^2}} = \dfrac{\cancel{x + y}}{\cancel{xy}} \cdot \dfrac{\cancel{x^2y^2}\,xy}{(y - x)\cancel{(y + x)}} = \dfrac{xy}{y - x}$ (A-5)

41. (A) $(-7) - (-5) = -7 + [-(-5)]$ Subtraction

(B) $5u + (3v + 2) = (3v + 2) + 5u$ Commutative (+)

(C) $(5m - 2)(2m + 3) = (5m - 2)2m + (5m - 2)3$ Distributive

(D) $9 \cdot (49) = (9 \cdot 4)y$ Associative ($\cdot$)

(E) $\dfrac{u}{-(v - w)} = -\dfrac{u}{v - w}$ Negatives

(F) $(x - y) + 0 = (x - y)$ Identity (+) (A-2)

42.

E M 45 25 20 10

(A) 90

(B) 45

(A-1)

43. $6\sqrt[5]{x^2} - 7\sqrt[4]{(x - 1)^3} = 6x^{2/5} - 7(x - 1)^{3/4}$ (A-5)

44. $2x^{1/2} - 3x^{2/3} = 2\sqrt{x} - 3\sqrt[3]{x^2}$ (A-5)

45. $\dfrac{4\sqrt{x} - 3}{2\sqrt{x}} = \dfrac{4x^{1/2}}{2x^{1/2}} - \dfrac{3}{2x^{1/2}} = 2 - \dfrac{3}{2}x^{-1/2}$ (A-5)

46. $\dfrac{3x}{\sqrt{3x}} = \dfrac{3x}{\sqrt{3x}} \cdot \dfrac{\sqrt{3x}}{\sqrt{3x}} = \dfrac{\cancel{3x}\sqrt{3x}}{\cancel{3x}} = \sqrt{3x}$ (A-5)

47. $\dfrac{x - 5}{\sqrt{x} - \sqrt{5}} = \dfrac{x - 5}{\sqrt{x} - \sqrt{5}} \cdot \dfrac{\sqrt{x} + \sqrt{5}}{\sqrt{x} + \sqrt{5}} = \dfrac{\cancel{(x - 5)}(\sqrt{x} + \sqrt{5})}{\cancel{x - 5}} = \sqrt{x} + \sqrt{5}$ (A-5)

48. $\dfrac{\sqrt{x - 5}}{x - 5} = \dfrac{\sqrt{x - 5}}{x - 5} \cdot \dfrac{\sqrt{x - 5}}{\sqrt{x - 5}} = \dfrac{\cancel{x - 5}}{\cancel{(x - 5)}\sqrt{x - 5}} = \dfrac{1}{\sqrt{x - 5}}$ (A-5)

49. $\dfrac{\sqrt{u + h} - \sqrt{u}}{h} = \dfrac{\sqrt{u + h} - \sqrt{u}}{h} \cdot \dfrac{\sqrt{u + h} + \sqrt{u}}{\sqrt{u + h} + \sqrt{u}}$

$= \dfrac{u + h - u}{h(\sqrt{u + h} + \sqrt{u})}$

$= \dfrac{\cancel{h}}{\cancel{h}(\sqrt{u + h} + \sqrt{u})} = \dfrac{1}{\sqrt{u + h} + \sqrt{u}}$ (A-5)

50. $\dfrac{x}{12} - \dfrac{x - 3}{3} = \dfrac{1}{2}$

Multiply each term by 12: $x - 4(x - 3) = 6$

$x - 4x + 12 = 6$

$-3x = 6 - 12$

$-3x = -6$

$x = 2$ (A-6)

51. $x^2 = 5x$

$x^2 - 5x = 0$ (solve by factoring)

$x(x - 5) = 0$

$x = 0$ or $x - 5 = 0$

$x = 5$ (A-7)

52. $3x^2 - 21 = 0$

$x^2 - 7 = 0$ (solve by the square root method)

$x^2 = 7$

$x = \pm\sqrt{7}$ (A-7)

53. $x^2 - x - 20 = 0$ (solve by factoring)

$(x - 5)(x + 4) = 0$

$x - 5 = 0$ or $x + 4 = 0$

$x = 5$ $\quad x = -4$ (A-7)

54. $2x = 3 + \frac{1}{x}$ $(x \neq 0)$

$2x^2 = 3x + 1$ (multiply by x)

$2x^2 - 3x - 1 = 0$

$$x = \frac{-(-3) \pm \sqrt{(-3)^2 - (4)(2)(-1)}}{2(2)}$$

$$= \frac{3 \pm \sqrt{9 + 8}}{4} = \frac{3 \pm \sqrt{17}}{4}$$ (A-7)

55. $2(x + 4) > 5x - 4$

$2x + 8 > 5x - 4$

$2x - 5x > -4 - 8$

$-3x > -12$ (Divide both sides by -3 and reverse the inequality)

$x < 4$ or $(-\infty, 4)$

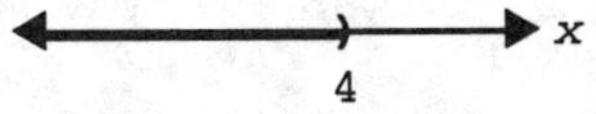

(A-6)

56. $1 - \frac{x - 3}{3} \leq \frac{1}{2}$

Multiply both sides of the inequality by 6. We do not reverse the direction of the inequality, since $6 > 0$.

$6 - 2(x - 3) \leq 3$

$6 - 2x + 6 \leq 3$

$-2x \leq 3 - 12$

$-2x \leq -9$

Divide both sides by -2 and reverse the direction of the inequality, since $-2 < 0$.

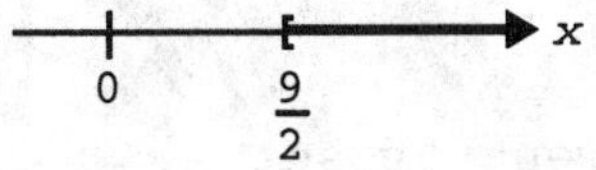

$x \geq \frac{9}{2}$ or $\left[\frac{9}{2}, \infty\right)$ (A-6)

57. $-2 \leq \frac{x}{2} - 3 < 3$

$-2 + 3 \leq \frac{x}{2} < 3 + 3$

$1 \leq \frac{x}{2} < 6$

$2 \leq x < 12$ or $[2, 12)$

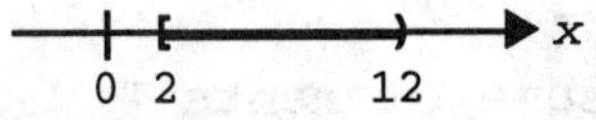

(A-6)

58. $2x - 3y = 6$
$-3y = -2x + 6$
$y = \frac{2}{3}x - 2$ (A-6)

59. $xy - y = 3$
$y(x - 1) = 3$
$y = \frac{3}{x - 1}$ (A-6)

60. True. $A \cap B$ is always a subset of both A and B, i.e. $A \cap B \subset A$ and $A \cap B \subset B$. Therefore, if $A = A \cap B$, then $A \subset B$.

61. The GDP per person is given by:

$$\frac{5,951,000,000,000}{255,100,000} = \frac{5.951 \times 10^{12}}{2.551 \times 10^{8}} \approx 2.3328 \times 10^{4} = \$23,328$$ (A-4)

62. Let x = the amount invested at 8%. Then $60,000 - x$ = amount invested at 14%. The interest on $60,000 at 12% for one year is:
$0.12(60,000) = 7200$
Thus, we want:
$0.08x + 0.14(60,000 - x) = 7200$
$0.08x + 8400 - 0.14x = 7200$
$-0.06x = -1200$
$x = 20,000$
Therefore, $20,000 should be invested at 8% and $40,000 should be invested at 14%. (A-6)

63. Let x = number of tapes produced
Cost: $C = 12x + 72,000$
Revenue: $R = 30x$
To find the break-even point, set $R = C$.
$30x = 12x + 72,000$
$18x = 72,000$
$x = 4000$
Thus, 4000 tapes must be sold for the producer to break even. (A-6)

EXERCISE A-1

Things to remember:

1. $a \in A$ means "a is an element of set A."
2. $a \notin A$ means "a is not an element of set A."
3. $\varnothing$ means "the empty set" or "null set."
4. $S = \{x \mid P(x)\}$ means "S is the set of all x such that $P(x)$ is true."
5. $A \subset B$ means "A is a subset of B."
6. $A = B$ means "A and B have exactly the same elements."

7. $A \not\subset B$ means "A is not a subset of B."

8. $A \neq B$ means "A and B do not have exactly the same elements."

9. $A \cup B = A$ union $B = \{x \mid x \in A \text{ or } x \in B\}$.

10. $A \cap B = A$ intersection $B = \{x \mid x \in A \text{ and } x \in B\}$.

11. $A' =$ complement of $A = \{x \in U \mid x \notin A\}$, where U is a universal set.

---

1. T **3.** T **5.** T **7.** T **9.** $\{1, 3, 5\} \cup \{2, 3, 4\} = \{1, 2, 3, 4, 5\}$

11. $\{1, 3, 4\} \cap \{2, 3, 4\} = \{3, 4\}$

13. $\{1, 5, 9\} \cap \{3, 4, 6, 8\} = \varnothing$

15. $\{x \mid x - 2 = 0\}$
$x - 2 = 0$ is true for $x = 2$.
Hence, $\{x \mid x - 2 = 0\} = \{2\}$.

17. $x^2 = 49$ is true for $x = 7$ and -7.
Hence, $\{x \mid x^2 = 49\} = \{-7, 7\}$.

19. $\{x \mid x$ is an odd number between 1 and 9 inclusive$\} = \{1, 3, 5, 7, 9\}$.

21. $U = \{1, 2, 3, 4, 5\}$; $A = \{2, 3, 4\}$. Then $A' = \{1, 5\}$.

23. From the Venn diagram, A has 40 elements.

25. A' has 60 elements.

27. $A \cup B$ has 60 elements (35 + 5 + 20).

29. $A' \cap B$ has 20 elements (common elements between A' and B.)

31. $(A \cap B)'$ has 95 elements.
[Note: $A \cap B$ has 5 elements.]

33. $A' \cap B'$ has 40 elements.

35. (A) $\{x \mid x \in R \text{ or } x \in T\}$.
$= R \cup T$ ("or" translated as $\cup$, union)
$= \{1, 2, 3, 4\} \cup \{2, 4, 6\}$
$= \{1, 2, 3, 4, 6\}$

(B) $R \cup T = \{1, 2, 3, 4, 6\}$

37. $Q \cap R = \{2, 4, 6\} \cap \{3, 4, 5, 6\}$
$= \{4, 6\}$
$P \cup (Q \cap R) = \{1, 2, 3, 4\} \cup \{4, 6\}$
$= \{1, 2, 3, 4, 6\}$

39. Yes. $A \cup B = B$ can be represented by the Venn diagram

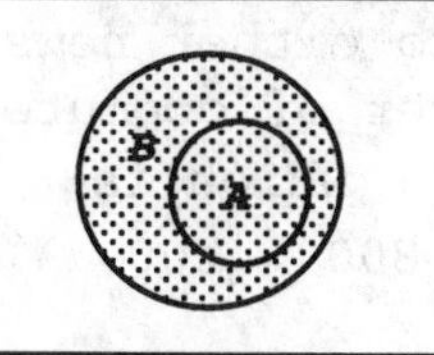

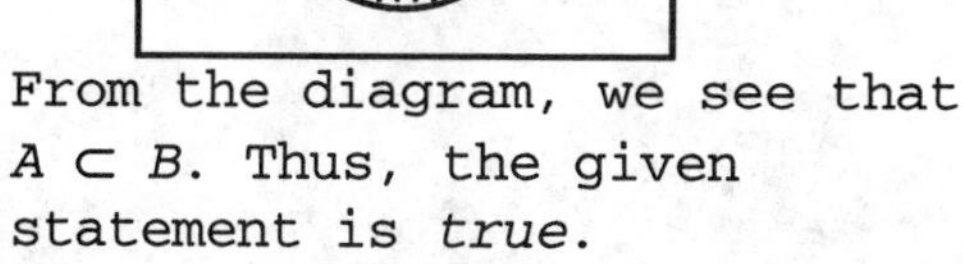
From the diagram, we see that $A \subset B$. Thus, the given statement is *true*.

41. Yes. The given statement is always *true*. To understand this, see the following Venn diagram.

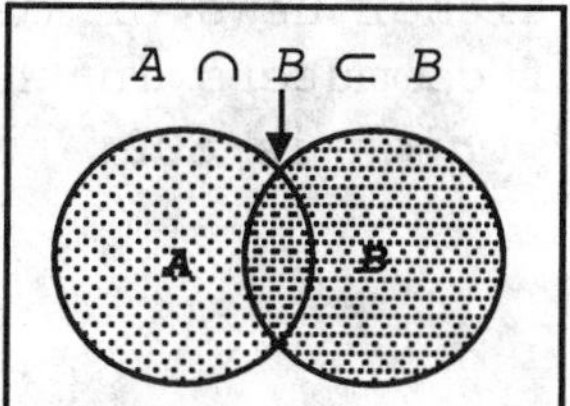

43. Yes. The given statement is *true*. To understand this, see the following Venn diagram.

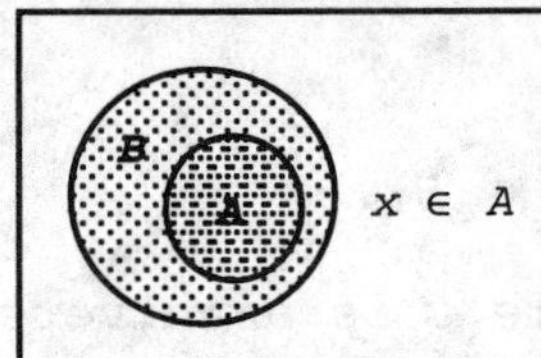

From the diagram, we conclude that $x \in B$.

45. (A) Set $\{a\}$ has two subsets: $\{a\}$ and $\emptyset$

(B) Set $\{a, b\}$ has four subsets: $\{a, b\}$, $\emptyset$, $\{a\}$, $\{b\}$

(C) Set $\{a, b, c\}$ has eight subsets: $\{a, b, c\}$, $\emptyset$, $\{a\}$, $\{b\}$, $\{c\}$, $\{a, b\}$, $\{a, c\}$, $\{b, c\}$

Parts (A), (B), and (C) suggest the following formula:

The number of subsets in a set with n elements $= 2^n$.

(D) $\{a, b, c, d\}$

Set of subsets. $A = \{\emptyset, \{a\}, \{b\}, \{c\}, \{d\}, \{a, b\}, \{a, c\}, \{a, d\}, \{b, c\}, \{b, d\}, \{c, d\}, \{a, b, c\}, \{a, b, d\}, \{a, c, d\}, \{b, c, d\}, \{a, b, c, d\}\}$

$n(A) = 16 = 2^4$

Based on parts (A)—(D), the number of subsets of a set with n elements is 2^n.

47. The Venn diagram that corresponds to the given information is shown at the right. We can see that $N \cup M$ has $300 + 300 + 200 = 800$ students.

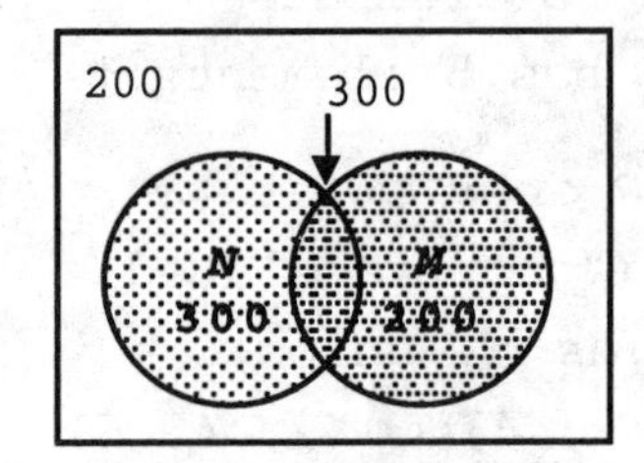

49. $(N \cup M)'$ has 200 students [because $N \cup M$ has 800 students and $(N \cup M)'$ has $1000 - 800 = 200$].

51. $N' \cap M$ has 200 students.

53. The number of commuters who listen to either news or music = number of commuters in the set $M \cup N$, which is 800.

55. The number of commuters who do not listen to either news or music = number of commuters in set $(N \cup M)'$, which is $1000 - 800 = 200$.

57. The number of commuters who listen to music but not news = number of commuters in the set $N' \cap M$, which is 200.

59. The six two-person subsets that can be formed from the given set $\{P, V_1, V_2, V_3\}$ are:

$\{P, V_1\}$ $\{P, V_3\}$ $\{V_1, V_3\}$
$\{P, V_2\}$ $\{V_1, V_2\}$ $\{V_2, V_3\}$

61. From the given Venn diagram $A \cap Rh = \{A+, AB+\}$

63. Again, from the given Venn diagram: $A \cup Rh = \{A-, A+, B+, AB-, AB+, O+\}$

65. From the given Venn diagram: $(A \cup B)' = \{O+, O-\}$

67. $A' \cap B = \{B-, B+\}$

69. Statement (2): for every $a, b \in C$, aRb and bRa means that everyone in the clique relates to one another.

EXERCISE A-2

Things to remember:

1. THE SET OF REAL NUMBERS

SYMBOL	NAME	DESCRIPTION	EXAMPLES
N	Natural numbers	Counting numbers (also called positive integers)	1, 2, 3, ...
Z	Integers	Natural numbers, their negatives, and 0	... -2, -1, 0, 1, 2, ...
Q	Rational numbers	Any number that can be represented as $\frac{a}{b}$, where a and b are integers and $b \neq 0$. Decimal representations are repeating or terminating.	-4; 0; 1; 25; $\frac{-3}{5}$; $\frac{2}{3}$; 3.67; $-0.333\overline{3}$; $5.2727\overline{27}$
I	Irrational numbers	Any number with a decimal representation that is nonrepeating and non-terminating.	$\sqrt{2}$; π; $\sqrt[3]{7}$; 1.414213...; 2.718281828...
R	Real numbers	Rationals and irrationals	

2. BASIC PROPERTIES OF THE SET OF REAL NUMBERS

Let a, b, and c be arbitrary elements in the set of real numbers R.

ADDITION PROPERTIES

ASSOCIATIVE: $(a + b) + c = a + (b + c)$

COMMUTATIVE: $a + b = b + a$

IDENTITY: 0 is the additive identity; that is, $0 + a = a + 0$ for all a in R, and 0 is the only element in R with this property.

INVERSE: For each a in R, $-a$ is its unique additive inverse; that is, $a + (-a) = (-a) + a = 0$, and $-a$ is the only element in R relative to a with this property.

MULTIPLICATION PROPERTIES

ASSOCIATIVE: $(ab)c = a(bc)$

COMMUTATIVE: $ab = ba$

IDENTITY: 1 is the multiplicative identity; that is, $1a = a1 = a$ for all a in R, and 1 is the only element in R with this property.

INVERSE: For each a in R, $a \neq 0$, $\frac{1}{a}$ is its unique multiplicative inverse; that is, $a\left(\frac{1}{a}\right) = \left(\frac{1}{a}\right)a = 1$, and $\frac{1}{a}$ is the only element in R relative to a with this property.

DISTRIBUTIVE PROPERTIES

$$a(b + c) = ab + ac$$

$$(a + b)c = ac + bc$$

3. SUBTRACTION AND DIVISION

For all real numbers a and b.

SUBTRACTION: $a - b = a + (-b)$

$$7 - (-5) = 7 + [-(-5)] = 7 + 5 = 12$$

DIVISION: $a \div b = a\left(\frac{1}{b}\right),\ b \neq 0$

$$9 \div 4 = 9\left(\frac{1}{4}\right) = \frac{9}{4}$$

NOTE: 0 can never be used as a divisor!

4. PROPERTIES OF NEGATIVES

For all real numbers a and b.

a. $-(-a) = a$

b. $(-a)b = -(ab) = a(-b) = -ab$

c. $(-a)(-b) = ab$

d. $(-1)a = -a$

e. $\frac{-a}{b} = -\frac{a}{b} = \frac{a}{-b}$, $b \neq 0$

f. $\frac{-a}{-b} = -\frac{-a}{b} = -\frac{a}{-b} = \frac{a}{b}$, $b \neq 0$

5. ZERO PROPERTIES

For all real numbers a and b.

a. $a \cdot 0 = 0$

b. $ab = 0$ if and only if $a = 0$ or $b = 0$ (or both)

6. FRACTION PROPERTIES

For all real numbers a, b, c, d, and k (division by 0 excluded).

a. $\frac{a}{b} = \frac{c}{d}$ if and only if $ad = bc$

b. $\frac{ka}{kb} = \frac{a}{b}$

c. $\frac{a}{b} \cdot \frac{c}{d} = \frac{ac}{bd}$

d. $\frac{a}{b} \div \frac{c}{d} = \frac{a}{b} \cdot \frac{d}{c}$

e. $\frac{a}{b} + \frac{c}{b} = \frac{a + c}{b}$

f. $\frac{a}{b} - \frac{c}{b} = \frac{a - c}{b}$

g. $\frac{a}{b} + \frac{c}{d} = \frac{ad + bc}{bd}$

h. $\frac{a}{b} - \frac{c}{d} = \frac{ad - bc}{bd}$

---

1. $uv = vu$ **3.** $3 + (7 + y) = (3 + 7) + y$ **5.** $1(u + v) = u + v$

7. T; Associative property of multiplication

9. T; Distributive property **11.** T; Definition of subtraction

13. T; Commutative property of addition **15.** T; Property of negatives

17. T; Multiplicative inverse property **19.** T; Property of negatives

21. F; $\frac{a}{b} + \frac{c}{d} = \frac{ad + bc}{bd}$ **23.** T; Distributive property

25. T; Zero property

27. No. For example: $2\left(\frac{1}{2}\right) = 1$. In general $a\left(\frac{1}{a}\right) = 1$ whenever $a \neq 0$.

29. (A) False. For example, -3 is an integer but not a natural number.

(B) True

(C) True. For example, for any natural number n, $n = \frac{n}{1}$.

31. $\sqrt{2}$, $\sqrt{3}$, ...; in general, the square root of any rational number that is not a perfect square; π, e.

33. (A) $8 \in N, Z, Q, R$ (B) $\sqrt{2} \in R$

(C) $-1.414 = -\frac{1414}{1000} \in Q, R$ (D) $\frac{-5}{2} \in Q, R$

35. (A) True. This is the associative property of addition.

(B) False. For example, $(3 - 7) - 4 = -4 - 4 = -8$
$\neq 3 - (7 - 4) = 3 - 3 = 0.$

(C) True. This is the associative property of multiplication.

(D) False. For example, $(12 \div 4) \div 2 = 3 \div 2 = \frac{3}{2}$
$\neq 12 \div (4 \div 2) = 12 \div 2 = 6.$

37.
$$\begin{aligned} C &= 0.090909\ldots \\ 100C &= 9.090909\ldots \\ 100C - C &= (9.090909\ldots) - (0.090909\ldots) \\ 99C &= 9 \\ C &= \frac{9}{99} = \frac{1}{11} \end{aligned}$$

EXERCISE A-3

Things to remember:

1. NATURAL NUMBER* EXPONENT

For n a natural number and b any real number,

$$b^n = b \cdot b \cdot \cdots \cdot b, \; n \text{ factors of } b.$$

For example, $2^3 = 2 \cdot 2 \cdot 2$ (= 8),
$3^5 = 3 \cdot 3 \cdot 3 \cdot 3 \cdot 3$ (= 243).

In the expression b^n, n is called the EXPONENT or POWER, and b is called the BASE.

* The natural numbers are the counting numbers: 1, 2, 3, 4, ...

2. FIRST PROPERTY OF EXPONENTS

For any natural numbers m and n, and any real number b,

$$b^m \cdot b^n = b^{m+n}.$$

For example, $3^3 \cdot 3^4 = 3^{3+4} = 3^7$.

3. POLYNOMIALS

a. A POLYNOMIAL IN ONE VARIABLE x is constructed by adding or subtracting constants and terms of the form ax^n, where a is a real number, called the COEFFICIENT of the term, and n is a natural number.

b. A POLYNOMIAL IN TWO VARIABLES x AND y is constructed by adding or subtracting constants and terms of the form ax^my^n, bx^k, cy^j, where a, b, and c are real numbers, called COEFFICIENTS and m, n, k, and j are natural numbers.

c. Polynomials in more than two variables are defined similarly.

d. A polynomial with only one term is called a MONOMIAL.
A polynomial with two terms is called a BINOMIAL.
A polynomial with three terms is called a TRINOMIAL.

4. DEGREE OF A POLYNOMIAL

a. A term of the form ax^n, $a \neq 0$, has degree n. A term of the form ax^my^n, $a \neq 0$, has degree $m + n$. A nonzero constant has degree 0.

b. The DEGREE OF A POLYNOMIAL is the degree of the nonzero term with the highest degree. For example, $3x^4 + \sqrt{2}x^3 - 2x + 7$ has degree 4; $2x^3y^2 - 3x^2y + 7x^4 - 5y^3 + 6$ has degree 5; the polynomial 4 has degree 0.

c. The constant 0 is a polynomial but it is not assigned a degree.

5. Two terms in a polynomial are called LIKE TERMS if they have exactly the same variable factors raised to the same powers. For example, in

$$7x^5y^2 - 3x^3y + 2x + 4x^3y - 1,$$

$-3x^3y$ and $4x^3y$ are like terms.

6. To multiply two polynomials, multiply each term of one by each term of the other, and then combine like terms.

7. SPECIAL PRODUCTS

a. $(a - b)(a + b) = a^2 - b^2$

b. $(a + b)^2 = a^2 + 2ab + b^2$

c. $(a - b)^2 = a^2 - 2ab + b^2$

1. The term of highest degree in $x^3 + 2x^2 - x + 3$ is x^3 and the degree of this term is 3.

3. $$\begin{aligned}(2x^2 - x + 2) + (x^3 + 2x^2 - x + 3) &= x^3 + 2x^2 + 2x^2 - x - x + 2 + 3\\ &= x^3 + 4x^2 - 2x + 5\end{aligned}$$

5. $$\begin{aligned}(x^3 + 2x^2 - x + 3) - (2x^2 - x + 2) &= x^3 + 2x^2 - x + 3 - 2x^2 + x - 2\\ &= x^3 + 1\end{aligned}$$

7. Using a vertical arrangement:
$$\begin{array}{rrrrrr} & & x^3 & + 2x^2 & - x & + 3\\ & & & 2x^2 & - x & + 2\\ \hline 2x^5 & + 4x^4 & - 2x^3 & + 6x^2 & & \\ & - x^4 & - 2x^3 & + x^2 & - 3x & \\ & & 2x^3 & + 4x^2 & - 2x & + 6\\ \hline 2x^5 & + 3x^4 & - 2x^3 & + 11x^2 & - 5x & + 6\end{array}$$

9. $$\begin{aligned}2(u - 1) - (3u + 2) - 2(2u - 3) &= 2u - 2 - 3u - 2 - 4u + 6\\ &= -5u + 2\end{aligned}$$

11. $$\begin{aligned}4a - 2a[5 - 3(a + 2)] &= 4a - 2a[5 - 3a - 6]\\ &= 4a - 2a[-3a - 1]\\ &= 4a + 6a^2 + 2a\\ &= 6a^2 + 6a\end{aligned}$$

13. $(a + b)(a - b) = a^2 - b^2$ (Special product 7a)

15. $$\begin{aligned}(3x - 5)(2x + 1) &= 6x^2 + 3x - 10x - 5\\ &= 6x^2 - 7x - 5\end{aligned}$$

17. $$\begin{aligned}(2x - 3y)(x + 2y) &= 2x^2 + 4xy - 3xy - 6y^2\\ &= 2x^2 + xy - 6y^2\end{aligned}$$

19. $(3y + 2)(3y - 2) = (3y)^2 - 2^2 = 9y^2 - 4$ (Special product 7a)

21. $$\begin{aligned}(3m + 7n)(2m - 5n) &= 6m^2 - 15mn + 14mn - 35n^2\\ &= 6m^2 - mn - 35n^2\end{aligned}$$

23. $(4m + 3n)(4m - 3n) = 16m^2 - 9n^2$

25. $(3u + 4v)^2 = 9u^2 + 24uv + 16v^2$ (Special product 7b)

27. $$\begin{aligned}(a - b)(a^2 + ab + b^2) &= a(a^2 + ab + b^2) - b(a^2 + ab + b^2)\\ &= a^3 + a^2b + ab^2 - a^2b - ab^2 - b^3\\ &= a^3 - b^3\end{aligned}$$

29. $(4x + 3y)^2 = 16x^2 + 24xy + 9y^2$

31. $$\begin{aligned}m - \{m - [m - (m - 1)]\} &= m - \{m - [m - m + 1]\}\\ &= m - \{m - 1\}\\ &= m - m + 1\\ &= 1\end{aligned}$$

33. $(x^2 - 2xy + y^2)(x^2 + 2xy + y^2) = (x - y)^2(x + y)^2$
$= [(x - y)(x + y)]^2$
$= [x^2 - y^2]^2$
$= x^4 - 2x^2y^2 + y^4$

35. $(3a - b)(3a + b) - (2a - 3b)^2 = (9a^2 - b^2) - (4a^2 - 12ab + 9b^2)$
$= 9a^2 - b^2 - 4a^2 + 12ab - 9b^2$
$= 5a^2 + 12ab - 10b^2$

37. $(m - 2)^2 - (m - 2)(m + 2) = m^2 - 4m + 4 - [m^2 - 4]$
$= m^2 - 4m + 4 - m^2 + 4$
$= -4m + 8$

39. $(x - 2y)(2x + y) - (x + 2y)(2x - y)$
$= 2x^2 - 4xy + xy - 2y^2 - [2x^2 + 4xy - xy - 2y^2]$
$= 2x^2 - 3xy - 2y^2 - 2x^2 - 3xy + 2y^2$
$= -6xy$

41. $(u + v)^3 = (u + v)(u + v)^2 = (u + v)(u^2 + 2uv + v^2)$
$= u^3 + 3u^2v + 3uv^2 + v^3$

43. $(x - 2y)^3 = (x - 2y)(x - 2y)^2 = (x - 2y)(x^2 - 4xy + 4y^2)$
$= x(x^2 - 4xy + 4y^2) - 2y(x^2 - 4xy + 4y^2)$
$= x^3 - 4x^2y + 4xy^2 - 2x^2y + 8xy^2 - 8y^3$
$= x^3 - 6x^2y + 12xy^2 - 8y^3$

45. $[(2x^2 - 4xy + y^2) + (3xy - y^2)] - [(x^2 - 2xy - y^2) + (-x^2 + 3xy - 2y^2)]$
$= [2x^2 - xy] - [xy - 3y^2] = 2x^2 - 2xy + 3y^2$

47. $(2x - 1)^3 - 2(2x - 1)^2 + 3(2x - 1) + 7$
$= (2x - 1)(2x - 1)^2 - 2[4x^2 - 4x + 1] + 6x - 3 + 7$
$= (2x - 1)(4x^2 - 4x + 1) - 8x^2 + 8x - 2 + 6x + 4$
$= 8x^3 - 12x^2 + 6x - 1 - 8x^2 + 14x + 2$
$= 8x^3 - 20x^2 + 20x + 1$

49. $2\{(x - 3)(x^2 - 2x + 1) - x[3 - x(x - 2)]\}$
$= 2\{x^3 - 5x^2 + 7x - 3 - x[3 - x^2 + 2x]\}$
$= 2\{x^3 - 5x^2 + 7x - 3 + x^3 - 2x^2 - 3x\}$
$= 2\{2x^3 - 7x^2 + 4x - 3\}$
$= 4x^3 - 14x^2 + 8x - 6$

51. $m + n$

53. Given two polynomials, one with degree m and the other of degree n, their product will have degree $m + n$ regardless of the relationship between m and n.

55. Since $(a + b)^2 = a^2 + 2ab + b^2$, $(a + b)^2 = a^2 + b^2$ only when $2ab = 0$; that is, only when $a = 0$ or $b = 0$.

57. Let x = amount invested at 9%.
Then $10,000 - x$ = amount invested at 12%.
The total annual income I is:

$$I = 0.09x + 0.12(10,000 - x)$$
$$= 1,200 - 0.03x$$

59. Let x = number of tickets at $10.
Then $3x$ = number of tickets at $30 and $4,000 - x - 3x = 4,000 - 4x$ = number of tickets at $50.
The total receipts R are:

$$R = 10x + 30(3x) + 50(4,000 - 4x)$$
$$= 10x + 90x + 200,000 - 200x = 200,000 - 100x$$

61. Let x = number of kilograms of food A.
Then $10 - x$ = number of kilograms of food B.
The total number of kilograms F of fat in the final food mix is:

$$F = 0.02x + 0.06(10 - x)$$
$$= 0.6 - 0.04x$$

EXERCISE A-4

Things to remember:

1. FACTORED FORMS

   A polynomial is in FACTORED FORM if it is written as the product of two or more polynomials. A polynomial with integer coefficients is FACTORED COMPLETELY if each factor cannot be expressed as the product of two or more polynomials with integer coefficients, other than itself and 1.

2. METHODS

   a. Factor out all factors common to all terms, if they are present.

   b. Try grouping terms.

   c. ac-Test for polynomials of the form

   $$ax^2 + bx + c \quad \text{or} \quad ax^2 + bxy + cy^2$$

   If the product ac has two integer factors p and q whose sum is the coefficient b of the middle term, i.e., if integers p and q exist so that

   $$pq = ac \quad \text{and} \quad p + q = b$$

   then the polynomials have first-degree factors with integer coefficients. If no such integers exist then the polynomials will not have first-degree factors with integer coefficients; the polynomials are *not factorable.*

3. SPECIAL FACTORING FORMULAS

a. $u^2 + 2uv + v^2 = (u + v)^2$ Perfect square
b. $u^2 - 2uv + v^2 = (u - v)^2$ Perfect square
c. $u^2 - v^2 = (u - v)(u + v)$ Difference of squares
d. $u^3 - v^3 = (u - v)(u^2 + uv + v^2)$ Difference of cubes
e. $u^3 + v^3 = (u + v)(u^2 - uv + v^2)$ Sum of cubes

---

1. $3m^2$ is a common factor: $6m^4 - 9m^3 - 3m^2 = 3m^2(2m^2 - 3m - 1)$

3. $2uv$ is a common factor: $8u^3v - 6u^2v^2 + 4uv^3 = 2uv(4u^2 - 3uv + 2v^2)$

5. $(2m - 3)$ is a common factor: $7m(2m - 3) + 5(2m - 3) = (7m + 5)(2m - 3)$

7. $(3c + d)$ is a common factor: $a(3c + d) - 4b(3c + d)$
$= (a - 4b)(3c + d)$

9. $2x^2 - x + 4x - 2 = (2x^2 - x) + (4x - 2)$
$= x(2x - 1) + 2(2x - 1)$
$= (2x - 1)(x + 2)$

11. $3y^2 - 3y + 2y - 2 = (3y^2 - 3y) + (2y - 2)$
$= 3y(y - 1) + 2(y - 1)$
$= (y - 1)(3y + 2)$

13. $2x^2 + 8x - x - 4 = (2x^2 + 8x) - (x + 4)$
$= 2x(x + 4) - (x + 4)$
$= (x + 4)(2x - 1)$

15. $wy - wz + xy - xz = (wy - wz) + (xy - xz)$
$= w(y - z) + x(y - z)$
$= (y - z)(w + x)$

or $wy - wz + xy - xz = (wy + xy) - (wz + xz)$
$= y(w + x) - z(w + x)$
$= (w + x)(y - z)$

17. $am - bn - bm + an = (am - bm) + (an - bn)$
$= m(a - b) + n(a - b)$
$= (a - b)(m + n)$

or $am - bn - bm + an = (am + an) - (bm + bn)$
$= a(m + n) - b(m + n)$
$= (m + n)(a - b)$

19. $3y^2 - y - 2$

$a = 3,\ b = -1,\ c = -2$

Step 1. Use the ac-test to test for factorability

$ac = (3)(-2) = -6$

pq
(1)(-6)
(-1)(6)
[(2)(-3)]
(-2)(3)

Note that $2 + (-3) = -1 = b$. Thus, $3y^2 - y - 2$ has first-degree factors with integer coefficients.

Step 2. Split the middle term using $b = p + q$ and factor by grouping.

$-1 = -3 + 2$

$$3y^2 - y - 2 = 3y^2 - 3y + 2y - 2 = (3y^2 - 3y) + (2y - 2)$$
$$= 3y(y - 1) + 2(y - 1)$$
$$= (y - 1)(3y + 2)$$

21. $u^2 - 2uv - 15v^2$

$a = 1,\ b = -2,\ c = -15$

Step 1. Use the ac-test

$ac = 1(-15) = -15$

pq
(1)(-15)
(-1)(15)
(3)(-5)
(-3)(5)

Note that $3 + (-5) = -2 = b$. Thus $u^2 - 2uv - 15v^2$ has first-degree factors with integer coefficients.

Step 2. Factor by grouping

$-2 = 3 + (-5)$

$$u^2 + 3uv - 5uv - 15v^2 = (u^2 + 3uv) - (5uv + 15v^2)$$
$$= u(u + 3v) - 5v(u + 3v)$$
$$= (u + 3v)(u - 5v)$$

23. $m^2 - 6m - 3$

$a = 1,\ b = -6,\ c = -3$

Step 1. Use the ac-test

$ac = (1)(-3) = -3$

pq
(1)(-3)
(-1)(3)

None of the factors add up to $-6 = b$. Thus, this polynomial is *not factorable*.

25. $w^2x^2 - y^2 = (wx - y)(wx + y)$ (difference of squares)

27. $9m^2 - 6mn + n^2 = (3m - n)^2$ (perfect square)

29. $y^2 + 16$

$a = 1,\ b = 0,\ c = 16$

Step 1. Use the ac-test

$ac = (1)(16)$

pq
(1)(16)
(-1)(-16)
(2)(8)
(-2)(-8)
(4)(4)
(-4)(-4)

None of the factors add up to $0 = b$. Thus this polynomial is *not factorable*.

31. $4z^2 - 28z + 48 = 4(z^2 - 7z + 12) = 4(z - 3)(z - 4)$

33. $2x^4 - 24x^3 + 40x^2 = 2x^2(x^2 - 12x + 20) = 2x^2(x - 2)(x - 10)$

35. $4xy^2 - 12xy + 9x = x(4y^2 - 12y + 9) = x(2y - 3)^2$

37. $6m^2 - mn - 12n^2 = (2m - 3n)(3m + 4n)$

39. $4u^3v - uv^3 = uv(4u^2 - v^2) = uv[(2u)^2 - v^2] = uv(2u - v)(2u + v)$

41. $2x^3 - 2x^2 + 8x = 2x(x^2 - x + 4)$ [Note: $x^2 - x + 4$ is *not factorable*.]

43. $r^3 - t^3 = (r - t)(r^2 + rt + t^2)$ (difference of cubes)

45. $a^3 + 1 = (a + 1)(a^2 - a + 1)$ (sum of cubes)

47. $(x + 2)^2 - 9y^2 = [(x + 2) - 3y][(x + 2) + 3y]$
$= (x + 2 - 3y)(x + 2 + 3y)$

49. $5u^2 + 4uv - 2v^2$ is *not factorable*.

51. $6(x - y)^2 + 23(x - y) - 4 = [6(x - y) - 1][(x - y) + 4]$
$= (6x - 6y - 1)(x - y + 4)$

53. $y^4 - 3y^2 - 4 = (y^2)^2 - 3y^2 - 4 = (y^2 - 4)(y^2 + 1)$
$= (y - 2)(y + 2)(y^2 + 1)$

55. $27a^2 + a^5b^3 = a^2(27 + a^3b^3) = a^2[3^3 + (ab)^3]$
$= a^2(3 + ab)(9 - 3ab + a^2b^2)$

EXERCISE A-5

Things to remember:

1. FUNDAMENTAL PROPERTY OF FRACTIONS

   If a, b, and k are real numbers with $b, k \neq 0$, then
   $\frac{ka}{kb} = \frac{a}{b}$.

   A fraction is in LOWEST TERMS if the numerator and denominator have no common factors other than 1 or -1.

2. MULTIPLICATION AND DIVISION

   For a, b, c, and d real numbers:

   a. $\frac{a}{b} \cdot \frac{c}{d} = \frac{ac}{bd}$, $b, d \neq 0$

   b. $\frac{a}{b} \div \frac{c}{d} = \frac{\frac{a}{b}}{\frac{c}{d}} = \frac{a}{b} \cdot \frac{d}{c}$, $b, c, d \neq 0$

   The same procedures are used to multiply or divide two rational expressions.

3. ADDITION AND SUBTRACTION

For a, b, and c real numbers:

a. $\dfrac{a}{b} + \dfrac{c}{b} = \dfrac{a + c}{b}$, $b \neq 0$

b. $\dfrac{a}{b} - \dfrac{c}{b} = \dfrac{a - c}{b}$, $b \neq 0$

The same procedures are used to add or subtract two rational expressions (with the same denominator).

4. THE LEAST COMMON DENOMINATOR (LCD)

The LCD of two or more rational expressions is found as follows:

a. Factor each denominator completely, including integer factors.

b. Identify each different factor from all the denominators.

c. Form a product using each different factor to the highest power that occurs in any one denominator. This product is the LCD.

The least common denominator is used to add or subtract rational expressions having different denominators.

---

1. $\dfrac{d^5}{3a} \div \left(\dfrac{d^2}{6a^2} \cdot \dfrac{a}{4d^3}\right) = \dfrac{d^5}{3a} \div \left(\dfrac{\cancel{a}\cancel{d^2}}{24\cancel{a^2}\cancel{d^3}}\right) = \dfrac{d^5}{3a} \div \dfrac{1}{24ad} = \dfrac{d^5}{3\cancel{a}} \cdot \dfrac{\cancel{24}^{\,8}\cancel{a}d}{1} = 8d^6$

(with a and d remaining below the cancelled a^2 and d^3)

3. $\dfrac{x^2}{12} + \dfrac{x}{18} - \dfrac{1}{30} = \dfrac{15x^2}{180} + \dfrac{10x}{180} - \dfrac{6}{180}$

$= \dfrac{15x^2 + 10x - 6}{180}$

We find the LCD of 12, 18, 30: $12 = 2^2 \cdot 3$, $18 = 2 \cdot 3^2$, $30 = 2 \cdot 3 \cdot 5$

Thus, LCD $= 2^2 \cdot 3^2 \cdot 5 = 180$.

5. $\dfrac{4m - 3}{18m^3} + \dfrac{3}{4m} - \dfrac{2m - 1}{6m^2}$

$= \dfrac{2(4m - 3)}{36m^3} + \dfrac{3(9m^2)}{36m^3} - \dfrac{6m(2m - 1)}{36m^3}$

$= \dfrac{8m - 6 + 27m^2 - 6m(2m - 1)}{36m^3}$

$= \dfrac{8m - 6 + 27m^2 - 12m^2 + 6m}{36m^3} = \dfrac{15m^2 + 14m - 6}{36m^3}$

Find the LCD of $18m^3$, $4m$, $6m^2$: $18m^3 = 2 \cdot 3^2 m^3$, $4m = 2^2 m$, $6m^2 = 2 \cdot 3m^2$

Thus, LCD $= 36m^3$.

7. $\dfrac{x^2 - 9}{x^2 - 3x} \div (x^2 - x - 12) = \dfrac{\cancel{(x - 3)}\cancel{(x + 3)}}{x\cancel{(x - 3)}} \cdot \dfrac{1}{(x - 4)\cancel{(x + 3)}} = \dfrac{1}{x(x - 4)}$

9. $\dfrac{2}{x} - \dfrac{1}{x - 3} = \dfrac{2(x - 3)}{x(x - 3)} - \dfrac{x}{x(x - 3)}$ LCD $= x(x - 3)$

$= \dfrac{2x - 6 - x}{x(x - 3)} = \dfrac{x - 6}{x(x - 3)}$

11. $\dfrac{3}{x^2 - 1} - \dfrac{2}{x^2 - 2x + 1} = \dfrac{3}{(x - 1)(x + 1)} - \dfrac{2}{(x - 1)^2}$ LCD $= (x - 1)^2(x + 1)$

$= \dfrac{3(x - 1)}{(x - 1)^2(x + 1)} - \dfrac{2(x + 1)}{(x - 1)^2(x + 1)}$

$= \dfrac{3x - 3 - 2(x + 1)}{(x - 1)^2(x + 1)} = \dfrac{3x - 3 - 2x - 2}{(x - 1)^2(x + 1)}$

$= \dfrac{x - 5}{(x - 1)^2(x + 1)}$

13. $\dfrac{x + 1}{x - 1} - 1 = \dfrac{x + 1}{x - 1} - \dfrac{x - 1}{x - 1} = \dfrac{x + 1 - (x - 1)}{x - 1} = \dfrac{2}{x - 1}$

15. $\dfrac{3}{a - 1} - \dfrac{2}{1 - a} = \dfrac{3}{a - 1} - \dfrac{-2}{-(1 - a)} = \dfrac{3}{a - 1} + \dfrac{2}{a - 1} = \dfrac{5}{a - 1}$

17. $\dfrac{2x}{x^2 - 16} - \dfrac{x - 4}{x^2 + 4x} = \dfrac{2x}{(x - 4)(x + 4)} - \dfrac{x - 4}{x(x + 4)}$ LCD $= x(x - 4)(x + 4)$

$= \dfrac{2x(x) - (x - 4)(x - 4)}{x(x - 4)(x + 4)}$

$= \dfrac{2x^2 - (x^2 - 8x + 16)}{x(x - 4)(x + 4)}$

$= \dfrac{x^2 + 8x - 16}{x(x - 4)(x + 4)}$

19. $\dfrac{x^2}{x^2 + 2x + 1} + \dfrac{x - 1}{3x + 3} - \dfrac{1}{6} = \dfrac{x^2}{(x + 1)^2} + \dfrac{x - 1}{3(x + 1)} - \dfrac{1}{6}$

LCD $= 6(x + 1)^2$

$= \dfrac{6x^2}{6(x + 1)^2} + \dfrac{2(x + 1)(x - 1)}{6(x + 1)^2} - \dfrac{(x + 1)^2}{6(x + 1)^2}$

$= \dfrac{6x^2 + 2(x^2 - 1) - (x^2 + 2x + 1)}{6(x + 1)^2}$

$= \dfrac{7x^2 - 2x - 3}{6(x + 1)^2}$

21. $\dfrac{2 - x}{2x + x^2} \cdot \dfrac{x^2 + 4x + 4}{x^2 - 4} = \dfrac{-\cancel{(x - 2)}}{x\cancel{(x + 2)}} \cdot \dfrac{\cancel{(x + 2)^2}}{\cancel{(x + 2)}\cancel{(x - 2)}} = -\dfrac{1}{x}$

23. $\frac{c+2}{5c-5} - \frac{c-2}{3c-3} + \frac{c}{1-c} = \frac{c+2}{5(c-1)} - \frac{c-2}{3(c-1)} - \frac{c}{c-1}$

LCD $= 15(c-1)$

$$= \frac{3(c+2)}{15(c-1)} - \frac{5(c-2)}{15(c-1)} - \frac{15c}{15(c-1)}$$

$$= \frac{3c + 6 - 5c + 10 - 15c}{15(c-1)} = \frac{-17c + 16}{15(c-1)}$$

25. $\frac{1 + \frac{3}{x}}{x - \frac{9}{x}} = \frac{\frac{x+3}{x}}{\frac{x^2-9}{x}} = \frac{x+3}{x} \cdot \frac{x}{x^2-9} = \frac{\cancel{x+3}}{\cancel{x}} \cdot \frac{\cancel{x}}{\cancel{(x+3)}(x-3)} = \frac{1}{x-3}$

27. $\frac{\frac{1}{2(x+h)} - \frac{1}{2x}}{h} = \left(\frac{1}{2(x+h)} - \frac{1}{2x}\right) \div \frac{h}{1}$

$$= \frac{x - x - h}{2x(x+h)} \cdot \frac{1}{h}$$

$$= \frac{-h}{2x(x+h)h} = \frac{-1}{2x(x+h)}$$

29. $\frac{\frac{x}{y} - 2 + \frac{y}{x}}{\frac{x}{y} - \frac{y}{x}} = \frac{\frac{x^2 - 2xy + y^2}{xy}}{\frac{x^2 - y^2}{xy}} = \frac{\overset{(x-y)}{\cancel{(x-y)^2}}}{\cancel{xy}} \cdot \frac{\cancel{xy}}{\cancel{(x-y)}(x+y)} = \frac{x-y}{x+y}$

31. (A) $\frac{x^2 + 4x + 3}{x + 3} = x + 4$: Incorrect

(B) $\frac{x^2 + 4x + 3}{x + 3} = \frac{\cancel{(x+3)}(x+1)}{\cancel{x+3}} = x + 1 \quad (x \neq -3)$

33. (A) $\frac{(x+h)^2 - x^2}{h} = 2x + 1$: Incorrect

(B) $\frac{(x+h)^2 - x^2}{h} = \frac{x^2 + 2xh + h^2 - x^2}{h} = \frac{2xh + h^2}{h} = \frac{\cancel{h}(2x+h)}{\cancel{h}} = 2x + h \ (h \neq 0)$

35. (A) $\frac{x^2 - 3x}{x^2 - 2x - 3} + x - 3 = 1$: Incorrect

(B) $\frac{x^2 - 3x}{x^2 - 2x - 3} + x - 3 = \frac{x\cancel{(x-3)}}{\cancel{(x-3)}(x+1)} + x - 3$

$$= \frac{x}{x+1} + x - 3$$

$$= \frac{x + (x-3)(x+1)}{x+1} = \frac{x^2 - x - 3}{x+1}$$

37. (A) $\frac{2x^2}{x^2 - 4} - \frac{x}{x-2} = \frac{x}{x+2}$: Correct

$\frac{2x^2}{x^2 - 4} - \frac{x}{x-2} = \frac{2x^2}{(x-2)(x+2)} - \frac{x}{x-2}$

$$= \frac{2x^2 - x(x + 2)}{(x - 2)(x + 2)}$$

$$= \frac{x^2 - 2x}{(x - 2)(x + 2)} = \frac{x\cancel{(x - 2)}}{\cancel{(x - 2)}(x + 2)} = \frac{x}{x + 2}$$

39. $$\frac{\frac{1}{3(x + h)^2} - \frac{1}{3x^2}}{h} = \left[\frac{1}{3(x + h)^2} - \frac{1}{3x^2}\right] \div \frac{h}{1}$$

$$= \frac{x^2 - (x + h)^2}{3x^2(x + h)^2} \cdot \frac{1}{h}$$

$$= \frac{x^2 - (x^2 + 2xh + h^2)}{3x^2(x + h)^2 h}$$

$$= \frac{-2xh - h^2}{3x^2(x + h)^2 h}$$

$$= \frac{-\cancel{h}(2x + h)}{3x^2(x + h)^2 \cancel{h}}$$

$$= -\frac{(2x + h)}{3x^2(x + h)^2} = \frac{-2x - h}{3x^2(x + h)^2}$$

41. $$1 - \frac{1}{1 - \frac{1}{1 - \frac{1}{x}}} = 1 - \frac{1}{1 - \frac{1}{\frac{x - 1}{x}}} = 1 - \frac{1}{1 - \frac{x}{x - 1}}$$

$$= 1 - \frac{1}{\frac{(x - 1) - x}{x - 1}} = 1 - \frac{1}{\frac{-1}{x - 1}} = 1 + x - 1 = x$$

EXERCISE A-6

Things to remember:

1. DEFINITION OF a^n, where n is an integer and a is a real number:

   a. For n a positive integer,
   $$a^n = a \cdot a \cdot \cdots \cdot a, \ n \text{ factors of } a.$$

   b. For $n = 0$,
   $$a^0 = 1, \ a \neq 0, \ 0^0 \text{ is not defined.}$$

   c. For n a negative integer,
   $$a^n = \frac{1}{a^{-n}}, \ a \neq 0.$$

   [Note: If n is negative, then $-n$ is positive.]

2. PROPERTIES OF EXPONENTS

GIVEN: n and m are integers and a and b are real numbers.

a. $a^m a^n = a^{m+n}$ $\qquad a^8 a^{-3} = a^{8+(-3)} = a^5$

b. $(a^n)^m = a^{mn}$ $\qquad (a^{-2})^3 = a^{3(-2)} = a^{-6}$

c. $(ab)^m = a^m b^m$ $\qquad (ab)^{-2} = a^{-2}b^{-2}$

d. $\left(\frac{a}{b}\right)^m = \frac{a^m}{b^m}$, $b \neq 0$ $\qquad \left(\frac{a}{b}\right)^5 = \frac{a^5}{b^5}$

e. $\frac{a^m}{a^n} = a^{m-n} = \frac{1}{a^{n-m}}$, $a \neq 0$ $\qquad \frac{a^{-3}}{a^7} = \frac{1}{a^{7-(-3)}} = \frac{1}{a^{10}}$

3. SCIENTIFIC NOTATION

Let r be any real number. Then r can be expressed as the product of a number between 1 and 10 and an integer power of 10, that is, r can be written

$$r = a \times 10^n,\ 1 \le a < 10,\ n \text{ an integer, } a \text{ in decimal form}$$

A number expressed in this form is said to be in scientific notation.

Examples:

$7 = 7 \times 10^0$ $\qquad 0.5 = 5 \times 10^{-1}$

$67 = 6.7 \times 10$ $\qquad 0.45 = 4.5 \times 10^{-1}$

$580 = 5.8 \times 10^2$ $\qquad 0.0032 = 3.2 \times 10^{-3}$

$43{,}000 = 4.3 \times 10^4$ $\qquad 0.000\,045 = 4.5 \times 10^{-5}$

---

1. $2x^{-9} = \frac{2}{x^9}$

3. $\frac{3}{2w^{-7}} = \frac{3w^7}{2}$

5. $2x^{-8}x^5 = 2x^{-8+5} = 2x^{-3} = \frac{2}{x^3}$

7. $\frac{w^{-8}}{w^{-3}} = \frac{1}{w^{-3+8}} = \frac{1}{w^5}$

9. $5v^8v^{-8} = 5v^{8-8} = 5v^0 = 5 \cdot 1 = 5$

11. $(a^{-3})^2 = a^{-6} = \frac{1}{a^6}$

13. $(x^6y^{-3})^{-2} = x^{-12}y^6 = \frac{y^6}{x^{12}}$

15. $82{,}300{,}000{,}000 = 8.23 \times 10^{10}$

17. $0.783 = 7.83 \times 10^{-1}$

19. $0.000\,034 = 3.4 \times 10^{-5}$

21. $4 \times 10^4 = 40{,}000$

23. $7 \times 10^{-3} = 0.007$

25. $6.171 \times 10^7 = 61{,}710{,}000$

27. $8.08 \times 10^{-4} = 0.000\,808$

29. $(22 + 31)^0 = (53)^0 = 1$

31. $\dfrac{10^{-3} \times 10^{4}}{10^{-11} \times 10^{-2}} = \dfrac{10^{-3+4}}{10^{-11-2}} = \dfrac{10^{1}}{10^{-13}} = 10^{1+13} = 10^{14}$

33. $(5x^2y^{-3})^{-2} = 5^{-2}x^{-4}y^{6} = \dfrac{y^6}{5^2x^4} = \dfrac{y^6}{25x^4}$

35. $\dfrac{8 \times 10^{-3}}{2 \times 10^{-5}} = \dfrac{8}{2} \times \dfrac{10^{-3}}{10^{-5}} = 4 \times 10^{-3+5} = 4 \times 10^{2}$

37. $\dfrac{8x^{-3}y^{-1}}{6x^2y^{-4}} = \dfrac{4y^{-1+4}}{3x^{2+3}} = \dfrac{4y^3}{3x^5}$

39. $\dfrac{7x^5 - x^2}{4x^5} = \dfrac{7x^5}{4x^5} - \dfrac{x^2}{4x^5} = \dfrac{7}{4} - \dfrac{1}{4x^3} = \dfrac{7}{4} - \dfrac{1}{4}x^{-3}$

41. $\dfrac{3x^4 - 4x^2 - 1}{4x^3} = \dfrac{3x^4}{4x^3} - \dfrac{4x^2}{4x^3} - \dfrac{1}{4x^3} = \dfrac{3}{4}x - x^{-1} - \dfrac{1}{4}x^{-3}$

43. $\dfrac{3x^2(x - 1)^2 - 2x^3(x - 1)}{(x - 1)^4} = \dfrac{x^2(x - 1)[3(x - 1) - 2x]}{(x - 1)^4} = \dfrac{x^2(x - 3)}{(x - 1)^3}$

45. $2x^{-2}(x - 1) - 2x^{-3}(x - 1)^2 = \dfrac{2(x - 1)}{x^2} - \dfrac{2(x - 1)^2}{x^3}$

$= \dfrac{2x(x - 1) - 2(x - 1)^2}{x^3}$

$= \dfrac{2(x - 1)[x - (x - 1)]}{x^3}$

$= \dfrac{2(x - 1)}{x^3}$

47. $\dfrac{9{,}600{,}000{,}000}{(1{,}600{,}000)(0.000\ 000\ 25)} = \dfrac{9.6 \times 10^9}{(1.6 \times 10^6)(2.5 \times 10^{-7})} = \dfrac{9.6 \times 10^9}{1.6(2.5) \times 10^{6-7}}$

$= \dfrac{9.6 \times 10^9}{4.0 \times 10^{-1}} = 2.4 \times 10^{9+1} = 2.4 \times 10^{10}$

$= 24{,}000{,}000{,}000$

49. $\dfrac{(1{,}250{,}000)(0.000\ 38)}{0.0152} = \dfrac{(1.25 \times 10^6)(3.8 \times 10^{-4})}{1.52 \times 10^{-2}} = \dfrac{1.25(3.8) \times 10^{6-4}}{1.52 \times 10^{-2}}$

$= 3.125 \times 10^4 = 31{,}250$

51. 2^{3^2}; 64
But, $2^{3^2} = 2^9 = 512$

53. $a^m a^0 = a^{m+0} = a^m$
Therefore, $a^m a^0 = a^m$ which implies $a^0 = 1$.

55. $\dfrac{u + v}{u^{-1} + v^{-1}} = \dfrac{u + v}{\frac{1}{u} + \frac{1}{v}} = \dfrac{u + v}{\frac{v + u}{uv}} = (u + v) \cdot \dfrac{uv}{v + u} = uv$

57. $\dfrac{b^{-2} - c^{-2}}{b^{-3} - c^{-3}} = \dfrac{\frac{1}{b^2} - \frac{1}{c^2}}{\frac{1}{b^3} - \frac{1}{c^3}} = \dfrac{\frac{c^2 - b^2}{b^2c^2}}{\frac{c^3 - b^3}{b^3c^3}} = \dfrac{\cancel{(c - b)}(c + b)}{\cancel{b^2c^2}} \cdot \dfrac{\overset{bc}{\cancel{b^3c^3}}}{\cancel{(c - b)}(c^2 + cb + b^2)}$

$= \dfrac{bc(c + b)}{c^2 + cb + b^2}$

59. (A) $213{,}701{,}000{,}000 = 2.13701 \times 10^{11}$

(B) $95{,}862{,}000{,}000 = 9.5862 \times 10^{10}$

$\dfrac{2.13701 \times 10^{11}}{9.5682 \times 10^{10}} = \dfrac{2.13701}{9.5862} \times 10 \approx 2.2293$

(C) $\dfrac{9.5862 \times 10^{10}}{2.13701 \times 10^{11}} = \dfrac{9.5862}{2.13701} \times 10^{-1} \approx 0.4486$

61. (A) $\dfrac{4{,}065{,}000{,}000{,}000}{255{,}100{,}000} = \dfrac{4.065 \times 10^{12}}{2.551 \times 10^{8}} \approx 1.5935 \times 10^{4} = \$15{,}935$

(B) $\dfrac{292{,}300{,}000{,}000}{255{,}100{,}000} = \dfrac{2.923 \times 10^{11}}{2.551 \times 10^{8}} \approx 1.146 \times 10^{3} = \$1{,}146$

(C) $\dfrac{292{,}300{,}000{,}000}{4{,}065{,}000{,}000{,}000} = \dfrac{2.923 \times 10^{11}}{4.065 \times 10^{12}} \approx 0.719 \times 10^{-1} = 0.0719$ or 7.19%

63. (A) $9 \text{ ppm} = \dfrac{9}{1{,}000{,}000} = \dfrac{9}{10^6} = 9 \times 10^{-6}$ (B) $0.000\,009$ (C) 0.0009%

65. $\dfrac{757.5}{100{,}000} \times 255{,}100{,}000 = \dfrac{7.575 \times 10^2}{10^5} \cdot 2.551 \times 10^8$

$= \dfrac{19.323825 \times 10^{10}}{10^5}$

$= 19.323825 \times 10^5$

$= 1{,}932{,}382.5$

To the nearest thousand, there were 1,932,000 violent crimes committed in 1992.

Things to remember:

1. *n*th ROOT

Let b be a real number. For any natural number n,

r is an nth ROOT of b if $r^n = b$

If n is odd, then b has exactly one real nth root.
If n is even, and $b < 0$, then b has NO real nth roots.
If n is even, and $b > 0$, then b has two real nth roots; if r is an nth root, then $-r$ is also an nth root.
0 is an nth root of 0 for all n

2. NOTATION

Let b be a real number and let $n > 1$ be a natural number. If n is odd, then the nth root of b is denoted

$b^{1/n}$ or $\sqrt[n]{b}$

If n is even and $b > 0$, then the PRINCIPAL nth ROOT OF b is the positive nth root; the principal nth root is denoted

$b^{1/n}$ or $\sqrt[n]{b}$

In the $\sqrt[n]{b}$ notation, the symbol $\sqrt{\ }$ is called a RADICAL, n is the INDEX of the radical and b is called the RADICAND.

3. RATIONAL EXPONENTS

If m and n are natural numbers without common prime factors, b is a real number, and b is nonnegative when b is even, then

$$b^{m/n} = \begin{cases} (b^{1/n})^m = (\sqrt[n]{b})^m \\ (b^m)^{1/n} = \sqrt[n]{b^m} \end{cases}$$

and $b^{-m/n} = \dfrac{1}{b^{m/n}}$, $b \neq 0$

The two definitions of $b^{m/n}$ are equivalent under the indicated restrictions on m, n, and b.

4. PROPERTIES OF RADICALS

If m and n are natural numbers greater than 1 and x and y are positive real numbers, then

a. $\sqrt[n]{x^n} = x$ $\qquad \sqrt[3]{x^3} = x$

b. $\sqrt[n]{xy} = \sqrt[n]{x}\sqrt[n]{y}$ $\qquad \sqrt[5]{xy} = \sqrt[5]{x}\sqrt[5]{y}$

c. $\sqrt[n]{\dfrac{x}{y}} = \dfrac{\sqrt[n]{x}}{\sqrt[n]{y}}$ $\qquad \sqrt[4]{\dfrac{x}{y}} = \dfrac{\sqrt[4]{x}}{\sqrt[4]{y}}$

1. $6x^{3/5} = 6\sqrt[5]{x^3}$

3. $(4xy^3)^{2/5} = \sqrt[5]{(4xy^3)^2}$

5. $(x^2 + y^2)^{1/2} = \sqrt{x^2 + y^2}$
[Note: $\sqrt{x^2 + y^2} \neq x + y$.]

7. $5\sqrt[4]{x^3} = 5x^{3/4}$

9. $\sqrt[5]{(2x^2y)^3} = (2x^2y)^{3/5}$

11. $\sqrt[3]{x} + \sqrt[3]{y} = x^{1/3} + y^{1/3}$

13. $25^{1/2} = (5^2)^{1/2} = 5$

15. $16^{3/2} = (4^2)^{3/2} = 4^3 = 64$

17. $-36^{1/2} = -(6^2)^{1/2} = -6$

19. $(-36)^{1/2}$ is not a rational number
-36 does not have a real square root; $(-36)^{1/2}$ is not a real number.

21. $\left(\frac{4}{25}\right)^{3/2} = \left(\left(\frac{2}{5}\right)^2\right)^{3/2} = \left(\frac{2}{5}\right)^3 = \frac{2^3}{5^3} = \frac{8}{125}$

23. $9^{-3/2} = (3^2)^{-3/2} = 3^{-3} = \frac{1}{3^3} = \frac{1}{27}$

25. $x^{4/5}x^{-2/5} = x^{4/5-2/5} = x^{2/5}$

27. $\frac{m^{2/3}}{m^{-1/3}} = m^{2/3-(-1/3)} = m^1 = m$

29. $(8x^3y^{-6})^{1/3} = (2^3x^3y^{-6})^{1/3} = 2^{3/3}x^{3/3}y^{-6/3} = 2xy^{-2} = \frac{2x}{y^2}$

31. $\left(\frac{4x^{-2}}{y^4}\right)^{-1/2} = \left(\frac{2^2x^{-2}}{y^4}\right)^{-1/2} = \frac{2^{2(-1/2)}x^{-2(-1/2)}}{y^{4(-1/2)}} = \frac{2^{-1}x^1}{y^{-2}} = \frac{xy^2}{2}$

33. $\frac{8x^{-1/3}}{12x^{1/4}} = \frac{2}{3x^{1/4+1/3}} = \frac{2}{3x^{7/12}}$

35. $\sqrt[5]{(2x + 3)^5} = [(2x + 3)^5]^{1/5} = 2x + 3$

37. $\sqrt{18x^3}\sqrt{2x^3} = \sqrt{36x^6} = (6^2x^6)^{1/2} = (6^2)^{1/2}(6^2)^{1/2}(x^6)^{1/2} = 6x^3$

39. $\frac{\sqrt{6x}\sqrt{10}}{\sqrt{15x}} = \sqrt{\frac{60x}{15x}} = \sqrt{4} = 2$

41. $3x^{3/4}(4x^{1/4} - 2x^8) = 12x^{3/4+1/4} - 6x^{3/4+8}$
$= 12x - 6x^{3/4+32/4} = 12x - 6x^{35/4}$

43. $(3u^{1/2} - v^{1/2})(u^{1/2} - 4v^{1/2}) = 3u - 12u^{1/2}v^{1/2} - u^{1/2}v^{1/2} + 4v$
$= 3u - 13u^{1/2}v^{1/2} + 4v$

45. $(5m^{1/2} + n^{1/2})(5m^{1/2} - n^{1/2}) = (5m^{1/2})^2 - (n^{1/2})^2 = 25m - n$

47. $(3x^{1/2} - y^{1/2})^2 = (3x^{1/2})^2 - 6x^{1/2}y^{1/2} + (y^{1/2})^2 = 9x - 6x^{1/2}y^{1/2} + y$

49. $\frac{\sqrt[3]{x^2} + 2}{2\sqrt[3]{x}} = \frac{x^{2/3} + 2}{2x^{1/3}} = \frac{x^{2/3}}{2x^{1/3}} + \frac{2}{2x^{1/3}} = \frac{1}{2}x^{1/3} + \frac{1}{x^{1/3}} = \frac{1}{2}x^{1/3} + x^{-1/3}$

51. $\dfrac{2\sqrt[4]{x^3} + 3\sqrt[3]{x}}{3x} = \dfrac{2x^{3/4} + 3x^{1/3}}{3x} = \dfrac{2x^{3/4}}{3x} + \dfrac{3x^{1/3}}{3x}$

$= \dfrac{2}{3}x^{3/4-1} + x^{1/3-1} = \dfrac{2}{3}x^{-1/4} + x^{-2/3}$

53. $\dfrac{2\sqrt[3]{x} - \sqrt{x}}{4\sqrt{x}} = \dfrac{2x^{1/3} - x^{1/2}}{4x^{1/2}} = \dfrac{2x^{1/3}}{4x^{1/2}} - \dfrac{x^{1/2}}{4x^{1/2}} = \dfrac{1}{2}x^{1/3-1/2} - \dfrac{1}{4} = \dfrac{1}{2}x^{-1/6} - \dfrac{1}{4}$

55. $\dfrac{12mn^2}{\sqrt{3mn}} = \dfrac{12mn^2}{\sqrt{3mn}} \cdot \dfrac{\sqrt{3mn}}{\sqrt{3mn}} = \dfrac{12mn^2\sqrt{3mn}}{3mn} = 4n\sqrt{3mn}$

57. $\dfrac{2}{\sqrt{x-2}} = \dfrac{2}{\sqrt{x-2}} \cdot \dfrac{\sqrt{x-2}}{\sqrt{x-2}} = \dfrac{2\sqrt{x-2}}{x-2}$

59. $\dfrac{7(x-y)^2}{\sqrt{x} - \sqrt{y}} = \dfrac{7(x-y)^2}{\sqrt{x} - \sqrt{y}} \cdot \dfrac{\sqrt{x} + \sqrt{y}}{\sqrt{x} + \sqrt{y}} = \dfrac{7(x-y)^2(\sqrt{x} + \sqrt{y})}{x - y}$

$= 7(x - y)(\sqrt{x} + \sqrt{y})$

61. $\dfrac{\sqrt{5xy}}{5x^2y^2} = \dfrac{\sqrt{5xy}}{5x^2y^2} \cdot \dfrac{\sqrt{5xy}}{\sqrt{5xy}} = \dfrac{5xy}{5x^2y^2\sqrt{5xy}} = \dfrac{1}{xy\sqrt{5xy}}$

63. $\dfrac{\sqrt{x+h} - \sqrt{x}}{h} = \dfrac{\sqrt{x+h} - \sqrt{x}}{h} \cdot \dfrac{\sqrt{x+h} + \sqrt{x}}{\sqrt{x+h} + \sqrt{x}}$

$= \dfrac{x + h - x}{h(\sqrt{x+h} + \sqrt{x})} = \dfrac{h}{h(\sqrt{x+h} + \sqrt{x})}$

$= \dfrac{1}{\sqrt{x+h} + \sqrt{x}}$

65. $\dfrac{\sqrt{t} - \sqrt{x}}{t - x} = \dfrac{\sqrt{t} - \sqrt{x}}{t - x} \cdot \dfrac{\sqrt{t} + \sqrt{x}}{\sqrt{t} + \sqrt{x}} = \dfrac{t - x}{(t - x)(\sqrt{t} + \sqrt{x})} = \dfrac{1}{\sqrt{t} + \sqrt{x}}$

67. $(x + y)^{1/2} \stackrel{?}{=} x^{1/2} + y^{1/2}$

Let $x = y = 1$. Then

$(1 + 1)^{1/2} = 2^{1/2} = \sqrt{2} \approx 1.414$

$1^{1/2} + 1^{1/2} = \sqrt{1} + \sqrt{1} = 1 + 1 = 2;\ \sqrt{2} \neq 2$

69. $(x + y)^{1/3} \stackrel{?}{=} \dfrac{1}{(x + y)^3}$

Let $x = y = 4$. Then

$(4 + 4)^{1/3} = 8^{1/3} = \sqrt[3]{8} = 2$

$\dfrac{1}{(4 + 4)^3} = \dfrac{1}{8^3} = \dfrac{1}{512}$

71. $-\dfrac{1}{2}(x - 2)(x + 3)^{-3/2} + (x + 3)^{-1/2} = \dfrac{-(x - 2)}{2(x + 3)^{3/2}} + \dfrac{1}{(x + 3)^{1/2}}$

$= \dfrac{-x + 2 + 2(x + 3)}{2(x + 3)^{3/2}}$

$= \dfrac{x + 8}{2(x + 3)^{3/2}}$

73. $$\frac{(x-1)^{1/2} - x\left(\frac{1}{2}\right)(x-1)^{-1/2}}{x-1} = \frac{(x-1)^{1/2} - \frac{x}{2(x-1)^{1/2}}}{x-1}$$

$$= \frac{\frac{2(x-1)^{1/2}(x-1)^{1/2}}{2(x-1)^{1/2}} - \frac{x}{2(x-1)^{1/2}}}{x-1}$$

$$= \frac{\frac{2(x-1) - x}{2(x-1)^{1/2}}}{x-1} = \frac{x-2}{2(x-1)^{3/2}}$$

75. $$\frac{(x+2)^{2/3} - x\left(\frac{2}{3}\right)(x+2)^{-1/3}}{(x+2)^{4/3}} = \frac{(x+2)^{2/3} - \frac{2x}{3(x+2)^{1/3}}}{(x+2)^{4/3}}$$

$$= \frac{\frac{3(x+2)^{1/3}(x+2)^{2/3}}{3(x+2)^{1/3}} - \frac{2x}{3(x+2)^{1/3}}}{(x+2)^{4/3}}$$

$$= \frac{\frac{3(x+2) - 2x}{3(x+2)^{1/3}}}{(x+2)^{4/3}} = \frac{x+6}{3(x+2)^{5/3}}$$

77. $22^{3/2} = 22^{1.5} \approx 103.2$ or $22^{3/2} = \sqrt{(22)^3} = \sqrt{10,648} \approx 103.2$

79. $827^{-3/8} = \frac{1}{827^{3/8}} = \frac{1}{827^{0.375}} \approx \frac{1}{12.42} \approx 0.0805$

81. $37.09^{7/3} \approx 37.09^{2.3333} \approx 4,588$

83. (A) $\sqrt{3} + \sqrt{5} \approx 1.732 + 2.236 = 3.968$

(B) $\sqrt{2+\sqrt{3}} + \sqrt{2-\sqrt{3}} \approx 2.449$

(C) $1 + \sqrt{3} \approx 2.732$

(D) $\sqrt[3]{10 + 6\sqrt{3}} \approx 2.732$

(E) $\sqrt{8 + \sqrt{60}} \approx 3.968$

(F) $\sqrt{6} \approx 2.449$

(A) and (E) have the same value:

$$(\sqrt{3} + \sqrt{5})^2 = 3 + 2\sqrt{3}\sqrt{5} + 5 = 8 + 2\sqrt{15}$$

$$[\sqrt{8 + \sqrt{60}}\,]^2 = 8 + \sqrt{4 \cdot 15} = 8 + 2\sqrt{15}$$

(B) and (F) have the same value.

$$(\sqrt{2+\sqrt{3}} + \sqrt{2-\sqrt{3}})^2 = 2 + \sqrt{3} + 2\sqrt{2+\sqrt{3}}\sqrt{2-\sqrt{3}} + 2 - \sqrt{3}$$

$$= 4 + 2\sqrt{4-3} = 4 + 2 = 6$$

$(\sqrt{6})^2 = 6.$

(C) and (D) have the same value.

$$\begin{aligned}(1 + \sqrt{3})^3 &= (1 + \sqrt{3})^2(1 + \sqrt{3}) \\ &= (1 + 2\sqrt{3} + 3)(1 + \sqrt{3}) \\ &= (4 + 2\sqrt{3})(1 + \sqrt{3}) \\ &= 4 + 6\sqrt{3} + 6 = 10 + 6\sqrt{3}\end{aligned}$$

$$\left(\sqrt[3]{10 + 6\sqrt{3}}\right)^3 = 10 + 6\sqrt{3}$$

EXERCISE A-8

Things to remember:

1. FIRST DEGREE, OR LINEAR, EQUATIONS AND INEQUALITIES

   A FIRST DEGREE, or LINEAR, EQUATION in one variable x is an equation that can be written in the form

   $ax + b = 0,$ STANDARD FORM

   where a and b are constants and $a \neq 0$. If the equality symbol = is replaced by $<$, $>$, $\leq$, or $\geq$, then the resulting expression is called a FIRST DEGREE, or LINEAR, INEQUALITY.

2. SOLUTIONS

   A SOLUTION OF AN EQUATION (or inequality) involving a single variable is a number that when substituted for the variable makes the equation (or inequality) true. The set of all solutions is called the SOLUTION SET. To SOLVE AN EQUATION (or inequality) we mean that we determine the solution set. Two equations (or inequalities) are EQUIVALENT if they have the same solution set.

3. EQUALITY PROPERTIES

   An equivalent equation will result if:

   a) The same quantity is added to or subtracted from each side of a given equation.

   b) Each side of a given equation is multiplied by or divided by the same nonzero quantity.

4. INEQUALITY PROPERTIES

   An equivalent inequality will result and the SENSE WILL REMAIN THE SAME if each side of the original inequality:

   a) Has the same real number added to or subtracted from it.

   b) Is multiplied or divided by the same positive number.

   An equivalent inequality will result and the SENSE WILL REVERSE if each side of the original inequality:

   c) Is multiplied or divided by the same negative number.

   NOTE: Multiplication and division by 0 is not permitted.

5. The double inequality $a \le x \le b$ means that $a \le x$ and $x \le b$. Other variations, as well as a useful interval notation, are indicated in the following table.

Interval Notation	Inequality Notation	Line Graph
$[a, b]$	$a \le x \le b$	
$[a, b)$	$a \le x < b$	
$(a, b]$	$a < x \le b$	
(a, b)	$a < x < b$	
$(-\infty, a]$	$x \le a$	
$(-\infty, a)$	$x < a$	
$[b, \infty)$	$x \ge b$	
(b, ∞)	$x > b$	

[Note: An endpoint on a line graph has a square bracket through it if it is included in the inequality and a parenthesis through it if it is not.]

1.
$$\begin{aligned} 2m + 9 &= 5m - 6 \\ 2m + 9 - 9 &= 5m - 6 - 9 && \text{[using 3(a)]} \\ 2m &= 5m - 15 \\ 2m - 5m &= 5m - 15 - 5m && \text{[using 3(a)]} \\ -3m &= -15 \\ \frac{-3m}{-3} &= \frac{-15}{-3} && \text{[using 3(b)]} \\ m &= 5 \end{aligned}$$

3.
$$\begin{aligned} x + 5 &< -4 \\ x + 5 - 5 &< -4 - 5 && \text{[using 4(a)]} \\ x &< -9 \end{aligned}$$

5.
$$\begin{aligned} -3x &\ge -12 \\ \frac{-3x}{-3} &\le \frac{-12}{-3} && \text{[using 4(c)]} \\ x &\le 4 \end{aligned}$$

7.
$$\begin{aligned} -4x - 7 &> 5 \\ -4x &> 5 + 7 \\ -4x &> 12 \\ x &< -3 \end{aligned}$$

Graph of $x < -3$ is:

9.
$$\begin{aligned} 2 \le x &+ 3 \le 5 \\ 2 - 3 \le x &\le 5 - 3 \\ -1 \le x &\le 2 \end{aligned}$$

Graph of $-1 \le x \le 2$ is:

11. $\frac{y}{7} - 1 = \frac{1}{7}$

Multiply both sides of the equation by 7. We obtain:

$y - 7 = 1$ [using 3(b)]

$y = 8$

13. $\frac{x}{3} > -2$

Multiply both sides of the inequality by 3. We obtain:

$x > -6$ [using 4(b)]

15. $\frac{y}{3} = 4 - \frac{y}{6}$

Multiply both sides of the equation by 6. We obtain:

$2y = 24 - y$

$3y = 24$

$y = 8$

17. $10x + 25(x - 3) = 275$

$10x + 25x - 75 = 275$

$35x = 275 + 75$

$35x = 350$

$x = \frac{350}{35}$

$x = 10$

19. $3 - y \le 4(y - 3)$

$3 - y \le 4y - 12$

$-5y \le -15$

$y \ge 3$

[Note: Division by a negative number, -3.]

21. $\frac{x}{5} - \frac{x}{6} = \frac{6}{5}$

Multiply both sides of the equation by 30. We obtain:

$6x - 5x = 36$

$x = 36$

23. $\frac{m}{5} - 3 < \frac{3}{5} - m$

Multiply both sides of the inequality by 5. We obtain:

$m - 15 < 3 - 5m$

$6m < 18$

$m < 3$

25. $0.1(x - 7) + 0.05x = 0.8$

$0.1x - 0.7 + 0.05x = 0.8$

$0.15x = 1.5$

$x = \frac{1.5}{0.15}$

$x = 10$

27. $2 \le 3x - 7 < 14$

$7 + 2 \le 3x < 14 + 7$

$9 \le 3x < 21$

$3 \le x < 7$

Graph of $3 \le x < 7$ is:

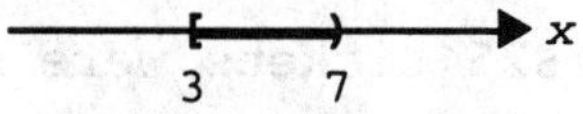

29. $-4 \le \frac{9}{5}C + 32 \le 68$

$-36 \le \frac{9}{5}C \le 36$

$-36\left(\frac{5}{9}\right) \le C \le 36\left(\frac{5}{9}\right)$

$-20 \le C \le 20$

Graph of $-20 \le C \le 20$ is:

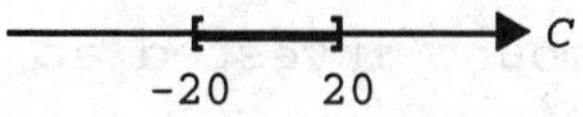

31. $3x - 4y = 12$

$3x = 12 + 4y$

$3x - 12 = 4y$

$y = \frac{1}{4}(3x - 12)$

$y = \frac{3}{4}x - 3$

33. $Ax + By = C$

$By = C - Ax$

$y = \frac{C}{B} - \frac{Ax}{B}, \ B \ne 0$

or $y = -\left(\frac{A}{B}\right)x + \frac{C}{B}$

35.
$$F = \frac{9}{5}C + 32$$
$$\frac{9}{5}C + 32 = F$$
$$\frac{9}{5}C = F - 32$$
$$C = \frac{5}{9}(F - 32)$$

37.
$$A = Bm - Bn$$
$$A = B(m - n)$$
$$B = \frac{A}{m - n}$$

39.
$$-3 \le 4 - 7x < 18$$
$$-3 - 4 \le -7x < 18 - 4$$
$$-7 \le -7x < 14.$$

Dividing by -7, and recalling <u>4</u>(c), we have

$1 \ge x > -2$ or $-2 < x \le 1$

The graph is: 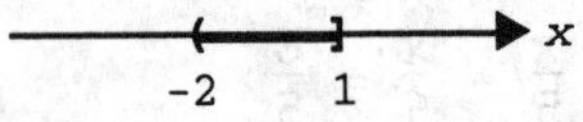

41. (A) $ab > 0$; $a > 0$ <u>and</u> $b > 0$, or $a < 0$ <u>and</u> $b < 0$

(B) $ab < 0$; $a > 0$ <u>and</u> $b < 0$, or $a < 0$ <u>and</u> $b > 0$

(C) $\frac{a}{b} > 0$; $a > 0$ <u>and</u> $b > 0$, or $a < 0$ <u>and</u> $b < 0$

(D) $\frac{a}{b} < 0$; $a > 0$ <u>and</u> $b < 0$, or $a < 0$ <u>and</u> $b > 0$

43. (A) If $a - b = 2$, then $a > b$.

(B) If $c - d = -1$, then $c < d$ ($d - c = 1$ so $d > c$).

45. Let $a, b > 0$. If $\frac{b}{a} > 1$, then $b > a$ so $a - b < 0$; $a - b$ is negative.

47. Let x = number of \$15 tickets. Then the number of \$25 tickets = $8000 - x$. Now,
$$15x + 25(8000 - x) = 165{,}000$$
$$15x + 200{,}000 - 25x = 165{,}000$$
$$-10x = -35{,}000$$
$$x = 3{,}500$$

Thus, 3,500 \$15 tickets and 8,000 - 3,500 = 4,500 \$25 tickets were sold.

49. Let x = the amount invested at 10%. Then $12{,}000 - x$ is the amount invested at 15%.

Required total yield = 12% of \$12,000 = $0.12\cdot 12{,}000$ = \$1,440. Thus,
$$0.10x + 0.15(12{,}000 - x) = 0.12\cdot 12{,}000$$
$$10x + 15(12{,}000 - x) = 12\cdot 12{,}000 \quad \text{(multiply both sides by 100)}$$
$$10x + 180{,}000 - 15x = 144{,}000$$
$$-5x = -36{,}000$$
$$x = \$7{,}200$$
Thus, we get \$7,200 invested at 10% and 12,000 - 7,200 = \$4,800 invested at 15%.

51. Let x be the price of the car in 1992. Then

$$\frac{x}{5,000} = \frac{140.3}{38.8} \quad \text{(refer to Table 2, Example 9)}$$

$$x = 5,000 \cdot \frac{140.3}{38.8} = 18,079.90$$

To the nearest dollar, the car would sell for $18,080.

53. Let x = number of books produced. Then

Costs: $C = 1.60x + 55,000$

Revenue: $R = 11x$

To find the break-even point, set $R = C$:

$$11x = 1.60x + 55,000$$
$$9.40x = 55,000$$
$$x = 5851.06383$$

Thus, 5851 books will have to be sold for the publisher to break even.

55. Let x = number of books produced.

Costs: $C(x) = 55,000 + 2.10x$

Revenue: $R(x) = 11x$

(A) The obvious strategy is to raise the sales price of the book.

(B) To find the break-even point, set $R(x) = C(x)$:

$$11x = 55,000 + 2.10x$$
$$8.90x = 55,000$$
$$x = 6179.78$$

The company must sell more than 6180 books.

(C) From Problem 53, the production level at the break-even point is: 5,851 books. At this production level, the costs are

$$C(5,851) = 55,000 + 2.10(5,851) = \$67,287.10$$

If p is the new price of the book, then we need

$$5851p = 67,287.10$$

and $p \approx 11.50$

The company should increase the price at least $0.50 (50 cents).

57. Let x = the number of rainbow trout in the lake. Then,

$$\frac{x}{200} = \frac{200}{8} \quad \text{(since proportions are the same)}$$

$$x = \frac{200}{8}(200)$$

$$x = 5,000$$

59. $\text{IQ} = \frac{\text{Mental age}}{\text{Chronological age}}(100)$

$$\frac{\text{Mental age}}{9}(100) = 140$$

$$\text{Mental age} = \frac{140}{100}(9)$$
$$= 12.6 \text{ years}$$

EXERCISE A-9

Things to remember:

1. A quadratic equation in one variable is an equation of the form

   (A) $ax^2 + bx + c = 0$,

   where x is a variable and a, b, and c are constants, $a \neq 0$.

2. Quadratic equations of the form $ax^2 + c = 0$ can be solved by the SQUARE ROOT METHOD. The solutions are:

   $x = \pm\sqrt{\frac{-c}{a}}$ provided $\frac{-c}{a} \geq 0$;

   otherwise, the equation has no real solutions.

3. If the left side of the quadratic equation (A) can be FACTORED,

   $ax^2 + bx + c = (px + q)(rx + s)$,

   then the solutions of (A) are

   $x = \frac{-q}{p}$ or $x = \frac{-s}{r}$.

4. The solutions of (A) are given by the QUADRATIC FORMULA:

   $$x = \frac{-b \pm \sqrt{b^2 - 4ac}}{2a}$$

   The quantity $b^2 - 4ac$ under the radical is called the DISCRIMINANT and:

   (i) (A) has two real solutions if $b^2 - 4ac > 0$;
   (ii) (A) has one real solution if $b^2 - 4ac = 0$;
   (iii) (A) has no real solution if $b^2 - 4ac < 0$.

5. FACTORABILITY THEOREM

   The second-degree polynomial, $ax^2 + bx + c$, with integer coefficients, can be expressed as the product of two first-degree polynomials with integer coefficients if and only if $\sqrt{b^2 - 4ac}$ is an integer.

6. FACTOR THEOREM

   If r_1 and r_2 are solutions of $ax^2 + bx + c = 0$, then $ax^2 + bx + c = a(x - r_1)(x - r_2)$.

---

1. $2x^2 - 22 = 0$
$x^2 - 11 = 0$
$x^2 = 11$
$x = \pm\sqrt{11}$

3. $(x - 1)^2 = 4$
$x - 1 = \pm\sqrt{4} = \pm 2$
$x = 1 \pm 2 = -1$ or 3

5. $2u^2 - 8u - 24 = 0$
$u^2 - 4u - 12 = 0$
$(u - 6)(u + 2) = 0$
$u - 6 = 0$ or $u + 2 = 0$
$u = 6$ or $u = -2$

7. $x^2 = 2x$
$x^2 - 2x = 0$
$x(x - 2) = 0$
$x = 0$ or $x - 2 = 0$
$x = 2$

9. $x^2 - 6x - 3 = 0$

$$x = \frac{-b \pm \sqrt{b^2 - 4ac}}{2a}, \quad a = 1,\ b = -6,\ c = -3$$

$$= \frac{-(-6) \pm \sqrt{(-6)^2 - 4(1)(-3)}}{2(1)}$$

$$= \frac{6 \pm \sqrt{48}}{2} = \frac{6 \pm 4\sqrt{3}}{2} = 3 \pm 2\sqrt{3}$$

11. $3u^2 + 12u + 6 = 0$

Since 3 is a factor of each coefficient, divide both sides by 3.

$u^2 + 4u + 2 = 0$

$$u = \frac{-b \pm \sqrt{b^2 - 4ac}}{2a}, \quad a = 1,\ b = 4,\ c = 2$$

$$= \frac{-4 \pm \sqrt{4^2 - 4(1)(2)}}{2(1)} = \frac{-4 \pm \sqrt{8}}{2} = \frac{-4 \pm 2\sqrt{2}}{2} = -2 \pm \sqrt{2}$$

13. $2x^2 = 4x$

$x^2 = 2x$ (divide both sides by 2)

$x^2 - 2x = 0$ (solve by factoring)

$x(x - 2) = 0$

$x = 0$ or $x - 2 = 0$

$x = 2$

15. $4u^2 - 9 = 0$ (solve by square root method)

$4u^2 = 9$

$u^2 = \frac{9}{4}$

$u = \pm\sqrt{\frac{9}{4}} = \pm\frac{3}{2}$

17. $8x^2 + 20x = 12$

$8x^2 + 20x - 12 = 0$

$2x^2 + 5x - 3 = 0$

$(x + 3)(2x - 1) = 0$

$x + 3 = 0$ or $2x - 1 = 0$

$x = -3$ or $2x = 1$

$x = \frac{1}{2}$

19. $x^2 = 1 - x$

$x^2 + x - 1 = 0$

$$x = \frac{-b \pm \sqrt{b^2 - 4ac}}{2a}, \quad a = 1,\ b = 1,\ c = -1$$

$$= \frac{-1 \pm \sqrt{(1)^2 - 4(1)(-1)}}{2(1)} = \frac{-1 \pm \sqrt{5}}{2}$$

21. $2x^2 = 6x - 3$

$2x^2 - 6x + 3 = 0$

$$x = \frac{-b \pm \sqrt{b^2 - 4ac}}{2a}, \quad a = 2,\ b = -6,\ c = 3$$

$$= \frac{-(-6) \pm \sqrt{(-6)^2 - 4(2)(3)}}{2(2)} = \frac{6 \pm \sqrt{12}}{4} = \frac{6 \pm 2\sqrt{3}}{4} = \frac{3 \pm \sqrt{3}}{2}$$

23. $y^2 - 4y = -8$

$y^2 - 4y + 8 = 0$

$$y = \frac{-b \pm \sqrt{b^2 - 4ac}}{2a}, \; a, = 1, \; b = -4, \; c = 8$$

$$= \frac{-(-4) \pm \sqrt{(-4)^2 - 4(1)(8)}}{2(1)} = \frac{4 \pm \sqrt{-16}}{2}$$

Since $\sqrt{-16}$ is not a real number, there are no real solutions.

25. $(x + 4)^2 = 11$

$x + 4 = \pm\sqrt{11}$

$x = -4 \pm \sqrt{11}$

27. $\frac{3}{p} = p$

$p^2 = 3$

$p = \pm\sqrt{3}$

29. $2 - \frac{2}{m^2} = \frac{3}{m}$

$2m^2 - 2 = 3m$

$2m^2 - 3m - 2 = 0$

$(2m + 1)(m - 2) = 0$

$m = -\frac{1}{2}, 2$

31. $x^2 + 40x - 84$

Step 1. Test for factorability

$\sqrt{b^2 - 4ac} = \sqrt{(40)^2 - 4(1)(-84)} = \sqrt{1936} = 44$

Since the result is an integer, the polynomial has first-degree factors with integer coefficients.

Step 2. Use the factor theorem

$x^2 + 40x - 84 = 0$

$x = \frac{-40 \pm 44}{2} = 2, -42$ (by the quadratic formula)

Thus, $x^2 + 40x - 84 = (x - 2)(x - [-42]) = (x - 2)(x + 42)$

33. $x^2 - 32x + 144$

Step 1. Test for factorability

$\sqrt{b^2 - 4ac} = \sqrt{(-32)^2 - 4(1)(144)} = \sqrt{448} \approx 21.166$

Since this is not an integer, the polynomial is not factorable.

35. $2x^2 + 15x - 108$

Step 1. Test for factorability

$\sqrt{b^2 - 4ac} = \sqrt{(15)^2 - 4(2)(-108)} = \sqrt{1089} = 33$

Thus, the polynomial has first-degree factors with integer coefficients.

Step 2. Use the factor theorem

$2x^2 + 15x - 108$

$x = \frac{-15 \pm 33}{4} = \frac{9}{2}, -12$

Thus, $2x^2 + 15x - 108 = 2\left(x - \frac{9}{2}\right)(x - [-12]) = (2x - 9)(x + 12)$

37. $4x^2 + 241x - 434$

Step 1. Test for factorability

$\sqrt{b^2 - 4ac} = \sqrt{(241)^2 - 4(4)(-434)} = \sqrt{65025} = 255$

Thus, the polynomial has first-degree factors with integer coefficients.

Step 2. Use the factor theorem

$4x^2 + 241x - 434$

$$x = \frac{-241 \pm 255}{8} = \frac{14}{8}, -\frac{496}{8} \text{ or } \frac{7}{4}, -62$$

Thus, $4x^2 + 241x - 434 = 4\left(x - \frac{7}{4}\right)(x + 62) = (4x - 7)(x + 62)$

39. $A = P(1 + r)^2$

$(1 + r)^2 = \frac{A}{P}$

$1 + r = \sqrt{\frac{A}{P}}$

$r = \sqrt{\frac{A}{P}} - 1$

41. $x^2 + 4x + C = 0$

The discriminant is: $16 - 4c$

(A) If $16 - 4c > 0$, i.e., if $c < 4$, then the equation has two distinct real roots.

(B) If $16 - 4c = 0$, i.e., if $c = 4$, then the equation has one real double root.

(C) If $16 - 4c < 0$, i.e., if $c > 4$, then there are no real roots.

43. Setting the supply equation equal to the demand equation, we have

$$\frac{x}{450} + \frac{1}{2} = \frac{6,300}{x}$$

$$\frac{1}{450}x^2 + \frac{1}{2}x = 6,300$$

$$x^2 + 225x - 2,835,000 = 0$$

$$x = \frac{-225 \pm \sqrt{(225)^2 - 4(1)(-2,835,000)}}{2} \quad \text{(quadratic formula)}$$

$$= \frac{-225 \pm \sqrt{11,390,625}}{2}$$

$$= \frac{-225 \pm 3375}{2}$$

$= 1,575$ units

Note, we discard the negative root since a negative number of units cannot be produced or sold. Substituting $x = 1,575$ into either equation (we use the demand equation), we get

$$p = \frac{6,300}{1,575} = 4$$

Supply equals demand at \$4 per unit.

45. $A = P(1 + r)^2 = P(1 + 2r + r^2) = Pr^2 + 2Pr + P$

Let $A = 144$ and $P = 100$. Then,

$100r^2 + 200r + 100 = 144$

$100r^2 + 200r - 44 = 0$

Using the quadratic formula,

$$r = \frac{-200 \pm \sqrt{(200)^2 - 4(100)(-44)}}{200}$$

$$= \frac{-200 \pm 240}{200} = -2.2,\ 0.20$$

Since $r > 0$, we have $r = 0.20$ or 20%.

47. $v^2 = 64h$

For $h = 1$, $v^2 = 64(1) = 64$. Therefore, $v = 8$ ft/sec.

For $h = 0.5$, $v^2 = 64(0.5) = 32$. Therefore, $v = \sqrt{32} = 4\sqrt{2} \approx 5.66$ ft/sec.

APPENDIX B SPECIAL TOPICS

EXERCISE B-1

Things to remember:

1. SEQUENCES

   A SEQUENCE is a function whose domain is a set of successive integers. If the domain of a given sequence is a finite set, then the sequence is called a FINITE SEQUENCE; otherwise, the sequence is an INFINITE SEQUENCE. In general, unless stated to the contrary or the context specifies otherwise, the domain of a sequence will be understood to be the set N of natural numbers.

2. NOTATION FOR SEQUENCES

   Rather than function notation $f(n)$, n in the domain of a given sequence f, subscript notation a_n is normally used to denote the value in the range corresponding to n, and the sequence itself is denoted $\{a_n\}$ rather than f or $f(n)$. The elements in the range, a_n, are called the TERMS of the sequence; a_1 is the first term, a_2 is the second term, and a_n is the nth term or general term.

3. SERIES

   Given a sequence $\{a_n\}$. The sum of the terms of the sequence, $a_1 + a_2 + a_3 + \cdots$ is called a SERIES. If the sequence is finite, the corresponding series is a FINITE SERIES; if the sequence is infinite, then the corresponding series is an INFINITE SERIES. Only finite series are considered in this section.

4. NOTATION FOR SERIES

   Series are represented using SUMMATION NOTATION.
   If $\{a_k\}$, $k = 1, 2, \ldots, n$ is a finite sequence, then the series

   $$a_1 + a_2 + a_3 + \cdots + a_n$$

   is denoted

   $$\sum_{k=1}^{n} a_k.$$

   The symbol Σ is called the SUMMATION SIGN and k is called the SUMMING INDEX.

5. ARITHMETIC MEAN

   If $\{a_k\}$, $k = 1, 2, \ldots, n$, is a finite sequence, then the ARITHMETIC MEAN $\overline{a}$ of the sequence is defined as

   $$\overline{a} = \frac{1}{n}\sum_{k=1}^{n} x_k.$$

1. $a_n = 2n + 3;\ a_1 = 2\cdot 1 + 3 = 5$

$a_2 = 2\cdot 2 + 3 = 7$

$a_3 = 2\cdot 3 + 3 = 9$

$a_4 = 2\cdot 4 + 3 = 11$

3. $a_n = \frac{n+2}{n+1};\ a_1 = \frac{1+2}{1+1} = \frac{3}{2}$

$a_2 = \frac{2+2}{2+1} = \frac{4}{3}$

$a_3 = \frac{3+2}{3+1} = \frac{5}{4}$

$a_4 = \frac{4+2}{4+1} = \frac{6}{5}$

5. $a_n = (-3)^{n+1};\ a_1 = (-3)^{1+1} = (-3)^2 = 9$

$a_2 = (-3)^{2+1} = (-3)^3 = -27$

$a_3 = (-3)^{3+1} = (-3)^4 = 81$

$a_4 = (-3)^{4+1} = (-3)^5 = -243$

7. $a_n = 2n + 3;\ a_{10} = 2\cdot 10 + 3 = 23$

9. $a_n = \frac{n+2}{n+1};\ a_{99} = \frac{99+2}{99+1} = \frac{101}{100}$

11. $\sum_{k=1}^{6} k = 1 + 2 + 3 + 4 + 5 + 6 = 21$

13. $\sum_{k=4}^{7} (2k - 3) = (2\cdot 4 - 3) + (2\cdot 5 - 3) + (2\cdot 6 - 3) + (2\cdot 7 - 3)$

$= 5 + 7 + 9 + 11 = 32$

15. $\sum_{k=0}^{3} \frac{1}{10^k} = \frac{1}{10^0} + \frac{1}{10^1} + \frac{1}{10^2} + \frac{1}{10^3} = 1 + \frac{1}{10} + \frac{1}{100} + \frac{1}{1000} = \frac{1111}{1000} = 1.111$

17. $a_1 = 5,\ a_2 = 4,\ a_3 = 2,\ a_4 = 1,\ a_5 = 6$. Here $n = 5$ and the arithmetic mean is given by:

$\bar{a} = \frac{1}{5}\sum_{k=1}^{5} a_k = \frac{1}{5}(5 + 4 + 2 + 1 + 6) = \frac{18}{5} = 3.6$

19. $a_1 = 96,\ a_2 = 65,\ a_3 = 82,\ a_4 = 74,\ a_5 = 91,\ a_6 = 88,\ a_7 = 87,\ a_8 = 91,$ $a_9 = 77$, and $a_{10} = 74$. Here $n = 10$ and the arithmetic mean is given by:

$\bar{a} = \frac{1}{10}\sum_{k=1}^{10} a_k = \frac{1}{10}(96 + 65 + 82 + 74 + 91 + 88 + 87 + 91 + 77 + 74)$

$= \frac{825}{10} = 82.5$

21. $a_n = \frac{(-1)^{n+1}}{2^n};\ a_1 = \frac{(-1)^2}{2^1} = \frac{1}{2}$

$a_2 = \frac{(-1)^3}{2^2} = -\frac{1}{4}$

$a_3 = \frac{(-1)^4}{2^3} = \frac{1}{8}$

$a_4 = \frac{(-1)^5}{2^4} = -\frac{1}{16}$

$a_5 = \frac{(-1)^6}{2^5} = \frac{1}{32}$

23. $a_n = n[1 + (-1)^n]$; $a_1 = 1[1 + (-1)^1] = 0$

$a_2 = 2[1 + (-1)^2] = 4$

$a_3 = 3[1 + (-1)^3] = 0$

$a_4 = 4[1 + (-1)^4] = 8$

$a_5 = 5[1 + (-1)^5] = 0$

25. $a_n = \left(-\frac{3}{2}\right)^{n-1}$; $a_1 = \left(-\frac{3}{2}\right)^0 = 1$

$a_2 = \left(-\frac{3}{2}\right)^1 = -\frac{3}{2}$

$a_3 = \left(-\frac{3}{2}\right)^2 = \frac{9}{4}$

$a_4 = \left(-\frac{3}{2}\right)^3 = -\frac{27}{8}$

$a_5 = \left(-\frac{3}{2}\right)^4 = \frac{81}{16}$

27. Given -2, -1, 0, 1, ... The sequence is the set of successive integers beginning with -2. Thus, $a_n = n - 3$, $n = 1, 2, 3, \ldots$.

29. Given 4, 8, 12, 16, ... The sequence is the set of positive integer multiples of 4. Thus, $a_n = 4n$, $n = 1, 2, 3, \ldots$.

31. Given $\frac{1}{2}, \frac{3}{4}, \frac{5}{6}, \frac{7}{8}, \ldots$ The sequence is the set of fractions whose numerators are the odd positive integers and whose denominators are the even positive integers. Thus,

$a_n = \frac{2n - 1}{2n}$, $n = 1, 2, 3, \ldots$.

33. Given 1, -2, 3, -4, ... The sequence consists of the positive integers with alternating signs. Thus,

$a_n = (-1)^{n+1}n$, $n = 1, 2, 3, \ldots$.

35. Given 1, -3, 5, -7, ... The sequence consists of the odd positive integers with alternating signs. Thus,

$a_n = (-1)^{n+1}(2n - 1)$, $n = 1, 2, 3, \ldots$.

37. Given $1, \frac{2}{5}, \frac{4}{25}, \frac{8}{125}, \ldots$ The sequence consists of the nonnegative integral powers of $\frac{2}{5}$. Thus,

$a_n = \left(\frac{2}{5}\right)^{n-1}$, $n = 1, 2, 3, \ldots$.

39. Given $x, x^2, x^3, x^4, \ldots$ The sequence is the set of positive integral powers of x. Thus, $a_n = x^n$, $n = 1, 2, 3, \ldots$.

41. Given $x, -x^3, x^5, -x^7, \ldots$ The sequence is the set of positive odd integral powers of x with alternating signs. Thus,

$a_n = (-1)^{n+1}x^{2n-1}, n = 1, 2, 3, \ldots$.

43. $$\sum_{k=1}^{5} (-1)^{k+1}(2k - 1)^2 = (-1)^2(2\cdot1 - 1)^2 + (-1)^3(2\cdot2 - 1)^2 + (-1)^4(2\cdot3 - 1)^2 + (-1)^5(2\cdot4 - 1)^2 + (-1)^6(2\cdot5 - 1)^2$$
$$= 1 - 9 + 25 - 49 + 81$$

45. $$\sum_{k=2}^{5} \frac{2^k}{2k + 3} = \frac{2^2}{2\cdot2 + 3} + \frac{2^3}{2\cdot3 + 3} + \frac{2^4}{2\cdot4 + 3} + \frac{2^5}{2\cdot5 + 3}$$
$$= \frac{4}{7} + \frac{8}{9} + \frac{16}{11} + \frac{32}{13}$$

47. $$\sum_{k=1}^{5} x^{k-1} = x^0 + x^1 + x^2 + x^3 + x^4 = 1 + x + x^2 + x^3 + x^4$$

49. $$\sum_{k=0}^{4} \frac{(-1)^k x^{2k+1}}{2k + 1} = \frac{(-1)^0 x}{2\cdot0 + 1} + \frac{(-1)x^3}{2\cdot1 + 1} + \frac{(-1)^2 x^5}{2\cdot2 + 1} + \frac{(-1)^3 x^7}{2\cdot3 + 1} + \frac{(-1)^4 x^9}{2\cdot4 + 1}$$
$$= x - \frac{x^3}{3} + \frac{x^5}{5} - \frac{x^7}{7} + \frac{x^9}{9}$$

51. (A) $2 + 3 + 4 + 5 + 6 = \sum_{k=1}^{5} (k + 1)$ (B) $2 + 3 + 4 + 5 + 6 = \sum_{j=0}^{4} (j + 2)$

53. (A) $1 - \frac{1}{2} + \frac{1}{3} - \frac{1}{4} = \sum_{k=1}^{4} \frac{(-1)^{k+1}}{k}$ (B) $1 - \frac{1}{2} + \frac{1}{3} - \frac{1}{4} = \sum_{j=0}^{3} \frac{(-1)^j}{j + 1}$

55. $$2 + \frac{3}{2} + \frac{4}{3} + \ldots + \frac{n + 1}{n} = \sum_{k=1}^{n} \frac{k + 1}{k}$$

57. $$\frac{1}{2} - \frac{1}{4} + \frac{1}{8} - \ldots + \frac{(-1)^{n+1}}{2^n} = \sum_{k=1}^{n} \frac{(-1)^{k+1}}{2^k}$$

59. $a_1 = 2$ and $a_n = 3a_{n-1} + 2$ for $n \geq 2$.

$a_1 = 2$
$a_2 = 3\cdot a_1 + 2 = 3\cdot2 + 2 = 8$
$a_3 = 3\cdot a_2 + 2 = 3\cdot8 + 2 = 26$
$a_4 = 3\cdot a_3 + 2 = 3\cdot26 + 2 = 80$
$a_5 = 3\cdot a_4 + 2 = 3\cdot80 + 2 = 242$

61. $a_1 = 1$ and $a_n = 2a_{n-1}$ for $n \geq 2$.

$a_1 = 1$
$a_2 = 2\cdot a_1 = 2\cdot1 = 2$
$a_3 = 2\cdot a_2 = 2\cdot2 = 4$
$a_4 = 2\cdot a_3 = 2\cdot4 = 8$
$a_5 = 2\cdot a_4 = 2\cdot8 = 16$

63. In $a_1 = \frac{A}{2}$, $a_n = \frac{1}{2}\left(a_{n-1} + \frac{A}{a_{n-1}}\right)$, $n \geq 2$, let $A = 2$. Then:

$$a_1 = \frac{2}{2} = 1$$
$$a_2 = \frac{1}{2}\left(a_1 + \frac{A}{a_1}\right) = \frac{1}{2}(1 + 2) = \frac{3}{2}$$
$$a_3 = \frac{1}{2}\left(a_2 + \frac{A}{a_2}\right) = \frac{1}{2}\left(\frac{3}{2} + \frac{2}{3/2}\right) = \frac{1}{2}\left(\frac{3}{2} + \frac{4}{3}\right) = \frac{17}{12}$$

$$a_4 = \frac{1}{2}\left(a_3 + \frac{A}{a_3}\right) = \frac{1}{2}\left(\frac{17}{12} + \frac{2}{17/12}\right) = \frac{1}{2}\left(\frac{17}{12} + \frac{24}{17}\right) = \frac{577}{408} \approx 1.414216$$

and $\sqrt{2} \approx 1.414214$

EXERCISE B-2

Things to remember:

1. A sequence of numbers $a_1, a_2, a_3, \ldots, a_n, \ldots$, is called an ARITHMETIC SEQUENCE if there is constant d, called the COMMON DIFFERENCE, such that

   $$a_n - a_{n-1} = d,$$

   that is,

   $$a_n = a_{n-1} + d$$

   for all $n > 1$.

2. A sequence of numbers $a_1, a_2, a_3, \ldots, a_n, \ldots$, is called a GEOMETRIC SEQUENCE if there exists a nonzero constant r, called the COMMON RATIO, such that

   $$\frac{a_n}{a_{n-1}} = r,$$

   that is,

   $$a_n = ra_{n-1}$$

   for all $n > 1$.

3. *n*TH TERM OF AN ARITHMETIC SEQUENCE

   If $\{a_n\}$ is an arithmetic sequence with common difference d, then

   $$a_n = a_1 + (n - 1)d$$

   for all $n > 1$.

4. *n*TH TERM OF A GEOMETRIC SEQUENCE

   If $\{a_n\}$ is a geometric sequence with common ratio r, then

   $$a_n = a_1r^{n-1}$$

   for all $n > 1$.

5. SUM FORMULAS FOR FINITE ARITHMETIC SERIES

   The sum S_n of the first n terms of an arithmetic series $a_1 + a_2 + a_3 + \ldots + a_n$ with common difference d, is given by

   (a) $S_n = \frac{n}{2}[2a_1 + (n - 1)d]$ (First Form)

or by

(b) $S_n = \frac{n}{2}(a_1 + a_n)$. (Second Form)

6. SUM FORMULAS FOR FINITE GEOMETRIC SERIES

The sum S_n of the first n terms of a geometric series $a_1 + a_2 + a_3 + a_n$ with common ratio r, is given by:

$$S_n = \frac{a_1(r^n - 1)}{r - 1}, \quad r \neq 1, \quad \text{(First Form)}$$

or by

$$S_n = \frac{ra_n - a_1}{r - 1}, \quad r \neq 1. \quad \text{(Second Form)}$$

7. SUM OF AN INFINITE GEOMETRIC SERIES

If $a_1 + a_2 + a_3 + \dots + a_n + \dots$, is an infinite geometric series with common ratio r having the property $-1 < r < 1$, then the sum S_∞ is defined to be:

$$S_\infty = \frac{a_1}{1 - r}.$$

---

1. (A) $-11, -16, -21, \dots$
This is an arithmetic sequence with common difference $d = -5$; $a_4 = -26$, $a_5 = -31$.

(B) $2, -4, 8, \dots$
This is a geometric sequence with common ratio $r = -2$; $a_4 = -16$, $a_5 = 32$.

(C) $1, 4, 9, \dots$
This is neither an arithmetic sequence $(4 - 1 \neq 9 - 4)$ nor a geometric sequence $\left(\frac{4}{1} \neq \frac{9}{4}\right)$.

(D) $\frac{1}{2}, \frac{1}{6}, \frac{1}{18}, \dots$
This is a geometric sequence with common ratio $r = \frac{1}{3}$; $a_4 = \frac{1}{54}$, $a_5 = \frac{1}{162}$.

3. $a_2 = a_1 + d = 7 + 4 = 11$
$a_3 = a_2 + d = 11 + 4 = 15$ (using 1)

5. $a_{21} = a_1 + (21 - 1)d = 2 + 20 \cdot 4 = 82$ (using 2)
$S_{31} = \frac{31}{2}[2a_1 + (31 - 1)d] = \frac{31}{2}[2 \cdot 2 + 30 \cdot 4] = \frac{31}{2} \cdot 124 = 1922$ [using 3(a)]

7. Using 3(b), $S_{20} = \frac{20}{2}(a_1 + a_{20}) = 10(18 + 75) = 930$

9. $a_2 = a_1 r = 3(-2) = -6$
$a_3 = a_2 r = -6(-2) = 12$
$a_4 = a_3 r = 12(-2) = -24$ (using 4)

11. Using 6, $S_7 = \frac{-3\cdot 729 - 1}{-3 - 1} = \frac{-2188}{-4} = 547.$

13. Using 5, $a_{10} = 100(1.08)^9 = 199.90.$

15. Using 5, $200 = 100r^8$. Thus, $r^8 = 2$ and $r = \sqrt[8]{2} \approx 1.09.$

17. Using 6, $S_{10} = \frac{500[(0.6)^{10} - 1]}{0.6 - 1} \approx 1242,$

$$S_\infty = \frac{500}{1 - 0.6} = 1250.$$

19. $S_{41} = \sum_{k=1}^{41} 3k + 3$. The sequence of terms is an arithmetic sequence.
Therefore,

$$S_{41} = \frac{41}{2}(a_1 + a_{41}) = \frac{41}{2}(6 + 126) = \frac{41}{2}(132) = 41(66) = 2{,}706$$

21. $S_8 = \sum_{k=1}^{8} (-2)^{k-1}$. The sequence of terms is a geometric sequence with common ratio $r = -2$ and $a_1 = (-2)^0 = 1.$

$$S_8 = \frac{1[(-2)^8 - 1]}{-2 - 1} = \frac{256 - 1}{-3} = -85$$

23. Let $a_1 = 13$, $d = 2$. Then, using 2, we can find n:

$67 = 13 + (n - 1)2$ or $2(n - 1) = 54$
$n - 1 = 27$
$n = 28$

Therefore, using 3(b), $S_{28} = \frac{28}{2}[13 + 67] = 14\cdot 80 = 1120.$

25. (A) $2 + 4 + 8 + \cdots$. Since $r = \frac{4}{2} = \frac{8}{4} = \cdots = 2$ and $|2| = 2 > 1$, the sum does not exist.

(B) $2, -\frac{1}{2}, \frac{1}{8}, \ldots$. In this case, $r = \frac{-1/2}{2} = \frac{1/8}{-1/2} = \cdots = -\frac{1}{4}.$
Since $|r| < 1$,

$$S_\infty = \frac{2}{1 - (-1/4)} = \frac{2}{5/4} = \frac{8}{5} = 1.6.$$

27. $f(1) = -1$, $f(2) = 1$, $f(3) = 3$, $\ldots$. This is an arithmetic progression with $a_1 = -1$, $d = 2$. Thus, using 3(a),

$$f(1) + f(2) + f(3) + \cdots + f(50) = \frac{50}{2}[2(-1) + 49\cdot 2] = 25\cdot 96 = 2400$$

29. $f(1) = \frac{1}{2}$, $f(2) = \left(\frac{1}{2}\right)^2 = \frac{1}{4}$, $f(3) = \left(\frac{1}{2}\right)^3 = \frac{1}{8}$, $\ldots$. This is a geometric progression with $a_1 = \frac{1}{2}$ and $r = \frac{1}{2}$. Thus, using 6:

$$f(1) + f(2) + \cdots + f(10) = S_{10} = \frac{\frac{1}{2}\left[\left(\frac{1}{2}\right)^{10} - 1\right]}{\frac{1}{2} - 1} \approx 0.999$$

31. Consider the arithmetic progression with $a_1 = 1$, $d = 2$. This progression is the sequence of odd positive integers. Now, using 3(a), the sum of the first n odd positive integers is:

$$S_n = \frac{n}{2}[2\cdot 1 + (n - 1)2] = \frac{n}{2}(2 + 2n - 2) = \frac{n}{2}\cdot 2n = n^2$$

33. Consider the time line:

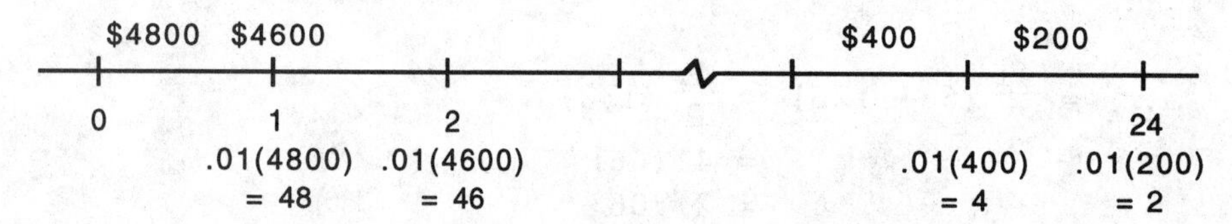

The total cost of the loan is $2 + 4 + 6 + \cdots + 46 + 48$. The terms form an arithmetic progression with $n = 24$, $a_1 = 2$, and $a_{24} = 48$. Thus, using 3(b):

$$S_{24} = \frac{24}{2}(2 + 48) = 24\cdot 25 = \$600$$

35. This is a geometric progression with $a_1 = 3{,}500{,}000$ and $r = 0.7$. Thus, using 7:

$$S_\infty = \frac{3{,}500{,}000}{1 - 0.7} \approx \$11{,}670{,}000$$

37.

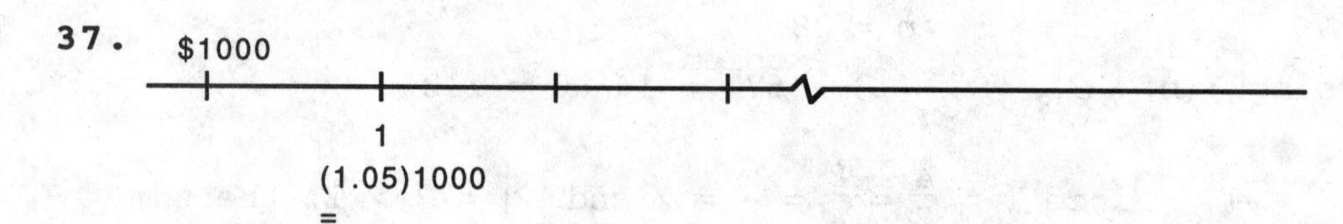

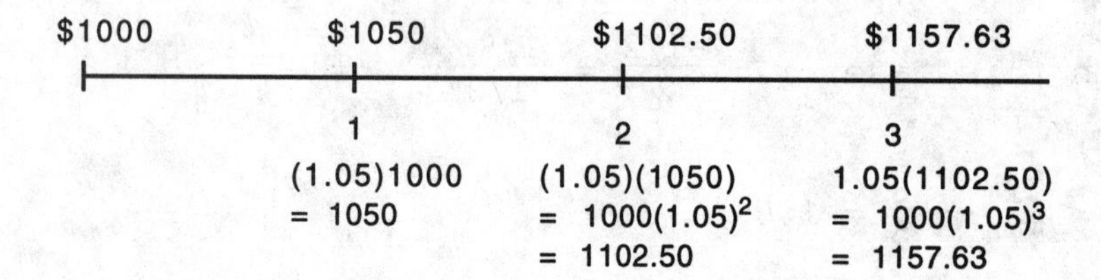

In general, after n years, the amount A_n in the account is:

$$A_n = 1000(1.05)^n$$

Thus, $A_{10} = 1000(1.05)^{10} \approx \$1628.89$

and $A_{20} = 1000(1.05)^{20} \approx \$2653.30$

EXERCISE B-3

Things to remember:

1. If n is a positive integer, then n FACTORIAL, denoted $n!$, is the product of the integers from 1 to n; that is,

   $$n! = n \cdot (n - 1) \cdot \ldots \cdot 3 \cdot 2 \cdot 1 = n(n - 1)!$$

   Also, $1! = 1$ and $0! = 1$.

2. If n and r are nonnegative integers and $r \le n$, then:

   $$C_{n,r} = \frac{n!}{r!(n - r)!}$$

3. BINOMIAL THEOREM

   For all natural numbers n:

   $$(a + b)^n = C_{n,0}a^n + C_{n,1}a^{n-1}b + C_{n,2}a^{n-2}b^2 + \cdots + C_{n,n-1}ab^{n-1} + C_{n,n}b^n.$$

1. $6! = 6 \cdot 5 \cdot 4 \cdot 3 \cdot 2 \cdot 1 = 720$

3. $\frac{10!}{9!} = \frac{10 \cdot 9!}{9!} = 10$

5. $\frac{12!}{9!} = \frac{12 \cdot 11 \cdot 10 \cdot 9!}{9!} = 1320$

7. $\frac{5!}{2!3!} = \frac{5 \cdot 4 \cdot 3!}{2 \cdot 1 \cdot 3!} = 10$

9. $\frac{6!}{5!(6 - 5)!} = \frac{6 \cdot 5!}{5!1!} = 6$

11. $\frac{20!}{3!17!} = \frac{20 \cdot 19 \cdot 18 \cdot 17!}{3!17!} = \frac{20 \cdot 19 \cdot 18}{3 \cdot 2 \cdot 1} = 1140$

13. $C_{5,3} = \frac{5!}{3!(5 - 3)!} = \frac{5!}{3!2!} = 10$ (see Problem 7)

15. $C_{6,5} = \frac{6!}{5!(6 - 5)!} = 6$ (see Problem 9)

17. $C_{5,0} = \frac{5!}{0!(5 - 0)!} = \frac{5!}{1 \cdot 5!} = 1$

19. $C_{18,15} = \frac{18!}{15!(18 - 15)!} = \frac{18 \cdot 17 \cdot 16 \cdot 15!}{15!3!} = \frac{18 \cdot 17 \cdot 16}{3 \cdot 2 \cdot 1} = 816$

21. Using 3,

$$(a + b)^4 = C_{4,0}a^4 + C_{4,1}a^3b + C_{4,2}a^2b^2 + C_{4,3}ab^3 + C_{4,4}b^4$$
$$= a^4 + 4a^3b + 6a^2b^2 + 4ab^3 + b^4$$

23. Using 3,

$$(x - 1)^6 = [x + (-1)]^6$$
$$= C_{6,0}x^6 + C_{6,1}x^5(-1) + C_{6,2}x^4(-1)^2 + C_{6,3}x^3(-1)^3$$
$$+ C_{6,4}x^2(-1)^4 + C_{6,5}x(-1)^5 + C_{6,6}(-1)^6$$
$$= x^6 - 6x^5 + 15x^4 - 20x^3 + 15x^2 - 6x + 1$$

25. $(2a - b)^5 = [2a + (-b)]^5$

$$= C_{5,0}(2a)^5 + C_{5,1}(2a)^4(-b) + C_{5,2}(2a)^3(-b)^2 + C_{5,3}(2a)^2(-b)^3$$
$$+ C_{5,4}(2a)(-b)^4 + C_{5,5}(-b)^5$$
$$= 32a^5 - 80a^4b + 80a^3b^2 - 40a^2b^3 + 10ab^4 - b^5$$

27. The fifth term in the expansion of $(x - 1)^{18}$ is:

$$C_{18,4}x^{14}(-1)^4 = \frac{18\cdot17\cdot16\cdot15}{4\cdot3\cdot2\cdot1}x^{14} = 3060x^{14}$$

29. The seventh term in the expansion of $(p + q)^{15}$ is:

$$C_{15,6}p^9q^6 = \frac{15\cdot14\cdot13\cdot12\cdot11\cdot10}{6\cdot5\cdot4\cdot3\cdot2\cdot1}p^9q^6 = 5005p^9q^6$$

31. The eleventh term in the expansion of $(2x + y)^{12}$ is:

$$C_{12,10}(2x)^2y^{10} = \frac{12\cdot11}{2\cdot1}4x^2y^{10} = 264x^2y^{10}$$

33. $C_{n,0} = \dfrac{n!}{0!(n - 0)!} = \dfrac{n!}{1\cdot n!} = 1 \qquad C_{n,n} = \dfrac{n!}{n!(n - n)!} = \dfrac{n!}{n!0!} = 1$

35. The next two rows are:

1 5 10 10 5 1 and 1 6 15 20 15 6 1,

respectively. These are the coefficients in the binomial expansions of $(a + b)^5$ and $(a + b)^6$.

Explorations in Finite Mathematics Software

This software enhances the understanding and visualization of many of the topics covered in the textbook. The self-documented software consists of two programs called FINITE1 and FINITE2.

Part 1 (Execute by entering FINITE1)

Part 1 covers finite math topics involving linear equations, linear inequalities, and matrices. In particular, it contains routines to perform Gaussian elimination, matrix operations, matrix inversion, simplex method, graphical solution of linear programming problems, the method of least-squares, and regular and absorbing Markov chains. The software follows the conventions of the text.

Part 2 (Execute by entering FINITE2)

Part 2 contains routines for Venn diagrams, Venn diagram counting problems, Venn diagram probability problems, computation of combinations, permutations, and factorials, statistical analysis of data, Galton board, binomial distribution, areas under normal curve, simple interest, compound interest, loan analysis, annuity analysis, and finance tables.

Requirements

The software runs on *any* IBM compatible computer having at least 384 K of memory and a graphics adapter, that is, CGA, EGA, MCGA, VGA, or Hercules.

Technical Support

If you have questions about the software, contact David I. Schneider by e-mail at dis@math.umd.edu.

Matrices On Diskette

Matrices appearing in examples from the text have been saved on the diskette with either suggestive names or names of the form CxSxEx. For instance, the matrix appearing in Chapter 8, Section 3, Example 7 has the name C8S3E7. The list of saved matrices appropriate to the current routine is displayed on the screen whenever you request "Load a matrix saved on disk."

To Invoke Explorations in Finite Mathematics from DOS

1. Place the diskette in a drive, say drive A.
2. Type A: and press the Enter key.
3. Type CD \FINITE and press the Enter key.
4. Type FINITE1 or FINITE2 and press the Enter key.

(*Note:* If a Hercules adapter is used, the program MSHERC.COM must be run before FINITE1 or FINITE2 is entered.)

To Invoke Explorations in Finite Mathematics from Windows 3.1

1. Place the diskette in a drive, say drive A.
2. From Program Manager, double-click on the DOS icon in the Main program group. Or, exit Windows.
3. Type A: and press the Enter key.
4. Type CD \FINITE and press the Enter key.
5. Type FINITE1 or FINITE2 and press the Enter key.

To Invoke Explorations in Finite Mathematics from Windows 95 (or later version)

1. Place the diskette in a drive, say drive A.
2. Click the Start button, point to Programs, and click MS-DOS Prompt.
3. If the DOS window does not fill the screen, hold down the Alt key and press the Enter key to enlarge the window.
4. Type A: and press the Enter key.
5. Type CD \FINITE and press the Enter key.
6. Type FINITE1 or FINITE2 and press the Enter key.

YOU SHOULD CAREFULLY READ THE FOLLOWING TERMS AND CONDITIONS BEFORE OPENING THIS DISKETTE PACKAGE. OPENING THIS DISKETTE PACKAGE INDICATES YOUR ACCEPTANCE OF THESE TERMS AND CONDITIONS. IF YOU DO NOT AGREE WITH THEM, YOU SHOULD PROMPTLY RETURN THE PACKAGE UNOPENED, AND YOUR MONEY WILL BE REFUNDED.

IT IS A VIOLATION OF COPYRIGHT LAWS TO MAKE A COPY OF THE ACCOMPANYING SOFTWARE EXCEPT FOR BACKUP PURPOSES TO GUARD AGAINST ACCIDENTAL LOSS OR DAMAGE.

Prentice-Hall, Inc. provides this program and licenses its use. You assume responsibility for the selection of the program to achieve your intended results, and for the installation, use, and results obtained from the program. This license extends only to use of the program in the United States or countries in which the program is marketed by duly authorized distributors.

LICENSE

You may:

a. use the program;
b. copy the program into any machine-readable form without limit;

LIMITED WARRANTY

THE PROGRAM IS PROVIDED "AS IS" WITHOUT WARRANTY OF ANY KIND, EITHER EXPRESSED OR IMPLIED, INCLUDING, BUT NOT LIMITED TO, THE IMPLIED WARRANTIES OF MERCHANTABILITY AND FITNESS FOR A PARTICULAR PURPOSE. THE ENTIRE RISK AS TO THE QUALITY AND PERFORMANCE OF THE PROGRAM IS WITH YOU. SHOULD THE PROGRAM PROVE DEFECTIVE, YOU (AND NOT PRENTICE-HALL, INC. OR ANY AUTHORIZED DISTRIBUTOR) ASSUME THE ENTIRE COST OF ALL NECESSARY SERVICING, REPAIR, OR CORRECTION.

SOME STATES DO NOT ALLOW THE EXCLUSION OF IMPLIED WARRANTIES, SO THE ABOVE EXCLUSION MAY NOT APPLY TO YOU. THIS WARRANTY GIVES YOU SPECIFIC LEGAL RIGHTS AND YOU MAY ALSO HAVE OTHER RIGHTS THAT VARY FROM STATE TO STATE.

Prentice-Hall, Inc. does not warrant that the functions contained in the program will meet your requirements or that the operation of the program will be uninterrupted or error free.

However, Prentice-Hall, Inc., warrants the diskette(s) on which the program is furnished to be free from defects in materials and workmanship under normal use for a period of ninety (90) days from the date of delivery to you s evidenced by a copy of your receipt.

LIMITATIONS OF REMEDIES

Prentice-Hall's entire liability and your exclusive remedy shall be:

1. the replacement of any diskette not meeting Prentice-Hall's "Limited Warranty" and that is returned to Prentice-Hall with a copy of your purchase order, or
2. if Prentice-Hall is unable to deliver a replacement diskette or cassette that is free of defects in materials or workmanship, you may terminate this Agreement by returning the program, and your money will be refunded.

IN NO EVENT WILL PRENTICE-HALL BE LIABLE TO YOU FOR ANY DAMAGES, INCLUDING ANY LOST PROFITS, LOST SAVINGS, OR OTHER INCIDENTAL OR CONSEQUENTIAL DAMAGES ARISING OUT OF THE USE OR INABILITY TO USE SUCH PROGRAM EVEN IF PRENTICE-HALL, OR AN AUTHORIZED DISTRIBUTOR HAS BEEN ADVISED OF THE POSSIBILITY OF SUCH DAMAGES, OR FOR ANY CLAIM BY ANY OTHER PARTY.

SOME STATES DO NOT ALLOW THE LIMITATION OR EXCLUSION OF LIABILITY FOR INCIDENTAL OR CONSEQUENTIAL DAMAGES, SO THE ABOVE LIMITATION OR EXCLUSION MAY NOT APPLY TO YOU.

GENERAL

You may not sublicense, assign, or transfer the license or the program except as expressly provided in this Agreement. Any attempt otherwise to sublicense, assign, or transfer any of the rights, duties, or obligations hereunder is void.

This Agreement will be governed by the laws of the State of New York.

Should you have any questions concerning this Agreement, you may contact Prentice-Hall, Inc., by writing to:

Prentice Hall
College Division
Upper Saddle River, NJ 07458

Should you have any questions concerning technical support you may write to:

IPS Publishing, Inc.
12606NE 95th Street, C-110
Vancouver, WA 98682
for technical support: 1-800-933-8878

YOU ACKNOWLEDGE THAT YOU HAVE READ THIS AGREEMENT, UNDERSTAND IT, AND AGREE TO BE BOUND BY ITS TERMS AND CONDITIONS. YOU FURTHER AGREE THAT IT IS THE COMPLETE AND EXCLUSIVE STATEMENT OF THE AGREEMENT BETWEEN US THAT SUPERSEDES ANY PROPOSAL OR PRIOR AGREEMENT, ORAL OR WRITTEN, AND ANY OTHER COMMUNICATIONS BETWEEN US RELATING TO THE SUBJECT MATTER OF THIS AGREEMENT.